NUCLEATION IN CONDENSED MATTER

APPLICATIONS IN MATERIALS AND BIOLOGY

PERGAMON MATERIALS SERIES

Vol. 1 **CALPHAD** by N. Saunders and A. P. Miodownik
Vol. 2 **Non-Equilibrium Processing of Materials** edited by C. Suryanarayana
Vol. 3 **Wettability at High Temperatures** by N. Eustathopoulos, M. G. Nicholas and B. Drevet
Vol. 4 **Structural Biological Materials** edited by M. Elices
Vol. 5 **The Coming of Materials Science** by R. W. Cahn
Vol. 6 **Multinuclear Solid-State NMR of Inorganic Materials** by K. J. D. MacKenzie and M. E. Smith
Vol. 7 **Underneath the Bragg Peaks: Structural Analysis of Complex Materials** by T. Egami and S. J. L. Billinge
Vol. 8 **Thermally Activated Mechanisms in Crystal Plasticity** by D. Caillard and J. L. Martin
Vol. 9 **The Local Chemical Analysis of Materials** by J. W. Martin
Vol. 10 **Metastable Solids from Undercooled Melts** by D. M. Herlach, P. Galenko and D. Holland-Moritz
Vol. 11 **Thermo-Mechanical Processing of Metallic Materials** by B. Verlinden, J. Driver, I. Samajdar and R. D. Doherty
Vol. 12 **Phase Transformations: Examples from Titanium and Zirconium Alloys** by S. Banerjee and P. Mukhopadhyay
Vol. 13 **Intermetallic Chemistry** by R. Ferro and A. Saccone
Vol. 14 **Self-Diffusion and Impurity Diffusion in Pure Metals: Handbook of Experimental Data** by G. Neumann and C. Tuijn

NUCLEATION IN CONDENSED MATTER

APPLICATIONS IN MATERIALS AND BIOLOGY

By

K. F. KELTON
Washington University in St. Louis, USA

A. L. GREER
University of Cambridge, UK

ELSEVIER

Amsterdam • Boston • Heidelberg • London • New York • Oxford
Paris • San Diego • San Francisco • Singapore • Sydney • Tokyo
Pergamon is an imprint of Elsevier

Pergamon is an imprint of Elsevier
Linacre House, Jordan Hill, Oxford OX2 8DP, UK
Radarweg 29, PO Box 211, 1000 AE Amsterdam, The Netherlands

British Library Cataloguing in Publication Data
A catalogue record for this book is available from the British Library

Library of Congress Cataloging-in-Publication Data
A catalog record for this book is available from the Library of Congress

ISBN: 978-0-08-042147-6

For information on all Pergamon publications
visit our website at books.elsevier.com

Printed and bound in the United Kingdom

Transferred to Digital Print 2010

The authors direct the reader to the companion site for this work, which includes colour versions of some of the figures (those with 'color online' noted in the text) and an updated sheet of errata. The site is available at URL http://www.elsevierdirect.com/companion.jsp?ISBN=9780080421476

This book is dedicated to the memory of

DAVID TURNBULL
(1915–2007)

for his leadership in research, directing the authors towards the study of nucleation

ROBERT CAHN
(1924–2007)

for his guidance and encouragement, driving onward the writing of this book

CONTENTS

PREFACE

Nucleation, for example involving the spontaneous appearance of small regions of a new phase in an old phase, is the usual starting process for transformations in a material. While the study of nucleation is an old one, stretching back to pioneering work by Fahrenheit in the early 1700's, it remains a very active field, on which several books have just recently appeared. Some overlap of the material covered in this book with that covered in those is unavoidable. However, we hope that our presentation is distinctive in covering a wide range of topics, and in taking the chosen topics up to the frontier of current work. The key theories of nucleation are presented and critiqued in light of experimental tests. Coupled processes that often underlie nucleation are discussed, as is the important practical theme of nucleation control. Given the number of unanswered questions in the field and the rapid progress being made, some sections of this book will soon appear dated; we feel, nevertheless, that it is useful to have a survey of the present state of the art. We have tried to provide the reader with comprehensive citations of original sources. This has involved much effort, starting from reference data that were often patchy or wrong. We now have close sympathy with the editors of the *Catalogue of Scientific Papers 1800–1863*, published by the Royal Society of London in 1867, who bemoaned the lack of complete data in tracing authors and publications: "None but those who have been engaged in a task of this kind can form any idea of the difficulty occasioned by such omissions."

This book aims to provide a comprehensive coverage of nucleation in condensed matter, and includes some topics (such as nucleation in biology, medicine, food and drink) that fall outside the usual focus in materials science and physics. We are, however, conscious of such omissions as the nucleation of magnetic and other ordered domains, and nucleation in vapors, which has practical application in atmospheric processes that are of increasing environmental concern. A discussion of these and other topics was originally planned, but it became clear that their inclusion would increase the length of this book beyond a reasonable limit. The study of nucleation in vapors, for example, could easily be the focus of another book of similar length. Inevitably, the choice of topics reflects the background and prejudices of the authors, whose own research has focused on the nucleation of crystals in liquids and glasses, but we hope that other important topics are fairly represented.

Having completed the book, we are struck by the evident lack of communication even within the field of nucleation studies. To cite just one example of many: line tension has been well studied for heterogeneous nucleation of condensation from a vapor, but has been largely ignored for crystallization of a liquid. We hope that in many ways, not least bridging such communication gaps, this book can stimulate further advances in nucleation studies.

It is a pleasure to dedicate the book to two particularly distinguished scholars. A glance even at the name index of this book will show that the influence of David Turnbull on studies of nucleation in condensed matter was pervasive and continues to be so. We both had the privilege of working in Turnbull's laboratory at Harvard, and our scientific collaboration stems from that time. In some sense this is the book that we wish Turnbull had written. Although he might have done it better, we have been able to take the story further on in time. This book would never have come about had it not been for the continual encouragement of Robert Cahn in the other Cambridge, acting principally in his capacity as inaugural editor of the *Pergamon Materials Series*. His influence on us, by his own example, in the art of good writing has been immense, even if we fail to reach his standards.

In our research and in writing this book, we have benefited from interactions with many colleagues around the world. We thank them for discussion and their shared insights into many of the new ideas that are expressed here. ALG thanks the many graduate students who contributed to his education on nucleation while they worked in his group. We thank Professors Lev Gelb and Michael Ogilvie, Washington University, for a critical reading of Chapters 4 and 10, and Linda Coffin, Emily Kelton, and many of KFK's graduate students, for their editorial assistance and comments on chapters at various stages. KFK particularly thanks his wife, Emily, and sons Franklin and James for their support during the many years spent in the preparation of this book. Of course, despite all of the input from others, the responsibility for any errors and omissions lies with the authors. The companion site for this work: http://www.elsevierdirect.com/companion.jsp?ISBN=9780080421476 includes color versions of some of the figures and an updated sheet of errata.

Ken Kelton
Lindsay Greer
October 2009

SYMBOLS

Chapter and section numbers are cited where a symbol has different meanings in different sections.

Symbol	Definition
	Prefixes
δ	incremental change
Δ	change or difference
	Subscripts
0	initial value
am	austenite/martensite
amor	amorphous
appl	applied
A	atoms; Avogadro
b	bubble
bu	break-up
B	Boltzmann, boundary, Bragg
c	coagulation; critical; crystallization (of a single polymer chain)
cap	spherical cap
cf	composition fluctuation; coupled-flux
ch	ccp/hcp
cl	cluster
class	classical
clus	cluster
col	columnar
core	dislocation core
crys	crystalline nucleus
ct	center
cyl	cylindrical
C	Curie
CNT	classical nucleation theory
d	droplet
disk	disk
D	free diffusion on lattice
DF	density-functional
DIT	diffuse-interface theory

eff	effective
emit	dislocation emission
eq	equilibrium
es	electrostatic
EM	electromigration
f	fusion
fg	free-growth
flu	fluctuation (of order or composition)
g	gas
gb	grain-boundary
gr	gram-atomic
grad	in a composition gradient
GLCH	Ginzburg-Landau Cahn-Hilliard
het	heterogeneous
hom	homogeneous
H	atomic hydrogen
He	helium
i	interfacial
inc	incubation
int	interstitial
ion	ion
IA	interfacial attachment
K	Kashchiev
Kn	Kauzmann
ℓ	dislocation line
l	liquid
lat	lateral (ledge movement on a crystal surface)
lf	linked-flux
lim	limiting
liq	liquidus
loop	dislocation loop
LJ	Lennard-Jones
m	medium (initial phase); melting
max	maximum
min	minimum
M	magnetic, molar
MWDA	modified weighted-density approximation
N	nucleation, nucleant phase
NC	nonclassical
Nuc	nucleant substrate or area
O	oxygen
pore	pore or bubble
pt	thermal plateau

P	nucleant patch; Poisson
PDFA	perturbative density-functional approximation
q	value to which a sample is quenched
r	reduced (i.e. normalized)
rel	relaxation
rept	reptation
res	resolved
s	shear; solid; start; first stem (of a polymer chain); supersaturation
sol	solidification; solutal
st	steady-state
step	surface step
strain	strain
sup	supersaturation
S	surface, per unit area
SCCT	self-consistent classical theory
SDFA	semi-empirical density-functional approximation
SF	stacking fault
th	thermal
us	unstable stacking
v	vapor
vac	vacancy
void	void
V	per unit volume
VW	Volmer-Weber
w	critical condition for wetting
WD	work done
α	original phase (α)
$\alpha\gamma$	in α phase in contact with γ phase; pertaining to the interface between α and γ phases
β	new phase (β)
γ	product phase (precipitate or interfacial reaction product)
κ	dielectric
	Superscripts
0	initial
+	forward
−	backward
′	final value
*	critical value (corresponding to the unstable equilibrium of a critical nucleus)
at	per atom
A	of A atoms

B	of B atoms
crit	critical
d	droplet
eq	equilibrium
l	liquid
m	medium (initial phase)
s	solid
st	steady-state
WDA	weighted-density approximation
z	for zero flux

Main Symbols

a	geometrical factor [Ch. 2, §8] ∣ activity [Ch. 2, §11] ∣ the ratio $\Delta g_{sl}/\Delta g_{il}$ [Ch. 4, §3] ∣ capillary length $(= (2\sigma/k_B T)\bar{v}C_\infty)$ [Ch. 4, §5] ∣ number of atoms/molecules of species A [Ch. 5, §1; Ch. 7, §2] ∣ lattice parameter, interatomic spacing [Ch. 4, §4; Ch. 9, §2; Ch. 14, §5; Ch. 15, §3; Ch. 16, §4]
a'	geometrical factor including the entropy of fusion per unit volume $(a' = a\,\Delta s_f^{-2})$ [Ch. 2, §8; Ch. 16, §4]
A	interfacial area [Ch. 2, §2] ∣ portion of A^* that is independent of the atomic mobility [Ch. 9, §2]
A	chemical species A
A	diagonalized matrix of **K**
A^*	kinetic (dynamical) pre-factor for nucleation rate (usually $\mathrm{mol}^{-1}\,\mathrm{s}^{-1}$)
$A(n)$	drift coefficient, $(k^-(n) - k^+(n))$ [Ch. 2, §10]
b	elements of matrix that diagonalizes **K** [Ch. 3, §3] ∣ $\sqrt{1-a}$, with a as in Ch. 4, Eq. (15) ∣ length of Burgers vector of dislocation [Ch. 6, §3; Ch. 12, §2; Ch. 14, §5] ∣ number of atoms/molecules of species B [Ch. 5, §1; Ch. 7, §2] ∣ thickness of a layer of chains (polymer crystal) [Ch. 11, §6] ∣ lattice parameter [Ch. 16, §4]
$\boldsymbol{b}$	Burgers vector
B	chemical species B
B	matrix of elements b diagonalizing **K**
B	applied magnetic field [Ch. 10, §3]
$B(n)$	drift coefficient, $\frac{1}{2}(k^-(n) + k^+(n))$ [Ch. 2, §10]
B_j	coefficient describing the interaction of j molecules [Ch. 4, §2]
B_J	the Brillouin function [Ch. 7, Eq. (21)]
c_p	specific heat at constant pressure
C	constant

C	solute concentration: C_∞ equilibrium C at the surface of an infinitely large particle; $\bar{C}$ average C; C^{i}_{α} concentration of solute in the α phase next to a precipitate; $C_{\gamma\alpha}$ concentration of A atoms in the product phase γ in contact with phase α
d	dimensionality of system [Ch. 4, §5] | average grain diameter in a polycrystalline phase [Ch. 6, §2; Ch. 13, §5]
D	diffusion coefficient ($m^2\,s^{-1}$): D' effective D in size (n) space; $\tilde{D}$ interdiffusivity; D^*_{A} tracer diffusivity of A atoms
e	charge on an electron (1.602192×10^{-19} C)
e	natural logarithm base (2.71828)
Δe	elastic strain energy per unit volume [Ch. 14, §3]
E	applied electric field [Ch. 6, §4; Ch. 15, §5] | Young modulus [Ch. 9, §2] | energy: E_{ch} strain energy (per mole) associated with the ccp-to-hcp transformation [Ch. 12, §4]; E_{loop} total line energy of a dislocation half-loop [Ch. 12, §2]; E_n energy of the nth state of the system [Ch. 10, §2]; E_{step} total energy of a surface step on a thin film [Ch. 12, §2]; E_{strain} uniform strain energy of a thin film [Ch. 12, §2]
E_1	single atom or molecule: E_n cluster of n atoms or molecules
f	Helmholtz free energy per unit volume: f_0 for a uniform system [Ch. 4, §4] | catalytic factor for heterogeneous nucleation: $f(\phi)$ as a function of contact angle ϕ; $f(\phi, X)$ as a function of contact angle ϕ and normalized nucleant size X [Ch. 6, §2] | order parameter: f_{bcc} for bcc; f_{ccp} for ccp; f_{liq} for the liquid [Ch. 10, §5]
F	Helmholtz free energy (usually per mole) [Ch. 2, §2]: F_{ex} particle–particle interaction contribution to F [Ch. 4, §4]; F_{id} ideal contribution to F [Ch. 4, §4)]; F_M magnetic contribution to F [Ch. 7, §5]; F^{at}_{M} energy per atom due to magnetic ordering [Ch. 7, §5]
$\bar{F}(n,s)$	Laplace transform of $F(n,t)$, $\mathscr{L}(F(n,t))$
$F(n,t)$	$N(n,t)/N^{eq}(n)$ [Ch. 2, §10; Ch. 3, §3; Ch. 3, §4]: $F^{st}(n)$ steady-state solution of $F(n,t)$; $\Delta F^{st}(n) = F^{st}(n+1) - F^{st}(n)$
g	Gibbs free energy per unit volume (free-energy density): g_0 for a system of uniform density [Ch. 4, §4]
$\mathfrak{g}$	surface coupling
g_L	Landé g-factor
G	Gibbs free energy (usually per mole)
$\mathfrak{G}$	applied strain-energy release rate [Ch. 12, §2]: $\mathfrak{G}_{emit}$ critical rate for dislocation emission at a crack tip | temperature gradient at solidification front [Ch. 13, §2]

$G(n, \rho)$	correction for interfacial rates in coupled-flux nucleation that takes account of entropy change in the shell and original phase with the incorporation of an atom into the cluster interface
h	enthalpy per unit volume ∣ height of a spherical cap [Ch. 6, §2], normalized field, the functional analog to ln s [Ch. 10, §3] ∣ Planck constant (6.626×10^{-34} J s) [Ch. 11, §6; Ch. 14, §3] ∣ thickness of thin film [Ch. 12, §2]
$\mathfrak{h}$	surface field
$\hbar$	Planck constant divided by 2π
H	enthalpy (usually per mole) ∣ external applied magnetic field [Ch. 7, §5]: H_{appl} external applied; H_{T} total
$\mathscr{H}$	the Hamiltonian, $= \sum_i \dot{q}_i p_i - \mathscr{L}$ equal to the total energy
H_γ	constant for the thickening rate of the γ product phase: $H_{\gamma\alpha}$ when affected by the atomic fluxes in the α phase
I	nucleation rate or flux (usually $\text{mol}^{-1}\,\text{s}^{-1}$)
j	flux of A atoms (shorthand for j^{A})
$\mathbf{j}$	probability current; field flux
J	total angular momentum [Ch. 7, Eq. (17)]
$J(n, t)$	flux of clusters past size n as a function of time [Ch. 2, §10]
k	rate constant [Ch. 10, §7]: k_{D} for free diffusion on a lattice; k_{IA} for interfacial attachment ∣ solute partition coefficient [Ch. 5, §4; Ch. 13, §2]: k^{eq} equilibrium value of coefficient [Ch. 5, §4]
k^+	forward rate constant
k^-	backward rate constant
k_{B}	Boltzmann constant (1.38062×10^{-23} J K^{-1})
K	bulk modulus
$\bar{\mathbf{K}}$	tridiagonal rate-constant matrix
l^*	a dimensionless edge-length of a cube-shaped critical cluster
ℓ	thickness of lamella (polymer crystal)
L	length of nucleus along a dislocation line [Ch. 6, §3] ∣ orbital angular momentum [Ch. 7, §5] ∣ edge length of the first Brillouin zone for the allowed sites in reciprocal space [Ch. 9, §2] ∣ coherence length of lattice (polymer crystal) [Ch. 11, §6]
$\mathscr{L}$	the Lagrangian: kinetic energy minus potential energy
m	mass: m, m_1 of a single molecule; m_{A} of a molecule of A; m_{e} of an electron ∣ Avrami exponent [Ch. 8, §5; Ch. 9, §2] ∣ liquidus slope (K at.%$^{-1}$ or K wt.%$^{-1}$) [Ch. 13, §2]
m_J	projection of the total angular momentum J onto the direction of the external magnetic field H

M	order parameter: M_c value of M for which $\omega_l(M) = \omega_s(M)$ [Ch. 4, §4] ∣ total magnetization [Ch. 7, §5] ∣ number of independent generalized coordinates [Ch. 10, §2]
$\mathbf{M}$	generalized transport mobility matrix
$\mathrm{M_s}$	martensite start temperature
n	number: of atoms or molecules in a cluster [Ch. 1, §2; Ch. 2, §2; Ch. 6, §4]; of methylene units in a polyethylene chain [Ch. 11, §6]; of planes in a stacking fault [Ch. 12, §4]; of molecules per unit volume [Ch. 14, §3]; of vacant lattice sites [Ch. 14, §5]
n^*	number of atoms or molecules in the critical nucleus
N	number (usually per mole): of clusters [Ch. 2, §2], of molecules [Ch. 4, §2], of droplets [Ch. 11, §2]; N_0 of single molecules in the original phase [Ch. 2, §2; Ch. 9, §2; Ch. 10, §3], of nucleant particles per unit volume [Ch. 13, §2]; N_1 of single molecules [Ch. 2, §4; Ch. 6, §1]; N_{amor} of still-uncrystallized (amorphous) droplets [Ch. 11, §2]; N_B of single molecules in contact with unit area of a boundary [Ch. 6, §2]; N_p of particles in Ostwald ripening [Ch. 4, §5]; N_S of excess surface atoms [Ch. 2, §2], of single molecules in contact with substrate per unit area of original phase [Ch. 6, §2]; N_{sol} of solute molecules per unit volume in the initial phase [Ch. 5, §5]; N_V of single molecules per unit volume [Ch. 6, §2; Ch. 9, §2], of atomic sites per unit volume [Ch. 14, §5]; N_{void} of voids per unit volume [Ch. 14, §5]
$\mathbb{N}$	time-dependent cluster population matrix
$N(n)$	cluster size distribution
N_A	Avogadro number (6.022142×10^{23} mol^{-1})
O	number of possible attachment sites: $O(a, b)$ on a cluster of a molecules of A and b molecules of B; $O(n)$ on a cluster of n atoms or molecules
O^*	total surface area of a critical cluster, $O(n^*)$
p	pressure
$\boldsymbol{p}$	vector momentum of a molecule
P	probability
$\mathscr{P}_{He}$	production rate of helium atoms under irradiation
q	single-molecule partition function [Ch. 4, §4] ∣ scattering vector ($= 4\pi \sin\theta/\lambda$) [Ch. 7, §1]
q_i	generalized coordinates: $\dot{q}_i (= \mathrm{d}q_i/\mathrm{d}t)$ generalized velocities [Ch. 10, §2]
$\boldsymbol{q}_n$	reciprocal-lattice vectors
Q	canonical partition function [Ch. 4, §2; Ch. 7, §5; Ch. 10, §2] ∣ strength of sources and sinks for solute diffusion [Ch. 4, §5] ∣

growth-restriction factor [Ch. 13, Eq. (2)] activation energy: Q_{rept} for reptational diffusion [Ch. 11, §6]; Q_0 for dislocation motion without applied stress [Ch. 12, §4]; Q_s for dislocation motion under applied stress [Ch. 12, §4]

r radius of curvature of an interface; radius of a bubble, cluster, crystal, nucleus, particle, pore; distance from center of a cluster | ratio of interfacial attachment rates (k_B^+/k_A^+) [Ch. 5, §4]

$\boldsymbol{r}$ position vector; vector from a site to its nearest-neighbor sites

r^* critical radius for nucleation

R universal gas constant (8.31434 J K^{-1} mol^{-1}) | lateral radius of/on a nucleant substrate: R_{Nuc} of planar particle face; R_p of a patch | R radius: of inscribing sphere [Ch. 6, Eq. (37)] of dislocation loop [Ch. 12, §2] | measured sample resistance [Ch. 9, §2]

$\boldsymbol{R}_i$ vectors specifying real-space lattice sites

s supersaturation | entropy per unit volume

$\mathfrak{s}$ stress: $\mathfrak{s}_{res}$ resolved shear stress on glide plane; $\mathfrak{s}_\mu$ minimum, temperature-independent resistance to dislocation motion

s_i spin state ($s_i = \pm 1$)

S entropy (usually per mole): S_M^{at} entropy per atom due to magnetic ordering [Ch. 7, §5]; ΔS_M^{at} contribution of magnetic ordering to the difference in S_M^{at} between liquid and solid phases [Ch. 7, §5] | total spin angular momentum [Ch. 7, §5]

$S(a,b)$ surface area of a cluster containing a molecules of A and b molecules of B [Ch. 5, §4]

$S(q)$ X-ray structure factor as a function of scattering vector q [Ch. 4, §4]

t time

T temperature (typically in absolute units)

ΔT supercooling ($= T_m - T$): ΔT_{max} maximum supercooling for solidification ($= T_m - T_{min}$) [Ch. 7, §2]; ΔT_N onset supercooling for nucleation [Ch. 13, §2]; ΔT_r reduced supercooling ($= \Delta T/T_m$ or $\Delta T/T_{liq}$) [Ch. 7, §2; Ch. 11, §2] | superheat ($= T_{max} - T_m$) [Ch. 14, §3]

T_g glass-transition temperature: T_{rg} ($= T_g/T_m$) reduced glass-transition temperature

T_m equilibrium melting temperature (typically in absolute units)

u the ratio n/n^* [Ch. 3, §4] | cluster growth rate [Ch. 8, §5]

U energy, internal energy of system | energy of a dislocation [Ch. 6, §3]: U_B a component of the strain energy; U_{core}

	energy per unit length of the core ∣ U_{es} electrostatic energy [Ch. 6, §4] ∣ U_S energy per unit area by which metastable surface layer exceeds stable surface layer [Ch. 6, §2]
$\langle U \rangle$	average potential energy, [Ch. 10, Eq. (4)]
$\bar{v}$	atomic or molecular volume ∣ partial molar volume [Ch. 5, §2]
v^*	volume of critical cluster [Ch. 5, §2] ∣ activation volume for dislocation nucleation [Ch. 12, §2]
$\mathbf{v}$	characteristic vectors: $\mathbf{v}_{cl}$ for the cluster; $\mathbf{v}_{liq}$ for the equilibrated liquid; $\mathbf{v}_{bcc}$ for the bcc phase; $\mathbf{v}_{ccp}$ for the ccp phase
V	volume of sample/system
w	number of available states
W	work of cluster formation
W^*	work of forming a critical cluster/nucleus (can also be represented as $W(n^*)$)
$W'(n)$	$W(n) - W(1)$ [Ch. 2, §2]
$W_{t,t'}(\xi \rightarrow \xi')$	transition probability [Ch. 4, §5]
$\delta W(n)$	the work of formation for a cluster of $n+1$ molecules less that of a cluster of n molecules
x	volume fraction transformed [Ch. 8, §3] ∣ number of gas atoms [Ch. 14, §5] ∣ a distance coordinate [Ch. 15, §5]
X	mole fraction ∣ normalized radius of nucleant substrate $(= R_{Nuc}/r)$ [Ch. 6, §2]
y	distance from a dislocation line [Ch. 6, §3] ∣ a spatial coordinate [Ch. 15, §4]
z	height of pill-box embryo [Ch. 6, §2] ∣ lattice coordination number [Ch. 9, §2]
z^*_A	effective charge on A atoms within the product phase γ
Z	Zeldovich factor [Ch. 2, Eq. (50)] [Ch. 2, §7; Ch. 9, §2; Ch. 14, §5] ∣ partition function [Ch. 7, Eq. (17)] ∣ constant in expression [Ch. 15, Eq. (26)] for the electromigration contribution to the intermixing flux of A and B in the product phase γ [Ch. 15, §5]
Z_N	configurational integral [Ch. 4, §2]
$\alpha(n, \rho)$	rate at which solute atoms diffuse into the shell surrounding a cluster [Ch. 5, §5]
$\alpha(\phi)$	geometrical factor for the curved surface area of spherical-cap embryos [Ch. 6, §2]
α_{es}	electrostatic energy factor [Ch. 6, Eq. (51)]
α_s	supersaturation ratio for dissolved gas $(= C_g/C_0)$ [Ch. 13, Eq. (12); Ch. 16, Eq. (3)]
β	inverse thermal energy $(= (k_B T)^{-1})$ [Ch. 4, §2] ∣ coefficient for the free-growth supercooling [Ch. 6, Eq. (37)] ∣

proportionality factor relating the number of silicon interstitials to the number of oxygen atoms consumed [Ch. 9, §3] | pre-factor in rate equation, governed by reputation [Ch. 11, §6]

$\beta(n, \rho)$ rate at which solute atoms diffuse out of the shell surrounding a cluster

γ unbiased molecular jump frequency (at cluster interface) [Ch. 2, §4; Ch. 5, §4; Ch. 15, §3]: γ_A for species A [Ch. 5, §4] | proportionality constant [Ch. 3, §4] | frequency of atomic motion on dislocation loop [Ch. 12, §2] | energy per unit area of void-matrix interface [Ch. 14, §5]

γ_I energy per unit area of an intrinsic stacking fault

$\gamma^+(n)$ rate of single-molecule addition at an interface site for a cluster containing n molecules

$\gamma^-(n-1)$ rate of single-molecule loss at an interface site for a cluster containing $n-1$ molecules

γ_0^+ rate constant for secondary nucleation (forward reaction from zero to one stem attached on a polymer crystal) [Ch. 11, §6]

γ_V attempt frequency per unit volume [Ch. 12, §2]

Γ gamma function [Ch. 4, §5] | line tension where an interphase interface meets a nucleant substrate [Ch. 6, §2] | free energy (per unit area) of an interface (parallel to close-packed planes): Γ_{am} between an austenite matrix and a martensite embryo; Γ_{ch} between a ccp matrix and an hcp embryo; Γ_{SF} of a stacking fault [Ch. 12, §2; Ch. 12, §4]

Γ_{eff} an effective mobility [Ch. 4, Eq. (39)]

Γ_{us} unstable stacking energy [Ch. 12, §2]

δ width of interphase interface or grain boundary | $\equiv (\bar{v}_{Co} - \bar{v}_{CuCo})/\bar{v}_{CuCo}$ [Ch. 9, §2]

Δ width (in range of n) of the near-critical region in which the work of cluster formation is within k_BT of W^* [Ch. 2, §9] | dimensionless free-energy difference between liquid and solid ($= \Delta g_{sl}/\rho_s k_B T_l$) [Ch. 4, §4] | excess solute concentration ($\bar{C} - C_\infty$) [Ch. 4, §5] | dimensionless parameter describing the driving force for nucleation: Δ_{cyl} of a cylinder on a dislocation line [Ch. 6, Eq. (46)]; Δ_{gb} on grain boundaries [Ch. 6, Eq. (51)] [Ch. 6, §3; Ch. 6, §5]

ε depth of potential minimum [Ch. 4, §4] | general component of the strain tensor [Ch. 6, §3]

$\varepsilon(n)$ normalization constant determined for each n from the condition that $\sum_{\rho=0}^{\rho_{max}} P(\rho) = 1$ [Ch. 5, §5]

ε_0 permittivity of free space

ε_κ	dielectric constant
ς	geometric factor, $(36\pi)^{1/3}$ for spherical clusters [Ch. 7, §2; Ch. 8, §9] ǀ particle volume fraction in a colloidal suspension [Ch. 7, §4]: ς_{cl} within a crystalline cluster; ς_{fl} within a shear-molten fluid; ς_l at the liquidus; ς_s at the solidus
ζ_E	Euler constant (= 0.5772) [Ch. 3, §4] ǀ number of degrees of freedom [Ch. 4, §5]
η	fractional density change on freezing [Ch. 4, §4] ǀ geometrical factor depending on cluster shape [Ch. 5, §2] ǀ liquid viscosity [Ch. 8, §6]
θ	time lag (or induction time) for nucleation
θ_B	Bragg angle for X-ray scattering
ϑ	cooling rate ($= \lvert dT/dt \rvert$)
κ	effective kinetic constant [Ch. 8, §5; Ch. 9, §2] ǀ gradient-energy coefficient [Ch. 9, §2] ǀ constant describing the frictional force, per monomer, hindering reptation [Ch. 11, §6] ǀ constant in expression for the size distribution $N(n)$ [Ch. 12, §4] ǀ rate constant : κ_γ for thickening of a γ-phase layer; $\kappa_{\gamma\alpha}$ for the motion of the $\gamma-\alpha$ interface [Ch. 15, §5]
κ_D	dynamical pre-factor [Ch. 4, §5]
λ	atomic or molecular jump distance or diameter ǀ eigenvectors of diagonalized matrix **A** [Ch. 3, §3] ǀ thermal de Broglie wavelength ($= (h^2/2\pi m k_B T)^{1/2}$) [Ch. 4, §2] ǀ curvature of $\omega(M)$: λ_l for the liquid phase; λ_s for the solid phase [Ch. 4, §4] ǀ X-ray wavelength [Ch. 7, §1] ǀ constant relating a component of the total magnetic field to the magnetization [Ch. 7, §5]
Λ	line tension (energy per unit length) of a dislocation
μ	chemical potential (free energy per atom): $\mu^0_{A,1}$ of pure liquid A
μ_B	Bohr magneton
μ_s	elastic shear modulus
ν	unbiased attempt frequency, Debye frequency
ν_P	Poisson ratio
$\aleph$	number of spins per unit volume
ξ	random variable [Ch. 2, §1] ǀ position in phase space [Ch. 4, §5] ǀ a particle configuration of the system [Ch. 10, §2] ǀ composition-dependent correction factor to ensure the correct limit of diffusion-limited growth for large clusters [Ch. 5, §5] ǀ a positive constant in [Ch. 15, Eq. (3)]

Ξ grand canonical partition function

ρ number density (number per unit volume) of molecules [Ch. 4, §2]

ρ number of solute molecules in the nearest-neighbor shell around a cluster [Ch. 5, §5]: $\rho_{max}(n)$ maximum number of atom sites in the nearest-neighbor shell of a cluster containing n atoms

ρ density (mol m^{-3}) [Ch. 14, §2] | ρ_A density of atoms (mol m^{-2}) in a close-packed plane [Ch. 12, §4]

σ interfacial energy per unit area (J m^{-2}) | surface energy (surface tension) of a liquid [Ch. 13, §6; Ch. 16, §3; Ch. 17, §4]

σ_{ls}^{M} gram-atomic liquid solid interfacial free energy (kJ/m^2) $\sigma_{ls}^{M} = \sigma_{ls} V_M^{2/3} N_A^{1/3}$ [Ch. 7, §2]

τ transient time for nucleation

τ^* transient time at the critical size

τ_0 characteristic time required to attempt a spin flip on a lattice [Ch. 10, §2]

υ velocity: of an interface, growth velocity; υ_{lat} of a ledge (polymer crystal)

ϕ contact angle: ϕ_∞ of planar interface at a nucleant surface (with no influence of line tension) [Ch. 6, §2] | angle between glide plane and crack plane for dislocation emission at a crack tip [Ch. 12, §2]

$\phi(r_s, t)$ normalized distribution of particle size [Ch. 4, §5]

Φ potential: $\Phi(a, b)$ generalized chemical potential [Ch. 5, Eq. (28)]; $\Phi(n)$ kinetic potential for clusters of size n [Ch. 2, §5]; Φ_{LJ} Lennard-Jones potential

χ number of nuclei, or nucleation events leading to grains (usually per mole): χ_S on the mold wall during solidification; χ_V in the bulk during solidification, or per unit volume [Ch. 7, §2; Ch. 13, §3]

$\tilde{\chi}$ number of sites: $\tilde{\chi}_S$ per unit area for surface nucleation; $\tilde{\chi}_V$ per unit volume for heterogeneous nucleation throughout the volume

X $\equiv \varpi(T)^3 \Delta g^{-2}\, T^{-1}$

Ψ interfacial contribution to the work of cluster formation [Ch. 2, §11; Ch. 5, §2] | weighting function [Ch. 4, §2]

ω intracluster interaction energy [Ch. 4, §2] | local dimensionless free energy for a uniform system [Ch. 4, §4] | lattice-misorientation twist angle [Ch. 12, §3] | jump frequency of a vacancy [Ch. 14, §5]

$\varpi(T)$ $= (W_{NC}^*/W_{CNT}^*)^{1/3}$

Ω	grand (or Landau) potential [Ch. 2, §2; Ch. 4, §2] ∣ number of configurations [Ch. 5, §5] ∣ Ω_0, in nucleation prefactor, a measure of the volume of the saddle point in configurational space [Ch. 4, §5]

ABBREVIATIONS AND ACRONYMS

Acronym	Meaning	Section where first used
a	amorphous	Chapter 15, Section 2
A	Al_2O_3	Chapter 8, Section 6
AFP	antifreeze protein	Chapter 16, Section 2
at.%	atomic percentage	
B	BaO or B_2O_3	Chapter 8, Section 6
bcc	body-centered cubic	Chapter 4, Section 5
BFZA	bright-field zone-axis	Chapter 9, Section 2
BMG	bulk metallic glass	Chapter 8, Section 2
BSE	bovine spongiform encephalopathy	Chapter 16, Section 6
C	CaO	Chapter 8, Section 6
CA	cellular automaton	Chapter 13, Section 2
CaOx	calcium oxalate	Chapter 16, Section 6
ccp	cubic close-packed	Chapter 4, Section 4
CET	columnar-to-equiaxed transition	Chapter 13, Section 2
CET	complete erasure time	Chapter 14, Section 2
CNT	classical nucleation theory	Chapter 4, Section 1
DC	direct-chill	Chapter 13, Section 3
DF	density-functional	Chapter 4, Section 4
DFA	density-functional approximation	Chapter 7, Section 3
DFT	density-functional theory	Chapter 4, Section 4
DIT	diffuse-interface theory	Chapter 4, Section 3
dpa	displacements per atom	Chapter 14, Section 5
DSC	differential scanning calorimetry	Chapter 7, Section 2
DTA	differential thermal analysis	Chapter 7, Section 2
EML	electromagnetic levitation	Chapter 7, Section 2
ESL	electrostatic levitation	Chapter 7, Section 2
ETM	early transition metal	Chapter 14, Section 2
fcc	face-centered cubic	Chapter 14, Section 5
FE	finite element	Chapter 13, Section 2
FFI	fatal familial insomnia	Chapter 16, Section 6

FIM	field-ion microscopy	Chapter 14, Section 4
FSE	feline spongiform encephalopathy	Chapter 16, Section 6
GLCH	Cahn–Hilliard model with Ginzburg–Landau free energy	Chapter 8, Section 6
GP	Guinier-Preston; Guinier-Preston zones	Chapter 14, Section 4
GSS	Gerstmann-Straussler-Scheinker disease	Chapter 16, Section 6
hcp	hexagonal close-packed; hexagonal close packing	Chapter 6, Section 2.6
HREM	high-resolution electron microscopy	Chapter 9, Section 2
Ht	the protein huntingtin	Chapter 16, Section 6
HV	high-vacuum	Chapter 7, Section 2
i	icosahedral	Chapter 7, Section 2
INA	ice-nucleating agent	Chapter 16, Section 2
ISRO	icosahedral short-range order	Chapter 8, Section 8
JMAK	Johnson–Mehl–Avrami–Kolmogorov	Chapter 8, Section 3
L	left hand	
L	Li_2O	Chapter 8, Section 6
L-H	Lauritzen–Hoffman	Chapter 11, Section 6
LJ	Lennard-Jones	Chapter 10, Section 3
Ln	lanthanide	Chapter 13, Section 5
LSW	Lifshitz–Slyozov–Wagner	Chapter 1, Section 3
LTM	late transition metal	Chapter 14, Section 2
m	metastable	Chapter 1, Section 2
M	MgO	Chapter 8, Section 10
MCS	Monte Carlo steps	Chapter 9, Section 2
MD	molecular dynamics	Chapter 10, Section 2
ML	metastable limit	Chapter 16, Section 5
M_s	martensite start temperature	Chapter 12, Section 4
MWDA	modified-weighted density-functional approximation	Chapter 4, Section 4
N	Na_2O	Chapter 8, Section 6
NASA	National Aeronautics and Space Administration	Chapter 7, Section 2
NMR	nuclear magnetic resonance	Chapter 7, Section 2
PC-RAM	phase-change random access memory	Chapter 14, Section 2

PDFA	perturbative density-functional approximation	Chapter 4, Section 4
PFZ	precipitate-free zone	Chapter 14, Section 4
ppt	parts per thousand	Chapter 13, Section 3
PrP	prion protein	Chapter 16, Section 6
PrP^C	cellular form of prion protein	Chapter 16, Section 6
PrP^{Sc}	scrapie form of prion protein	Chapter 16, Section 6
R	right-hand	
RE	rare earth	Chapter 14, Section 2
rf	radio frequency	Chapter 7, Section 2
S	SiO_2	Chapter 8, Section 6
SAM	self-assembled monolayer	Chapter 16, Section 4
SANS	small-angle neutron scattering	Chapter 9, Section 2
SAXS	small-angle X-ray scattering	Chapter 8, Section 6
SCCT	self-consistent classical nucleation theory	Chapter 7, Section 3
sCJD	sporadic Creutzfeldt–Jacob disease	Chapter 16, Section 6
SDFA	semiempirical density-functional approximation	Chapter 4, Section 4
SEM	scanning electron microscopy	Chapter 8, Section 5
SFT	stacking-fault tetrahedron	Chapter 14, Section 5
SIBM	strain-induced grain-boundary migration	Chapter 12, Section 3 (Figure 11)
SRO	short-range order	Chapter 8, Section 8
SSAR	solid-state amorphization reaction	Chapter 15, Section 2
TD	thermal donor	Chapter 9, Section 3
TEM	transmission electron microscopy	Chapter 7, Section 2
TM	transition metal	Chapter 13, Section 5
TTT	time–temperature–transformation	Chapter 14, Section 3
u	unstable	Chapter 1, Section 2
UHV	ultra-high vacuum	Chapter 7, Section 2
ULSI	ultra-large-scale integration	Chapter 9, Section 3
vCJD	variant Creutzfeldt–Jacob disease	Chapter 16, Section 6
VLSI	very-large-scale integration	Chapter 9, Section 3
WDA	weighted density-functional approximation	Chapter 4, Section 4
wt%	percentage by weight	
XPS	X-ray photoelectron spectroscopy	Chapter 8, Section 6
Z	ZrO_2	Chapter 8, Section 10

2D	two-dimensional	
3D	three-dimensional	
3DAP	three-dimensional atom probe	Chapter 9, Section 1

CHAPTER 1

Introduction

Contents

1. WHAT IS NUCLEATION?

The word *nucleus*, defined by the Oxford English Dictionary as "the central and most important part of an object, movement, or group, forming the basis for its activity and growth" [1], was introduced into English usage early in the eighteenth century, derived from the Latin for *kernel* or *inner part*. In the nineteenth century, it was adopted to describe a small region of a new phase appearing during a phase change such as melting or freezing. Such *phase transitions* (also known, particularly in metallurgy, as *phase transformations*) are ubiquitous in the natural world and are hence important in a very wide range of scientific disciplines, including astrophysics, metallurgy, materials science, electronic engineering, atmospheric physics, mineralogy, chemical engineering, biology, food science, and medicine.

Statistical fluctuations generate nuclei through the transient appearance and disappearance of small regions of a new phase within an original (or "parent") phase. The lifetime of a fluctuation is related to its size. Only when a "critical" size is exceeded is the dissolution probability small enough for the fluctuation to evolve into a macroscopic region of the new phase. The *stochastic* behavior of shrinkage and growth is consistent with the existence of a barrier to the phase transition — the *nucleation barrier*. Familiar examples of nucleation-limited phase transformations are the condensation of vapors, the crystallization of liquids, and precipitation in liquids and solids. Nucleation can also be central to less familiar processes, such as phase separation, magnetic domain formation,

Pergamon Materials Series, Volume 15
ISSN 1470-1804, DOI 10.1016/S1470-1804(09)01501-6

dislocation formation, star and galaxy formation, and even the emergence of matter in the primordial universe. A growing realization, discussed in the later chapters of this book, is that nucleation can be important even in biological processes. In whatever field, the nucleation mechanism can be complex and often proceeds in stages; it may involve coupled phase transitions, coupled kinetic fluxes, and coupled fluctuations in different order parameters that characterize the original and new phases.

Sometimes the term *nucleation* is applied loosely to any appearance of a new phase. In many of these cases, the precise definition of a nucleation process is not satisfied. For example, the new phase may grow by irreversible aggregation, rather than by stochastic formation and dissolution. Some processes can mimic nucleation arbitrarily closely. On holding at an elevated temperature, a dispersion of one solid phase in another may show coarsening in which larger particles of the dispersed phase grow while smaller particles shrink and disappear. In this way, the dispersion evolves to contain fewer larger particles with a smaller total interfacial area between the two phases. This coarsening, often called *Ostwald ripening,* does not involve nucleation, although a critical size exists, above which particles grow and below which they shrink. While this is analogous to a critical nucleus, it is so large that molecular-scale fluctuations are insignificant; growth and shrinkage are by-and-large deterministic. On the other hand, fluctuations might play a role if the particles were sufficiently fine, such as nanocrystallites grown in an amorphous phase.

In other cases, nucleation is essential, but is not the controlling factor in initiating the overall transformation. For nucleation occurring on catalyst surfaces, for example, subsequent growth may be characterized by a competition (likely to be mediated by heat or solute) in which only some nuclei can participate in the overall transformation. Microstructural development is then controlled largely by growth. Nucleation can be observed only within a limited window of thermodynamic and kinetic conditions. It may be, then, that in some cases nucleation plays a hidden but still central role. In such cases, details of the phase transformation mechanisms and of the role of nucleation may be very difficult or impossible to determine.

Since nucleation is the initial step in many phase transformations, the ability to control that process is often of significant practical concern. It is the key, for example, to the production of desired microstructures that are tailored to technological needs and it is fundamental for the survival of some biological organisms (Chapter 16). The pattern shown in Figure 1 is a decorative example of nucleation control, where the potter has controlled conditions to produce a pleasing pattern of crystalline areas in an otherwise amorphous glaze. Each crystalline area grew from a distinct

Fig. 1 A vase coated with a glaze that has been partially crystallized during firing. A distribution of crystal sizes is observed, with the larger crystallites having nucleated first; growth has led to crystal impingement. For further examples, see Ref [2]. (Photo courtesy of Brian Barber.)

initiation event; the pattern is determined by the sequence of events and by the growth of those nuclei to consume the amorphous phase.

The phenomenological evidence for nucleation, the development of theoretical models, measurement methods, the use of the resulting data in evaluating nucleation models, and examples of nucleation control in practical situations are the themes developed in Chapters 2–17 of this book. Before beginning that discussion, however, it is useful to examine briefly the experimental observations and theoretical understanding that form the basis of our knowledge of nucleation processes.

2. HISTORICAL BACKGROUND

2.1 Experimental observations

In 1721, Fahrenheit discovered a tendency for water cooled to below its freezing temperature (a *supercooled* or *undercooled* liquid) to resist the formation of the crystalline phase, ice. In 1724, he reported the results of a systematic set of experiments in which sealed containers of boiled water were set outside on winter evenings when the temperature of the surroundings was lower than the freezing point of the water (thirty-two degrees at standard atmospheric pressure in the units of his new temperature scale, i.e. 32°F) [3]. Surprisingly, he found that the water remained a fluid, even when left outside overnight in an air temperature of 15°F. When small ice particles were introduced into the supercooled water, however, crystallization followed immediately, with the temperature of the ice–water mixture rising to 32°F. Further, he reported that when carrying a glass vial of the supercooled water from his bedchamber to a nearby room, he stumbled on the stairs connecting the rooms, agitating the liquid, and it immediately crystallized.

These observations were reproduced and extended to other liquids by Triewald, Musschenbroek, Brugmanns, and Mairan [4, 5], Lowitz [6], and others; a critical review was written as early as 1775 (see [7]). The amount of supercooling can be considerable. For water, Fahrenheit reported a maximum supercooling of 17°F ($\sim$8 K). In 1820, Kaemtz [8] achieved a supercooling of 19 K and in a series of papers from 1844 to 1847, Regnault reported 32.8 K, a value not equaled or exceeded until almost a century later. Gay-Lussac demonstrated the generality of supercooling and confirmed Fahrenheit's report that mechanical vibration could induce crystallization in supercooled liquids [9, 10].

The degree of supercooling was often variable. Schröder, von Dusch, and Violette recognized that this was due largely to airborne particles and particles residing in the containers [11–13]. Eliminating these particles improved the reproducibility of the observed supercooling [14, 15]. The structure and chemistry of the particles were identified as critical factors for catalyzing crystallization. Lowitz observed that the introduction of small particles of the primary crystallizing phase into supercooled liquids readily initiated crystallization, while the introduction of particles of unrelated phases often had little influence [6]. Ostwald demonstrated that the effectiveness of nucleation was out of proportion with the amount of catalyzing particles, the introduction of less than one part in 10^9 being sufficient to trigger crystallization in aqueous solutions of sodium chlorate [16].

While deep supercooling was readily observed in water and a number of organic liquids, even by the middle of the twentieth century it had not been observed in metallic liquids. This was often taken as evidence that the structures of the metallic liquid and the primary crystallizing phase were very similar. Turnbull convincingly demonstrated that this was not the reason for the absence of supercooling; the problem was that impurity catalysis of nucleation is easy in metallic liquids. If the influence of catalytic sites was minimized, in his case by dispersing the liquid into small droplets so that a significant fraction of the liquid contained no sites (Chapter 7, Section 2.1.1), a significant supercooling could be achieved [17]. Liquid mercury provides a dramatic example of this; it can be supercooled by over 25% of its melting point before crystallization (Figure 2).

An enormous number of studies have now been made. Collectively these demonstrate that supercooling is a common property of all liquids, regardless of the nature of their chemical bonding (i.e., metallic, organic, ionic, etc.). These observations unequivocally demonstrate the existence of a barrier to the formation of the new phase. They also reveal the existence of two types of nucleation, *homogeneous nucleation*, which as we shall see corresponds to spatially and temporally independent fluctuations of regions of the original phase (Chapters 2–5), and a second type, *heterogeneous nucleation*, which is catalyzed at specific sites (Chapter 6).

In the great majority of cases, the study of nucleation focuses on its kinetics, since the critical nuclei themselves are too small to be observable. Recently, however, nucleation has been directly observed on a microscopic scale. The first key study concerned crystallization of the protein apoferritin [18]. The slow timescale for crystallization and the

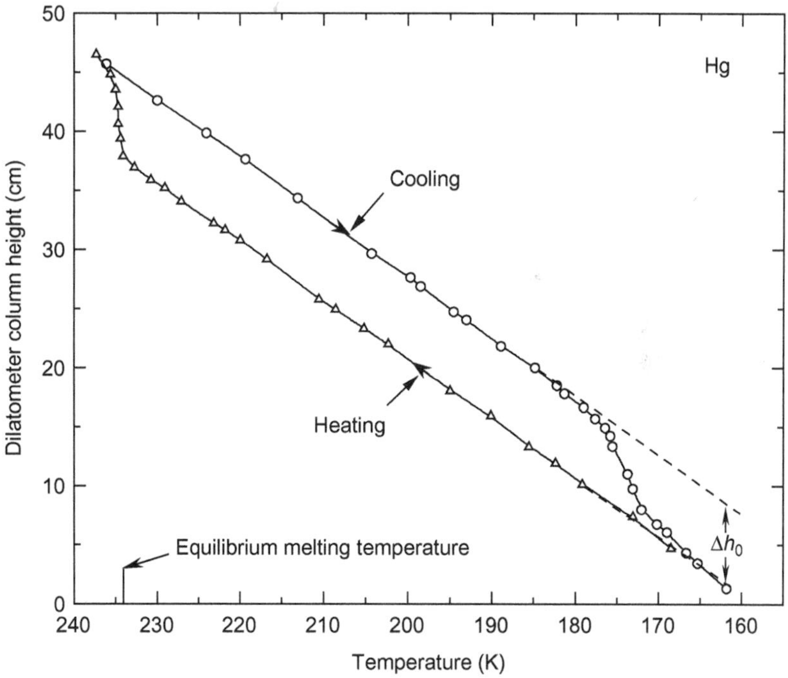

Fig. 2 Dilatometric measurements of an emulsion of liquid mercury droplets in a carrier liquid as a function of temperature demonstrating the large supercooling possible in many liquids; Δh_0 is the change in column height in the dilatometer that corresponds to the increase in density of the mercury due to crystallization (Reprinted with permission from Ref. [17], copyright (1952), American Institute of Physics.)

large size of the molecules allowed their positions in the crystal phase to be determined using atomic force microscopy. Yau and Vekilov found that crystals containing fewer than 20–50 molecules tended to dissolve, while those containing more molecules tended to grow. Similar behavior was also observed in the nucleation of crystalline phases in concentrated colloidal suspensions [19], using fast laser-scanning confocal microscopy to obtain a three-dimensional measurement of the positions of the colloidal particles. The samples were "melted" by agitation. To measure homogeneous nucleation from the "melt," attention was focused on regions of the sample that were 8–15 atomic diameters away from the container wall and heterogeneous nucleation was minimized by coating the walls with smaller particles before adding the colloidal system. As shown in Figure 3(a, b), crystal-like regions (shown by the dark spheres) formed and dissolved in the liquid phase (light spheres). Only when they contained more than about 20 particles did they grow into

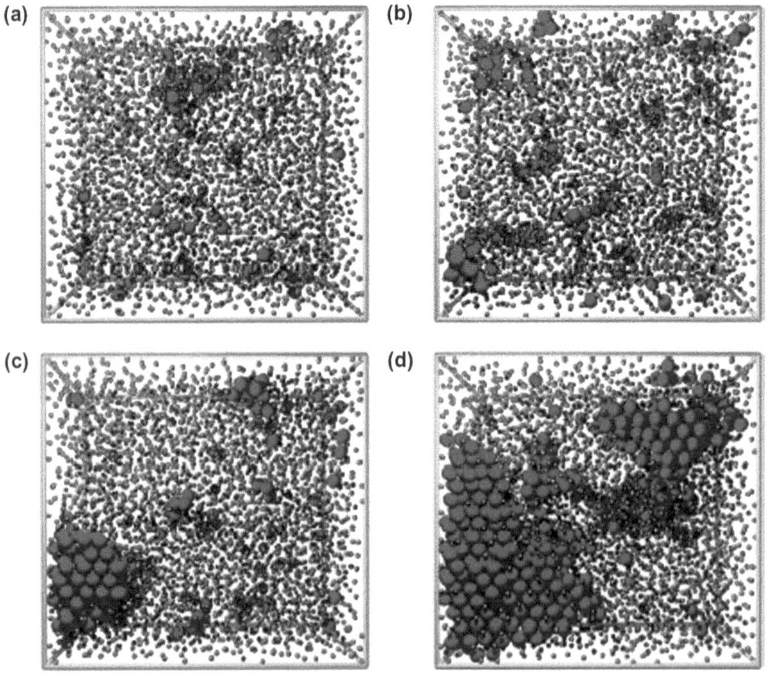

Fig. 3 Scanning confocal microscopy observations of particle positions during the crystallization of a colloid of polymethyl methacrylate spheres that were stabilized with poly-12-hydroxystearic acid. Particles in a crystal-like configuration are dark, while those in the liquid-like phase are light. Time after shear melting: (a) 20 min; (b) 43 min; (c) 66 min; (d) 89 min. (From Ref. [19], reprinted with permission from AAAS.)

large crystallites (Figure 3c, d). These studies show directly that small regions of the new phase do not uniformly grow. Instead, they are continually forming, disappearing, and reforming, continuing to grow only when they exceed a critical size; such is the stochastic nature of true nucleation.

Despite significant improvements in experimental techniques, it remains difficult to study nucleation processes outside a limited range of physical conditions. Within the past 20 to 30 years, it has become increasingly common to study nucleation processes in idealized systems using computer simulation methods (Chapter 10). Due to limited computer resources, the earliest simulations did little more than indicate that under favorable conditions "nucleation occurs." This has now changed, and computer simulations are providing new information on the kinetic evolution of clusters, and the volume and interfacial structures of nuclei. Even with the enormous advances in computing power, however, studies are restricted to numbers of atoms or molecules that are tiny compared to real systems, and to timescales much shorter than those for experimental measurements. That nucleation is "observed" in a computer model so restricted in size and time indicates that the driving force for nucleation is often much higher than in real systems.

2.2 Theoretical development

A system is generally said to be in equilibrium (*stable* or *metastable*, see Figure 4 and the following discussion) when its thermodynamic properties do not change with time. This corresponds to the condition that the first derivative of the free energy with respect to the appropriate variables be zero. A phase-space illustration of this is given in Figure 4, showing the free energy as a function of generalized coordinates that describe the transformation from state A of a system to state B. The sign of the coordinates is taken as positive if they follow the phase-space direction from the original to the new state. In this schematic, A is a *metastable equilibrium* state for which the free energy has a local minimum. For such a state, the system is stable to small fluctuations in the coordinates, but will eventually evolve (possibly after an extremely long time) to a more stable state. B is a *stable equilibrium* state, corresponding to a global minimum in the free energy. As the system makes its transition from A to B, the free energy increases to a point where it is a local maximum. In that *unstable equilibrium* state, the system is unstable to fluctuations in the coordinates; hence, the state is short-lived. For metastable and stable equilibrium states the second derivative of the free energy with respect to all coordinates must be greater than zero; at least one of the derivatives is less than zero for unstable

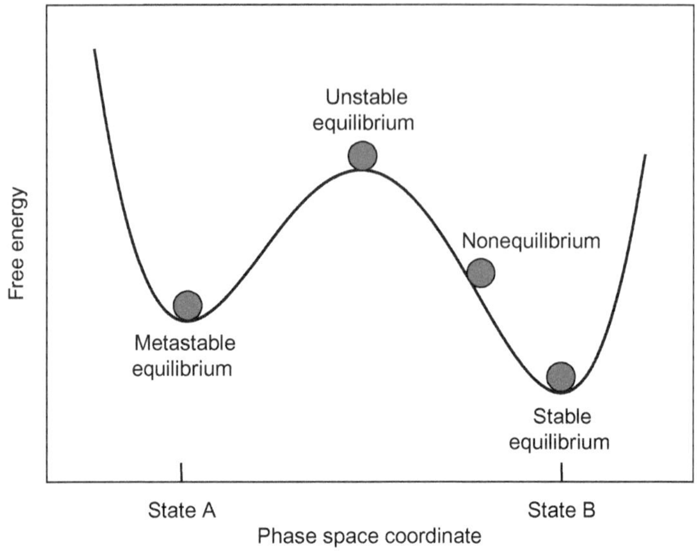

Fig. 4 Schematic illustration of the evolution of the system from a metastable equilibrium state (State A), through a state of unstable equilibrium and finally to a state of stable equilibrium (State B).

equilibrium states. During the transition between equilibrium states, the system evolves through a series of *nonequilibrium* or *unstable* states in which the properties of the system are continuously changing.

The schematic diagram in Figure 4 maps onto the experimental observations for nucleation discussed in Section 2.1 of this chapter, in particular pointing to the existence of a barrier to the transformation as the system passes from A to B. The physical origin of this barrier is at the core of nucleation theory and is discussed in detail in following chapters. If the free energy continuously decreases with an increase in the coordinates (as is the case for the nonequilibrium state shown in Figure 4), there is no barrier for the continuing transformation. This could, for example, correspond to irreversible aggregation, distinct from nucleation, even though both processes generate new growth centers.

The range of thermodynamic parameters for phase coexistence is typically shown in an equilibrium phase diagram. An example for a binary system (i.e. containing two chemical components) is shown in Figure 5a, illustrating the equilibrium between two isostructural phases of different chemical compositions as a function of temperature. At high temperatures above the coexistence curve, shown here by a solid line, the two chemical species are completely miscible and only one phase exists. Below the coexistence curve, two phases of different chemical composition are in equilibrium and the single phase is

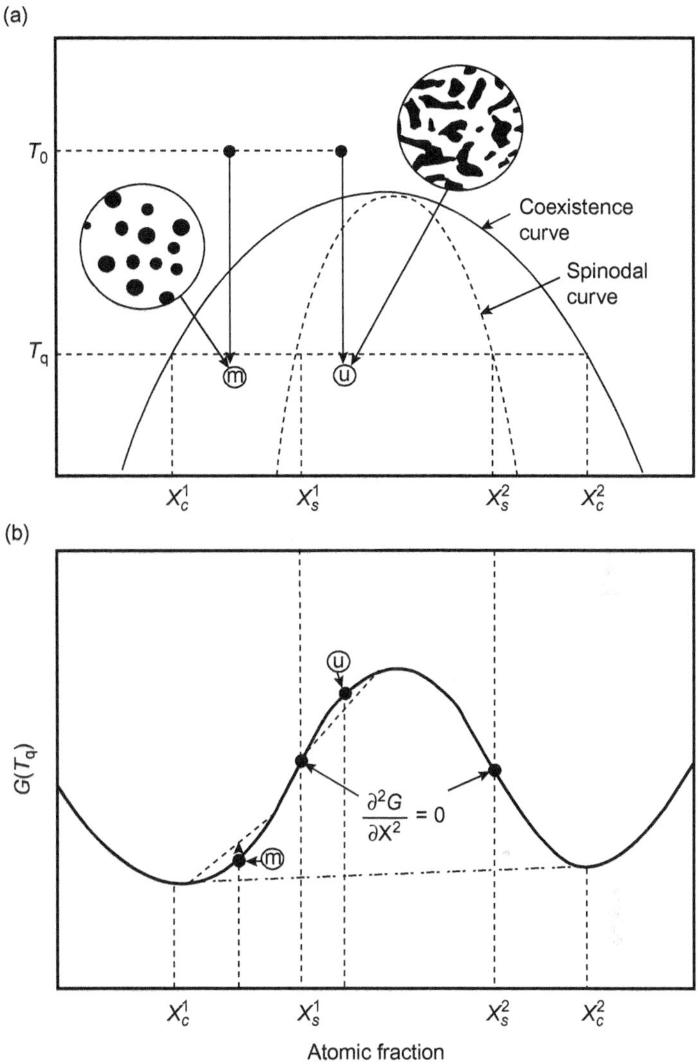

Fig. 5 (a) A phase diagram for a two-component system showing phase separation with schematic depictions of the morphologies obtained by quenching into the metastable (m) or unstable (u) regions. The coexistence and the classical chemical spinodal curves are indicated. (b) A diagram showing the relation of the boundaries of the spinodal and coexistence regions to the shape of the Gibbs free-energy curve. (Reprinted from Ref. [22], copyright (1991), with permission from Elsevier.)

metastable (indicated by "m" in Figure 5). Phase separation within the high-temperature phase is therefore expected upon cooling. However, above the spinodal curve (the dashed curve in Figure 5a), the free energy initially increases with spontaneous composition fluctuations in the single phase (Figure 5b) giving rise to an energetic barrier that stabilizes the system against phase separation ($d^2G/dX^2 > 0$, where G is the Gibbs free energy and X is the mole fraction of one component specifying the chemical composition). At the spinodal curve, the second derivative of the free energy changes sign and the system becomes unstable to concentration fluctuations, however small ($d^2G/dX^2 < 0$). The spinodal curve, then, marks the boundary of metastability; the system is in unstable equilibrium inside the region of the phase diagram bounded by the spinodal curve. (For a more extensive discussion of states of equilibrium and phase diagrams, see [20, 21].) It is important to note that even when the system contains the equilibrium fractions of the two phases, stable equilibrium has generally not been reached due to the small domain sizes of the phases. The free-energy contribution from the large total area of interface between the regions of the two phases increases the overall free energy, becoming a driving free energy for grain coarsening (discussed in more detail in Chapter 4).

Gibbs recognized significant differences between transformations in the metastable and unstable regions [23–25]. If the system represented in Figure 5 was quenched from a temperature T_0 above the coexistence curve to a temperature T_q in the metastable region, phase separation would proceed by nucleation and growth, characterized by large-amplitude chemical fluctuations over a small spatial extent, generally leading to the droplet morphology indicated schematically. If it was quenched to the same temperature in the unstable, or spinodal, region (u), the phase transition would occur by a spinodal mechanism that is characterized by long-range fluctuations of initially infinitesimal amplitude and would generally produce an interconnected structure. While such differences in phase morphology seem clear, they are insufficient for an unambiguous identification of the phase transformation mechanism. For example, an apparently interconnected structure could also result from the superposition of a high density of separately nucleated grains [26, 27]. Small-angle scattering experiments [28, 29] and three-dimensional atom probe (also called atom probe tomography, APT) studies (discussed in Chapters 9 and 14) can better identify spinodal transformations.

While applied here to chemical phase separation, Gibbs' ideas are readily extended to describe phase transitions that are characterized by other order parameters. Although the thermodynamic notion of a sharp division between nucleation and spinodal transformations is useful, it is not precisely valid since it fails to take account of the kinetic aspects of the processes. What is determined, then, is an operational, rather than an intrinsic, limit on metastability [30]. This is supported by field-theory calculations showing a gradual, rather than an abrupt, change in the dynamical behavior of a quenched system in the vicinity of the spinodal curve defined as in Figure 5b [31]. Further, Gibbs was primarily concerned with processes occurring near equilibrium where the probability of a significant number of fluctuations leading to the stable phase is infinitesimal, corresponding to a large barrier to the phase transition. When the system is quenched deeply into the metastable region, this barrier decreases until it becomes of the same order as the thermal energy (k_BT, where k_B is the Boltzmann constant and T is the temperature in absolute units), defining a limit on metastability for which fluctuations leading to the new phase become comparable in number to the equilibrium thermodynamic fluctuations in the original phase.

The significance of Gibbs' ideas on nucleation was largely ignored until 1926, when Volmer and Weber recognized the importance of kinetics and constructed the first complete theory of nucleation [32, 33]. Gibbs' formulation of the reversible work considered a cluster of radius r of the new phase containing n atoms or molecules. A work, $W(n)$, was used to calculate an equilibrium cluster size distribution for clusters smaller than a "critical cluster," characterized by a critical radius, r^*, and corresponding to a maximum in the required work of formation, W^*. The nucleation rate, I, was taken to be proportional to the Boltzmann-weighted probability of having a critical fluctuation,

$$I = A^* \exp\left(-\frac{W^*}{k_BT}\right) \tag{1}$$

The prefactor, A^*, contains kinetic factors that describe the rate of single atom or molecule addition to the cluster. Farkas [34] (working from a suggestion by Szilard) formulated a more detailed kinetic model for cluster evolution that became the basis for later treatments. Becker and Döring argued that an equilibrium distribution is inappropriate, suggesting instead a steady-state distribution, and obtaining an expression for the steady-state nucleation rate [35]. These approaches also yield expressions for the nucleation rate that have the form of Eq. (1), although the precise values for the kinetic prefactor are different. All these earlier treatments deal with the vapor-to-liquid phase transition, where the kinetic coefficients can be defined in terms of the molecular velocity distribution. Turnbull and Fisher extended this formalism to include the case of crystal nucleation from the liquid by redefining these kinetic coefficients as functions of the diffusion rate in the liquid phase [36]. This approach was readily extended to include nucleation processes in solids.

Collectively, these treatments comprise what is commonly known as the "classical theory of nucleation," a robust theory that is easily and widely used. It provides a ready explanation for steady-state and time-dependent nucleation processes [37] and the kinetic model can even be quantitatively predictive [38] (Chapters 2 and 3). However, there are significant problems with the classical theory as well, motivating the development of new formulations of both the thermodynamic and kinetic models (Chapter 4). The theoretical development reviewed in this section has focused on homogeneous nucleation; the important topic of heterogeneous nucleation is covered in Chapter 6. All the discussion so far has considered nucleation of phase changes in which the relevant species for kinetic analysis are atoms or molecules. Yet the concept of nucleation can also be applied in other cases. Examples are the nucleation of dislocations and the mediation of phase changes by movement of dislocations; these are analyzed in Chapter 12.

3. THE AIM AND PLAN OF THIS BOOK

The central aim of this book is to enable a reader, faced with a phenomenon in which nucleation appears to play a role, to determine whether nucleation is indeed important and to develop a quantitative and predictive description of the nucleation behavior. Therefore, we examine prominent nucleation models, tests of these models by comparing their predictions with experimental data, and some practical uses that emerge from a deeper understanding of nucleation processes. The examples have been chosen to be most useful to the targeted audience of experimental scientists working in the areas of soft and hard condensed-matter physics, materials science and related areas of biology and medicine.

The book is divided into three parts. In Part I, the classical theories of steady-state and time-dependent nucleation are developed. The classical theory is appropriate for cases in which the nucleation kinetics is governed by processes that occur at the interface between the original and new phases. Strictly, it is inapplicable for cases where other processes are governing, such as the long-range diffusion of solute that is important in solid-state precipitation. Ways in which the classical theory might be extended to take account of coupling between interfacial and long-range diffusion kinetic fluxes are discussed. These retain the flavor of the classical theory, keeping Gibbs' assumption of a sharp dividing surface between the original and new phase and a kinetic model based on the attachment/detachment of single atoms or molecules at the cluster surface. Both points are taken up in Part I, where phenomenological and density functional approaches for the general case of diffuse interfaces and other kinetic models are discussed. Often it is assumed that the nucleation, growth, and coarsening processes are well separated. While it has long been known that this assumption is invalid for many condensation processes, it is now becoming clear that it may also be invalid for precipitation processes when the nucleation rates are extremely high; examples include the nucleation and growth of nanocrystal phases in some metallic glasses (Chapter 14) and some solid-state precipitation processes (Chapter 9).

Different communities have adopted varied views of what constitutes a valid theory of nucleation. For example, physicists and chemists often favor density-functional theories, which are readily expressed in terms of differences in the order parameters of the original and new phases. Most often, they are concerned with steady-state nucleation processes. Materials scientists, on the other hand, overwhelmingly adopt the classical theory of nucleation, defined in terms of macroscopic, measurable quantities, and often applied to non-steady-state processes. There are efforts to adapt continuum models to describe nucleation fluxes. One approach is based on an extension of the Lifshitz–Slyozov–Wagner (LSW) theory, originally developed for diffusion-limited coarsening [39, 40]. Clearly, with such a large number of approaches, it is important to compare theoretical predictions with experimental data. This is the purpose of Part II, where nucleation processes in liquids, glasses, and crystals are examined in light of some of the theories discussed in Part I. Computer simulations are increasingly providing valuable new "experimental" insights into nucleation processes and are discussed in Part II as well. Space limitations and the focus of this book do not allow a discussion of all possible systems. A prominent omission, for example, is nucleation in the vapor phase. This is certainly an important area, but it is omitted in favor of more examples in condensed phases. Nevertheless, nucleation in the vapor phase is mentioned from time to time, notably in Chapters 2 and 4, as it is an important basis for analyses of the equilibrium cluster population in other cases.

A discussion of some practical uses of nucleation is provided in Part III; a central theme is that of nucleation control. Topics include a discussion of the introduction of particles that catalyze nucleation (heterogeneous nucleation) for grain refinement in the solidification of liquids of metallurgical importance and the tailoring of annealing treatments to produce conventional precipitation-hardened alloys, as well as micro- and nano-structured composites in ceramic and metallic glasses. Nucleation on defects, such as dislocations, is also discussed. Perhaps unusual is a discussion of the importance of nucleation in the food and drink industry, as well as the emerging view of the importance of nucleation processes in biology and medicine. Again, some topics have been omitted. Nucleation in thin-film deposition, for example, is not covered since there already exist numerous books and review articles on this topic, and since thin-film formation from the vapor phase is often dominated by growth rather than nucleation. Nucleation in the atmosphere is also an important topic of increasing environmental concern. It is not discussed here, however, due to the omission of the more fundamental discussion on nucleation from the vapor phase, which is needed as a basis for a discussion of nucleation control in the vapor.

In the following chapters, key phenomenological evidence for a stochastic nucleation process is presented, the theories developed to describe this process are discussed, and predictions from the theories are compared with nucleation data for condensed systems. A better understanding of nucleation processes has practical consequences, leading to improved microstructural control in materials and a deeper appreciation of biological processes that may also be of medical interest. The study of nucleation is an old subject, but recent discoveries have demonstrated that it is still exciting, with much left to learn.

REFERENCES

[1] *Oxford Dictionary of English*, Oxford University Press, Oxford (2005).
[2] D. Creber, *Crystalline Glazes*, A & C Black Ltd, London (1997), p. 96.
[3] D.G. Fahrenheit, Experimenta et observationes de congelatione aquae in vacuo factae, Phil. Trans. Roy. Soc. **39** (1724) 78–89.
[4] G.E. Fischer, Gehlers phys. Wörterbuch **1** (1789) 678.
[5] G.E. Fischer, Geschichte der Physik **5** (1804) 279.
[6] J.T. Lowitz, Aufsatz über das Krystallisiren der Salze, Crells Chemische Annalen **1** (1795) 6.
[7] C.A. Angell, Supercooled water, in: *Water, A Comprehensive Treatise, vol. 7: Water and Aqueous Solutions at Subzero Temperatures*, Ed. F. Franks, Plenum Press, New York (1982), pp. 1–81.
[8] W. Kaemtz, Traite de Meteorologie **1** (1820) 290.
[9] J.L. Gay-Lussac, De l'influence de la pression de l'air sur la cristallisation des sels, Ann. Chim. **87** (1813) 225–236.
[10] J.L. Gay-Lussac, Premier mémoire sur la dissolubilité des sels dans l'eau, Ann. Chim. Phys. **11** (1819) 296–315.
[11] S.A. von Dusch, Liebigs Annalen **89** (1853) 232.
[12] S.A. von Dusch, Liebigs Annalen **109** (1859) 35.
[13] C. Violette, Recherches sur la cause de la cristallisation des solutions salines sursaturées, Comptes Rend. **60** (1865) 831–833.
[14] D. Gernez, Sur la cristallisation des dissolutions salines sursaturées et sur la présence normale du sulfate de soude dans l'air, Comptes Rend. **60** (1865) 833–837.
[15] D. Gernez, Sur les causes d'erreur que présente l'étude des dissolutions sursaturées, Comptes Rend. **61** (1865) 71–73.
[16] W. Ostwald, Studien über die Bildung und Umwandlung fester Körper: 1 Abhandlung Übersättigung und Überkaltung, Z. Phys. Chem. **22** (1897) 289–330.
[17] D. Turnbull, Kinetics of solidification of supercooled liquid mercury droplets, J. Chem. Phys. **20** (1952) 411–424.
[18] S.-T. Yau, P.G. Vekilov, Quasi-planar nucleus structure in apoferritin crystallization, Nature **406** (2000) 494–497.
[19] U. Gasser, E.R. Weeks, A. Schofield, P.N. Pusey, D.A. Weitz, Real-space imaging of nucleation and growth in colloidal crystallization, Science **292** (2003) 258–262.

[20] M. Hillert, *Phase Equilibria, Phase Diagrams and Phase Transformations — Their Thermodynamic Basis*, Cambridge University Press, Cambridge (1998).
[21] R.A. Swalin, *Thermodynamics of Solids*, John Wiley & Sons, New York (1972).
[22] K.F. Kelton, Crystal nucleation in liquids and glasses, in: *Solid State Physics*, Eds. H. Ehrenreich, D. Turnbull, Academic Press, Boston (1991), pp. 75–178.
[23] J.W. Gibbs, On the equilibrium of heterogeneous substances, Trans. Connect. Acad. **3** (1878) 108–248.
[24] J.W. Gibbs, On the equilibrium of heterogeneous substances, Trans. Connect. Acad. **3** (1878) 343–524.
[25] J.W. Gibbs, *The Collected Works of J. Willard Gibbs*, Vol. I, II, Longmans, Green and Co., New York (1928).
[26] J.E. Hilliard, Spinodal decomposition, in: *Phase Transformations*, Ed. H.I. Aaronson, American Society for Metals, Metals Park, OH (1970), pp. 497–560.
[27] H. Herman, R.K. MacCrone, Comments on 'Separation of phases by spinodal decomposition in the systems Al_2O_3–Cr_2O_3 and Al_2O_3–Cr_2O_3–Fe_2O_3', J. Am. Ceram. Soc. **55** (1972) 50.
[28] V. Gerold, G. Kostorz, Small-angle scattering applications to materials science, J. Appl. Crystallogr. **11** (1978) 376–404.
[29] K. Hono, K.-I. Hirano, Early stages of decomposition of alloys (spinodal or nucleation), Phase Transitions **10** (1987) 223–255.
[30] J.S. Langer, Kinetics of metastable states, in: *Lecture Notes in Physics, vol. 132: Systems Far from Equilibrium*, Ed. L. Garrido, Springer-Verlag, Berlin (1980), pp. 12–47.
[31] J.D. Gunton, M. San Miguel, P.S. Sahni, The dynamics of first-order phase transitions, in: *Phase Transitions and Critical Phenomena* Eds. C. Domb, J.L. Lebowitz, vol. 8, Academic Press, New York (1983), pp. 267–482.
[32] M. Volmer, A. Weber, Nuclei formation in supersaturated states (transl.), Z. Phys. Chem. **119** (1926) 227–301.
[33] M. Volmer, *Kinetik der Phasenbildung*, Vol. 122, Steinkopff, Dresden (1939).
[34] L. Farkas, Keimbildungsgeschwindigkeit in übersättigten Dämpfen, Z. Phys. Chem. **125** (1927) 236–242.
[35] R. Becker, W. Döring, Kinetic treatment of grain-formation in super-saturated vapours, Ann. Physik **24** (1935) 719–752.
[36] D. Turnbull, J.C. Fisher, Rate of nucleation in condensed systems, J. Chem. Phys. **17** (1949) 71–73.
[37] K.F. Kelton, A.L. Greer, C.V. Thompson, Transient nucleation in condensed systems, J. Chem. Phys. **79** (1983) 6261–6276.
[38] K.F. Kelton, A.L. Greer, Test of classical nucleation theory in a condensed system, Phys. Rev. **B38** (1988) 10089–10092.
[39] I.M. Lifschitz, V.V. Slyozov, The kinetics of precipitation from super-saturated solid solutions, J. Phys. Chem. Solids **19** (1961) 35–50.
[40] C. Wagner, Theory of the ageing of precipitates by redissolution (Ostwald maturing), Z. Elektrochemie **65** (1961) 581–591.

PART I: Theory

CHAPTER 2

The Classical Theory

Contents

As discussed in Chapter 1, while it is often difficult to initiate first-order phase transitions, they generally proceed quickly once started. This barrier to the formation of the new phase, the nucleation barrier, is a central concept in all theories of nucleation. In this chapter and in Chapter 3, we present an overview of the most commonly used theory, the classical theory of nucleation.

1. THE NUCLEATION BARRIER

Phase transformations in the region of metastability (Chapter 1, Figure 3) are initiated within the original phase by the nucleation of small regions of

Pergamon Materials Series, Volume 15
ISSN 1470-1804, DOI 10.1016/S1470-1804(09)01502-8

the new phase, which then grow to macroscopic dimensions. Identification of an order parameter (or parameters) that best reflect the differences between the original and new phases is the first step in the construction of a theory of such phase transformations. In most cases, the order parameter is defined in terms of the densities, atomic structures, or chemical compositions of the original and new phases. Nucleation is characterized by large amplitude fluctuations of this order parameter. These fluctuations are localized and are stochastic[1] in both space and time. The probability that a fluctuation occurs is governed by thermodynamic conditions, specifically the minimum work required to create the fluctuation.

Generally the "fluctuation" is taken to be a cluster of a few atoms (or molecules) in the configuration of the new phase. As will be demonstrated in Section 2 of this chapter, the nucleation barrier arises from the energy penalty for creating an interface between this cluster and the original phase. If the phase transition is thermodynamically favored, sufficiently large clusters of the new phase must have a lower free energy than the same atoms retaining the configuration of the original phase. However, the atoms in the region of the interface between the original and new phases are in a higher energy state than they would have in the two macroscopic phases. For small clusters of the new phase, most of the atoms reside in the interfacial region, and the creation of such clusters therefore requires work. For a critical cluster, having radius r^* and containing n^* atoms, the work of cluster formation has a maximum W^*, which constitutes the barrier to nucleation. As clusters grow beyond the critical size, the fraction of atoms in the interfacial region decreases, as does the work of cluster formation. Clusters larger than r^* (or n^*) are therefore favored to grow to complete the phase transformation. The net number of clusters passing this critical size per unit time is the nucleation rate. These points are illustrated in Figure 1.

It should be pointed out that fluctuations are, of course, not limited to the metastable region; they also occur in systems in equilibrium. However, only under metastable conditions is there an extended range of stability for fluctuations combined with a characteristic energy barrier to give the distinctive phenomena of nucleation. The rate at which stable clusters of the new phase appear, the nucleation rate, is intimately related to the growth of these fluctuations in the metastable region.

The mathematical expression for the work of forming a cluster of n atoms, $W(n)$, is fundamental to the development of analytical expressions for the nucleation rate. It is derived in Section 2, according to the classical theory of nucleation. In following chapters, the fundamental assumptions of that theory are examined. Remarkably, as will become evident in

[1] Viewed most simply, a stochastic function depends explicitly on time, t, and position, $\mathbf{r}$, and on random variables, i.e. $f(t, \mathbf{r}, \xi_i, \xi_j)$. The random variables, ξ_i and ξ_j here, have a range of possible values, each with a defined probability that introduces a statistical uncertainty in $f(t, \mathbf{r})$.

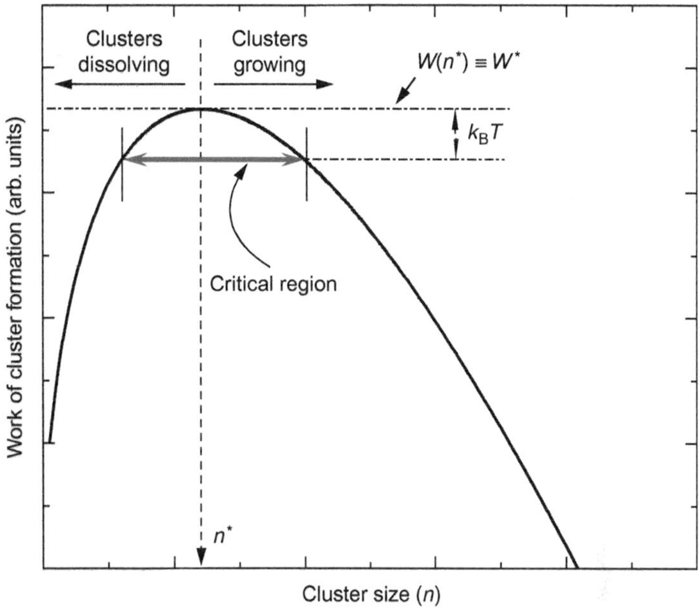

Fig. 1 The reversible work of formation of a cluster of n atoms, $W(n)$, as a function of n. The critical size, n^*, the critical work of formation, $W(n^*) = W^*$, and the critical region (within which $W(n^*) - W(n) \leq k_B T$) are indicated. Below n^* clusters are on average dissolving; above n^* they are on average growing.

Part II of this book, the classical theory is relatively robust. It can be successfully applied to a wide range of nucleation phenomena, and its simplicity and flexibility make it attractive.

We begin our discussion of the classical theory by focusing on thermodynamic issues, calculating the amount of work required to form small clusters of a new phase in the original phase. A kinetic model for the growth of these small clusters is essential for a quantitative understanding of nucleation. The classical model is introduced and used to calculate a nucleation rate, taken to be independent of time for stationary conditions of supersaturation. However, stationary conditions often do not apply. A growing number of experiments in condensed systems indicate that the nucleation rate can dip below, or even rise above, the steady-state value with time. A discussion of this *time-dependent* nucleation is the focus of Chapter 3.

2. THERMODYNAMICS OF CLUSTER FORMATION

If the free energy of a new phase is lower than that of the existing phase, a phase transition is possible, though the transformation might take some time if, for example, the kinetics is slow or the system is close to equilibrium. The thermodynamic driving force, which plays a central role in determining the nucleation of the new phase, is the difference in free energies of the original and new phases. For homogeneous nucleation, which occurs randomly in space and time, small clusters of the new phase arise spontaneously within the original metastable phase. These clusters are assumed to be assemblies of the fundamental units of the original phase (single atoms or molecules in most instances). The nucleation rate is proportional to the probability of formation of these small clusters, described using the theory of thermodynamic fluctuations as first proposed by Gibbs [1].

From statistical arguments, the entropy, S, is specified by the number of available states in the thermodynamic system, w (see, for example, Ref. [2]). A fluctuation in a small volume will be accompanied by a change in the entropy (ΔS) from the bulk (or average) value ($\overline{S}$), which reflects the change in the number of available states from that characteristic of $\overline{S}$ (i.e. from w_0 to w):

$$\Delta S = S - \overline{S} = k_B \ln \frac{w}{w_0}, \tag{1}$$

where k_B is the Boltzmann constant. When the homogeneous medium is equilibrated quickly (in a time shorter than that required for the formation of the fluctuation), the probability for the fluctuation to occur, P, is

$$P \propto \exp\left(\frac{\Delta S}{k_B}\right), \tag{2}$$

which can be expressed in terms of the minimum work required for the formation of the fluctuation, W_{min} [2]:

$$P \propto \exp\left(-\frac{W_{min}}{k_B T}\right), \tag{3}$$

where T is the temperature in absolute units.

Within the classical theory of nucleation, the relevant fluctuation is that which produces a cluster or droplet in unstable equilibrium with the original phase. To calculate the corresponding W_{min} required to produce this critical cluster, it is convenient to first calculate the energy change, U, corresponding to cluster formation. This is most easily illustrated for the nucleation of a liquid droplet in a gas phase.

Consider a droplet of the new phase forming spontaneously within the original phase by a reversible process (Figure 2). For simplicity, assume a single-component system in a strain-free environment held at constant temperature during the transition. Let the system, composed of the original and new phases, be enclosed by a frictionless piston/cylinder

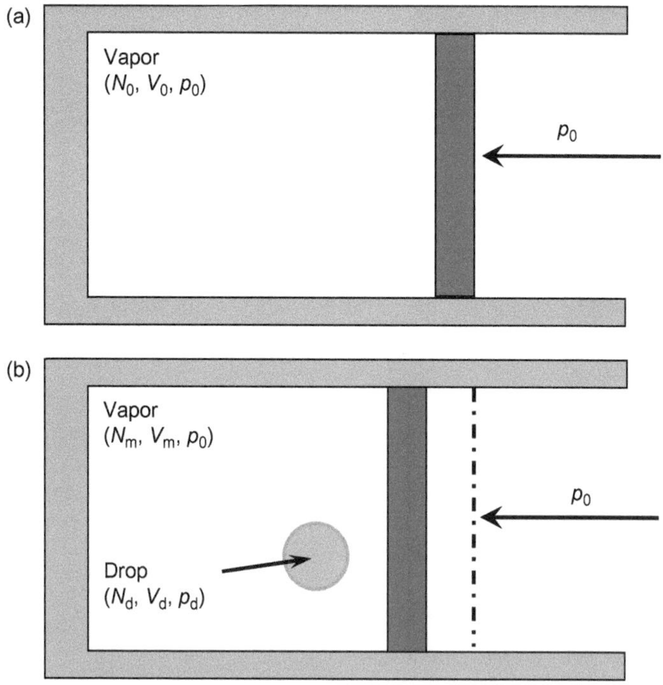

Fig. 2 Schematic illustration of the nucleation of a droplet in a supersaturated vapor held at constant pressure: (a) supersaturated vapor in a frictionless piston/cylinder with impermeable insulating walls to maintain an adiabatic system; (b) formation of drop of different density, with a corresponding decrease in the number of moles and volume in the supersaturated vapor.

assembly consisting of an impermeable adiabatic wall that allows work to be exchanged with the surroundings. Assume that the external pressure remains constant at a value, p_0. The pressure of the original phase with and without droplet formation is then also p_0. Let p_d be the pressure inside the droplet. Consider a volume, V_0, of the original phase and a final volume V' of the phase mixture.

With increasing distance from the center of the droplet, the order parameter changes from its value in the new phase to that in the original phase. For the case considered here, the density is a natural order parameter, since it differs significantly between the liquid and vapor phases. A gradual change over some distance, the interfacial region, is expected (Figure 3), making the choice of the dividing interface between the two phases ambiguous. In a one-component system, however, it is possible, and natural, to choose the interface so that the number of excess surface atoms, N_s, is zero (illustrated in Figure 3).

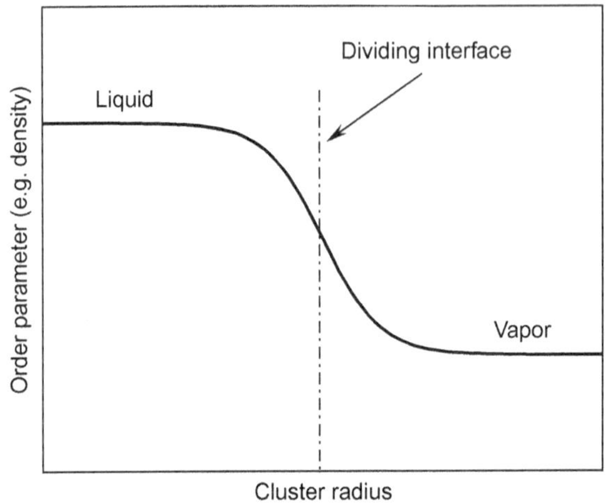

Fig. 3 Schematic illustration of the change in order parameter (here the density) as a function of distance from the center of the cluster growing into the original phase. The Gibbs dividing interface is indicated.

With these assumptions, the change in energy on forming the droplet is

$$\Delta U = U(S', V', N') - U(S_0, V_0, N_0) = \left(\left(-p_0 V_m + TS_m + \mu_m(p_0, T)N_m\right) + \left(-p_d V_d + TS_d + \mu_d(p_d, T)N_d + \sigma A\right)\right) - \left(-p_0 V_0 + TS_0 + \mu_m(p_0, T)N_0\right). \quad (4)$$

Here S', V', N' are the entropy and volume of the system, and number of molecules in the system after droplet formation and S_0, V_0, N_0 are the values before. The subscripts "m" and "d" refer to the droplet and medium (original phase) (see Figure 2), μ is the chemical potential and σA is the work required to create the interface between the original and new phases; A is the area of the interface between the droplet and vapor and σ is the interfacial free energy. Assuming that the number of atoms is conserved, $N_0 = N' = N_m + N_d$, and noting that the final volume, $V' = V_m + V_d$,

$$\Delta U = (-p_d + p_0)V_d - p_0 \Delta V + T\Delta S + \sigma A + N_d\left(\mu_d(p_d, T) - \mu_m(p_0, T)\right), \quad (5)$$

where $\Delta V = V' - V_0$. Using this value for the change in energy, the change in Gibbs free energy can be constructed,

$$\Delta G = \Delta U + p_0 \Delta V - T\Delta S = (-p_d + p_0)V_d + N_d\left(\mu_d(p_d, T) - \mu_m(p_0, T)\right) + \sigma A. \quad (6)$$

It has been assumed that droplet formation occurs reversibly with no heat exchange through the piston walls, giving $S_0 = S_m + S_d$. Note that the chemical potential for the droplet is evaluated at the internal droplet pressure, which is larger than the outside pressure by an amount given by the Laplace equation [3]

$$p_d - p_0 = \frac{2\sigma}{r}, \tag{7}$$

where r is the radius of the droplet. Combining the Gibbs–Duhem equation for a single-component system,

$$S\mathrm{d}T - V\mathrm{d}p + N\mathrm{d}\mu = 0, \tag{8}$$

with Eq. (7), it readily follows that at constant temperature,

$$N_d\left(\mu_d(p_d(r), T) - \mu_d(p_0, T)\right) = V_d(p_d - p_0) = \frac{2\sigma V_d}{r}, \tag{9}$$

giving

$$\Delta G = N_d\left(\mu_d(p_0, T) - \mu_m(p_0, T)\right) + \sigma A = N_d\Delta\mu + \sigma A, \tag{10}$$

where $\Delta\mu$ is the change in chemical potential for a single molecule on moving from the original phase to the new phase. For a reversible fluctuation at constant pressure leading to the appearance of a cluster containing n molecules, the change in Gibbs free energy, equal to the minimum work of cluster formation ($W_{min} \equiv W(n)$), is then

$$W(n) = n\Delta\mu + \sigma A. \tag{11}$$

The first term in Eq. (11) is a volume term, reflecting the "strength" of the thermodynamic driving free energy; the second term is the energy penalty for the creation of an interface. If the phase transition is favored, $\Delta\mu$ is negative; the interfacial free energy is always positive. Obviously, if the volume of the cluster scales with n, the interfacial area scales with $n^{2/3}$. The surface term then is dominant in the small cluster limit, since it approaches zero less rapidly than the volume term. For a sufficiently large cluster size, however, the volume term dominates, driving the work of cluster formation to negative values. This relation, then, predicts the variation of $W(n)$ with cluster size shown in Figure 1. It demonstrates that the barrier to the phase transformation is the creation of the interface between the cluster of the new phase and the original phase.

For transitions where V and N are constant, but the pressure is allowed to vary, the appropriate thermodynamic potential is not the Gibbs free energy, but the Helmholtz free energy, F. If the number is also allowed to vary, the Landau potential, Ω, must be used. In all cases, the equilibrium cluster size distribution (number of clusters per mole that contain n molecules), $N^{eq}(n)$, follows readily from Eq. (3):

$$N^{eq}(n) = N_A \exp\left(-\frac{W(n)}{k_B T}\right), \tag{12}$$

where N_A is the Avogadro number and $W(n)$ is computed using the appropriate thermodynamic potential.

Because the equilibrium number density drops off rapidly with increasing n, for all practical purposes, the equilibrium number of single molecules should equal the initial supply, N_A. This is not, however, predicted by Eq. (12) if the value for $W(1)$ is computed from Eq. (11). It can only be true if $W(1)$ is zero. As will be argued in Section 5 of this chapter, $W(n)$ in Eq. (12) should be replaced by $W'(n) = W(n) - W(1)$. For large clusters, of course, this correction term for the single-molecule state is inconsequential.

For illustration, consider spherical clusters with the same composition as the original phase, formed by fluctuations in the homogeneous original phase. These are reasonably good assumptions for the vapor-to-liquid or liquid-to-crystal transformation; the following discussion will therefore be particularized to those cases. Many possible complicating effects are left out of this simple treatment. Stress effects arising during transformations within the solid state, for example, are ignored here, but are discussed in Chapters 9 and 14.

A critical weakness of the classical theory of nucleation is the use of parameters derived from properties measured for macroscopic phases to describe small clusters. The interfacial region between the original and new phases is generally described by a geometrical interface with a specific free energy, σ, per unit area. The value of σ for a curved boundary, however, is relatively independent of the position chosen for the interface only when the radius of curvature is much larger than the width of the transition region, which is not the case for small clusters. Further, σ is also a function of the curvature [4], which is large for small clusters. Estimates of σ obtained from an analysis of nucleation data will therefore bear little resemblance to the values defined for large droplets or for a planar interface. Further, the density in the transition region is intermediate between that of the two phases, implying that the mean free energy and the number of atoms associated with a growing cluster are not precise concepts. These and related problems are discussed in more detail in following chapters, particularly Chapter 4.

Assuming constant pressure, the reversible work for the formation of a spherical cluster containing n atoms or molecules with an isotropic interfacial energy is (see Eq. (11))

$$W(n) = n\Delta\mu + (36\pi)^{1/3}\bar{v}^{2/3}n^{2/3}\sigma, \tag{13}$$

where $\bar{v}$ is the molecular volume.[2] For supercooled liquids and glasses, $\Delta\mu$ depends on the magnitude of the supercooling, $\Delta T = T_m - T$, where

[2] The second term in Eq. (13) is obtained for spherical clusters by expressing the surface free energy, $4\pi r^2\sigma$ in terms of the number of clusters, using $n\bar{v} = (4/3)\pi r^3$. Pre-factors different from $(36\pi)^{1/3}$ are obtained for other cluster geometries.

T_m is the equilibrium melting point, and T is the temperature, both in absolute units. Above T_m, $\Delta\mu$ for crystallization is positive; below T_m, it is negative. The dependence of $W(n)$ on n shown in Figure 1 was computed from Eq. (13). As already discussed, $W(n)$ at first increases monotonically with increasing cluster size, but eventually decreases for large clusters, where the volume free energy decrease on phase transition is dominant. The maximum work as a function of cluster size, denoted by $W(n^*)$ (or often W^*), is readily found by solving $(\mathrm{d}W(n)/\mathrm{d}n)_n = 0$, giving

$$W(n^*) = \frac{16\pi}{3}\frac{\sigma^3}{(\Delta g)^2}. \tag{14}$$

This critical required work for nucleation corresponds to a critical cluster size, n^*, given by

$$n^* = \frac{32\pi}{3\bar{v}}\frac{\sigma^3}{|\Delta g|^3}. \tag{15}$$

Here the Gibbs free energy difference per unit volume has been introduced for convenience of notation, $\Delta g = \Delta\mu/\bar{v}$. Clusters of size n^* are in unstable equilibrium; those smaller than n^* (often called embryos) tend to shrink while clusters larger than this value will on average grow. In the crudest sense, then, nucleation is simply the production of post-critical clusters, called nuclei.

If the interfacial energy is anisotropic, at constant pressure

$$W(n) = n\Delta\mu + \sum_i A_i\sigma_i = n\Delta\mu + A\bar{\sigma}, \tag{16}$$

where A_i and σ_i are the areas and interfacial energies of the cluster facets. The "effective" interfacial energy, $\bar{\sigma}$, is defined as

$$\bar{\sigma} = \frac{\sum_i A_i\sigma_i}{\sum_i A_i}. \tag{17}$$

The basic form for $W(n)$ remains unchanged, though the constant for the $n^{2/3}$ term will be different due to the different cluster surface area. For continuously curved surfaces, the summation must be replaced by the integral of the interfacial energy over the surface. There will be an additional, more fundamental, correction as well. This formulation assumes compact clusters; a fractal morphology is likely for small clusters.

This discussion has adhered to a strictly atomistic picture for the cluster. It is often more convenient and general to express the properties of a cluster simply in terms of its radius. The work of cluster formation is then (for a spherical cluster)

$$W(r) = \frac{4\pi}{3}r^3\Delta g + 4\pi r^2\sigma. \tag{18}$$

Setting $(dW(r)/dr)_{r^*}$ equal to zero gives the critical radius for nucleation,

$$r^* = \frac{2\sigma}{|\Delta g|}. \tag{19}$$

This could also have been obtained simply from Eq. (15) by noting that $(4\pi/3)r^{*3} = n^*\bar{v}$. The critical work of cluster formation, $W(r^*)$, of course, remains the same (Eq. 14).

3. KINETIC MODEL FOR CLUSTER FORMATION

Volmer and Weber formulated the first kinetic model of nucleation [5], the central ideas of which have served as a basis for further development by Szilard [6], Farkas [7], Volmer [8, 9], Becker and Döring [10], Zeldovich [11], Frenkel [12, 13], Turnbull and Fisher [14], and others.

Volmer and Weber assumed that clusters of n molecules, E_n, grow or shrink slowly by the addition or loss of a single molecule, E_1, following a series of bimolecular reactions:

$$E_{n-1} + E_1 \underset{k^-(n)}{\overset{k^+(n-1)}{\rightleftarrows}} E_n,$$

$$E_n + E_1 \underset{k^-(n+1)}{\overset{k^+(n)}{\rightleftarrows}} E_{n+1}. \tag{20}$$

Here $k^+(n)$ is the rate of single-molecule addition to a cluster of size n and $k^-(n)$ is the rate of loss. It is implicitly assumed that reactions of clusters with dimers, trimers, etc., are too infrequent to be comparable with single-molecule attachment. The kinetic model is illustrated in Figure 4, showing the evolving cluster population density and the reactions involved. The time-dependent cluster size distribution, $N(n,t)$ is determined by solving a system of coupled differential equations of the form

$$\frac{\partial N(n,t)}{\partial t} = N(n-1,t)k^+(n-1) - N(n,t)\big(k^+(n) + k^-(n)\big) + N(n+1,t)k^-(n+1), \tag{21}$$

which has the form of a master equation. The nucleation rate past a cluster size n, $I(n, t)$, is the time-dependent flux of clusters past that size and is given by

$$I(n,t) = N(n,t)k^+(n) - N(n+1,t)k^-(n+1). \tag{22}$$

The nucleation rate has dimensions of inverse time and is proportional to the total number of molecules in the system. It is important to emphasize that the nucleation rate is in general a function of both the time and the

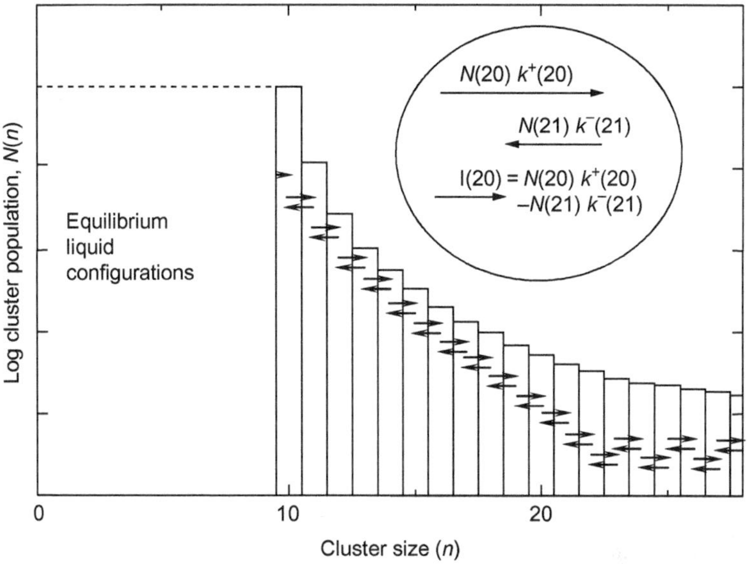

Fig. 4 A histogram of the cluster populations as a function of their size, n, (i.e. number of molecules, n) showing the fluxes that describe cluster growth. Below some lower limit (here $n = 10$), the clusters of the new phase are indistinguishable from equilibrium fluctuations in the liquid.

cluster size at which it is measured (Chapter 3). This is key to understanding many precipitation and nucleation processes in solids and glasses.

As mentioned, the heterophase fluctuations leading to transitory phase transformations of small regions of the original phase also occur in stable phases, leading to several conceptual problems in identifying the cluster size distribution that is appropriate for nucleation studies. Even if it were possible to follow the cluster evolution directly, it would prove difficult to decide which clusters have configurations that are distinct from equilibrium fluctuations in the original phase. For gas condensation or precipitation from a dilute solution, this is less problematic due to the much lower density of the original phase. It is more important, however, when modeling nucleation in condensed phases, such as the crystallization of a liquid or a glass, and is a problem central to the analysis of computer-generated nucleation data (Chapter 10). Any distinctions between nucleation-induced and equilibrium fluctuations are, in fact, artificial since, for the fluctuation approach to be meaningful, all configurations must be microstates accessible from the metastable equilibrium state of the original phase. For illustration, in Figure 4, the lower limit for clusters of the new phase is taken to contain 10 molecules. This is arguably valid, for example, in a ccp metal for which the smallest configuration that is clearly distinguishable from the range of configurations available in the liquid is that of two edge-sharing octahedra consisting of 10 atoms. The discovery of quasicrystals, condensed quasiperiodic phases with a non-crystallographic symmetry, has further exacerbated these conceptual difficulties (see Refs. [15, 16] for example). Since the most prevalent quasicrystal, the icosahedral phase, has short-range order that is similar to that in the liquid [17], there may be little distinction between those two phases on the cluster level. This is discussed in greater depth in Chapter 7.

4. COMPUTATION OF THE RATE CONSTANTS

The classical theory of nucleation was first developed and applied to the case of condensation from a supersaturated vapor. While vapor condensation is not discussed in this book, it is still useful to first examine how the kinetic constants are obtained for nucleation in that case. As already mentioned, in some ways, vapor nucleation is more straightforward than nucleation in condensed phases. For liquid or crystal nucleation from the vapor, the forward attachment rate for a single molecule to a cluster E_n, $k^+(n)$, is equal to the number of collisions per unit time between free molecules and the cluster multiplied by the probability that the molecule is incorporated into the cluster (called the sticking coefficient). Assuming that all are incorporated (sticking coefficient equal to 1), from the kinetic theory of gases and following Wu [18],

$$k^+(n) = \sqrt{\frac{k_B T}{2\pi m_1}}\, N(1)\,(36\pi\bar{v}^2)^{1/3} n^{2/3}(1+n^{-1/3})^2\sqrt{1+n^{-1}}, \tag{23}$$

where m_1 is the mass of the molecule and $N(1)$ is the population of single molecules. This differs slightly from the expression normally given, which, for an ideal gas, treats the cluster as having the shape of a flat surface rather than a sphere, and does not account for cluster translation:

$$k^+(n) = \frac{p}{\sqrt{2\pi m_1\, k_B T}}(36\,\pi\,\bar{v}^2)^{1/3} n^{2/3}. \tag{24}$$

The back rate, or evaporation rate, is more difficult to estimate, since there is no clear way to compute it from first principles. To obtain an expression, let us first consider a different, well-defined problem, that of chemical species A transforming to species B with a forward rate constant k_A^+. The reverse reaction of B going to A with a rate constant of k_B^- must, of course, also be taken into account. The rate of change of the fraction of species A is then

$$\frac{dX_A}{dt} = -k_A^+ X_A + k_B^- X_B, \tag{25}$$

where X_A and X_B are the mole fractions of A and B respectively. If the chemical reaction is a closed system, after some time X_A and X_B become constant. This is a condition of equilibrium in which there is no longer a net flux toward either A or B. Setting $dX_A/dt = 0$ for this case, the reverse rate can be readily obtained given the forward rate and the equilibrium mole fractions of A and B,

$$k_B^- = k_A^+ \frac{X_A^{eq}}{X_B^{eq}}. \tag{26}$$

A similar method, often called the constrained equilibrium hypothesis, or detailed balance, is generally followed in developing the back rates for nucleation. Assuming an equilibrium cluster distribution, there should be no net cluster flux, giving a zero nucleation rate at all cluster sizes. In analogy with Eq. (26), the back rate for cluster growth can be expressed in terms of the forward rate as

$$k^-(n+1) = k^+(n)\frac{N^{eq}(n)}{N^{eq}(n+1)}, \tag{27}$$

where the equilibrium cluster size distribution, $N^{eq}(n)$, is computed using Eq. (12). Since the ratios of the densities are taken, the problem of normalizing $W(n)$ to $W(1)$ can be ignored. Careful thought shows, however, that there are problems with the application of the equilibrium hypothesis to describe nucleation. For nucleation to occur, the original phase must be metastable. The constrained equilibrium hypothesis, however, requires that the new stable phase be held in equilibrium with the metastable phase, which is unphysical. Further, as computed from Eq. (12), the cluster size distribution in the equilibrium phase is not normalizable. It increases without bound for clusters larger than the critical size (Figure 6). As expected, an equilibrium size distribution does not exist for nucleation; it becomes merely a mathematical convenience for solving the problem. One approach to justify it is within the context of a new type of potential, a *kinetic potential*, defined within the kinetic model for nucleation. This is discussed in Section 5 of this chapter.

Excepting the difficulties with the constrained equilibrium hypothesis, the formulation of the rate constants for vapor condensation is straightforward within the framework of the kinetic theory of gases. The correct procedure for computing the rate constants in condensed phases is less obvious. The atoms joining the cluster of the new phase already reside in the interfacial region of the cluster and have nearly the same packing density and configuration as atoms within the cluster (particularly when compared with the large differences in these properties between atoms or molecules in the gas and condensed phases). Turnbull and Fisher assumed that with the addition or removal of a single atom or molecule to a cluster, the system passes through a configuration (the activated complex) that is higher in energy than either the original or new state (Figure 5). The forward and backward rate constants are calculated using transition state theory, which again relies on the constrained equilibrium hypothesis to construct the back rate,

$$
\begin{aligned}
k^{+}(n) &= O(n)\gamma \exp\left(-\frac{\delta W(n)}{2k_{\mathrm{B}}T}\right) = O(n)\gamma^{+}(n),\\
k^{-}(n+1) &= O(n)\gamma \exp\left(+\frac{\delta W(n)}{2k_{\mathrm{B}}T}\right) = O(n)\gamma^{-}(n+1);
\end{aligned}
\tag{28}
$$

where

$$\gamma^{+}(n) = \gamma \exp\left(-\frac{\delta W(n)}{2k_{\mathrm{B}}T}\right) \text{ and } \gamma^{-}(n+1) = \gamma \exp\left(+\frac{\delta W(n)}{2k_{\mathrm{B}}T}\right).$$

In Eq. (28), $\delta W(n)$ is the work of formation for a cluster of $n + 1$ molecules less that of a cluster containing n molecules; γ is an unbiased molecular jump frequency at the cluster interface, $\gamma^{+}(n)$ is the rate of single-molecule addition at an interface site for a cluster containing n molecules, and $\gamma^{-}(n + 1)$ is the rate of single-molecule loss from a cluster containing $n + 1$ molecules, $O(n)$ represents the number of possible attachment sites on a cluster of n molecules, approximately

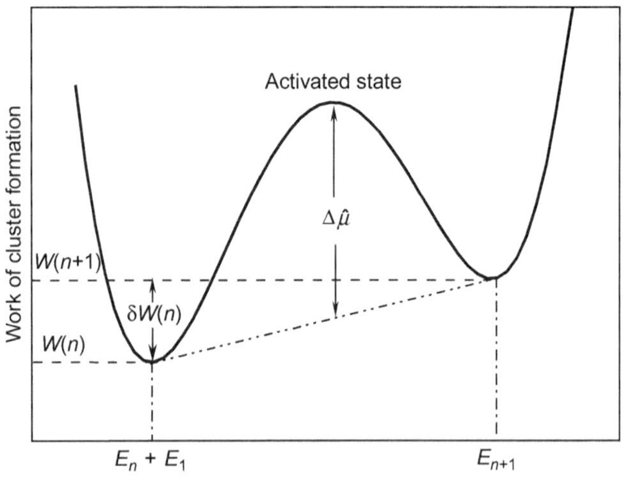

Fig. 5 Schematic illustration of the energy barrier in transition rate theory, showing the work of cluster formation as a function of position in configuration (or phase) space. The initial minimum in work corresponds to the assembly of a cluster of size n and an isolated single molecule; the second minimum corresponds to a cluster of size n + 1.

$4n^{2/3}$ for a spherical cluster. Expressed in terms of the attempt frequency, ν, the unbiased jump frequency is given by

$$\gamma = \nu \exp\left(-\frac{\Delta\hat{\mu}}{k_B T}\right), \tag{29}$$

where $\Delta\hat{\mu}$ is the difference between the energy of the activated state and the average energies of the initial and final states (Figure 5). The jump frequency is generally taken to be the same as that governing bulk diffusion, D,

$$\gamma = \frac{6D}{\lambda^2}, \tag{30}$$

where λ is the atomic jump distance. Since the structure near the interface is unlikely to be similar to that of the parent phase, this assumption is somewhat questionable; it is also likely that any scaling between the two jump frequencies should be dependent on cluster size. Interestingly, in at least one system that has been investigated thoroughly, the temperature dependencies of the two atomic mobilities are quite similar, and the same constant appears to describe the evolution of extremely small clusters, containing fewer than 100 molecules, and the growth of macroscopic ones [19].

5. KINETIC POTENTIAL — AN ALTERNATIVE TO THE CONSTRAINED EQUILIBRIUM HYPOTHESIS

The constrained equilibrium hypothesis leads to a relation between the forward and backward rate constants based on the existence of an equilibrium distribution of the form given by Eq. (12). Since this distribution is not physically realizable for nucleation processes, however, the derived rates become suspect. Several approaches have been suggested to get round these problems [20–32]. These, however, rely on ad-hoc assumptions for the cluster distribution, are too sensitive to the details of assumed potential interactions between atoms or molecules used to compute the cluster distribution to be of much practical importance, or are essentially equivalent to detailed balance arguments.

Recall that Eq. (12) arose from an assumption of cluster formation as an equilibrium fluctuation process and has led to the present difficulties. Another, more kinetic, approach to nucleation is possible, however. We can simply assume that there exists *some* hypothetical cluster distribution, not necessarily given by Eq. (12), which enforces the zero-flux hypothesis. As long as this distribution is viewed only as a consequence of this hypothesis, if it can be mathematically formulated it is useful, even though it is not physically realizable. Wu has recently explored the consequences of this approach [18]. It is briefly summarized here.

Taking Eq. (22) as the starting point, it must be possible to construct a mathematical distribution, $N^z(n)$, which forces the condition of a zero flux, i.e. $I(n, t) = 0$ for all n. This is satisfied if

$$k^-(n+1) = k^+(n)\frac{N^z(n)}{N^z(n+1)}. \tag{31}$$

Although this has the same form as Eq. (27), the distribution that leads to this condition is not yet known. Since we are avoiding the equilibrium assumption, Eq. (12) cannot be assumed. We can instead construct the cluster size distribution that forces this condition in terms of the single-molecule density and the set of rate constants. To show this, simply write the zero-flux condition for each cluster size and solve for the cluster density as,

$$\begin{aligned}
N^z(2) &= N^z(1)\frac{k^+(1)}{k^-(2)} \\
N^z(3) &= N^z(2)\frac{k^+(2)}{k^-(3)} = N^z(1)\frac{k^+(1)}{k^-(2)}\frac{k^+(2)}{k^-(3)} \\
N^z(4) &= N^z(3)\frac{k^+(3)}{k^-(4)} = N^z(1)\frac{k^+(1)}{k^-(2)}\frac{k^+(2)}{k^-(3)}\frac{k^+(3)}{k^-(4)} \\
&\vdots
\end{aligned} \tag{32}$$

For any cluster size, then,

$$N^z(n+1) = N^z(1)\prod_{i=1}^{n}\frac{k^+(i)}{k^-(i+1)}. \tag{33}$$

It is useful to define a new potential that will allow us to compute this distribution in the manner followed to obtain Eq. (12). This kinetic potential, $\Phi(n)$, originates not in the equilibrium calculation of the work of cluster formation, but in the kinetic equations of the classical nucleation theory. We require that

$$N^z(n) = N^z(1)\exp\left(-\frac{\Phi(n)}{k_B T}\right). \tag{34}$$

This is true if

$$\Phi(n) = -k_B T \ \ln\prod_{i=1}^{n-1}\frac{k^+(i)}{k^-(i+1)} = -k_B T\sum_{i=1}^{n-1}\ln\left(\frac{k^+(i)}{k^-(i+1)}\right). \tag{35}$$

As before, the rates can be defined from the differences in the kinetic potential,

$$-k_B T \ \ln\frac{k^+(n)}{k^-(n+1)} = \Phi(n+1) - \Phi(n). \tag{36}$$

The distribution computed from Eq. (34) is the same as given by Eq. (12) provided that $N^z(1)$ is set to the number of single molecules in the ensemble and $\Phi(n)$ is equal to $W(n)$, the reversible work of cluster formation, affording a different statement of the constrained equilibrium hypothesis. This still suffers, of course, from the problems with the original equilibrium hypothesis. When a nucleating system is in metastable equilibrium, the cluster size distribution is predicted to diverge at large cluster sizes, which is an unphysical prediction. The derivation of Eq. (34) by kinetic arguments, however, provides a way out. Within an equilibrium view, a probability is assigned for a fluctuation leading to a cluster of any size based on the value of the work of cluster formation relative to the energy of isolated molecules. No insight is provided for how the cluster might actually form. The kinetic model, on the other hand, views cluster formation through a succession of steps, each leading to a cluster of slightly larger size. To achieve an extremely large size, which equilibrium arguments would lead us to expect (*incorrectly*) is extremely favorable, the kinetic argument requires that clusters first grow through sizes that are not highly probable. As we shall see, when steady-state kinetics is considered, the cluster size density goes to zero, not infinity, in the limit of large cluster size.

Before leaving this section, it is useful to point out one important use for the kinetic potential. The expression that we derived earlier for the work of cluster formation (Eq. 12) leads to conceptual difficulties when applied blindly to small clusters, predicting that a finite amount of work is required to form a single-molecule "cluster." This clearly makes no sense; the problem arises from a failure to properly reference the work of cluster formation to the single molecule. Wu discusses this at length, giving a historical slant to the solution [18]. Using the kinetic potential and a rigorous approach to the capillary approximation, he derived the correctly referenced, self-consistent, form for the work of cluster formation. For spherical clusters,

$$W(n) = (n-1)\Delta\mu + (36\pi)^{1/3}\bar{v}^{2/3}(n^{2/3}-1)\sigma, \tag{37}$$

which correctly predicts zero work for the formation of a single-molecule cluster.

6. NUMERICAL EXPLORATION OF THE CONSEQUENCES OF THE KINETIC MODEL FOR NUCLEATION

Having developed expressions for the rate of cluster formation, we could immediately move to the development of analytical expressions for the nucleation rate. Instead, we first explore the behavior of the cluster distribution as prescribed by the kinetic model developed in Section 3 of this chapter, best accomplished by a numerical solution of those equations. Numerical approaches will be used several times in this book to illustrate nucleation behavior, to determine the validity of fundamental assumptions made for analytical solutions, and to test those solutions. They are extended readily to model a large number of experimental situations, including arbitrary initial cluster size distributions, non-isothermal annealing treatments, glass formation, and nucleation of a phase with a composition that is different from the original phase. Also, numerical treatments allow a quantitative analysis of time-dependent nucleation data (Chapter 3) that are often obtained under experimental conditions that are too complicated to analyze analytically.

A finite-difference (or Euler) method [33] is most easily used to obtain a solution of the coupled differential equations underlying the kinetic model of the classical theory of nucleation. Time is divided into a large number of small intervals, δt, and the number of clusters of size n at the end of the interval, $N(n, t + \delta t)$ is computed using

$$N(n,t+\delta t) = N(n,t) + \delta t \, \frac{\partial N(n,t)}{\partial t}, \tag{38}$$

where $\partial N(n, t)/\partial t$ is given by Eq. (21). It is necessary to choose arbitrary upper and lower limits for these calculations. The consequences of these limits are examined in Chapter 3, when discussing the time-dependent nucleation rate; they have little impact on the discussion within this chapter. More important at this stage, the differential equations for nucleation are stiff, i.e. the rates for clusters of different size in the distribution can vary widely. Therefore, to maintain the stability of the solution using Euler's method, the time increments, δt, must be kept small and safeguards introduced in the computer code to constantly check for signs of divergence and make appropriate adjustments to the time increments [34].

Cluster size distributions calculated from Eq. (38) as a function of time are shown in Figure 6. The equilibrium size distribution, computed from Eq. (12), is also shown. For this illustration, parameters were chosen that were appropriate for the precipitation of oxygen-rich clusters in single-crystal silicon [35, 36] (see Chapter 9, Section 3). For simplicity, the initial distribution was taken to be an ensemble of single atoms and strain effects were ignored. The definition of the precipitate cluster can be reasonably extended to dimers, avoiding the difficulties mentioned when treating liquid/crystal transitions, i.e. of distinguishing between a crystal cluster and an equilibrium liquid fluctuation.

With time, the population of n-size clusters increases. Eventually, however, after long annealing times, the cluster size distribution takes on a constant, or steady-state, value, $N^{st}(n)$. Except for extremely long annealing times, when the original number of single molecules is

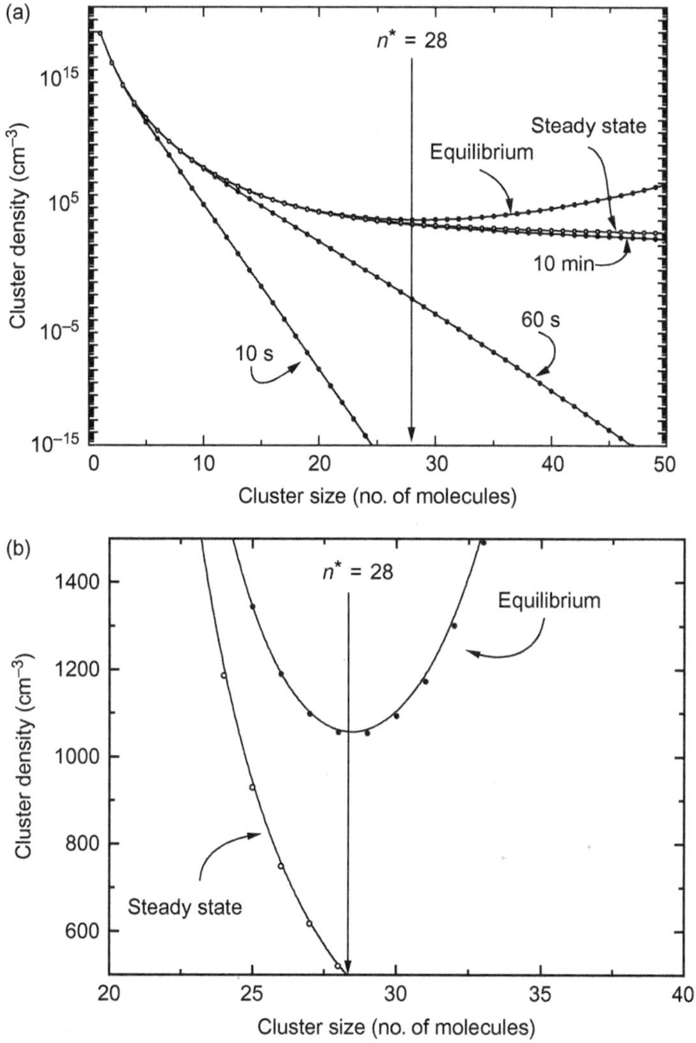

Fig. 6 The calculated time-dependent cluster size distribution as a function of cluster size using parameters appropriate for oxide precipitation in single-crystal silicon at a temperature of 800°C (Chapter 9) (See Chapter 3 for discussion of time-dependent nucleation). The initial oxygen concentration used is $8.0 \times 10^{23}\,\mathrm{m}^{-3}$; the equilibrium concentration at that temperature is $3.2 \times 10^{22}\,\mathrm{m}^{-3}$. (a) The calculated cluster population as a function of cluster size for different annealing times at 800°C, the steady-state distribution and the equilibrium distribution. (b) The steady-state and equilibrium distributions near the critical size, demonstrating that $N^{st}(n^*) = N^{eq}(n^*)/2$.

depleted, this time-invariant distribution is maintained. For small clusters, it has a similar value to the equilibrium distribution, both approaching the same value for single-molecule clusters. Unlike the equilibrium distribution, however, which rises without bound as the cluster size approaches infinity, the steady-state distribution goes to zero in that limit, similar to the form proposed in Section 5 for the hypothetical distribution derived using the kinetic potential. A close examination shows that at the critical size, $N^{st}(n^*) = N^{eq}(n^*)/2$.

Most approaches to nucleation focus on the critical size. We have already emphasized the importance of this in the development of the work of cluster formation. Based on the steady-state distribution shown in Figure 6, however, it might be questioned why this particular size is so important. This will become clearer with the discussions in the later part of this chapter and the next; from both a kinetic and a thermodynamic point of view it remains a special size in the problem. This is best seen by looking at the relation between the forward and backward rates shown in Figure 7 as a function of cluster size. The critical size is that point where the forward and backward rates are equal, consistent with the thermodynamic notion of the unstable equilibrium of critical-sized clusters. For clusters smaller than the critical size, the back rate is larger, indicating that on average they are dissolving, while for clusters larger than the critical size, the forward rate is larger, indicating cluster growth.

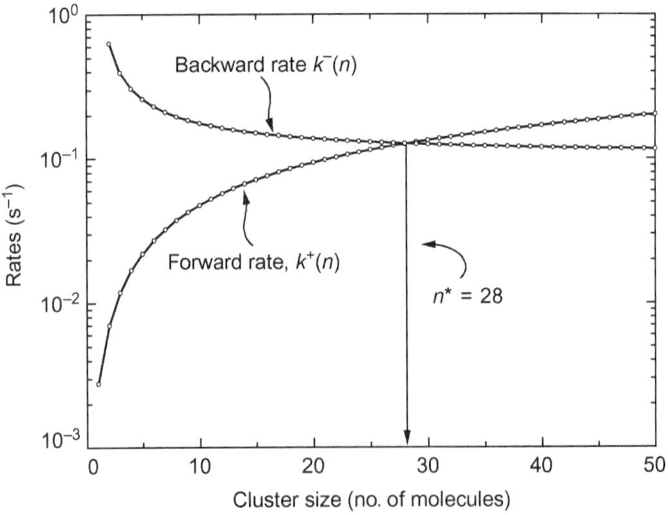

Fig. 7 The forward and backward rates used to compute the time-dependent cluster size distributions in Figure 6. They become equal for clusters of the critical size.

The insight gained from the behavior of the cluster distribution from this numerical analysis is now used to develop analytical expressions for the nucleation rate.

7. STEADY-STATE HOMOGENEOUS NUCLEATION — DISCRETE CLUSTER MODEL

Volmer and Weber, the first to formulate the kinetic problem of nucleation, chose to base their development on the equilibrium cluster distribution [5]. Although the unphysical nature of this distribution for nucleation phenomena has been discussed at length, their simple estimate of the nucleation rate remains useful for quickly identifying the most relevant physical parameters. Realizing that the cluster population diverged at large cluster sizes, they chose simply to remove the bothersome clusters from consideration. In their model, any clusters that made it to the critical size grew very quickly to macroscopic sizes, leaving the distribution, and had no further influence on the nucleation behavior. Their modified equilibrium distribution is then

$$\begin{aligned} n<n^* \quad & N(n) = N^{\mathrm{eq}}(n) = N_{\mathrm{A}} \exp\left(-\frac{W(n)}{k_{\mathrm{B}}T}\right) \\ n>n^* \quad & N(n) = 0. \end{aligned} \tag{39}$$

Because there is no back-flux from clusters larger than n^*, the nucleation rate I^{VW} (now constant in time) is easily evaluated, using Eq. (22) to give

$$I^{\mathrm{VW}} = N^{\mathrm{eq}}(n^*)k^+(n^*) = k^+(n^*)N_{\mathrm{A}} \exp\left(-\frac{W(n^*)}{k_{\mathrm{B}}T}\right). \tag{40}$$

Within the classical-theory assumptions that fluctuations are actual clusters, that the kinetics are interface-controlled, and that capillarity[3] holds, this simple equation correctly describes the functional dependence of the nucleation rate on the most relevant physical parameters, i.e. the work of cluster formation and the atomic mobility, derived from the diffusion coefficient in condensed systems. It also correctly predicts that the rate scales linearly with the size of the system.

As demonstrated in Figure 6, however, the equilibrium distribution is not the one achieved from cluster evolution dictated by the kinetic equations of the classical theory of nucleation, even for $n<n^*$. Instead, it

[3] For nucleation, *capillarity* means that the cluster of the new phase can be viewed as having a sharp surface.

eventually approaches a steady-state distribution. Becker and Döring [10] argued that this had to be the case (without the benefit of the numerical solution) and proposed that the correct form for the time-independent nucleation rate, the steady-state rate, I^{st}, must be derived from the steady-state distribution, $N^{st}(n)$,

$$I^{st} = N^{st}(n)k^{+}(n) - N^{st}(n+1)k^{-}(n+1). \tag{41}$$

For a true steady-state condition, I^{st} should be independent of time and the same for all cluster sizes. A time-dependent nucleation rate is often observed experimentally; the origins of this are discussed in Chapter 3. To maintain the steady-state condition in the model, the correct number of molecules must be added to the original phase for each nucleus removed, to maintain a constant level of single molecules. In practice, the number of molecules involved in forming the nuclei is sufficiently small that the single-molecule depletion has a negligible effect in the early stages of transformation where nucleation is often most important.

Adopting the constrained equilibrium hypothesis, Eq. (27) can be used to derive the backward rate constant in terms of the equilibrium cluster size distribution and the forward rate constant; Eq. (41) can then be written as

$$I^{st} = N^{eq}(n)k^{+}(n)\left(\frac{N^{st}(n)}{N^{eq}(n)} - \frac{N^{st}(n+1)}{N^{eq}(n+1)}\right). \tag{42}$$

The boundary conditions for the steady-state distribution are readily determined from an examination of Figure 6,

$$\begin{gathered} N^{st}(n) \to N^{eq}(n) \text{ as } n \to 0 \\ \text{and} \\ N^{st}(n) \to 0 \text{ as } n \to \infty. \end{gathered} \tag{43}$$

As an approximation to these limits, lower and upper bounds on the cluster size, $\tilde{u}$ and $\tilde{v}$ respectively, are chosen such that for $n \leq \tilde{u}$, $N^{st}(n) = N^{eq}(n)$ and for $n \geq \tilde{v}$, $N^{st}(n) = 0$. Fortunately, the solution does not depend strongly on the values chosen for $\tilde{u}$ and $\tilde{v}$, provided that the energies corresponding to these cluster sizes are at least $k_B T$ lower than the energy at the critical size, lying outside the critical region (Figure 1). Summing Eq. (42) for all values of n between $\tilde{u}$ and $\tilde{v}$ and using the stated boundary conditions,

$$\sum_{n=\tilde{u}}^{\tilde{v}} \frac{I^{st}}{N^{eq}(n)k^{+}(n)} = \sum_{n=\tilde{u}}^{\tilde{v}} \left(\frac{N^{st}(n)}{N^{eq}(n)} - \frac{N^{st}(n+1)}{N^{eq}(n+1)}\right) = \frac{N^{st}(\tilde{u})}{N^{eq}(\tilde{u})} - \frac{N^{st}(\tilde{v}+1)}{N^{eq}(\tilde{v}+1)} = 1. \tag{44}$$

The steady-state nucleation rate is then

$$I^{st} = \left(\sum_{n=\tilde{u}}^{\tilde{v}} \frac{1}{N^{eq}(n) k^+(n)} \right)^{-1}. \tag{45}$$

Since $1/N^{eq}(n)$ has a maximum at n^*, those terms of the sum nearest n^* will make the greatest contribution. Further, since the cluster-size dependence of the rate constants is much weaker than that of $N^{eq}(n)$, little error is introduced if $k^+(n)$ is replaced by the forward rate constant at the critical size $k^+(n^*)$, giving (using Eq. (28))

$$I^{st} = \gamma^+(n^*) O(n^*) \left(\sum_{n=\tilde{u}}^{\tilde{v}} \frac{1}{N^{eq}(n)} \right)^{-1}. \tag{46}$$

An approximate analytical expression for I^{st} is obtained by replacing $W(n)$ by the first two non-zero terms in a Taylor expansion about n^*,

$$W(n) = W(n^*) + \frac{\hat{\xi}^2}{2!} \left(\frac{\partial^2 W(n)}{\partial n^2} \right)_{n^*}, \tag{47}$$

defining $\hat{\xi} = n - n^*$, taking $N^{eq}(n)$ to be a continuous function of n, and replacing the sum by an integral

$$\sum_{n=\tilde{u}}^{\tilde{v}} \frac{1}{N^{eq}(n)} = \frac{1}{N_A} \exp \frac{W(n^*)}{k_B T} \int_{\tilde{u}-n^*}^{\tilde{v}-n^*} d\hat{\xi} \exp\left(-\frac{W(n^*)\hat{\xi}^2}{3n^{*2} k_B T} \right). \tag{48}$$

The limits of the integral may be extended to $\pm\infty$, because of the strong maximum of $1/N^{eq}(n^*)$, giving the error integral which can be solved readily,

$$\sum_{n=\tilde{u}}^{\tilde{v}} \frac{1}{N^{eq}(n)} \simeq \frac{1}{N_A} \exp\left(\frac{W(n^*)}{k_B T} \right) \left(\frac{3\pi k_B T}{W(n^*)} \right)^{1/2} n^*. \tag{49}$$

The homogeneous steady-state nucleation rate for spherical clusters is then

$$I^{st} = N^{eq}(n^*) \gamma^+(n^*) O(n^*) \left(\frac{|\Delta\mu|}{6\pi k_B T n^*} \right)^{1/2} = N^{eq}(n^*) k^+(n^*) Z, \tag{50}$$

which differs from the Volmer and Weber expression (Eq. 40) by only a multiplicative factor Z called the Zeldovich factor,

$$Z = \left(\frac{|\Delta\mu|}{6\pi k_B T n^*} \right)^{1/2}. \tag{51}$$

In most cases, $0.01 \leq Z \leq 0.1$. Given I^{st}, the steady-state distribution is easily obtained

$$I^{st}\sum_{m=n}^{\tilde{v}}\frac{1}{N^{eq}(m)k^{+}(m)}=\frac{N^{st}(n)}{N^{eq}(n)}, \tag{52}$$

where $N^{st}(n) = 0$ for $n \geq \tilde{v}$. This distribution is identical to that shown in Figure 6, which was obtained from a numerical solution of the coupled differential equations. None of the approximations made in this section was required for the numerical calculation, indicating that those approximations introduce little error in the derivation. Further, it is easily shown that $N^{st}(n^*) = N^{eq}(n^*)/2$, in agreement with that observation from the numerical solution.

8. ESTIMATE OF THE STEADY-STATE NUCLEATION RATE IN A CONDENSED SYSTEM

Assuming nucleation in a condensed system with rate constants given by Eq. (28), and assuming that the atomic mobility scales with the bulk diffusion coefficient, the steady-state nucleation rate per mole, given in Eq. (50), can be written as

$$I^{st}=\frac{24Dn^{*2/3}N_A}{\lambda^2}\left(\frac{|\Delta\mu|}{6\pi k_B T n^*}\right)^{1/2}\exp\left(-\frac{W(n^*)}{k_B T}\right). \tag{53}$$

As a concrete example, let us examine the predicted temperature dependence of I^{st} for a supercooled liquid. Substituting the value for $W(n^*)$, from Eq. (14),

$$I^{st}\propto D\exp\left(-a\frac{\sigma^3}{T\,(\Delta g)^2}\right)\propto D\exp\left(-a'\frac{\sigma^3}{T\,\Delta T^2}\right). \tag{54}$$

The free energy is assumed to scale with the amount of supercooling of the liquid ($\Delta g = \Delta s_f\,(T - T_m) = \Delta s_f \Delta T$, where Δs_f is the entropy of fusion per unit volume). The cluster geometry is reflected in the constants ($a = 16\pi/3k_B$ for a spherical cluster); $a' = a\Delta s_f^{-2}$. Equation (53) is plotted in Figure 8a, using parameters appropriate to the nucleation of the crystal phase in lithium disilicate glass [37]. The nucleation rate increases sharply with increasing supercooling, yet decreases again at low temperatures, due to the decreasing atomic mobility, described by the diffusion coefficient. For comparison, the Volmer–Weber nucleation rate (Eq. 40) and that computed numerically (from Eq. 45) are also shown. All show the same general behavior with temperature though, as expected, the Volmer–Weber rate is larger than either the numerically computed or the Becker–Döring rates. There is little difference between the Becker–Döring nucleation rates and the exact ones computed numerically.

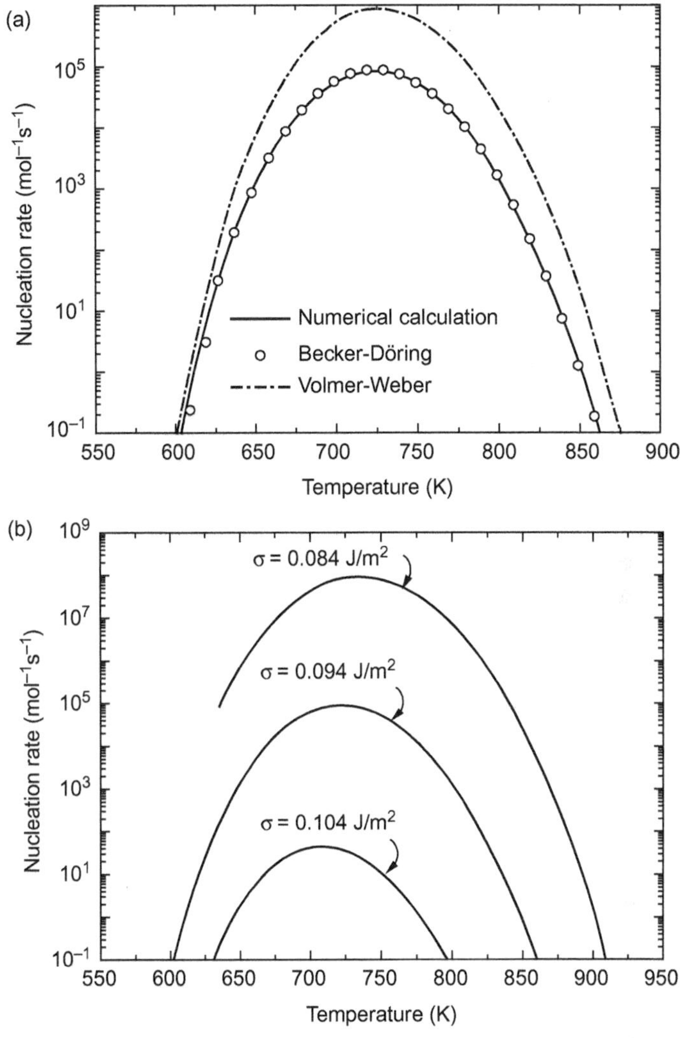

Fig. 8 The computed steady-state nucleation rates as a function of temperature using parameters appropriate to the crystallization of lithium disilicate glass (Chapter 8). (a) A comparison of the calculated steady-state rates from the Volmer–Weber (Eq. (40)) and Becker–Döring (Eq. (53)) solutions, and a numerical computation of the sum in Eq. (45). (b) The calculated rates from Eq. (45) showing the strong dependence on the interfacial free energy.

Only in the crystallization of silicate glasses has the nucleation rate been measured over a sufficiently wide temperature range to explore the behavior shown in Figure 8a; the experimental data are in good agreement with these predictions (Chapter 8). The predicted sharp increase in nucleation rate with decreasing temperature leads to the concept of a maximum supercooling for liquids (Chapter 7) and is in agreement with the qualitative discussion of a barrier given at the beginning of this chapter. Finally, it should be noted that the nucleation rate is extremely sensitive to the value of the interfacial energy between the original and new phases, σ. A variation of only a few percent in σ can alter the nucleation rate by orders of magnitude (Figure 8b). This is often perceived as a weakness of the theory since, as will be discussed several times, σ is an effective parameter that is but poorly known except from nucleation studies.

Eq. (53) has the form

$$I^{st} = A^* \exp\left(-\frac{W(n^*)}{k_B T}\right), \tag{55}$$

where $W(n^*)$ is the height of the nucleation barrier, now identified with the work of formation for a critical cluster of the new phase. The nucleation rate is proportional to the thermodynamic probability of having a fluctuation leading to the formation of a critical cluster, governed by the height of the nucleation barrier, and a dynamical factor, A^* incorporating the rate at which the cluster grows. The value of the prefactor is at the heart of much of the controversy over the validity of the classical nucleation theory. Its value determined from experimental data is often orders of magnitude larger than that predicted by theory. This is examined in detail in Chapter 4 of this book; the discrepancies appear to be a consequence of the assumption made in the classical theory that the width of the interface between the original phase and clusters of the new phase is very small.

9. ZELDOVICH–FRENKEL EQUATION — CONTINUOUS CLUSTER MODEL

Nucleation can be viewed as one example of a class of processes with discrete states that evolve with time. Many such processes are best described by a master equation that relates the quantities of interest to transition probabilities between the different states. The coupled differential equations describing cluster evolution (Eq. (21)) are one example of a master equation. Even for the simple case of steady-state nucleation, approximate analytical solutions can be obtained from this master equation only by making various approximations and truncations; the more general time-dependent behavior is difficult to obtain. As first shown by Zeldovich [11], the steady-state and time-dependent nucleation rates can be computed by constructing a Fokker–Planck equation that approximates the master equation, often called the Zeldovich–Frenkel equation. From Eqs. (21) and (22) the change in the density of clusters of size "n" is related to the difference in the forward fluxes promoting clusters to size "n" and size "$n + 1$"

$$\frac{\partial N(n,t)}{\partial t} = I(n-1,t) - I(n,t). \tag{56}$$

Computing the flux from Eq. (22) and using Eq. (27) to relate the forward and backward rate constants,

$$I(n,t) = k^{+}(n)N^{\mathrm{eq}}(n)\left(\frac{N(n,t)}{N^{\mathrm{eq}}(n)} - \frac{N(n+1,t)}{N^{\mathrm{eq}}(n+1)}\right). \tag{57}$$

Taking the cluster distribution to the continuous limit,

$$I(n,t) = -k^{+}(n)N^{\mathrm{eq}}(n)\frac{\partial}{\partial n}\left(\frac{N(n,t)}{N^{\mathrm{eq}}(n)}\right). \tag{58}$$

Finally, combining Eqs. (56) and (58), and again taking the continuous limit, the Zeldovich–Frenkel equation for nucleation is obtained:

$$\frac{\partial N(n,t)}{\partial t} = \frac{\partial}{\partial n}\left(k^{+}(n)N^{\mathrm{eq}}(n)\frac{\partial}{\partial n}\left(\frac{N(n,t)}{N^{\mathrm{eq}}(n)}\right)\right). \tag{59}$$

Substituting for $N^{\mathrm{eq}}(n)$ from Eq. (12), this can be rewritten as

$$\frac{\partial N(n,t)}{\partial t} = \frac{\partial}{\partial n}\left(k^{+}(n)\frac{\partial N(n,t)}{\partial n}\right) + \frac{1}{k_{\mathrm{B}}T}\frac{\partial}{\partial n}\left(k^{+}(n)N(n,t)\frac{\partial W(n)}{\partial n}\right), \tag{60}$$

which has the form of a diffusion equation where $k^{+}(n)$ and $N(n, t)$ are analogs of the diffusion coefficient and concentration respectively. Clearly, the flux must also contain both diffusion and drift components,

$$I(n,t) = -k^{+}(n)\frac{\partial N(n,t)}{\partial n} + N(n,t)\frac{dn}{dt}, \tag{61}$$

where $\mathrm{d}n/\mathrm{d}t$, the drift velocity in size space, is also the macroscopic growth velocity, describing the growth of a single nucleus when fluctuations are neglected:

$$\frac{\mathrm{d}n}{\mathrm{d}t} = -\frac{k^{+}(n)}{k_{\mathrm{B}}T}\frac{\partial W(n)}{\partial n}. \tag{62}$$

The diffusion component is most important near the critical size; for clusters much larger than n^{*}, it contributes little and the nuclei grow with a rate,

$$I(n > n^{*}, t) \rightarrow \frac{\mathrm{d}n}{\mathrm{d}t}N(n,t). \tag{63}$$

For large cluster sizes Eq. (63) provides a boundary condition equivalent to that assumed in the derivation of the steady-state solution (i.e. $N(n, t)\to 0$ as $t\to\infty$). Following Ref. [11] an identical procedure to that used to obtain Eqs. (45) and (52), the steady-state nucleation rate and population are given by

$$\frac{1}{I^{\mathrm{st}}} = \int_0^\infty \frac{\mathrm{d}n}{k^+(n)N^{\mathrm{eq}}(n)} = \frac{1}{k^+(n^*)N^{\mathrm{eq}}(n^*)} \int_0^\infty \mathrm{d}n \exp\left(\frac{(n-n^*)^2}{2k_\mathrm{B}T}\left(\frac{\partial^2 W(n)}{\partial n^2}\right)_{n^*}\right) \tag{64}$$

and

$$\frac{N^{\mathrm{st}}(n)}{N^{\mathrm{eq}}(n)} = I^{\mathrm{st}} \int_n^\infty \frac{\mathrm{d}m}{k^+(m)N^{\mathrm{eq}}(m)} \tag{65}$$

respectively. Assuming spherical nuclei, setting all of the forward rate constants equal to $k^+(n^*)$, describing $W(n)$ by the first two non-zero terms in a Taylor expansion about n^*, as shown, and extending the integration to even bounds about n^* (i.e. $\pm\infty$), the steady-state rate given by Eq. (50) is obtained. For more general cases, I^{st} can be written as

$$I^{\mathrm{st}} = \frac{\Delta}{2\tau\sqrt{\pi}} N^{\mathrm{eq}}(n^*) \tag{66}$$

where Δ is the width of the near-critical region,

$$\frac{1}{\Delta^2} = -\frac{1}{2k_\mathrm{B}T}\left.\frac{\partial^2 W(n)}{\partial n^2}\right|_{n^*}, \tag{67}$$

defined as the range of cluster sizes for which $W(n)$ is within $k_\mathrm{B}T$ of that of the critical size (Figure 1). The relaxation time, τ, is the average time required for a cluster to diffuse through the critical region,

$$\tau^{-1} = \left.\frac{\mathrm{d}}{\mathrm{d}n}\left(\frac{\mathrm{d}n}{\mathrm{d}t}\right)\right|_{n^*} = \frac{2k^+(n^*)}{\Delta^2}. \tag{68}$$

It is easy to show that Eq. (66) reduces to Eq. (50) for spherical nuclei with $W(n)$ given by Eq. (13).

It should come as no surprise that the solution to the Zeldovich–Frenkel equation is identical to that obtained in the Becker–Döring treatment since a set of coupled linear differential equations can always be replaced by one partial differential equation. Why then bother with the Zeldovich approach? As we shall see in the next chapter, the real strength of the Zeldovich-Frenkel equation is in formulating analytical approximations to the time-dependent nucleation rate. The time required to reach the steady-state rate, for example, is governed to a large extent by the diffusion through the critical region, parameterized by τ given in Eq. (68). While the Zeldovich–Frenkel equation has been criticized and alternative master-equations have been proposed (Section 10 of this chapter), it remains important from an historical perspective, it is relatively easy to solve, it provides valuable physical insight into the underlying dynamics of time-dependent nucleation, and it does yield approximate analytical expressions that can describe most nucleation phenomena within experimental error.

10. ALTERNATIVE MASTER EQUATIONS

It is possible to view nucleation as one of a class of processes that have discrete states that evolve with time. Master equations for such processes relate the quantities of interest to the transitions between these states. The coupled differential equations of the Becker–Döring treatment are one example. In another, investigated by Shizgal and Barrett [38], the time-dependent cluster size distribution, $N(n, t)$, is obtained from an infinite-order differential equation with continuous n, called the forward Kramers–Moyal equation [39, 40]:

$$\frac{\partial N(n,t)}{\partial t} = \sum_{j=1}^{\infty} \frac{1}{j!} \frac{\partial^j}{\partial n^j} \Big(\big(k^-(n) + (-1)^j k^+(n)\big) N(n,t) \Big). \tag{69}$$

A truncation after two terms leads to a class of Fokker–Planck equations that approximately describes the process. The Zeldovich–Frenkel equation is one such equation, founded on particular choices for the drift and diffusion terms and the way in which the discrete cluster density is taken to the continuous limit. Recently, several have criticized those choices [18, 41, 42]. Following Shizgal and Barrett,

$$\frac{\partial N(n,t)}{\partial t} = \frac{\partial}{\partial n} \big(A(n)N(n,t)\big) + \frac{\partial^2}{\partial n^2} \big(B(n)N(n,t)\big), \tag{70}$$

where n is taken to be a continuous variable, and $A(n)$ and $B(n)$ are given by

$$A(n) = k^-(n) - k^+(n) \qquad B(n) = \frac{k^-(n) + k^+(n)}{2}. \tag{71}$$

$A(n)$ can be viewed as a drift coefficient and $B(n)$ as a diffusion coefficient; in general both are functions of the cluster size. By continuity,

$$\frac{\partial N(n,t)}{\partial t} = -\frac{\partial J(n,t)}{\partial n}, \tag{72}$$

so the flux, $J(n, t)$, equivalent to the nucleation rate at a particular size, is given by

$$J(n,t) = -\left(A(n)N(n,t) + \frac{\partial}{\partial n} B(n)N(n,t) \right). \tag{73}$$

Since the flux vanishes in equilibrium, this can be used to determine the density of clusters, $N^{eq}(n)$. Defining $F(n, t) = N(n, t)/N^{eq}(n)$, the Fokker–Planck equation can be expressed as

$$\frac{\partial F(n,t)}{\partial t} = \frac{1}{N^{eq}(n)}\frac{\partial}{\partial n}\left(B(n)N^{eq}(n)\frac{\partial F(n,t)}{\partial n}\right). \tag{74}$$

Due to the truncation, the equilibrium distribution, $N^{eq}(n)$ obtained from Eq. (74) differs from that obtained from the original discrete equation. If the correct value for $N^{eq}(n)$ is used, Eqs. (70) and (74) retain the same mathematical form, but $A(n)$ and $B(n)$ are no longer given by Eq. (71). The central problem is then to make the proper choice for the values of $A(n)$ and $B(n)$. The Zeldovich–Frenkel equation, for example, results when $B(n) = k^+(n)$ and $N^{eq}(n) = N_A \exp(-W(n)/k_B T)$.

Shizgal and Barrett [38] considered two possible sets of values for $A(n)$ and $B(n)$, and found that both choices gave better agreement with the numerical solution than did the Zeldovich–Frenkel equation. Rabin and Gitterman [43] argued that $A(n)$ and $B(n)$ are completely determined near critical points by dynamics. They argued that it is essential to distinguish between slow and fast critical quenches, since the steady-state population values will be maintained by a larger fraction of clusters during slower quenches. This was verified by a numerical simulation of nucleation in an Ising model [44].

We shall see in Chapter 3 that based on the discrete rate equations the value of the quenched-in cluster distribution can have a large effect on the time-dependent nucleation behavior. In Chapter 8, we will also see that predictions of the time-dependent nucleation behavior from a numerical solution to the discrete kinetic equations are in good agreement with experimental data, indicating that these are at least approximately valid. By directly solving these, then, the problems discussed in this section may be avoided.

11. THE NUCLEATION THEOREM

Gibbs first noticed that his thermodynamic formulation of nucleation led to a simple relation between the critical work and the critical size [1] (see also Nielsen [45]). This is readily found by taking the derivative of $W(n^*)$ (Eq. 14) with respect to $\Delta\mu$ and comparing with Eq. (15),

$$\frac{dW(n^*)}{d|\Delta\mu|} = \frac{d}{d|\Delta\mu|}\left(\frac{16\,\pi}{3}\frac{\sigma^3}{|\Delta g|^2}\right) = -\left(\frac{32\,\pi}{3\bar{v}}\frac{\sigma^3}{|\Delta g|^3}\right) = -n^*. \tag{75}$$

While particular to Gibbs' choice for a sharp dividing surface between the cluster and the original phase, it is straightforward to derive an expression that is valid for any choice for the surface and volume regions of the cluster, including the more realistic cases that will be considered in Chapter 4. Following Kashchiev [46], assume that the work of cluster formation can be written as

$$W(n, \Delta\mu) = -n\,\Delta\mu + \Psi(n, \Delta\mu), \tag{76}$$

where the dependence of W on $\Delta\mu$ has been written explicitly; $\Psi(n, \Delta\mu)$ is the excess free-energy contribution from the interfacial region (Note that in this section $\Delta\mu$ represents the *magnitude* of the change in chemical potential.). While this still assumes that the work of cluster formation can be expressed as a surface and volume term, it is more general than Eq. (11), since no dividing interface has been explicitly assumed. Taking n to be continuous and $\Psi(n, \Delta\mu)$ and $W(n, \Delta\mu)$ to be differentiable functions,

$$\left.\frac{\partial W(n, \Delta\mu)}{\partial n}\right|_{n^*} = -\Delta\mu + \left.\frac{\partial \Psi(n, \Delta\mu)}{\partial n}\right|_{n^*} = 0. \tag{77}$$

Differentiating the critical work of formation, $W(n^*, \Delta\mu) = -n^*\Delta\mu + \Psi(n^*, \Delta\mu)$,

$$\frac{\partial W(n^*, \Delta\mu)}{\partial(\Delta\mu)} = -n^* + \frac{\partial \Psi(n^*, \Delta\mu)}{\partial(\Delta\mu)} + \left(-\Delta\mu + \left.\frac{\partial \Psi(n, \Delta\mu)}{\partial n}\right|_{n^*}\right)\frac{\mathrm{d}n^*}{\mathrm{d}(\Delta\mu)}, \tag{78}$$

and using Eq. (77),

$$\frac{\partial W(n^*, \Delta\mu)}{\partial(\Delta\mu)} = -n^* + \frac{\partial \Psi(n^*, \Delta\mu)}{\partial(\Delta\mu)}. \tag{79}$$

This general relation between $W(n^*, \Delta\mu)$, n^*, and $\Delta\mu$ is called the nucleation theorem. Justification for the phenomenological derivation reproduced here has been given from more general statistical–mechanical [47, 48] and thermodynamic [49] treatments and generalized to multicomponent cases [50]. Kinetic extensions to the nucleation theorem, based on the law of mass action and detailed balance, have recently appeared [51]. The nucleation theorem allows a relatively unambiguous assessment of the size of the critical size from experimental data in some cases, and it has led to scaling relations for the critical nucleus in both classical and non-classical theories [52]. Unfortunately, as discussed later in this section, it is of limited use for many of the processes that are of most importance in materials processing, such as the nucleation of condensed phases from liquids and glasses and solid-state transformations.

The key result from this chapter is that the steady-state nucleation rate has the form $I^{\mathrm{st}} = A^*\exp(-W^*/k_{\mathrm{B}}T)$ where A^*, the dynamical prefactor, is a function of the driving free energy, interfacial free energy, viscosity and temperature. It follows, then, that

$$\frac{\partial(k_{\mathrm{B}}T\ln I^{\mathrm{st}})}{\partial(\Delta\mu)} = \frac{\partial(k_{\mathrm{B}}T\ln A^*)}{\partial(\Delta\mu)} - \frac{\partial W(n^*, \Delta\mu)}{\partial(\Delta\mu)}. \tag{80}$$

From Eq. (79),

$$\frac{\partial\,(k_{\mathrm{B}}T\ln I^{\mathrm{st}})}{\partial(\Delta\mu)} = \frac{\partial(k_{\mathrm{B}}T\ln A^{*})}{\partial(\Delta\mu)} + n^{*} - \left(\frac{\partial\Psi(n^{*},\Delta\mu)}{\partial(\Delta\mu)}\right). \tag{81}$$

If $\Psi(n^{*}, \Delta\mu)$ is a weak function of the driving free energy, then its derivative is approximately zero, giving

$$n^{*} \cong \frac{\partial\,(k_{\mathrm{B}}T\ln I^{\mathrm{st}})}{\partial(\Delta\mu)} - \frac{\partial\,(k_{\mathrm{B}}T\ln A^{*})}{\partial(\Delta\mu)}, \tag{82}$$

a simple expression relating the number of atoms in the critical size to the measured steady-state nucleation rate, which is independent of the particular model used to describe nucleation.

Nucleation processes in a vapor are normally measured at constant temperature, as a function of the supersaturation, $s = \ln(p/p^{\mathrm{eq}})$, where p^{eq} is the equilibrium vapor pressure and p is the actual vapor pressure. For vapor nucleation, $A^{*} \propto \exp(2s)$ and $\Delta\mu = k_{\mathrm{B}}Ts$ [46], and it follows from Eq. (82) that the critical size for homogeneous nucleation is computed directly from the slope of a plot of the logarithm of the nucleation rate as a function of the logarithm of the vapor pressure,

$$n^{*} = \frac{\partial\,\ln(I^{\mathrm{st}})}{\partial s} - 2 = \frac{\partial\,\ln(I^{\mathrm{st}})}{\partial\ln p} - 2. \tag{83}$$

Nucleation processes from solution are also normally studied at constant temperature and as a function of the supersaturation, now expressed in terms of the actual and equilibrium activities (a and a^{eq} respectively) of the dissolved species, $s = \ln(a/a^{\mathrm{eq}})$, leading to an almost identical expression for the critical size,

$$n^{*} = \frac{\partial\,\ln(I^{\mathrm{st}})}{\partial s} - 1 = \frac{\partial\,\ln(I^{\mathrm{st}})}{\partial\ln a} - 1; \tag{84}$$

the numerical factor differs because in these cases $A^{*} \propto \exp(s)$ [46].

Eq. (82) yields accurate estimates for the critical size in condensation and precipitation processes because the prefactor depends only weakly on the experimental variable, the supersaturation. This is not the case for nucleation in crystallization processes, either from the melt or solid. In these cases, $\Delta\mu$ is a function of the actual temperature, T, and the difference between the transformation temperature (e.g. melting temperature), ΔT, and T. Typically $\Delta\mu$ is taken to be proportional to ΔT, e.g. $\Delta\mu = (\Delta S_{\mathrm{f}}/N_{\mathrm{A}})\Delta T$ for crystallization from the liquid, for example, where ΔS_{f} is the entropy of fusion per mole. Here, the temperature is the experimental variable. Since A^{*} must contain a dynamical factor describing the atomic mobility, it is also a very strong function of the temperature in such cases, $A^{*} = C\exp(-\Delta H/N_{\mathrm{A}}k_{\mathrm{B}}T)$ where, for example,

ΔH corresponds to the activation enthalpy per mole for diffusion (see Eq. 53). From Eq. (82),

$$n^* = -\frac{N_A k_B}{\Delta S_f}\left(\frac{\partial (T \ln I^{st})}{\partial T} - \ln C\right). \tag{85}$$

To obtain an estimate of n^*, it is essential to have a good value for C; unfortunately ln C can have variable large and negative values, between 10 and 100, of the same order as the first term, making estimates from the nucleation theorem suspect. While it is possible to improve this uncertainty by measurements of the induction time for time-dependent nucleation (Chapter 3), Eq. (82) should be used with caution when describing nucleation of crystallization.

12. SUMMARY

In this chapter, the theoretical framework for the most commonly used theory of nucleation, the classical theory, has been outlined. This development forms the basis for much of the discussion in the remainder of this book. Some key points are:

- The probability of obtaining a cluster of n molecules is related to the work of cluster formation, which, following Gibbs, can be approximately expressed by a volume and a surface term:

  $$W(n) = n\Delta\mu + \sigma A.$$

  Here $\Delta\mu$ is the change in free energy per molecule upon phase change, σ is the interfacial free energy between the new and original phase and A is the surface area of the cluster.
- The critical cluster, containing n^* molecules, is that cluster size for which $W(n)$ is a maximum, $W(n^*) = W^*$.

  $$n^* = \frac{32\pi}{3\bar{v}}\frac{\sigma^3}{|\Delta g|^3}$$

  for spherical clusters, corresponding to a critical radius

  $$r^* = \frac{2\sigma}{|\Delta g|},$$

  and a critical work of cluster formation,

  $$W^* = \frac{16\pi}{3}\frac{\sigma^3}{(\Delta g)^2}.$$

  The free-energy change is expressed here per unit volume, $\Delta g = \Delta\mu/\bar{v}$, and $\bar{v}$ is the molecular volume.

- Clusters smaller than n^* will *on average* shrink while those larger than n^* will *on average* grow.
- An "equilibrium" distribution of clusters as a function of size can be defined

$$N^{\rm eq}(n) = N_{\rm A}\exp\left(-\frac{W(n)}{k_{\rm B}T}\right),$$

although this is not a physically realizable distribution in nucleation processes.
- The time evolution of the cluster size distribution is specified by a set of coupled linear differential equations that describe the absorption and desorption of molecules as a function of cluster size. This kinetic picture underlies time-dependent nucleation phenomena, which will be explored in the next chapter.
- In most cases, a time-independent nucleation rate, i.e. the steady-state rate, is attained eventually. It has the general form,

$$I^{\rm st} = A^*\exp\left(-\frac{W^*}{k_{\rm B}T}\right).$$

This identifies the relevant parameters for nucleation, the work of formation for the critical cluster, $W^* = W(n^*)$, and the dynamical prefactor, which, in condensed systems, is proportional to the diffusion coefficient for nucleation. Assuming the kinetic model of the classical theory of nucleation, the steady-state rate can be expressed as

$$I^{\rm st} = N^{\rm eq}(n^*)\gamma^+(n^*)O(n^*)\left(\frac{|\Delta\mu|}{6\pi k_{\rm B}Tn^*}\right)^{1/2} = N^{\rm eq}(n^*)k^+(n^*)Z,$$

where Z is the Zeldovich factor, generally having a value between 0.01 and 0.1, $\gamma^+(n^*)$ is the rate of single-molecule addition at an interface site for a cluster containing n molecules (see Eq. (28)), and $O(n^*)$ is the number of attachment sites on a critical cluster.
- There exists a general relation between W^*, n^* and $\Delta\mu$, called the nucleation theorem:

$$\frac{\partial W(n^*,\Delta\mu)}{\partial(\Delta\mu)} = -n^* + \frac{\partial\Psi(n^*,\Delta\mu)}{\partial(\Delta\mu)},$$

where $\Psi(n^*, \Delta\mu)$ describes the contributions from the interfacial region when the work of cluster formation is written as $W(n^*, \Delta\mu) = -n^*\Delta\mu + \Psi(n^*, \Delta\mu)$. (Note that here $\Delta\mu$ represents the *magnitude* of the change in chemical potential.) For a more detailed discussion of the nucleation theorem, see Ref. [53].

REFERENCES

[1] J.W. Gibbs, *Scientific Papers*, Vol. I, II, Longmans, London (1906).
[2] L.D. Landau, E.M. Lifshitz, *Statistical Physics*, Oxford, Pergamon (1969) pp. 22–28 & pp. 471–475.

[3] C.H.P. Lupis, *Chemical Thermodynamics of Materials*, North Holland, New York (1983) pp. 352–353.
[4] R.C. Tolman, The effect of droplet size on surface tension, J. Chem. Phys. **17** (1949) 333–337.
[5] M. Volmer, A. Weber, Keimbildung in übersättigten Gebilden, Z. Phys. Chem. **119** (1926) 277–301.
[6] L. Szilard, referred to in Farkas [7]
[7] L. Farkas, Keimbildungsgeschwindigkeit in übersättigten Dämpfen, Z. Phys. Chem. **125** (1927) 236–242.
[8] M. Volmer, Particle formation and particle action as a special case of heterogeneous catalysis, Z. Elektrochem. Angew. Phys. Chem. **35** (1929) 555–561.
[9] M. Volmer, *Kinetik der Phasenbildung*, Vol. 122, Steinkopff, Dresden (1939).
[10] R. Becker, W. Döring, Kinetic treatment of grain-formation in super-saturated vapours, Ann. Phys. **24** (1935) 719–752.
[11] J.B. Zeldovich, On the theory of new phase formation; cavitation, Acta Physiochimica U.R.S.S. **18** (1943) 1–22.
[12] J. Frenkel, Statistical theory of condensation phenomena, J. Chem. Phys. **7** (1939) 200–201.
[13] J. Frenkel, *Kinetic Theory of Liquids*, Clarendon, Oxford (1946).
[14] D. Turnbull, J.C. Fisher, Rate of nucleation in condensed systems, J. Chem. Phys. **17** (1949) 71–73.
[15] J.C. Holzer, K.F. Kelton, The structural relations between amorphous, icosahedral, and crystalline phases, in: *Crystal-Quasicrystal Transitions*, Eds. M. Yacaman, M. Torres, Elsevier, Amsterdam, (1993), pp. 103–114.
[16] K.F. Kelton, Quasicrystals and Related Structures, in: *Intermetallic Compounds*, Eds. J.H. Westbrook, R.L. Fleischer, John Wiley, New York, (1995), pp. 229–268.
[17] K.F. Kelton, G.W. Lee, A.K. Gangopadhyay, R.W. Hyers, T. Rathz, J. Rogers, M.B. Robinson, D. Robinson, First x-ray scattering studies on electrostatically-levitated metallic liquids — demonstrated influence of local icosahedral order on the nucleation barrier of ordered phases, Phys. Rev. Lett. **90** (2003); 195504/1–4.
[18] D.T. Wu, Nucleation theory, in: Eds. H. Ehrenreich, F. Spaepen, Solid State Physics, Vol. 50, Academic, Boston, (1997), pp. 37–187.
[19] K.F. Kelton, M.C. Weinberg, Calculation of macroscopic growth rates from nucleation data, J. Non-Cryst. Solids **180** (1994) 17–24.
[20] F.C. Frank, M. Tosi, On the theory of polymer crystallization, Proc. Roy. Soc. Lond. A **263** (1961) 323–339.
[21] B. Lewis, V. Halpern, Surface diffusion capture in nucleation theory, J. Cryst. Growth **33** (1976) 39–52.
[22] C.H. Yang, H. Qui, Theory of homogenous nucleation: a chemical kinetic view, J. Chem. Phys. **84** (1986) 416–423.
[23] J.L. Katz, H. Wiedersich, Nucleation theory without Maxwell demons, J. Coll. Interf. Sci. **61** (1977) 351–355.
[24] J.L. Katz, F. Spaepen, A kinetic approach to nucleation in condensed systems, Philos. Mag. B. **37** (1978) 137–148.
[25] J.L. Katz, M.D. Donohue, A kinetic approach to homogeneous nucleation theory, Adv. Chem. Phys. **40** (1979) 137–155.
[26] J. Frenkel, Statistical theory of condensation phenomena, J. Phys. **1** (1939) 243–246.
[27] E. Ruckenstein, B. Nowakowski, A kinetic theory of nucleation in liquids, J. Coll. Interface Sci. **137** (1990) 583–592.
[28] B. Nowakowski, E. Ruckenstein, A kinetic approach to the theory of nucleation in gases, J. Chem. Phys. **94** (1991) 1397–1402.
[29] B. Nowakowski, E. Ruckenstein, Homogeneous nucleation in gases: a three-dimensional Fokker-Planck equation for evaporation from clusters, J. Chem. Phys. **94** (1991) 8487–8492.

[30] B. Nowakowski, E. Ruckenstein, A kinetic treatment of heterogeneous nucleation, J. Phys. Chem. **96** (1992) 2313–2316.

[31] S.H. Bauer, D.J. Frurip, Homogeneous nucleation in metal vapors. 5. A self-consistent kinetic model, J. Phys. Chem. **81** (1977) 1015–1024.

[32] C.F. Wilcox, S.H. Bauer, Estimation of homogeneous nucleation flux via a kinetic model, J. Chem. Phys. **94** (1991) 8302–8309.

[33] W.H. Press, S.A. Teukolsky, W.T. Vetterling, B.P. Flannery, *Numerical Recipes in Fortran*, Cambridge University Press, Cambridge (1992).

[34] K.F. Kelton, A.L. Greer, C.V. Thompson, Transient nucleation in condensed systems, J. Chem. Phys. **79** (1983) 6261–6276.

[35] K.F. Kelton, R. Falster, D. Gambaro, M. Olmo, M. Cornara, P.F. Wei, Oxygen precipitation in silicon: Experimental studies and theoretical investigations within the classical theory of nucleation, J. Appl. Phys. **85** (1999) 8097–8111.

[36] P.F. Wei, K.F. Kelton, R. Falster, Coupled-flux nucleation modeling of oxygen precipitation in silicon, J. Appl. Phys. **88** (2000) 5062–5070.

[37] A.L. Greer, K.F. Kelton, Nucleation in lithium disilicate glass: A test of classical theory by quantitative modeling, J. Am. Ceram. Soc. **74** (1991) 1015–1022.

[38] B. Shizgal, J.C. Barrett, Time dependent nucleation, J. Chem. Phys. **91** (1989) 6505–6518.

[39] H.A. Kramers, Brownian motion in field of force and diffusion model of chemical reactions, Physica **7** (1940) 284–304.

[40] J.E. Moyal, Stochastic processes and statistical physics, J. R. Stat. Soc. Lond. **B11** (1949) 150–210

[41] F.C. Goodrich, Nucleation rates and the kinetics of particle growth. I. The pure birth process, Proc. Roy. Soc. Lond. A **277** (1964) 155–166.

[42] F.C. Goodrich, Nucleation rates and the kinetics of particle growth. II. The birth and death process, Proc. Roy. Soc. Lond. A **277** (1964) 167–182.

[43] Y. Rabin, M. Gitterman, Time-dependent effects in nucleation near the critical point, Phys. Rev. A **29** (1984) 1496–1505.

[44] I. Edrei, M. Gitterman, Fast and slow quenches in nucleation: comparison of the theory with experiment and numerical simulations, Phys. Rev. A **33** (1986) 2821–2824.

[45] A.E. Nielsen, Kinetics of precipitation, *International Series of Monographs on Analytical Chemistry*. Vol. 18, Macmillan, New York (1964).

[46] D. Kashchiev, On the relation between nucleation work, nucleus size, and nucleation rate, J. Chem. Phys. **76** (1982) 5098–5102.

[47] Y. Viisanen, R. Strey, H. Reiss, Homogeneous nucleation rates for water, J. Chem. Phys. **99** (1993) 4680–4692.

[48] R. Strey, Y. Viisanen, Measurement of the molecular content of binary nuclei. Use of the nucleation rate surface for ethanol-hexanol, J. Chem. Phys. **99** (1993) 4693–4704.

[49] Y. Viisanen, R. Strey, A. Laaksonen, M. Kulmala, Measurement of the molecular content of binary nuclei. II. Use of the nucleation rate surface for water-ethanol, J. Chem. Phys. **100** (1994) 6062–6072.

[50] D.W. Oxtoby, D. Kashchiev, A general relation between the nucleation work and the size of the nucleus in multicomponent nucleation, J. Chem. Phys. **100** (1994) 7665–7671.

[51] R. McGraw, D.T. Wu, Kinetic extensions of the nucleation theorem, J. Chem. Phys. **118** (2003) 9337–9347.

[52] R. McGraw, A. Laaksonen, Scaling properties of the critical nucleus in classical and molecular-based theories of vapor-liquid nucleation, Phys. Rev. Lett. **76** (1996) 2754–2757.

[53] D. Kashchiev, *Nucleation — Basic Theory with Applications*, Butterworth-Heinemann, Oxford (2000).

CHAPTER 3

Time-Dependent Effects within the Classical Theory

Contents

1. QUALITATIVE DISCUSSION OF TIME-DEPENDENT NUCLEATION

In a limited number of cases, particularly for the silicate glasses, the nucleation rate can be measured directly by counting the number of nuclei produced per unit amount of substance (e.g., volume, mole, etc.) as a function of time (see Chapter 8). Frequently, the measured rates change with time, and hence are not described by the steady-state rates derived in Chapter 2. Such time-dependent, or *transient*, nucleation rates are common in condensed phases, having been observed in a wide range of glass-forming systems (Chapter 8), for example, and playing a central role in some solid-state precipitation processes (Chapter 9). Transient nucleation can be important for metastable phase formation

Pergamon Materials Series, Volume 15
ISSN 1470-1804, DOI 10.1016/S1470-1804(09)01503-X

and stability (Chapter 8). In this chapter, we examine transient nucleation within the bounds of the classical theory of nucleation.

The number of nuclei, χ, produced as a function of time, t, is equal to the integral of the time-dependent nucleation rate, $I(t)$,

$$\chi(t) = \int_0^t I(t)\, dt. \tag{1}$$

Were the nucleation rate steady-state, a plot of the number of nuclei produced as a function of time should yield a straight line with slope equal to the steady-state value, I^{st}. As shown in Figure 1, however, a nonlinear behavior is frequently observed. In this figure, showing the rate of production of nuclei of the cubic crystal phase in lithium-disilicate glass, the nucleation rate (given by the local slope) is initially low, but increases with time to I^{st}. For long annealing times, $\chi(t)$ is approximately

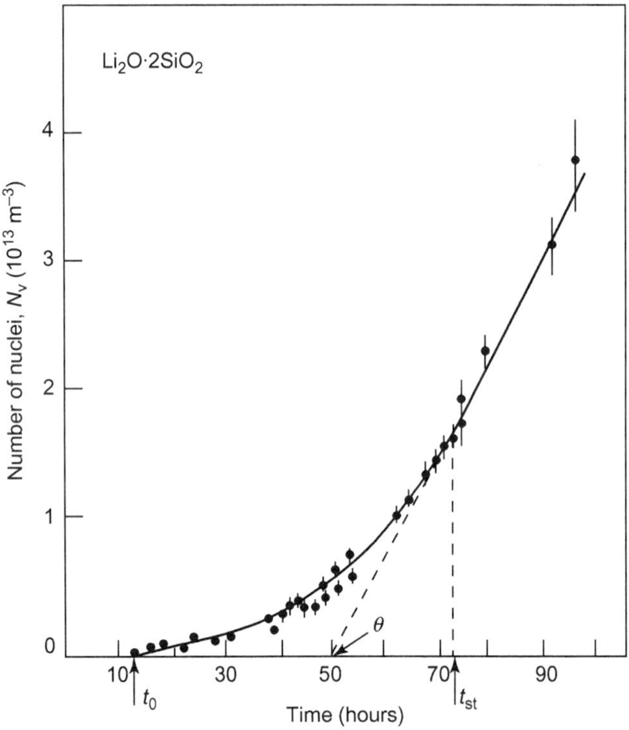

Fig. 1 The number of nuclei produced as a function of time at 703 K for lithium-disilicate glass. The time for onset of nucleation t_0, the time to steady-state t_{st}, and the induction time, θ, are indicated. (Reprinted from Ref. [1], copyright (1981), with permission from Elsevier.)

$$\chi(t) = I^{st}(t - \theta), \quad \text{for } t \gg \theta, \tag{2}$$

where θ is an *effective time-lag*, also called an *induction time*. Experimentally, θ is obtained by extrapolating the number of nuclei produced as a function of time in the steady-state regime to the time axis (Figure 1). The induction time, the time before measurable nucleation is observed, t_0, and the time to reach steady state, t_{st}, are also indicated in this figure. For nucleation from the vapor or liquid, the transient period is short compared with the period of observation, and is therefore mostly of little concern.

To understand the origin of time-dependent nucleation, it is useful to return to the discussion presented in Chapter 2 of the evolution of the cluster size distribution with time. Assuming an initial distribution of single molecules, a numerical solution to the coupled differential equations describing cluster growth (Chapter 2, Eq. (21)) showed that the population of larger clusters increases with annealing time, eventually approaching the steady-state cluster size distribution at that temperature (Chapter 2, Figure 6a). Consequently, the nucleation rate calculated at a particular cluster size will also increase with increasing temperature, since it is expected to scale with the cluster population (Chapter 2, Eqs. (40) and (50)).

The nucleation rate in the glass shown in Figure 1 of this chapter is time-dependent for similar reasons. Here, rather than starting with a distribution of single molecules, the cluster size distribution is more similar to one at a higher temperature. During the glass quench, the temperature is decreased more rapidly than relaxation to the appropriate cluster size distribution can accommodate. The non-steady-state nucleation rate observed in Figure 1, then, reflects the time required for the evolution of the initial cluster size distribution to the steady-state one at the annealing temperature.

For illustration, Figure 2 shows the steady-state cluster size distributions calculated from Eq. (52) of Chapter 2 for two different temperatures, using parameters that are appropriate for the crystallization of lithium-disilicate glass (see Chapter 8). Assume that the glass was held at a temperature, $T_1 = 550°\text{C}$, for example, for a sufficiently long time to achieve the steady-state cluster size distribution. If the temperature is abruptly lowered to $T_2 = 390°\text{C}$, the cluster size distribution will immediately begin to evolve to the steady-state distribution appropriate for that temperature, following the reaction kinetics described by Eqs. (20) and (21) of Chapter 2. Because a finite amount of time is required to achieve the new cluster size distribution, however, the nucleation rate cannot instantly assume its steady-state value at T_2. Based on the steady-state calculation, and accounting for the instantaneous change in critical

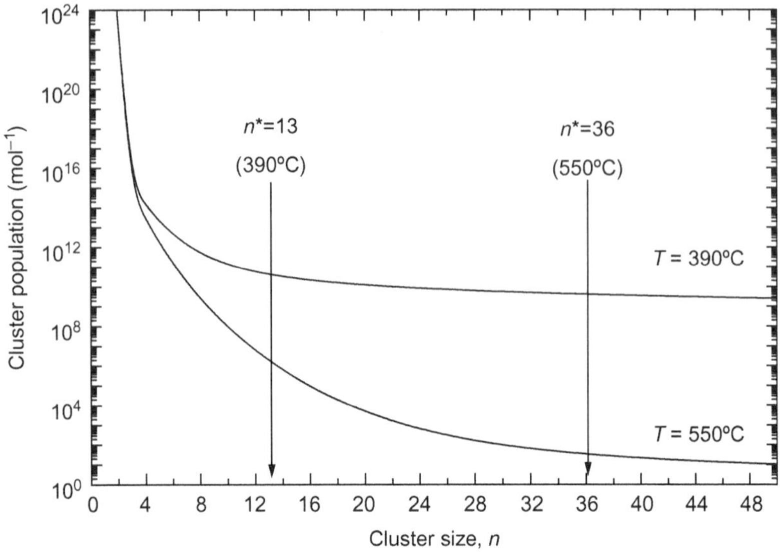

Fig. 2 The steady-state cluster populations at 390°C and 550°C, calculated using thermodynamic and kinetic parameters that are appropriate for crystallization of lithium-disilicate glass (Chapter 8). The critical sizes at the two temperatures are indicated. (Reprinted from Ref. [22], copyright (1991), with permission from Elsevier.)

size, the nucleation rate at T_2 should be proportional to the time-dependent critical cluster population at T_2 multiplied by the forward jump rate at T_2, all evaluated at $n^*(T_2)$, the critical cluster size at T_2:

$$I\left(n^*(T_2), T_2, t\right) \propto N\left(n^*(T_2), T_2, t\right) \, k^+\left(n^*(T_2), T_2\right). \tag{3}$$

Since the cluster population at $n^*(T_2)$ is initially lower than the appropriate steady-state population, the nucleation rate is initially significantly smaller than expected. As the cluster size distribution evolves toward the steady-state one for T_2, the nucleation rate rises to its steady-state value. This is precisely the behavior deduced from Figure 1. Were the temperature change reversed, so that T_2 were larger than T_1, $n^*(T_2)$ would initially be too large, and the nucleation rate would decrease with time. This behavior has also been observed (see Ref. [2] for example). Since the observed time-dependent nucleation rate reflects the cluster dynamics at much smaller cluster sizes, near and even below the critical size, it becomes possible, then, to probe that regime. In Chapter 8, it will be shown that by studying the nucleation behavior following appropriately chosen annealing treatments, the validity of the kinetic model for nucleation can be evaluated.

2. NUMERICAL ANALYSIS OF TIME-DEPENDENT NUCLEATION

Initial investigations of time-dependent nucleation were focused on nucleation from the vapor phase. The prime interest was in the formation of droplets in supersonic flows, such as occur in jet engines, which were under active development at the time. All estimates of the transient time for nucleation were based on expressions derived from the Zeldovich–Frenkel (ZF) equation developed in Chapter 2 (Section 9). Experimental and theoretical work showed that the transient times were so short in the cases of interest as be of little practical importance. However, there was renewed interest in the 1970s and 1980s, when detailed nucleation studies of metallic and silicate glasses began to reveal the behavior shown in Figure 1. This has led to a reevaluation of the earlier analytical expressions developed for time-dependent nucleation, casting doubt on some of the key assumptions made in their derivation. Further, as already mentioned in Chapter 2 (Section 10), even the fundamental equation on which most of these studies depend—the ZF equation—has recently been called into question.

Numerical calculations are free from the many, largely untestable, approximations made to obtain analytical solutions. They can also be extended to model realistic cases, including arbitrary initial cluster size distributions, nonisothermal annealing conditions, nucleation of a new phase with a composition different from that of the original phase, and heterogeneous nucleation. Following the approach used in Chapter 2, numerical calculations will be used to motivate a discussion of time-dependent nucleation; details of the numerical treatment used can be found in Ref. [3]. Attention will focus on the evolution of the cluster size distribution with annealing, which underlies all time-dependent nucleation phenomena, and on the evaluation of a few representative analytical expressions for the time-dependent nucleation rate and induction time that can be used for the analysis of experimental data.

2.1 Single-molecule initial distribution

Consider first the simplest initial distribution, containing only single molecules, and assume parameters that are similar to those for lithium-disilicate glass (see Chapter 8 for more detail) at an annealing temperature of 450°C. As shown in Chapter 2, Figure 6a, the cluster size distribution changes with annealing, with the population of larger clusters increasing and eventually reaching the time-independent steady-state distribution, N^{st}. The nucleation rate at a given cluster size increases accordingly. Figure 3a shows the calculated nucleation rate at the critical size, n^* for this annealing temperature. It is initially

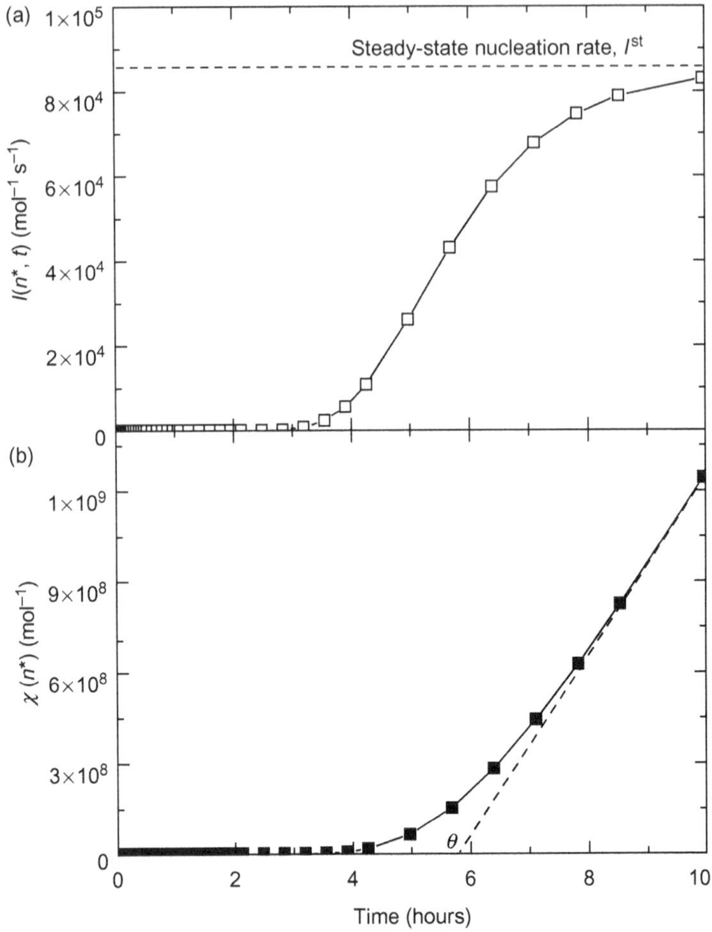

Fig. 3 The calculated time-dependent nucleation rate (a) and number of nuclei (b) at 450°C, measured for $n^* = 18$. The steady-state nucleation rate I^{st}, and the induction time θ, are indicated. Calculations were made using parameters appropriate for nucleation in lithium-disilicate glass (see Chapter 8).

low, reflecting the small cluster population at n^*. It increases sigmoidally with annealing time to approach the constant, steady-state, nucleation rate. As shown in Figure 3b, this is reflected in $\chi(t)$, the number of clusters larger than n^* (i.e., nuclei) appearing as a function of time (see Eq. (1)). After a long annealing time, $\chi(t)$ increases linearly with time; the slope of $\chi(t)$ is then equal to the steady-state rate. An extrapolation to the time axis gives a nonzero intercept, the induction time, θ. This predicted behavior is identical in character to that experimentally observed (Figure 1).

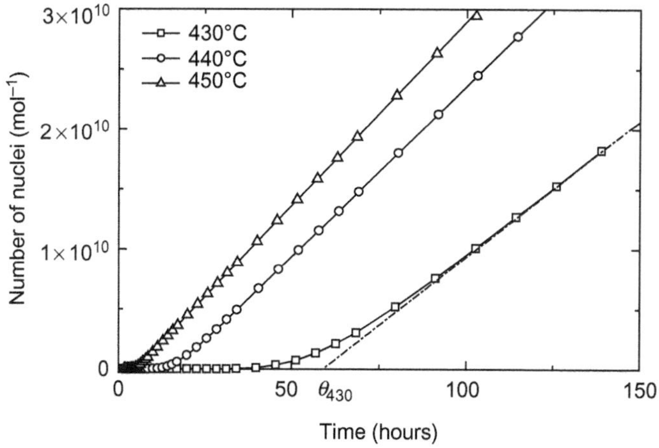

Fig. 4 The calculated number of nuclei for three temperatures, measured at $n = 52$, the critical size at 600°C. The induction time at 430°C, θ_{430}, is shown for illustration. Calculations were made using parameters appropriate for crystal nucleation in lithium-disilicate glass (see Chapter 8).

Figure 4 shows the calculated number of nuclei as a function of time for different annealing temperatures. For simplicity, since the critical size is also a function of temperature, the nucleation rate reported here is measured for the same size at all temperatures (i.e., $n^*_{600°C} = 52$). The shapes of the curves are independent of temperature, although the slope of the linear portions of the curves and the induction times are different. The different slopes are not surprising; they simply reflect the temperature dependence of the steady-state nucleation rate, discussed in Chapter 2.

The induction time has a strong temperature dependence (Figure 5a). There is little correlation between θ and the thermodynamic parameters of nucleation, illustrated by the weaker dependence on the work of critical cluster formation, $W^*(T)$, in Figure 5b. In contrast, θ scales very closely with the atomic mobility, calculated from the diffusion coefficient (Figure 5a).

Figure 6 shows the time-dependent behavior of the cluster size distribution, $N(n,t)$, the nucleation rate, $I(n,t)$, and the rate of change of the cluster population, $dN(n, t)/dt$, for a range of cluster sizes above and below the critical size. For all cluster sizes, the population increases monotonically with annealing time to approach the steady-state population at the annealing temperature (Figure 6a). Although the forward flux measured at any cluster size (the nucleation rate) must also approach the steady-state value for long annealing times, the behavior at intermediate times differs (Figure 6b). At and above n^*, $I(n, t)$ increases

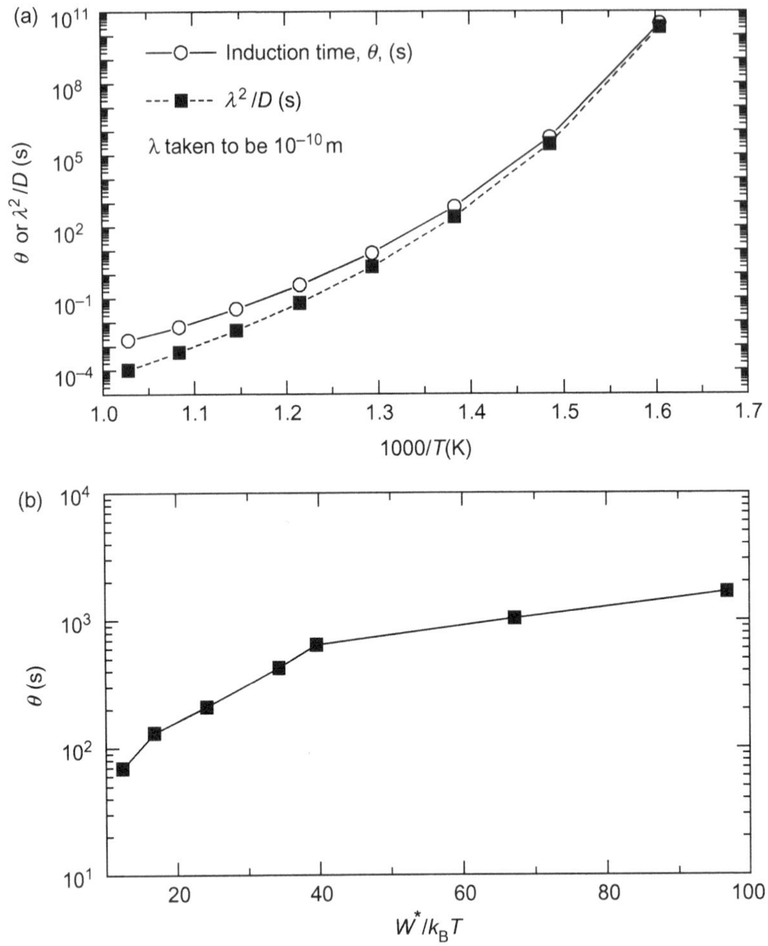

Fig. 5 A comparison between the induction time θ, the average atomic mobility ($\approx\lambda^2/D$, where D is the diffusion coefficient and λ is a jump distance), and $W(n^*)$, the critical work of cluster formation. Calculations were made using parameters appropriate for crystal nucleation in lithium-disilicate glass (see Chapter 8).

monotonically toward the same steady-state rate. For cluster sizes less than n^*, however, it first overshoots the steady-state rate before approaching it from the top side for long annealing times. The maximum is of greater magnitude and occurs at shorter times for smaller values of n. For all cluster sizes, $\mathrm{d}N(n,t)/\mathrm{d}t$ exhibits a maximum (Figure 6c). For smaller cluster sizes, the maximum is larger and occurs at shorter times. Also, for $n<n^*$, the curves in Figure 6c show an inflection point after the maximum.

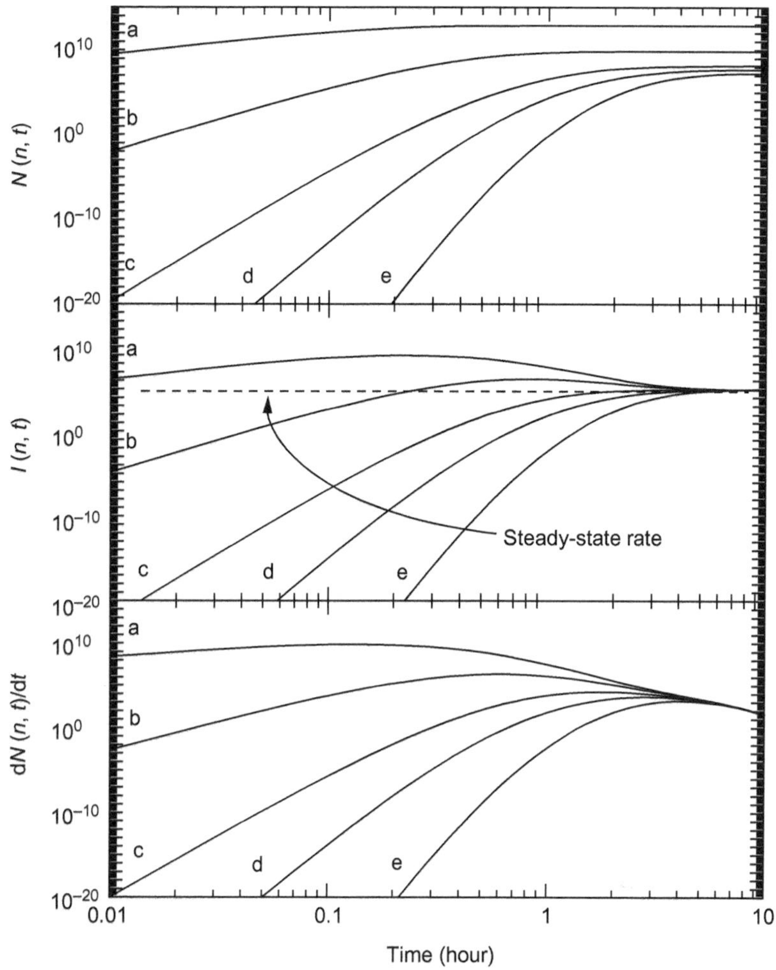

Fig. 6 (a) The cluster population, $N(n, t)$, (b) the forward flux, $I(n, t)$, and (c) the rate of change of cluster population, $dN(n,t)/dt$ as a function of times for selected cluster sizes: a, $n = 20$; b, $n = 31$; c, $n^* = 38$; d, $n = 49$; e, $n = 64$. Calculations were made for 450°C assuming parameters appropriate for nucleation in lithium-disilicate glass (see Chapter 8).

2.2 Effect of preexisting cluster distributions

In a real system, it is likely that there is a more complicated initial cluster size distribution, determined by the thermal history of the sample. Figure 7 contrasts the influence of three different initial (but still idealized) cluster size distributions on the transient nucleation behavior: (1) an initial single-molecule distribution; (2) a steady-state distribution characteristic of a temperature higher than the annealing temperature of 450°C; and (3) a steady-state distribution from a temperature that is lower than the annealing temperature. Again, these numerical calculations were made using parameters appropriate for the crystallization of lithium-disilicate glass (Chapter 8). The calculated nucleation rates for the three distributions are shown in Figure 7a; the corresponding numbers of critical nuclei produced are shown in Figure 7b. The single-molecule and high-temperature initial distributions predict similar changes with annealing for both the nucleation rate and the number of nuclei generated. The smaller time-lag for the high-temperature distribution simply reflects the nonzero initial cluster population. Very different behavior is observed for the low-temperature initial distribution. Based on the steady-state arguments made earlier in this chapter, the population of clusters of a given size becomes larger as the temperature is lowered, reflecting the greater driving free energy for the transformation. When the temperature is subsequently raised, the initial nucleation rate can be much higher than expected due to this large

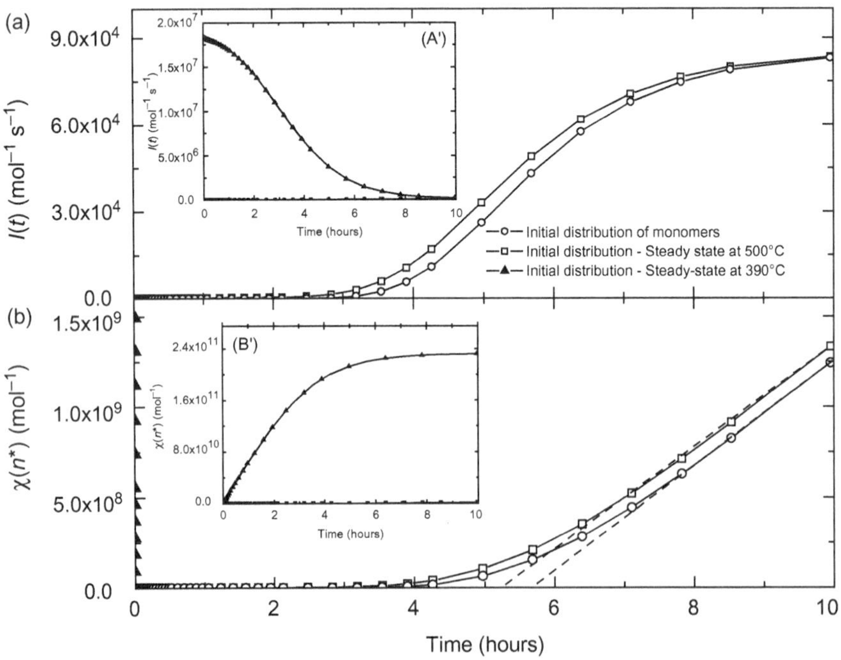

Fig. 7 The nucleation rate (a) and the number of nuclei produced (b) with annealing time at 450°C, assuming an initial distribution of single molecules, the steady-state distribution at 500°C, and the steady-state distribution at 390°C. The results for the initial steady-state distribution at 390°C are shown in the inset. Calculations were made using parameters appropriate for nucleation in lithium-disilicate glass (see Chapter 8).

cluster population (Figure 7a inset). With annealing, the cluster population decreases to the steady-state value at the higher temperature, and the nucleation rate drops to the appropriate steady-state value.

Clearly, the initial distribution can have a profound effect on the nucleation behavior and must be considered for a quantitative analysis. It is a weakness of almost all existing expressions for transient nucleation that they are derived for an initial distribution of single molecules. At best, they can be used for qualitative insight only, though frequently that is sufficient.

3. ANALYTICAL SOLUTIONS TO THE DISCRETE COUPLED DIFFERENTIAL EQUATIONS

Numerical treatments are required if accurate information is needed for complicated annealing treatments. A simple analytical expression for the transient nucleation rate and the induction time would be sufficient in most cases, however. Such expressions have traditionally been developed from the ZF equation (Chapter 2, Eq. (59)), which, as already mentioned in Chapter 2, has been recently questioned. Further, questionable assumptions are frequently made to solve the partial differential equation, casting some doubt on the treatments and their results [4]. A second approach focuses on the discrete Becker-Döring equations. A brief discussion of these approaches and the expressions obtained is provided in Sections 3 and 4.

The most straightforward approach to a solution of the discrete coupled equations is the derivation of an expression for the time-dependent cluster population. Once obtained, the nucleation rate, transient time, and all other parameters of interest are readily found. Frequently, however, this approach involves more intense numerical calculation than for the direct numerical treatment already discussed. The second approach focuses on obtaining an expression for the induction time, which is the directly measured quantity. Both approaches are briefly examined in this section.

3.1 Time-dependent cluster population

The coupled set of linear differential equations describing the rate of change of the cluster population (Chapter 2, Eq. (21)) can be written in matrix form as (considering only the derivative with respect to time),

$$\frac{d\mathbb{N}}{dt} = \bar{\mathbf{K}}\,\mathbb{N}, \tag{4}$$

where $\mathbb{N}$ is the time-dependent cluster population matrix

$$\mathbb{N} = \begin{bmatrix} N(\tilde{u}, t) \\ N(\tilde{u}+1, t) \\ \vdots \\ N(\tilde{v}, t) \end{bmatrix} \tag{5}$$

and $\bar{\mathbf{K}}$ is a tridiagonal rate-constant matrix,

$$\bar{\mathrm{K}} = \begin{pmatrix} -k^{+}(\tilde{u}) & k^{-}(\tilde{u}+1) & 0 & 0 & \cdots\cdots & 0 \\ k^{+}(\tilde{u}) & -[k^{-}(\tilde{u}+1)-k^{+}(\tilde{u}+1)] & k^{-}(\tilde{u}+2) & 0 & \cdots\cdots & 0 \\ 0 & k^{+}(\tilde{u}+1) & -[k^{-}(\tilde{u}+2)-k^{+}(\tilde{u}+2)] & k^{-}(\tilde{u}+3) & \cdots\cdots & 0 \\ \cdot & \cdot & & & & \\ \cdot & \cdot & & & & \\ \cdot & \cdot & & & & \\ 0 & 0 & \cdots\cdots & & k^{+}(\tilde{v}-1) & -[k^{-}(\tilde{v})+k^{+}(\tilde{v})] \end{pmatrix} \tag{6}$$

Here, ũ and ṽ are the lower and upper limits, respectively, on the number of atoms in the clusters (see Chapter 2, Section 7). This seems a natural starting point for analytical treatments. The time-dependent population can be obtained directly by diagonalizing the rate-constant matrix [3]. The cluster population as a function of time in terms of the initial distribution is then

$$N(n, t) = \sum_{m,j=\tilde{u}}^{\tilde{v}} N_0(j, 0)\, b_{nm} b_{mj}^{-1}\, \exp(\lambda_m t), \tag{7}$$

where the bs are the elements of the matrix $\mathbf{B}$ that diagonalizes $\bar{\mathbf{K}}$ (i.e., $\mathbf{A} = \mathbf{B}^{-1}\, \bar{\mathbf{K}}\, \mathbf{B}$), the λs are the eigenvectors of the diagonalized matrix, $\mathbf{A}$, and $N_0(j,0)$ is the initial density of clusters containing j molecules. This expression for the cluster population is exact. It can be applied to any isothermal, multistate rate problem with any initial cluster population, when only transitions between neighboring states are allowed. The method could be extended to determine the time-dependent cluster population for transformations involving a changing driving free energy, such as in nonisothermal transformations, or precipitation processes with a changing supersaturation, by discretizing the driving free energy as a function of time. The generally large size of the rate coefficient matrix, however, makes the diagonalization difficult, even in the case of constant driving free energy. To model nucleation under other conditions requires the complete diagonalization of $\bar{\mathrm{K}}$ for each discrete change in driving free energy, making the procedure impractical with reasonable computer resources. Practically, then, this approach is of less use than the numerical methods discussed previously.

3.2 Calculation of the induction time

The second method focuses attention on the induction time for nucleation, θ, the experimentally observable quantity that is often of most interest (see Figure 1 of this chapter for example). The numerical results discussed earlier show that the induction time and the nucleation rate are both functions of the cluster size at which they are measured. An expression for $\theta(n)$ is obtained when Eq. (1), the general expression for the number of nuclei produced as a function of time, is combined with Eq. (2), which gives the time-dependent number in the asymptotic limit:

$$\theta(n) = t - \int_0^t \frac{I(n,t)}{I^{st}}\,\mathrm{d}t = \int_0^t \left(1 - \frac{I(n,t)}{I^{st}}\right)\mathrm{d}t \Rightarrow \int_0^\infty \left(1 - \frac{I(n,t)}{I^{st}}\right)\mathrm{d}t. \quad (8)$$

The final expression follows since the time-dependent nucleation rate approaches its steady-state value in the asymptotic limit. The induction time is, therefore, a function of the complete time history of the nucleation rate. Several solutions to this equation have been offered [5–10]. The simplest approach, proposed recently by Wu [9], is reproduced here. Were the integrand, Y, a perfect differential, that is,

$$\frac{\mathrm{d}Y(n,t)}{\mathrm{d}t} = 1 - \frac{I(n,t)}{I^{st}}, \quad (9)$$

the solution for the induction time would be trivial:

$$\theta(n) = Y(n,\infty) - Y(n,0) = Y^{st}(n) - Y(n,0), \quad (10)$$

where $Y^{st}(n)$ represents the steady-state (or infinite-time) value.

As discussed in Chapter 2, from the Becker–Döring solution to the steady-state rate using the principle of detailed balance, the flux past a cluster of size n can be written as

$$I(n,t) = k^+(n)N^{eq}(n)\left(\frac{N(n,t)}{N^{eq}(n)} - \frac{N(n+1,t)}{N^{eq}(n+1)}\right) = -k^+(n)N^{eq}(n)\Delta F(n,t), \quad (11)$$

where $F(n,t) = N(n,t)/N^{eq}(n)$ and $\Delta F(n,t) = F(n+1,t) - F(n,t)$. (Note that in Ref. [9] quantities are referenced to the mid-point, $n+1/2$, so that there is no confusion that the flux is from n to $n+1$; the different notation here is adopted to be consistent with that used in Chapter 2). The same time-independent boundary conditions used in Chapter 2 to obtain the value for the steady-state nucleation rate are again assumed

$$F(\tilde{u},t) = 1$$
$$F(\tilde{v},t) = 0. \quad (12)$$

The rate of change of the cluster population in this notation is then (from Eqs. (21) and (22) of Chapter 2)

$$\frac{\mathrm{d}N(n,t)}{\mathrm{d}t} = I(n-1,t) - I(n,t). \tag{13}$$

As discussed before, the minimum and maximum cluster sizes should be chosen to be as near to 1 and as large as possible respectively. Employing manipulations like those used in Chapter 2 to obtain an expression for the steady-state nucleation rate, Eq. (13) is summed from cluster size p to n, giving

$$\frac{\mathrm{d}}{\mathrm{d}t}\sum_{r=p+1}^{n} N(r,t) = I(p,t) - I(n,t) = -k^{+}(p)N^{\mathrm{eq}}(p)\,\Delta F(p,t) - I(n,t). \tag{14}$$

For steady state,

$$I^{\mathrm{st}} = -k^{+}(n)\,N^{\mathrm{eq}}(n)\Delta F^{\mathrm{st}}(n) = -k^{+}(n)N^{\mathrm{eq}}(n)\big(F^{\mathrm{st}}(n+1) - F^{\mathrm{st}}(n)\big), \tag{15}$$

valid for any value of n. Dividing both sides of Eq. (14) by $(k^{+}(p)\,N^{\mathrm{eq}}(p))$, and taking the steady-state rate from Eq. (15), Eq. (14) can be rewritten as

$$-\frac{1}{I^{\mathrm{st}}}\frac{\mathrm{d}}{\mathrm{d}t}\left(\Delta F^{\mathrm{st}}(p)\sum_{r=p+1}^{n} N(r,t)\right) = -\Delta F(p,t) + \Delta F^{\mathrm{st}}(p)\frac{I(n,t)}{I^{\mathrm{st}}}. \tag{16}$$

Summing Eq. (16) over p, over the allowed values of cluster size in the ensemble, $\tilde{\mathrm{u}}$ to $\tilde{\mathrm{v}}-1$, and using the boundary conditions in Eq. (12), Eq. (16) can be rewritten as

$$\begin{aligned} &-\left(\frac{1}{I^{\mathrm{st}}}\right)\frac{\mathrm{d}}{\mathrm{d}t}\left(\sum_{p=\tilde{\mathrm{u}}}^{\tilde{\mathrm{v}}-1}\Delta F^{\mathrm{st}}(p)\sum_{r=p+1}^{n} N(r,t)\right) \\ &= F(n_{\tilde{\mathrm{u}}}) - F(n_{\tilde{\mathrm{v}}}) + (F^{\mathrm{st}}(n_{\tilde{\mathrm{u}}}) - F^{\mathrm{st}}(n_{\tilde{\mathrm{v}}}))\frac{I(n,t)}{I^{\mathrm{st}}} \\ &= 1 - \frac{I(n,t)}{I^{\mathrm{st}}}. \end{aligned} \tag{17}$$

Using the principle of summation by parts (analogous to integration by parts), Wu [4, 9] showed that Eq. (17) could be recast into the form

$$Y(n,t) = 1 - \frac{I(n,t)}{I^{\mathrm{st}}} = \frac{1}{I^{\mathrm{st}}}\frac{\mathrm{d}}{\mathrm{d}t}\left(\sum_{p=\tilde{\mathrm{u}}+1}^{n} N(p,t) - \sum_{p=\tilde{\mathrm{u}}+1}^{\tilde{\mathrm{v}}-1} F^{\mathrm{st}}(p)N(p,t)\right), \tag{18}$$

which is the exact differential that we needed to obtain the induction time. From Eq. (10) the induction time at any cluster size is given by

$$\theta(n) = \frac{1}{I^{\mathrm{st}}}\left(\sum_{p=\tilde{\mathrm{u}}+1}^{n}(N^{\mathrm{st}}(p) - N(p,0)) - \sum_{p=\tilde{\mathrm{u}}+1}^{\tilde{\mathrm{v}}-1}\frac{N^{\mathrm{st}}(p)}{N^{\mathrm{eq}}(p)}(N^{\mathrm{st}}(p) - N(p,0))\right), \tag{19}$$

where $N(p,0)$ is the initial density of clusters of size p (i.e., the initial cluster distribution). This expression is exact. It requires only that the initial cluster size distribution and the steady-state one (given in Chapter 2, Eq. (52)) be known.

The induction time, θ, is only the zeroth moment of the time-dependent nucleation flux. The time-dependent rate could be expressed exactly given the steady-state nucleation rate, and the complete set of moments [11],

$$M_k(n) = \int_0^\infty t^k \left(1 - \frac{I(n,t)}{I^{st}}\right) dt, \tag{20}$$

where $M_k(n)$ is the kth moment about zero time of the function $(1 - I(n,t)/I^{st})$. A similar approach was first followed by Hile [7] and Shizgal and Barrett [8]. For cases where the nucleation rate increases monotonically with time, $I(n, t)$ can be characterized by I^{st}, θ, and the reduced moment, $M_r(n)$, which describes the time breadth of the transient region [4, 12]. The value of $M_r(n)$ is of considerable interest and controversy, but will not be discussed here (see Refs. [12–14] for more information).

4. ANALYTICAL SOLUTIONS OF THE ZELDOVICH–FRENKEL EQUATION

As already discussed (Chapter 2, Section 9), the ZF equation has traditionally been the starting point for calculations of both steady-state and time-dependent nucleation. For that reason, and because solutions lead to extremely simple expressions for the induction time that can be used to evaluate quickly the potential importance of transient effects, these are discussed here. Because it is difficult to obtain closed-form solutions to the ZF equation, many approximate solutions have been proposed, based on different methods for separating the *diffusion* region, near the critical size, from the *drift* regions at other cluster sizes, and generally assuming a parabolic form for the barrier (as used to obtain the steady-state nucleation rates in Chapter 2). Further, it is frequently assumed (incorrectly) that the cluster population evolution underlying time-dependent nucleation behavior can be described by a single relaxation time. An overview of the derivation of these expressions is provided in this section. The limits of validity are discussed in Section 5.

4.1 Single relaxation time

Recall that the ZF equation (Chapter 2, Eq. (59)) has the appearance of a diffusion equation, where $k^{+}(n)$ and $N(n, t)$ are analogs of the diffusion coefficient and concentration respectively. The flux in size space, $I(n, t)$, also has two components, corresponding to the change in the cluster size distribution with cluster size and dn/dt, the drift velocity in size space (Chapter 2, Eq. (61)). Now imagine a situation where a distribution of clusters is allowed to equilibrate under a small supersaturation, s_1, giving the steady-state distribution, $N^{st}(n, s_1)$. If the supersaturation is suddenly increased to s_2, it is reasonable to expect that the steady-state population can be attained up to some cluster size, $\hat{n}$. For clusters larger than $\hat{n}$, however, $N(n, s_2) \approx N^{st}(n, s_1)$. The time required to establish the steady-state population that is appropriate for s_2 at these larger cluster sizes is the average time required for a cluster to grow from $\hat{n}$ to n. The diffusion of clusters in size space is most important near the critical size; the drift velocity term becomes dominant at larger cluster sizes. Focusing our attention on cluster motion in the critical region (i.e. n near n^*), then, it is reasonable to expect that the time-dependent cluster population will be analogous to the concentration in the problem of one-dimensional atomic diffusion under a concentration gradient, that is (evaluated at n^*),

$$N(n^*, t) = N^{st}(n) \exp\left(-\frac{(n - \hat{n})^2}{4D't}\right), \tag{21}$$

where D' is an effective "diffusion coefficient" in size space. Because the flux is proportional to the cluster population, it is reasonable to expect a similar form for the time-dependent nucleation rate,

$$I(n^*, t) = I^{st} \exp\left(-\frac{(n^* - \hat{n})^2}{4D't}\right). \tag{22}$$

For diffusion in the one-dimensional size space, $(n^* - \hat{n}) = \sqrt{D'\hat{t}}$, where $\hat{t}$ is the time required to diffuse through the critical region, that is, the transient time τ. The time-dependent nucleation rate is, therefore, expected to have the form,

$$I(n, t) = I^{st} \exp\left(-\gamma \frac{\tau}{t}\right), \tag{23}$$

where γ is a proportionality constant. From Eq. (1), the total number of nuclei is

$$\chi(t) = \int_0^t I(n, t')\,dt' = \int_0^t I^{st} \exp\left(-\gamma \frac{\tau}{t'}\right) dt'. \tag{24}$$

It is important to note that the transient time is related to, but not equal to, the induction time θ.

The value of D' depends strongly on the cluster size. Below n^*, the net forward rate is small; above n^* it becomes increasingly rapid, dominated by the drift term. If the rate-limiting step is cluster diffusion through the critical region, it is reasonable to take $D' \approx k^+(n^*)$, where $k^+(n^*)$ is the rate of single-molecule addition to the critical cluster (see Chapter 2, Section 3). Assuming that there is no initial distribution, $\hat{n} \approx 0$, the time-dependent nucleation rate is obtained,

$$I(n,t) \simeq I^{\mathrm{st}} \exp\left(-\frac{(n^*)^2}{4k^+(n^*)t}\right). \tag{25}$$

Because the backward cluster flux is small in the initial stages of cluster relaxation, when the cluster size population at larger sizes is approximately zero, the assumptions made are reasonable for short times. Further, although Eq. (25), originally proposed by Zeldovich [15], is probably the most widely used, comparisons with numerical solutions demonstrate that the form of $I(n,t)$ and the predicted transient times are incorrect [3]. Several alternative expressions have been proposed (see Ref. [3] for a review). An expression obtained by Kashchiev [16], which is based on the assumption of a single relaxation time, gives good agreement with numerical calculations over a reasonable range of parameter space [3]. It is an easily used, reasonably accurate method for analyzing experimental time-dependent nucleation data in condensed systems.

4.2 Kashchiev treatment

Assuming the ZF equation (Chapter 2, Eqs. (59) and (60)) and choosing an initial distribution of single molecules, the boundary conditions on the cluster population are

$$\begin{aligned} &N(n,0) = 0, \qquad \text{for } n>1 \\ &N(1,t) = N^{\mathrm{eq}}(1), \\ &\underbrace{\lim}_{n\to\infty} N(n,t) \to 0. \end{aligned} \tag{26}$$

Following the development of the Becker–Döring equation for the steady-state nucleation rate, the time-dependent forward flux at a cluster size, n, is conveniently written as (see Chapter 2, Eq. (58))

$$I(n,t) = -k^+(n)N^{\mathrm{eq}}(n)\frac{\partial}{\partial n}\left(\frac{N(n,t)}{N^{\mathrm{eq}}(n)}\right) = -k^+(n^*)N^{\mathrm{eq}}(n)\frac{\partial F(n,t)}{\partial n}, \tag{27}$$

where, as before, $F(n,t) = N(n,t)/N^{eq}(n)$. By continuity (see Chapter 2, Eq. (59)),

$$\frac{\partial F(n,t)}{\partial t} = \frac{1}{N^{eq}(n)} \frac{\partial}{\partial n} \left(k^+(n) N^{eq}(n) \frac{\partial F(n,t)}{\partial n} \right). \tag{28}$$

If $F(n,t)$ is expressed as the sum of the steady-state solution, $F^{st}(n)$, and the deviation from steady state, $\psi(n, t)$

$$F(n,t) = F^{st}(n) + \psi(n,t), \tag{29}$$

two separate differential equations are obtained from Eq. (28), one for the steady-state solution and one for the time-dependent deviation. Assuming that $\psi(n, t)$ is separable into the product of a function of n alone and one of only t , $\psi(n,t) = f(n)g(t)$, the time-dependent equation becomes

$$\frac{1}{g(t)} \frac{dg(t)}{dt} = \frac{1}{N^{eq}(n) f(n)} \frac{d}{dn} \left(k^+(n) N^{eq}(n) \frac{df(n)}{dn} \right). \tag{30}$$

Because the left side of Eq. (30) is a function of time only and the right-hand side is a function of size alone, both must be equal to a constant, $-\lambda$. The function $g(t)$ is then a decaying exponential in time, and the size-dependent function, $f(n)$, is given by the solutions of the resulting Sturm–Liouville equation. The deviation is then

$$\psi(n,t) = \sum_{k=1}^{\infty} A_k f_k(n)\, e^{-\lambda_k t}, \tag{31}$$

where $f_k(n)$ and λ_k are the eigenfunctions and eigenvalues, respectively, of the Sturm–Liouville equation (see Ref. [16]).

Following the development in Chapter 2 for the steady-state nucleation rate, the reversible work of cluster formation is expanded about the critical size, keeping only the first two nonzero terms. Both the steady-state and time-dependent solutions can be expressed in terms of two different functions of n, $u_1(n)$ and $u_2(n)$, which are themselves dependent on the error function,

$$\varphi(\xi) = \text{erf}(\xi) = \frac{2}{\sqrt{\pi}} \int_0^{\xi} \exp(-y^2)\, dy, \tag{32}$$

where

$$\xi = (n - n^*) \sqrt{-\frac{1}{2k_B T} \left(\frac{\partial^2 W(n)}{\partial n^2} \right)_{n=n^*}} = \frac{(n - n^*)}{\Delta}. \tag{33}$$

As discussed in Chapter 2, Δ is the width (in units of cluster size) of the critical region, for which the work of cluster formation is within $k_B T$ of

$W(n^*)$ (Chapter 2, Figure 1). The absolute value of $\varphi(\tilde{\xi})$ differs from unity only when $|\hat{\xi}|<2$, or equivalently, $|n-n^*|/\Delta<2$. A straightforward manipulation shows that $f(n)$ is zero outside the critical region, demonstrating that cluster evolution within the critical region dominates the observed transient nucleation behavior. This important conclusion supports some of the assumptions originally made by Zeldovich and underlies the rest of Kashchiev's derivation. It is also strictly incorrect, since additional time is required for clusters to grow larger than n^*, where nucleation rates are actually measured; further, a more complete analysis demonstrates that more than one characteristic time is needed to describe the relaxation of the cluster distribution to the steady-state value. These points are discussed further in Sections 4.3 and 4.4.

Several additional assumptions are made to solve the differential equation; $k^+(n)$ is set to $k^+(n^*)$, for all n, and $u_1(n)$ and $u_2(n)$ are assumed to be linear functions of n in the critical region, making their second derivatives zero. As discussed in Chapter 2, $N^{st}(n^*) = N^{eq}(n^*)/2$, giving $F^{st}(n^*) = 1/2$; the steady-state population is taken to vary linearly about this value through the critical region. The solution for $F(n, t)$ in the small neighborhood around the critical size is then,

$$\begin{aligned} F(n,t) = {} & \frac{1}{2} - \frac{n-n^*}{\Delta\sqrt{\pi}} \\ & + \sum_{n=1}^{\infty} \frac{1}{n\pi}\left(\cos(n\pi) - 1 - \frac{4}{\sqrt{\pi}}(1+\cos n\pi)\right) \\ & \times \sin\left(\frac{n\pi(n-n^*+2\Delta)}{4\Delta}\right) \times \exp\left(\frac{n^2\pi^2k^+(n^*)t}{16\Delta^2}\right). \end{aligned} \tag{34}$$

From Eq. (34) and a summation identity discussed in Ref. [16], a universal scaling relation is predicted, involving only the steady-state nucleation rate, I^{st}, and the transient time according to Kashchiev, τ_K

$$\frac{I(n^*,t)}{I^{st}} = 1 + 2\sum_{m=1}^{\infty}(-1)^m \exp\left(-\frac{m^2t}{\tau_K}\right), \tag{35}$$

where

$$\tau_K = -\frac{24k_BTn^*}{\pi^2k^+(n^*)\Delta\mu} = \frac{4}{\pi^3k^+(n^*)Z^2} \tag{36}$$

and Z is the Zeldovich factor (Chapter 2, Eq. (51)).

4.3 Asymptotic solution

Asymptotic methods have been used to obtain solutions of the ZF equation that avoid the restrictive assumptions of the Kashchiev treatment [17–19]. In one approach, assuming a high nucleation barrier ($W^* \gg k_B T$) and letting $F(n,t) = N(n,t)/N^{eq}(n)$, a Laplace transform is taken of the ZF equation to obtain the following ordinary differential equation [19],

$$\frac{\hat{\Delta}^2}{2}\frac{d}{du}\frac{k^+(n)}{k^+(n^*)}\frac{d\bar{F}(n,s)}{du} + \frac{du}{dt}\tau\frac{d\bar{F}(n,s)}{du} - \vartheta\bar{F}(n,s) = -\tau F(u,0), \qquad (37)$$

where

$$\bar{F}(n,s) = \mathscr{L}(F(n,t)) = \int_0^\infty \exp(-st)F(n,t)\,dt,$$

$$u = n/n^*, \qquad \hat{\Delta}^{-2} = -\frac{1}{2k_B T}\left(\frac{\partial^2 W}{\partial u^2}\right)_{n^*},$$

$$\tau^{-1} = \frac{d}{du}\left(\frac{du}{dt}\right)_{n^*}, \quad \text{and} \quad \vartheta = s\tau.$$

Note that $\hat{\Delta}$ is closely related to the width of the critical region, Δ, which we have discussed earlier in this chapter and in Chapter 2; $\hat{\Delta} = \Delta/n^*$, where n^* is the number of molecules in the critical cluster. Linearly independent solutions for Eq. (37) are obtained in three different regions of cluster size, that near the critical size and two on either side. The width of the near-critical region is small, $\Delta \approx n^*/\sqrt{W^*/k_B T}$, while the subcritical and postcritical regions are larger ($(n^* - n) \gg \Delta$ and $(n - n^*) \gg \Delta$, respectively). By forcing the solutions to match asymptotically, to lowest order, the flux in the regions of physical interest is found and the critical flux $I(n^*, t)$ is obtained by taking the inverse Laplace transform for $u = 1$,

$$I(n^*, t) = -I^{st}\alpha(t)\int_0^\infty \frac{\partial F(u,0)}{\partial u}\exp\left[-z^2(u)\left(\alpha^2(t) - 1\right)\right]du, \qquad (38)$$

where

$$\alpha(t) = \left(1 - \exp\left(-\frac{2t}{\tau}\right)\right)^{-1/2}$$

$$z(u) = -\exp\left(\int_{1-\hat{\Delta}}^{u}\frac{du}{\tau(du/dt) + i0}\right) \qquad (39)$$

($i0$ means that the real axis, u, is approached from above in the integral evaluation). For the case where the initial cluster size distribution is close to the steady-state distribution, $F(u,0) \approx \frac{1}{2}\mathrm{erfc}(z(u))$, where erfc is the complementary error function, $\mathrm{erfc}(x) = 1 - \mathrm{erf}(x)$ (see Eq. (32)). Assuming no preexisting nuclei, this case leads (from Eq. (39)) to

$$I(n^*, t) = I^{st}\alpha(t)\exp\left(-\beta(t)\right)$$
$$\beta(t) = (\alpha^2(t) - 1)\exp\left((1 - \hat{\Delta})\frac{2t_\theta}{\hat{\tau}}\right), \tag{40}$$

where t_θ is a characteristic time related to the induction time. The most important result from this analysis is the emergence of two different relaxation times instead of one as obtained in Section 4.2, a short time, $\tau/2$ and a long time $t_\theta(1 - \hat{\Delta}) \approx \tau\ln(\hat{\Delta}^{-1})$ [17, 18]. After $\alpha(t)$ has relaxed to its steady-state value of unity, there still exists a large time interval $\tau \ll t \ll t_\theta(1 - \hat{\Delta})$ where $\beta(t)$ has reached its asymptotic form $\beta(t) = \exp(2(t_\theta(1 - \hat{\Delta}) - t))$, but the ratio of $I(n,t)/I^{st}$ is still far from unity. These results arise from the different kinetic rates in the diffusion and drift regions. They are in conflict with a key assumption within the Kashchiev treatment, that the distribution differs from the steady-state one only in the critical region.

As discussed in Chapter 2, clusters that sufficiently exceed the critical size are biased toward continued growth. In this cluster size regime, the time-dependent nucleation rate is given by [18]

$$I(n, t) = I^{st}\exp\left(-\exp\left(-\frac{t - t_{\mathrm{rel}}(n)}{\tau}\right)\right), \tag{41}$$

where $t_{\mathrm{rel}}(n)$ is a size-dependent "relaxation time," defined as the time when the nucleation rate has reached $1/e$ of its steady-state value. The induction time, $\theta(n)$, can be related to $t_{\mathrm{rel}}(n)$, $\theta(n) = t_{\mathrm{rel}}(n) + \zeta_E\tau$, where ζ_E is Euler's constant (0.5772...). The quantity most often measured in nucleation experiments is the number of nuclei with size exceeding the observation size n, $\chi(n,t)$. Assuming a realistic model for cluster growth and a large nucleation barrier, the integration of Eq. (41), gives

$$\chi(n, t) = \tau I^{st}E_1\left(\exp\left(-\frac{t - t_{\mathrm{rel}}(n)}{\tau}\right)\right), \tag{42}$$

where E_1 is the exponential integral [20]. For long times this has the form,

$$\chi(n, t) = I^{st}(t - \theta), \quad \theta = t_{\mathrm{rel}} + \zeta_E\tau, \tag{43}$$

where the induction time, θ, is the intercept of the extrapolation of the linear long-time behavior with the time axis (Figure 1). Knowing θ, the transient time, τ, (the time required to diffuse through the critical region) can be determined from Eq. (42),

$$\tau = \frac{\chi(n, \theta)}{I^{st}E_1(e^{-\zeta_E})} = \frac{2.03\chi(n, \theta)}{I^{st}}. \tag{44}$$

Given I^{st} and the number of nuclei generated per unit time at the critical size, the transient time, τ can be computed. Acoording to [17], the Kashchiev value, τ_K, from

$$\chi(n^*,\theta) = 2I^{st}\tau_K \sum_{m=1}^{\infty} \frac{(-1)^{m+1}}{m^2} \exp\left(-\frac{\pi^2 m^2}{6}\right)\tau \approx 0.39\, I^{st}\tau_K\,, \tag{45}$$

is greater by more than 50% from the values obtained from Eq. (42). The prediction of two dominant time-scales for relaxation in that treatment, instead of the single one predicted by all previous analyses, can result in complex relaxation phenomena, which are observed, both experimentally and in numerical treatment (Chapter 8).

4.4 Calculation of the induction time

As discussed in Section 4.2, it is often more convenient to focus attention on the induction time alone. Extending Eq. (19) to the continuous case of the ZF equation, the induction time for a cluster of size n is

$$\theta(n) = \frac{1}{I^{st}}\left(\int_0^n \left(N^{st}(n') - N(n',0)\right) dn' - \int_0^\infty \frac{N^{st}(n')}{N^{eq}(n')}\left(N^{st}(n') - N(n',0)\right) dn'\right), \tag{46}$$

where $N(n', 0)$ is the initial cluster size distribution. Assuming a parabolic expansion of the work of cluster formation around the critical size, and evaluating the steady-state nucleation rate and cluster population [21],

$$I^{st} = \frac{1}{\Delta\sqrt{\pi}} k^+(n^*) N^{eq}(n^*), \quad \frac{N^{st}(n)}{N^{eq}(n)} \approx \frac{1}{2}\text{erfc}\left(\frac{n-n^*}{\Delta}\right), \tag{47}$$

leading to an expression for the induction time at the critical size,

$$\theta(n^*) = \frac{\zeta_E \tau}{2} + \tau \ln\left(\frac{n^*}{\Delta}\right) \tag{48}$$

and one for calculating the induction time at any other size

$$\theta(n) = \theta(n^*) + \frac{\zeta_E \tau}{2} + \tau \ln\left(\frac{n-n^*}{\Delta/2}\right), \tag{49}$$

where τ, the maximum relaxation time (often taken to be the time required for a cluster to diffuse through the critical region), is given by

$$\tau = \left(\frac{d}{dn}\left(\frac{dn}{dt}\right)_{n^*}\right)^{-1} = \frac{\Delta^2}{2k^+(n^*)}, \tag{50}$$

which is equivalent to the definition in Eq. (37). Wu arrived at essentially the same expressions [9].

These equations are valid when n^* is somewhat larger than the width of the critical region, Δ. They predict that the transient time, τ, depends on the measured incubation time. The ratio, τ/θ, depends on thermodynamic parameters and external conditions such as supersaturation or supercooling.

Applying this approach to particular nucleation models, Shneidman and Weinberg [21] showed that the ratio of induction to transient time for the Turnbull–Fisher model for nucleation, most relevant for nucleation in condensed phases and discussed in Chapter 2, should be described well by

$$\frac{\theta(n)}{\tau} = \hat{\xi} + \ln \hat{\xi} + \ln \frac{6W^*}{k_B T} + \zeta_E - 1, \tag{51}$$

where

$$\hat{\xi} = \left(\frac{n}{n^*}\right)^{1/3} - 1. \tag{52}$$

Equations (49) and (51) allow the induction time at any cluster size to be computed, given the value at one cluster size. They can, therefore, be used to estimate the induction time near the critical size from the induction time measured at a different size.

5. LIMITS OF APPLICABILITY OF SELECTED ANALYTICAL EXPRESSIONS

Given the large number of analytical expressions, it is natural to question which, if any, are sufficiently sound to provide an accurate estimate of the transient nucleation behavior. The numerical solutions used throughout Chapters 2 and 3 can inform us on this point. Figure 5 shows that the dominant contribution to the magnitude of the induction time comes from the atomic mobility, presumably reflected by the diffusion coefficient. This is true for all the treatments considered. Differences between them arise from their different dependences on thermodynamic parameters.

Here, the expressions of Kashchiev (Eq. (36)), Shneidman (Eq. (48)), and Shneidman and Weinberg (Eq. (51)) are examined; a brief evaluation of other popular expressions can be found in Ref. [3]. The work W^* required to form a critical cluster is the most useful measure of the thermodynamic parameters. Nucleation studies are generally made in the range $25k_BT < W^* < 60k_BT$; $W^*/k_BT \approx 40$ at the peak nucleation rate in lithium-disilicate glass for example (Chapter 8). To span any conceivably

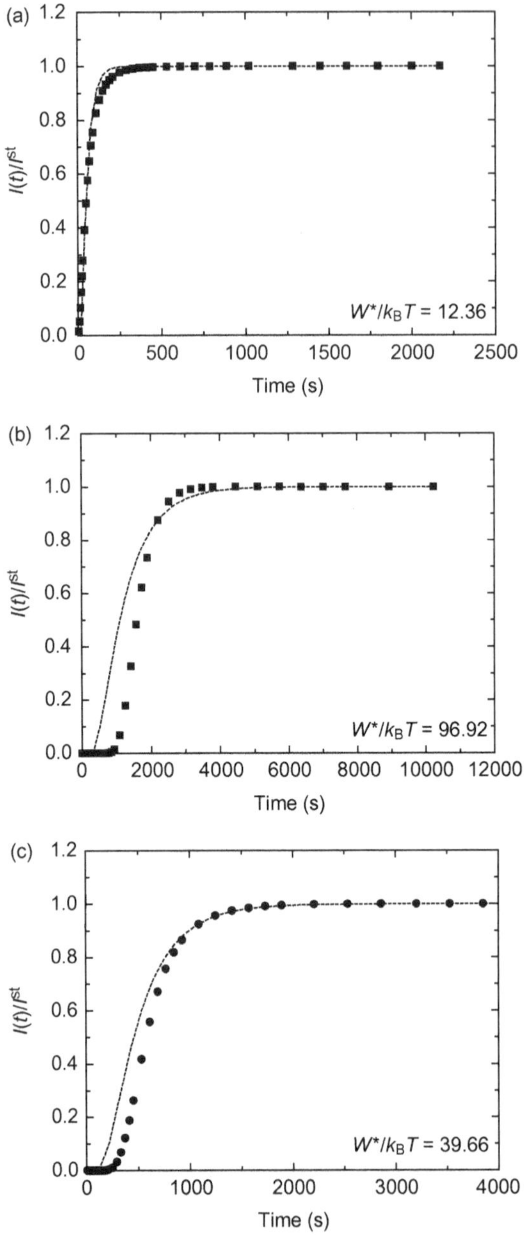

Fig. 8 The fit of the numerical data for the time-dependent nucleation rate to Eq. (35). Fits are shown for a range of values of W^*, the critical work of cluster formation: (a) $W^*/k_BT = 12.4$; (b) $W^*/k_BT = 40.0$; (c) $W^*/k_BT = 96.9$. Calculations were made using parameters appropriate for nucleation in lithium-disilicate glass (see Chapter 8); W^* was varied by changing the entropy of fusion.

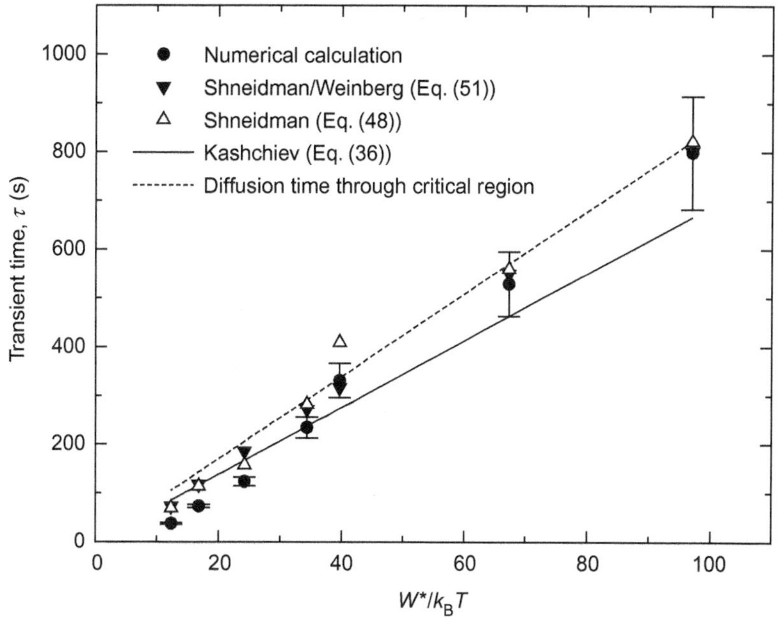

Fig. 9 The transient time as a function of W^*, the work of formation of the critical cluster. Transient times calculated from a numerical solution of the coupled differential equations of nucleation (by fitting the nucleation rates to Eq. (35)) are compared with values predicted from expressions due to Kashchiev (Eq. (36)), Shneidman (Eq. (48)), and Shneidman-Weinberg (Eq. (51)). Calculations were made using parameters appropriate for nucleation in lithium-disilicate glass (See Chapter 8); W^* was varied by changing the entropy of fusion. Errors are indicated for the values obtained from the numerical calculation.

reasonable experimental case, then, we examine the range $12k_BT \leq W^* \leq 98k_BT$, take the diffusion coefficient to be similar to that for lithium-disilicate glass, and assume an initial single-molecule distribution. To compare directly with analytical expressions, the time-dependent nucleation rates are evaluated at the critical size for the W^* used.

The numerical transient times are determined by fitting the calculated nucleation rates to Eq. (35). The fits are reasonable (Figure 8), but not perfect; they become worse as W^* increases. Equation (35) should then be viewed only as an approximate functional form for time-dependent nucleation, although it is adequate for use with measured data, given normal experimental error. Figure 9 compares these calculated transient times with those predicted from Eq. (36) (by fitting the numerically computed nucleation rates to Eq. (35)), and those computed from the induction times measured from the numerical calculations using Eqs. (48)

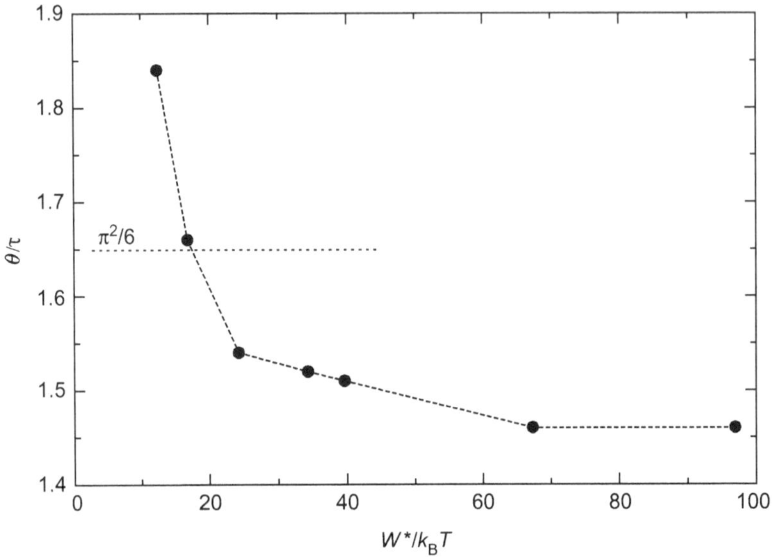

Fig. 10 The calculated ratio of the induction and transient times for nucleation as a function of W^*, the work of formation of the critical cluster. The constant value for this ratio, $\pi^2/6$, predicted by Kashchiev is indicated. Calculations were made using parameters appropriate for nucleation in lithium-disilicate glass; W^* was varied by changing the entropy of fusion.

and (51). The time required for a cluster to "diffuse" through the critical region (Eq. (50)) is also shown. Over this extended range, all transient times show a similar dependence on W^*. The transient times derived from the numerically computed induction times, using Eqs. (48) and (51), are in good agreement with the times required for diffusion through the critical region (Eq. (50)). These agree well with the numerical values for high values of W^* ($W^*/k_BT > 40$); however, they rise slightly above the numerical data for smaller values of W^*. The agreement with the Kashchiev expression, Eq. (36), is worse; predictions fall below the numerical data for $W^*/k_BT > 40$ and rise above the data for $W^*/k_BT < 35$.

Kashchiev determined that the ratio of the induction time to the transient time is a constant, equal to $\pi^2/6$ [16]. The ratio determined from the numerical calculations (Figure 10) is instead a strong function of W^*. It is equal to the Kashchiev value only when the value of W^* is near $20k_BT$.

The induction time, θ, is a more convenient parameter for comparison between experimental and numerically calculated data. An exact expression for θ, which takes into account the initial cluster size distribution, is given in Eq. (19). That expression can be used even when the initial cluster size distribution is not a single-molecule distribution.

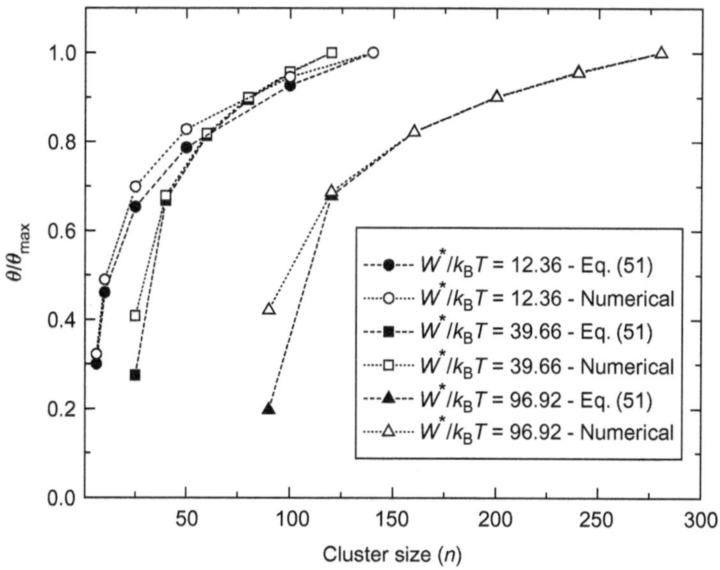

Fig. 11 The ratio of induction times for different cluster sizes to the value at the maximum cluster size. Values calculated from a numerical solution to the coupled differential equations of nucleation (fitting the nucleation rates to Eq. (35)) and those predicted from Eq. (51) are shown for three different values of W^*, the work of formation of the critical cluster. These calculations were made using parameters appropriate for nucleation in lithium-disilicate glass; W^* was varied by changing the entropy of fusion.

Unfortunately, because θ is computed from the differences of sums of large numbers, limited numerical precision makes its application difficult. In many cases, it is easier to numerically solve the coupled differential equations directly.

As will be discussed in detail in Chapter 8, the experimental induction time is generally determined for clusters that are substantially larger than the critical size. As indicated in the last section, Shneidman and Weinberg's expression, Eq. (51), gives θ/τ at any cluster size. Predictions made from Eq. (51) are compared with values computed numerically in Figure 11. Three values for W^* are considered; they span the total range considered in Figures 9 and 10. Since the transient time is independent of cluster size, Eq. (51) can be tested by comparing the ratios θ/θ_{max} computed numerically and from Eq. (51). Here, θ_{max} is the induction time at the maximum cluster size considered in the numerical calculations. The values for this ratio calculated from Eq. (51) agree well with the numerical data for large clusters. The agreement becomes worse as the cluster size approaches the critical size for each W^*. In all cases, however, the agreement is good for $n > 1.5n^*$.

From these studies, it may be concluded that for cases where the initial cluster size distribution is reasonably close to an ensemble of single molecules, such as in a rapidly quenched glass, the isothermal transient time can be estimated using any of the expressions presented in Sections 4.2 to 4.4. Also, Eq. (51) gives a reasonable estimate of the induction time at any cluster size if it is known for one cluster size. If more quantitative agreement is required, if the initial distribution is not well approximated by an ensemble of single molecules, or if the external parameters are changing during the nucleation, a numerical approach is likely required.

6. SUMMARY

In this chapter, we have examined the origins of non-steady-state nucleation within the classical theory. Some useful analytical expressions for time-dependent nucleation in simple situations were presented. Key points are as follows:

- For many cases, nucleation rates differ from steady-state values in the early stages of nucleation.
- The induction time, θ, is an experimental measure of the time required for the nucleation rate to approach steady state. It is determined by extrapolating the number of nuclei produced as a function of time in the steady-state regime to the time axis.
- The relaxation of the inherited cluster size distribution to steady state at the annealing temperature is the origin of the time-dependent nucleation rate.
- The orders of magnitude of θ and τ are set by the atomic mobility (diffusion coefficient); thermodynamic parameters have a smaller influence.
- The transient time, τ, is a measure of the rate of relaxation of the cluster size distribution toward steady state. It is the maximum relaxation time and generally defined to be the time required for a cluster to "diffuse" through the critical region (Eq. (50)).
- The expressions of Kashchiev (Eq. (36)), Shneidman (Eq. (48)), or Shneidman and Weinberg (Eq. (51)) can be used to estimate θ and τ when the initial distribution is essentially an ensemble of single molecules.
- Time-dependent nucleation behavior depends strongly on the initial cluster size distribution. It can be quite complex, in some cases causing the nucleation rate to increase above the steady-state value before re-approaching it later (Chapter 8).

- Quantitative studies of time-dependent nucleation under experimentally realistic conditions generally require numerical solutions of the coupled rate equations central to the classical theory.

REFERENCES

[1] V.M. Fokin, A.M. Kalinina, V.N. Filipovich, Nucleation in silicate glasses and effect of preliminary heat treatment on it, J. Cryst. Growth **52** (1981) 115–121.

[2] A.M. Kalinina, V.N. Filipovich, V.M. Fokin, Stationary and non-stationary crystal nucleation rate in a glass of $2Na_2O.CaO.3SiO_2$ stoichiometric composition, J. Non-Cryst. Solids **38–39** (1980) 723–728.

[3] K.F. Kelton, A.L. Greer, C.V. Thompson, Transient nucleation in condensed systems, J. Chem. Phys. **79** (1983) 6261–6276.

[4] D.T. Wu, Nucleation theory, in: *Solid State Physics*, Eds. H. Ehrenreich, F. Spaepen, Academic Press, Boston (1996), pp. 37–187.

[5] R.P. Andres, M. Boudart, Time lag in multistate kinetics: Nucleation, J. Chem. Phys. **42** (1965) 2057–2064.

[6] H.L. Frisch, C.C. Carlier, Time lag in nucleation, J. Chem. Phys. **54** (1971) 4326–4330.

[7] L.R. Hile, PhD Thesis, Princeton University (1970).

[8] B. Shizgal, J.C. Barrett, Time dependent nucleation, J. Chem. Phys. **91** (1989) 6505–6518.

[9] D.T. Wu, The time lag in nucleation theory, J. Chem. Phys. **97** (1992) 2644–2650.

[10] V.A. Shneidman, M.C. Weinberg, Transient nucleation induction time from the birth-death equations, J. Chem. Phys. **97** (1992) 3629–3638.

[11] N. Miloshev, On a new approach for finding the non-steady state kinetic rate of phase formation, Atmos. Res. **28** (1992) 173–183.

[12] L.S. Bartell, D.T. Wu, On the reduced moment in the transient regime of homogeneous nucleation, J. Chem. Phys. **127** (2007) 164508/1–6.

[13] V.A. Shneidman, Asymptotic relations between time-lag and higher moments of transient nucleation flux, J. Chem. Phys. **119** (2003) 12487–12491.

[14] D. Kashchiev, Moments of the rate of nonstationary nucleation, J. Chem. Phys. **122** (2005) 114506/1–6.

[15] J.B. Zeldovich, On the theory of new phase formation: Cavitation, Acta Physiochim. U.R.S.S. **18** (1943) 1–22.

[16] D. Kashchiev, Solution of the non-steady state problem in nucleation kinetics, Surf. Sci. **14** (1969) 209–220.

[17] V.A. Shneidman, Size-distribution of new-phase particles in nonstationary condensation of a supercooled gas, Sov. Phys. Tech. Phys. **32** (1987) 76–81.

[18] V.A. Shneidman, Establishment for a steady-state nucleation regime. Theory and comparison with experimental data for glasses, Sov. Phys. Tech. Phys. **33** (1988) 1338–1342.

[19] V.A. Shneidman, Transient critical flux in nucleation theory, Phys. Rev. A, **44** (1991) 2609–2611.

[20] M. Abramowitz, I. Stegun, *Handbook of Mathematical Functions with Formulas, Graphs and Mathematical Tables*, National Bureau of Standards, Applied Mathematics Series, US Government Printing Office, Washington (1970).

[21] V.A. Shneidman, M.C. Weinberg, Induction time in transient nucleation theory, J. Chem. Phys. **97** (1992) 3621–3628.

[22] K.F. Kelton, Crystal nucleation in liquids and glasses, in: *Solid State Physics*, Eds. H. Ehrenreich, D. Turnbull, Academic Press, Boston (1991), pp. 75–178.

CHAPTER 4

Beyond the Classical Theory

Contents

1. INTRODUCTION

As will be demonstrated in Parts II and III of this book, *classical nucleation theory* (CNT) is robust, relatively easy to use, and capable of handling a wide range of nucleation phenomena. By treating nucleation in terms of the dynamics of microscopic clusters, it explains the origins of the nucleation barrier and time-dependent nucleation rates. The model is not flawless, however. It is based on an extremely simplified view of cluster properties. Capillarity (the assumption that the interface between a cluster of the new phase and the original phase is sharp) is assumed for all clusters so that small clusters containing only a few atoms are treated as if they were macroscopic droplets. The expressions for the nucleation barrier and the nucleation rate, both calculated originally for vapor condensation, are taken to be universally applicable for all nucleation-based first-order phase transitions. This disregards any impact of the atomic structures of the original and new phases. As we shall see in later

Pergamon Materials Series, Volume 15
ISSN 1470-1804, DOI 10.1016/S1470-1804(09)01504-1

chapters, CNT frequently fails to quantitatively explain experimental data. Further, it predicts a nonzero barrier to the phase transformation in all cases, failing to account for spinodal transformations in the unstable regions of the free energy (discussed in Chapter 1).

In this chapter some of the approaches going beyond CNT are discussed. A statistical–mechanical derivation of the work of cluster formation is discussed and a density-functional approach that can take account of the atomic order in the original and new phases is outlined. The density-functional approach presented is primarily relevant for amorphous-to-crystal transitions (i.e., liquid/crystal or glass/crystal). Typically, density-functional models are more complicated than CNT and depend on parameters that are often unavailable, making them difficult to use for a quantitative comparison with experimental data. However, the increasing sophistication of experimental measurements and the technological need for better nucleation control will continue to drive the search for more accurate models.

2. A STATISTICAL–MECHANICAL TREATMENT OF CLUSTER FORMATION

The work of cluster formation is at the heart of the difficulties with CNT. It was derived in Chapter 2 for vapor condensation using thermodynamic arguments; now it is useful to examine a statistical–mechanical derivation. Of the several approaches taken [1, 2], the sketch given here follows closely that of ten Wolde and Frenkel [3].

Consider a gas of N molecules in a volume V that is placed in contact with a particle reservoir at temperature T and chemical potential μ. The grand canonical partition function (see McQuarrie [4] and Goodstein [5] for a discussion of statistical mechanics and partition functions) is:

$$\Xi(V,T,\mu) = \sum_{N=0}^{\infty} Q(N,V,T)(\mathrm{e}^{\beta\mu})^N, \tag{1}$$

where $Q(N, V, T)$ is the canonical partition function

$$Q(N,V,T) = \sum_j \mathrm{e}^{-\beta E_j} \to \frac{1}{h^{3N}\,N!}\int \mathrm{e}^{-\beta\mathscr{H}(N)}\mathbf{dr}_1 \cdots \mathbf{dr}_N\,\mathbf{dp}_1 \cdots \mathbf{dp}_N, \tag{2}$$

in which $\mathbf{r}$ and $\mathbf{p}$ are the position and momentum, respectively, for each molecule, h is the Planck constant, and $\beta = (k_B T)^{-1}$ (where k_B is the Boltzmann constant and T is the temperature in absolute units). The index j in Eq. (2) enumerates all possible states (i.e., all possible shapes, positions, and internal configurations of the clusters with energy E_j). $\mathscr{H}(N)$ is the N-body Hamiltonian for interacting particles (with the kinetic and interaction components assumed to be separable):

$$\mathscr{H}(N) = \frac{1}{2m} \sum_{i=1}^{N} \mathbf{p}_i \cdot \mathbf{p}_i + U(\mathbf{r}_1, \ldots, \mathbf{r}_N). \tag{3}$$

By integrating over the kinetic energy (momentum) contribution, we obtain

$$Q(V, N, T) = \frac{1}{\lambda^{3N} N!} \int \mathrm{d}\mathbf{r}_1 \cdots \mathrm{d}\mathbf{r}_N \mathrm{e}^{-\beta U(\mathbf{r}_1 \cdots \mathbf{r}_N)} = \frac{1}{\lambda^{3N} N!} Z_N, \tag{4}$$

where Z_N is the configuration integral and $\lambda = \left(h^2/2\pi m k_\mathrm{B} T\right)^{1/2}$is the thermal de Broglie wavelength, where m is the molecular mass. In infinite dilution, the interactions between gas molecules can be ignored (internal energy $U = 0$), and $Z_N = V^N$; the resulting canonical partition function leads directly to the ideal-gas equation of state. For more dense gases, the virial equation of state takes account of molecular interactions as a power-series expansion of the number density $\rho = (N/V)$, $p/k_\mathrm{B}\mathrm{T} = \rho + B_2(T)\rho^2 + B_3(T)\rho^3 + \cdots + B_j(T)\rho^j + \cdots$ (where p is the pressure). The coefficients, B_j, can be derived within statistical mechanics by considering the interactions of j molecules in a volume V (see Ref. [4]). The N-body problem is then reduced to a series of one-body, two-body,…, j-body, problems and the dense gas is viewed as a distribution of clusters containing different numbers of molecules.

For vapor condensation, some molecules are associated with the original gas phase and some are associated with the new liquid phase. It is reasonable to describe the liquid in terms of clusters as well, leading to two cluster distributions. Defining which clusters are associated with which phase is not trivial; this will be discussed more in Chapter 10. For now, we assume that an operational criterion can be formulated. Following ten Wolde and Frenkel, and using the notation in Chapter 2, Section 2, let N_d be the number of molecules that are in liquid clusters (d for droplets) and N_m be the number in the original gas phase (m for medium) so that $N = N_\mathrm{m} + N_\mathrm{d}$. Since the potential interaction depends on how many particles are in the two phases, and noting that there are $N!/(N_\mathrm{d}!N_\mathrm{m}!)$ ways of selecting N_d and N_m particles from the total number of particles, the grand canonical ensemble of this two-phase mixture can be written as

$$\begin{aligned} \Xi(V, T, \mu) = & \sum_{N_\mathrm{d}=0}^{\infty} \frac{\mathrm{e}^{\beta\mu N_\mathrm{d}}}{\lambda^{3N_\mathrm{d}} N_\mathrm{d}!} \sum_{N_\mathrm{m}=0}^{\infty} \frac{\mathrm{e}^{\beta\mu N_\mathrm{m}}}{\lambda^{3N_\mathrm{m}} N_\mathrm{m}!} \int \Psi(\mathbf{r}_1^\mathrm{d}, \ldots, \mathbf{r}_{N_\mathrm{d}}^\mathrm{d}, \mathbf{r}_1^\mathrm{m}, \ldots, \mathbf{r}_{N_\mathrm{m}}^\mathrm{m}) \\ & \times \mathrm{e}^{-\beta U(\mathbf{r}_1^\mathrm{d}, \ldots, \mathbf{r}_{N_\mathrm{d}}^\mathrm{d}, \mathbf{r}_1^\mathrm{m}, \ldots, \mathbf{r}_{N_\mathrm{m}}^\mathrm{m})} \mathrm{d}\mathbf{r}_1^\mathrm{d} \cdots \mathrm{d}\mathbf{r}_{N_\mathrm{d}}^\mathrm{d} \mathrm{d}\mathbf{r}_1^\mathrm{m} \cdots \mathrm{d}\mathbf{r}_{N_\mathrm{m}}^\mathrm{m}, \end{aligned} \tag{5}$$

where $\Psi(\mathbf{r}_1^\mathrm{d}, \ldots, \mathbf{r}_{N_\mathrm{d}}^\mathrm{d}, \mathbf{r}_1^\mathrm{m}, \ldots, \mathbf{r}_{N_\mathrm{m}}^\mathrm{m})$ is a weighting function that is one when the number of particles in liquid configurations is equal to N_d and zero otherwise. The steps to a solution of Eq. (5) are involved and not repeated here. ten Wolde and Frenkel show (see Appendix A of reference [3]) that by expressing $\Psi(\mathbf{r}_1^{\mathrm{d}},\ldots,\mathbf{r}_{N_{\mathrm{d}}}^{\mathrm{d}},\mathbf{r}_1^{\mathrm{m}},\ldots,\mathbf{r}_{N_{\mathrm{m}}}^{\mathrm{m}})$ in terms of weighting functions that describe liquid clusters containing different numbers of molecules, n, and ignoring cluster–cluster interactions, Eq. (5) can be expressed in terms of these clusters and their interaction with the vapor phase,

$$\begin{aligned}\Xi(V,T,\mu) &= \exp(-\beta\omega^{(0)}) \\ &\quad \times \exp\left(\sum_n \left(\mathrm{e}^{\beta\mu n}\left(\frac{Vn^3}{\lambda^{3n}n!}\int \mathrm{e}^{-\beta\omega^{(n)}(\mathbf{r}'_1\cdots\mathbf{r}'_{n-1};\mu)}\mathrm{d}\mathbf{r}'_1\cdots d\mathbf{r}'_{n-1}\right)\right)\right) \\ &= \exp(-\beta\omega^{(0)})\exp\left(\sum_n Z_n\mathrm{e}^{\beta\mu n}\right), \end{aligned} \tag{6}$$

where Z_n is for a liquid cluster containing n atoms. The primed coordinates indicate that the positions are measured relative to the center of mass of the cluster. $\omega^{(n)}(\mathbf{r}'_1\cdots\mathbf{r}'_{n-1};\mu)$ is the intracluster interaction energy for a cluster of size n and $\omega^{(0)}$ is a constant representing the potential in the absence of clusters. This result can be used to obtain an expression for the cluster size distribution.

Since $pV = k_{\mathrm{B}}T\ln\Xi$, the grand (or Landau) potential, $\Omega = -pV$, is

$$\Omega(T,\mu) = \omega^{(0)} - k_{\mathrm{B}}T\sum_n Z_n\mathrm{e}^{\beta\mu n}, \tag{7}$$

yielding the thermodynamic description of the cluster ensemble. The average total number of particles is

$$N = k_{\mathrm{B}}T\left(\frac{\partial \ln \Xi(N,T,V)}{\partial\mu}\right)_{V,T} = \sum_n nZ_n\mathrm{e}^{\beta\mu n}. \tag{8}$$

Since (i) $N = \sum_{n=1}^{\infty} nN(n)$ (where $N(n)$ is the ensemble average number of clusters that contain n atoms or molecules) and (ii) defining the Helmholtz free energy of the cluster of size n as $F(n) = -k_{\mathrm{B}}T \ln Z_n$, Eq. (8) gives

$$N(n) = \mathrm{e}^{-(F(n)-n\mu)/k_{\mathrm{B}}T} = \mathrm{e}^{-(\Delta F/k_{\mathrm{B}}T)} \tag{9}$$

using $\beta^{-1} = k_{\mathrm{B}}T$. By expressing $N(n)$ in terms of a probability, $P(n)$ as $P(n) = N(n)/N$, and defining an intensive Gibbs free energy of the cluster (work of cluster formation, $W(n)$) in terms of $P(n)$, $W(n) = -k_{\mathrm{B}}T\ln(P(n))$, we justify the form obtained in Chapter 2 (Eq. 12), that is,

$$N(n) = N\mathrm{e}^{-W(n)/k_{\mathrm{B}}T}. \tag{10}$$

It should be noted that $W(n)$ generally contains terms in n and $n^{2/3}$ (plus terms of higher order in n) [6] so that to lowest order it will have the form of Eq. (11) in Chapter 2. However, the explicit concept of a surface was not introduced in ten Wolde and Frenkel's treatment. The definition of a cluster is incorporated into the definition of mean force.

This treatment has focused on the definition of a cluster, the work of cluster formation, and the cluster size distribution for the case of vapor condensation. It supports several points that were assumed in the thermodynamic treatment in Chapter 2. However, the assumptions made here are inappropriate for liquid-to-solid transitions or solid-to-solid transitions. Unlike the vapor, the liquid phase cannot be described by a sequence of independent n-body interactions. Cluster–cluster interactions cannot be ignored. In these cases the structures of the original phases are much more ordered than in the gas and can have a significant impact on the nucleation barrier. A density-functional formalism will be introduced in Section 4 to address some of these questions. Before that, however, we examine a phenomenological model that helps to focus two key issues more clearly, that is, interfacial width and ordering.

3. DIFFUSE-INTERFACE THEORY

Computer simulations [7–10], density-functional calculations [11–14], and discrete studies [15, 16] show that the width of the interface between the original phase and clusters of the new phase can be a significant fraction of the cluster radius. This is illustrated schematically in Figure 1a for a crystal cluster growing from a liquid. Within the cluster, the density oscillates with the periodicity of the crystal and approaches a constant value in the liquid far from the crystal; the liquid is ordered near the cluster. Such a broad interfacial region is in conflict with a central assumption of CNT and has significant consequences for the calculation of the work of cluster formation.

It is useful to first examine a phenomenological thermodynamic approach, *diffuse-interface theory* (DIT), proposed independently by Spaepen [17] and Gránásy [18–20], which is physically appealing and can be readily fit to nucleation data using parameters that are similar to those used in CNT. It leads to a natural explanation of a positive temperature dependence of the interfacial free energies for liquid–crystal and glass–crystal phase transformations, claimed from fits of CNT to nucleation data for liquids (Chapter 7) and glasses (Chapter 8). Finally, it provides a good introduction to concepts that will be taken up again when we discuss the density-functional approach.

For simplicity, we consider the case of a spherical crystal cluster in an isotropic amorphous (liquid or glass) medium. Then the radial variation of $g(r)$, where r is measured from the center of the cluster, can represent the free-energy density (free energy per unit volume) for this discussion. As illustrated in Figure 1b, $g(r)$ increases, above that in the

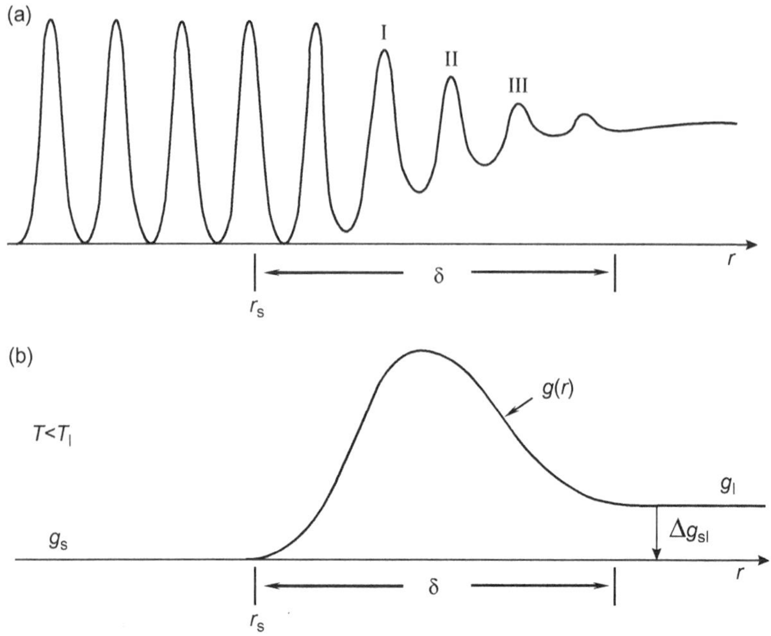

Fig. 1 (a) The atomic density on traversing the interface between the liquid and solid; (b) the corresponding free-energy density. The interfacial region extends from r_s to $r_s + \delta$, where r_s is the radius of the solid cluster. (Reprinted from Ref. [17], copyright (1994), with permission from Elsevier.)

liquid, g_l, and the solid, g_s, phases due to the interface (the interfacial free energy is positive, equal to the excess energy per unit area from the introduction of an interface). Assuming constant pressure (as was the case for the CNT derivation in Chapter 2), the reversible work of formation for a cluster of the new phase having a radius of r_s and an interfacial width δ is,

$$W(r_s) = 4\pi \int_0^{\infty} (g(r) - g_l) r^2 \mathrm{d}r. \tag{11}$$

From Figure 1, $g(r) = g_s$ for $r < r_s$; $g(r) = g_l$ for $r > r_s + \delta$. As for CNT (Chapter 2), the critical cluster radius for nucleation r_s^* is determined by the condition

$$\left.\frac{\mathrm{d}W(r_s)}{\mathrm{d}r_s}\right|_{r_s = r_s^*} = 0. \tag{12}$$

Given a functional form for $g(r)$, the problem would be solved. A general solution has been proposed in terms of moments for the free energy [17]. However, for pedagogical reasons, a simpler treatment [17] is presented here.

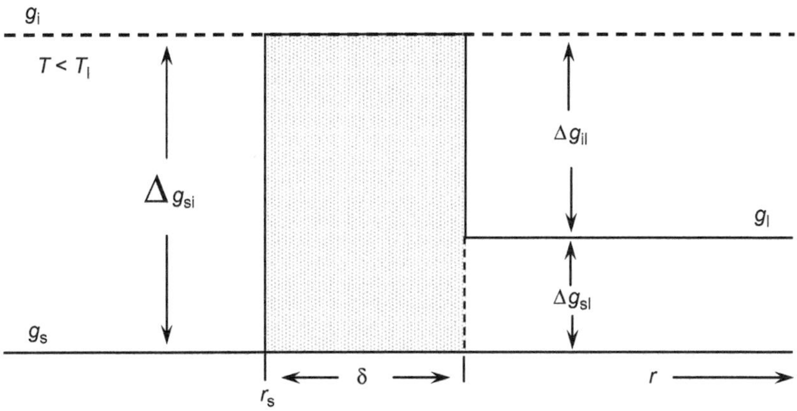

Fig. 2 Simplified view of the free-energy density, assuming a step-function approximation to the more realistic free-energy change shown in Figure 1. The magnitudes of the free-energy-density differences are indicated. (Adapted from Ref. [17], copyright (1994), with permission from Elsevier.)

Assume that the free-energy density can be approximated by a step function, as shown in Figure 2. Let the free energy of the interface be g_i, which is assumed to be constant between r_s and $r_s + \delta$. Below the melting temperature, the reversible work of cluster formation is,

$$W(r_s) = 4\pi \int_0^{r_s} (g_s - g_l) r^2 \mathrm{d}r + 4\pi \int_{r_s}^{r_s+\delta} (g_i - g_l) r^2 \mathrm{d}r. \quad (13)$$

For this simple case, the work of cluster formation, measured relative to the free energy of the liquid, is

$$W(r_s) = \frac{4\pi}{3} r_s^3 \Delta g_{sl} + \frac{4\pi}{3}\left((r_s + \delta)^3 - r_s^3\right)\Delta g_{il}, \quad (14)$$

where to maintain notational consistency with Chapter 2, we have defined $\Delta g_{sl} = g_s - g_l$ and $\Delta g_{il} = g_i - g_l$ as the free-energy differences per unit volume between the liquid and solid phases and the liquid and interface, respectively. (Note that $\Delta g_{sl} < 0$.) Defining $a = \Delta g_{sl}/\Delta g_{il}$, we get

$$W(r_s) = \frac{4\pi}{3} \Delta g_{il} (a r_s^3 + 3 r_s^2 \delta + 3 r_s \delta^2 + \delta^3). \quad (15)$$

Computing the critical radius from $(\mathrm{d}W/\mathrm{d}r_s)|_{r_s^*} = 0$,

$$\frac{4\pi}{3} \Delta g_{il} (3 a r_s^2 + 6 r_s \delta + 3\delta^2)_{r_s^*} = 0, \quad (16)$$

giving

$$r_s^* = -\frac{\delta}{a} \pm \frac{\delta}{a}\sqrt{1-a} = -\frac{\delta}{a}(1 \mp \sqrt{1-a}). \quad (17)$$

Choosing the positive sign for the second term so that r_s^* goes to infinity as the driving free energy (Δg_{sl}) goes to zero,

$$r_s^* = -\frac{\delta}{1-b} \quad \text{where} \quad b^2 = 1-a. \tag{18}$$

Using this value in Eq. (15), the critical work of nucleation is

$$W^* = \frac{4\pi}{3}\Delta g_{il}\delta^3 \frac{b^2}{(1-b)^2}. \tag{19}$$

The interfacial free energy, σ, is rigorously defined for a planar interface in equilibrium; for a curved diffuse interface, the definition is less clear. However, Eq. (13) of Chapter 2, developed for a sharp interface, can be taken as a defining equation for σ. By equating the two expressions developed for W^* thus far (Eq. (14) in Chapter 2 and Eq. (19) in this chapter) an expression for σ is obtained,

$$\sigma = \Delta g_{il}\delta\left(\frac{ab}{2(1-b)}\right)^{2/3}. \tag{20}$$

Substituting for a and b, and taking $\Delta g_{is} = g_i - g_s$ to be the free-energy difference per unit volume between solid and interface (Figure 2),

$$\sigma = \delta(\Delta g_{il}\Delta g_{is}\Delta g_{i,av})^{1/3}, \tag{21}$$

where

$$\Delta g_{i,av} = \frac{1}{4}\left(\sqrt{\Delta g_{il}} + \sqrt{\Delta g_{is}}\right)^2 = \frac{1}{2}\left(\frac{1}{2}(\Delta g_{il} + \Delta g_{is}) + (\Delta g_{il}\Delta g_{is})^{1/2}\right). \tag{22}$$

$\Delta g_{i,av}$ is then the arithmetic mean of the arithmetic and geometric averages of Δg_{il} and Δg_{is}. For small supercoolings, the driving free energy, Δg_{sl}, is small. Since $|\Delta g_{is}| = |\Delta g_{il}| + |\Delta g_{sl}|$ (see Figure 2), we can assume that $\Delta g_{is} \approx \Delta g_{il}$. To leading order, the arithmetic and geometric means are then the same, $(\Delta g_{il} + \Delta g_{is}) \approx 2(\Delta g_{il}\Delta g_{is})^{1/2}$ and σ is approximately (from Eq. (21)),

$$\sigma \approx \frac{\delta}{2}(\Delta g_{il} + \Delta g_{is}) = \delta\left(\Delta g_{il} - \frac{\Delta g_{sl}}{2}\right), \tag{23}$$

(remembering that the sign of Δg_{sl} is negative). Since $(\partial g/\partial T)_P = -s$, where s is the entropy (per unit volume),

$$\left(\frac{\partial \Delta g_{il}}{\partial T}\right)_P = -(s_i - s_l) \text{ and } \left(\frac{\partial \Delta g_{sl}}{\partial T}\right)_P = -(s_s - s_l) = \Delta s_f, \tag{24}$$

where s_l, s_s, and s_i are the entropies of the liquid, solid, and interface, and Δs_f is the entropy of fusion. Assuming that the interfacial width is independent of temperature, the temperature dependence of the interfacial free energy is, from Eq. (23),

$$\frac{d\sigma}{dT} \approx \delta\left((s_l - s_i) - \frac{\Delta s_f}{2}\right). \tag{25}$$

As we shall see in Chapters 7 and 8, CNT-based analyses of experimental nucleation measurements in liquids and glasses indicate that $d\sigma/dT > 0$. This is opposite to the behavior expected for a free energy ($= h - Ts$, where h is the enthalpy and s is the entropy) and opposite to that observed for vapor condensation. From Eq. (25), this requires that $s_i < s_l - (\Delta s_f/2)$ equivalent to $s_i < (s_l + s_s)/2$, demonstrating that the amorphous phase is more ordered at the cluster interface.

Using the expression for σ in Eq. (23) to compute the radius of a critical cluster for CNT, r^*_{CNT}, (i.e., inserting Eq. (23) into Eq. (19) of Chapter 2)

$$r^*_{\text{CNT}} = \frac{2\sigma}{|\Delta g_{sl}|} = \delta\left(1 - \frac{2}{a}\right). \tag{26}$$

Subtracting the radius of a critical cluster in the DIT (Eq. (18)), in a form that is expanded to first order in a, shows that the CNT cluster interface (Gibbsian or equimolar dividing surface) lies exactly in the middle of the interfacial layer,

$$r^*_{\text{CNT}} - r^*_s \approx \frac{\delta}{2}. \tag{27}$$

The interfacial free energy for CNT is the excess free energy computed with respect to this interface. In a similar way, the nucleation rate in the DIT is obtained by substituting the work of cluster formation (Eq. (19)) for the CNT value in Eqs. (53) or (55) of Chapter 2.

The DIT is a simple extension of CNT, bringing in to lowest order the importance of the atomic structures of the original and new phases. It shows that the unexpected temperature dependence of the interfacial free energy obtained from fits of CNT to nucleation data for liquids and glasses (Chapters 7 and 8) arises from a failure to take account of the ordering of the amorphous phase near the crystal clusters. However, the DIT is an ad-hoc theory, resting on unproven assumptions (such as the temperature independence of the thickness of the interface). It is not readily extended to take into account the influence of the structural details of the ordered phases on the nucleation rate, or the coupling of other phase transitions to the nucleation barrier (see Chapter 7). Next, a more fundamental approach is discussed.

4. DENSITY-FUNCTIONAL THEORY

Density-functional theories (DFT), based on an order-parameter description of the phase transition, allow a more formal treatment of nucleation that is intermediate between the macroscopic thermodynamic description of CNT and the microscopic-based computer simulation methods

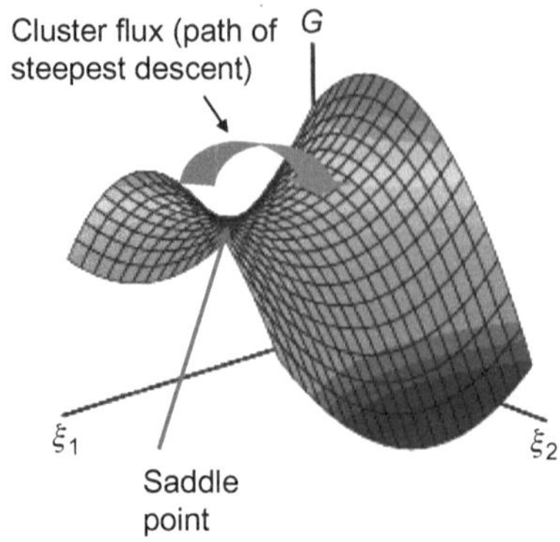

Fig. 3 The free energy, as a function of two arbitrary phase-space variables, in the saddle transition region between metastable and stable configurations.

that will be discussed in Chapter 10. DFTs can be extended to handle a wide range of order parameters and a coupling between phase transitions, even of different order. As in CNT, nucleation proceeds by the decay of a metastable state to a stable state by a localized fluctuation. The critical cluster corresponds to a particular saddle-point configuration of the free energy (shown schematically in Figure 3 for two arbitrary phase-space variables, ξ_1 and ξ_2, see also Ref. [21]). Field-theoretic methods like DFT provide information about the free-energy barrier and factors that control the width of the saddle region, which determine the time that the "cluster" spends there, and hence the nucleation rate.

4.1 Basic formalism

A spatially inhomogeneous density underlies the thermodynamic treatment in Section 3. Since the density varies continuously through the interface between the cluster and the original phase, it could be taken as an order parameter for a DFT description of the phase transition, expressing the free energy in terms of the density. By minimizing this free energy, using the techniques of variational calculus, the density profile around the nucleating cluster can be obtained, allowing $g(r)$ to be calculated rather than assumed, as was the case in the DIT.

Central to density-functional theories is the concept of a *functional*. In contrast to a *function*, which acts on one or more variables, a *functional* acts on an entire function. For example, assume that the number density, $\rho(\mathbf{r})$ (i.e., the number of particles per unit volume as a function of position) is an appropriate order parameter for the phase transition. For a uniform system, the dependence of the Gibbs free energy on ρ could be represented as a function, $g(\rho)$. For an inhomogeneous density, however, the free energy is a functional, $G[\rho]$, which acts on the function $\rho(\mathbf{r})$. (Note that square brackets are used to indicate a functional operation.) If the functional is *local*, the value for the total free energy can be computed by integrating the contributions from all points in space over the volume of the system,

$$G[\rho] = \int_V g(\rho(\mathbf{r}))\mathrm{d}\mathbf{r}, \tag{28}$$

For a *nonlocal functional*, the value is not simply the sum of the local contributions but must take account of factors such as density–density correlations. An example of a nonlocal function will be considered in Section 4.

Since $G = F + pV$, where F is the Helmholtz free energy at constant pressure, p is the pressure, and V is the volume,

$$G[\rho] = \int_V \left(f(\rho(\mathbf{r})) + p\right)\mathrm{d}\mathbf{r}, \tag{29}$$

where $f(\rho(\mathbf{r}))$ is the Helmholtz free energy density as a function of position. The work (also at constant pressure) required to form a fluctuation containing N molecules in the configuration of the new phase, from those atoms in the original phase is

$$W = G[\rho] - N\mu, \tag{30}$$

where μ is the chemical potential in the original phase. Combining Eqs. (28) and (30), gives

$$W[\rho] = \int_V \left(g(\rho(\mathbf{r})) - \mu\rho(\mathbf{r})\right)\mathrm{d}\mathbf{r}, \tag{31}$$

which is equivalent to Eq. (11). In terms of the Helmholtz free-energy density (using Eq. (29)),

$$W[\rho] = \int_V \left(\left(f(\rho(\mathbf{r})) - \mu\rho(\mathbf{r})\right) + p\right)\mathrm{d}\mathbf{r}. \tag{32}$$

As already emphasized, Eqs. (31) and (32) are general relations, not relying on any assumed dividing surface between the original and new phases and not introducing an interfacial free energy. Further, while the number density was used for illustration, the functional forms also hold for more general order parameters.

A critical fluctuation in the density, ρ^*, is required to "nucleate" the final phase, corresponding to the saddle point in Figure 3. This is analogous to the critical size within CNT, and again the focus is on computing the work required to form this critical fluctuation. The critical nucleus at the saddle point corresponds to a stationary state; the variation of the work with respect to the order parameter is equal to zero,

$$(\delta W[\rho]/\delta\rho)_{\rho=\rho^*} = 0. \tag{33}$$

This is equivalent to the requirement in CNT that W be a maximum at n^* (see Figure 1 in Chapter 2). The problem is then reduced to solving

Eq. (33), which requires that the functional $G[\rho]$ be known. The simplest assumption is that g is a function of ρ and its gradient, $g = g(\rho(r), \nabla\rho(r))$, which is appropriate for the case where the density and the free energy vary slowly with position (Section 4.2).

Consider first a simple one-dimensional case, that is, $g = g(\rho, \mathrm{d}\rho/\mathrm{d}x)$, where the explicit spatial dependence of the density has been dropped for notational ease. From Eq. (31), $W = \int \mathrm{d}x \psi(\rho, \mathrm{d}\rho/\mathrm{d}x)$, where $\psi(\rho, \mathrm{d}\rho/\mathrm{d}x) \equiv g(\rho, \mathrm{d}\rho/\mathrm{d}x) - \mu\rho$. The variational condition in Eq. (33) corresponds to

$$\delta W = \int \mathrm{d}x\left(\frac{\partial\psi}{\partial\rho}\delta\rho - \frac{\partial\psi}{\partial(\partial\rho/\partial x)}\delta\left(\frac{\partial\rho}{\partial x}\right)\right) = 0 \quad \text{for} \quad \rho = \rho^*, \tag{34}$$

giving the Euler–Lagrange equation [22]

$$\left(\frac{\partial\psi}{\partial\rho} - \frac{\mathrm{d}}{\mathrm{d}x}\frac{\partial\psi}{\partial(\partial\rho/\partial x)}\right)_{\rho=\rho^*} = 0. \tag{35}$$

Following Kashchiev [63], Eq. (35) is readily generalized to three dimensions to give

$$\left(\frac{\partial}{\partial\rho}(g(\rho, \nabla\rho) - \mu\rho) - \sum_i \frac{\partial^2}{\partial x_i \partial\rho'_i}(g(\rho, \nabla\rho) - \mu\rho)\right)_{\rho=\rho^*} = 0, \tag{36}$$

where x_i and ρ'_i are the orthogonal components of $\mathbf{r}$ and $\nabla\rho$ respectively. Since ρ and ρ' are independent, Eq. (36) reduces to

$$\left(\frac{\partial g(\rho, \nabla\rho)}{\partial\rho} - \sum_i \frac{\partial^2 g(\rho, \nabla\rho)}{\partial x_i \partial\rho'_i}\right)_{\rho=\rho^*} = \mu. \tag{37}$$

Using the chain rule for partial derivatives, this becomes ($\rho''_{i,j} = \partial^2\rho/\partial x_i \partial x_j$)

$$\left(\frac{\partial g(\rho, \nabla\rho)}{\partial\rho} - \sum_i \frac{\partial^2 g(\rho, \nabla\rho)}{\partial x_i \partial\rho'_i} - \sum_i \rho'_i \frac{\partial^2 g(\rho, \nabla\rho)}{\partial\rho\partial\rho'_i} - \sum_{i,j} \rho''_{i,j} \frac{\partial^2 g(\rho, \nabla\rho)}{\partial\rho'_i \partial\rho'_j}\right)_{\rho=\rho^*} = \mu. \tag{38}$$

Equation (38) is solved subject to the appropriate boundary conditions to obtain the critical density, $\rho^*(\boldsymbol{r})$. It should be noted that this equation is equivalent to chemical equilibrium between the critical cluster of the new phase and the original phase (i.e., $\mu^* = \mu$). The nucleation barrier W^* is obtained by solving Eq. (31) using $\rho^*(\boldsymbol{r})$.

These density-functional relations are similar to those from CNT, although this tends to be obscured by the more elaborate formalism. Table 1 compares the key equations of the two methods.

Up to now, we have assumed that the density is the appropriate order parameter for the phase transition. This may not be true for cases such as the liquid–crystal transition, where the densities of the two phases are close. The order parameter may also be a function of time, t, as well as position, r. In that case, the time evolution is typically described by a Landau–Ginzburg equation (see Ref. [23]). For a time- and position-dependent order parameter (such as structural order), $M(r, t)$, characterizing nucleation from a liquid (an elemental liquid in this case), is

$$\frac{\partial M(\mathbf{r},t)}{\partial t} = -\frac{\Gamma_{\mathrm{eff}}}{\rho_{\mathrm{s}} k_{\mathrm{B}} T_{\mathrm{m}}} \frac{\delta W[M]}{\delta M}, \tag{39}$$

where ρ_s is the density of the bulk solid, T_m is the equilibrium melting temperature and Γ_{eff} is an effective mobility that characterizes the change in order parameter with time; $\delta W[M]/\delta M$ denotes a functional derivative. Within the framework of nonequilibrium thermodynamics, Eq. (39) describes an order-parameter "flux" as a linear function of the applied force, or driving free energy, given by the derivative of W with respect to M. It is an example of a phase-field equation, with $M(r, t)$ representing the local phase. Phase-field kinetic equations guarantee that the free energy decreases with increasing time [24].

The predictions from a density-functional calculation of nucleation depend on the order parameters chosen to describe the phase transition and the approximations that are made to obtain solutions for their spatial dependence. While the results obtained differ in detail, they all indicate that, in agreement with computer simulations and the prediction of the DIT, the interfaces between clusters of the new phase and the original phase are diffuse rather than sharp, leading to a different work of cluster formation and a different "critical size" for nucleation than expected from CNT. In the following sections, we give an overview of the types of approximations made and work through a simple case where only one order parameter is assumed. In Chapters 7 and 8, predictions from select density-functional calculations will be compared with experimental data and contrasted with predictions from CNT.

Table 1 Key equations for classical theory (CNT) and density-functional theory

Classical Theory	Density-Functional Theory
$W_n = n\delta\mu + (36\pi)^{1/3}\bar{v}^{2/3}n^{2/3}\sigma$	$W[\rho] = \int_V (g(\rho, \nabla\rho) - \mu\rho) \mathrm{d}\, r$, where $\rho \equiv \rho(\mathbf{r})$
$\left.\frac{\mathrm{d}\, W_n}{\mathrm{d}\, n}\right\vert_{n=n^*} = 0$	$(\delta W[\rho]/\delta\rho)_{\rho=\rho^*} = 0$
$W^* = W(n^*)$	$W^* = W(\rho^*)$
$\mu^* = \mu$ (μ is the chemical potential of the original phase)	$\left[\frac{\partial g(\rho,\nabla\rho)}{\partial\rho} - \sum_i \frac{\partial^2 g(\rho,\nabla\rho)}{\partial x_i \partial\rho'} - \sum_i \rho'_i \frac{\partial^2 g(\rho,\nabla\rho)}{\partial\rho\partial\rho'_i} - \sum_{i,j} \rho''_{i,j} \frac{\partial^2 g(\rho,\nabla\rho)}{\partial\rho'_i\partial\rho'_j}\right]_{\rho=\rho^*} = \mu.$

4.2 Square-gradient approximation

Consider a one-dimensional example, taking the density, ρ, as the relevant order parameter. Assume that the free energy is a continuous function of the local density and its derivatives (independent variables), and that $d\rho/dx$ (also written as ρ') is small compared with the reciprocal of intermolecular distances. Expanding around the free energy g_0 for a system of uniform density $\bar{\rho}$,

$$g(\rho, \rho', \rho'', \ldots) = g_0(\bar{\rho}) + (\rho - \bar{\rho})\left(\frac{\partial g}{\partial \rho}\right)_{\bar{\rho}} + \rho'\left(\frac{\partial g}{\partial \rho'}\right)_{\bar{\rho}} + \rho''\left(\frac{\partial g}{\partial \rho''}\right)_{\bar{\rho}} + \cdots$$
$$+ \frac{1}{2}(\rho - \bar{\rho})^2\left(\frac{\partial^2 g}{\partial \rho^2}\right)_{\bar{\rho}} + \frac{1}{2}(\rho')^2\left(\frac{\partial^2 g}{\partial (\rho')^2}\right)_{\bar{\rho}} + \cdots \quad (40)$$

The density varies smoothly between the solid and the liquid phases (Figure 4a); the dividing surface (at x_s) is chosen so that the sum of the excess density is zero (the Gibbs dividing surface). The contributions from the first derivative, $d\rho/dx$, are significant only near x_s, about which they are approximately symmetric (Figure 4b). Hence, they do not contribute to first order to the integrated free energy, $G[\rho]$. This symmetry also causes terms of $(\rho-\bar{\rho})^m$ to vanish under integration, when m is odd. By absorbing the even terms of $(\rho-\bar{\rho})^m$ into g_0 and truncating the series to lowest order, Eq. (40) reduces to [25]

$$g(\rho, \rho', \rho'', \ldots) = g_0 + \rho''\left(\frac{\partial g}{\partial \rho''}\right)_{\bar{\rho}} + \frac{1}{2}(\rho')^2\left(\frac{\partial^2 g}{\partial (\rho')^2}\right)_{\bar{\rho}} + \cdots$$
$$= g_0 + \kappa_1(\partial^2\rho/\partial x^2) + \kappa_2(\partial\rho/\partial x)^2 + \cdots. \quad (41)$$

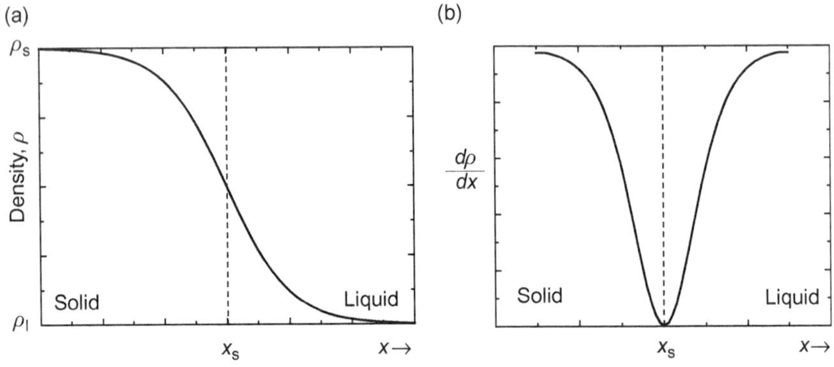

Fig. 4 (a) The variation in density through the interface (located at x_s) from the solid (density ρ_s) to the liquid (density ρ_l); (b) derivative of the density as a function of position.

The same form for the free-energy density is obtained in three dimensions

$$g(\rho, \nabla\rho, \nabla^2\rho, \ldots) = g_0 + \kappa_1 \nabla^2\rho + \kappa_2(\nabla\rho)^2 + \cdots, \tag{42}$$

where

$$\kappa_1 = \left(\frac{\partial g}{\partial \nabla^2 \rho}\right)_{\bar{\rho}} \quad \text{and} \quad \kappa_2 = \left(\frac{\partial^2 g}{\partial (|\nabla\rho|)^2}\right)_{\bar{\rho}}, \tag{43}$$

and $(\nabla\rho)^2 = \nabla\rho \cdot \nabla\rho$. Using Eq. (42), the Gibbs free-energy functional, $G[\rho]$, is obtained by integrating over the total volume

$$G[\rho] = \int_V g(\rho, \nabla\rho, \nabla^2\rho, \ldots)\mathrm{d}\,\mathbf{r} = \int_V \left(g_0 + \kappa_1 \nabla^2\rho + \kappa_2(\nabla\rho)^2 + \cdots\right) \mathrm{d}\mathbf{r}. \tag{44}$$

Applying the divergence theorem, we get

$$\int_V \kappa_1(\nabla^2\rho)\mathrm{d}\mathbf{r} = -\int_V \frac{\mathrm{d}\,\kappa_1}{\mathrm{d}\,\rho}(\nabla\rho)^2\mathrm{d}\mathbf{r} + \int_S (\kappa_1 \nabla\rho \cdot \hat{n})\,\mathrm{d}\mathbf{s}, \tag{45}$$

where $\hat{n}$ is a unit vector normal to the boundary of integration. The boundary can always be chosen so that $\nabla\rho \cdot \hat{\boldsymbol{n}}$ is zero at the external surface, causing the second integral to vanish. Following Cahn and Hilliard [25], using Eq. (45) to eliminate the terms in the second derivative of the density in Eq. (44), the Gibbs free-energy functional is

$$G[\rho] = \int_V \left(g_0 + \kappa(\nabla\rho)^2 + \cdots\right) \mathrm{d}\mathbf{r}, \tag{46}$$

where

$$\kappa = -\frac{\mathrm{d}\kappa_1}{\mathrm{d}\rho} + \kappa_2, \tag{47}$$

and the free-energy density is

$$g(\rho(\mathbf{r})) = g_0 + \kappa(\nabla\rho)^2 + \cdots. \tag{48}$$

The first term is the free-energy density for a homogeneous system that is well defined in the metastable and unstable regions. The gradient in the second term increases the free energy with spatial variations in density, thus favoring a homogeneous system.

Inserting Eq. (48) into Eq. (37) or (38) gives for the critical fluctuation [25, 26]

$$\left(\frac{\partial g_0}{\partial \rho} - \frac{\partial \kappa}{\partial \rho}(\nabla\rho)^2 - 2\kappa(\nabla^2\rho)\right)_{\rho=\rho^*} = \mu. \tag{49}$$

For a spherically symmetric nucleus, the density is a function only of the radial distance, and $(\nabla\rho)^2 = (\mathrm{d}\rho/\mathrm{d}\,r)^2$ and $\nabla^2\rho = (2/r)(\mathrm{d}\,\rho/\mathrm{d}\,r) + \mathrm{d}^2\rho/\mathrm{d}r^2$, giving (from Eq. (49)),

$$\left(2\kappa\left(\frac{\mathrm{d}^2\rho}{\mathrm{d}\,r^2}\right)+\frac{4\kappa}{r}\frac{\mathrm{d}\rho}{\mathrm{d}r}+\frac{\partial\kappa}{\partial\rho}\left(\frac{\mathrm{d}\rho}{\mathrm{d}r}\right)^2\right)_{\rho^*} = \left(\frac{\partial g_0}{\partial\rho}\right)_{\rho^*} - \mu, \tag{50}$$

subject to the boundary conditions,

$$\frac{\mathrm{d}\,\rho}{\mathrm{d}\,r} = 0 \quad \text{at} \quad r = 0 \quad \text{and} \quad r = \infty$$
$$\text{and} \quad \rho = \bar{\rho} \quad \text{at} \quad r = \infty. \tag{51}$$

These equations must be solved numerically to obtain the density profile and the work of cluster formation.

4.3 Single-order-parameter description of nucleation of a crystal from a liquid

A single-order-parameter analysis, often called a semi-empirical density-functional approximation (SDFA), is used to illustrate the application of DFT to nucleation processes. It has the advantage that it can be solved exactly, and the parameters entering the theory can be deduced from bulk properties, allowing predictions to be compared with experimental data. There exist several SDFA formulations of nucleation [27–29]; the example presented here is from Bagdassarian and Oxtoby [30].

Consider the case of crystal nucleation in a liquid. Define a single, nonconserved structural order parameter, $M(\mathbf{r}, t)$ that is a function of both space and time, so that $M(\mathbf{r}, t) = 0$ in the *bulk* liquid and $M(\mathbf{r}, t) = M_s$ in the solid, where M_s is a positive number. For the *bulk* solid, $M_s = 1$. Assuming the square-gradient approximation (Section 4.2), the scaled work of cluster formation can be expressed in terms of M, as

$$\frac{W_{\mathrm{SDFA}}[M]}{\rho_s k_B T_m} = \int_V \left(\omega(M) + \frac{1}{2}K_M^2|\nabla M|^2\right)\mathrm{d}\,\mathbf{r}, \tag{52}$$

where ρ_s is the density of the solid (note, density differences between liquid and solid are ignored in this treatment) and ω is a local dimensionless free energy for a uniform system that has a degree of order M. Note that to simplify the notation, $M(\mathbf{r}, t)$ is written as M; the temporal and spatial dependences are implicit. K_M^2 is the coefficient of the square-gradient term (Eq. (46)) divided by $\rho_s k_B T_m$; K_m is a measure of the correlation length for the order parameter. From Eq. (39), taking the functional derivative of Eq. (52), the temporal and spatial evolutions of M are described by

$$\frac{\partial M}{\partial t} = -\Gamma_{\text{eff}}\left(\frac{\mathrm{d}\,\omega(M)}{\mathrm{d}M} - K_M^2\nabla^2 M\right). \tag{53}$$

As a lowest-order approximation, assume a double-well construction for which $\omega(M)$ has a parabolic variation with M. Since it is a free energy, ω will assume the lower of the liquid or solid values for any M,

$$\omega(M) = \min\left(\frac{\lambda_l}{2}M^2; \frac{\lambda_s}{2}(M - M_s)^2 + \Delta\right), \tag{54}$$

where λ_l is the curvature of the $\omega(M)$ curve for the liquid phase, λ_s is the curvature for the solid phase, and Δ is a dimensionless free-energy difference between the liquid and solid that scales with the supercooling ($\Delta = \Delta g_{sl}/\rho_s k_B T_m$). As shown in Figure 5, $\omega_l(M)$ for the liquid phase intersects with $\omega_s(M)$ of the solid phase at $M = M_c$. The location of M_c is a function of λ_l, λ_s and Δ, increasing with increasing Δ (corresponding to an increasing supercooling). From Eq. (54),

$$\Delta = \frac{1}{2}\lambda_l M_c^2 - \frac{1}{2}\lambda_s(M_c - M_s)^2. \tag{55}$$

$\omega(M)$ changes discontinuously on going through M_c. For $M<M_c$, it follows the curve $\omega_l(M)$ for the liquid phase and for $M>M_c$, it follows that of the solid phase, $\omega_s(M)$ (in both cases shown by the dark solid portions of the curves).

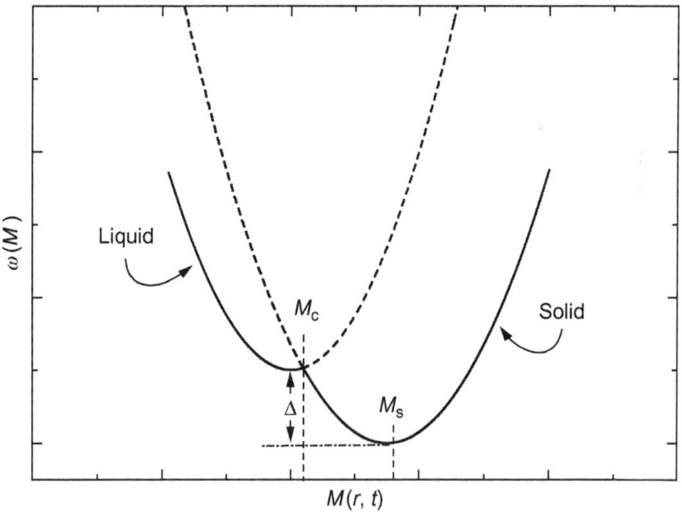

Fig. 5 Noninteracting work of cluster formation, $\omega(M)$ (i.e. local Landau potential) as a function of the structural order parameter, $M(r, t)$ for the liquid and solid phases. The dark line shows the lowest value of $\omega(M)$ in the two branches on either side of the special value of the order parameter, M_c.

Since $M \to M_s$ in the center of a solid nucleus (with $M \to 1$ in the center of a macroscopic solid cluster) and $M \to 0$ as $r \to \infty$, M can be used to describe the profile of the interfacial region between the solid clusters and the liquid phase. The cluster radius, r_s, is the boundary between the cluster and the liquid; for $r < r_s$, $M > M_c$ (the solid-like region) and for $r > r_s$, $M < M_c$ (the liquid-like region). The critical nucleus (radius r_s^*) corresponds to a stationary profile ($\delta W[M]/\delta M = 0$, from Eq. (39) when $\partial M/\partial t = 0$). Assuming that the critical nuclei are spherically symmetric, Eq. (53) yields

$$\begin{aligned} \frac{d^2 M}{d r^2} + \frac{2}{r}\frac{d M}{d r} - \frac{\lambda_s}{K_M^2}(M - M_s) = 0 \quad & \text{for} \quad r < r_s^* \quad (M > M_c) \\ \frac{d^2 M}{d r^2} + \frac{2}{r}\frac{d M}{d r} - \frac{\lambda_l}{K_M^2} M = 0 \quad & \text{for} \quad r > r_s^* \quad (M < M_c), \end{aligned} \tag{56}$$

where M is now only a function of r. In the solid-like region, $dM/dr \to 0$ as $r \to 0$, in the liquid-like region $M \to 0$ as $r \to \infty$, and at the liquid/solid boundary ($r = r_s^*$) the solid and liquid order parameters and their derivatives with respect to r are equal. Since these differential equations have Yukawa-like solutions, $\propto e^{-\alpha r}/r$, a reasonable form for the order parameter that satisfies these boundary conditions is (see Ref. [30])

$$\begin{aligned} M &= M_s + \frac{r_s^*(M_s - M_c)}{r} \frac{\exp(-r\sqrt{\lambda_s}/K_M) - \exp(r\sqrt{\lambda_s}/K_M)}{\exp(r_s^*\sqrt{\lambda_s}/K_M) - \exp(-r_s^*\sqrt{\lambda_s}/K_M)} \quad (r < r_s^*) \\ M &= \frac{r_s^* M_c}{r} \exp\left(-\frac{(r - r_s^*)\sqrt{\lambda_l}}{K_M} \right) \quad (r > r_s^*). \end{aligned} \tag{57}$$

The correlation length for M is then $\Lambda = K_M/\sqrt{\lambda_s}$. By requiring that the order parameters in the liquid and solid be equal at r_s^*, these equations lead to the following expression,

$$\frac{M_c}{M_s} = \frac{1 - (K_M/r_s^*\sqrt{\lambda_s}) \tanh(r_s^*\sqrt{\lambda_s}/K_M)}{1 + \sqrt{(\lambda_l/\lambda_s)} \tanh(r_s^*\sqrt{\lambda_s}/K_M)}. \tag{58}$$

Finding the value of M_c from Eq. (58) for a critical size r_s^*, and determining Δ from Eq. (55), the order-parameter profile as a function of distance from the cluster center can be calculated from Eq. (57). The work of formation for a spherical critical cluster within this SDFA, W_{SDFA}^* is (from Eq. (52))

$$\frac{W_{SDFA}^*}{\rho_s k_B T_m} = 4\pi \int_0^\infty \left(\omega(M) + \frac{1}{2} K_M^2 \left(\frac{d M}{d r} \right)^2 \right) r^2 d r; \tag{59}$$

the appropriate value of M from Eq. (57) is chosen depending on the value of the integration variable, r.

For the critical nucleus, $\partial M/\partial t = 0$ and Eq. (53) becomes

$$\frac{\mathrm{d}\,\omega(M)}{\mathrm{d}M} = \frac{K_M^2}{r^2}\frac{\mathrm{d}}{\mathrm{d}\,r}\left(r^2\frac{\mathrm{d}\,M}{\mathrm{d}\,r}\right). \tag{60}$$

Integrating the second term in Eq. (59) by parts and assuming that M goes to zero as $r \to \infty$ more rapidly than r becomes large,

$$\frac{W^*_{\mathrm{SDFA}}}{\rho_s k_B T_m} = 2\pi \int_0^\infty \left[2r^2\omega(M) - K_M^2 M \frac{\mathrm{d}}{\mathrm{d}\,r}\left(r^2\frac{\mathrm{d}\,M}{\mathrm{d}\,r}\right)\right]\mathrm{d}\,r. \tag{61}$$

Using Eq. (60),

$$\frac{W^*_{\mathrm{SDFA}}}{\rho_s k_B T_m} = 2\pi\left(\int_0^{r_s^*}\left(2\omega(M) - M\frac{\mathrm{d}\omega(M)}{\mathrm{d}M}\right)r^2\mathrm{d}r + \int_{r_s^*}^\infty\left(2\omega(M) - M\frac{d\omega(M)}{dM}\right)r^2\mathrm{d}r\right), \tag{62}$$

where the range of the integral has been broken into the liquid-like ($r > r_s^*$) and solid-like ($r < r_s^*$) regions. From Eq. (54), only the first integral ($0 \le r \le r_s^*$) contributes to the free energy, giving

$$\frac{W^*_{\mathrm{SDFA}}}{\rho_s k_B T_m} = \frac{4\pi(r_s^*)^3\Delta}{3} + 2\pi\lambda_s M_s \int_0^{r_s^*}(M_s - M)r^2\mathrm{d}r. \tag{63}$$

Combining this with Eqs. (57) and (58), the work of formation for the critical cluster can be written as

$$\frac{W^*_{\mathrm{SDFA}}}{\rho_s k_B T_m} = \frac{4\pi(r_s^*)^3\Delta}{3} + 2\pi M_s M_c r_s^*\left(1 + \frac{r_s^*\sqrt{\lambda_l}}{K_M}\right)K_M^2. \tag{64}$$

The CNT work of cluster formation (Eq. (18) of Chapter 2) can be written as

$$\frac{W^*_{\mathrm{CNT}}}{\rho_s k_B T_m} = \frac{4\pi(r^*_{\mathrm{CNT}})^3\Delta}{3} + 4\pi(r^*_{\mathrm{CNT}})^2\sigma', \tag{65}$$

where r^*_{CNT} is the critical radius in CNT and $\sigma' = \sigma/\rho_s k_B T_m$, the scaled interfacial free energy. When $\Delta \to 0$, corresponding to near equilibrium between the liquid and solid phases, the critical cluster size becomes very large ($r_s^* \to \infty$). In that limit, $r_s^* \to r^*_{\mathrm{CNT}}$ and $W^*_{\mathrm{SDFA}} \to W^*_{\mathrm{CNT}}$, giving (from Eqs. (64) and (65)),

$$\sigma' = \frac{M_s M_c}{2r_s^*}\left(1 + \frac{r_s^*\sqrt{\lambda_l}}{K_M}\right)K_M^2. \tag{66}$$

Also in that limit, Eq. (58) becomes

$$\frac{M_c}{M_s} \approx \frac{1 - (K_M/r_s^*\sqrt{\lambda_s})\left(1 - 2\exp(-2r_s^*\sqrt{\lambda_s}/K_M)\right)}{1 + \sqrt{\lambda_l/\lambda_s}\left(1 - 2\exp(-2r_s^*\sqrt{\lambda_s}/K_M)\right)} \approx \frac{\sqrt{\lambda_s}}{\sqrt{\lambda_l} + \sqrt{\lambda_s}}\left(1 - \frac{K_M}{r_s^*\sqrt{\lambda_s}}\right), \tag{67}$$

since the exponential argument may be ignored. It also follows directly that

$$1-\frac{M_c}{M_s}\approx\frac{\sqrt{\lambda_l}}{\sqrt{\lambda_l}+\sqrt{\lambda_s}}\left(1+\frac{K_M}{r_s^*\sqrt{\lambda_l}}\right). \tag{68}$$

Substituting Eqs. (67) and (68) into Eq. (66), and ignoring terms of orders $(1/r_s^*)$ and $(1/r_s^{*2})$, we obtain an expression for the interfacial free energy in the large r limit,

$$\sigma'=\frac{K_M M_s^2}{2}\left(\frac{\sqrt{\lambda_l\lambda_s}}{\sqrt{\lambda_l}+\sqrt{\lambda_s}}\right)\Rightarrow\sigma=\frac{\rho_s k_B T_m K_M M_s^2}{2}\left(\frac{\sqrt{\lambda_l\lambda_s}}{\sqrt{\lambda_l}+\sqrt{\lambda_s}}\right). \tag{69}$$

It is useful to scale the key parameters to obtain dimensionless quantities

$$\tilde{r}\equiv r\frac{\sqrt{\lambda_s}}{K_M},\quad \tilde{r}_s^*\equiv r_s^*\frac{\sqrt{\lambda_s}}{K_M},\quad \tilde{r}_{CNT}^*\equiv r_{CNT}^*\frac{\sqrt{\lambda_s}}{K_M},\quad \tilde{\rho}_s\equiv\rho_s\frac{K_M^3}{\lambda_s^{3/2}},\quad \tilde{\Delta}\equiv\frac{2\Delta}{\lambda_s}. \tag{70}$$

In terms of such dimensionless quantities, the work of cluster formation for the classical critical cluster is

$$\frac{W_C^*}{\tilde{\rho}_s k_B T_m \lambda_s}=\frac{8\pi M_s^6}{3\tilde{\Delta}^2}\left(\frac{\beta}{1+\beta}\right)^3, \tag{71}$$

corresponding to a classical critical radius of

$$\tilde{r}_{CNT}^*=-\frac{2M_s}{\tilde{\Delta}}\left(\frac{\beta}{1+\beta}\right), \tag{72}$$

where $\beta=\sqrt{\lambda_l/\lambda_s}$. The scaled work of cluster formation for the DFT becomes (from Eq. (64) with the scaled variables given in Eq. (70))

$$\frac{W_{SDFA}^*}{\tilde{\rho}_s k_B T_m \lambda_s}=\frac{2\pi}{3}\tilde{\Delta}(\tilde{r}_s^*)^3+2\pi M_s M_c \tilde{r}_s^*(1+\beta\tilde{r}_s^*). \tag{73}$$

We will examine how well this expression for the work fits experimental data for crystal nucleation in liquids and glasses in Chapters 7 and 8. But for now, let us examine the profile of the order parameter. In terms of the scaled parameters, Eqs. (57) and (58) become

$$\begin{aligned} M&=M_s+\frac{\tilde{r}_s^*(M_s-M_c)}{\tilde{r}}\frac{e^{-\tilde{r}}-e^{\tilde{r}}}{e^{\tilde{r}_s^*}-e^{-\tilde{r}_s^*}} && (r<r_s^*)\\ M&=\frac{\tilde{r}_s^* M_c}{\tilde{r}}\exp\left(-\beta(\tilde{r}-\tilde{r}_s^*)\right) && (r>r_s^*), \end{aligned} \tag{74}$$

and Eq. (58) becomes

$$\frac{M_c}{M_s}=\frac{1-\left(\tanh(\tilde{r}_s^*)/\tilde{r}_s^*\right)}{1+\beta\tanh(\tilde{r}_s^*)}. \tag{75}$$

The relation between $\tilde{\Delta}$ and M_c (Eq. (55)) is

$$\tilde{\Delta} = \beta^2 M_c^2 - (M_c - M_s)^2. \tag{76}$$

Figure 6 shows a plot of the order parameter, M, for three different values for $\tilde{\Delta}$; the parameters chosen for these calculations are listed in the figure. An insert shows the profile for a larger value for the scaled critical radius ($\tilde{r}_s^* = 20$), corresponding to $\tilde{\Delta} = -0.05$. It is not straightforward to make a comparison between the scaled parameter, $\tilde{\Delta}$, and the degree to which the nucleating system is out of equilibrium; but clearly an increase in the magnitude of $\tilde{\Delta}$ corresponds to a higher driving free energy for the transformation.

It is clear from the curves in Figure 6 that at all supercooling values, the interface between the liquid and the solid cluster is diffuse; the order parameter, M drops smoothly from the value in the solid to that in the liquid ($M = 0$). The profile sharpens dramatically relative to the cluster radius, however, as the critical size becomes large,

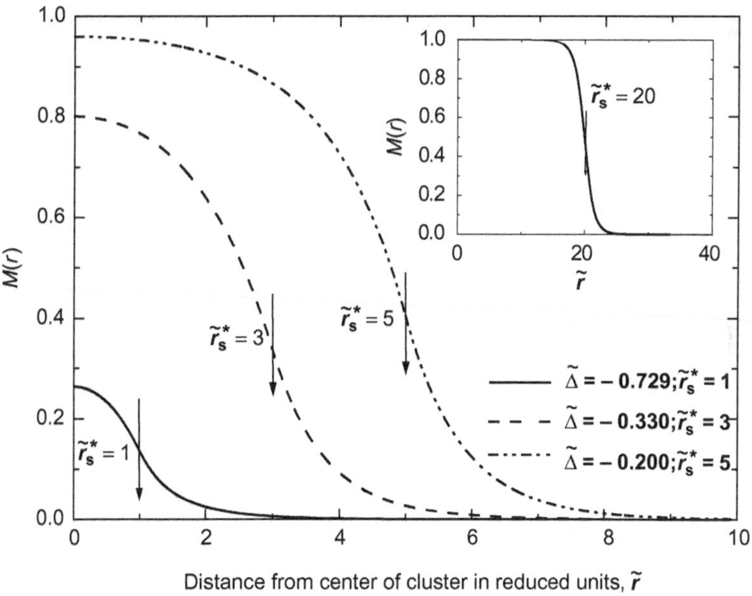

Fig. 6 Order parameter M as a function of the scaled distance from the center of the critical cluster, computed from Eq. (73), taking $M_s = 1$, $\lambda_s = \lambda_l = \lambda$, and $\beta = 1$. Profiles for three different values of $\tilde{\Delta}$ are shown with the corresponding values of the reduced critical size $\tilde{r}_s^*$. The inset demonstrates a sharpening of the interfacial profile relative to the cluster size with decreasing driving free energy for the transformation. (Reprinted with permission from Ref. [30], copyright (1994), American Institute of Physics.)

corresponding to transformations under conditions closer to equilibrium. Note that the actual (unscaled) interfacial width remains relatively constant, supporting this assumption made in the discussion of the DIT (Section 3).

Table 2 shows the interfacial width Δr_{int} as a function of supercooling; Δr_{int} was arbitrarily chosen to correspond to the range of r spanning $0.9M_{\max}$ to $0.1M_{\max}$. (As the critical size increases, it becomes easier to make a less arbitrary definition of the cluster width, but this crude estimate will serve to illustrate the point that we wish to make.) The physical estimates for the critical radius, r^*, and the interfacial width, were made by assuming that $\lambda = 4$ and $K_M = 0.34$ nm (see Ref. [31]). With increasing critical cluster size, the interfacial width approaches a constant value that is less significant relative to the cluster radius, becoming more consistent with the assumptions of CNT.

The maximum magnitude of the order parameter also serves to divide regions where CNT might be valid from those where it is not. The value for M in the macroscopic solid phase is equal to one. For large critical clusters (such as $\tilde{r}_s^* = 20$ in the inset of Figure 6, this value is reached in most of the "solid" cluster, as assumed in CNT calculations. For larger driving free energies, however, or smaller critical cluster sizes, this value is never attained, even at the cluster center. For transformations taking place far from equilibrium, then, it is fundamentally incorrect to use thermodynamic parameters taken from the macroscopic solid to calculate the work of cluster formation, an assumption typically made for nucleation calculations.

Figure 7 compares the classical and density-functional predictions for the reduced critical size as a function of the driving free energy, using the same value of β as for Figure 6. For CNT, $\tilde{r}_{\text{CNT}}^*$ is computed from Eq. (72). For the DFT, Eq. (76) is first solved for the value of M_c corresponding to a given $\tilde{\Delta}$; using this, Eq. (75) is solved iteratively to obtain $\tilde{r}_s^*$. For small driving free energies, the critical radius is the same as that predicted by CNT. As the driving free energy becomes greater, corresponding to a larger supercooling in liquids, for example, the

Table 2 Interfacial width as a function of supercooling

$\tilde{\Delta}$	$\tilde{r}_s^*$	$\Delta\tilde{r}_{\text{int}}$	r_s^* (nm)	Δr_{int} (nm)
−0.050	20	3.50	3.40	0.595
−0.200	5	3.10	0.85	0.527
−0.330	3	2.60	0.51	0.442
−0.729	1	1.45	0.17	0.247

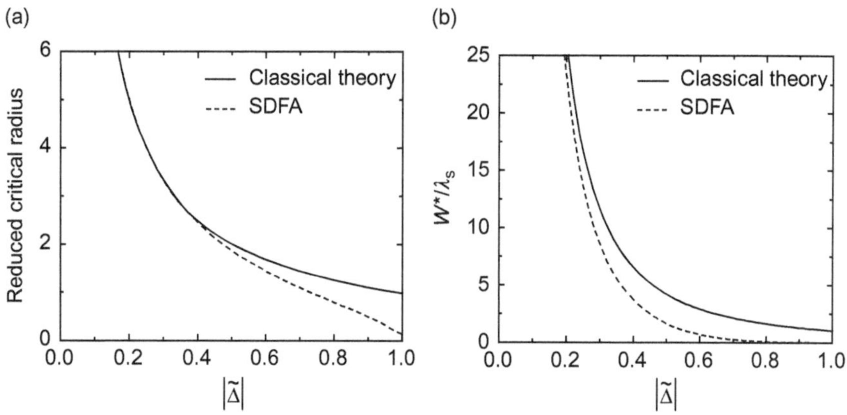

Fig. 7 (a) The scaled critical size as a function of the magnitude of the driving free energy, computed assuming $M_s = 1$, $\lambda_s = \lambda_l = \lambda$, and $\beta = 1$ (from Eq. (71)). (b) Scaled work of cluster formation, computed from Eq. (72). The scaled values for CNT were computed using Eqs. (14) and (19) of Chapter 2. (Reprinted with permission from Ref. [30], copyright (1994), American Institute of Physics.)

value computed from this density-functional analysis falls below that expected from CNT. Correspondingly, the work of cluster formation falls below that predicted by CNT as the driving free energy increases (Figure 7b). In qualitative agreement with calculations from more sophisticated density-functional treatments, these results once again indicate that the thermodynamic model of CNT is valid only for transitions that occur very near to equilibrium. The decrease of W^*_{SDFA} to zero is a failing of this particular model, since there is no point at which the liquid becomes unstable relative to the solid. Nonetheless, it does indicate that a properly constructed density-functional model could describe the transition from a nucleation-and-growth mechanism to a spinodal transformation, which the CNT cannot do.

4.4 More sophisticated density-functional approaches

In addition to its pedagogical usefulness, the single-order parameter treatment presented in Section 4.3 has the significant advantages that an analytical solution can be obtained and that predictions are readily compared with experimental data (Chapters 8 and 9). However, while a single structural order parameter may provide a reasonable description of the vapor–liquid interface, it would seem that a quantitative description of the liquid–crystal interface requires more sophisticated treatments. Since a comprehensive discussion of advanced density-functional approaches is beyond the scope of this book, only a brief discussion is presented here. A more detailed discussion can be found elsewhere [32].

4.4.1 Perturbative density-functional approximation (PDFA)

Perturbation approaches for treating liquid–crystal interfaces were pioneered by Ramakrishnan and Yussouff [33] and Haymet and Oxtoby [12]. Ignoring possible interactions with external fields, the Helmholtz free energy is divided into two contributions, an exact term that describes a noninteracting "ideal" system, F_{id} and an excess contribution that takes account of the particle–particle interactions, F_{ex}

$$F[\rho] = F_{\text{id}}[\rho] + F_{\text{ex}}[\rho]. \tag{77}$$

As already mentioned in Section 2, for structureless free particles in the absence of interactions (see Ref. [4] for example),

$$F_{\text{id}} = -k_{\text{B}}T \ln Q = -k_{\text{B}}T \ln \frac{q^N}{N!} = -k_{\text{B}}T \ln \frac{(V/\lambda^3)^N}{N!}, \tag{78}$$

where Q is as in Eq. (2), q is the single-molecule partition function, and λ is the thermal de Broglie wavelength. In terms of the density (see Ref. [12]),

$$F_{\text{id}} = k_{\text{B}}T \int_{\text{V}} \rho_0(\mathbf{r}) \Big(\ln \left(\Lambda \rho_0(\mathbf{r})\right) - 1 \Big) \, \mathrm{d}\mathbf{r}, \tag{79}$$

where $\Lambda = \lambda^3$. The excess free energy will also be assumed to be a local functional of the density, $F_{\text{ex}}[\rho] = \int f(\mathbf{r}, \rho) \mathrm{d}\,\mathbf{r}$.

In principle, an expression for $F_{\text{ex}}[\rho]$ can be obtained by expanding around the homogeneous liquid density, taking account of all the n-body correlations in the liquid, but practical considerations force a low-order truncation. Expanding to second order, and again assuming the square-gradient approximation (Section 4.2), F_{ex} can be written in terms of the Ornstein-Zernike direct correlation function of the liquid, $c^{(2)}$ [12, 64],

$$F_{\text{ex}}[\rho] = F_{\text{ex}}[\rho_1] - c^{(1)}(\rho_1) \int \Delta\rho(\mathbf{r}) \mathrm{d}\,\mathbf{r} - \frac{1}{2} \iint c^{(2)}(\mathbf{r}' - \mathbf{r}, \rho_1) \Delta\rho(\mathbf{r}) \Delta\rho(\mathbf{r}') \mathrm{d}\,\mathbf{r}\, \mathrm{d}\,\mathbf{r}', \tag{80}$$

where $-c^{(1)}(\rho_1)$ is the excess chemical potential of the liquid, ρ_1 is the density of the homogeneous liquid and $\Delta\rho(\mathbf{r}) = \rho_{\text{s}}(\mathbf{r}) - \rho_1$; ρ_{s} is the average density of the solid. A significant virtue of this formulation is that the Fourier components of the two-particle correlation function are related to the structure factor, $S(q)$, which can be obtained from X-ray or neutron diffraction measurements. This method is then particularly attractive for studying nucleation in complex liquids, where accurate interatomic potentials are not known.

An expression for the critical work of cluster formation, in terms that will be used in Chapters 8 and 9 for an analysis of experimental nucleation data in liquids and glasses, is [33, 65],

$$W^*_{\mathrm{PDFA}} = 4\pi\rho_1 k_\mathrm{B}T \int_0^\infty r^2 \left\{ \Delta\omega(\eta(r),\ m_i(r)) - \frac{c_0''}{4}\left(\frac{\mathrm{d}\,\eta(r)}{\mathrm{d}\,r}\right)^2 - \sum_i \frac{c_i''}{12}\left(\frac{\mathrm{d}\,m_i(r)}{\mathrm{d}\,r}\right)^2 \right\} \mathrm{d}\,r, \tag{81}$$

where η is the fractional density change on freezing, $\Delta\omega$ is the excess local grand potential density, $\Delta\omega = (c_0 - 1)\eta + (1/2)c_0\eta^2 + (1/2)\sum_i c_i m_i^2$, and the c_i are the Fourier coefficients in the direct correlation function; m_i are the Fourier components of the crystal density (e.g. Eq. (82)), and $c_i'' = \mathrm{d}^2c/\mathrm{d}\,q^2$ is evaluated at the ith reciprocal lattice vector, q_i.

4.4.2 Nonperturbative DFT: weighted density- (WDA) and modified-weighted density- (MWDA) functional approximations

As is clear from the last example, crystal nucleation from a liquid should take account of the periodic density fluctuations in the solid,

$$\rho(\boldsymbol{r}) = \rho_0 + \rho_\mathrm{s} \sum_i m_i \mathrm{e}^{i\boldsymbol{q}_i\cdot\mathbf{r}}, \tag{82}$$

where ρ_0 is the average density (varying from the density of the liquid to that of the solid), and $\boldsymbol{q}_i$ are the reciprocal lattice vectors. The Fourier coefficients, m_i, are the order parameters for the liquid-to-crystal phase transition. Within a perturbation approach, it might be assumed that an approximate solution could be obtained by including only some large, but finite number of the Fourier components [34, 35] or by picking only those components corresponding to the reciprocal wave vectors that are close to the position of the primary peak in $S(q)$ [28, 36]. However, low-order truncations generally give rise to unphysical results, such as a negative density in some locations. Nonperturbative approaches are required to solve such problems; the most commonly used are the WDA and MWDA.

In the WDA ([37–39] and the references therein), the excess free energy of the system is a function of a smoothed (or weighted) spatially invariant density, $\bar{\rho}$

$$F_{\mathrm{ex}}^{\mathrm{WDA}}[\rho] = \int f_0(\bar{\rho}(\mathbf{r}))\rho(\mathbf{r})\mathrm{d}\,\mathbf{r}, \tag{83}$$

where f_0 is the free energy per unit volume for a homogeneous system evaluated at a weighted density $\bar{\rho}(\mathbf{r})$, given by

$$\bar{\rho}(\mathbf{r}) = \int w(\mathbf{r} - \mathbf{r}'; \bar{\rho}(\mathbf{r}))\rho(\mathbf{r}')\mathrm{d}\mathbf{r}'. \tag{84}$$

For a weighting function that is a delta function, the local density-functional expression for the free energy is obtained (Eq. (28)). The weighting function w is chosen so that (i) $\int w(\mathbf{r}-\mathbf{r}';\bar{\rho}(\mathbf{r}))\mathrm{d}\mathbf{r}'=1$ and (ii) the second functional derivative of $F_{\mathrm{ex}}^{\mathrm{WDA}}$ is equal to the two-particle direct correlation function for a homogeneous fluid $c_0^{(2)}$,

$$\lim_{\rho\to\rho_0}\frac{\delta^2 F_{\mathrm{ex}}^{\mathrm{WDA}}[\rho]}{\delta\rho(\mathbf{r})\delta\rho(\mathbf{r}')}=-k_{\mathrm{B}}Tc_0^{(2)}(\mathbf{r}-\mathbf{r}';\rho_0), \tag{85}$$

where ρ_0 is the average density (or density of a uniform system).

In the WDA approximation, then, all the information in the two-particle correlation is preserved in the expression for the free energy. Curtin and Ashcroft showed that the condition in Eq. (85) led to a differential equation in Fourier space for the density dependence of the weighting function that could be solved iteratively on a grid in Fourier and density space.

MWDA maintains the advantages of WDA, but is computationally simpler [40]. In contrast with WDA, the MWDA proposes an expression for the free-energy density for a homogeneous liquid, $f_0[\hat{\rho}]=F_{\mathrm{ex}}^{\mathrm{MWDA}}[\rho]/N$, where N is the number of particles and $\hat{\rho}$ is a position-independent weighted density defined by

$$\hat{\rho}=\frac{1}{N}\int\rho(\mathbf{r})\mathrm{d}\,\mathbf{r}\int\tilde{w}(\mathbf{r}-\mathbf{r}';\hat{\rho})\rho(\mathbf{r}')\mathrm{d}\mathbf{r}', \tag{86}$$

where $\tilde{w}(\mathbf{r}-\mathbf{r}';\hat{\rho})$ is the new weighting function, different from that used in WDA. The Fourier transform, $\tilde{w}(q)$, is proportional to the Fourier transform of the Ornstein–Zernike two-particle direct correlation function.

4.4.3 Nucleation in a Lennard-Jones liquid

As mentioned in the last section, in principle nucleation of a crystal from a liquid should take account of the infinite number of order parameters in the Fourier expansion of the periodic density, which is not possible. The following description of nucleation in a Lennard-Jones liquid avoids this complexity [39]. Since the Lennard-Jones potential favors a cubic close-packed (ccp) crystal, the solid-phase density is represented by a sum of Gaussians centered on ccp lattice sites,

$$\rho(\mathbf{r})=2a_0^3\rho_0\left(\frac{\alpha}{\pi}\right)^{3/2}\sum_i\exp\Big(-\alpha(\mathbf{r}-\mathbf{R}_i)\cdot(\mathbf{r}-\mathbf{R}_i)\Big), \tag{87}$$

where a_0 is the lattice constant, $\mathbf{R}_i$ are the real-space lattice sites, α sets the width of the Gaussian, and ρ_0 is the average density. Equation (87) can be expressed as a Fourier series with coefficients, m_i

$$m_i=\mathrm{e}^{-q_i^2/4\alpha}, \tag{88}$$

where the q_is are the reciprocal lattice vectors. If α is small, indicating a localization in q-space, the atoms are delocalized in real space, corresponding to the case of a liquid; correspondingly, m_i approaches zero. If α is large, indicating a delocalization in q-space, the atoms are localized in real space, corresponding to a solid; in this case m_i approaches one. The limits on m_i agree with those required for a good order parameter. Only m_1 is needed to specify all of the other Fourier components, however, as is easily seen by taking the logarithm of Eq. (88) and comparing terms of different q_i,

$$m_i = m_1^{(q_i/q_1)^2}. \tag{89}$$

Specification of the crystal phase, therefore, requires only two order parameters: the average density ρ_0 and the structural order parameter m_1. If we allow these parameters to be a function of position and to have the necessary range of values, they can be used to describe both the crystal and liquid phases. The work of cluster formation is the difference in the grand canonical (Landau) potential, $\Omega[\rho(\mathbf{r})]$; $\rho(\mathbf{r})$ is the *microscopic* density, which is a functional of both $\rho_0(\mathbf{r})$ and $m_i(\mathbf{r})$. As discussed in Section 4.3, the critical cluster corresponds to the extremum of $\Omega[\mathbf{r}]$, with respect to those two parameters,

$$\frac{\partial\Omega}{\partial\rho_0(r)} = 0 \quad \text{and} \quad \frac{\partial\Omega}{\partial m_1(r)} = 0; \text{ evaluated at } r^*. \tag{90}$$

The Landau potential can be computed from the local grand canonical potential $w[\rho_0(\mathbf{r}), m_1(\mathbf{r})]$, which can be obtained from the local Helmholtz free-energy functional per unit volume, $f[\rho_0(\mathbf{r}), m_1(\mathbf{r})]$,

$$w\left[\rho_0(\mathbf{r}), m_1(\mathbf{r})\right] = f\left[\rho_0(\mathbf{r}), m_1(\mathbf{r})\right] - \mu\rho_0(\mathbf{r}). \tag{91}$$

The Helmholtz free energy can be written as the sum of an ideal term,

$$F_{\mathrm{id}} = k_B T \int_V \rho_0(\mathbf{r})\Big(\ln\left(\rho_0(\mathbf{r})\Lambda\right) - 1\Big)\,\mathrm{d}\mathbf{r}, \tag{92}$$

(where $\Lambda = \lambda^3$ (λ is the thermal de Broglie wavelength)) and an excess term, $F_{\mathrm{ex}}[\rho_0(\mathbf{r}), m_1(\mathbf{r})]$, that accounts for the interactions between molecules. For the Lennard-Jones (LJ) potential,

$$\Phi_{\mathrm{LJ}}(r) = 4\varepsilon\left(\left(\frac{\sigma}{r}\right)^{12} - \left(\frac{\sigma}{r}\right)^{6}\right), \tag{93}$$

(where ε is the depth of the minimum in the potential occurring at $\sim 1.122\sigma$), $F_{\mathrm{ex}}[\rho_0(\mathbf{r}), m_1(\mathbf{r})]$ is written as the sum of a hard-sphere repulsive interaction (the positive term in Φ_{LJ}), and a short-range attractive component (the negative term in Φ_{LJ}). Using MWDA (Section 4.4.2) and the square-gradient approximation, the work of cluster formation is determined,

$$W^{*}_{\mathrm{MWDA}} = 4\pi\rho_s k_B T_m \int_0^{\infty} r^2 \left\{ \begin{array}{l} \omega[\rho_0(r), m_1(\mathbf{r})] - \omega_{\mathrm{m}} + \frac{1}{2}K^2_{m_1 m_1}\left(\frac{\partial m_1(r)}{\partial r}\right)^2 \\ + \frac{1}{2\rho_S^2}K^2_{\rho_0\rho_0}\left(\frac{\partial \rho_0(r)}{\partial r}\right)^2 + \frac{1}{\rho_s}K^2_{m_1\rho_0}\left(\frac{\partial m_1(r)}{\partial r}\right)\left(\frac{\partial \rho_0(r)}{\partial r}\right) \end{array} \right\} dr \quad (94)$$

where ω_{m} is the local grand canonical potential of the metastable liquid. Equation (94) is analogous to Eq. (52), except with two order parameters. The correlation lengths, $K^2_{m_1 m_1}$, $K^2_{\rho_0\rho_0}$ and $K^2_{m_1\rho_0}$, are functionals of $\rho_0(\mathbf{r})$ and $m_1(\mathbf{r})$, and are related to $c^{(2)}$, the direct correlation function of the liquid. Equation (94) will be used in Chapters 7 and 8 in the discussion of experimental nucleation data in liquids and glasses.

5. NONCLASSICAL FORMULATIONS OF NUCLEATION KINETICS

The discussions in the previous sections of this chapter have focused on the work of cluster formation, addressing the inadequacies in the thermodynamic model of nucleation assumed within CNT. However, the validity of the kinetic model developed in Chapter 2, which is essentially based on chemical kinetics concepts, is also of concern. The definition of the forward and backward rates, particularly for diffusion-controlled nucleation processes in condensed phases (Chapter 5), and the detailed-balance arguments on which the theory rests (Chapter 2), are both questionable. A very brief discussion of alternatives to the rate-equation approach of CNT is provided in this section. For more detailed discussions of these issues, see Refs. [41–43].

5.1 Cluster theory

One approach is to treat the microscopic dynamics in terms of $P(\xi, t)$, the probability of the system being at a particular point in phase space (position ξ at time t). Following Binder and Stauffer [43], the time evolution of $P(\xi, t)$ is described by a master equation:

$$\frac{\partial P(\xi,t)}{\partial t} = -\int_0^t \sum_{\xi'} W_{t,t'}(\xi \to \xi')P(\xi,t')\mathrm{d}t' + \int_0^t \sum_{\xi} W_{t,t'}(\xi' \to \xi)P(\xi',t')\,\mathrm{d}t', \quad (95)$$

where $W_{t,t'}(\xi \to \xi')$ is a transition probability [44, 45]. Assuming that the evolution of the system depends only on the instantaneous state, the transition probabilities satisfy the detailed-balance condition,

$$W(\xi \to \xi')\exp\left(\frac{-\mathscr{H}(\xi)}{k_B T}\right) = W(\xi' \to \xi)\exp\left(\frac{-\mathscr{H}(\xi')}{k_B T}\right), \tag{96}$$

where $\mathscr{H}(\xi)$ is the Hamiltonian. A "cluster" is defined by 'relevant' coordinates that are of most practical interest for the nucleation process. These coordinates may include the cluster volume and surface area, as in CNT, but may also include other quantities that are relevant in some cases, such as the internal states of the cluster. It is assumed that $P(x,t)$ is sharply peaked at the mean values of the relevant coordinates.

Equation (95) can be recast to describe the cluster evolution. The time-dependent number of clusters having the same values of ξ is of most interest, analogous to the time-dependent cluster size population that underlies the nucleation rate within CNT. Binder and Stauffer obtain a general relation describing the time evolution of multicoordinate clusters (see Eq. (3.16) in Ref. [43]) that describes both nucleation and coagulation. Assuming that the number of molecules in a cluster is the only relevant cluster coordinate, their key result can be recast into the notation used in Chapter 2,

$$\begin{aligned}\frac{\partial N(n,t)}{\partial t} &= \frac{\partial}{\partial n}\left(k_n^+ \frac{\partial N(n,t)}{\partial n}\right) + \frac{1}{k_B T}\frac{\partial}{\partial n}\left(k_n^+ N(n,t)\frac{\partial W(n)}{\partial n}\right) \\ &+ \frac{1}{2}\int_{n_c}^{n} W(n-n',n')\frac{N(n',t)\,N(n-n',t)}{N^{\mathrm{eq}}(n')\,N^{\mathrm{eq}}(n-n')}\,\mathrm{d}n' \\ &- \frac{N(n,t)}{N^{eq}(n)}\int_{n_c}^{\infty} W(n,n')\frac{N(n',t)}{N^{\mathrm{eq}}(n')}\,\mathrm{d}n',\end{aligned} \tag{97}$$

where n_c is a lower cutoff for coagulation, in the region of the critical size. The first two terms of Eq. (97) are identical to Eq. (60) of Chapter 2, the Zeldovich–Frenkel equation describing the stochastic growth and dissolution of clusters in CNT. The third and fourth terms describe the coagulation of particles larger than the critical size (see Ref. [46]). If coagulation is not important, for example in the early stages of the phase transformation, and if the dominant coordinate is cluster size, the kinetic formulation of CNT is then reasonable.

5.2 Field-theory approach

The change in focus from the atomistic view of CNT to a generalized-order-parameter field description in the density-functional calculations discussed in Section 4 suggests another way to handle the kinetic aspects of nucleation. Like the cluster-based approach discussed in Section 5.1, these field-theoretic approaches also lend some support for the kinetic model of CNT, yielding an equation for the steady-state rate that has the same form as the CNT result, albeit with a different interpretation of the exponential prefactor.

As an example, taking the density field, $\rho(\mathbf{r})$ as the appropriate order parameter to describe the phase transition, within the theory of nonequilibrium thermodynamics there is a simple linear relation between the thermodynamic forces and the field flux for conserved-order-parameter dynamics,

$$\frac{\partial \rho(\mathbf{r})}{\partial t} = -\nabla \cdot j(\mathbf{r})$$
$$j(\mathbf{r}) = -\mathbf{M}\,\nabla\left(\frac{\partial F[\rho(\mathbf{r})]}{\partial \rho(\mathbf{r})}\right), \tag{98}$$

where $F[\rho(\mathbf{r})]$ is a free-energy functional, here taken as the Helmholtz free energy, $j(\mathbf{r})$ is a probability current, and $\mathbf{M}$ is a generalized transport mobility matrix. These equations do not take account of the fluctuations (e.g., in cluster size and shape) that are at the heart of nucleation. However, they can form the basis of a correct model if we introduce a Langevin force that fluctuates on a time scale that is short compared to the time scale for the molecular rearrangements accompanying nucleation (see Refs. [41, 47, 48], for example). As in our previous discussions, the metastable state decays via a locally unstable fluctuation (critical fluctuation). It is often convenient to introduce a new set of variables, η_i, $i = 1, 2, \ldots \zeta$, to describe the ζ degrees of freedom of the metastable system, with time-dependent currents and probability density that are a function of these variables, $J_i(\{\eta\},t)$ and $\widehat{\rho}(\{\eta\},t)$. With these changes, and the introduction of fluctuation terms, $k_B T(\partial \widehat{\rho}/\partial \eta)$, whose statistical properties correspond to those of a heat bath at temperature T, we obtain new kinetic equations for nucleation

$$\frac{\partial \widehat{\rho}}{\partial t} = -\sum_{i=1}^{\zeta} \frac{\partial J_i}{\partial \eta_i}$$
$$J_i = -\sum_{j=1}^{\zeta} M_{ij}\left(\frac{\partial F\{\eta\}}{\partial \eta_j}\widehat{\rho} + k_B T\,\frac{\partial \widehat{\rho}}{\partial \eta_j}\right). \tag{99}$$

As in previous discussions, the equilibrium probability density has the form,

$$\widehat{\rho}^{\,\mathrm{eq}}\{\eta\} \propto \exp\left(-\frac{F\{\eta\}}{k_B T}\right). \tag{100}$$

The metastable and stable states have configurations that minimize F and, hence, maximize $\widehat{\rho}^{\,\mathrm{eq}}$. Near equilibrium, the nucleation rate is given by the probability flow rate across the saddle point, which can be written as [48]

$$I = \kappa_D \Omega_0 \exp\left(-\frac{W^*}{k_B T}\right), \tag{101}$$

which is identical to the form predicted for the steady-state nucleation rate by CNT (see Eq. (55) of Chapter 2). The prefactor Ω_0 is a measure of the volume of the saddle-point region in configurational space; it is a generalization of the Zeldovich factor that was discussed in Chapter 2 (Eq. (51)). The leading terms of Ω_0 are the volumes of the system, measured as the cube of the correlation length, and factors that account for distortions of the droplet from a symmetric minimum-energy configuration. The dynamical prefactor, κ_D, describes the exponential growth rate of the unstable mode.

To apply the field-theoretic description of nucleation to particular situations, one must make the proper choice of the statistical variables, $\{\eta_i\}$, and the conjugate thermodynamic forces. For condensation from the vapor, Langer and Turski [49, 50] pointed out that a hydrodynamic description is most appropriate, since condensation (and nucleation from a liquid or solid) is characterized by a semimacroscopic density fluctuation involving large numbers of molecules. Grant and Gunton [51] followed this approach to describe nucleation of the solid from the liquid, making assumptions that should be reasonable for simple body-centered cubic (bcc) metals such as the alkali metals, and assuming that the nucleation is limited by conduction of heat away from the interface. In cases such as the crystallization of glasses (Chapter 8), however, interfacial and diffusive kinetics, rather than heat transport from the interface, dictate the nucleation kinetics.

While this approach is more rigorous than the phenomenological rate-theory approach of CNT (Chapter 2), the forms of the nucleation rate equations are, as mentioned, identical (Eqs. (55) of Chapter 2 and (101) of this chapter). A test of these formulations requires experiments that can differentiate between contributions to the preterms. At present, we know of no studies that have been able to do this.

5.3 Nucleation and coarsening processes

The comparison of experimental data with model predictions, the subject of the second part of this book, shows that measurable nucleation occurs over a very narrow range in supercooling or supersaturation. It is difficult to measure only the nucleation step; what is generally measured is the time before a significant volume fraction has transformed. A theory is needed, therefore, to describe the complete evolution of the phase transition from the nucleation stage through growth to the final stage of Ostwald ripening, where larger particles grow at the expense of smaller ones, reducing the total interfacial free energy, but without further transformation of the initial phase. For interface-controlled transformations, CNT can be directly extended to account for particle growth [52]; extensions that account for nucleation and growth and eventual coarsening in diffusion-controlled transformations are discussed in Chapter 5.

Langer and Schwartz were the first to formulate a comprehensive mean-field model for nucleation and growth of droplets in a metastable phase-separating near-critical fluid [53]. For small supersaturations, they found a significant time delay before nucleation appeared to commence after quenching into the metastable region, suggesting that time was needed for nuclei to grow to measurable size. For large supersaturations nucleation occurred quickly and growth was slow. When droplet coarsening was taken into account, the overall transformation rate was much smaller than expected based on nucleation considerations alone, accounting for the strong differences that had been observed previously between experimental measurements of nucleation and predictions from CNT [54–56]. The possibility of nucleation and growth with simultaneous coarsening is generally assumed to be negligible for studies of phase transformations in condensed phases such as liquids or glasses. However, it can occur in some metallic glasses that have extremely high crystal nucleation rates ($\sim 10^{20}$–$10^{22}\,\mathrm{m}^{-3}\,\mathrm{s}^{-1}$) at temperatures near or below the glass transition temperatures, where crystal growth rates are very small (Chapter 8).

The classic treatment of coarsening is due to Lifshitz and Slyozov, and Wagner (LSW) [57, 58]. It is valid for the late stages of coarsening of a system in which the volume fraction of the minority phase is vanishingly small. A clear and concise summary of the LSW theory of coarsening is found in the introduction of Ref. [59] and followed here.

The thermodynamic driving force for coarsening is the size-dependent chemical potential of the droplets or particles of the new phase, which changes the equilibrium concentration solute in the matrix adjacent to a particle of radius r_s from that adjacent to a particle of infinite radius according to the Gibbs–Thomson equation

$$C(r)|_{r_s} = C_\infty\left(1 + \frac{a}{r_s}\right), \tag{102}$$

where a is the capillary length, equal to $(2\sigma/k_B T)\bar{v}C_\infty$, $\bar{v}$ is the atomic volume, and C_∞ is the equilibrium solute concentration at the surface of an infinitely large particle of the new phase. In the LSW treatment, this becomes one of the boundary conditions for the solution of the diffusion equation for the concentration around a particle in the steady-state limit, $\nabla^2 C(r) = 0$ (since $\partial C/\partial t = 0$ in steady state), which determines the flow of material between the particles. Far from the particle, the concentration approaches the average solute concentration in the system, $\bar{C}$,

$$\lim_{r\to\infty} C(r) = \bar{C}. \tag{103}$$

If there is no nucleation, $\bar{C}$ is a constant. Assuming that the particles are spherical, solute conservation between the original and new phases requires

$$\frac{d}{dt}\left(\frac{4\pi}{3}r_s^3\right) = 4\pi r_s^2 D \frac{dC(r)}{dr}\bigg|_{r=r_s}, \tag{104}$$

where D is the diffusion coefficient, giving

$$\frac{dr_s}{dt} = D\frac{dC(r)}{dr}\bigg|_{r=r_s} = \frac{D}{r_s}\left(\Delta(t) - \frac{a}{r_s}\right), \tag{105}$$

where $\Delta(t) = \bar{C} - C_\infty$. There exists, then, a critical radius for coarsening, $r_s^* = a/\Delta$; particles for which $r_s < r_s^*$ will shrink while those with $r_s > r_s^*$ will grow.

A properly normalized distribution of particle sizes, $\phi(r_s, t)$, is assumed so that

$$\int \phi(r_s, t)dr_s = N(t), \tag{106}$$

where $N(t)$ is the time-dependent number of particles. If nucleation has ceased, $\phi(r_s, t)$ satisfies the continuity equation,

$$\frac{\partial\phi(r_s,t)}{\partial t} + \frac{\partial}{\partial r_s}\left(\frac{dr_s}{dt}\phi(r_s,t)\right) = 0. \tag{107}$$

Conservation of solute requires

$$\Delta(t) + \frac{4\pi}{3}\int_0^\infty r_s^3\phi(r_s,t)d\,r_s = C_0, \tag{108}$$

where C_0 is the initial solute concentration. The solutions to these equations are well known; in particular the predicted asymptotic growth rate of the average droplet radius is proportional to $t^{1/3}$,

$$\bar{r}_s(t) = \left(\frac{4aDt}{9}\right)^{1/3}. \tag{109}$$

The power-law growth and dynamical scaling first revealed in the LSW theory are now viewed as key characteristics of first-order phase transitions [60]. Experiments indicate that interactions among particles and the spatial locations of the particles during nucleation and growth are important [61]. Sagui and Grant have developed a model that combines steady-state nucleation with a modified LSW theory, taking account of these correlations between particles [62]. Instead of the simple diffusion equation in the LSW model ($\nabla^2 C(r) = 0$), a multiple-droplet diffusion equation must be solved

$$\frac{\partial C(r,t)}{\partial t} - \nabla^2 C(r,t) = -a \sum_{i=1}^{N_p(t)} Q_i \delta(r - r_i), \tag{110}$$

where $a = d\,\pi^{d/2}/\Gamma((d/2)+1)$, Γ is the gamma function, and d is the dimensionality of the system. The sum is taken over the particle number, $N_p(t)$. Depending on their size, the particles are either sources or sinks for solute diffusion; Q represents their strength. Nuclei appear based on the nucleation rate, $I(t)$, which is the integral over the rates for a distribution of particle sizes

$$I(t) = \int_0^\infty I(r_s, t) \mathrm{d}\, r_s. \tag{111}$$

A Gaussian form is assumed for the distribution of nucleation rates, $I(r_s, t)$. The general form for the homogenous nucleation rate (Eq. (101)) is used, with the work of cluster formation described by surface and volume terms (as in Eq. (11) of Chapter 2). Although steady-state homogenous nucleation is assumed, the phase transformation will cause the supersaturation to decrease, decreasing the driving free energy for nucleation and causing the nucleation rate to decrease with increasing time.

Figure 8 shows the results of a numerical integration of these equations in two dimensions; the general features should be the same for a three-dimensional case. A smooth transition between the stages of the phase transition is observed with the appearance of nuclei (Figure 8a, b) followed by growth (Figure 8c), and finally coarsening (Figure 8d). Changes in the solute field are particularly interesting. During nucleation and growth, the solute concentration is depleted in those regions containing the highest density of particles. The finite diffusion coefficient in the original phase also causes a shift in the local concentration field around the particles during the coarsening stage. As observed, the solute concentration is enriched around those particles with $r_s < r_s^*$ since they are dissolving, and depleted around those with $r_s > r_s^*$, which are growing.

This analysis is based on the assumption of a steady-state nucleation rate; the time-dependent nucleation due to cluster evolution as discussed in Chapter 3 is not considered. While a time-dependent nucleation rate that was calculated within CNT could be grafted onto the LSW model, a more fundamental difficulty is that this model does not take into account the coupling between interfacial attachment and long-range diffusion, which could be very important for precipitation and other solid-state transformations. As will be discussed in Chapter 5, this coupling of two stochastic fluxes leads to an evolving solute field around the clusters that appears to have many similarities with that observed from coarsening.

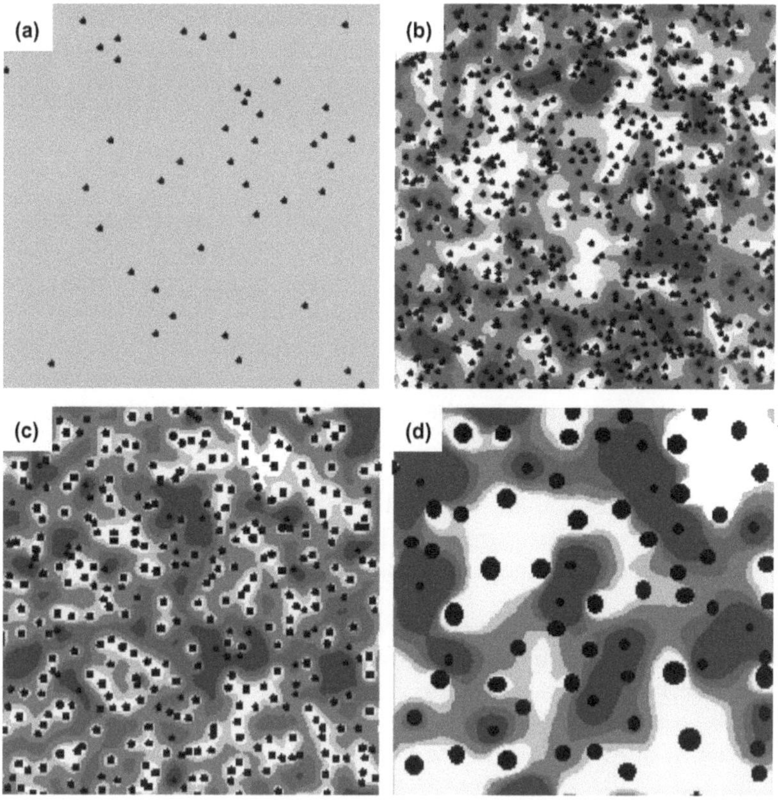

Fig. 8 Time evolution of a first-order phase transition in 2D, showing nucleation (a) and (b), growth (c) and coarsening (d). The gray scale in the background represents the solute concentration, with dark indicating the highest, and white the lowest, solute concentration. (Reprinted figure from Ref. [62], copyright (1999), American Physical Society.)

6. SUMMARY

In this chapter, we have focused primarily on corrections to the thermodynamic model of nucleation theory. We have demonstrated that many of the problems stated at the beginning of this chapter, and which we will explore in later chapters, can be reduced or eliminated by calculating the work of cluster formation, $W(n)$, from a density-functional approach, instead of the Gibbs model assumed in CNT. DFT indicates that the cluster/parent phase interface is broad, that the properties of small clusters are not the same as those of macroscopic pieces of the new phase, and that unless the phase transformation occurs near equilibrium, the classical-theory calculation of the work of cluster formation is incorrect.

Close to equilibrium, a critical fluctuation of the new phase resembles the critical nucleus in CNT in the following ways (see for example [26]):

1. The order parameter (such as the density) at the center of the nucleus approaches the equilibrium value for the new phase;
2. The specific energy associated with the interface approaches that of a flat interface;
3. The size of the fluctuation (corresponding to the cluster radius) is very large;
4. The thickness of the interface is small compared with the size of the fluctuation;
5. The work of formation for the fluctuation is in reasonable agreement with predictions from CNT.

Near equilibrium, then, where the critical fluctuation and the work required to form it are large, CNT is valid. As the departure from equilibrium becomes larger, significantly different behavior emerges:

1. The order parameter in the center of the nucleus less resembles that of the new phase and approaches that of the original phase;
2. The interface between a region of the new phase and the original phase becomes diffuse;
3. The work of formation for the fluctuation vanishes.

The first point demonstrates that the use of macroscopic thermodynamic parameters to describe nucleation is questionable, particularly when nucleation occurs far from equilibrium. The last point indicates that when the driving free energy becomes sufficiently large, a nucleation-based transformation changes continuously to a spinodal transformation, where small regions of the new phase are in unstable equilibrium with the original phase.

A key point is that the thermodynamic basis of CNT is strictly valid only for transformations occurring close to equilibrium. Its use outside that regime is questionable.

The kinetic model of CNT does not take into account coagulation or simultaneous coarsening and nucleation. Further, it is inappropriate for cases where the compositions of the original and new phases are different and when long-range diffusional fluxes become important. Formulations based on the analysis of LSW were briefly discussed. However, these require that the solute concentration at the interface of the cluster be at equilibrium. An extension of the rate-theory approach of CNT to consider the cases other than this is discussed in Chapter 5.

REFERENCES

[1] H. Reiss, R.K. Bowles, Some fundamental statistical mechanical relations concerning physical clusters of interest to nucleation theory, J. Chem. Phys. **111** (1999) 7501–7504.
[2] I.J. Ford, Statistical mechanics of nucleation: a review, Proc. Inst. Mech. Eng. **218**(part C), (2004) 883–899.
[3] P.R. ten Wolde, D. Frenkel, Computer simulation study of gas-liquid nucleation in a Lennard-Jones system, J. Chem. Phys. **109** (1998) 9901–9918.
[4] D.A. McQuarrie, *Statistical Mechanics*, Harper & Row, New York (1976).
[5] D.L. Goodstein, *States of Matter*, Prentice-Hall, Inc, Englewood Cliffs, NJ (1975).
[6] R.K. Pathria, *Statistical Mechanics*, Butterworth-Heinemann, Oxford (1996).
[7] J.Q. Broughton, A. Bonnissent, F.F. Abraham, The fcc (1 1 1) and (1 0 0) crystal-melt interfaces: a comparison by molecular dynamics simulation, J. Chem. Phys. **74** (1981) 4029–4039.
[8] J.Q. Broughton, G.H. Gilmer, Molecular dynamics of the crystal fluid interface. 5. Structure and dynamics of crystal–melt systems, J. Chem. Phys. **84** (1986) 5749–5758.
[9] W.J. Ma, J.R. Banavar, J. Koplik, A molecular dynamics study of freezing in a confined geometry, J. Chem. Phys. **97** (1992) 485–493.
[10] B.B. Laird, A.D.J. Haymet, The crystal–liquid interface of a body-centered-cubic-forming substance: Computer simulations of the r^{-6} potential, J. Chem. Phys. **91** (1989) 3638–3646.
[11] W.A. Curtin, Density-functional theory of the solid–liquid interface, Phys. Rev. Lett. **59** (1987) 1228–1231.
[12] A.D.J. Haymet, D.W. Oxtoby, A molecular theory for the solid–liquid interface, J. Chem. Phys. **74** (1981) 2559–2565.
[13] D.W. Oxtoby, A.D.J. Haymet, A molecular theory of the solid–liquid interface. II. Study of bcc crystal–melt interfaces, J. Chem. Phys. **76** (1982) 6262–6272.
[14] D.W. Oxtoby, in: *Fundamentals of Inhomogeneous Fluids*, Ed. D. Henderson, Marcel Dekker, New York (1992), p. 407.
[15] F. Spaepen, A structural model for the solid–liquid interface in monatomic systems, Acta Metall. **23** (1975) 729–743.
[16] D.R. Nelson, F. Spaepen, Polytetrahedral order in condensed matter, in: *Solid State Physics*, Eds. H. Ehrenreich, D. Turnbull, Academic Press, Boston (1989), pp. 1–90.
[17] F. Spaepen, Homogeneous nucleation and the temperature dependence of the crystal–melt interfacial tension, in: *Solid State Physics*, Eds. H. Ehrenreich, D. Turnbull, Academic Press, New York (1994), pp. 1–32.
[18] L. Gránásy, Diffuse interface theory of nucleation, J. Non-Cryst. Sol. **162** (1993) 301–303.
[19] L. Gránásy, Diffuse interface theory for homogeneous vapor condensation, J. Chem. Phys. **104** (1996) 5188–5198.
[20] L. Gránásy, Diffuse interface model of crystal nucleation, J. Non-Cryst. Sol. **219** (1997) 49–56.
[21] K.F. Kelton, Crystal nucleation in liquids and glasses, in: *Solid State Physics*, Eds. H. Ehrenreich, D. Turnbull, Academic Press, New York (1991), pp. 75–178.
[22] H. Goldstein, *Classical Mechanics*, Addison-Wesley, Reading, MA (1950).
[23] N. Goldenfeld, *Lectures on Phase Transitions and the Renormalization Group*, Addison-Wesley, Boston (1992).
[24] O. Penrose, P.C. Fife, Thermodynamically consistent models of phase-field type for the kinetics of phase transitions, Phys. D. **43** (1990) 44–62.
[25] J.W. Cahn, J.E. Hilliard, Free energy of a nonuniform system I. Interfacial free energy, J. Chem. Phys. **28** (1958) 258–267.
[26] J.W. Cahn, J.E. Hilliard, Free energy of a nonuniform system. III. Nucleation in a two-component incompressible fluid, J. Chem. Phys. **31** (1959) 688–699.

[27] W.H. Shih, Z.Q. Wang, X.C. Zeng, D. Stroud, Ginzburg–Landau theory for the solid–liquid interface of bcc elements, Phys. Rev. A. **35** (1987) 2611–2618.
[28] M. Iwamatsu, K. Horii, Application of a one-order parameter density functional model of crystal-melt transition to nucleation and steady-state kinetics: crystal-melt asymmetry, Mater. Sci. Eng. A. **A226–A228** (1997) 99–103.
[29] L. Gránásy, T. Pusztai, Diffuse interface analysis of crystal nucleation in hard-sphere liquid, J. Chem. Phys. **117** (2002) 10121–10124.
[30] C.K. Bagdassarian, D.W. Oxtoby, Crystal nucleation and growth from the undercooled liquid: A nonclassical piecewise parabolic free-energy model, J. Chem. Phys. **100** (1994) 2139–2148.
[31] D.W. Oxtoby, P.R. Harrowell, The effect of density change on crystal growth rates from the melt, J. Chem. Phys. **96** (1992) 3834–3843.
[32] D.W. Oxtoby, Crystallization of liquids: A density functional approach, in: *Liquids, Freezing and Glass Transition*, Eds. J.P. Hansen, D. Levesque, J. Zinn-Justin, Elsevier, Amsterdam (1991), pp. 145–191.
[33] T.V. Ramakrishnan, M. Yussouff, First-principles order-parameter theory of freezing, Phys. Rev. B. **19** (1979) 2775–2794.
[34] A.D.J. Haymet, Freezing and interfaces: Density functional theories in two and three dimensions, Prog. Solid State Chem. **17** (1986) 1–32.
[35] B.B. Laird, J.D. McCoy, A.D.J. Haymet, Density functional theory of freezing: Analysis of crystal density, J. Chem. Phys. **87** (1987) 5449–5456.
[36] P. Harrowell, D.W. Oxtoby, A molecular theory of crystal nucleation from the melt, J. Chem. Phys. **80** (1984) 1639–1646.
[37] W.A. Curtin, N.W. Ashcroft, Weighted-density-functional theory of inhomogeneous liquids and the freezing transition, Phys. Rev. A **32** (1985) 2909–2919.
[38] Y.C. Shen, D.W. Oxtoby, Density functional theory of crystal growth: Lennard-Jones fluids, J. Chem. Phys. **104** (1996) 4233–4242.
[39] Y.C. Shen, D.W. Oxtoby, Nucleation of Lennard-Jones fluids: A density functional approach, J. Chem. Phys. **105** (1996) 6517–6524.
[40] A.R. Denton, N.W. Ashcroft, Modified weighted-density-functional theory of nonuniform classical liquids, Phys. Rev. A **39** (1989) 4701–4708.
[41] J.S. Langer, Theory of the condensation point, Ann. Phys. **41** (1967) 108–157.
[42] J.S. Langer, Statistical theory of the decay of metastable states, Ann. Phys. **54** (1969) 258–275.
[43] K. Binder, D. Stauffer, Statistical theory of nucleation, condensation and coagulation, Adv. Phys. **25** (1976) 343–396.
[44] L. van Hove, The approach to equilibrium in quantum statistics — A perturbation treatment to general order, Physica **23** (1957) 441–480.
[45] R. Balescu, J. Wallenborn, On the structure of the time-evolution process in many body systems, Physica **54** (1971) 477–503.
[46] S.K. Friedlander, C.S. Wang, The self-preserving particle size distribution for coagulation by Brownian motion, J. Colloid Inter. Sci. **22** (1966) 126–132.
[47] J.S. Langer, Statistical theory of the decay of metastable states, Ann. Phys. **54** (1969) 258–275.
[48] J.S. Langer, Metastable states, Physica **73** (1974) 61–72.
[49] J.S. Langer, L.A. Turski, Hydrodynamic model of the condensation of a vapor near its critical point, Phys. Rev. A **8** (1973) 3230–3243.
[50] L.A. Turski, J.S. Langer, Dynamics of a diffuse liquid–vapor interface, Phys. Rev. A **22** (1980) 2189–2195.
[51] M. Grant, J.D. Gunton, Theory for the nucleation of a crystalline droplet from the melt, Phys. Rev. B **32** (1985) 7299–7307.

[52] K.F. Kelton, M.C. Weinberg, Calculation of macroscopic growth rates from nucleation data, J. Non-Cryst. Solids **180** (1994) 17–24.
[53] J.S. Langer, A.J. Schwartz, Kinetics of nucleation in near-critical fluids, Phys. Rev. A **21** (1980) 948–958.
[54] B.E. Sundquist, R.A. Oriiani, Homogeneous nucleation in a miscibility gap system. A critical test of nucleation theory, J. Chem. Phys. **36** (1962) 2604–2615.
[55] R.B. Heady, J.W. Cahn, Experimental test of classical nucleation theory in a liquid–liquid miscibility gap system, J. Chem. Phys. **58** (1973) 896–910.
[56] J.S. Huang, W.I. Goldburg, M.R. Moldover, Observation of anomalously large supercooling in carbon dioxide, Phys. Rev. Lett. **34** (1975) 639–642.
[57] I.M. Lifshitz, V.V. Slyozov, The kinetics of precipitation from super-saturated solid solutions, Phys. Chem. Solids **19** (1961) 35–50.
[58] C. Wagner, Theory of the ageing of precipitates by redissolution (Ostwald maturing), Z. Elektrochem. **65** (1961) 581–591.
[59] J.H. Yao, K.R. Elder, H. Guo, M. Grant, Theory and simulation of Ostwald ripening, Phys. Rev. B **47** (1993) 14110–14125.
[60] J.D. Gunton, M.S. Miguel, P.S. Sahni, *Phase Transitions and Critical Phenomena*, Vol. 8, Academic Press, London (1983).
[61] P.W. Voorhees, R.J. Schaefer, In situ observation of particle motion and diffusion interactions during coarsening, Acta Metall. **35** (1987) 327–339.
[62] C. Sagui, M. Grant, Theory of nucleation and growth during phase separation, Phys. Rev. E **59** (1999) 4175–4187.
[63] D. Kashchiev, *Nucleation – Basic Theory with Applications*, Butterworth-Heinemann, Oxford (2000) pp. 97–112.
[64] W.A. Curtin, Density-functional theory of crystal-melt interfaces, Phys. Rev. E **39** (1989) 6775–6791.
[65] L. Gránásy, P.F. James, Nucleation in oxide glasses: comparison of theory and experiment, Proc. Roy. Soc. Lond. A **454** (1998) 1745–1766.

CHAPTER 5

Multi-Component Systems

Contents

1. INTRODUCTION

With the inclusion of time-dependent effects (Chapter 3), the classical theory provides a reasonable model for nucleation in a one-component phase. However, the common case of nucleation in a multi-component phase is more complicated. For example, the composition of the nuclei can depend on their size, and the composition of the regions surrounding these nuclei can vary in a complicated way that reflects the stochastic nature of cluster growth during nucleation.

Pergamon Materials Series, Volume 15
ISSN 1470-1804, DOI 10.1016/S1470-1804(09)01505-3

We expect the steady-state nucleation rate to have the familiar form:

$$I^{st} = A^* \exp\left(-\frac{W^*}{k_B T}\right), \tag{1}$$

where W^* is the reversible work of formation for a critical cluster, containing n^* molecules, composed of a molecules of A, b molecules of B, c molecules of C, etc.; A^* is a composition-dependent prefactor, k_B is the Boltzmann constant, and T is the temperature in absolute units. In Section 2, the calculation of the work of cluster formation is extended to the multi-component case, following a treatment by Oxtoby and Kashchiev [1]. The new phase is assumed to be incompressible, making their approach useful for modeling nucleation in vapor-to-liquid and liquid-to-solid phase transitions in multi-component phases. In subsequent sections, we examine some kinetic models for nucleation in multi-component phases for interface-limited processes (Section 4) and when interface and long-range diffusion processes are competitive (Section 5). A common method for treating diffusion-dominated nucleation and coarsening was discussed in Chapter 4 (Section 5.3). As for the one-component case, we expect time-dependent nucleation to be important, which is treated in detail in Section 5.

2. WORK OF CLUSTER FORMATION

Following the approach used in Chapter 2 (Section 2.2), consider the case of condensation, corresponding to the reversible formation of a cluster of liquid (droplet, d) within a gas phase (medium, m) that now contains multiple atomic species. Assume a constant temperature, T, and a constant pressure on the medium, $p_m = p_0$, during the phase transition. As for the one-component case, a dividing interface defines the boundary between the cluster interior and the medium (see Chapter 2, Figure 3). The change in Gibbs free energy (ΔG) due to cluster formation can be expressed in terms of the excess free-energy contributions from the interfacial region Ψ, the difference in chemical potentials for the ith species in the original ($\mu_{m,i}$) and new ($\mu_{d,i}$) phases, and the pressure–volume work,

$$\Delta G = (p_0 - p_d)V_d + \sum_i (\mu_{d,i} - \mu_{m,i})n_{d,i} + \sum_i (\mu_{s,i} - \mu_{m,i})n_{s,i} + \Psi\left(V_d, \{\mu_{m,i}\}\right); \tag{2}$$

$\{\mu_{m,i}\}$ indicates the set of chemical potentials for species i. V_d is the volume of the droplet phase, p_d is the pressure within the droplet, $\mu_{s,i}$ is the chemical potential of the molecules in the interface, $n_{d,i}$ is the number of molecules of species i in the new phase cluster, and $n_{s,i}$ is the number of excess molecules in the interface. Setting the derivatives of ΔG with respect to $n_{d,i}$, $n_{s,i}$, and V_d separately equal to zero gives the conditions of chemical and mechanical equilibrium for the critical cluster

$$\mu_{m,i} = \mu^*_{d,i} = \mu^*_{s,i} \quad \text{(for all } i\text{)}$$
$$p^*_d = p_0 + \frac{\partial \Psi^*}{\partial V^*_d} \,. \tag{3}$$

The condition on pressure is the analog of the Laplace equation used for the derivation of Eq. (10) in Chapter 2, although a constant interfacial free-energy is not assumed here. Combining Eqs. (2) and (3), the work of critical cluster formation, W^*, is

$$W^* = \Delta G\left(V^*_d, \{\mu_{m,i}\}\right) = (p_0 - p^*_d)V^*_d + \Psi^*\left(V^*_d, \{\mu_{m,i}\}\right), \tag{4}$$

where V^*_d is the volume of a critical cluster of the droplet phase. The Gibbs–Duhem relations for the molecules in the droplet phase, the initial phase, and the interface are [2]

$$V^*_m dp_0 = \sum_i n^*_{m,i} d\mu_{m,i}, \quad V^*_d dp^*_d = \sum_i n^*_{d,i} d\mu^*_{d,i}, \quad d\Psi^* = -\sum_i n^*_{s,i} d\mu^*_{s,i}, \tag{5}$$

where V_m is the volume of the original phase in the two-phase mixture ($V = V_m + V_d$, where V is the total volume of the system). The asterisk superscripts imply that the quantities are evaluated when the droplet is a critical cluster. Expressing the volume of the droplet of the final phase in terms of the partial molar volumes,

$$V_d = \sum_i n_{d,i}\,\bar{v}_{d,i} \quad \text{where} \quad \bar{v}_{d,i} = \left(\frac{\partial V_d}{\partial n_{d,i}}\right)_{T,P_d,n_{d,j(j\neq i)}}, \tag{6}$$

and using the second relation in Eq. (5), we can write,

$$\sum_i n^*_{d,i}\,(\bar{v}^*_{d,i} dp^*_d - d\mu^*_{d,i}) = 0, \tag{7}$$

where $n^*_{d,i}$ is the number of molecules and $\bar{v}^*_d$ is the partial molar volume of species i in the critical cluster. For a nontrivial solution,

$$\bar{v}^*_{d,i} dp^*_d = d\mu^*_{d,i} \ \text{(for all } i\text{)}. \tag{8}$$

At constant temperature and assuming an incompressible final phase, the integral of Eq. (8) gives

$$\bar{v}^*_{d,i}(p^*_d - p_0) = \mu^*_{d,i}(T, p^*_d) - \mu^*_{d,i}(T, p_0) \equiv -\Delta\mu^*_i. \tag{9}$$

Using Eqs. (4), (6) and (9), the work of formation for a critical cluster containing more than one component can be written as

$$W^* = \sum_i n^*_{d,i}\Delta\mu^*_i + \Psi^*\left(V^*_d, \{\mu_{m,i}\}\right). \tag{10}$$

Often the driving free energies for each species and the excess free-energy are not known. Following the one-component case, the driving free-energy is assigned an effective value and the free energy is taken to be the surface free-energy for a planar interface, giving

$$W^* = n^* \Delta\mu' + \eta n^{*2/3} \bar{v}^{2/3} \sigma', \tag{11}$$

where n^* is the total number of molecules of all types in the critical cluster, $\Delta\mu'$ is an *effective* driving free-energy, σ' is an *effective* surface free-energy, $\bar{v}$ is the average atomic volume, and η is a geometric factor dependent on the shape of the cluster, equal to $(36\pi)^{1/3}$ for spherical clusters. As we saw in Chapter 4 for the one-component case, the assignment of a sharp cluster interface, with a constant interfacial free-energy (the capillarity approximation), is incorrect for small clusters. Problems with the interface are exacerbated in multi-component phases, where the chemical compositions of the interface and cluster interior are functions of the cluster size.

In Section 2.11 we showed that in a one-component system, the derivative of W^* with respect to $\Delta\mu$ led to the nucleation theorem. It is useful to examine this for a multi-component system. Taking the derivatives of W^* with respect to the chemical potential of each of the species i in the original phase, while holding the others fixed,

$$\begin{aligned}\frac{\partial W^*}{\partial \mu_{m,i}} &= V_d^* \frac{\partial (p_0 - p_d^*)}{\partial \mu_{m,i}} + (p_0 - p_d^*) \frac{\partial V_d^*}{\partial \mu_{m,i}} \\ &\quad + \frac{\partial \Psi^*}{\partial V_d^*} \frac{\partial V_d^*}{\partial \mu_{m,i}} + \frac{\partial \Psi^*}{\partial \mu_{m,i}}.\end{aligned} \tag{12}$$

From mechanical equilibrium (Eq. (3)), this reduces to

$$\frac{\partial W^*}{\partial \mu_{m,i}} = V_d^* \frac{\partial (p_0 - p_d^*)}{\partial \mu_{m,i}} + \frac{\partial \Psi^*}{\partial \mu_{m,i}}. \tag{13}$$

Using the Gibbs–Duhem relations (Eq. (5)) and the equality between the chemical potentials (Eq. (3)), we obtain

$$\frac{\partial W^*}{\partial \mu_{m,i}} = V_d^* \left(\frac{n_{m,i}^*}{V_m} - \frac{n_{d,i}^*}{V_d^*} \right) - n_{s,i}^*. \tag{14}$$

Defining the critical number density, $\rho^* = n^*/v^*$, where v^* is the volume of the critical cluster, gives

$$\frac{\partial W^*}{\partial \mu_{m,i}} = -\left(n_{s,i}^* + n_{d,i}^* \left(1 - \frac{\rho_{m,i}}{\rho_{d,i}^*} \right) \right) = -\Delta n_i^*, \tag{15}$$

where Δn_i^* is the excess number of molecules of type i in the nucleus over those in the same volume of the original phase, that is, the excess number in the critical cluster. The number of molecules of each species, i, in the critical size is then a function of the number in the interior of the critical droplet, $n^*_{\mathrm{d},i}$, and those in the interface, $n^*_{\mathrm{s},i}$. If the number density of the initial phase is small relative to that of the critical-sized droplet (as is the case for nucleation from a dilute vapor), the experimental critical cluster size is simply the sum of these for all species,

$$n^* = \sum_i (n^*_{\mathrm{d},i} + n^*_{\mathrm{s},i}). \tag{16}$$

From Eq. (1),

$$\frac{\partial W^*}{\partial \mu_{\mathrm{m},i}} = -\frac{\partial (k_{\mathrm{B}}T \ln I^{st})}{\partial \mu_{\mathrm{m},i}} + \frac{\partial (k_{\mathrm{B}}T \ln A^*)}{\partial \mu_{\mathrm{m},i}}. \tag{17}$$

The slope of the experimental steady-state nucleation rate with respect to the driving free-energy for the phase transformation can then be used to compute the number of molecules of each species in the critical cluster. Equation (17) holds for any nucleation process that satisfies Eq. (1), regardless of whether it is classical, atomistic, homogeneous, heterogeneous, diffusion-controlled, interface-controlled, etc. For nucleation at constant temperature in a supersaturated solution, the dependence of A^* on the supersaturation is weak, giving an expression for the excess number of molecules of type i in the critical nucleus,

$$\Delta n^*_i = -\left(\frac{\partial W^*}{\partial \mu_{\mathrm{m},i}}\right)_T = \left(\frac{\partial (k_{\mathrm{B}}T \ln I^{\mathrm{st}})}{\partial \mu_{\mathrm{m},i}}\right)_T - \varepsilon, \tag{18}$$

where ε is a constant between 0 and 1 [1]. An analysis of crystal nucleation rates from the melt is more complicated, since these are typically inferred from maximum supercooling measurements, obtained under nonisothermal conditions at constant pressure (Chapter 7). Here, the difference between the nucleation and the liquidus temperatures defines the driving free-energy; a derivative with respect to $\mu_{\mathrm{m},i}$ becomes a derivative with respect to temperature. Also, the contribution of the prefactor, A^*, can no longer be ignored, since it is itself generally a strong function of temperature (as was discussed in Section 2.11 for nucleation in the one-component case).

3. MULTI-COMPONENT KINETIC MODELS FOR NUCLEATION (GENERAL CONSIDERATIONS)

As for the case of nucleation in a one-component system (Chapter 2), the A^* (Eq. (1)) must be determined from a kinetic analysis. We have already discussed some of the problems with the definition of the work of cluster formation in a multi-component system; many more questions arise when we consider the kinetic aspects of nucleation. The construction of the appropriate cluster distribution, the importance of interface versus diffusive fluxes, and even the definition of a nucleation rate become significant problems.

Consider the kinetic development of a cluster with a composition that is different from that of the initial phase. Unlike the one-component case, the nucleation and growth of this cluster will involve both the attachment of molecules at the interface and the transport of material in the original phase to the interfacial region. If these processes are competitive, they must be coupled in some way; the stochastic nature of cluster evolution near the critical size makes this much more difficult than for growth. The rates must be defined carefully. Interfacial rates will likely be different for the different atomic species. If long-range diffusion is dominant, the rates must be defined so that the growth rate of large clusters reduces to the expression for diffusion-limited growth [3]. Time-dependent composition changes in the regions of the original phase near each cluster must be considered. The composition of the clusters themselves might be time- as well as size-dependent. These effects cannot be ignored; in Chapter 3, we saw that thermal and processing history could have a profound impact on subsequent nucleation behavior.

A complete treatment of time-dependent nucleation that takes all these features into account may not be possible. Fortunately, in many cases either the interfacial or long-range-diffusion fluxes dominate. As we shall see, however, in the next two sections where we discuss these two simplest cases, the formulation of a model that captures the essence of the physical process can be difficult. To simplify the development, we will limit the discussions to nucleation in a binary system.

4. INTERFACE-LIMITED NUCLEATION

Expressions for the nucleation rate are developed by following a classical-theory-based approach. The case where atomic diffusion in the original phase is sufficiently fast that interfacial processes are rate-limiting is considered in this section. The case where atomic diffusion becomes more important is taken up in Section 5.

4.1 Kinetic model

Consider first the case where the long-range diffusion rate is sufficiently fast that interfacial attachment becomes the rate-limiting step. This is equivalent to the assumption that the composition of the regions near the developing clusters is time-independent, staying the same as the original phase. Since the volume of the new phase is small in the nucleation stage of the transformation, it can be assumed that the overall composition of the original phase remains constant as well. As already discussed, for multi-component nucleation the cluster distribution cannot be defined by the number of molecules only, but must also include the chemical composition of the cluster. Define $N(a, b, t)$ to be the number of clusters (cluster population) having a units of A and b units of B (where a unit might be an atom, or a molecule). As indicated, the cluster population is in general a function of time, t. A cluster can increase or decrease in size by either adding or losing a number of A or B units. Following the classical theory, we assume that the most probable step involves only one A or B unit at a time, giving rate kinetics of the form,

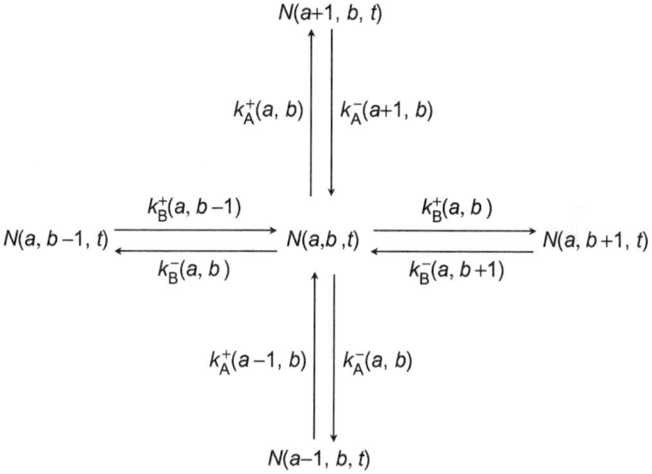

(19)

Here $k_A^+(a,b)$ and $k_B^+(a,b)$ are the rates at which A or B molecules, respectively, are added to a cluster containing a molecules of A and b molecules of B (cluster size, $n = a + b$); $k_A^-(a,b)$ and $k_B^-(a,b)$ are the rates at which A or B molecules, respectively, leave that cluster. Cluster evolution can then be viewed as a movement on a two-dimensional lattice with the points on the lattice corresponding to different cluster compositions (Figure 1). Developing clusters hop through the lattice one step at a time, in either a or b directions. Note that the population development for a one-component system is equivalent to cluster motion along a linear lattice. From Eq. (19), the rate of change of the cluster distribution is given by the solutions to a set of coupled differential equations of the form,

$$\begin{aligned}\frac{\partial N(a,\ b,\ t)}{\partial t} &= N(a-1,\ b,\ t)k_A^+(a-1,\ b) + N(a,\ b-1,\ t)k_B^+(a,\ b-1)\\ &+N(a+1,\ b, t)k_A^-(a+1,\ b) + N(a,\ b+1,\ t)k_B^-(a,\ b+1)\\ &-N(a,\ b,\ t)\left(k_A^+(a,\ b) + k_B^+(a,\ b) + k_A^-(a,\ b) + k_B^-(a,\ b)\right).\end{aligned} \tag{20}$$

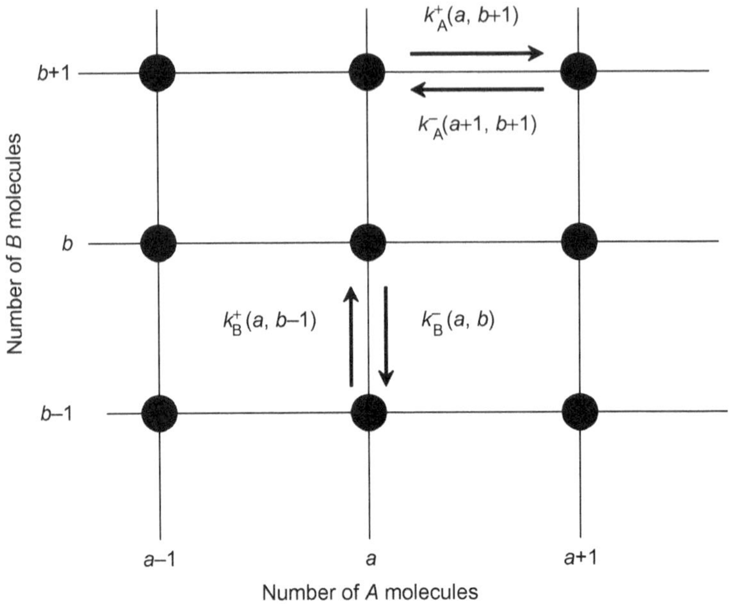

Fig. 1 Schematic illustration of the rate processes underlying interface-limited nucleation in a binary system. Cluster evolution is viewed as a movement on a two-dimensional lattice, where the lattice points correspond to the cluster composition.

As in Chapter 2, the precise definition of the rates depends on the physical case under consideration. For gas condensation,

$$k^+_{\rm A}(a,b) = \sqrt{\frac{k_{\rm B}T}{2\pi m_{\rm A}}} N_{\rm A}(1)\ (36\pi)^{1/3}(a\bar{v}_{\rm A} + b\bar{v}_{\rm B})^{2/3}$$
$$\times \left(1 + \left(\frac{\bar{v}_{\rm A}}{a\bar{v}_{\rm A} + b\bar{v}_{\rm B}}\right)^{1/3}\right)^2 \left(1 + \left(\frac{m_{\rm A}}{am_{\rm A} + bm_{\rm B}}\right)\right)^{1/2}$$

and

$$k^+_{\rm B}(a,b) = \sqrt{\frac{k_{\rm B}T}{2\pi m_{\rm B}}} N_{\rm B}\ (1)(36\pi)^{1/3}(a\bar{v}_{\rm A} + b\bar{v}_{\rm B})^{2/3}$$
$$\times \left(1 + \left(\frac{\bar{v}_{\rm B}}{a\bar{v}_{\rm A} + b\bar{v}_{\rm B}}\right)^{1/3}\right)^2 \left(1 + \left(\frac{m_{\rm B}}{am_{\rm A} + bm_{\rm B}}\right)\right)^{1/2} \qquad (21)$$

where $N_{\rm A}(1)$ and $N_{\rm B}(1)$ are the population of single molecules, $m_{\rm A}$ and $m_{\rm B}$ are the molecular masses, and $\bar{v}_{\rm A}$and $\bar{v}_{\rm B}$ are the atomic volumes of

A and B, respectively [4]. For crystallization from the liquid or glass, or solid-state transformations (which are interface- rather than diffusion-limited), the generalizations of the Turnbull–Fisher rates are

$$k_A^+(a,b) = O(a,b)\gamma_A \exp\left(-\frac{\delta W(a)}{2k_BT}\right) = 6O(a,b)\frac{D_A}{\lambda_A^2}\exp\left(-\frac{W(a+1,b)-W(a,b)}{2k_BT}\right)$$
and
$$k_B^+(a,b) = O(a,b)\gamma_B \exp\left(-\frac{\delta W(b)}{2k_BT}\right) = 6O(a,b)\frac{D_B}{\lambda_B^2}\exp\left(-\frac{W(a,b+1)-W(a,b)}{2k_BT}\right), \tag{22}$$

where $O(a, b)$ represents the number of attachment sites on a cluster containing n $(=a + b)$ molecules. For spherical clusters $O(a, b)$ is proportional to $((a\bar{v}_A + b\bar{v}_B)/\bar{v})^{2/3}$, where $\bar{v}$ is an average atomic volume. If the atomic volumes of A and B are nearly equal, $O(a,b)$ becomes proportional to $(a + b)^{2/3}$ or $n^{2/3}$ as in the one-component case (Section 4 of Chapter 2). For species A and B respectively, γ_A and γ_B are the unbiased interfacial attachment rates, D_A and D_B are the diffusion coefficients at the interface (assuming as in Chapter 2 that the interfacial mobility is the same as that for diffusion in the original phase), and λ_A and λ_B are the jump distances.

As for the one-component case, the backward rates are computed by assuming detailed balance, using the equilibrium distribution, $N^{eq}(a, b)$, corresponding to zero cluster flux,

$$k_A^-(a+1,b) = k_A^+(a,b)\frac{N^{eq}(a,b)}{N^{eq}(a+1,b)}$$
$$k_B^-(a,b+1) = k_B^+(a,b)\frac{N^{eq}(a,b)}{N^{eq}(a,b+1)}. \tag{23}$$

The nucleation flux (in general time-dependent) has two components, $I_A(a, b, t)$ and $I_B(a, b, t)$, corresponding to the flux of clusters of composition (a, b) that become clusters of composition $(a + 1, b)$ or $(a, b + 1)$, respectively:

$$I_A(a,b,t) = k_A^+(a,b)N(a,b,t) - k_A^-(a+1,b)N(a+1,b,t)$$
$$I_B(a,b,t) = k_B^+(a,b)N(a,b,t) - k_B^-(a,b+1)N(a,b+1,t). \tag{24}$$

Combining Eqs. (23) and (24),

$$I_A(a,b,t) = N^{eq}(a,b)k_A^+(a,b)\left(\frac{N(a,b,t)}{N^{eq}(a,b)} - \frac{N(a+1,b,t)}{N^{eq}(a+1,b)}\right)$$
$$I_B(a,b,t) = N^{eq}(a,b)k_B^+(a,b)\left(\frac{N(a,b,t)}{N^{eq}(a,b)} - \frac{N(a,b+1,t)}{N^{eq}(a,b+1)}\right). \tag{25}$$

4.2 Constrained-equilibrium (zero-flux) distribution

The calculation of the zero-flux distribution is again a central problem. It might appear natural to assume [5]

$$N^{eq}(a,b) = \left(N_A(1) + N_B(1)\right)\exp\left(-\frac{W(a,b)}{k_BT}\right). \tag{26}$$

However, this choice has been disputed on thermodynamic grounds [6, 7]. Similar problems were raised in the definition of the equilibrium distribution for nucleation in a one-component system (Chapter 2, Section 5). As for that case, we solve the problem by assuming that the equilibrium distribution can be expressed in terms of a chemical potential, $\Phi(a, b)$ defined relative to the single-molecule state of either species [4]

$$\begin{aligned} N^{eq}(a,b) &= N_A(1)\exp\left(-\frac{\Phi_A(a,b)}{k_BT}\right) \\ &\text{or} \\ N^{eq}(a,b) &= N_B(1)\exp\left(-\frac{\Phi_B(a,b)}{k_BT}\right). \end{aligned} \tag{27}$$

These two distributions must be the same; the two kinetic potentials can, therefore, differ from each other only by a function of the relative single-molecule concentrations of the two species (easily solved from Eq. (27)). This allows the specific kinetic potentials to be expressed in terms of a generalized potential, $\Phi(a, b)$,

$$\begin{aligned} \Phi_A(a,b) &= \Phi(a,b) - \Phi(1,0) \\ \Phi_B(a,b) &= \Phi(a,b) - \Phi(0,1), \end{aligned} \tag{28}$$

where

$$\Phi(0,1) - \Phi(1,0) = -k_BT\ln\left(\frac{N_A(1)}{N_B(1)}\right). \tag{29}$$

The equilibrium distribution is then defined as

$$N^{eq}(a,b) = C\exp\left(-\frac{\Phi(a,b)}{k_BT}\right), \tag{30}$$

where C is a constant. As for the one-component case, $\Phi(a, b)$ is defined to be zero in the original phase of mixed molecules, which can be expressed as

$$N_A(1)\Phi(1,0) + N_B(1)\Phi(0,1) = 0, \tag{31}$$

where the two pure molecular concentrations are defined using Eq. (30)

$$\begin{aligned} N_A(1) &= C\exp\left(-\frac{\Phi(1,0)}{k_BT}\right) \\ N_B(1) &= C\exp\left(-\frac{\Phi(0,1)}{k_BT}\right). \end{aligned} \tag{32}$$

Solving Eqs. (31) and (32) gives

$$C = \left(N_A(1)\right)^{N_A(1)/(N_A(1)+N_B(1))} \times \left(N_B(1)\right)^{N_B(1)/(N_A(1)+N_B(1))}, \qquad (33)$$

which is significantly more complicated than our originally assumed form, Eq. (26). By the constrained-equilibrium hypothesis, $\Phi(a, b)$ is equal to W^*, defined earlier (Section 3).

4.3 Numerical treatment and comparison with analytical prediction

As for the cases in previous chapters, it is useful to determine the behavior expected from simple numerical solutions for multi-component nucleation, before discussing the analytical approaches. When the rates of interfacial cluster attachment for the different species do not differ widely, the coupled differential equations describing the nucleation kinetics (Eqs. (19) and (20)) can be solved with the finite-difference method used for nucleation in a one-component system (Chapter 2) [8–10]). For illustration, consider nucleation from an idealized binary supercooled liquid, or glass, so that strain influences can be ignored. For simplicity, we approximate the excess free-energy (Eqs. (2) and (10)) by the product of the surface area of the cluster and an effective interfacial free-energy (which can in principle be a function of the cluster composition). The work of cluster formation is then

$$W(a, b) = a(\mu_{A,s} - \mu_{A,l}) + b(\mu_{B,s} - \mu_{B,l}) + S(a, b)\sigma(a, b), \qquad (34)$$

where $\mu_{A,s}$ and $\mu_{B,s}$ are the chemical potentials of A and B, respectively, in the solid phase (s) and $\mu_{A,l}$ and $\mu_{B,l}$ are the chemical potentials in the liquid phase (l). $S(a, b)$ is the surface area of a cluster containing a molecules of A and b molecules of B, and $\sigma(a, b)$ is the interfacial free-energy. The liquid-phase chemical potentials are

$$\begin{aligned} \mu_{A,l} &= \mu^0_{A,l} + k_B T \ln X_{A,l} \\ \mu_{B,l} &= \mu^0_{B,l} + k_B T \ln X_{B,l} = \mu^0_{B,l} + k_B T \ln(1 - X_{A,l}), \end{aligned} \qquad (35)$$

where $\mu^0_{A,l}$and $\mu^0_{B,l}$ are the chemical potentials of pure liquid A and B, and $X_{A,l}$ and $X_{B,l}$ are the fractions of A and B, respectively, in the liquid phase. The second term in Eq. (35) arises from the entropy of mixing and the use of Stirling's approximation [11]. These expressions are valid for the chemical potentials for the liquid phase since the number of molecules is large; they are incorrect for the cluster, however, where the number of molecules is small [8]. To account for this, the work of cluster formation should be

$$\begin{aligned} W(a, b) &= a(\mu_{A,s} - \mu^0_{A,l}) + b(\mu_{B,s} - \mu^0_{B,l}) \\ &- k_B T\left(a \ln X_{A,l} + b\, ln(1 - X_{A,l})\right) + k_B T\, ln\left(\frac{(a+b)!}{a!b!}\right) + S(a, b)\sigma(a, b). \end{aligned} \qquad (36)$$

The materials parameters used are artificial, adapted from those for crystallization in $Li_2O \cdot 2SiO_2$ glass [8–10]. While these parameters do not correspond to an actual physical system, relevant points regarding nucleation in a typical binary system can be discovered by their use. Calculations show that principal features of the time-dependent nucleation rate are similar to those found in one-component systems [9, 10]. Assuming an initial state of disassociated molecules, the nucleation rises from zero to approach a stationary, steady-state rate with increasing annealing time, with the same functional form as for one-component nucleation (see Chapter 3, Figure 3). As for those calculations in a one-component system, the induction time for nucleation and the steady-state rate depend on the fundamental kinetic and thermodynamic parameters and the annealing conditions.

Figure 2 shows the contours for the reversible work of cluster formation computed from Eq. (36), taking the liquidus temperature to

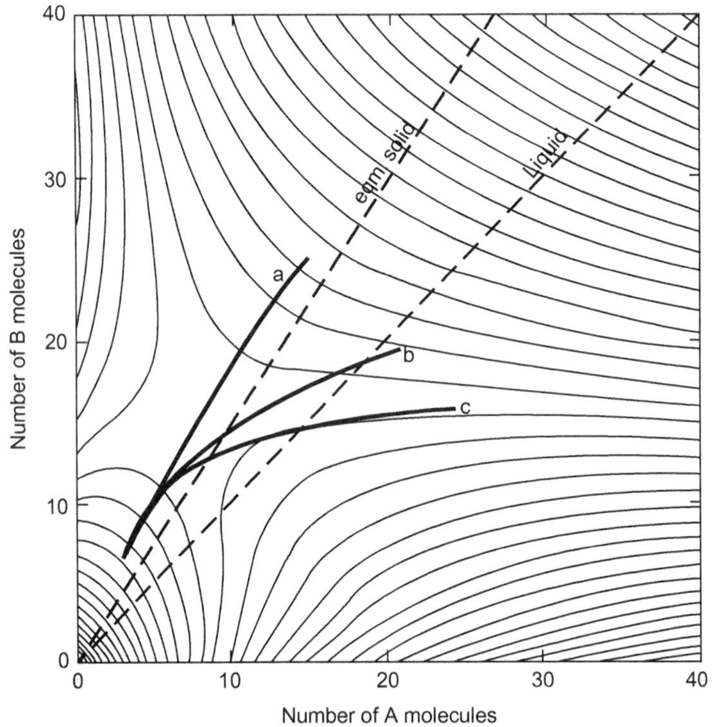

Fig. 2 Contours of the reversible work of cluster formation in a binary system, calculated from Eq. (36) with the parameters given in Section 4.3. The liquidus temperature is 1300 K, the entropy of melting for the two components is 40 J K^{-1} mol^{-1}, and the partition coefficient, k^{eq}, is 0.8. (From Ref. [12], with permission.)

be 1300 K, the entropy of melting for the two components as $40\,\mathrm{J\,K^{-1}\,mol^{-1}}$, and assuming a partition coefficient k^{eq} of 0.8 ($k^{eq} = X_{A,s}/X_{A,l}$), where $X_{A,s}$ and $X_{A,l}$ are the mole fractions of A in the solid and liquid phases, respectively. The mole fraction in the glass is 0.5, shown by the dashed line in the figure (liquid). The equilibrium composition of the solid (consistent with $k^{eq} = 0.8$) is also indicated (eqm solid). Because k^{eq} is not equal to one, the contours are not symmetric about $a = b$.

Taking the molecular volume and interfacial free energy to be independent of composition, the computed steady-state cluster distribution when components A and B are given the same interfacial mobility is shown in Figure 3a. The critical size is 23 molecules. The maximum population passes through the saddle region, following the maximum driving force composition. This is clear from Figure 2, where curve a represents the locus of points of highest population density in Figure 3a (i.e., following the ridge of the profile). These results appear to support a common assumption, made in the next section (Section 4.4), that the most important contributions to the nucleation rate are for processes proceeding along this line. It is important to realize, however, that the interfacial mobilities of the species were assumed to be the same for this calculation. In practice, they can differ by several orders of magnitude.

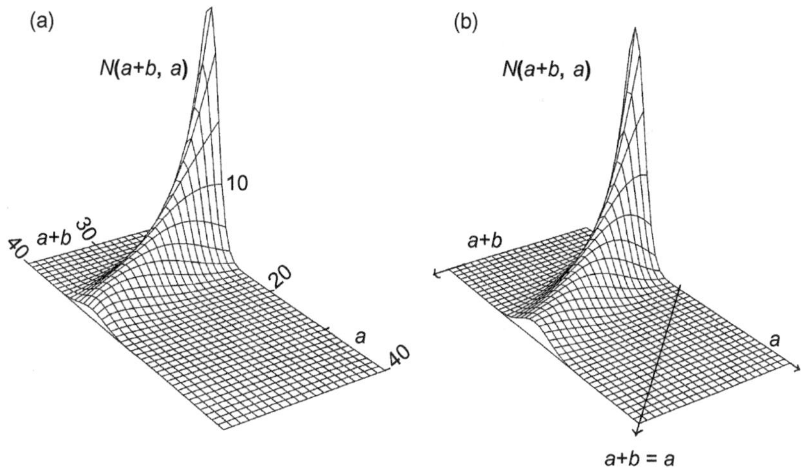

Fig. 3 (a) The steady-state cluster distribution computed with the parameters mentioned and assuming that A and B have the same interfacial mobility; (b) the computed distribution when the mobility of B is reduced by an order of magnitude below that of A. The critical size in both cases contains 23 molecules. (From Ref. [12], with permission.)

The influence of a difference in the interfacial mobilities of the two species is illustrated in Figure 3b, where the mobility of B has been reduced to an order of magnitude below that of A. The distribution obtained is significantly different. The composition of the maximum cluster population is shifted from that corresponding to the maximum driving force composition toward one richer in the species with the higher attachment mobility. The locus of points of highest population density for this case is shown as curve *b* in Figure 2. For these calculations, the deviation in composition from the saddle region is small until the cluster approaches the critical size ($n^* = 23$). Larger clusters, however, depart radically in composition, becoming more A-rich. Decreasing the mobility of B still further (by another order of magnitude) produces clusters that are even more A-rich, and causes the deviation from the maximum driving force composition to occur earlier. Calculations indicate that the mean chemical composition of the clusters approaches the saddle-point composition only for clusters smaller than about ($n^*/2$), with the sign of the deviation determined primarily by the heat of fusion of the pure components [10].

Before further discussion, it is useful to develop an analytical expression for the steady-state rate. In Section 4.5, we compare predictions from that expression with the numerical results.

4.4 Steepest-descent solution for nucleation kinetics

The nucleation rate is frequently obtained from the cluster transition rate through the saddle region of the free-energy surface, in the direction of the steepest descent (see Chapter 4, Figure 3). As illustrated in the last section, however, this is invalid in all but the simplest cases. Nevertheless, it illustrates some key features of interface-limited nucleation in a multi-component system.

The cluster flux in (a, b) space is shown schematically in Figure 4. Following the Zeldovich treatment for nucleation in a one-component system (Section 2.9), it is useful to rewrite the discrete differential equations for binary nucleation (Eq. (25)) as partial derivatives of continuous variables. Assuming large clusters,

$$\begin{aligned} I_{\mathrm{A}}(a,b,t) &= -N^{\mathrm{eq}}(a,b)k_{\mathrm{A}}^{+}(a,b)\left(\frac{\partial}{\partial a}\left(\frac{N(a,b,t)}{N^{\mathrm{e}}(a,b)}\right)\right)_{b} \\ I_{\mathrm{B}}(a,b,t) &= -N^{\mathrm{eq}}(a,b)k_{\mathrm{B}}^{+}(a,b)\left(\frac{\partial}{\partial b}\left(\frac{N(a,b,t)}{N^{\mathrm{e}}(a,b)}\right)\right)_{a}. \end{aligned} \tag{37}$$

The composition of the critical cluster, (a^*, b^*), by definition is a point of unstable equilibrium. Attention is focused on passage through the

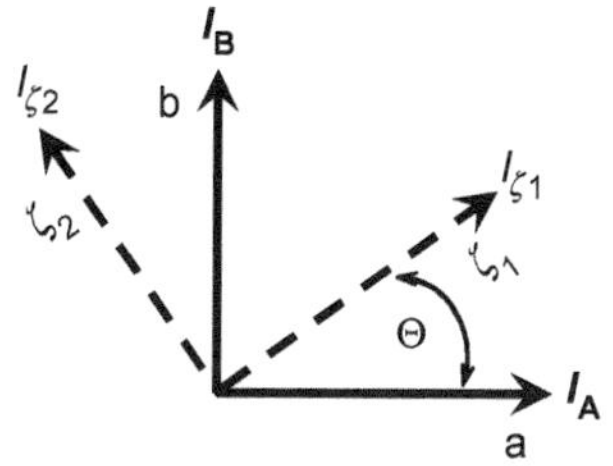

Fig. 4 Schematic illustration of the reorientation of the axis from that representing the cluster composition (a, b), to (ζ_1, ζ_2) so that ζ_1 points along direction of steepest decent through the saddle region.

saddle region near n^*. Following Reiss [5], it is convenient to introduce a new coordinate system (ζ_1, ζ_2) that is rotated from (a, b) by an angle Θ (Figure 4) so that ζ_1 points along the streamline direction through the saddle region, that is, in the direction of the steepest descent. The coordinates in the original coordinate system are easily related to those in the new coordinate system by the rotation matrix,

$$\begin{pmatrix} a \\ b \end{pmatrix} = \begin{pmatrix} \cos\Theta & -\sin\Theta \\ \sin\Theta & \cos\Theta \end{pmatrix} \begin{pmatrix} \zeta_1 \\ \zeta_2 \end{pmatrix}. \tag{38}$$

From continuity,

$$-\left(\frac{\partial I_{\zeta_1}}{\partial \zeta_1} + \frac{\partial I_{\zeta_2}}{\partial \zeta_2}\right) = \frac{\partial N(\zeta_1, \zeta_2)}{\partial t}, \tag{39}$$

where I_{ζ_1} and I_{ζ_2} are the components of the nucleation flux along ζ_1 and ζ_2, respectively. We now seek an expression for the steady-state nucleation rate. In that case, the distribution is constant with time so that $\partial N(\zeta_1, \zeta_2)/\partial t = \partial N^{st}(\zeta_1, \zeta_2)/\partial t = 0$, where $N^{st}(\zeta_1, \zeta_2)$ is the steady-state cluster population. By construction, I_{ζ_1} points in the direction of the principal nucleation rate, I, allowing I_{ζ_2} to be ignored, giving $\partial I^{st}_{\zeta_1}/\partial \zeta_1 = 0$; the principal nucleation rate is, therefore, a function of ζ_2 only, $I^{st}_{\zeta_1} = I^{st}(\zeta_2)$. The total flux past a particular cluster size in the saddle region is obtained by integrating over ζ_2. For the steady-state nucleation rate at the critical size (a^*, b^*),

$$I^{st} = I^{st}(a^*, b^*) = \int_{-\infty}^{\infty} I^{st}(\zeta_2) d\zeta_2. \tag{40}$$

The components along the direction of A and B are easily related to $I^{st}(\zeta_2)$

$$I^{st}_A = I^{st}(\zeta_2)\cos\Theta \quad \text{and} \quad I^{st}_B = I^{st}(\zeta_2)\sin\Theta. \tag{41}$$

To compute I^{st}_A and I^{st}_B from Eq. (41), the partial derivatives with respect to a and b must be expressed with respect to the rotated coordinates. With the rotation matrix,

$$\begin{aligned}\frac{\partial}{\partial a}\left(\frac{N^{\rm st}(a,b)}{N^{\rm eq}(a,b)}\right)_b &= \frac{\partial}{\partial \zeta_1}\left(\frac{N^{\rm st}(\zeta_1,\zeta_2)}{N^{\rm eq}(\zeta_1,\zeta_2)}\right)\cos\Theta - \frac{\partial}{\partial \zeta_2}\left(\frac{N^{\rm st}(\zeta_1,\zeta_2)}{N^{\rm eq}(\zeta_1,\zeta_2)}\right)\sin\Theta \\ \frac{\partial}{\partial b}\left(\frac{N^{\rm st}(a,b)}{N^{\rm eq}(a,b)}\right)_a &= \frac{\partial}{\partial \zeta_1}\left(\frac{N^{\rm st}(\zeta_1,\zeta_2)}{N^{\rm eq}(\zeta_1,\zeta_2)}\right)\sin\Theta + \frac{\partial}{\partial \zeta_2}\left(\frac{N^{\rm st}(\zeta_1,\zeta_2)}{N^{\rm eq}(\zeta_1,\zeta_2)}\right)\cos\Theta.\end{aligned} \tag{42}$$

Combining Eqs. (37) and (42),

$$\begin{aligned}I_{\rm A}^{\rm st} &= -N^{\rm eq}(\zeta_1,\zeta_2)k_{\rm A}^{+}\left(\frac{\partial}{\partial \zeta_1}\left(\frac{N^{\rm st}(\zeta_1,\zeta_2)}{N^{\rm eq}(\zeta_1,\zeta_2)}\right)\cos\Theta - \frac{\partial}{\partial \zeta_2}\left(\frac{N^{\rm st}(\zeta_1,\zeta_2)}{N^{\rm eq}(\zeta_1,\zeta_2)}\right)\sin\Theta\right) \\ I_{\rm B}^{\rm st} &= -N^{\rm eq}(\zeta_1,\zeta_2)k_{\rm B}^{+}\left(\frac{\partial}{\partial \zeta_1}\left(\frac{N^{\rm st}(\zeta_1,\zeta_2)}{N^{\rm eq}(\zeta_1,\zeta_2)}\right)\sin\Theta + \frac{\partial}{\partial \zeta_2}\left(\frac{N^{\rm st}(\zeta_1,\zeta_2)}{N^{\rm eq}(\zeta_1,\zeta_2)}\right)\cos\Theta\right),\end{aligned} \tag{43}$$

where the arguments of the rate constants have been omitted for simplicity of notation. From Eq. (41), $I_{\rm B}^{\rm st}/I_{\rm A}^{\rm st} = \tan\theta$. With this, Eq. (43) gives

$$\begin{aligned}&(k_{\rm A}^{+} - k_{\rm B}^{+})\sin\Theta\cos\Theta\left(\frac{\partial}{\partial \zeta_1}\left(\frac{N^{\rm st}(\zeta_1,\zeta_2)}{N^{\rm eq}(\zeta_1,\zeta_2)}\right)\right) = \\ &(k_{\rm A}^{+}\sin^2\Theta + k_{\rm B}^{+}\cos^2\Theta)\left(\frac{\partial}{\partial \zeta_2}\left(\frac{N^{\rm st}(\zeta_1,\zeta_2)}{N^{\rm eq}(\zeta_1,\zeta_2)}\right)\right).\end{aligned} \tag{44}$$

Combining Eqs. (41), (43), and (44), we obtain

$$I^{\rm st}(\zeta_2) = \frac{I_{\rm A}^{\rm st}}{\cos\Theta} = -k_{\rm A}^{+}k_{\rm B}^{+}N^{\rm eq}(\zeta_1,\zeta_2)\frac{\frac{\partial}{\partial \zeta_1}\left(\frac{N^{\rm st}(\zeta_1,\zeta_2)}{N^{\rm eq}(\zeta_1,\zeta_2)}\right)}{(k_{\rm A}^{+}\tan^2\Theta + k_{\rm B}^{+})\cos^2\Theta}. \tag{45}$$

Rearranging Eq. (45), using the identity $1 + \tan^2\Theta = \sec^2\Theta$, and integrating over $d\zeta_1$,

$$\int_0^{\infty}\frac{\partial}{\partial \zeta_1}\left(\frac{N^{\rm st}(\zeta_1,\zeta_2)}{N^{\rm eq}(\zeta_1,\zeta_2)}\right)d\zeta_1 = -\int_0^{\infty}\frac{I^{\rm st}(\zeta_2)(k_{\rm A}^{+}\tan^2\Theta + k_{\rm B}^{+})d\zeta_1}{k_{\rm A}^{+}k_{\rm B}^{+}(1+\tan^2\Theta)N^{\rm eq}(\zeta_1,\zeta_2)}. \tag{46}$$

As for the one-component case, the major contribution to this integral arises from regions near the critical cluster size. Assuming, then, that the interfacial attachment rates, $k_{\rm A}^{+}$ and $k_{\rm B}^{+}$, may be replaced by the rates for the critical cluster, $k_{\rm A^*}^{+}$ and $k_{\rm B^*}^{+}$, and taking the same boundary conditions on the distribution as for the one-component case, $N^{\rm st}(\zeta_1,\zeta_2)/N^{\rm eq}(\zeta_1,\zeta_2) \to 1$ as $\zeta_1 \to 0$ and $N^{\rm st}(\zeta_1,\zeta_2)/N^{\rm eq}(\zeta_1,\zeta_2) \to 0$ as $\zeta_1 \to \infty$,

$$I^{\rm st}(\zeta_2) = \frac{k_{\rm A^*}^{+}k_{\rm B^*}^{+}(1+\tan^2\Theta)}{k_{\rm B^*}^{+} + k_{\rm A^*}^{+}\tan^2\Theta}\left(\int_0^{\infty}\frac{d\zeta_1}{N^{\rm eq}(\zeta_1,\zeta_2)}\right)^{-1}. \tag{47}$$

Except for the changed prefactor, this now has the form of the expression for the steady-state nucleation rate in a one-component system (Eqs. (46) and (64) in Chapter 2). As in that case, $N^{eq}(\zeta_1, \zeta_2) = C\ \exp(-W(\zeta_1, \zeta_2)/k_B T)$, where C is a constant; $1/N^{eq}(\zeta_1, \zeta_2)$ is, therefore, sharply peaked in the ζ_1 direction near the critical size (ζ_1^*, ζ_2^*), allowing a quadratic expansion of the work of cluster formation about that point,

$$
\begin{aligned}
W(\zeta_1, \zeta_2) \approx W(\zeta_1^*, \zeta_2^*) &+ \frac{1}{2}\left(\frac{\partial^2 W(\zeta_1, \zeta_2)}{\partial^2 \zeta_1}\right)_{\zeta_1^*,\, \zeta_2^*} (\zeta_1 - \zeta_1^*)^2 \\
&+ \frac{1}{2}\left(\frac{\partial^2 W(\zeta_1, \zeta_2)}{\partial \zeta_1 \partial \zeta_2}\right)_{\zeta_1^*,\, \zeta_2^*} (\zeta_1 - \zeta_1^*)(\zeta_2 - \zeta_2^*) \\
&+ \frac{1}{2}\left(\frac{\partial^2 W(\zeta_1, \zeta_2)}{\partial^2 \zeta_2}\right)_{\zeta_1^*,\, \zeta_2^*} (\zeta_2 - \zeta_2^*)^2. \qquad (48)
\end{aligned}
$$

Terms in $(\zeta_i - \zeta_i^*)$ vanish, since $W(\zeta_1^*, \zeta_2^*)$ is an extremum. The cross-term in Eq. (48) is also small compared with the other two derivatives. Introducing the notation, $\hat{\xi}_1 = \zeta_1 - \zeta_1^*$ and $\hat{\xi}_2 = \zeta_2 - \zeta_2^*$, and defining,

$$
P = \frac{1}{2}\left(\frac{\partial^2 W}{\partial^2 \zeta_1^2}\right)_{\zeta_1^*,\, \zeta_2^*} \quad \text{and} \quad Q = \frac{1}{2}\left(\frac{\partial^2 W}{\partial^2 \zeta_2^2}\right)_{\zeta_1^*,\, \zeta_2^*}, \qquad (49)
$$

we obtain

$$
\begin{aligned}
\int_0^\infty d\zeta_1/N^{eq}(\zeta_1, \zeta_2) &\approx \frac{1}{C}\exp\left(\frac{W^* + Q\hat{\xi}_2^2}{k_B T}\right) \times \int_{-\infty}^{\infty} \exp\left(-\frac{P\hat{\xi}_1^2}{k_B T}\right) d\hat{\xi}_1, \\
&\approx \frac{1}{C}\sqrt{\frac{\pi k_B T}{P}} \exp\left(\frac{W^* + Q\hat{\xi}_2^2}{k_B T}\right),
\end{aligned} \qquad (50)
$$

where P and Q represent the mutually orthogonal curvatures in the work at the saddle point. As for the one-component case, the extension to a lower limit of negative infinity is justified by the small contribution made by clusters with ζ_1 much smaller than the critical value. From Eq. (47), the principal steady-state nucleation rate as a function of ζ_2 is, therefore,

$$
I^{st}(\zeta_2) \approx C\frac{k_{A^*}^+ k_{B^*}^+ (1 + \tan^2\Theta)}{k_{B^*}^+ + k_{A^*}^+ \tan^2\Theta} \sqrt{\frac{P}{\pi k_B T}} \exp\left(-\frac{W^* + Q\hat{\xi}_2^2}{k_B T}\right). \qquad (51)
$$

To obtain the total steady-state rate, $I^{st}(\zeta_2)$ must be integrated over ζ_2 (Eq. (40)),

$$I^{st} \approx C\frac{k^+_{A^*}k^+_{B^*}(1+\tan^2\Theta)}{k^+_{B^*}+k^+_{A^*}\tan^2\Theta}\sqrt{\frac{P}{\pi k_B T}}\exp\left(-\frac{W^*}{k_B T}\right)\times\int_{-\infty}^{\infty}\exp\left(-\frac{Q\hat{\xi}_2^2}{k_B T}\right)d\hat{\xi}_2, \quad (52)$$

finally giving,

$$I^{st} \approx C\frac{k^+_{A^*}k^+_{B^*}(1+\tan^2\Theta)}{k^+_{B^*}+k^+_{A^*}\tan^2\Theta}\sqrt{\frac{P}{Q}}\exp\left(-\frac{W^*}{k_B T}\right). \quad (53)$$

Equation (53) has the form

$$I^{st} = k^+_{eff}(a^*, b^*)\sqrt{\frac{P}{Q}}\ N^{eq}(a^*, b^*), \quad (54)$$

where $k^+_{eff}(a^*, b^*)$ is an effective forward rate constant at the critical size. This is similar to the expression derived for polymorphic nucleation (Eq. (50) in Chapter 2), consisting of a weighted forward rate constant, a factor analogous to the Zeldovich factor, and the equilibrium number of critical clusters. When Q, the curvature in the direction of ζ_2, is large, the pass through the saddle region is narrow with steep sides. As expected, Eq. (54) predicts that the nucleation flux is lowered in this case. P represents the negative curvature in the direction of primary flow, ζ_1. A large value of P represents a narrower barrier and results in a higher nucleation rate (Eq. (54)). The approach presented in this section has been generalized to treat nucleation in multi-component systems [13, 14].

4.5 How good is this expression for I^{st}?

A key assumption in the development of Eq. (53) is that the chemical composition of the evolving clusters follows the ridge through the saddle region of $W(a, b)$. As discussed earlier, this appears to be satisfied only when the interfacial attachment mobilities are equal. In that case, the predictions of Eq. (53) are in good agreement with the numerical results presented in Section 4.3 (Figure 5a). Since the cluster evolution deviates from the path through the ridge of the saddle region when the interfacial attachment mobilities of the two species become different (Section 4.3), Eq. (53) is expected to fail in that case. In contradiction, it continues to give a reasonable prediction of the numerical steady-state rate (Figure 5b). This is likely a fortuitous result of the parameter set chosen, possibly due to compensating kinetic and thermodynamic effects [12], where an overestimate in the nucleation rate arising from invalid values for W, P, and Q is partially offset by

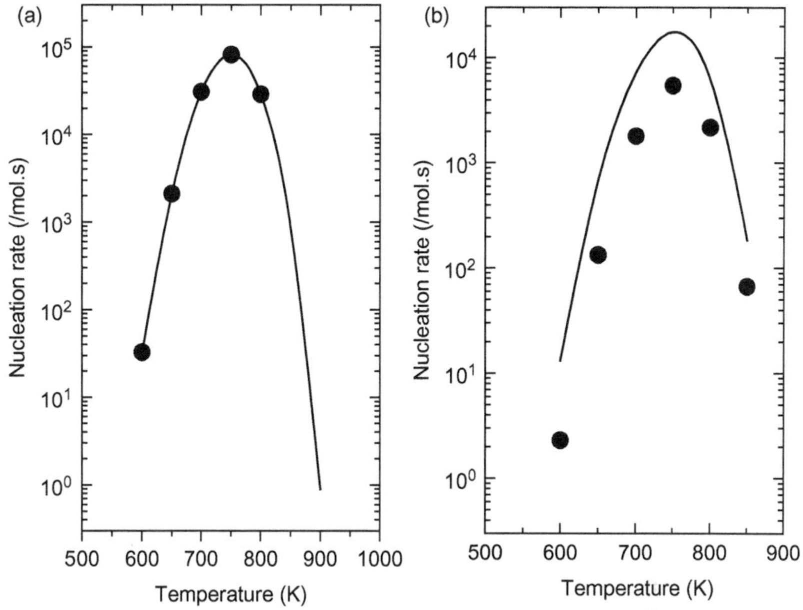

Fig. 5 (a) Steady-state nucleation rates as a function of temperature in a binary system, using the parameters listed in Section 4.3 and assuming that A and B have the same interfacial mobility; (b) steady-state nucleation rates as a function of temperature when the mobility of B is reduced by an order of magnitude below that of A. The data points are the values computed numerically; the solid line is the prediction from Eq. (53). (Data from Ref. [12].)

the kinetic coefficients, which were computed assuming that the mobilities of the two species are equal. In support of this, numerical [10] and experimental [15] studies have shown dramatic disagreement with predictions from Eq. (53).

4.6 Toward more correct analytical treatments of the nucleation rate

The limited validity of Eq. (53) and the lack of an expression for the time-dependent nucleation rate in a multi-component system are significant problems. The development of less restrictive solutions for interface-controlled, multi-component nucleation is beyond the scope of this book. However, it is useful to make reference to some of these treatments.

The development of Eq. (53) relies on two key assumptions:

(a) The nucleation rate is dominated by the cluster flux through the saddle point of the surface of the work of cluster formation, $W(a, b)$;

(b) The direction of the flux is constant in the vicinity of the saddle region and corresponds to the direction of steepest descent.

Binder and Stauffer [16, 17] eliminated the second assumption, obtaining a different expression for the steady-state nucleation rate,

$$I^{st} \approx C\frac{k_{A^*}^+ k_{B^*}^+ (1+\tan^2\Theta)}{k_{B^*}^+ + k_{A^*}^+ \tan^2\Theta}\Lambda \exp\left(-\frac{W^*}{k_B T}\right). \tag{55}$$

This has precisely the form Eq. (53), but differs in practice. The ratio of the interfacial attachment rates, $r = k_B^+/k_A^+$, is used to determine the angle of rotation, Θ, of the direction of the cluster flux from the original coordinates. Also, the term of the ratio of curvatures, $\sqrt{P/Q}$ in Eq. (53), is replaced by a more complicated expression, Λ (defined as follows). Expressed in terms of the original cluster coordinates (a, b), Θ and Λ are given by

$$\Lambda = -\frac{(\Delta_{aa}\cos^2\Theta - \Delta_{bb}\sin^2\Theta - 2\Delta_{ab})}{\sqrt{2\Delta_{ab}^2 - \Delta_{aa}\Delta_{bb}}}, \text{ where}$$

$$\Delta_{aa} = \frac{1}{2}\frac{\partial^2 W(a,b)}{\partial a^2}, \quad \Delta_{bb} = \frac{1}{2}\frac{\partial^2 W(a,b)}{\partial b^2}, \quad \Delta_{ab} = \Delta_{ba} = \frac{1}{2}\frac{\partial^2 W(a,b)}{\partial a \partial b}$$

$$\tan\Theta = s + \sqrt{s^2 + r}, \text{with} \quad s = -\frac{(\Delta_{aa}/\Delta_{ab} - r(\Delta_{bb}/\Delta_{ab}))}{2}. \tag{56}$$

Temkin and Shevelev [18] demonstrated that the one-component solution cannot be recovered if assumption (a) is made. They offer a more complete solution for the condensation of a binary vapor mixture where the two components have different kinetic activities. They abandon detailed balance in the derivation of their rates, however, making the solution questionable (and certainly difficult) for application to actual physical processes.

Were it possible to map nucleation in multi-component systems onto nucleation in a one-component system, the solution could be obtained easily [4, 9]. Wu has offered the most complete solution for binary nucleation by this approach, obtaining closed-form solutions for the steady-state nucleation rate, the cluster density, and the induction time [4]. Wu's approach can also be extended to incorporate more than two components.

5. COUPLED INTERFACE/DIFFUSION NUCLEATION

The assumption made in the previous sections, of a fixed composition in the original phase adjacent to the developing clusters of the new phase, is likely correct for nucleation from the vapor and some liquid-to-solid transitions. It is often incorrect, however, for nucleation of a phase of different composition from the original phase for crystals and glasses, where the diffusion rates are much slower. This can have profound consequences for the nucleation behavior. If, for example, an embryo becomes critical in unstable equilibrium with the original phase of average composition, the depletion of the original phase immediately surrounding the cluster will locally shift the driving free energy and hence the critical size. Further, the dissolution rate as a function of cluster size will lead to a size-dependent cluster-neighborhood composition in the original phase. To take account of these and similar effects, a variable describing the shifts of the composition of the original phase as a function of cluster growth must be introduced. The most correct treatment, where the compositions of both the cluster and the original phase dynamically change, is at present intractable. In this section, then, we follow a simpler approach, fixing the cluster composition and allowing the composition of the original phase to change with time.

Even with these assumptions, the problem is extremely difficult to solve. The interfacial attachment kinetics for each cluster must be coupled with its unique variable long-range diffusion field, which is a function of cluster size and sample history. A simplified model is, therefore, adopted [19–21]. Three regions of interest are defined: the cluster, the original phase in the neighborhood of the cluster, and the remainder of the original phase (Figure 6). Taking precipitation as an example, the cluster distribution now is a function of the cluster size, n,

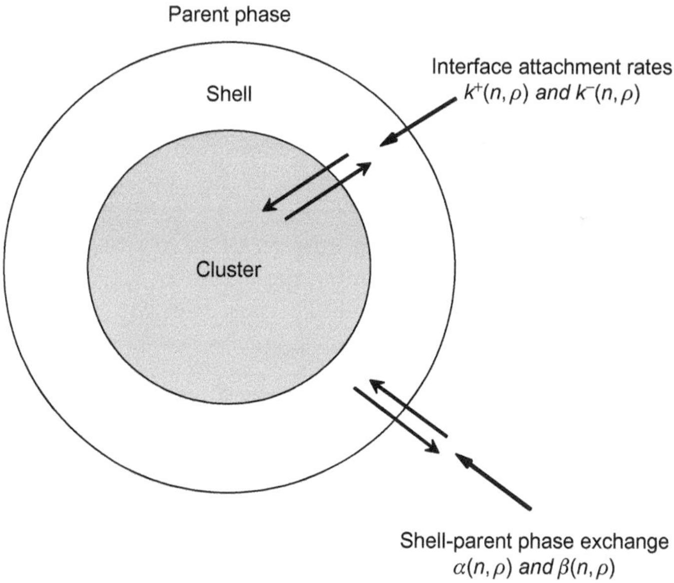

Fig. 6 Schematic illustration of the coupled-flux model for nucleation, showing the interfacial and diffusive rates. (Reprinted from Ref. [21], copyright (2000), with permission from Elsevier.)

and the number of solute molecules, ρ, in the nearest-neighbor shell to the cluster. As within the classical theory, clusters grow by single-molecule incorporation from the nearest-neighbor shell. The continuous diffusive fluxes in the original phase to the shell region are approximated by a set of discrete jumps, between the shell and the original phase. To ensure the correct diffusion time-scale, these rates can be adjusted [21] by equating the calculated growth rates for large clusters and the macroscopic diffusion-limited growth rate [3]. Recent computer simulations have demonstrated that this model, though crude, captures many essential features of the coupled fluxes [22].

5.1 Thermodynamic considerations

For illustration, consider the precipitation of a pure solute from a solid solution. Assume that the equilibrium cluster distribution in this coupled-flux model for nucleation has the following form:

$$N^{\mathrm{eq}}(n,\rho) = N_{\mathrm{sol}} \exp\left(-\frac{W(n)}{k_{\mathrm{B}}T}\right) P(\rho), \tag{57}$$

where N_{sol} is the number of solute molecules per unit volume in the initial phase, $W(n)$ is the work of cluster formation, and $P(\rho)$ is the normalized probability for having ρ solute molecules in the nearest-neighbor shell around the cluster. For simplicity, take the work of formation of clusters of pure solute to be the same as for the classical theory (Eq. (11) in Chapter 2). The computation of $P(\rho)$ requires a knowledge of the difference in the solute atom free-energy for an atom randomly placed in the original phase and when localized to a site in the cluster neighborhood. If the interaction enthalpy is the same for these two cases, a free-energy difference will arise only from differences in the configurational entropy, related to the number of configurations, Ω, in the usual way, that is, $S = k \ln \Omega$. Denoting the maximum number of atom sites in the nearest-neighbor shell of a cluster containing n molecules as $\rho_{\max}(n)$, the number of configurations for a shell containing ρ molecules, Ω_ρ, is

$$\Omega_\rho = \frac{(\rho_{\max}(n))!}{\rho!(\rho_{\max}(n)-\rho)!}. \tag{58}$$

Assuming that the N_{sol} molecules are distributed among the total number of sites per unit volume (N_{sites}) (i.e., $N_{\mathrm{sol}}/N_{\mathrm{sites}}$ is the atom fraction), the number of configurations in a unit volume of the original phase is

$$\Omega_{\mathrm{m}} = \frac{N_{\mathrm{sites}}!}{N_{\mathrm{sol}}!(N_{\mathrm{sites}} - N_{\mathrm{sol}})!}. \tag{59}$$

The probability for a fluctuation, arising from an exchange of a single-molecule, is equal to the exponential of the sum of the entropy differences in the shell and original phase that results from the exchange. From Eqs. (58) and (59) (see Ref. [21] for more detail),

$$P(\rho) \propto \frac{(\rho_{\max}(n))!N_{\mathrm{sol}}!(N_{\mathrm{sites}} - N_{\mathrm{sol}})!}{\rho!(\rho_{\max}(n) - \rho)!(N_{\mathrm{sol}} - \rho)!(N_{\mathrm{sites}} - N_{\mathrm{sol}} + \rho)!}. \tag{60}$$

For clusters sizes important for nucleation, $N_{\mathrm{sol}} \gg \rho$. Assuming for simplicity that the initial phase is also dilute in solute, Eq. (60) can be simplified using Stirling's approximation [11]

$$P(\rho) = \varepsilon(n)\frac{(\rho_{\max}(n))!}{\rho!(\rho_{\max}(n) - \rho)!}\left(\frac{N_{\mathrm{sol}}}{N_{\mathrm{sites}} - N_{\mathrm{sol}}}\right)^{\rho}, \tag{61}$$

where $\varepsilon(n)$ is a normalization constant determined for each n from the condition that $\sum_{\rho=0}^{\rho_{\max}} P(\rho) = 1$.

5.2 Kinetic model for nucleation from a dilute solution

As discussed in Chapters 2 and 3, in the classical theory clusters are assumed to evolve slowly in size, n, by a series of bimolecular reactions that involve the gain or loss of a single atom or molecule. For the coupled-flux model, assuming that the numbers of molecules in the clusters and their nearest-neighbor shells change by the gain or loss of one molecule at a time, the rate constants must be described within a two-dimensional (n, ρ) space (Figure 7).

The staircase appearance of the figure is a consequence of the assumption that only the nearest-neighbor shell is considered; the upper limit on ρ increases with cluster size, n, approximately as $4n^{2/3}$. In analogy to Eq. (19) for interface-limited nucleation in a multi-component system, the fluxes dictating the kinetics of cluster growth can be expressed as

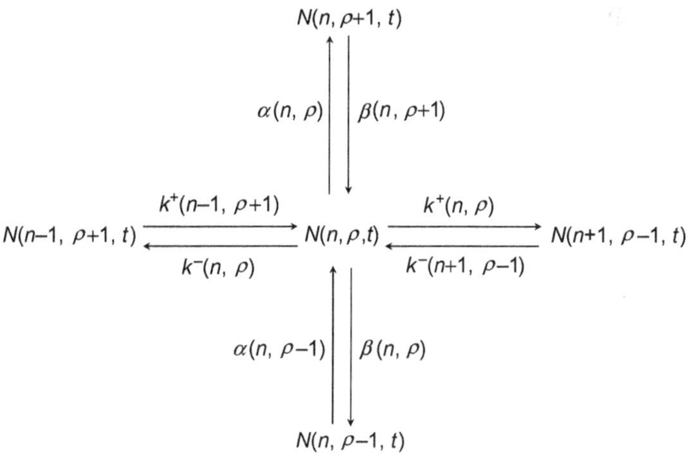

(62)

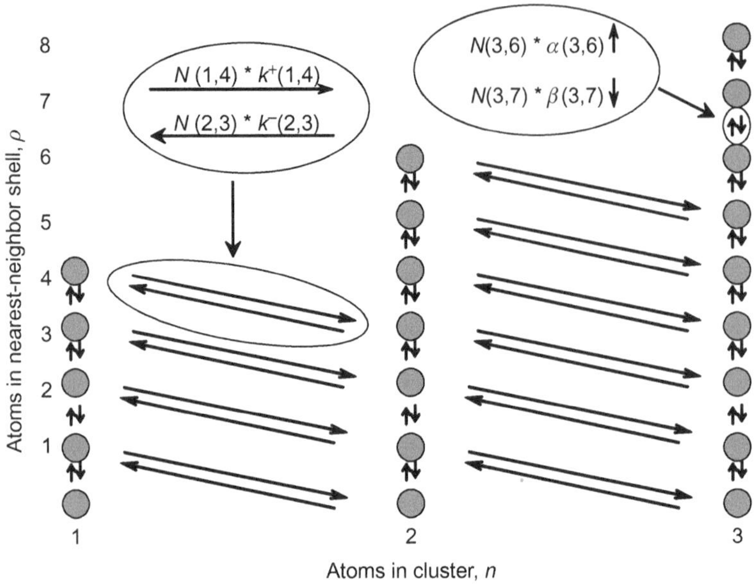

Fig. 7 Schematic illustration of the rate processes for the coupled-flux model for nucleation. (Reprinted from Ref. [21], copyright (2000), with permission from Elsevier.)

Here $\alpha(n, \rho)$ is the rate at which solute molecules diffuse into the shell of surrounding material, $\beta(n, \rho)$ is the rate at which they diffuse out, $k^+(n, \rho)$ is the rate of attachment of solute molecules from the shell onto the cluster, and $k^-(n, \rho)$ is the rate of detachment from the cluster. As for polymorphic transformations, clusters increase or decrease in size by the single attachments of molecules, but the supply of these is limited by the rate at which molecules arrive in the shell of surrounding material. The coupled differential equations that describe the rate of change of the cluster distribution have the following form:

$$\begin{aligned}\frac{\partial N(n,\rho)}{\partial t} &= \alpha(n,\rho-1)N(n,\rho-1,t) - \Big(\alpha(n,\rho)+\beta(n,\rho)\Big)N(n,\rho,t)\\ &\quad + \beta(n,\rho+1)N(n,\rho+1,t) + k^+(n-1,\rho+1)N(n-1,\rho+1,t)\\ &\quad + k^-(n+1,\rho-1)N(n+1,\rho-1,t) - \Big(k^+(n,\rho)+k^-(n,\rho)\Big)N(n,\rho,t).\end{aligned} \tag{63}$$

The rate at which an atom leaves the nearest-neighbor shell to return to the original phase should be proportional to $\rho D/\lambda^2$, where D is the diffusion coefficient for the solute in the solution phase and λ is the jump distance. Assuming that detailed balance holds for the exchange between the cluster shell and the original phase,

$$\alpha(n,\rho-1)N^{\text{eq}}(n,\rho-1) = \beta(n,\rho)N^{\text{eq}}(n,\rho). \tag{64}$$

Taking the equilibrium distribution given in Eq. (57) and assuming that the rates are symmetric,

$$\alpha(n,\rho-1) = \xi\rho\frac{D}{\lambda^2}\left(\frac{\rho_{\max}(n)-\rho+1}{\rho}\right)^{1/2}\left(\frac{N_{\mathrm{sol}}}{N_{\mathrm{sites}}-N_{\mathrm{sol}}}\right)^{1/2}$$
$$\beta(n,\rho) = \xi\rho\frac{D}{\lambda^2}\left(\frac{\rho_{\max}(n)-\rho+1}{\rho}\right)^{-1/2}\left(\frac{N_{\mathrm{sol}}}{N_{\mathrm{sites}}-N_{\mathrm{sol}}}\right)^{-1/2}. \tag{65}$$

The constant ξ has been introduced to take into account that molecules do not immediately become equivalent to the initial randomly dispersed state when they make the transition outside of the nearest-neighbor shell; ξ decreases with decreasing solute concentration in the initial phase, becoming very small in the limit of large dilution. By matching the growth rate of large clusters to the known diffusion-limited growth rate, ξ can be evaluated [21],

$$\xi = \frac{(4\pi)^{2/3}}{2}(3\bar{v})^{1/3}\lambda^2(N_{\mathrm{sites}}X_\infty)^{1/2}, \tag{66}$$

where X_∞ is the solute-atom fraction, $(N_{\mathrm{sol}}/N_{\mathrm{sites}})$ in the original phase, far from the growing cluster, and $\bar{v}$ is the solute molecular volume. It is reasonable to assume that the forward rate for interfacial attachment should also be proportional to the number of molecules and to the attempt frequency,

$$k^+(n,\rho) = \rho\left(\frac{6D_{\mathrm{i}}}{\lambda^2}\right), \tag{67}$$

where D_{i} is an effective diffusion coefficient governing interfacial attachment (i.e., a measure of the interfacial attachment mobility). Again assuming detailed balance at the cluster/shell interface,

$$k^+(n,\rho)N^{\mathrm{eq}}(n,\rho) = k^-(n+1,\rho-1)N^{\mathrm{eq}}(n+1,\rho-1). \tag{68}$$

Computing $N^{\mathrm{eq}}(n,\rho)$ from Eqs. (57) and (61) and assuming that the rates are symmetric,

$$k^+(n,\rho) = \rho\frac{6D_{\mathrm{i}}}{\lambda^2}\exp\left(-\frac{\delta W(n)}{2k_{\mathrm{B}}T}\right)G(n,\rho)$$
$$k^-(n+1,\rho-1) = \rho\frac{6D_{\mathrm{i}}}{\lambda^2}\exp\left(+\frac{\delta W(n)}{2k_{\mathrm{B}}T}\right)\left(\frac{1}{G(n,\rho)}\right). \tag{69}$$

Here, as for the classical theory, $\delta W(n)$ is the work required to form a cluster of size $n + 1$ less than that required to form a cluster of size n (i.e., $\delta W(n) = W(n+1) - W(n)$). Interestingly, these interfacial rates are the same as those from the classical theory, modified by a correction factor, $G(n, \rho)$, that takes into account the changes in entropy in the shell and original phase with the attachment of an atom to the cluster interface,

$$G(n,\rho)=\left(\frac{\varepsilon(n)}{\varepsilon(n+1)}\frac{\left(\rho_{\max}(n)\right)!}{\left(\rho_{\max}(n+1)\right)!\rho}\frac{1}{}\frac{\left(\rho_{\max}(n+1)-\rho+1\right)!}{\left(\rho_{\max}(n+1)-\rho\right)!}\left(\frac{N_{\text{sol}}}{N_{\text{sites}}-N_{\text{sol}}}\right)\right)^{-1/2}. \quad (70)$$

5.3 Numerical solution to kinetic model

The differential equations describing coupled-flux nucleation are "stiff" due to the large differences in the magnitudes of the rate constants across the two-dimensional space. This makes it difficult to solve them by using the explicit approaches (i.e., Eulerian and Runge-Kutta) employed in Chapters 2 and 3 for the classical theory. Implicit methods [23] provide a solution to this problem, allowing estimates of the time-dependent rate to be made with reasonable computing resources [21]. Here, we explore some of the key results from those calculations using parameters that are reasonable for precipitation and crystallization processes in metals and semiconductors (see Chapter 9, Sections 2 and 3), assuming an initial distribution of uniformly distributed single atoms or molecules. In Section 4, we saw that while the magnitude of the nucleation rate and the cluster composition in the multi-component nucleation were functions of the interfacial attachment rates for the different species, the overall behavior of the time-dependent nucleation rate was similar to that predicted by the classical theory for a one-component system. The fundamentally different rates for coupled-flux nucleation might lead one to expect radically different nucleation behavior in that case. To show the perhaps surprising similarity to classical-theory predictions, and to emphasize those features that are different, we present a slightly more extended discussion of the numerical results than was provided in Section 4.

5.3.1 Comparison with classical-theory predictions

Like the classical theory of nucleation, the predicted ratio of the equilibrium to steady-state distributions (summed over ρ to yield densities that are a function of n only) is nearly 0.5 (Section 2.6). For an initial distribution of isolated single molecules, the coupled-flux model predicts that the nucleation rate is initially low, reflecting a small cluster population at n^*. With increasing time, the nucleation rate increases in a sigmoidal fashion, and eventually approaches a constant, steady-state rate. This was predicted by the classical theory of nucleation for similar initial conditions (see Chapter 3, Figure 3). The time-dependent density of clusters larger than n^* (i.e., nuclei) reflects this transient behavior in the same way as the classical theory. Initially, the density of nuclei is low, consistent with a small nucleation rate. For long annealing times, the number of nuclei grows linearly with time with a slope equal to the steady-state coupled-flux nucleation rate, I_{cf}^{st}. An extrapolation of this linear region to the time axis gives a non-zero intercept, θ_{cf}, the coupled-flux induction time for nucleation. As for the classical theory, the precise time-dependent nucleation behavior is a strong function of the pre-existing cluster distribution [21].

Although the functional form is identical, the calculated induction time is much longer than predicted by the classical theory (Figure 8). The steady-state coupled-flux nucleation rate has the form $I_{cf}^{st} = A^* \exp(-W^*/k_B T)$, reflecting the same dependence on W^* as the classical theory due to the same thermodynamics for interfacial attachment in the two models. The magnitude of the steady-state rate is much smaller, however (Figure 8). This and the larger induction times reflect the increased importance of long-range diffusion for setting the nucleation time scales in the coupled-flux model. An important result from experimental and theoretical studies of the classical theory of nucleation is that I_{cf}^{st} and θ_{cf} are useful scaling parameters. This remains true within the coupled-flux model (Figure 9).

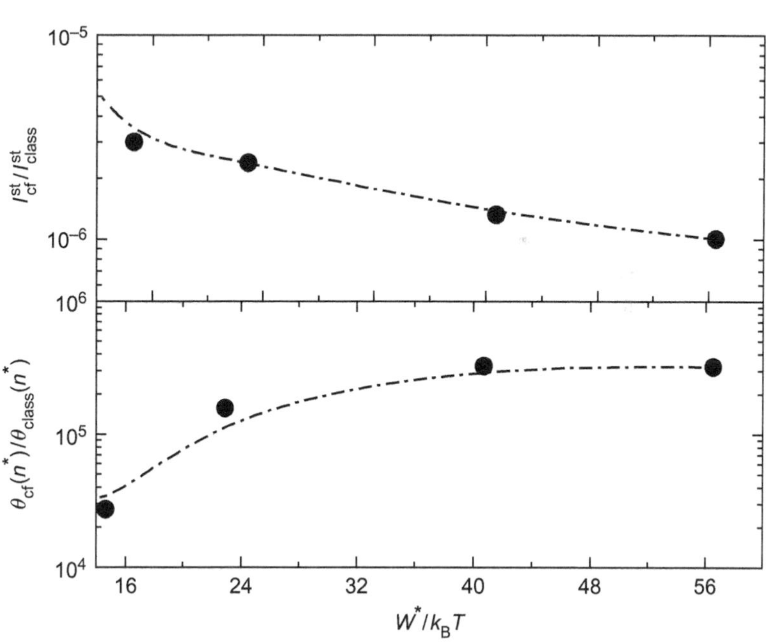

Fig. 8 Comparison of the ratios of the linked flux to classical theory values for the steady-state nucleation rates $I_{cf}^{st}/I_{class}^{st}$ and the induction times $\theta_{cf}(n^*)/\theta_{class}(n^*)$, computed from the numerical solutions to the coupled-flux equations. The dotted line is a guide for the eye, showing the trend of the numerical data. (Reprinted from Ref. [21], copyright (2000), with permission from Elsevier.)

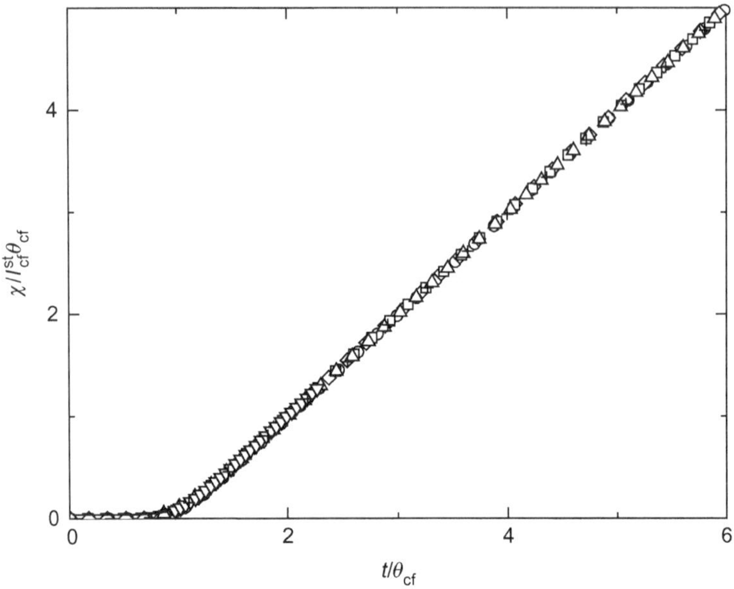

Fig. 9 The effective number of nuclei produced as a function of time per unit volume, χ, scaled to the steady-state nucleation rate multiplied by the induction time $\chi/I^{st}_{cf}\theta_{cf}$ as a function of the annealing time scaled to the induction time t/θ_{cf}. The steady-state nucleation rate varies by almost seven orders of magnitude (9×10^{19} to $5.7 \times 10^{15}\,m^{-3}\,s^{-1}$), with a corresponding change of the induction time from 11.5 to 24 h. (Reprinted from Ref. [21], copyright (2000), with permission from Elsevier.)

5.3.2 Enhanced solute concentration near subcritical clusters

It is well known that the solute concentration near a growing solute phase should be less than the average concentration in the original phase. A striking consequence of coupling the interfacial and diffusive fluxes is a predicted enhanced solute concentration in the neighborhood of subcritical clusters. This is shown in Figure 10, where the average solute-atom concentration in the cluster neighborhood is plotted as a function of cluster size, n. The average concentration is defined by

$$\langle C\rangle_n = \frac{\langle\rho\rangle_n}{\text{shell volume}} \approx \frac{\langle\rho\rangle_n}{(4\pi)^{1/3}(3n\bar{v})^{2/3}\lambda}, \tag{71}$$

where

$$\langle\rho\rangle_n = \frac{\sum_{\rho=0}^{\rho_{\max}(n)} \rho N(n,\rho)}{\sum_{\rho=0}^{\rho_{\max}(n)} N(n,\rho)}, \tag{72}$$

where $\bar{v}$ is the atomic volume and λ is the average jump distance. The data in Figure 10 show that after annealing for a time sufficiently greater

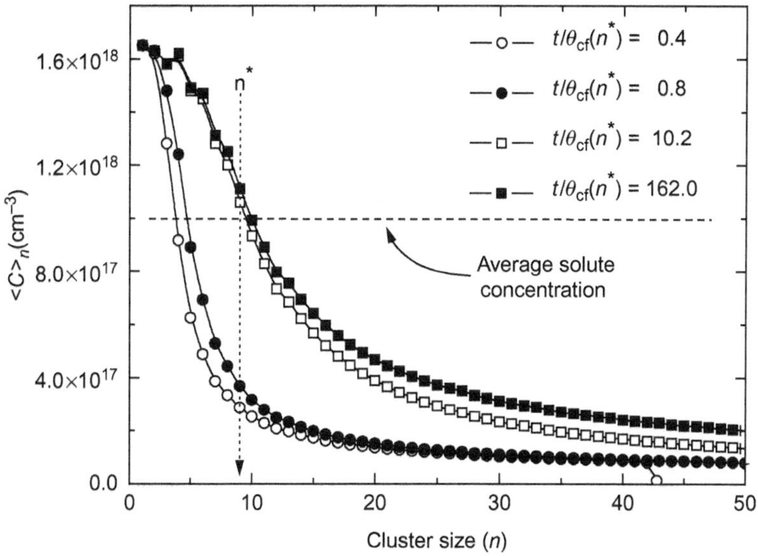

Fig. 10 Effective concentration, $\langle C \rangle_n$, of the original phase in the nearest-neighbor shell (computed from Eqs. (71) and (72) of the clusters of size n and different annealing times (scaled relative to the induction time at the critical size). The neighborhood concentration for small clusters near n^* is enhanced above the average sample concentration (shown as a dotted line). (Reprinted from Ref. [21], copyright (2000), with permission from Elsevier.)

than $\theta_{cf}(n^*)$, the solute concentration in the nearest-neighbor shell rises above the average sample concentration for clusters smaller than the critical cluster. The concentration falls below the average value in the neighborhoods of large clusters, as expected for diffusion-limited growth.

5.3.3 Effects of relative interfacial/diffusive mobilities and solute concentration

The magnitude of effect of the introduction of diffusion in the original phase on the nucleation behavior depends strongly on the ratio of the atomic mobilities for interfacial attachment and long-range diffusion, $\xi D/D_i$, where ξD is a measure of the effective diffusion mobility into the neighborhood shell (D is the effective diffusion coefficient for solute in the original phase and ξ is defined in Eq. (66)) and as mentioned before, D_i is the effective diffusion coefficient governing interfacial attachment. This is demonstrated in Figure 11. For $\xi D < D_i$, the coupled-flux nucleation rate scales linearly with the ratio of mobilities. For higher diffusion mobilities, however, the rate approaches an asymptote, equal to the classical theory rate scaled by the concentration (expressed here as

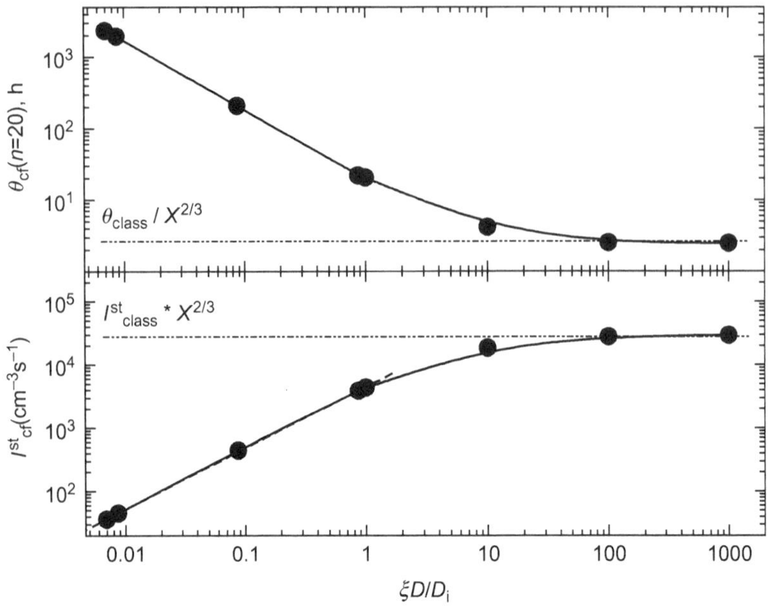

Fig. 11 The steady-state nucleation rates and induction times (evaluated at $n = 20$) computed from the coupled-flux differential equations as a function of the ratio of the atomic mobilities for interfacial and long-range diffusion $\xi D/D_i$. D is the diffusion coefficient in the original phase, D_i is the effective diffusion coefficient governing interfacial attachment, and ξ is defined in Eq. (66). The atom concentration (atomic fraction), X, is defined by N_{sol}/N_{sites}, where N_{sol} is the number of solute molecules and N_{sites} is the number of possible solute sites, both per unit volume. The critical size is eight in these calculations. (Reprinted from Ref. [21], copyright (2000), with permission from Elsevier.)

atom fraction) to the two-thirds power (i.e., $X^{2/3}$, where $X = N_{sol}/N_{sites}$). The induction time, computed within the coupled-flux model scales inversely with the long-range diffusion coefficient when it is less than the interfacial diffusion coefficient. For faster diffusion, it approaches a similar asymptote to that for the steady-state rate, again equal to the induction time calculated within the classical theory, scaled by the inverse two-thirds power of the concentration.

This asymptotic behavior might at first appear puzzling. It is a consequence of the composition dependence of the rates of interfacial attachment, $k^+(n^*)$, which scale with $n^{2/3}$, reflecting the number of nearest-neighbor molecules that can attach to the interface. The number of nearest-neighbor molecules should scale with $X^{2/3}$ when the diffusion mobility becomes high, since all possible sites in the original phase become equally likely to be populated. Within the classical theory,

$I^{st}_{class} \approx k^+(n^*)N^{eq}(n^*)$; the asymptotic behavior then simply reflects the $X^{2/3}$ dependence of $k^+(n^*)$ (Chapter 2, Section 4). For high diffusion rates, the induction time for coupled-flux nucleation approaches $\theta_{class}X^{2/3}$, again reflecting the $X^{2/3}$ dependence of $k^+(n^*)$ ($\theta = \pi^2\tau/6 \propto 1/k^+(n^*)$ (Eq. (36) in Chapter 3).

Given the behavior in the classical regime, it is useful to explore the concentration dependence in the range where diffusion effects become important. As shown in Figure 12, the steady-state rate predicted by the coupled-flux model increases as X^2 when the concentration is dilute ($X < 0.1$). When the concentration dependence for $k^+(n^*)$ is included, the classical theory predicts that steady-state rate should scale as $X^{5/3}$ ($I^{st}_{class} \approx k^+(n^*)N^{eq}(n^*)$). The composition dependence of the induction time for nucleation in the coupled-flux model decreases with increasing measurement cluster size; $\theta_{cf}(n^*) \propto X^{-5/3}$, decreasing to X^{-1} for $\theta_{cf}(4n^*)$. As already discussed, $\theta(n^*) \propto X^{-2/3}$ within the classical theory. A coupled-flux approach, then, predicts a moderately stronger concentration dependence of the nucleation properties than can be argued within the classical theory, signaling a possible test for the approach. While no experimental verification of the coupled-flux

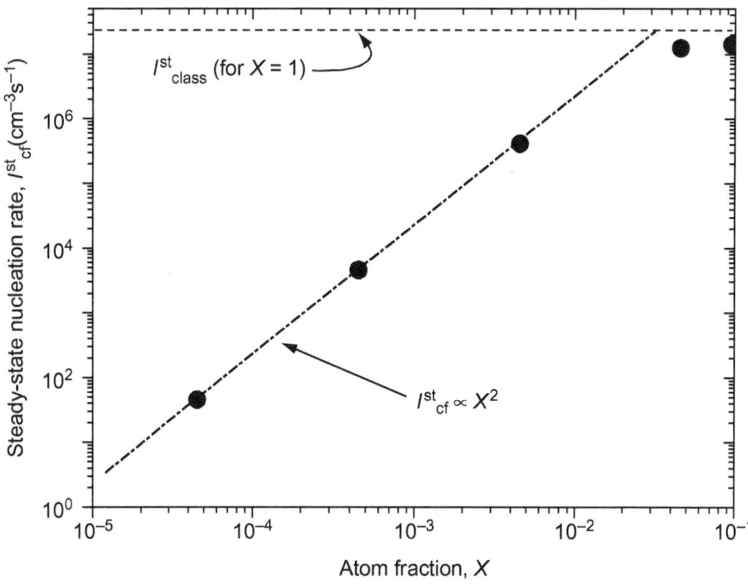

Fig. 12 The steady-state nucleation rate calculated within the coupled-flux model as a function of the concentration of the initial phase, keeping W^* constant. The concentration dependence of ξ is calculated from Eq. (66). (Reprinted from Ref. [24], copyright (2000), with permission from Elsevier.)

model yet exists, it has been confirmed recently by computer simulation [22].

5.4 Analytical solution

In 1968, Russell proposed a model for the time-dependent and steady-state nucleation rates in coupled-flux (or in Russell's terms, *linked-flux*) nucleation [19]. To our knowledge, this remains the only analytical treatment of this problem. Although his development shares features with interface-limited nucleation in a multi-component system (Section 4), the solution of the coupled kinetic equations is more difficult. The development of the partial differential equation describing nucleation is presented here; Ref. [19] should be consulted for more detail, including the approximations made to obtain a solution.

Russell chose a Gaussian distribution for the number of molecules in the shell region to construct the equilibrium cluster distribution. As in Section 5.1, let ρ be the number of solute molecules in the nearest-neighbor shell around a cluster containing n molecules. Assuming that fluctuations in ρ are small compared with the average value, $\bar{\rho}$, the equilibrium distribution is

$$N^{\mathrm{eq}}(n,\rho) = N_0 \exp\left(-\frac{W(n)}{k_{\mathrm{B}}T}\right)(2\pi\langle\Delta\rho\rangle^2)^{-1/2}\exp\left(-\frac{(\rho-\bar{\rho})^2}{2\langle\Delta\rho\rangle^2}\right), \tag{73}$$

where $\langle\Delta\rho\rangle$ is the standard deviation. This approach is fundamentally equivalent to that presented in Section 5.1. The calculation of the moments of the distribution requires physical input from an assumed microscopic model. Our consideration of entropic effects alone is only one of many possible choices that could be made to derive the parameters for the distribution in Eq. (73).

It should, however, be emphasized that a significant simplification was made when deriving the kinetic model presented in Section 5.2. In general, for precipitation of a phase of solute concentration C from an original phase of concentration C', the incorporation of an A atom into a cluster generates currents in both n and ρ space. The shell of nearest neighbors is expanded into the original phase to incorporate $1/C$ new sites, containing C'/C molecules of A (Figure 13). For nucleation from concentrated solutions, then, the depletion of A molecules in the shell with cluster growth is less than expected. For nucleation from a dilute solution, considered in the last two sections, C'/C approaches zero, allowing this effect to be ignored. This effect must be included, however, when considering nucleation from more concentrated solutions, and was included in Russell's treatment.

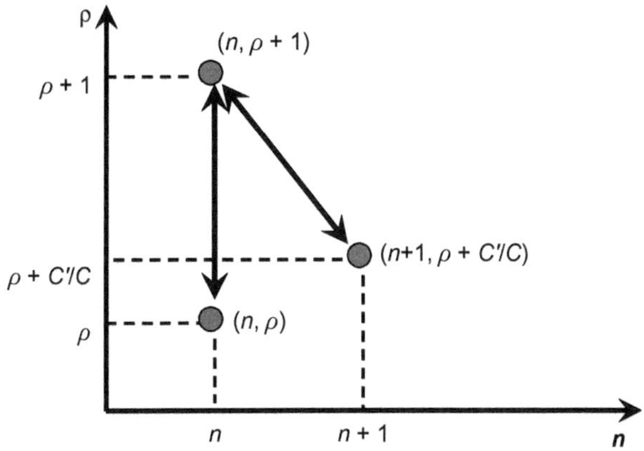

Fig. 13 Schematic illustration of coupled fluxes in phase space for the linked-flux model. The promotion of the cluster (n, ρ + 1) to size (n + 1) would appear to decrease the number of molecules in the nearest neighbor shell by 1 to ρ. The expansion of the cluster shell, however, partly compensates, giving instead the cluster (n + 1, ρ + C'/C). (Reprinted from Ref. [19], copyright (1968), with permission from Elsevier.)

The incorporation of an atom into the cluster results in fluxes in the (n, ρ) space, I_n and I_ρ

$$
\begin{aligned}
I_n &= k^+(n,\rho)N(n,\rho,t) - k^-\left(n+1,\rho-1+\frac{C'}{C}\right)N\left(n+1,\rho-1+\frac{C'}{C},t\right)\\
I_\rho &= k^+(n,\rho)\left(-1+\frac{C'}{C}\right)N(n,\rho,t) - k^-\left(n+1,\rho-1+\frac{C'}{C}\right)\times\\
&\quad\left(-1+\frac{C'}{C}\right)N\left(n+1,\rho-1+\frac{C'}{C},t\right),
\end{aligned}
\tag{74}
$$

where the notation reflects that the expansion of the interface upon atom incorporation leads to less than a loss of one atom in the cluster neighborhood. Using detailed balance and assuming that the ratio $N(n,\rho,t)/N^{eq}(n,\rho)$ varies slowly, the fluxes can be written as

$$
\begin{aligned}
I_n &= -k^+(n,\rho)N^{eq}(n,\rho)\frac{\partial}{\partial n}\left(\frac{N(n,\rho,t)}{N^{eq}(n,\rho)}\right)\\
&\quad+\left(\frac{C-C'}{C}\right)k^+(n,\rho)N^{eq}(n,\rho)\frac{\partial}{\partial \rho}\left(\frac{N(n,\rho,t)}{N^{eq}(n,\rho)}\right)\\
I_\rho &= \left(\frac{C-C'}{C}\right)k^+(n,\rho)N^{eq}(n,\rho)\frac{\partial}{\partial n}\left(\frac{N(n,\rho,t)}{N^{eq}(n,\rho)}\right)\\
&\quad-\left(\frac{C-C'}{C}\right)^2 k^+(n,\rho)N^{eq}(n,\rho)\frac{\partial}{\partial \rho}\left(\frac{N(n,\rho,t)}{N^{eq}(n,\rho)}\right).
\end{aligned}
\tag{75}
$$

As before, we can write the forward rate constant as (see Eq. (69), where the factor $G(n, \rho)$ is not included because of the different statistical form assumed for the equilibrium distribution)

$$k^{+}(n, \rho) = \rho \frac{6D_{\mathrm{i}}}{\lambda^2} \exp\left(-\frac{\delta W(n)}{2k_{\mathrm{B}}T}\right), \tag{76}$$

where D_{i} is again the effective diffusion coefficient describing the interfacial mobility. The flux between the matrix and shell is a diffusive flux given by

$$I''_{\rho} = -\beta(n, \rho)N(n, \rho, t) - \alpha(n, \rho - 1)N(n, \rho - 1, t), \tag{77}$$

where $\beta(n, \rho)$ is the rate at which molecules move from the shell to the original phase, which, following discussions in previous sections, should be defined as

$$\beta(n, \rho) = \frac{\xi D \rho}{\lambda^2}. \tag{78}$$

For a binary system, the interdiffusion coefficient is $D = 2D_{\mathrm{A}}D_{\mathrm{B}}/(D_{\mathrm{A}} + D_{\mathrm{B}})$, where D_{A} and D_{B} are the diffusion coefficients of species A and B, respectively, in the original phase. Assuming detailed balance and passing to the continuum limit, Eq. (77) becomes

$$I''_{\rho} = -\beta(n, \rho)N^{\mathrm{eq}}(n, \rho)\frac{\partial}{\partial \rho}\left(\frac{N(n, \rho, t)}{N^{\mathrm{eq}}(n, \rho)}\right). \tag{79}$$

Combining Eqs. (75) and (79) gives the total fluxes in (n, ρ) space due to interfacial attachment and long-range diffusion

$$\begin{aligned}
I_n &= -k^{+}(n, \rho)N^{\mathrm{eq}}(n, \rho)\frac{\partial}{\partial n}\left(\frac{N(n, \rho, t)}{N^{\mathrm{eq}}(n, \rho)}\right) \\
&\quad + k^{+}(n, \rho)\delta N^{\mathrm{eq}}(n, \rho)\frac{\partial}{\partial \rho}\left(\frac{N(n, \rho, t)}{N^{\mathrm{eq}}(n, \rho)}\right) \\
I_{\rho} &= k^{+}(n, \rho)\delta N^{\mathrm{eq}}(n, \rho)\frac{\partial}{\partial n}\left(\frac{N(n, \rho, t)}{N^{\mathrm{eq}}(n, \rho)}\right) \\
&\quad - (\delta^2 + \zeta^2)k^{+}(n, \rho)N^{\mathrm{eq}}(n, \rho)\frac{\partial}{\partial \rho}\left(\frac{N(n, \rho, t)}{N^{\mathrm{eq}}(n, \rho)}\right),
\end{aligned} \tag{80}$$

where

$$\zeta^2 = \frac{\beta(n, \rho)}{k^{+}(n, \rho)} \quad \text{and} \quad \delta = \left(\frac{C - C'}{C}\right). \tag{81}$$

Carrying out the differentiation and assuming Eq. (73) shows that these fluxes correspond to diffusion in an external field; the off-diagonal terms are equal, as expected from Onsager's theorem, validating the choice of the fluxes. Analogous to Eq. (39), by continuity,

$$-\left(\frac{\partial I_n}{\partial n}+\frac{\partial I_\rho}{\partial \rho}\right)=\frac{\partial N(n,\rho,t)}{\partial t}. \tag{82}$$

Using this, the differential equation to be solved for $N(n,\rho,t)$ is

$$\begin{aligned}\frac{\partial N(n,\rho,t)}{\partial t}&=k^+(n,\rho)N^{\text{eq}}(n,\rho)\left(\frac{\partial^2}{\partial n^2}\left(\frac{N(n,\rho,t)}{N^{\text{eq}}(n,\rho)}\right)-2\delta\frac{\partial^2}{\partial n\partial\rho}\left(\frac{N(n,\rho,t)}{N^{\text{eq}}(n,\rho)}\right)\right.\\&\quad\left.+(\delta^2+\zeta^2)\frac{\partial^2}{\partial\rho^2}\left(\frac{N(n,\rho,t)}{N^{\text{eq}}(n,\rho)}\right)\right)\\&\quad+\frac{\partial}{\partial n}\left(\frac{N(n,\rho,t)}{N^{\text{eq}}(n,\rho)}\right)\left(\frac{\partial\big(k^+(n,\rho)N^{\text{eq}}(n,\rho)\big)}{\partial n}-\delta\frac{\partial\big(k^+(n,\rho)N^{\text{eq}}(n,\rho)\big)}{\partial\rho}\right)\\&\quad+\frac{\partial}{\partial\rho}\left(\frac{N(n,\rho,t)}{N^{\text{eq}}(n,\rho)}\right)\\&\quad\times\left((\delta^2+\zeta^2)\frac{\partial\big(k^+(n,\rho)N^{\text{eq}}(n,\rho)\big)}{\partial\rho}-\delta\frac{\partial\big(k^+(n,\rho)N^{\text{eq}}(n,\rho)\big)}{\partial n}\right),\end{aligned} \tag{83}$$

which is a second-order, pseudolinear, elliptic partial differential equation. In the derivation of Eq. (83), the ratio $\zeta^2\equiv\zeta^2(n,\rho)=\beta(n,\rho)/k^+(n,\rho)$ (Eq. (81)) was taken to be independent of ρ. As for the case of interface-limited binary nucleation (Section 4), Eq. (83) is most easily solved by introducing a linear transformation to a new coordinate system oriented along the ridge through the saddle region, which assumes that the cluster moves along the direction of steepest descents. The solution is not reproduced here; the interested reader is referred to Ref. [19]. Within these assumptions, the steady-state nucleation rate obtained is

$$I^{\text{st}}_{\text{cf}}=\frac{\zeta^2(n^*)}{\Delta'\left(\delta^2+\zeta^2(n^*)\right)}k^+(n^*)N^{\text{eq}}(n^*)\approx\frac{\zeta^2(n^*)}{\left(\delta^2+\zeta^2(n^*)\right)}I^{\text{st}}_{\text{class}}, \tag{84}$$

where $I^{\text{st}}_{\text{class}}$ is the steady-state nucleation rate from the classical theory and Δ' is the width of the critical region in (n,ρ) space, given by

$$\frac{1}{2}\left(\frac{\Delta'}{2}\right)^2\left(\frac{\partial^2 W(n,\rho)}{\partial n^2}\right)_{n=n^*}=-k_{\text{B}}T. \tag{85}$$

Here $1/\Delta'$ is essentially the Zeldovich factor (Eq. (51) in Chapter 2); this was used to obtain the final expression in Eq. (84) ($I^{\text{st}}_{\text{class}}=Zk^+(n^*)N^{\text{eq}}(n^*)$). Using the principle of time reversibility to calculate the time required for a cluster at the critical size to dissolve to a monomer (assuming that this is the time required to grow to the critical size), assuming spherical clusters and taking the conditions of observable nucleation, where $W^* \approx 60k_BT$, the transient time for nucleation at the critical size is

$$\tau_{cf} = \frac{n^{*2}\left(\delta^2 + \zeta^2(n^*)\right)}{10\beta(n^*)}. \tag{86}$$

5.5 Comparison with numerical calculations

Because the rates in these equations are functions of both the cluster size and the number of molecules in the cluster neighborhood, it is unclear how to interpret $\beta(n^*)$, $k^+(n^*)$, and $\zeta^2(n^*)$ in Eqs. (84) and (86). For the dilute case considered in Section 5.2 ($\delta \approx 1$) [21],

$$\begin{aligned} I_{\mathrm{cf}}^{\mathrm{st}} &\approx \frac{I_{\mathrm{class}}^{\mathrm{st}}}{\left(1 + \dfrac{k^+(n^*,1)}{\beta(n^*,1)}\right)} \\ \tau_{\mathrm{cf}} &\approx \left(1 + \frac{k^+(n^*,1)}{\beta(n^*,1)}\right)\tau_{\mathrm{class.}} \end{aligned} \tag{87}$$

The expression for the transient time was obtained by noting that $\tau_{\mathrm{class}}(n^*)$ is often estimated by $n^{*2}/Ck^+(n^*)$, where C is a constant between 1 and 4 (see [19]). A better approximation would result if one of the expressions for $\tau_{\mathrm{class}}(n^*)$ discussed in Chapter 3 was used. Predictions from Eq. (87) are in poor agreement with calculated results for precipitation from a dilute solution [21]. Better agreement is obtained by scaling the classical theory results by the ratio of $\alpha(n^*,0)$ to $k^+(n^*,1)$

$$\begin{aligned} I_{\mathrm{cf}}^{\mathrm{st}} &\approx \frac{\alpha(n^*,0)}{k^+(n^*,1)} I_{class}^{st} \\ \tau_{\mathrm{cf}} &\approx \frac{k^+(n^*,1)}{\alpha(n^*,0)}\tau_{\mathrm{class.}} \end{aligned} \tag{88}$$

These expressions can be justified by assuming a serial rate process, as in interface/diffusion-limited growth [21].

6. SUMMARY

In this chapter, the practically important case of nucleation in a multi-component system has been examined, both in the interface-limited and diffusion-limited regimes. The introduction of the additional degree of freedom greatly complicated the formulation and solution of the kinetic models. Nevertheless, by again examining numerical solutions and simple analytical models, insight can be gained into some key aspects of this complex, but practically important, type of nucleation.

- As for nucleation in a one-component system, the steady-state nucleation rate has the form,

$$I^{st} = A^* \exp\left(-\frac{W^*}{k_B T}\right),$$

where W^* is the reversible work of formation for the critical cluster, composed of a molecules of A, b molecules of B, c molecules of C, etc., and A^* is the concentration-dependent prefactor.

Thermodynamic points

- W^* is given by

$$W^* = \sum_i n^*_{d,i} \Delta\mu^*_i + \Psi^*\left(V^*_d, \{\mu_{m,i}\}\right),$$

where $\Delta\mu^*_i$ is the free-energy difference per atom for species i in the initial and final phase, $n^*_{d,i}$ is the number of molecules of species i in the critical cluster, and Ψ is an excess free energy that is a function of the volume of the critical cluster and the set of chemical potentials of species i in the final phase. If the driving free energy is taken to be an effective free energy and Ψ is approximated by the surface free energy for a planar interface,

$$W^* = n^* \Delta\mu' + \eta n^{*2/3} \bar{v}^{2/3} \sigma',$$

where n^* is the total number of molecules of all types in the critical cluster, $\Delta\mu'$ is an effective free energy of transformation, σ' is an effective surface free energy, and η is a geometric factor dependent on the shape of the cluster, equal to $(36\pi)^{1/3}$ for spherical clusters.
- As for nucleation in a one-component system, even with no kinetic model, the number of molecules of species i in the critical cluster can be obtained from the slope of the steady-state nucleation rate, I^{st}, with respect to the chemical potential in the initial phase. The excess number of molecules of type i in the critical nucleus, Δn^*_i, defined as the number of molecules in the critical cluster less those that would have occupied the same volume in the initial phase, is

$$\Delta n^*_i = -\left(\frac{\partial W^*}{\partial \mu_{m,i}}\right)_T = \left(\frac{\partial (k_B T \ln I^{st})}{\partial \mu_{m,i}}\right)_T - \varepsilon,$$

where ε is between 0 and 1.

Interface-limited nucleation

Nucleation in a two-component system was considered. Clusters of a molecules of A and b molecules of B form in a system initially consisting of N_A^0 and N_B^0 A and B molecules respectively. Key points include:

- An "equilibrium" distribution of clusters as a function of size can be defined,

$$N^{eq}(a,b) = C\exp\left(-\frac{W^*}{k_B T}\right),$$

  where

$$C = (N_A(1))^{N_A(1)/(N_A(1)+N_B(1))} \times (N_B(1))^{N_B(1)/(N_A(1)+N_B(1))},$$

  where N_A (1) and N_B (1) are the initial densities of A molecules and B molecules, respectively.
- Two key assumptions are typically made when seeking analytical expressions for the binary nucleation rate: (i) the nucleation rate is dominated by the cluster flux through the saddle point of the surface of the work of cluster formation, $W(a, b)$, and (ii) the direction of the flux is constant in the vicinity of the saddle region and corresponds to the direction of steepest descent. With these approximations, a commonly used expression for the steady-state rate was obtained,

$$I^{st} \approx C\frac{k_{A^*}^+ k_{B^*}^+ (1 + \tan^2\Theta)}{k_{B^*}^+ + k_{A^*}^+ \tan^2\Theta}\sqrt{\frac{P}{Q}}\exp\left(-\frac{W^*}{k_B T}\right),$$

  where

$$P = \frac{1}{2}\left(\frac{\partial W}{\partial^2 \zeta_1^2}\right)_{\zeta_1^*,\, \zeta_2^*}$$

  and

$$Q = \frac{1}{2}\left(\frac{\partial W}{\partial^2 \zeta_2^2}\right)_{\zeta_1^*,\, \zeta_2^*}.$$

  Θ Is the rotation angle from the initial coordinate system, (a, b), to a new one, (ζ_1, ζ_2), where the abscissa points in the direction of the dominant cluster flux (Figure 4). The rates are a function of the details of the phase transition (i.e., vapor → liquid, liquid → solid); for gas condensation see Eq. (21); for nucleation in condensed phases see Eq. (22).
- Numerical calculations showed that this expression for the steady-state rate works reasonably well if the two species have similar interfacial mobilities. The key assumptions underlying the derivation, however, are flawed. The composition of the cluster is not a function of thermodynamics alone; it is often richer in the more mobile species.

Coupled interface/diffusion nucleation

A simplified model was developed for nucleation in cases where long-range diffusion in the original phase, to the neighborhood of the developing clusters, becomes important. This is a common case for solid state transformations and precipitation processes. For illustration, the precipitation of pure A from a supersaturated solution was considered. Key results include the following:

- It is impossible to model nucleation in partitioning systems where diffusion is important, without taking into account the linking of the interfacial and diffusive fluxes.
- Time-dependent nucleation rates scale with the relevant mobility, that is, interfacial for rapid diffusion processes and the diffusion mobility for cases where long-range diffusion is the rate-limiting step.
- The nucleation rate is smaller and the induction is larger than calculated from the classical theory, sometimes by several orders of magnitude.
- Predictions from the coupled-flux model agree with those from the classical theory of nucleation when rate of cluster growth is limited by interfacial attachment.
- As in the classical theory, $\chi/I_{\mathrm{cf}}^{\mathrm{st}}\theta_{\mathrm{cf}}$ as a function of t/θ_{cf} is a universal curve, where χ is the number of nuclei as a function of time per unit volume, $I_{\mathrm{cf}}^{\mathrm{st}}$ is the steady-state rate, and θ_{cf} is the induction time.
- The solute concentration is always enhanced near a subcritical cluster, rising above the mean concentration in the original phase; it falls below the mean matrix concentration for large clusters as expected for diffusion-limited growth.
- The steady-state rates and the induction times for diffusion-controlled nucleation from a dilute solution can be estimated from the classical expressions (see Eq. (88)).

REFERENCES

[1] D.W. Oxtoby, D. Kashchiev, A general relation between the nucleation work and the size of the nucleus in multicomponent nucleation, J. Chem. Phys. **100** (1994) 7665–7671.

[2] R.A. Swalin, *Thermodynamics of Solids*, Wiley, New York (1972).

[3] F.S. Ham, Theory of diffusion-limited precipitation, J. Phys. Chem. Solids **6** (1958) 335–351.

[4] D.T. Wu, Nucleation theory, in: *Solid State Physics*, Eds. H. Ehrenreich, F. Spaepen, Academic Press, Boston (1996), pp. 37–187 .

[5] H. Reiss, The kinetics of phase transitions in binary systems, J. Chem. Phys. **18** (1950) 840–848.

[6] M. Kulmala, A. Laaksonen, S.L. Girschick, The self-consistency correction to homogeneous nucleation: extension to binary systems, J. Aerosol Sci. **23** (1992) 309–312.

[7] G. Wilemski, B.E. Wyslouzil, Binary nucleation kinetics. 1. self-consistent size distribution, J. Chem. Phys. **103** (1995) 1127–1136.
[8] A.L. Greer, P.V. Evans, R.G. Hamerton, D.K. Shangguan, K.F. Kelton, Numerical modelling of crystal nucleation in glasses, J. Cryst. Growth **99** (1990) 38–45.
[9] Z. Kozisek, P. Demo, Transient kinetics of binary nucleation, J. Cryst. Growth **132** (1993) 491–503.
[10] Z. Kozisek, P. Demo, Transient nucleation in binary ideal solution, J. Chem. Phys. **102** (1995) 7595–7601.
[11] F. Reiff, *Fundamentals of Statistical and Thermal Physics*, McGraw Hill, Singapore (1965, 1984).
[12] P.V. Evans, *Solidification of Metals and Alloys far from Equilibrium*. PhD thesis, Univ. of Cambridge (1988), p. 119.
[13] G. Wilemski, Binary nucleation. I. Theory applied to water–ethanol vapors, J. Chem. Phys. **62** (1975) 3763–3771.
[14] C.S. Kiang, R.D. Cadle, P. Hamil, V.A. Mohnen, G.K. Yue, Ternary nucleation applied to gas to particle conversion, J. Aerosol. Sci. **6** (1975) 465–474.
[15] H. Vehkamaki, P. Paatero, M. Kulmala, A. Laaksonen, Binary nucleation kinetics—a matrix method, J. Chem. Phys. **101** (1994) 9997–10002.
[16] K. Binder, D. Stauffer, Statistical theory of nucleation, condensation and coagulation, Adv. Phys. **25** (1976) 343–396.
[17] D. Stauffer, Kinetic theory of two-component ('heteromolecular') nucleation and condensation, J. Aerosol. Sci. **7** (1976) 319–333.
[18] D.E. Temkin, V.V. Shevelev, On the theory of nucleation in two-component systems, J. Cryst. Growth **52** (1981) 104–110.
[19] K.C. Russell, Linked flux analysis of nucleation in condensed phases, Acta Metall. **16** (1968) 761–769.
[20] K.F. Kelton, A new model for nucleation in bulk metallic glasses, Philos. Mag. Lett. **77** (1998) 337–343.
[21] K.F. Kelton, Time-dependent nucleation in partitioning transformations, Acta Mater. **48** (2000) 1967–1980.
[22] H. Diao, R. Salazar, K.F. Kelton, L.D. Gelb, Impact of diffusion on concentration profiles around near-critical nuclei and implications for theories of nucleation and growth, Acta Mater. **56** (2008) 2585–2591.
[23] W.H. Press, S.A. Teukolsky, W.T. Vetterling, B.P. Flannery, *Numerical Recipes in Fortran*, Cambridge Univ. Press, Cambridge (1992).
[24] K. F. Kelton, Kinetic model for nucleation in partitioning systems, *J. Non-Cryst. Sol.* **274** (2000) 147–154.

CHAPTER 6

Heterogeneous Nucleation

Contents

1. INTRODUCTION

In previous chapters, we have considered nucleation occurring within a single phase without inhomogeneities of structure or composition. Such nucleation can occur with equal probability in any part of the system, and is termed *homogeneous*. This is rare outside specially designed experiments; in most cases inhomogeneities play a role in catalyzing nucleation, which is then termed *heterogeneous*. Heterogeneous nucleation may occur under driving forces much less than that required for detectable homogeneous nucleation, and nucleation control normally implies control of some distribution of heterogeneities. Gases and to a great extent liquids are intrinsically devoid of microstructure, and the uniformity necessary to ensure that nucleation within them is homogeneous can in principle be realized with sufficient purity or cleanliness. Even in such cases, however, heterogeneous nucleation at container walls or a free surface may dominate. Also, a closer examination of liquids (Chapter 7) shows that spontaneous nonuniformities that develop in their structure may blur the distinction between homogeneous and heterogeneous nucleation. In real solids, the natural microstructure

Pergamon Materials Series, Volume 15
ISSN 1470-1804, DOI 10.1016/S1470-1804(09)01506-5

ensures that nucleation within them is almost always heterogeneous. In this chapter, we analyze how heterogeneities can act as favorable sites for nucleation. The following sections consider planar, linear and point heterogeneities in turn.

Just as for homogeneous nucleation (Chapter 2, Section 8), a steady-state nucleation rate I^{st} can be derived in the form

$$I^{st} = N(1)k^{+}(n^{*})Z \exp\left(\frac{-W^{*}_{het}}{k_{B}T}\right), \quad (1)$$

where $N(1)$ is the number of (atoms or) molecules, $k^{+}(n^{*})$ is the rate of molecule addition to a cluster of critical size (i.e. with n^{*} molecules), Z is the Zeldovich factor (Chapter 2, Eq. (51)), and W^{*}_{het} is the critical work for heterogeneous nucleation. In this chapter, we consider how each of these parameters is affected by the heterogeneity. We note, however, that the key factors are $N(1)$ and W^{*}_{het}, as for a wide range of geometries $k^{+}(n^{*})Z$ is approximately the molecular jump frequency. First, the nucleation rate itself has to be specified with care. For example, for nucleation on a substrate, it may be appropriate to define a rate per unit area of substrate; on the other hand, it may be more useful to define a rate per unit volume involving the rate per unit area and the total area of substrate dispersed through unit volume. Of the parameters controlling I^{st}, mostly the focus is on W^{*}_{het} because the importance of heterogeneous nucleation arises from this critical work being less than for homogeneous nucleation. However, the prefactor $N(1)$ is also important: for homogeneous nucleation, it is the total number of molecules per unit volume, whereas for heterogeneous nucleation it is the greatly reduced number in contact with the relevant heterogeneities. Analogous to the treatment in Chapter 3, Section 1 (Figure 1), we also consider transient times for heterogeneous nucleation, and find that these can be shorter or longer than in the homogeneous case. Throughout the chapter, the focus, for simplicity, is mostly on one-component systems; multicomponent systems can be treated in ways analogous to those used in Chapter 5.

2. NUCLEATION ON INTERFACES

Any interface is a heterogeneity of potential importance for nucleation. Examples include a free surface, the contact with the wall of a container, the surface of an embedded particle, an interphase boundary in a liquid or solid, and a grain or domain boundary in a solid. All such interfaces have positive free energies, and the effective elimination of part of the interface when an embryo is formed favors heterogeneous nucleation. In many cases, we are concerned with a transition, for example, from α phase to β phase, occurring in contact with a third phase that forms a nucleant substrate. We will see that the interactions with the substrate can be such as to remove completely the barrier to nucleation of the new phase. In Section 2.1 we consider some aspects of three-phase interactions before going on to treat nucleation specifically.

2.1 Wetting, line tension, and adsorption

Transitions at interfaces are of broad interest, in both theory and practice. They can be quite distinct from transitions exhibited by the same phases in the bulk, and they are relevant in areas as diverse as surface reconstruction, ferromagnetic ordering, and surface segregation [1]. The field of interface or surface phase transitions lies beyond the scope of this book, but it is necessary to treat some of the basics of the three-phase interactions that are involved.

To focus the discussion, we begin by considering a nucleant solid, denoted by N, which is inert (i.e., nonreacting) and has a nondeformable planar surface. Transitions at such a surface have mostly been considered for two coexisting fluids, often vapor and liquid. A particular, widely studied case is the condensation of a liquid onto a solid substrate from a saturated vapor. To preserve generality, however, the phases in contact with N we label as α and β. Interfacial energies can be defined between each pair of phases, $\sigma_{\alpha\beta}$, $\sigma_{\alpha N}$, and $\sigma_{\beta N}$, and we take these energies to be isotropic. When the inequality

$$\sigma_{\alpha\beta} \geq |\sigma_{\alpha N} - \sigma_{\beta N}| \tag{2}$$

is satisfied, both α and β phases can simultaneously be in contact with N. A planar α–β interface would meet the surface of N at the *contact angle* ϕ_∞ given by the Young equation [2]

$$\sigma_{\alpha\beta} \cos \phi_\infty = \sigma_{\alpha N} - \sigma_{\beta N}, \tag{3}$$

derived by balancing interface tensions resolved parallel to the substrate surface (Figure 1a). As pointed out by Gibbs [3], in equilibrium the more stable phase takes the form of a spherical cap (Figure 1b,c). The work of formation of such a cap, and the implications for nucleation of the phase of which the cap is composed, are considered in Section 2.2. We note in passing that the spherical cap is not the only possible shape for a phase formed at an interface. If the substrate is deformable, as when the phase is formed on the surface of a liquid [4] or at the interface between two liquids [5], a lenticular shape is expected; this may also form when a transformation is nucleated at grain boundaries within a solid phase (Section 2.6). Among other factors, anisotropy of interfacial energy may lead to different shapes (as seen later, Figure 22). Even for isotropic interfacial energies, the spherical cap cannot provide a good description of the shape of an

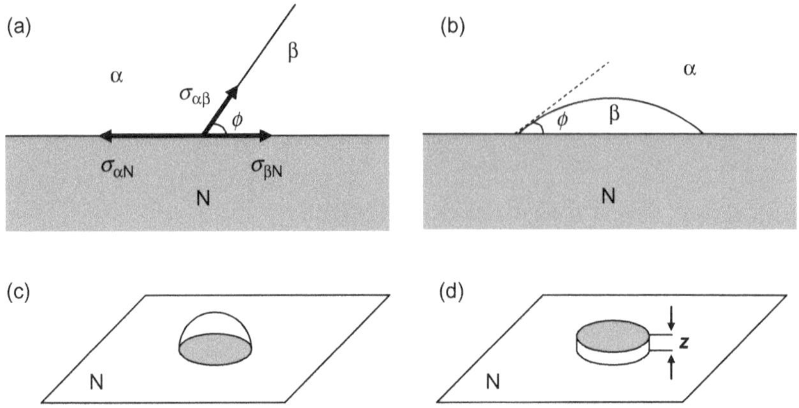

Fig. 1 (a) At the junction of a planar α-β interface with a planar surface of a substrate, the contact angle ϕ is determined by a balance of the interfacial energies according to the Young equation (Eq. (3)). (b, c) When β is the more stable phase, it adopts the classical *spherical-cap* geometry, where (in the absence of line tension) the contact angle is as in (a). (d) When ϕ is small, a *pill-box* shape is more appropriate than the spherical cap.

embryo of β if the contact angle is small; around the edge, a substantial fraction of the cap is, unphysically, less than one atomic diameter in height. Instead of a smooth increase in thickness at the edge of the cap, it is more appropriate to assume an abrupt step of height approximately one atomic diameter, giving an overall pill-box shape for the embryo (Figure 1d). For the moment, however, we consider only the spherical cap.

That the line along which three phases coexist may itself have an excess energy, was originally noted by Gibbs [3], but the effects of this *line tension* on the spherical-cap geometry were first considered by Gretz [6, 7]. For a spherical cap of lateral radius R, the line tension Γ alters the mechanical equilibrium so that the contact angle ϕ is given by

$$\sigma_{\alpha\beta} \cos \phi = \sigma_{\alpha N} - \sigma_{\beta N} - \frac{\Gamma}{R}. \tag{4}$$

Substituting from Eq. (3) we have

$$\cos \phi = \cos \phi_{\infty} - \frac{\Gamma}{\sigma_{\alpha\beta} R}, \tag{5}$$

showing that $\phi \to \phi_{\infty}$ as $R \to \infty$. Measurements of the contact angle of liquid caps, for example, by optical interferometric techniques [8], show that the effect of line tension is measurable for small R. The variation of ϕ with R is of the expected form (Figure 2a), and using Eq. (5), values of Γ can be extracted (Figure 2b). When Γ is positive, it increases ϕ, impeding

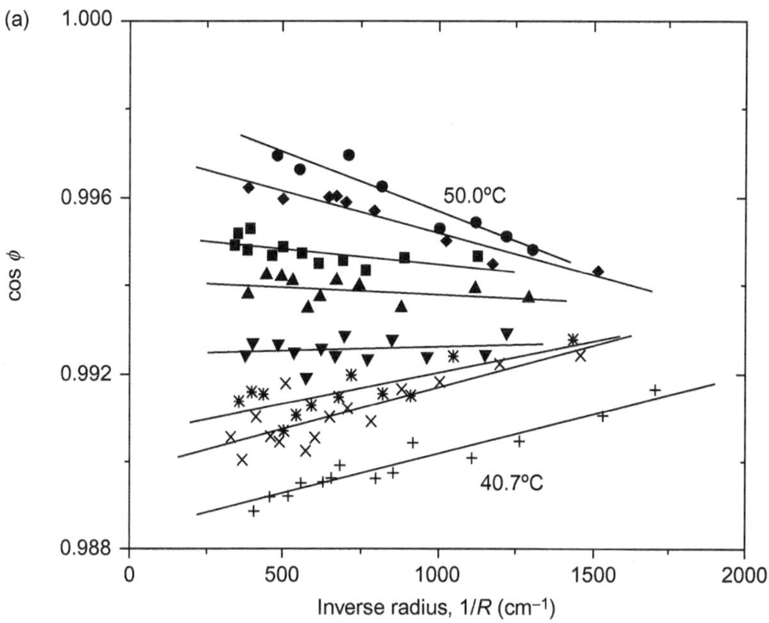

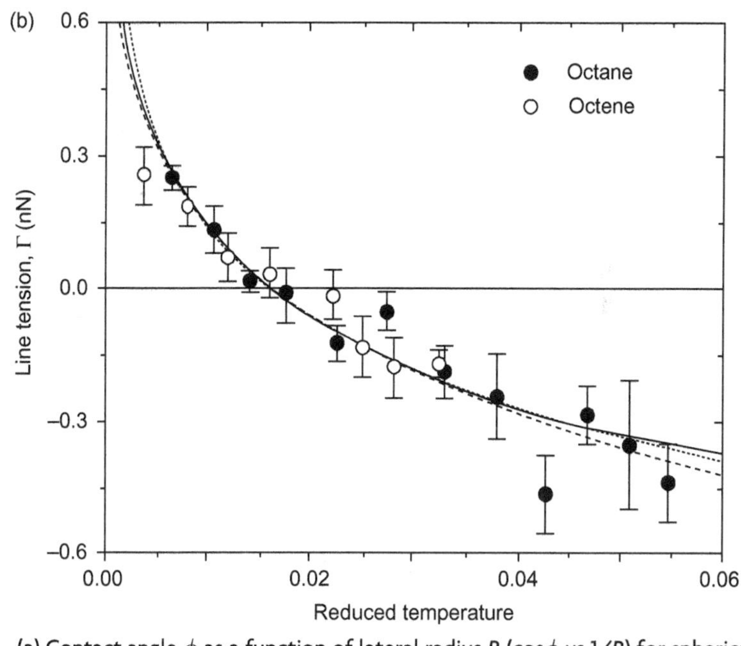

Fig. 2 (a) Contact angle ϕ as a function of lateral radius R ($\cos\phi$ vs $1/R$) for spherical-cap droplets of 1-octene on coated silicon at a range of temperatures between the values shown. (b) From data of the kind in (a), the line tension Γ for 1-octene and n-octane droplets can be obtained as a function of reduced temperature $(T_w - T)/T_w$, where T_w is the critical temperature for wetting. (Adapted with permission from Ref. [8], copyright (2001) by the American Physical Society.)

the spreading of the spherical cap. As the case normally considered is that of liquid caps coexisting with vapor in contact with the substrate, Γ is described as impeding the *wetting* of the surface. At small sizes, Γ may lead to the detachment of the cap from the surface, even for a phase that would have wet the surface on a macroscopic scale [9, 10].

When Eq. (2) is satisfied, the spherical cap (Figure 1) is the stable form for coexisting phases in contact with the substrate; this is described as *partial wetting*. When Eq. (2) is not satisfied, the mechanical balance represented by Eq. (3) cannot be established, and there is *complete wetting* of the substrate by either the α or the β phase. For example, if in principle $\sigma_{\alpha N} > (\sigma_{\alpha\beta} + \sigma_{\beta N})$ when the α and β phases coexist, then a layer of β always separates α and N, preventing their direct contact. In such a case, the effective value of $\sigma_{\alpha N}$ is given by [11]

$$\sigma_{\alpha N} = \sigma_{\alpha\beta} + \sigma_{\beta N}. \tag{6}$$

In the case of α and β phases forming a spinodal system, both $\sigma_{\alpha\beta}$ and $(\sigma_{\alpha N} - \sigma_{\beta N})$ tend to zero as the system is heated toward a critical temperature beyond which the phases merge (Chapter 1, Section 2.2, Figure 5). Cahn [11] suggested that, as $\sigma_{\alpha\beta}$ tends more strongly to zero, the system should in general show a transition from partial to complete wetting at a critical temperature T_w (used to normalize the temperature scale in Figure 2b). This *wetting transition* and related surface transitions have been very widely studied [1, 12]. Wetting transitions are normally first-order, but can be of higher order [12]. We consider briefly some aspects, focusing on first-order transitions in which nucleation plays a role.

In complete wetting, stabilization by the interfacial energies has the consequence that the layer of the wetting phase can continue to exist even when the phase would not be stable in bulk [11]. In that case, there is a *prewetting transition* in which the equilibrium state is a thin adsorbed layer below the transition temperature and thick wetting layer above it [1, 13]. The thickness of the wetting layer is determined by the diffuseness of the interfaces between the phases, and it diverges as conditions approach those for stability of the phase in bulk form. Each state (thin or thick) can persist metastably when the other is stable. There is a nucleation barrier for droplets of the thick layer to form when the thin layer is superheated above the transition temperature, or for pinholes of the thin layer to form when the thick layer is supercooled [13–16]; the barrier for the latter process is much higher than the other two [12]. We consider the work of formation of a circular disk of radius R of the stable layer from the metastable layer. The energy per unit area of the metastable layer exceeds that of the stable layer by U_S. The work W_{disk} of forming the disk is then

$$W_{disk} = -\pi R^2 U_S + 2\pi R\Gamma. \tag{7}$$

This is a two-dimensional analog of Eq. (18) in Chapter 2, and setting $\mathrm{d}\,W_{\mathrm{disk}}/\mathrm{d}\,R = 0$, the critical radius R^* is given by

$$R^* = \frac{\Gamma}{U_S}. \tag{8}$$

The origins of the line tension Γ are complex and lie outside our present treatment, but have been comprehensively reviewed [8, 17]. When the α–β interface of the spherical cap approaches the substrate, it is distorted over a length scale comparable to the interface width. Locally the interface curvature can change sign. Theory suggests that the line tension, which is a measure of the energy of the cap not accounted for by volume and interfacial terms, is very dependent on the order of the wetting transition and the range of intermolecular forces. A typical magnitude of Γ is in the range 10^{-12} to 10^{-10} N [18]. As shown in Figure 2b, Γ is temperature-dependent and can change sign. Typically for $T \ll T_w$ (where T_w is the temperature of the wetting transition), Γ is negative. For $T \leq T_w$, Γ is positive and it diverges as T_w is approached. As a result, nucleation becomes difficult and metastable surface states can be long-lived [13–16]. Interfacial energies can be curvature-dependent, and the treatment of Tolman [19] can be used to make a first-order correction for this. The possible curvature dependence of Γ can be treated in an analogous way [13], and its consequences have been explored [20].

For a pill-box-shaped embryo of lateral radius R and height z (Figure 1d), the total energy of this step is $2\pi R z \sigma_{\alpha\beta}$, and in the simplest case z can be taken to be λ, the atomic diameter [21]. For a number of cases of crystalline nuclei forming on a substrate, the energy appears to be 40–50% of this simple estimate [10, 22]. This edge energy is superficially similar to what would be obtained from a line tension of $\Gamma = \lambda\sigma_{\alpha\beta}$, but it is quite distinct. The edge energy is defined for a value of R different from that which would be obtained by fitting a single interfacial energy over the β-phase volume. The edge energy must always be positive, whereas (as already noted) the line energy may be positive or negative.

Summarizing this section, we have noted that with favorable interfacial energies, there can be wetting of a substrate by a phase, under conditions where the phase would not be stable in bulk. If we consider changing the conditions toward those for bulk stability, the formation of the wetting layer is likely to be by a first-order transition requiring nucleation, but the barrier to nucleation is low and the layer is almost certain to be formed before the phase would become stable in bulk. When the conditions for bulk stability are reached, the phase can then grow without any further nucleation barrier. Even in the absence of a wetting layer, a thin adsorbed layer can certainly affect the surface of the nucleant, in particular changing interfacial energies and thereby the contact angle. In some cases, an adsorbed layer may form with a structure that can act as a template for subsequent growth of the new phase on the substrate; specific cases will be explored later (e.g., Chapter 13, Section 3).

2.2 Spherical-cap model

We consider the heterogeneous nucleation of a new phase β, forming from an original phase α in contact with a nucleant substrate phase N. If Eq. (2) is not satisfied, then when both α and β are present, either of them wets N. If β wets N, then growth of that phase can proceed without a nucleation barrier. If α wets N, then β does not make contact with N, and N cannot act as a nucleant substrate for β. We are now concerned only with the case when Eq. (2) is satisfied. The new phase then forms on the substrate as a spherical cap (Figure 1b,c) whose work of formation we now derive. As for homogeneous nucleation (Chapter 2, Section 1), we first ignore any effects of strain (these effects, important in the solid state, are considered in, for example, Chapter 14, Section 3.3). The work of formation of the cap then depends on the free energies of the bulk, relaxed, original and new phases (taken to have the same number of molecules per unit volume) and on the interfacial energies, as in the homogeneous case, but also on the line tension discussed in Section 2.1. As in that section, we take the interfacial energies to be isotropic, and the substrate to have a planar, inert, nondeformable surface. We also take the original phase α and the new phase β to be of the same composition, the Gibbs free energy changing by Δg per unit volume on transformation from α to β. To begin with, we consider the case when the line tension is negligible. The spherical cap (Figure 1b) then has a contact angle ϕ set by the interfacial energies and the line tension, and given by Eq. (3).

Taking the radius of curvature of the α–β interface to be r, the volume of the spherical cap is $\pi(2 - 3\cos\phi + \cos^3\phi)r^3/3$, conveniently expressed as the volume of a sphere of radius r, multiplied by $f(\phi)$, where

$$f(\phi) = \frac{(2 - 3\cos\phi + \cos^3\phi)}{4}. \tag{9}$$

The area of the α–β interface is $2\pi(1-\cos\phi)r^2$, conveniently expressed as the surface area of a sphere of radius r, multiplied by $\alpha(\phi)$, where

$$\alpha(\phi) = \frac{(1 - \cos\phi)}{2}. \tag{10}$$

The area of the β–N interface is $\pi(r\sin\phi)^2$. When an embryo of β phase is formed in the shape of the spherical cap, α–β and β–N interfaces are created, but a portion of the α–N interface is destroyed. Combining the various terms and rearranging using Eq. (3), the work of forming an embryo of radius r (Figure 3) $W(r)$ is given by

$$W(r) = \left\{ \frac{4\pi}{3} r^3 \Delta g + 4\pi r^2 \sigma_{\alpha\beta} \right\} f(\phi), \tag{11}$$

where, strictly, $\phi = \phi_\infty$ as line tension is ignored. This work is identical to that for homogeneous nucleation, except that it is multiplied by the factor $f(\phi)$, scaling simply with the volume of the embryo. Setting $\mathrm{d}W/\mathrm{d}r = 0$, it is evident that the critical radius thus derived is identical to that for homogeneous nucleation. The critical radius r^*,

$$r^* = -\frac{2\sigma_{\alpha\beta}}{\Delta g}, \tag{12}$$

defines a curvature at which the α–β interface is in (unstable) equilibrium, and this is independent of whether the interface is part of a full sphere or of a spherical cap. The work of forming the critical heterogeneous nucleus W^*_{het},

$$W^*_{\text{het}} = W^*_{\text{hom}} f(\phi), \tag{13}$$

is related to that for the homogeneous case W^*_{hom} by the same factor $f(\phi)$ (plotted in Figure 4), which varies from 0 for $\phi = 0$ to 1 for $\phi = \pi$. The form of this variation makes clear that the presence of a nucleant substrate strongly favors nucleation, especially at low contact angles. The contact angle is a convenient measure of the intrinsic potency of a nucleant substrate, and effective values are often used as parameters even when the exact geometry of Figure 1b does not apply. The contact

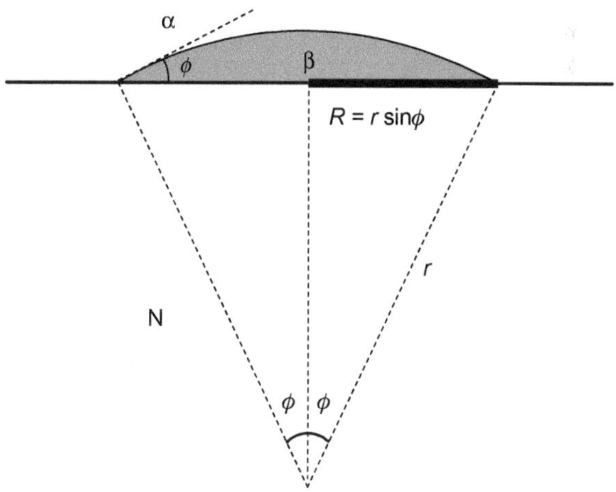

Fig. 3 A spherical-cap embryo has a lateral radius $R = r \sin \phi$, where r is the radius of curvature of the α–β interface.

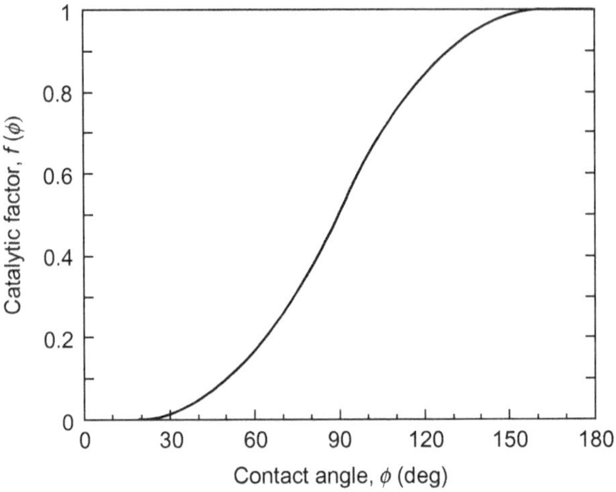

Fig. 4 The catalytic factor $f(\phi)$ for a spherical-cap embryo with contact angle ϕ (Eqs. (9) and (13)).

angle is strongly affected by chemical interactions at the nucleant surface. In the solid state, it can also be affected by interface structure, for example, by the degrees of lattice matching between the α and β phases and the substrate. Some specific examples of chemical and structural effects on ϕ will appear later in the book. A special case is when the substrate is the β phase itself; then $\sigma_{\alpha N} \equiv \sigma_{\alpha\beta}$ and $\sigma_{\beta N} = 0$, leading to $\phi = 0$ (Eq. 3). The substrate then acts as a *seed*, and there is clearly no nucleation barrier.

Homogeneous nucleation can occur with equal probability in any part of the system, whereas heterogeneous nucleation can occur only in contact with the substrate. The factor N_V for homogeneous nucleation must be replaced by N_S, the very much lower number of molecules in contact with the substrate. The value of N_S is the product of the number of sites per unit area of substrate and the total area of substrate per unit volume of the original phase. If, for example, the nucleant substrates are particles dispersed through the volume, then N_S depends on the surface area of the individual particles and their population density. A derivation analogous to that in Chapter 2, Section 7 shows that the Zeldovich factor for a spherical-cap embryo is modified by the factor $1/\sqrt{f(\phi)}$ [23]. The rate constant $k^+(n^*)$ is proportional to the number of sites on the surface of the critical nucleus at which transformation can occur, and this differs from case to case. For a transition between condensed phases, molecular transport across the α–β interface is relevant; $k^+(n^*)$ is then proportional to the α–β interfacial area and, therefore, to $\alpha(\phi)$. For condensation from the vapor, however, the relevant transport is likely to be of molecules already adsorbed on the substrate; for such a case $k^+(n^*)$ depends on the degree of adsorption of molecules on the substrate and on the surface diffusivity [24].

Taking the case of a transition between phases for which the relevant molecular transport is across the α–β interface, the steady-state nucleation rates for the heterogeneous and homogeneous cases are closely related:

$$I_{\text{het}}^{\text{st}} = A^* \frac{N_{\text{S}}}{N_{\text{V}}} \frac{\alpha(\phi)}{\sqrt{f(\phi)}} \exp\left\{ -\frac{W^* f(\phi)}{k_{\text{B}}T} \right\}, \tag{14}$$

where A^* and W^* have the same meanings as in Eq. (55) in Chapter 2. The term $\alpha(\phi)/\sqrt{f(\phi)}$ is close to unity for most angles, although it does increase significantly for angles less than 10°, diverging to infinity as ϕ approaches zero. The basic form of nucleation kinetics is, therefore, the same in homogeneous and heterogeneous cases. If the conditions are changed to progressively increase the driving force for the transformation from α to β, the nucleation rate shows a very sharp increase at some point, defining an effective onset for nucleation, analogous to that in the homogeneous case. The magnitude of the deviation from equilibrium necessary to reach this onset is the most direct measure of the potency of the nucleant substrates, more potent nucleants giving detectable nucleation onset at smaller deviations.

Time-dependent heterogeneous nucleation can be treated in the same way as for the homogeneous case. The treatments in Chapter 3 suggest that the transient time should be proportional to the number of molecules in the critical nucleus and inversely proportional to the number of sites on the surface of the critical nucleus at which there can be exchange of molecules between the original and new phases. Thus, at a given temperature, with constant Δg and $\sigma_{\alpha\beta}$ and varying only ϕ, the transient time τ should be proportional to the ratio of the volume to the area of the α–β interface, evaluated at the critical size. A full analysis [25] shows that this would be exact in a continuum; there are deviations from the continuum behavior when a numerical model is used to treat the exchange of discrete molecules between the α and β phases (Figure 5). The transient time for heterogeneous nucleation τ_{het} is then related to that for homogeneous nucleation τ_{hom} by

$$\tau_{\text{het}} = \tau_{\text{hom}} \frac{2 - 3\cos\phi + \cos^3\phi}{2(1 - \cos\phi)} = \tau_{\text{hom}} \frac{f(\phi)}{\alpha(\phi)}. \tag{15}$$

For $\phi = \pi/2$ or π, the transient time is identical with τ_{hom}. For $\pi/2 < \phi < \pi$, the transient time is greater than, but close to τ_{hom}. For $\phi < \pi/2$, τ_{het} decreases steadily with ϕ, down to zero at $\phi = 0$ (Figure 5). It is important to note that while the work of formation of a critical nucleus is reduced, perhaps dramatically, for nucleation on a substrate, the transient time may even be increased.

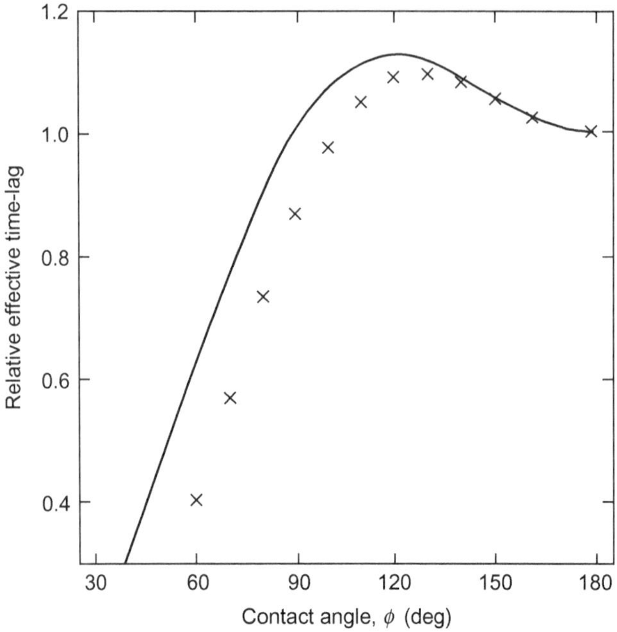

Fig. 5 The time-lag for transient heterogeneous nucleation (normalized with respect to that for homogeneous nucleation, $\phi = 180°$) as a function of the contact angle ϕ. The data points ($\times$) are from a full numerical calculation; the solid line is an analytical approximation taking the time-lag to scale with the volume to curved-interface ratio of the spherical-cap nucleus. (Reprinted from Ref. [25], copyright (1990), with permission from Elsevier.)

The preceding discussions of I^{st} and τ have been for one-component systems; they must be modified when transport of a solute is involved, as treated in Chapter 5.

We now turn to the effect of line tension on heterogeneous nucleation. The spherical cap makes a contact angle ϕ with the substrate given by Eq. (4), and its lateral radius $R = r \sin \phi$ (Figure 3). The work of formation of the cap has an extra term to include the line tension Γ:

$$W(r) = \frac{4\pi}{3} r^3 \Delta g f(\phi) + 4\pi r^2 \alpha(\phi) \sigma_{\alpha\beta} + \pi (r \sin \phi)^2 (\sigma_{\beta N} - \sigma_{\alpha N}) + 2\pi (r \sin \phi) \Gamma, \tag{16}$$

where $f(\phi)$ and $\alpha(\phi)$ are from Eqs. (9) and (10). Substituting from Eq. (4), we have

$$W(r) = \left(\frac{4\pi}{3} r^3 \Delta g + 4\pi r^2 \sigma_{\alpha\beta} \right) f(\phi) + \pi (r \sin \phi) \Gamma. \tag{17}$$

Superficially, this is very similar to Eq. (11), but it must be noted that ϕ is not only different in value from ϕ_{∞}, but also depends on $r \sin \phi$

(≡R in Eq. (5)). Because ϕ varies with r, differentiation of $W(r)$ with respect to r is not straightforward. Nevertheless, the critical condition can be found because Eq. (12) still applies for the α–β interface in unstable equilibrium at the critical condition. With r^* determined from Eq. (12) and with a known Γ, Eq. (5) yields a value for ϕ^*, the contact angle at the critical condition. With a positive Γ, at low enough r^* there is a regime in which Eq. (5) has no solution. This corresponds to the condition, noted in Section 2.1, where the line tension prevents the formation of a cap and causes detachment of the β phase from the substrate. Outside this regime, Eq. (5) has two solutions but, as shown by Navascués and Tarazona [26], the smaller value universally applies to the critical nucleus. Taking this value, the work of formation of the critical nucleus $W^*_{\rm het}$ is given by [7]

$$W^*_{\rm het} = W^*_{\rm hom}\, f(\phi^*) - \frac{2\pi \sin \phi^* \sigma_{\alpha\beta}\Gamma}{\Delta g}. \tag{18}$$

As noted in Section 2.1, Γ may be positive or negative. In the latter case, Eq. (5) always has a solution. In the limit of small r^*, there would be barrierless formation of small caps on the substrate, even for positive values of the free-energy change Δg. As Γ is negative only well below the temperature of the wetting transition, this barrierless regime is probably in the unphysical limit of r^* less than the molecular radius. For large r^*, independent of the sign of Γ, the values of contact angle and $W^*_{\rm het}$ converge on those that would apply in the absence of any line tension. The effect of line tension can thus dramatically affect the temperature dependence of nucleation behavior.

Figure 6, from the work of Navascués and Tarazona [26], shows measured critical supersaturations for nucleation of zinc condensation onto glass over a broad range of substrate temperature. The data cannot be matched by calculations based on $\Gamma = 0$ and a fixed contact angle. On the other hand, inclusion of a positive Γ ($\sim 4.2 \times 10^{-11}\,{\rm J\,m^{-1}}$, and decreasing slightly with temperature) gives an excellent fit to the data. This practical example illustrates the general point that self-consistent application of classical theory to heterogeneous nucleation on a substrate must include the line tension. If the line tension is ignored, which has often been the case, a breakdown of classical theory as r^* is reduced can appear sooner than is fundamentally justified (as explored further in the next section). Unfortunately, in the absence of measured values of Γ, it must be treated as an adjustable parameter.

2.3 Diffuse-interface modeling

As discussed in Chapter 4, the breakdown of the classical theory at small values of r^* can be viewed as a consequence of the diffuseness of the

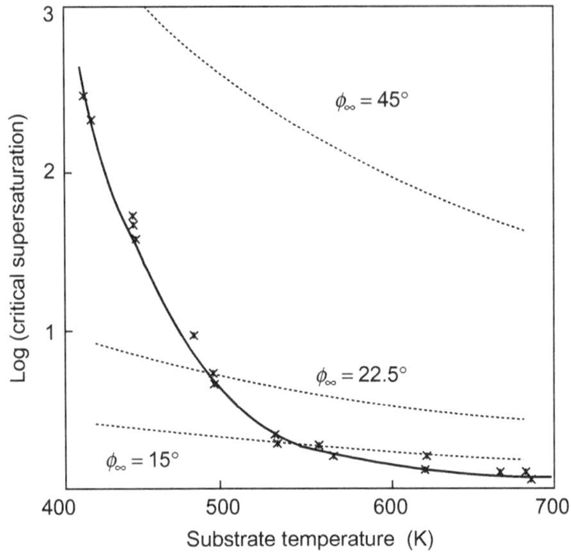

Fig. 6 The critical supersaturation for the condensation of zinc vapor onto glass as a function of the substrate temperature. The form of the data cannot be matched for any assumed contact angle ϕ_∞ by classical nucleation theory neglecting line tension (dashed lines). When line tension is included, a good fit can be obtained (solid line). (Reprinted with permission from Ref. [26], copyright (1981), American Institute of Physics.)

interface between the original and new phases. Diffuse interfaces can be well treated using density-functional modeling, and this has been applied to nucleation on a planar nondeformable substrate by Talanquer and Oxtoby [27]. They considered the condensation of a liquid from its vapor, a system with a critical temperature T_c. As such a system is heated to approach T_c, the vapor–liquid interface width diverges. The interaction with the substrate is described in terms of a surface field $\mathfrak{h}$ and surface coupling $\mathfrak{g}$ that accounts for possible enhancement of molecular interactions at the substrate. As $\mathfrak{h}$ is increased, substrate contact with the liquid is favored relative to contact with the vapor, and the contact angle ϕ decreases until there is complete wetting of the substrate by the liquid. Conversely, as $\mathfrak{h}$ is decreased, ϕ increases until there is complete coverage or "wetting" of the substrate by vapor, a condition that can be termed *complete drying*.

Modeling a van der Waals fluid, and using parameters appropriate for water, Talanquer and Oxtoby [27] made numerical comparisons of the predictions of classical nucleation theory without line tension, the classical theory including line tension, and density-functional theory (see Chapter 4). Figure 7 shows the ratio of predicted nucleation rates for temperatures far from the critical point. For homogeneous nucleation at

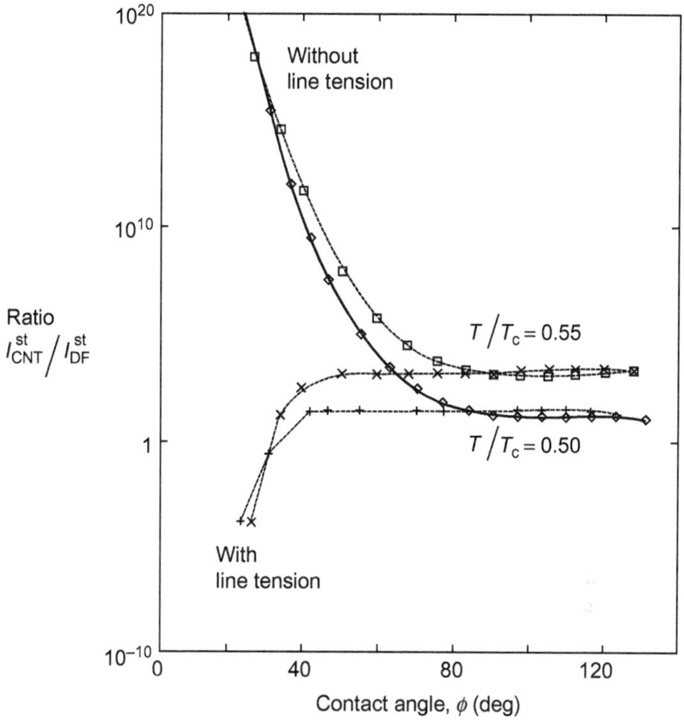

Fig. 7 Calculations of nucleation rates on a substrate for condensation of a van der Waals fluid. The parameters used are based on those for water. The steady-state rate I^{st}_{CNT} predicted by classical theory with and without line tension is compared with the predictions, I^{st}_{DF}, of a density-functional theory calculation over a range of contact angle for two reduced temperatures. (Reprinted with permission from Ref. [27], copyright (1996), American Institute of Physics.)

moderate driving force (supersaturation), the classical theory gives good predictions (i.e., close to the results of the density-functional theory); it slightly overestimates the nucleation rate, a problem that becomes worse as the temperature is raised toward T_c. This is the behavior seen in Figure 7 at large values of contact angle. As ϕ is decreased (by adjusting the surface field $\mathfrak{h}$) to less than 90°, the tendency to overestimate the nucleation rate becomes extreme for the classical theory neglecting line tension. When line tension is included, however, the classical theory continues to match the density-functional predictions until $\phi < 40°$.

From the treatment of Talanquer and Oxtoby [27], we can now summarize much of our coverage so far of nucleation on planar substrates. Figure 8 summarizes the behavior of a system (such as vapor–liquid) showing a critical point, as a function of supersaturation s and the surface field $\mathfrak{h}$ [27]. The behavior is shown at a fixed temperature, in this case $0.5T_c$. The figure shows the bulk wetting transition

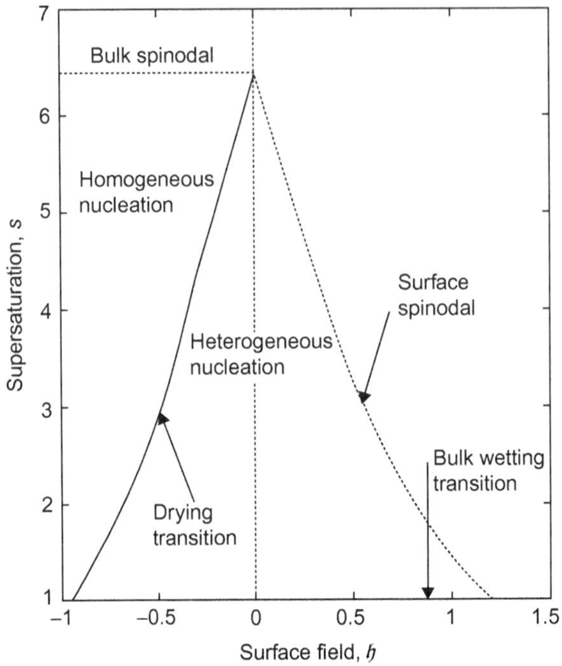

Fig. 8 The nucleation behavior for condensation of a van der Waals fluid (based on water) on a substrate. The temperature is 50% of the critical temperature T_c. More positive values of surface field $\mathfrak{h}$ correspond to smaller contact angle. The solid lines show first-order, discontinuous transitions. (Reprinted with permission from Ref. [27], copyright (1996), American Institute of Physics.)

(Section 2.1); for supersaturations $s<1$, this would continue as a prewetting transition. For $s>1$, as $\mathfrak{h}$ is decreased, ϕ increases until there is a sharp transition to complete drying (i.e., complete coverage or wetting of the substrate by the original phase, here taken to be vapor) accompanied by a discontinuous increase of ϕ to 2π. Beyond this transition, the substrate plays no role, and nucleation is homogeneous. As s is raised, homogeneous nucleation is progressively favored, and the barrier to nucleation goes to zero when the bulk spinodal is reached.

An increasing tendency to adsorption is described by increasing the surface field $\mathfrak{h}$, which gives a decreasing ϕ. The work of formation of heterogeneous nuclei on the substrate goes to zero at the surface spinodal [10]. For small s, the bulk wetting transition is reached before the surface spinodal. The wetting transition has to be nucleated. The nucleation barrier (Eq. (7)) arises from the line tension (positive in this region) and, as noted in Section 2.1, it may be significant.

The effects of diffuse interfaces can be treated by phase-field modeling, which is increasingly widely applied [28]. The background to the method was outlined in Chapter 4 in the discussion of density-functional approaches to nucleation. The phase-field approach can be applied to the wetting of a solid substrate by a liquid [29]. The generality of the approach is particularly attractive for complex cases with patterned substrates, a topic of emerging interest in nanotechnology [30]. Gránásy et al. [31] have applied phase-field modeling to the nucleation of the crystal phase from a pure melt in contact with a nucleating substrate. They specified a macroscopic equilibrium contact angle for crystal embryos on the substrate, and examined the nucleation behavior for two types of boundary conditions at the interface between the liquid and the substrate. In one case, the ordering in the liquid at the wall is taken to be negligible, and in this case the contact angle remains roughly constant even as, at large supercooling, the radius of the critical nucleus shrinks to become comparable with the crystal/liquid interface width (Figure 9a). In the other case, analogous to setting a higher value of the surface field in the treatment of Talanquer and Oxtoby [27], the phase field in the liquid at the wall is set to a constant value, simulating a substrate–liquid interaction inducing partial crystalline order in the liquid. In this case (Figure 9b) as the supercooling is increased, the contact angle decreases, tending to zero at a critical supercooling. The work of formation of the critical nucleus falls much more rapidly with increasing supercooling in the latter case (Figure 9c). The transition to complete "wetting" by the solid is a surface spinodal for freezing, analogous to that already discussed for liquid condensation from the vapor onto a substrate. The existence of such a surface spinodal for freezing remains to be verified by experiment or by atomistic simulations.

2.4 Substrate size and shape

In Section 2.2 we considered nucleation on a planar substrate of a linear dimension much larger than the radius of the critical nucleus r^*. In that case, the nucleation rate can be defined as the frequency of nucleation events per unit area of substrate, but this cannot be an appropriate definition when the size of the substrate surface, or the extent over which it can be considered flat, is comparable with r^*. In dealing with heterogeneous nucleation throughout the book, it will frequently be found that substrate size and shape have to be taken into account.

A suitable case to illustrate the main points is that of a spherical substrate, first analyzed thoroughly by Fletcher [32] and recently re-examined [23]. An embryo of β phase forms from an α phase in contact with a nucleant phase N in the form of a sphere of radius R_{Nuc} (Figure 10b). Considering the geometry, it is appropriate to normalize R_{Nuc} with respect to the radius of curvature r of the α–β interface, defining $X = R_{\mathrm{Nuc}}/r$. The embryo makes a contact angle ϕ with the substrate. To simplify the analysis, the line tension is ignored; ϕ is given

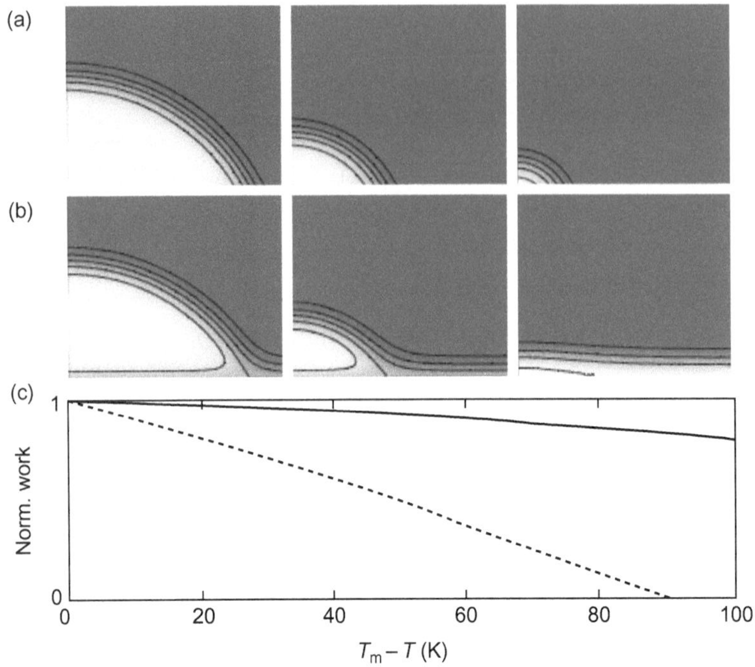

Fig. 9 Two-dimensional phase-field modeling of critical nuclei of a crystal phase nucleating from its pure melt on a substrate. The images (the bottom edge of which is the top of the substrate) show one half of the nuclei (image width 12 nm) at supercoolings of (L to R) 20, 40, and 90 K. The macroscopic equilibrium contact angle for the solid embryos on the substrate is 61.2°. In (a), there is negligible ordering in the liquid. In (b), the substrate induces partial crystalline order in the liquid and on cooling a surface spinodal for freezing is approached at a critical supercooling of 92.0 K. In (c), the work of formation of the nuclei (normalized with respect to the work in a classical model) is shown: solid line corresponding to (a), and dashed line corresponding to (b). (Reprinted with permission from Ref. [31], copyright (2007) by the American Physical Society.)

by Eq. (3), and is independent of r and X. As derived by Fletcher, the work of formation of the critical nucleus $W^*_{\text{het, sphere}}$ is related to that for homogeneous nucleation W^*_{hom} under the same conditions by

$$W^*_{\text{het, sphere}} = W^*_{\text{hom}} f(\phi, X), \tag{19}$$

where

$$f(\phi, X) = \frac{1}{2} + \frac{1}{2}\left(\frac{1 - X\cos\phi}{u}\right)^3 + \frac{X^3}{2}\left(2 - 3\left(\frac{X - \cos\phi}{u}\right) + \left(\frac{X - \cos\phi}{u}\right)^3\right)$$
$$+ \frac{3X^2\cos\phi}{2}\left(\left(\frac{X - \cos\phi}{u}\right) - 1\right) \tag{20}$$

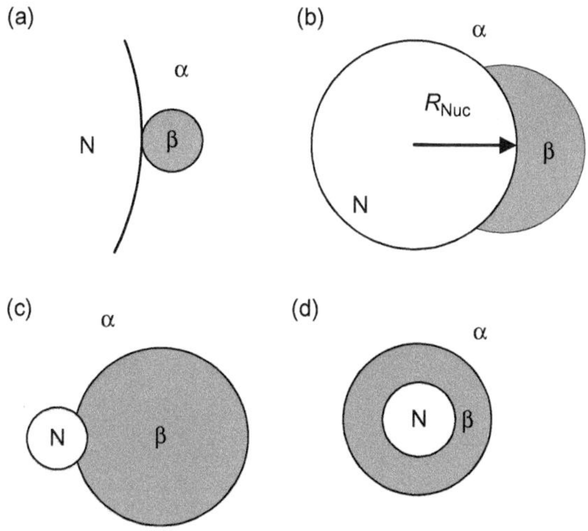

Fig. 10 Examples of nucleation on a spherical particle for different values of $X = R_{Nuc}/r$, where R_{Nuc} is the radius of the particle and r the radius of curvature of the interface between the embryo and the surrounding medium: (a) $X \gg 1$; (b) $X \geq 1$; (c) $X \ll 1$; (d) the special case of complete wetting.

and

$$u = (1 + X^2 - 2X \cos \phi)^{1/2}. \tag{21}$$

The catalytic factor $f(\phi, X)$ is plotted in Figure 11. For large X (as $R_{nuc} \to \infty$), the substrate is effectively planar (Figure 10a) and $f(\phi, X)$ tends to $f(\phi)$ as given in Eq. (9) for a planar substrate. For small X (Figure 10c), $f(\phi, X)$ tends to 1, as the work of formation is not much reduced by the presence of the nucleant substrate. Figure 11 shows straightforwardly that the catalytic factor does not deviate significantly from its value for a planar substrate until $X < 10$. It is reasonable to use the value for a planar substrate for $X > 10$, and to take the substrate to be ineffective as a nucleant for $X < 1$. A wide variety of substrate and embryo geometries have been considered [33], but qualitatively the results are not changed from those for spheres.

A special case arises when the new phase β wets the substrate, illustrated for a sphere in Figure 10d. In that case, the work of formation of the critical nucleus is defined only if $R_{Nuc} < r^*$, in which case the work of formation of an embryo of external radius r is given simply by

$$W(r) = \frac{4\pi}{3}\Delta g(r^3 - R_{nuc}^3) + 4\pi r^2 \sigma_{\alpha\beta} + 4\pi r^2(\sigma_{\beta N} - \sigma_{\alpha N}), \tag{22}$$

which, rearranged using Eq. (9) for wetting, gives

$$W(r) = \frac{4\pi}{3}\Delta g(r^3 - R_{nuc}^3) + 4\pi\sigma_{\alpha\beta}(r^2 - R_{nuc}^2). \tag{23}$$

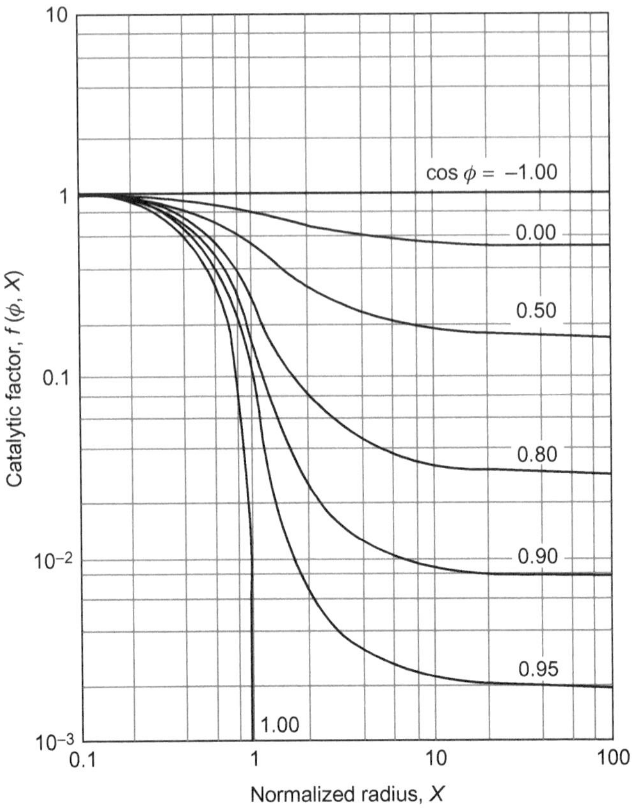

Fig. 11 The catalytic factor $f(\phi, X)$ (from Eqs. (20) and (21)) for heterogeneous nucleation of a spherical cap on a spherical nucleant particle of radius R_{Nuc}. The contact angle is ϕ and the normalized particle radius $X = R_{Nuc}/r$ where r is the radius of curvature of the α–β interface (Figure 3). (Reprinted with permission from Ref. [32], Copyright (1958), American Institute of Physics.)

The significance of such substrate wetting for nucleation under changing thermodynamic conditions is analyzed in Section 2.5.

So far, substrate-size effects have been analyzed for particles or droplets acting as the nucleant. One can also consider nucleant *patches* on a planar surface (Figure 12a). Within the patches (taken here to be circular), there is partial or complete wetting by the new β phase, while elsewhere there is complete wetting by the original α phase (or partial wetting by β but much weaker, i.e., at larger contact angle, than on the patches). Such patches have been invoked to interpret nucleation in solidification experiments [34] and are, for example, the form adopted by ice-nucleating agents on bacteria [35] (Chapter 16, Section 2.3). Figure 13 is from work where, on a planar substrate, an array of polar nucleant patches has been applied by microcontact printing onto an otherwise nonpolar surface; this array is

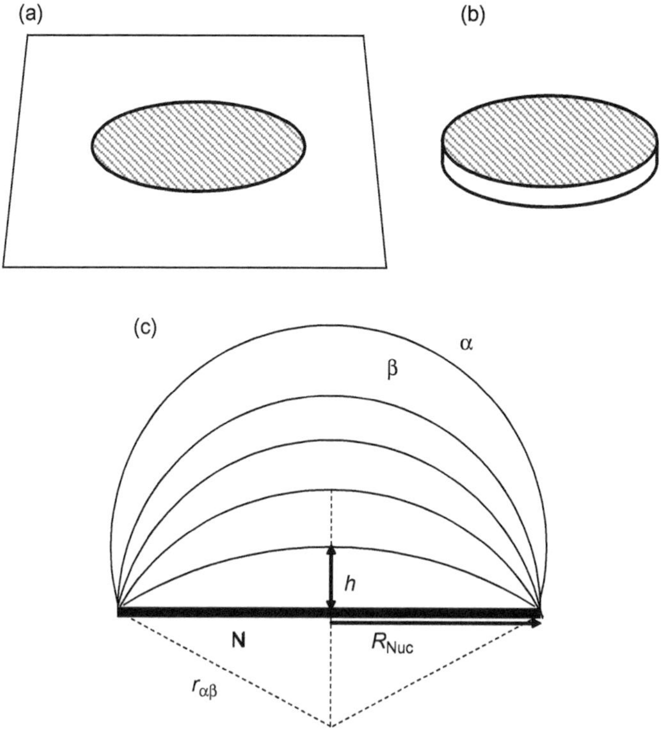

Fig. 12 Examples (shaded) of circular nucleant areas: (a) a surface patch, (b) the active face of a nucleant particle. The growth of β phase from such a nucleant area (c) involves an increase in the curvature of the α–β interface enabled by an increase in supercooling. The curvature is maximum when the α–β interface is hemispherical ($h = R_{Nuc}$) and there is free growth beyond that point. The onset of free growth as the supercooling is increased constitutes *athermal* heterogeneous nucleation.

then used as a template for the development of the new phase (in this case calcite, of interest in biomineralization [30]). A critical nucleus of the β phase can form in the usual spherical-cap geometry on a patch, provided the radius of the patch R_p exceeds the lateral radius $r^* \sin\phi$ of the cap (where ϕ is the contact angle). When $R_p < r^* \sin\phi$, the shape of the embryo is a spherical cap, but the included angle at the edge of the patch is not governed by Eq. (3) (since the same substrate surface does not continue outside the patch) and is greater than the contact angle that would apply if β were to partially wet the patch. Especially when ϕ is small (and therefore $R_p \ll r^*$), the work of nucleation from such patches is little reduced from that for the homogeneous case. Nevertheless, nucleation on such patches may be very significant, as explained in Section 2.5.

It is clear from the foregoing treatment of nucleation on a sphere that convexity of a nucleant substrate hinders its catalysis of heterogeneous nucleation. Conversely, concavity favors nucleation. Analogous to the behavior shown in Figure 11, the curvature of the substrate surface does not have a significant effect on its nucleation potency until the radius of curvature of the concavity is comparable with the critical radius for nucleation. Nevertheless, effects may be significant for solid substrates with defects. The examples of conical and cylindrical cavities are shown in Figure 14. Nucleation in such favored sites has been analyzed by Turnbull [36]. Of more interest than the conditions for true nucleation of β in the cavity is the preservation of the β phase within the cavity under thermodynamic conditions beyond its stability in bulk. In the cases sketched in Figure 14, the contact angle is such that the sign of curvature of the α–β interface is reversed from usual; thus the curvature, unlike that normally encountered in nucleation, acts to stabilize the β phase. In many cases (prominent examples are in solidification [36] (Chapter 13, Section 3.1) and bubble formation [37] (Chapter 16, Section 3; Chapter 17, Section 4)), what appears to be nucleation of β is in fact growth of preexisting β surviving in cavities. In Section 2.5 such cases are analyzed further.

There have been many studies of the effects of substrate geometry on the critical work of heterogeneous nucleation. For example the cases of wedge-shaped cavities and slit-like pores have been analyzed [38]. In many such cases, however, the nucleation kinetics at fixed supersaturation is of less interest than the behavior under increasing supersaturation, as discussed next.

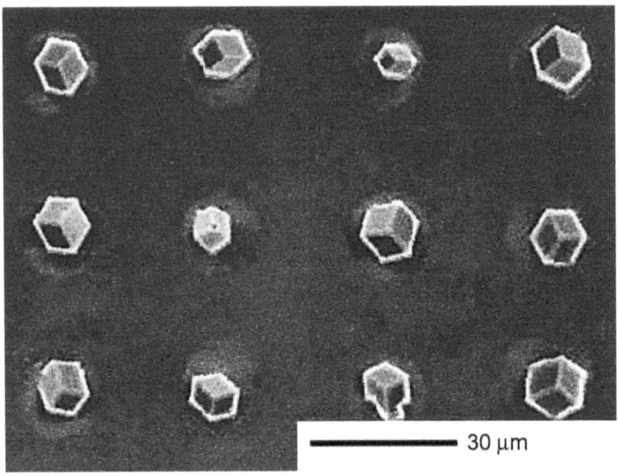

Fig. 13 A monolayer coating on a palladium substrate has been patterned to give an array of polar patches separated by nonpolar regions. The crystallization of calcite from solution is nucleated on the patches. (Reprinted by permission from Macmillan Publishers Ltd. *Nature* [30], copyright 1999.)

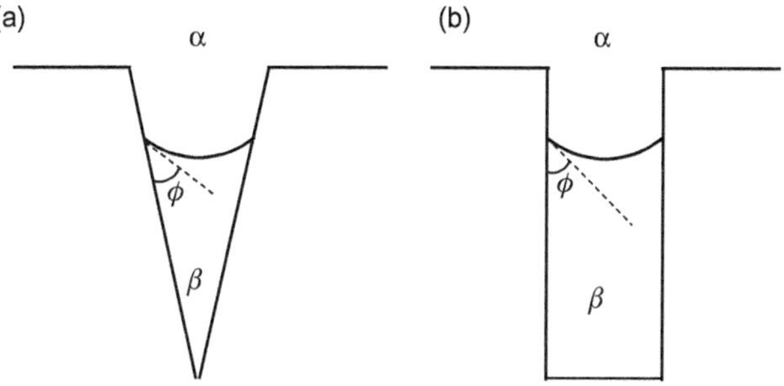

Fig. 14 Examples of (a) conical and (b) cylindrical cavities in a substrate N. The β phase is stabilized outside the conditions for its stability in bulk by the curvature of the α–β interface. The preserved phase may later act as nucleant seeds for transformation of α to β.

2.5 Athermal nucleation

The term *athermal nucleation* was coined by Fisher et al. [39] to describe the case when subcritical embryos are "automatically promoted to nuclei" when the critical size decreases and sweeps past their own size. The nucleation considered so far in this book can in contrast be considered to be *thermal* in that it requires thermal activation to surmount a critical work of formation. While thermal nucleation is a *stochastic* process that can establish a steady-state rate of production of nuclei under fixed conditions, athermal nucleation is a more *deterministic* process in which nuclei are produced only when the conditions are changed to shrink the critical size. The change in critical size inherent in athermal nucleation could arise in many ways, for example, by changing the supersaturation of a gas dissolved in a liquid (examples are given in Chapter 13, Section 6; Chapter 16, Section 3, and Chapter 17, Section 4), but in this section we focus on nucleation of a transformation by cooling. Athermal nucleation on cooling has been a standard way of analyzing the martensitic and related transformations in steels (Chapter 12, Section 4), but we will show that it has a wider relevance.

The rate of athermal nucleation is proportional to the cooling rate, and therefore likely to dominate at a high cooling rate [40]. The effect is difficult to analyze for *homogeneous* nucleation because of the transience of the size distribution of preexisting embryos. Nevertheless, the importance of athermal nucleation has been noted (e.g., for pulsed-laser melted thin films of silicon cooled at $\geq 10^{10}\,\mathrm{K\,s^{-1}}$), and for this case a nucleation-mechanism diagram has been used to identify the regimes of quench rate and supercooling in which thermal or athermal nucleation (of the crystalline phase in the liquid) would be dominant [41]. For rapid quenches, however, the quasi-steady-state approximation breaks down (according to principles set out in Chapter 3), and transient effects tend to reduce the number of nuclei. Shneidman has analyzed the competition between such transient effects and true athermal nucleation, which increases the number of nuclei when there is no further evolution of the embryo size distribution [42]. He notes that athermal homogeneous nucleation is significant only for the combination of extremely high cooling rate and large supercooling. In contrast, for *heterogeneous* nucleation, the athermal process is of wide relevance under typical processing conditions. This is because the heterogeneities in the system can define a stable population of embryos with a fixed size distribution.

We consider (based on the analysis in Ref. [43]) the same type of transformation as when treating spherical-cap nucleation (Section 2.2): β nucleates in α, where both are condensed phases of the same composition. The nucleation of the β phase is catalyzed by a nucleant area, for simplicity taken to be a plane circle, of radius R_{Nuc}. Such circular areas can represent active patches on a planar surface (Figure 12a) as discussed in Section 2.4, or the active faces of nucleant particles (Figure 12b, e.g., TiB_2 inoculant particles for the solidification of aluminum alloys, Chapter 13, Section 3). For potent nucleation, the initial formation of β phase may be so easy that it is not the effective barrier to transformation of the α phase. As pointed out in Section 2.1, adsorption on a surface may provide a template for growth of a new phase, or the formation of a wetting layer can stabilize a new phase before it would be stable in bulk. Even without such effects, classical heterogeneous nucleation with a small contact angle (Figure 3) can readily form β phase such that $r_{\alpha\beta} \gg R_{\mathrm{Nuc}}$. In that case, the initially formed β is not a *transformation nucleus*. The process by which it becomes a transformation nucleus is illustrated in Figure 12c. At a small supercooling, by adsorption, wetting, or heterogeneous nucleation, β forms on the substrate and spreads laterally to cover the nucleant area. Subsequent transformation occurs by increasing the height h of the spherical cap. As noted in Section 2.4, the angle at the edge of the cap is not a contact angle defined by a balance of interfacial tensions (Figure 1). As h increases, the radius of curvature $r_{\alpha\beta}$ of the α–β interface decreases, and growth must stop when $r_{\alpha\beta}$ has decreased to equal r^*, the critical nucleation radius for the ambient supercooling. The β phase is *dormant*, as it is not yet a nucleus for transformation that can spread throughout the α phase. The critical radius is inversely proportional to free-energy change Δg associated with the phase transformation. For crystallization of a liquid, Δg is governed by the supercooling ΔT, and given to a good approximation at small ΔT by

$$\Delta g = \Delta s \Delta T, \tag{24}$$

where Δs is the entropy change per unit volume on transforming from α to β. For the dormant cap $r_{\alpha\beta} = r^*$, and as the supercooling is increased, r^* decreases, permitting further growth. When r^* has decreased to equal R_{Nuc}, the β phase has the form of a hemisphere ($h = R_{\mathrm{Nuc}}$) and $r_{\alpha\beta}$ is at a minimum. At this point, the β phase is no longer dormant, since further growth causes a favorable increase in $r_{\alpha\beta}$ and free growth permits transformation of the entire α phase. The critical supercooling ΔT_{fg} for the onset of free growth is obtained from Eqs. (12) and (24):

$$\Delta T_{\mathrm{fg}} = -\frac{2\sigma_{\alpha\beta}}{\Delta s R_{\mathrm{nuc}}}. \tag{25}$$

The consequence of Eq. (25) is that, on cooling, the free growth of the β phase starts first on the largest nucleants in the system, and that it proceeds from successively smaller nucleants as cooling continues. The effective nucleation rate of β depends on the population and size distribution of the nucleants (as has been explored for cases of practical interest (Chapter 13, Section 3).

To analyze the relationship between athermal and thermal nucleation, we consider the work of formation of the cap of β phase. The radius of curvature of the α–β interface $r_{\alpha\beta}$ is related to the height of the cap h by

$$r_{\alpha\beta} = \frac{h}{2} + \frac{R_{\mathrm{nuc}}^2}{2h}. \tag{26}$$

The volume of the cap, V_{cap}

$$V_{\mathrm{cap}} = \pi\left(r_{\alpha\beta}h^2 - \frac{h^3}{3}\right) = \pi\left(\frac{R_{\mathrm{nuc}}^2 h}{2} + \frac{h^3}{6}\right), \tag{27}$$

and the area of the α–β interface $A_{\alpha\beta}$

$$A_{\alpha\beta} = 2\pi r_{\alpha\beta} h = \pi(h^2 + R_{\mathrm{nuc}}^2), \tag{28}$$

can be expressed in terms of h. (Note that Eqs. (9) and (10) do not apply, as there is no defined contact angle ϕ.) The work required to form a cap of solid (W_{cap}) has contributions from interfacial energies and from the free-energy change associated with the solidification of the volume V_{cap}. Consistent with there being a preexisting β phase, the reference point for energy ($W_{\mathrm{cap}} = 0$) is taken to be that of an infinitesimally thin layer of β coating the entire nucleant area. The only relevant interfacial energy is then $A_{\alpha\beta}\sigma_{\alpha\beta}$. The work of cap formation,

$$W_{\mathrm{cap}} = \sigma_{\alpha\beta}(A_{\alpha\beta} - \pi R_{\mathrm{nuc}}^2) - V_{\mathrm{cap}}\Delta s \Delta T, \tag{29}$$

by substituting from Eqs. (27) and (28), can be expressed as:

$$W_{\mathrm{cap}} = \sigma_{\alpha\beta}\pi h^2 - \Delta s \Delta T\left(\frac{\pi h^3}{6} + \frac{\pi R_{\mathrm{nuc}}^2 h}{2}\right). \tag{30}$$

The form of W_{cap} is best presented in terms of dimensionless quantities [43]. The dimensionless cap height is taken to be h/R_{Nuc}. The dimensionless supercooling is obtained by scaling with respect to the free-growth supercooling (Eq. (25)); thus, athermal nucleation occurs when the dimensionless supercooling $\Delta T/\Delta T_{fg} = 1$. A dimensionless work of formation can be obtained by normalizing with respect to $W^*_{hom,\,\Delta T_{fg}}$, the critical work for homogeneous nucleation at the free-growth supercooling ΔT_{fg}:

$$W^*_{hom,\,\Delta T_{fg}} = \frac{16\pi\sigma^3_{\alpha\beta}}{3\Delta s^2 \Delta T^2_{fg}} = \frac{4\pi\sigma_{\alpha\beta}R^2_{nuc}}{3}. \tag{31}$$

Rearranging Eq. (30) in terms of these dimensionless quantities gives

$$\frac{W_{cap}}{W^*_{hom,\,\Delta T_{fg}}} = -\frac{1}{4}\left(\frac{\Delta T}{\Delta T_{fg}}\right)\left(\frac{h}{R_{nuc}}\right)^3 + \frac{3}{4}\left(\frac{h}{R_{Nuc}}\right)^2 - \frac{3}{4}\frac{\Delta T}{\Delta T_{fg}}\frac{h}{R_{nuc}}. \tag{32}$$

The form of Eq. (32) is plotted in Figure 15 for five values of dimensionless supercooling. For $\Delta T < \Delta T_{fg}$, the work of cap formation passes through a minimum followed by a maximum as h/R_{Nuc} is

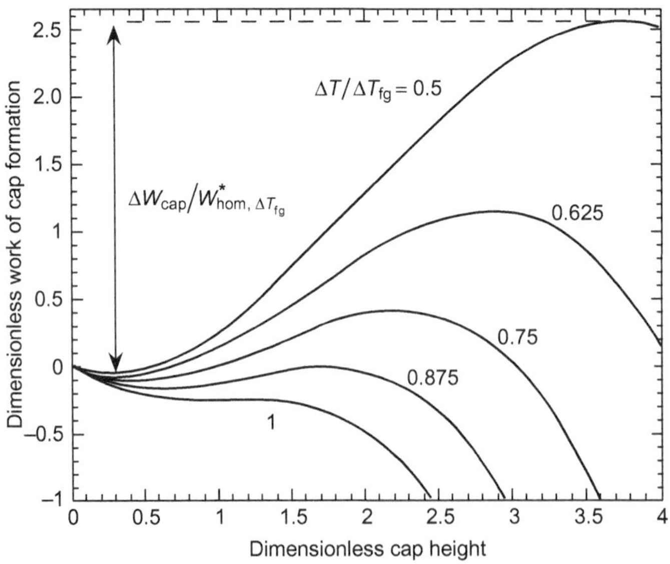

Fig. 15 Dimensionless work of formation ($\Delta W_{cap}/W^*_{hom,\,\Delta T_{fg}}$) of the β-phase cap as a function of dimensionless cap height (h/R_{nuc}) plotted for various values of dimensionless supercooling ($\Delta T/\Delta T_{fg}$) (Eq. (32)). The work of formation is normalized with respect to the critical work for homogeneous nucleation at the *free-growth* supercooling. The minima (maxima) in these energy curves represent metastable (unstable) equilibrium configurations. (Reprinted from Ref. [43], Copyright (2005), with permission from Elsevier.)

increased. These extrema occur at

$$\frac{h}{R_{\text{nuc}}} = \left(\frac{\Delta T_{\text{fg}}}{\Delta T} \pm \sqrt{\left(\frac{\Delta T_{\text{fg}}}{\Delta T} \right)^2 - 1} \right), \tag{33}$$

and represent conditions of equilibrium across the liquid/solid interface. At these points, the radius of curvature of that interface has the critical value r^* given by Eq. (12). Geometrically, the β-phase caps on the nucleant area are obtained by cutting a sphere of equilibrium curvature r^* with a plane such that the circle of intersection has a radius equal to R_{Nuc}. The cap of smaller volume (Figure 16a) is in metastable equilibrium; the larger cap (Figure 16b) is in unstable equilibrium (analogous to the state of the critical nucleus in homogeneous nucleation).

The energy difference between the minima and maxima on the curves in Figure 15 exists for $\Delta T/\Delta T_{\text{fg}} < 1$, and is the barrier for nucleation. As the dimensionless supercooling is increased, the barrier decreases. At $\Delta T/\Delta T_{\text{fg}} = 1$, the work of formation as a function of cap height no longer exhibits extrema; there is just the stationary point at

$$h = r_{\alpha\beta} = r^* = -\frac{2\sigma_{\alpha\beta}}{\Delta s \Delta T}, \tag{34}$$

when the β phase takes the form of a hemisphere and there is no barrier to free growth (Figure 15). Cooling through the condition $\Delta T/\Delta T_{\text{fg}} = 1$ gives athermal nucleation. At $\Delta T/\Delta T_{\text{fg}} < 1$, there could be thermal

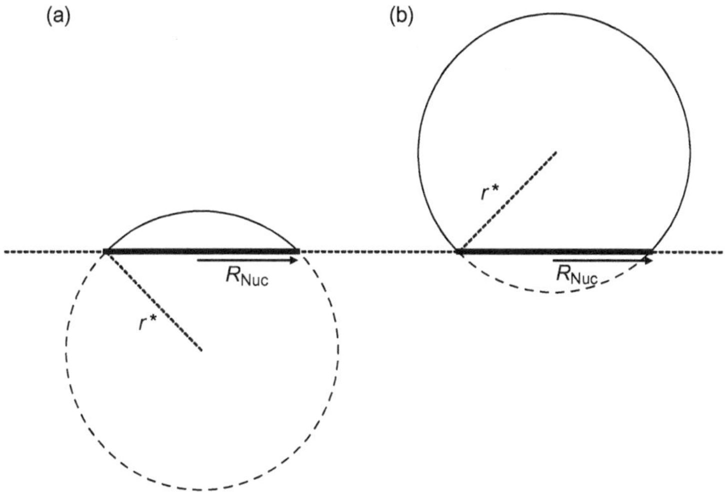

Fig. 16 The (a) dormant, metastable-equilibrium and (b) critical, unstable-equilibrium configurations of the cap of β phase for $\Delta T < \Delta T_{\text{fg}}$. The radius of curvature of the α–β interface in each case has the critical value r^*.

activation over the nucleation barrier [43], and we now assess the likelihood of this preempting athermal nucleation on cooling.

The initial, infinitesimally thin coating of β on the nucleant area grows naturally to the metastable and dormant condition shown in Figure 16a. From that condition, the critical work of thermal nucleation ΔW_{cap} is the difference in energy between the two extrema, which can be expressed in dimensionless terms as

$$\frac{\Delta W_{\mathrm{cap}}}{W^{*}_{\mathrm{hom},\,\Delta T_{\mathrm{fg}}}} = \left(\left(\frac{\Delta T_{\mathrm{fg}}}{\Delta T}\right)^{2} - 1\right)\sqrt{1 - \left(\frac{\Delta T}{\Delta T_{\mathrm{fg}}}\right)^{2}}. \qquad (35)$$

The ratio $\Delta W_{\mathrm{cap}}/W^{*}_{\mathrm{hom},\,\Delta T_{\mathrm{fg}}}$ tends to 1 (or 0) as $\Delta T/\Delta T_{\mathrm{fg}}$ tends to 0 (or 1). The critical work for thermal nucleation (Eq. (35)) is plotted (on a logarithmic scale, given its range of values) as a function of the dimensionless supercooling in Figure 17, and shows a sharp transition. For small $\Delta T/\Delta T_{\mathrm{fg}}$, the dimensionless work $\Delta W_{\mathrm{cap}}/W^{*}_{\mathrm{hom},\,\Delta T_{\mathrm{fg}}} \gg 1$, while for large $\Delta T/\Delta T_{\mathrm{fg}}$, the work $\Delta W_{\mathrm{cap}}/W^{*}_{\mathrm{hom},\,\Delta T_{\mathrm{fg}}} \ll 1$. The likelihood of thermal activation over this energy barrier depends on the ratio of the barrier height to the thermal energy $k_{\mathrm{B}}T$ (where k_{B} the Boltzmann

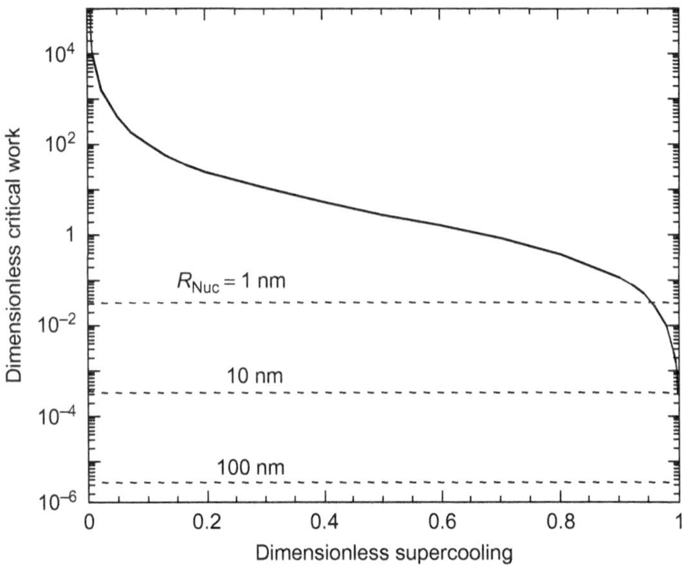

Fig. 17 The dimensionless critical work for free growth ($\Delta W_{\mathrm{cap}}/W^{*}_{\mathrm{hom},\,\Delta T_{\mathrm{fg}}}$) as a function of dimensionless supercooling ($\Delta T/\Delta T_{\mathrm{fg}}$) (Eq. (35)). The dashed lines indicate values of the critical work below which thermal activation of nucleation is likely to precede athermal nucleation on cooling. These values depend on the radius of the nucleant area (Eq. (36), with substitution for $T_{\mathrm{m}}/\sigma_{\alpha\beta}$ typical for metals). (Reprinted from Ref. [43], copyright (2005), with permission from Elsevier.)

constant and T is the temperature). Following the arguments of Feder et al. [44] for the random nature of molecule exchange with embryos of near-critical size, thermal activation becomes significant when the energy barrier $\Delta W_{\mathrm{cap}} \leq k_{\mathrm{B}}T$. Taking $\Delta W^*_{\mathrm{hom},\,\Delta T_{\mathrm{fg}}}$ from Eq. (31), the condition for significant thermal activation of nucleation, before the free-growth criterion is met, is [43]

$$\frac{\Delta W_{\mathrm{cap}}}{W^*_{\mathrm{hom},\,\Delta T_{\mathrm{fg}}}} \leq \frac{3k_{\mathrm{B}}T}{4\pi\sigma_{\alpha\beta}R^2_{\mathrm{nuc}}}. \tag{36}$$

The condition depends on the size of the nucleant area ($\propto R^2_{nuc}$) and the ratio $T/\sigma_{\alpha\beta}$. We now consider specifically the nucleation of metallic crystals from their melts. The supercoolings of interest are small, and we can take T to be the melting temperature T_{m}. For metallic elements, the liquid–solid interfacial energies $\sigma_{\alpha\beta}$ and the melting temperatures T_{m} are roughly proportional, with $\sigma_{\alpha\beta}/T_{\mathrm{m}} \approx 1.2 \times 10^{-4}\,\mathrm{J\,m^{-2}\,K^{-1}}$ (Chapter 7, Table 6). Substituting this ratio (which does not differ greatly for other classes of material) into Eq. (36), an approximate, but universal, link is established between $\Delta W_{\mathrm{cap}}/\Delta W^*_{\mathrm{hom},\,\Delta T_{\mathrm{fg}}}$ and R_{Nuc}. Figure 17 shows (as dashed horizontal contours) the values of $\Delta W_{\mathrm{cap}}/\Delta W^*_{\mathrm{hom},\,\Delta T_{\mathrm{fg}}}$ corresponding to $R_{\mathrm{Nuc}} = 1, 10$, and 100 nm. The inequality in Eq. (36) is satisfied for a given R_{Nuc} when the main curve in Figure 17 falls below the corresponding contour, and thermal nucleation is then likely. It can be seen that thermal nucleation is significant only at large dimensionless supercoolings when $\Delta T/\Delta T_{\mathrm{fg}}$ approaches one. It is more significant for smaller nucleant areas (as shown, e.g., in phase-field modeling [31]), but even for the smallest area considered, with $R_{\mathrm{Nuc}} = 1$ nm, thermal nucleation would be significant only for $\Delta T/\Delta T_{\mathrm{fg}} > 0.95$. In the real case of inoculation of liquid aluminum (Chapter 13, Section 3), the particles on which grain nucleation occurs have faces with R_{Nuc} typically not smaller than 1.5 μm. In such a case, thermal nucleation is significant only for $(1-\Delta T/T_{\mathrm{fg}}) < 5 \times 10^{-8}$. Thus it is fully justified to take the condition for grain initiation to be that given by Eq. (25). For micrometer-sized nucleant areas, thermal nucleation is negligible in advance of the free-growth condition (Eq. (25)) being met on cooling. Thus the effective nucleation of solid, at small supercooling on micrometer-sized nucleants, is a clear-cut example of an athermal process. It is governed by temperature change, and is completely deterministic, not stochastic [43].

The deterministic nature of athermal nucleation rules out any transient effects of the kind considered in Chapter 3. There is no nucleation rate under fixed thermodynamic conditions, and therefore no time-dependence of the rate. There are, however, kinetic effects on athermal nucleation, arising because the rapid motion of the metastable α–β interface that is required as the free-growth condition is approached may be hindered. The example of hindrance by solute partitioning in solidification has been analyzed and found to be not negligible [43].

So far, we have considered the nucleant areas to be plane circles; the interface between the nucleus and the original phase is then a spherical cap and is easily treated analytically. Numerical treatments can be applied to obtain the minimal-area interfaces for nuclei of given volumes forming on nucleant substrates of arbitrary shape [45]. Figure 18 shows nuclei at the onset of free growth formed on (a) a hexagon, (b) a square,

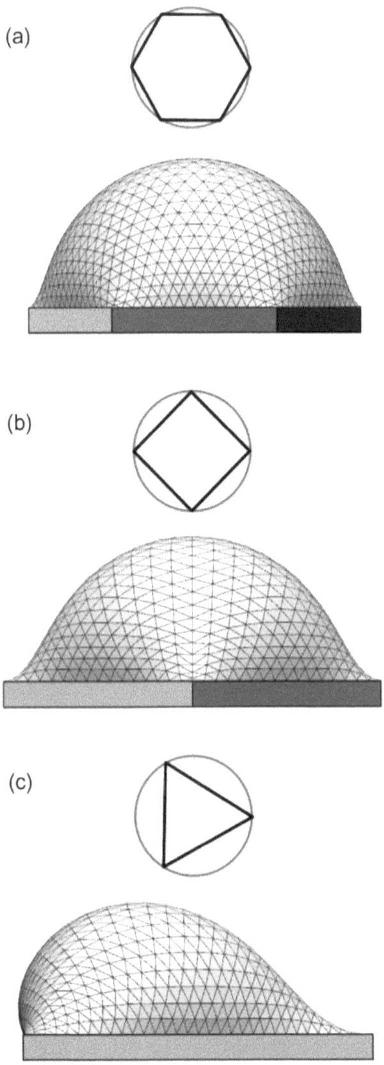

Fig. 18 The shape of the nucleus at the critical condition for the onset of free growth from polygonal substrates: (a) hexagon, (b) square, and (c) equilateral triangle. (From Ref. [45], with permission, Taylor & Francis Ltd. http://www.informaworld.com.)

and (c) an equilateral triangle. The free-growth supercooling (Eq. (25)) can be expressed more generally as

$$\Delta T_{fg} = -\frac{2\beta\sigma_{\alpha\beta}}{\Delta s R}, \tag{37}$$

where β is a coefficient that takes the value 1 for a plane circular substrate, but takes different values for other substrate shapes (planar or 3D). The characteristic linear dimension for substrates of arbitrary shape is taken to the radius R of the smallest sphere within which the shape can be inscribed. Values of β are given in Table 1. For planar substrates of given R, most notably for the equilateral triangle, the area is significantly less than that of the corresponding circle and β is significantly greater than 1. However, if substrates of equal area are considered, the free-growth supercooling on an equilateral triangle is only 5% greater than that on a circle. As shown by the comparison in Table 1, for polygonal substrates ΔT_{fg} can then reasonably be approximated by the value that would be obtained (from Eq. (25)) for a circular substrate of equal area.

While planar substrates are found in some cases of practical relevance (Chapter 13, Section 3), it is also important to treat nucleation on 3D substrates. An analytical treatment is straightforward for a spherical substrate with an adsorbed or wetting layer from which the β phase can grow. The work of formation of a nucleus in this case (Figure 10d) has already been given in Eq. (23). When plotted in dimensionless coordinates analogous to Figure 15, the curves show a maximum without a preceding minimum. The configuration of the dormant β phase is just the coating on the nucleant particle. Otherwise, the behavior is very similar to that in Figure 15, and ΔT_{fg} is the same as for a circular

Table 1 Values of β (to which the critical supercooling for the onset of free growth is proportional (Eq. (37)) for polygonal and polyhedral substrates

Substrate shape	Value of β in Eq. (37)	$\left(\frac{A(\text{polygon})}{A(\text{circle})}\right)_{\text{equal } R}$	$\left(\frac{\Delta T_{fg}(\text{polygon})}{\Delta T_{fg}(\text{circle})}\right)_{\text{equal area}}$
Circle	1	1	1
Hexagon	1.103±0.006	0.827	1.003±0.005
Square	1.270±0.012	0.637	1.013±0.010
Equilateral triangle	1.630±0.036	0.413	1.048±0.023
Sphere	1		
Cube	1.039±0.004		
Octahedron	1.094±0.006		

For the 2-D substrates, the relative areas for a given inscribing radius R and the relative free-growth supercoolings ΔT_{fg} for a given area are also given. (From Ref. [45], with permission, Taylor & Francis Ltd. http://www.informaworld.com.)

substrate of the same radius. The energy barrier for thermal nucleation decreases as $\Delta T/\Delta T_{fg}$ increases toward 1; it has a form very similar [46] to that shown in Figure 17. As for planar nucleant areas, stochastic behavior can dominate over deterministic athermal nucleation only for very small sizes. Numerical treatments of the α–β interface have been applied for β nuclei forming on polyhedral substrates [45]. Figure 19 shows the nucleant surfaces at the onset of free growth from a substrate in the form of (a) an octahedron and (b) a cube. As shown by the β values in Table 1, the free-growth supercooling on such shapes is slightly greater than it would be on the sphere within which they can be inscribed.

To conclude this section on athermal nucleation, we consider whether the phenomenon may correctly be regarded as *nucleation* at all. Dormant volumes of β phase formed on nucleant areas become transformation nuclei on cooling. In the athermal limit, the production of transformation

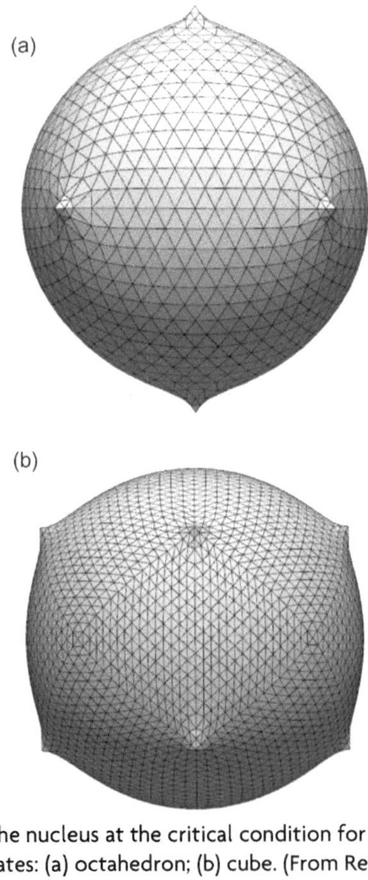

Fig. 19 The shape of the nucleus at the critical condition for the onset of free growth from polyhedral substrates: (a) octahedron; (b) cube. (From Ref. [45], with permission, Taylor & Francis Ltd. http://www.informaworld.com.)

nuclei is barrierless and deterministic, yet it is part of a continuum behavior in which the nucleation barrier shrinks as thermodynamic conditions change (Figure 15). Because it is part of a continuum involving the stochastic process of surmounting a barrier, the athermal limit can still reasonably be termed nucleation. The onset of homogeneous nucleation as conditions change is also because of a decreasing barrier. In contrast to heterogeneous nucleation, however, the prefactor $N(1)$ in Eq. (1) is so large that there is a detectable onset of nucleation while the barrier is non-negligible.

2.6 Interfaces in the solid state

The nature of the interface between one crystal and another depends on the structures, compositions, lattice parameters, and relative crystallographic orientations of the two crystals, as well as on the orientation of the interface between them. Even for grain boundaries, for which the first three factors are the same on each side of the boundary, there is a wide range of possible structures and associated energies and local atomic mobilities. The vast subject of crystal–crystal interfaces lies well beyond the scope of this book; an excellent discussion is given, for example, by Sutton and Balluffi [47]. For present purposes, we must note the basic types of interface. Interphase interfaces can be fully *coherent* when lattice planes are continuous through the interface. This is shown in Figure 20a for the simple case of a precipitate in a matrix with the same crystal structure. Where the lattice parameters in the two phases are different (as in Figure 20a), coherency can be maintained when the precipitate (or more generally the linear dimensions of the interface between the two phases) is small, but is likely to be lost as it grows. While coherency is most straightforward for phases with the same crystal structure, phases with different structures can still be coherent on particular crystallographic planes. When crystals have different structures, or even the same structure with a general misorientation, the interface between them is likely to be *incoherent*, that is, disordered, with no registry of the lattices across the interface (Figure 20b). Incoherent interfaces have much higher energy than coherent interfaces and, associated with this, the atomic mobility at incoherent interfaces is also much higher [48]. This means that the interface itself is more mobile and diffusion of solute atoms along the interface is faster. Many interfaces are *semicoherent*, with interfacial dislocations introduced into an otherwise coherent interface to lower the overall elastic strain energy (Figure 20c). Finally, it must be noted that many precipitate particles in a matrix are *partially coherent* in the sense that some faces are coherent, while others are not (Figure 20d, and examples in Chapter 14, Section 4.1).

Grain boundaries can be considered in a similar way. *Low-angle boundaries* (i.e., with small angles of crystallographic misorientation

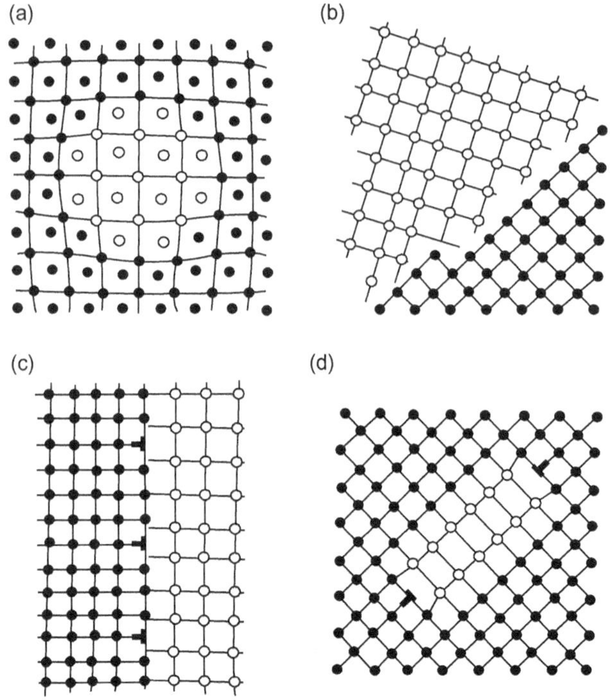

Fig. 20 Crystalline interphase interfaces: (a) coherent embedded particle; (b) incoherent; (c) semicoherent, showing interfacial dislocations; (d) partially coherent embedded particle, with some interfaces coherent, some semicoherent.

between neighboring grains) can be represented by arrays of dislocations in the boundary and have energies approximately proportional to the misorientation. *High-angle boundaries* have an energy that is high and approximately independent of misorientation, except near some special orientations where the energy plunges to low values.

For coherent interphase interfaces, interfacial energies have been estimated by summing structural and chemical contributions [49, 50]. The elastic strain energy of dislocation arrays provides a reasonable way of estimating the energies of low-angle grain boundaries. Otherwise, for incoherent interfaces or general high-angle grain boundaries, it is difficult to make simple predictions of interfacial energies. Values range from a few mJ m^{-2} for twin boundaries to approximately 500 mJ m^{-2} for incoherent interfaces, including high-angle grain boundaries [50]. An interface of high energy can be a potent nucleation catalyst because an embryo forming on the interface has its work of formation significantly reduced by the elimination of part of the original interface area. The coherency or incoherency of the interfaces between the embryo and the crystal on either side is also important, and not only because of the effect on interfacial energies. Incoherent interfaces are sinks for vacancies. Precipitation in alloys often occurs following a quench and in the presence of an excess vacancy concentration. Annihilation of vacancies at an incoherent boundary surrounding an embryo generates stresses that contribute an extra term (positive or negative, depending on the sign of the dilatational strain on transformation) to the driving free energy for the transformation [51].

It is clear that interfacial structures between crystals can be complex and interfacial energies may be highly *anisotropic* (dependent on orientation). In that case, it is no longer reasonable to take homogeneous nuclei to be spheres, or heterogeneous nuclei on planar substrates to be spherical caps. For a crystalline phase in isolation or embedded in a fluid, the Wulff construction can be used to find the equilibrium shape if the orientation dependence of the interfacial energy σ is known [52–54]. The use of this construction is illustrated in Figure 21a, showing a polar

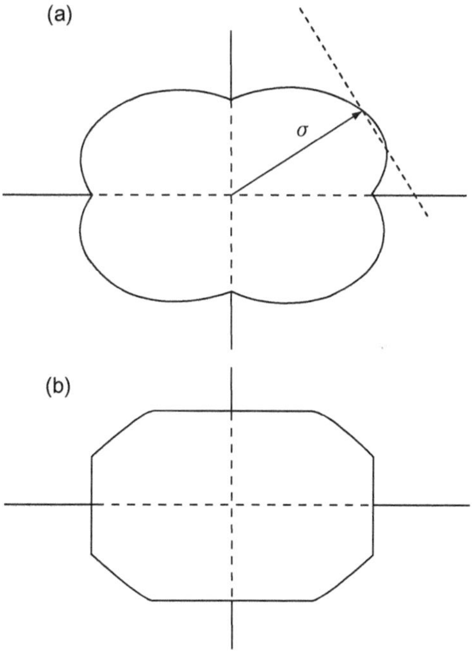

Fig. 21 The Wulff construction (shown here in two dimensions) for the equilibrium shape of a crystal *in vacuo* or in a fluid: (a) a polar plot of interfacial energy σ, showing low-energy cusps; (b) the resultant equilibrium shape, showing facets corresponding to the cusps, separated by nonfaceted interface. In (a), the dashed line represents the plane normal to the radial vector of length σ; the inner envelope of all such planes is the shape in (b). (Adapted from Ref. [50], copyright (1980), with permission from Elsevier.)

plot in which the distance from the origin to the surface in any direction is the magnitude of σ for an interfacial plane with that direction as its normal. Favored interfacial orientations then correspond to low-energy cusps. Taking a line from the origin, a plane perpendicular to the line is constructed through the point where the line intersects the σ surface. The inner envelope defined by all such planes defines the equilibrium shape. Interestingly, that shape (as shown in Figure 21b) may have both facets and curved surfaces. While the Wulff construction is easy to apply to find the equilibrium shape of a free surface or an interface in contact with a fluid, there is extra complexity when the external phase is crystalline, and therefore itself anisotropic. The complexity is still greater when the particle of interest (an embryo or nucleus in the present case) sits on a boundary between two crystals. Equilibrium shapes in this case can be derived using an extension of the Wulff approach. This topic has been treated by Cahn and Hoffmann [55, 56] and Lee and Aaronson [57, 58]. A compact summary of their conclusions has been presented by Russell [50].

The coherency of interfaces affects their mobility. Nuclei, however, are close to equilibrium and therefore unaffected by anisotropic interfacial mobilities. This is not true for growth, with the important consequence that growth morphologies are not a good guide to the shapes relevant in analyzing nucleation [59].

So far in this book, we have treated interfacial energies per unit area, and interfacial tensions (with the same units, $\mathrm{N\ m^{-1} \equiv J\ m^{-2}}$) as identical. This is true for interfaces between fluids, but for a solid phase at an interface it is important to distinguish between *interfacial energy* and *interfacial stress* (often termed the surface stress). The interfacial free energy is the reversible work to create unit area of fresh interface on the unstrained solid; the interfacial stress is the reversible work to extend, by elastic deformation of the solid, the area of an existing interface [60]. Often interfacial energies and stresses can have similar magnitudes, but they can even be of opposite sign [61]. The formation of an embryo involves creation of fresh interface, and the work of formation therefore directly involves the interfacial energy, not the interfacial stress. Nevertheless, the interfacial stress can affect the stress state within nuclei, with significant effects on nucleation kinetics in highly compressible systems [62] and on, for example, the stresses in thin films formed by the coalescence of islands [63].

To examine nucleation on a crystal–crystal interface, we consider first the simple geometry when all interfaces are incoherent and nonfaceted. An embryo of β forms at a disordered high-angle grain boundary in the α phase. It has no special orientation relationship with either grain; the two α–β interfaces then have similar energies $\sigma_{\alpha\beta}$, leading to a symmetrical embryo with a lenticular shape (Figure 22a). Representing the grain-boundary energy as $\sigma_{\alpha\alpha}$, the contact angle ϕ is given (ignoring any line

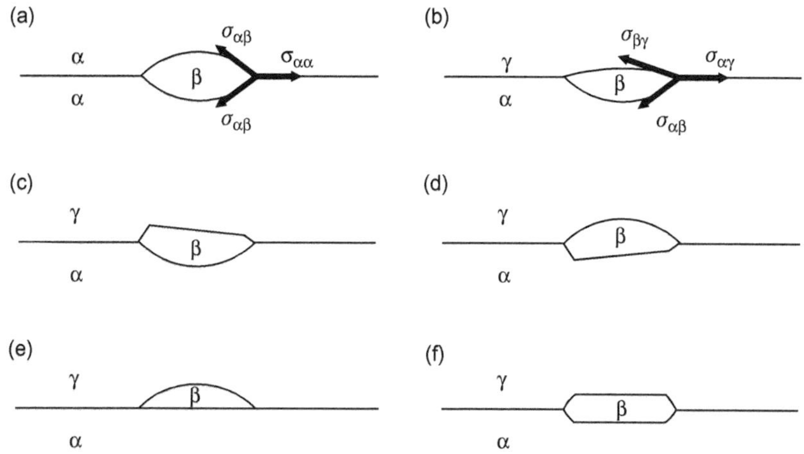

Fig. 22 Possible configurations of a β-phase embryo: (a) on an α-phase grain boundary with an incoherent interface with each grain; (b–f) on an α-γ interphase boundary, with (b) an incoherent interface with each phase, (c) a coherent facet with γ; (d) a coherent facet with α, (e) a coherent interface with α, with $\sigma_{\alpha\beta} < 0.5\sigma_{\alpha\gamma}$; and (f) on a coherent α–γ interface and coherent with both α and γ.

tension) by

$$\sigma_{\alpha\alpha} = 2\sigma_{\alpha\beta}\cos\phi. \tag{38}$$

The critical radius of curvature of the α–β interfaces is still given by Eq. (12) so that the work of formation of the lenticular nucleus is exactly twice that for a spherical cap (Eqs. (9) and (13)) with the same ϕ:

$$W^*_{\text{het}} = W^*_{\text{hom}} 2f(\phi). \tag{39}$$

The nucleation rate is given by an expression of the form in Eq. (1). As for the spherical cap (Section 2.2), the Zeldovich factor Z is that for the homogeneous case multiplied by the factor $1/\sqrt{f(\phi)}$ (Eq. (9)). When molecular transport to and from the embryo is predominantly through the two α–β interfaces, $Zk^+(n^*)$ is therefore that for the homogeneous case multiplied by $2\alpha(\phi)/\sqrt{f(\phi)}$, exactly twice the factor for the spherical cap Eq. (14). It is alternatively possible that the transport is predominantly along the grain boundary itself. This may well be the case for precipitation of solute-rich particles on grain boundaries, when the boundaries are fast transport paths for the solute [64]. Then $k^+(n^*)$ depends on the solute adsorption on the boundary [24], and the effective diffusivity in Chapter 2, Eq. (30) is the grain-boundary diffusivity. As shown by Russell [50], for nucleation of B-rich β phase from a matrix α containing a mole fraction X_α of B atoms, the factor $Zk^+(n^*)$ is given roughly by:

$$Zk^+(n^*) \approx \frac{D_{\text{gb}}X_\alpha}{\lambda^2}, \tag{40}$$

where D_{gb} is the grain-boundary diffusivity of the solute B, and λ is the atomic diameter. For deriving the nucleation rate per unit area of the boundary, the prefactor N_B, representing the number of molecules per unit area in contact with the boundary, is of order $2\lambda N_V$, where N_V is the number of molecules per unit volume in the system. It may also be appropriate to derive the nucleation rate per unit volume. In a polycrystal of average grain diameter d, the fraction of atoms in contact with a grain boundary is of order $10\lambda/d$ so that the prefactor $N_{V,\,gb}$ is given approximately by

$$N_{V,\,gb} = \frac{10N_V\lambda}{d}. \tag{41}$$

Further detail on nucleation kinetics at grain boundaries, including an analysis of transient times, has been given by Russell [65].

We now turn to consider embryo shapes less symmetric than that shown in Figure 22a. If the interface on which the β phase nucleates is between two phases (α and γ in Figure 22b), rather than between two grains of the same phase, then differing interfacial energies $\sigma_{\alpha\beta}$ and $\sigma_{\gamma\beta}$ imply an asymmetric lens shape to balance the interfacial tensions. Whether at an interphase interface or a grain boundary, the new phase β may have an orientation relationship with the crystal on one side of the boundary (Figure 22c), or on the other side (Figure 22d), permitting a low-energy facet to form at which the interface is coherent. The equilibrium shape of β can be derived using an extension of the Wulff procedure outlined earlier [55–58]. The work of formation of such an embryo, and the associated nucleation kinetics, have been considered in detail by Johnson et al. [66]. Lee and Aaronson [58] have shown that the work of formation of the critical nucleus increases as the misorientation angle between the facet and the boundary plane increases. As the energy per unit area of the facet is lowered relative to the other interfacial energies, it occupies more area; when the energy per unit area of the facet is $\leq 50\%$ of that of the grain boundary or interphase interface on which nucleation is centered, it is energetically favorable for the facet to lie in the plane of the interface [67], giving a simplified spherical-cap geometry (Figure 22e).

Examples of nucleation from metallurgical practice are given in Chapter 14, Section 4. These include cases of grain boundaries being favorable and unfavorable sites for nucleation.

Nucleation on interphase boundaries in most cases is similar to nucleation on grain boundaries, but it presents further possibilities. In early stages of precipitation, the phase that forms is often not the equilibrium one, but rather a metastable phase selected because it can be fully or partially coherent with the matrix, thereby reducing interfacial energy and the work of nucleation (precipitation sequences are considered further in Chapter 14, Section 4). The coherent interface between such a phase (γ in Figure 22f) and the matrix α is then a special site that permits the β phase nucleating on it to have coherent interfaces on both sides, a highly favorable configuration [66, 68, 69].

So far we have considered that interfaces may be favorable sites for nucleation because of their excess energy. However, the local structure at an interface, differing from the crystal or grain on either side, may resemble the structure of a phase about to nucleate. In that case, the interfacial configurations may act as a *template* for nucleation of the new phase. This possibility has been considered in molecular-dynamics simulations (see Chapter 10 for a discussion of such methods) by Palko et al. [70], in which templated nucleation of a hexagonal polymorph is predicted at symmetrical (310) tilt grain boundaries in cubic NaCl and LiCl. The thickness of the metastable phase formed at such a boundary is controlled by the mechanical strain normal to the boundary, and can include several atomic layers in which the rings characteristic of the hexagonal phase are evident (Figure 23).

The spatial arrangement of interfaces in the solid is also important. Grain boundaries, for example, exist as part of a grain structure, involving not only the boundaries themselves, but also grain edges (where three grains meet) and grain corners (where four grains meet). The consequences for nucleation can be most easily considered when the nucleating phase is always incoherent with the matrix phase. On an edge, an embryo is bounded by three surfaces of uniform curvature (Figure 24a). At a corner, such an embryo has a tetrahedral shape with four surfaces of uniform curvature (Figure 24b). In each case, the critical

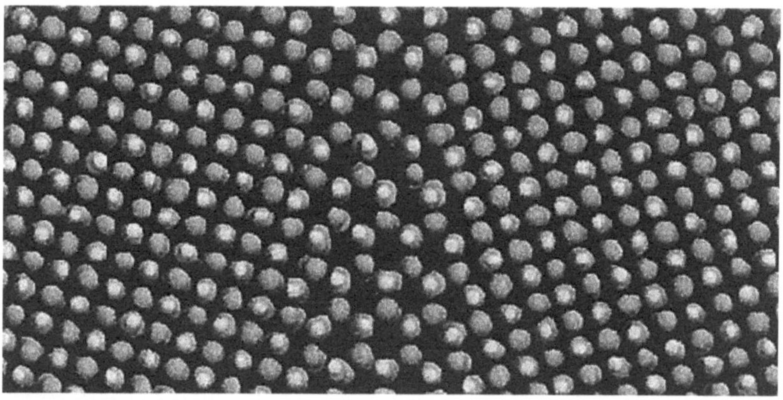

Fig. 23 The structure, predicted from molecular-dynamics simulations, of a symmetrical (310) tilt grain boundary in NaCl under a stress normal to the boundary. The six-membered rings on the boundary can act as a template for nucleation of a metastable hexagonal structure. (Reprinted in part with permission from Ref. [70], copyright (2004) American Chemical Society.)

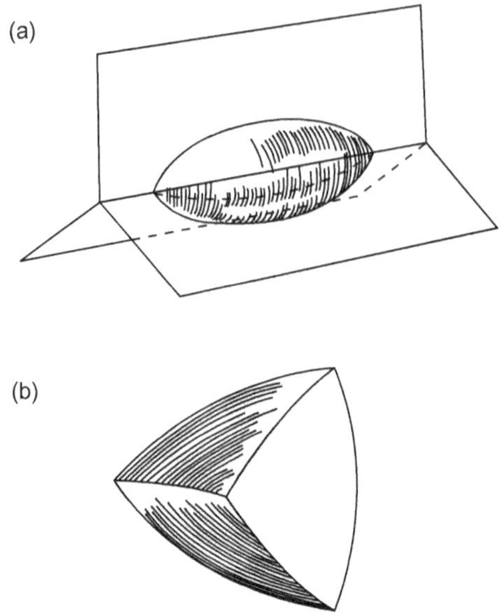

Fig. 24 The shape of a β-phase embryo having incoherent interfaces with the surrounding α-phase grains: (a) three-sided at a grain edge; (b) four-sided at a grain corner. (Reprinted from Ref. [50], copyright (1980), with permission from Elsevier.)

work of nucleation can be derived with appropriate geometrical factors that are functions of the contact angle ϕ [71]. For $0<\phi<\pi$, the work of forming a heterogeneous nucleus is less than that for the homogeneous case, an important contribution to the reduction coming from the energy of the eliminated area of the original boundary or boundaries. (And, strictly, although not considered in the classical treatment [71], the line and point energies of grain edges and corners should also be considered.) For a given contact angle, the work of formation of the critical nucleus decreases in the order: homogeneous nucleation, heterogeneous nucleation on a boundary, on an edge and at a corner. As will be shown in Section 5, this does not imply, however, that transformations in crystalline solids will predominantly be nucleated on grain corners.

Throughout Section 2, we have considered nucleation on interfaces that are in metastable equilibrium. The interface is not migrating, and there are no net diffusional fluxes across it. Nonequilibrium features such as these lie beyond the coverage of this chapter, but they are important in interfacial reactions and are treated in Chapter 15. Grain boundaries are considered again in Chapter 15, Section 2, where it is noted that reduction of grain-boundary area is not the only reason why they can be preferred sites for nucleation.

3. NUCLEATION ON DISLOCATIONS

In this chapter, we have already met (in Section 2.6) line defects in the form of grain edges, but these were not assigned a specific line energy. The line defects of overriding importance in crystalline solids are *dislocations*. A description of dislocations and their interactions is beyond the scope of this book; the reader is referred to standard texts [72, 73]. Some crystal–crystal phase transformations are mediated by dislocations on the interface between the two phases. In such cases, the interfacial dislocations can play a central role in the nucleation of the new phase; the example of the nucleation of martensite is covered in Chapter 12, Section 4.4. In the present section, we are concerned instead with dislocations within an original phase, and their role in catalyzing the nucleation of a new phase that does not form by a displacive or martensitic mechanism. An example is given in Figure 25; in this case gold-rich plate-like precipitates nucleate within an iron alloy only at dislocations lines because high elastic strain energy prohibits nucleation elsewhere [74]. Earlier reviews of this topic [50, 75] give more detail on some of the points covered next. The nucleation of dislocations is considered in Chapter 12, Section 2.

Dislocations within a phase lie on crystallographically defined slip planes and can be regarded as boundaries between regions on a given plane that have slipped relative to each other. The magnitude and direction of the

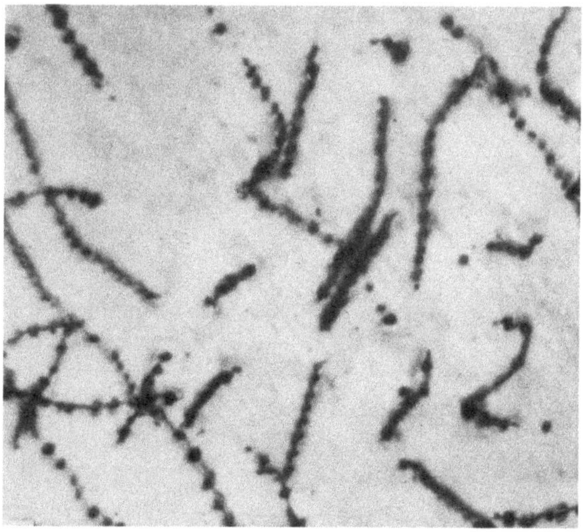

Fig. 25 Transmission electron micrograph of an α-Fe–1at%Au alloy. The supersaturated solid solution has been annealed at 500°C for 24 h to give precipitate particles on dislocations. (Reprinted from Ref. [74], copyright (1969), with permission from Marcel Dekker.)

relative slip are described by the Burgers vector; in the following treatment, we use only the magnitude b. The line vector of the dislocation and its Burgers vector both lie in the slip plane. The Burgers vector can lie at any angle to the line vector, but two special cases are distinguished: when the vectors are parallel, the dislocation is of *screw* type; when they are orthogonal, the dislocation is of *edge* type (Figure 26). The latter can be envisaged as the edge of a partial plane in the solid, and in that case near the dislocation line the strain has a net dilatational component that is compressive where the extra plane exists and tensile where it does not. In Figure 26a, these regions are respectively above and below the line, at A and B. However, the strain components around an edge dislocation are predominantly shear, and around a screw dislocation almost exclusively so. While the overall strain state is complex, all the components ε take the form

$$\varepsilon \propto \frac{b}{y}, \tag{42}$$

where y is the distance from the dislocation line. Equation (42), derived from linear elasticity theory, obviously cannot apply as $y \to 0$. A cylinder of small y is defined to be the *dislocation core*, a disordered region a few atomic spacings in diameter. Outside the core, Eq. (42) shows that the strains decay relatively slowly, and the main conclusions are that the total elastic strain energy in the matrix may be large and that effects can be felt at distances from the dislocation line much greater than the core radius. It is useful to consider that an isolated dislocation has an energy per unit length, the *self-energy*, but this is far from straightforward to evaluate. It is the sum of the energy of the core and of the strain energy in the matrix. To evaluate the latter, the inner and

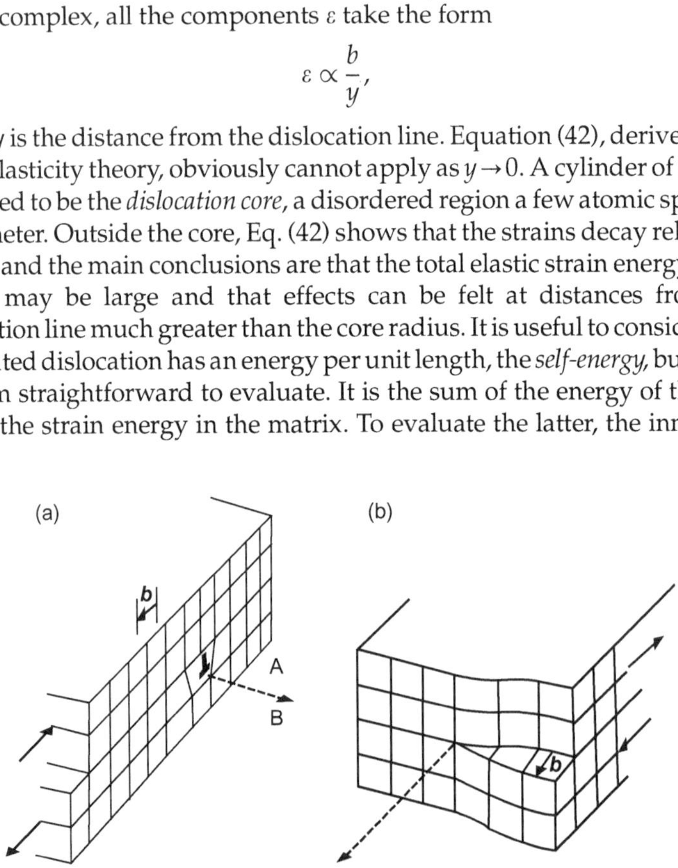

Fig. 26 Schematic illustrations of (a) an edge and (b) a screw dislocation in an idealized simple cubic lattice. The Burgers vector ***b*** and corresponding relative slip in the crystal are shown in each case. Above an edge dislocation at A, in region of the extra plane, the net dilatational strain is compressive; below the line at B, it is tensile. (Adapted from Ref. [50], copyright (1980), with permission from Elsevier.)

outer radii (values of y), between which the integration of the strain energy is performed, have to be chosen somewhat arbitrarily. Overall, however, the energy per unit length of a dislocation line, equivalent to a line tension Λ is given roughly by

$$\Lambda = \frac{\mu_s b^2}{2}, \tag{43}$$

where μ_s is the elastic shear modulus. The core energy is of order 10% of the total energy. It is clear that removal of a segment of dislocation line by forming a new phase would favor the nucleation of that phase. However, the strain fields are such that nucleation can also be favored near, but not on, a dislocation line. The nature of the interfaces between the original and new phases (incoherent, semicoherent, or coherent) is important. We first consider incoherent interfaces. A special case of an incoherent interface is that between a solid and its melt. Dislocations may play a crucial role in the internal nucleation of melting in crystalline solids [76], a topic covered in Chapter 14, Section 3.

We take the case of a cylinder of new phase β forming along a dislocation line in the original phase α. If the interface between the two phases is incoherent, then the lack of lattice registry means that the volume occupied by the new phase is completely free of the strains associated with the former dislocation. This case was analyzed by Cahn [77]. The work W_{cyl} to form a unit length of cylinder of radius r is

$$W_{cyl} = \pi r^2 \Delta g + 2\pi r \sigma_{\alpha\beta} - U_{core} - U_B \ln(r/r_{core}), \tag{44}$$

where Δg is the free-energy change (per unit volume) driving the transformation, $\sigma_{\alpha\beta}$ is the energy (per unit area) of the interface between the two phases, U_{core} is the core energy of the dislocation, and the last term is the elastic strain energy in a cylindrical shell between the outer radius of the core r_{core} and r. (If there was a volume change on forming the β phase, then the effect would be to decrease the magnitude of Δg by adding the associated strain energy as shown in Chapter 9.) The quantity U_B is $\mu_s b^2/4\pi$ for a screw dislocation and $\mu_s b^2/4\pi(1-\nu_P)$ for an edge dislocation, where ν_P is the Poisson ratio. The last two terms in Eq. (44) make up the energy of the dislocation line eliminated by the formation of the cylinder. Differentiating Eq. (44) with respect to r, and setting $dW_{cyl}/dr = 0$, the critical radius of the cylinder r^* is found to be [77]:

$$r^* = -\frac{\sigma_{\alpha\beta}}{\Delta g}(1 \pm (1 - \Delta_{cyl})^{1/2}), \tag{45}$$

where

$$\Delta_{cyl} = -\frac{2\Delta g U_B}{\pi \sigma_{\alpha\beta}^2}. \tag{46}$$

If $\Delta_{cyl}<1$, Eq. (45) can be solved, but there are two roots, corresponding to a minimum and maximum in W_{cyl} versus r (Figure 27, curve A). If $\Delta_{cyl}>1$, there is no solution for a critical radius r^* (Figure 27, curve B). The lack of any nucleation barrier in the latter case arises from the large negative contribution that the energy of the eliminated dislocation line makes to W_ℓ (Eq. 44). The forms of the curves in Figure 27 are similar to those in Figure 15, with Δ_{cyl} analogous to the reduced supercooling $\Delta T/\Delta T_{fg}$. As the driving force for transformation is increased, athermal nucleation occurs when $\Delta_{cyl}=1$. For $\Delta_{cyl}<1$, there is a cylinder, with radius r_0, of dormant β phase around the dislocation line (Figure 28). In this condition, thermal fluctuations can generate critical transformation nuclei, and Cahn [77] has shown that the rate of nucleation is significant only when $\Delta_{cyl}>0.4$–0.7. He showed that the critical nucleus would take the form of a bulge in the metastable cylinder of radius r_0 (Figure 28), and integrated Δ_{cyl} (Eq. 44) along the length L of the nucleus to obtain the dependence of its critical work of formation W^*_{cyl} on Δ_{cyl} (Figure 29). The high-energy region associated with a dislocation line has a radius much smaller than those of the circular areas considered in Section 2.5; consequently, thermal nucleation on dislocations can predominate over athermal nucleation, in contrast to the cases in Section 2.5.

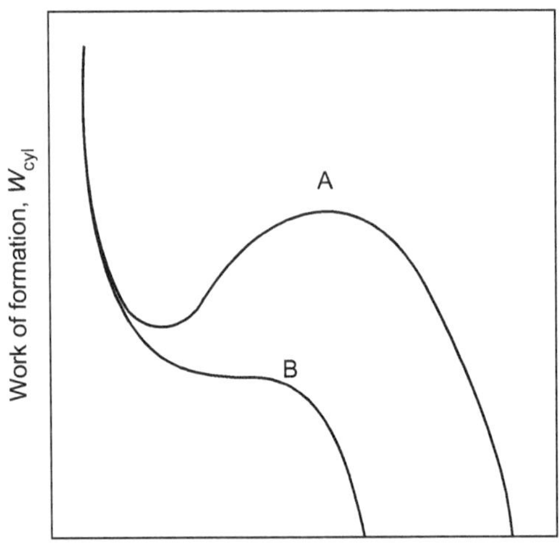

Fig. 27 The work of formation W_{cyl} of a cylinder of radius r of new phase β around a dislocation line in the original phase α. The α–β interface is incoherent and W_{cyl} is given by Eq. (44). When the driving force for the α to β transformation is low, there is an energy barrier for nucleation (A); when the driving force is high, there may be no barrier (B). (Adapted from Ref. [77], copyright (1957), with permission from Elsevier.)

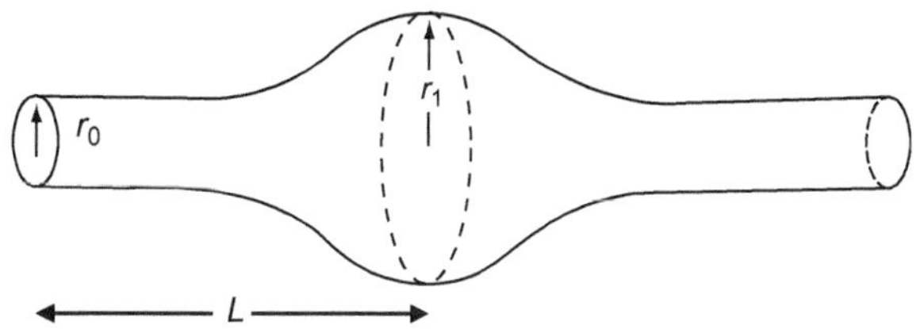

Fig. 28 A tube of new phase β formed along a dislocation line in original phase α has a metastable radius r_0 corresponding to the minimum in curve A in Figure 27. Thermal fluctuations over length L can cause the tube radius to reach the critical value r_1 corresponding to the maximum in curve A. This bulge in the tube is a critical nucleus for the new phase. (Adapted from [77], Copyright (1957), with permission from Elsevier.)

Cahn's analysis assumes that the strain field in the matrix outside the new phase is completely unaffected by the formation of the phase. A more thorough calculation, by Gomez-Ramirez and Pound [78], of the strain energy changes accompanying nucleation gives a work of formation very similar to that derived by Cahn. They showed that nuclei on screw dislocations are likely to be prolate ellipsoids of revolution with axial ratio of approximately 0.83. On edge dislocations, the shape is similar, but the cross-section (perpendicular to the dislocation line) is not circular as before, but heart shaped, arising from the tensile/compressive asymmetry around the extra partial plane.

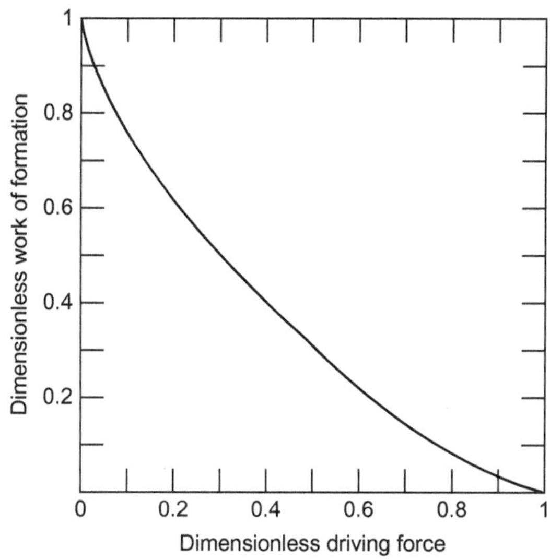

Fig. 29 The work of formation W^*_{cyl} of the critical nucleus shown in Figure 28, normalized with respect to the work for homogenous nucleation, as a function of the parameter Δ_{cyl} (Eq. (46)) that is proportional to the driving force for the transformation. (Reprinted from Ref. [77], copyright (1957), with permission from Elsevier.)

The analysis of Gomez-Ramirez and Pound does not predict a continuous cylinder, radius r_0, of metastable new phase along the dislocation line, but rather a string of discrete metastable embryos. These embryos can remain metastable even when Δg is positive, and they can act as athermal nuclei when the conditions are changed sufficiently rapidly to avoid the intervention of thermal nucleation [78, 79].

The rate of dislocation-catalyzed nucleation within a solid phase is the product of the length of dislocation line per unit volume (i.e., the dislocation density ρ_ℓ) and the nucleation rate per unit length of dislocation. Typical values of ρ_ℓ can vary widely, from 10^{10} m^{-2} for an annealed metal to 10^{16} m^{-2} for a heavily work-hardened metal. While similar densities can be found in ceramics, in carefully processed semiconductors ρ_ℓ may be effectively zero. Thus dislocation density may be a major factor in exercising control over nucleation in solids, and some examples feature in Chapter 14, Sections 4 and 5. The nucleation rate along a dislocation line is proportional to the number of atoms effectively in contact with the line. This is often taken to be of the order of five atoms per atomic diameter along the line [75]. Thus the first term in the prefactor for the nucleation rate per unit volume on dislocations is given by

$$N_{V,\ell} \approx \frac{5\rho_\ell}{\lambda}, \tag{47}$$

where λ is the atomic or molecular diameter. Russell [50] estimates that $N(1)$ for nucleation on dislocations in an annealed metal is typically 10^{-9} times that for homogeneous nucleation. The remaining part of the prefactor, $k^+(n^*)Z$ (Eq. (1)), is of the order of the atomic jump frequency, and it should be noted that for nucleation on dislocations, it may not be a good approximation to take this to be the jump frequency within the lattice of the original phase. In particular, if a solute is precipitating out of solution, the dislocation itself is expected to act as an effective fast transport path for the solute. The effective jump frequency is then that associated with diffusion along the dislocation line [50].

We now turn to consider the nucleation of coherent precipitate particles at dislocations. With lattice continuity between the matrix and the particle nucleating within it, a dislocation line would not be severed by a particle forming on it. However, since the self-energy is proportional to the shear modulus μ_s (Eq. (43)), a precipitate phase with μ_s lower than that of the matrix would still lower the energy of the dislocation. Treatments of this case [80, 81] assume that the strain field inside the embryo is identical to that inside the matrix it replaced. As pointed out by Christian [75], this is not a good approximation. Qualitatively it remains clear, however, that for coherent precipitates, dislocations are favored sites for nucleation of more compliant phases, but suppress the nucleation of elastically stiffer phases.

Apart from the effect of differing μ_s in original and new phases, dislocations can influence the nucleation of coherent precipitates because of the strain field associated with the precipitate. This has mostly been considered for purely dilatational transformation strains. As already noted, near an edge dislocation there are regions in which the matrix is in compression or tension. If a precipitate phase has a stress-free lattice parameter less than that of the matrix, the work of forming a critical nucleus is lowered in the region of the matrix under compression; conversely, nucleation of a precipitate phase with a lattice parameter larger than that of the matrix is favored in the tensile region of the matrix. In this way, nucleation is generally favored near the dislocation line, but not on it [80, 82, 83]. Larché [83] has shown that the catalytic effect of the dislocation line can be significant. Catalytic effects of this kind are restricted to edge dislocations. The shear strains around both edge and screw dislocations may, however, influence the nucleation of any phase when the transformation strain has a significant shear component. As pointed out by Christian [75], this would apply for nucleation of martensite (Chapter 12, Section 4). In general, it seems that for a new phase crystallographically coherent with the matrix in which it forms, dislocations can facilitate nucleation of coherent phases, but cannot reduce the nucleation barrier to zero (as may be true for the incoherent case, Figure 27, curve B) [75].

When one crystalline phase precipitates within another, it is often the case that some of the faces of the precipitate are coherent with the matrix, while others are incoherent. The implications of this at the nucleation stage have been considered qualitatively by Russell [50]. For example, a tetragonal phase forming in a cubic matrix might be coherent only on its {001} faces. In homogeneous nucleation, the nuclei are then expected to be disk-like, with large {001} faces and a narrow, incoherent periphery. If it is assumed that the *c* lattice parameter of the precipitate is larger than the lattice parameter of the matrix (i.e., the transformation strain involves a positive dilatation), then it would be favorable for the precipitate to nucleate in the tensile region near an edge dislocation. There is a further energetic benefit if part of the incoherent periphery of the particle lies along the dislocation line; in this way, a segment of line is, in effect, removed. The expected precipitate geometry is shown in Figure 30.

Last in this section, we note that dislocations may dissociate into partial dislocations separated by a *stacking fault* [72, 73]. In a cubic-closed packed metal, the stacking fault is a four-monolayer-thick ribbon of hexagonal close packing (hcp). This is a planar defect of low interfacial energy, but it can be a potent nucleant [50]. In particular, the local packing may act as a template for nucleation of an hcp phase, analogous to the effect discussed in Section 2.6 for grain boundaries (Figure 23).

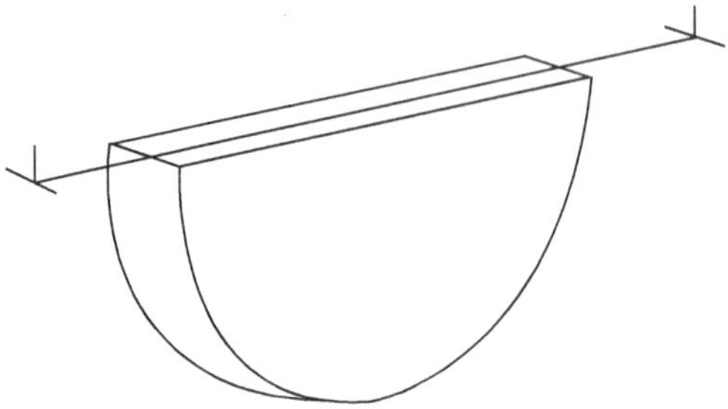

Fig. 30 A possible shape for a precipitate particle, semicoherent with the matrix, nucleating on an edge dislocation. The large faces of the particle are coherent with the matrix. The periphery is incoherent and part of it lies along the dislocation line. For the case shown, the dilatation strain is positive, and the precipitate forms in the tensile region below the line. (Reprinted from Ref. [50], copyright (1980), with permission from Elsevier.)

4. NUCLEATION ON ATOMIC-SCALE HETEROGENEITIES

Having surveyed nucleation on *planar* heterogeneities (Section 2) and *linear* heterogeneities (Section 3), we turn now to the role of *point* heterogeneities, taking these to be of atomic dimension. We have seen in Section 2.4 that a smaller nucleating substrate has a weaker catalytic effect for nucleation (Figure 11), and that the effect is expected to be negligible when the size of the substrate is less than that of the critical nucleus. It would then seem that there could be no catalytic effect of a heterogeneity with the size of one atom. However, we have seen in Section 3 that the atomic-scale discontinuity at a dislocation line gives rise to a long-range strain field, and that this is the basis for catalysis of nucleation. Analogous long-range effects permit atomic-scale point heterogeneities to be potent nucleation catalysts.

The role of strain fields can be discounted in fluids, but there is a strain field around a misfitting atom (or a vacancy) in a crystalline solid solution. This can be regarded as part of the driving force for solute atoms to cluster and ultimately form precipitates of a new phase. However, the energy of the strain fields around individual solute atoms is appropriately considered as part of the free energy of the solid solution [75] and is, therefore, already included in the chemical driving force Δg. Furthermore, the strain field around a misfitting atom has a zero dilatational component [75], and so does not bias the diffusional jumps of neighboring misfitting solute atoms. Thus elastic strain does not provide the long-range effect needed for nucleation catalysis by point heterogeneities.

That effect can, however, be provided by electric charge. It has long been recognized that condensation of supersaturated vapors is greatly accelerated in the presence of ions, an effect reviewed and analyzed by

Thomson [84]. Although condensation from a vapor lies outside the focus of this book, we now look further at the catalysis by ions because of the principles thereby illustrated. In addition, the condensation of water droplets on trails of ions caused by radiation (readily studied using a *cloud chamber* [85, 86]) is a phenomenon of importance in a wide range of fields from meteorology to particle physics. Thomson [84] noted that in the presence of an electric field, the electrostatic energy is lowered by the polarization of a dielectric medium. If the embryo is thus polarizable, its energy is reduced and nucleation is facilitated. We now consider this effect further for droplets nucleating in a supersaturated vapor. When a singly charged ion of radius r_{ion} is coated by a spherical shell with outer radius r, of a medium of dielectric constant ε_{κ}, the electrostatic energy U_{es} is changed by [87, 88]

$$\Delta U_{\text{es}} = \frac{e^2}{2}\left(1 - \frac{1}{\varepsilon_{\kappa}}\right)\left(\frac{1}{r} - \frac{1}{r_{\text{ion}}}\right), \tag{48}$$

where e is the electronic charge. The change ΔU_{es} is so negative for small $(r - r_{\text{ion}})$ that it is always favorable for the ion to be coated with a thin layer of the condensing phase (Figure 10d). The work of formation $W_{\text{ion}}(r)$ of a droplet of radius r centered on the ion is then given by Eq. (23); setting $R_{\text{Nuc}} = r_{\text{ion}}$ and including ΔU_{es}, we have

$$W_{\text{ion}}(r) = \frac{4\pi}{3}\Delta g(r^3 - r_{\text{ion}}^3) + 4\pi\sigma(r^2 - r_{\text{ion}}^2) + \frac{e^2}{2}\left(1 - \frac{1}{\varepsilon_{\kappa}}\right)\left(\frac{1}{r} - \frac{1}{r_{\text{ion}}}\right), \tag{49}$$

where Δg is the free-energy change per unit volume of condensed phase, and σ is the energy per unit area of the droplet surface. Figure 31 is a schematic plot of Eq. (49): at moderate supersaturation (curve A), the work of formation has a minimum at which the droplet is in metastable equilibrium with the vapor, as well as a maximum corresponding to unstable equilibrium at the critical nucleation radius r^*. The critical work of nucleation ΔW_{ion}^* is given by the difference in energy between the two equilibrium conditions. At larger supersaturation, the minimum and maximum converge, and the barrier to nucleation, may disappear (curve B). This form of behavior is analogous to those for nucleation on a limited area of substrate (Figure 15) and on a dislocation (Figure 27). In each case, the origin is the same: the heterogeneity provides an increasingly negative contribution to the work of formation at small embryo dimensions, while the free energy driving the transformation, scaling with the embryo volume, provides a dominant negative contribution at large sizes; in between, the positive interfacial energy makes the predominant contribution to $\mathrm{d}W_{\text{ion}}(r)/\mathrm{d}r$.

The detailed analysis of condensation kinetics lies beyond our scope. We assume that the catalytic effect of the ion lies in its reduction of the

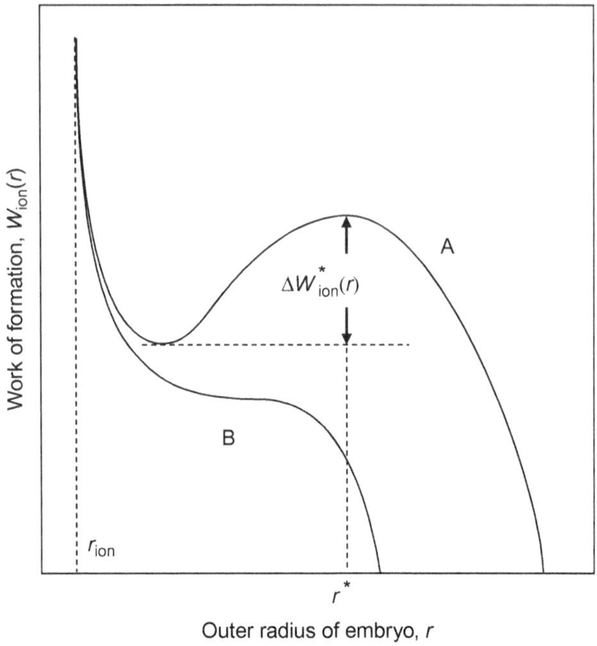

Fig. 31 For condensation from a vapor, the work of formation of an embryo on an ion of radius r_{ion} as a function of the outer radius r of the embryo. At moderate supersaturation (A), there is a barrier to nucleation at critical radius r^*, the critical work of nucleation $\Delta W^*_{ion}(r)$ being determined as shown. At higher supersaturation (B), the nucleation barrier can disappear.

work of nucleation (Eq. (49)), without any significant effect on the kinetics of molecular attachment to an embryo. The nucleation rate is given by an equation of the form of Eq. (1). The prefactors $k^+(n^*)$ and Z are little affected by the presence of the ion, and it follows that the nucleation transient time τ is also only weakly affected [88].

The analysis for condensation on an ion (Eq. (49)) would work equally well for nucleation on a larger nucleant particle, but the catalytic effect would decrease with increasing particle radius. The maximum contribution of electrostatic energy to lowering the work of nucleation (seen as the embryo radius $r \to \infty$) is inversely proportional to r_{ion} (from Eq. (48)). On the other hand, a nucleant particle could carry a larger charge than a single ion, and ΔU_{es} is proportional to the square of the net charge.

There is a striking feature of ion-catalyzed condensation that is not accounted for by the analysis presented so far: the catalytic effect can depend on the sign of the ionic charge. Ions of either sign catalyze the nucleation, but for many liquids one sign is much more effective than the opposite. For water, negative ions are by far the better catalysts; for ethanol, the reverse is the case. Thomson [84] reviewed these effects and explained them in terms of an electrified double layer at the droplet surfaces where, for *polar liquids*, the molecules are oriented to give a net dipole. For water (as can be demonstrated by positive charge acquisition on bubbling a gas through it) the dipole has the negative charge on the outside, and it is thus oriented parallel to the polarization induced by a negative ion at the droplet center. For a positive ion, the reversal of the natural surface dipole raises the surface energy, an effect that impedes nucleation. For ethanol, in contrast, the surface dipole has the positive charge on the outside, and the relative effectiveness of negative and positive ions is reversed.

In a cloud chamber, tracks of particles are revealed by condensation on the resulting trails of ions. In a *bubble chamber* (capable of faster operation than a cloud chamber), there is a similar phenomenon in which bubbles are nucleated along particle tracks in liquid hydrogen [86, 89]. We note briefly, however, that the nucleation mechanism in the latter case is completely different. In the cloud chamber, the catalytic effect of the ions relies on the condensed liquid having a higher dielectric constant than the original vapor. Evidently, in a bubble chamber the nucleating phase has the lower dielectric constant; as pointed out below (in considering the influence of external fields), the higher polarization of the surrounding liquid would *inhibit* the nucleation of a bubble on a point charge. It has been suggested that the presence of several charges on a bubble surface could aid its nucleation by their mutual repulsion, but this has not been found to give a satisfactory explanation of bubble formation. Rather, it appears that the stimulation of bubble formation along particle tracks is due to a heat spike locally increasing the superheating of the liquid [86].

Although outside the scope of our immediate focus on point heterogeneities, it is convenient to note here the effects of an externally applied electric field on nucleation. We consider an embryo of phase β forming homogeneously in a matrix of phase α. Analogous to the condensation case already considered, if β has a higher dielectric constant greater than α (i.e., $\varepsilon_{\kappa,\beta} > \varepsilon_{\kappa,\alpha}$), then the electrostatic energy of the system is lowered when the embryo forms, that is, the work of formation of the critical nucleus is reduced and nucleation is facilitated. Conversely, if $\varepsilon_{\kappa,\beta} < \varepsilon_{\kappa,\alpha}$ then nucleation is inhibited. As derived by Isard et al. [90] for nucleation in condensed phases, the work $W(n)$ of forming a spherical cluster of n molecules in an applied field E is

$$W(n) = (\Delta\mu - \alpha_{es}E^2)n + \sigma_{\alpha\beta}A_{\alpha\beta}, \tag{50}$$

where $\Delta\mu$ is the change in chemical potential on transformation in the absence of the field, and $\sigma_{\alpha\beta}$ and $A_{\alpha\beta}$ are the energy per unit area and the area of the α–β interface. Equation (50) is identical to Eq. (11) in

Chapter 2, but with the addition of the electrostatic energy per molecule $\alpha_{es}E^2$ in which α_{es} is given by [90]:

$$\alpha_{es} = \frac{3\varepsilon_0\varepsilon_\alpha(\varepsilon_{\kappa,\beta} - \varepsilon_{\kappa,\alpha})\bar{v}}{2(\varepsilon_{\kappa,\beta} + 2\varepsilon_{\kappa,\alpha})}, \tag{51}$$

where ε_0 is the permittivity of free space and $\bar{v}$ is the molecular volume. As expected, the driving force for the transformation can be increased or decreased when the field is applied, depending on the relative magnitudes of $\varepsilon_{\kappa,\alpha}$ and $\varepsilon_{\kappa,\beta}$. The change in driving force can readily be used to calculate the changes in critical nucleus size n^* (Eq. (15) in Chapter 2) and critical work of nucleation W^* (Eq. (14) in Chapter 2). The application of a field essentially leaves unaffected the prefactor A^* in Eq. (55) in Chapter 2 for the homogeneous nucleation rate.

In an applied field, the equilibrium shape of the embryo is no longer spherical. For example, when $\varepsilon_{\kappa,\beta} > \varepsilon_{\kappa,\alpha}$ favoring nucleation, the embryo is elongated parallel to the field. Such effects, however, require only a small correction to Eq. (51) [91].

For an externally applied field, the typical electrostatic energy per molecule is so small that the effects on nucleation are negligible. For example, for the condensation of water, we can take $\varepsilon_{\kappa,\alpha} = 1$ (for the vapor) and $\varepsilon_{\kappa,\alpha} = 80$. Then, a uniform field as high as 1 MV m^{-1} gives an increase in effective supersaturation of only 1 part in 10^7. This shows clearly that the concentration of charge at a point (i.e., an ion) is vital in obtaining a strong effect on nucleation. In line with expectation, experiments on condensed systems such as glass ceramics (Chapter 14, Section 2.1) have failed to show any acceleration of nucleation, even in high applied fields [90]. If the precipitating phase is metallic, for which $\varepsilon_{\kappa,\beta}$ can be taken to be infinity, then there is better chance of a detectable effect. There have been reports of nucleation being accelerated at low fields and even with $\varepsilon_{\kappa,\beta} < \varepsilon_{\kappa,\alpha}$ when it should be inhibited [92]. The origins of such effects are not clear, but they must lie outside the effects on the critical work W^* described in this section; perhaps they can be attributed to enhanced solute transport.

That an applied electric field typically has only a weak effect on phase stabilities holds also for applied *magnetic* fields. In the latter case, the absence of any equivalent for a point charge prohibits a strong effect on nucleation analogous to that for ions. It should be noted, however, that even a weak effect on phase stability can significantly affect nucleation, as discussed in Chapter 7, Section 5. Other effects of electric field are considered in Chapter 15, Section 5.4.

5. PATTERN AND COMPETITION

Behavior such as that shown in Figure 13 is deceptively simple in that the pattern of transformation is just the arrangement of heterogeneous nucleant sites. Such a one-to-one correspondence arises only under very special conditions. Normally there is a range of sites of different types, and there may be competition between these sites and with homogeneous nucleation. The nature of that competition and its influence on the pattern of nucleation are the focus of this section.

We turn first to the discussion in Section 2.6 concerning nucleation of β phase on grain faces, edges, and corners within an α-phase matrix. It was established that the work of formation of the critical nucleus is least at a corner and progressively greater at an edge or on a grain face. The products $k^+(n^*)Z$ are not greatly different for the different cases, but there is a very large difference in the numbers of molecules that can participate in the nucleation. For homogeneous nucleation, all the molecules can participate and the prefactor N_V is their number per unit volume. In a polycrystalline material with average grain diameter d and grain-boundary width δ, the number of participating molecules per unit volume is approximately $N_V(\delta/d)$ for nucleation on grain faces, $N_V(\delta/d)^2$ at grain edges, and $N_V(\delta/d)^3$ at grain corners. For different conditions, the dominant nucleation type (i.e., having the highest rate per unit volume) can be calculated. Figure 32 shows the regimes of predominant nucleation in a form due to Cahn [71]. The abscissa, $\sigma_{\alpha\alpha}/\sigma_{\alpha\beta}$, is a normalized grain-boundary energy, higher values of which imply a greater potency of grain boundaries (and edges and corners) as nucleation sites. The ordinate is a parameter Δ_{gb} defined by

$$\Delta_{gb} = \frac{\ln(d/\delta)}{W^*/k_B T}, \tag{52}$$

including the normalized grain diameter (d/δ) and the critical work for homogeneous nucleation W^*. For the range of conditions of interest, in which temperature changes affect W^* much more than $k_B T$, Δ_{gb} is a measure of the departure from equilibrium. Larger Δ_{gb} also corresponds to a larger grain size, for which nucleation on grain boundaries would be expected to be less dominant. Figure 32 shows that for small driving forces, the most favorable sites — corners — dominate the transformation. As the driving force is increased, the grain size increased, or $\sigma_{\alpha\alpha}/\sigma_{\alpha\beta}$ decreased, however, other less favorable sites dominate, and in the limit homogeneous nucleation dominates. Although this figure is for nucleation associated with grain boundaries, it illustrates important general characteristics of heterogeneous nucleation: that the catalytic potency of a site is not the only factor determining its significance; the driving force for the transformation and the population of such sites (which may depend on a microstructural length scale) are also important. The figure shows the predominant nucleation under constant driving force for transformation. In many cases, however, conditions are changed to progressively increase the driving force. The predominant type of nucleation may then

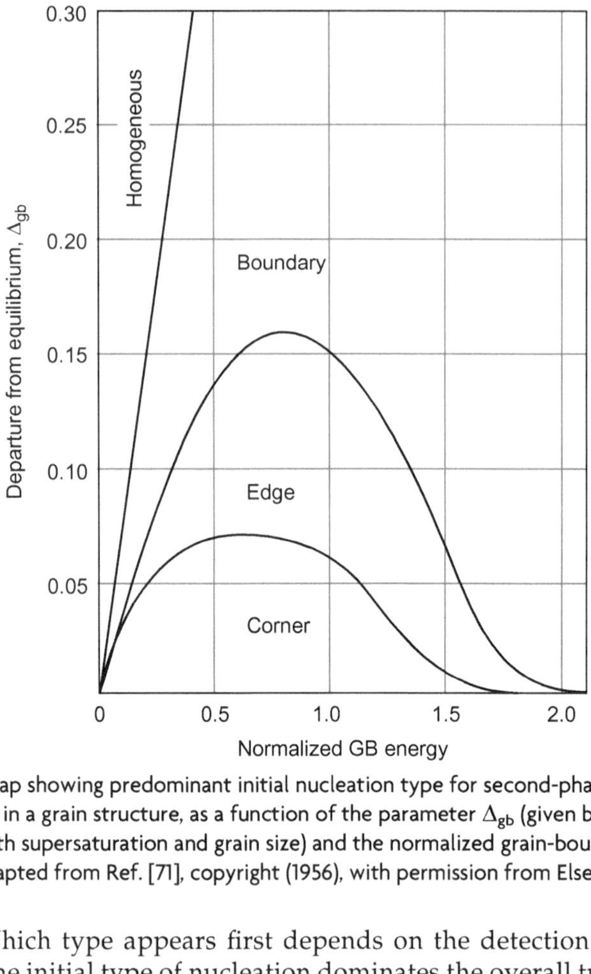

Fig. 32 A map showing predominant initial nucleation type for second-phase precipitation in a grain structure, as a function of the parameter Δ_{gb} (given by Eq. (52), increasing with supersaturation and grain size) and the normalized grain-boundary energy $\sigma_{\alpha\alpha}/\sigma_{\alpha\beta}$. (Adapted from Ref. [71], copyright (1956), with permission from Elsevier.)

change. Which type appears first depends on the detection limit [71]. Whether the initial type of nucleation dominates the overall transformation depends on the growth rate of the transforming regions.

Figure 32 is based on initial nucleation rates only; the effects of the transformation on the continuing nucleation are not accounted for. For steady-state *homogeneous* nucleation (and noting that our discussion is for polymorphic transformations showing no change in composition), the nucleation rate decreases with time in direct proportion to the volume fraction of the system remaining untransformed; nucleation may occur with equal probability in the untransformed volume. For *heterogeneous* nucleation, however, only molecules in contact with the relevant heterogeneities can participate in nucleation. This number can be reduced to zero by transformation of the sample around the heterogeneities; the heterogeneous nucleation rate is then zero even though the transformed fraction of the sample may be small. This *site saturation* is favored by a high nucleation rate, a low population of heterogeneities, and a high growth rate, and it can greatly affect the final microstructure. For example, in a polycrystalline material with a high driving force for transformation, copious nucleation on grain boundaries can give a distinctive microstructure in which the boundaries are coated with thin films of the new phase. On the other hand, in the same material with a lower driving force for transformation, nucleation is slower and the sporadic events lead to a microstructure essentially indistinguishable from the one that would be produced by random nucleation in the bulk. From the final microstructure, it is often difficult to deduce the original nucleation behavior, as discussed, for example, by Tong et al. [93].

The competition considered so far has been essentially geometrical. The probability of nucleation at a given site is influenced by nucleation at other sites only when growth of the new phase from other sites transforms the surrounding medium. This has been termed as *hard impingement*. There are, however, possibilities for longer-range interactions. The most important is in multicomponent systems when there is solute partitioning between the original α phase and the new β phase. This can be illustrated by considering in more detail the system shown in Figure 13, in which calcite nucleates from solution preferentially on a pattern of polar patches. The crystal growth on any patch depletes the surrounding region of calcium and carbonate ions. There is a characteristic length scale for this depletion that can be observed directly around an isolated patch (Figure 33a,b). The solute depletion around the patch reduces the local supersaturation below the value necessary for nucleation on the surrounding nonpolar substrate, leading to a region of the substrate that is free of additional nuclei. To achieve control of the kind shown in Figure 13, where only one type of site is operative, the array of sites must be sufficiently closely spaced that there is overlap (*soft impingement*) of their depletion zones (Figure 33c). One type of nucleation eliminates the other through competition for solute.

In this case, soft impingement of the solute diffusion fields around each nucleant patch does not affect the operation of the patches themselves. It could do so, however, if the patches were even more closely spaced. In Figure 13, the smallness of some of the calcite crystals implies that nucleation on those patches was nearly suppressed by solute depletion caused by growth on neighboring patches. A stronger example of this type of effect, involving transport of heat rather than solute, is discussed in detail in Chapter 13, Section 3. In solidification of a melt, the release of latent heat from the growing solid has an effect analogous to that of solute depletion in the previous example; the local rise in temperature reduces the probability of additional nucleation. Since thermal diffusivities are much greater than solutal diffusivities, the effective depletion zones can have radii greatly exceeding the distances between preferred nucleation sites.

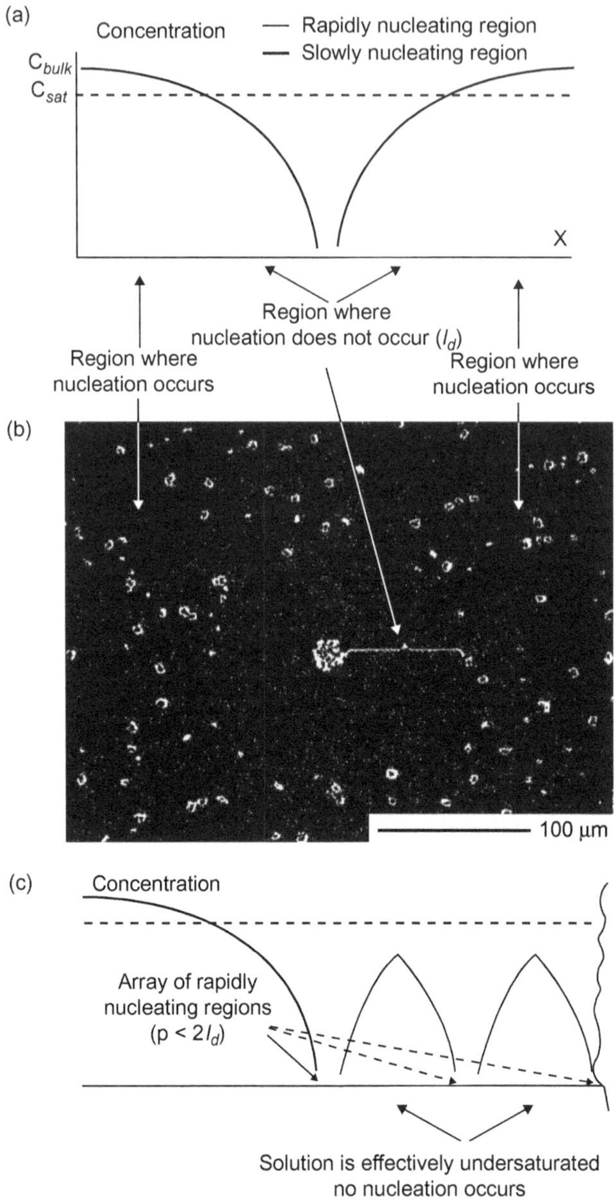

Fig. 33 Nucleation of calcite from solution on a patterned substrate: (a) growth of calcite on a nucleant patch depletes the surrounding substrate of solute; (b) solute depletion below the level required for nucleation on the main substrate leads to a nucleation-free zone around the nucleant patch; (c) an array of nucleant patches can prevent any nucleation other than on the patches themselves (Figure 13). (Reprinted by permission from Macmillan Publishers Ltd. *Nature* [30], copyright 1999.)

The result is that very few of the sites become active nuclei, the antithesis of the *site saturation* mentioned earlier in this section.

The soft-impingement effects noted here are of interest in determining how a given nucleation event may be influenced by others. The similar solute or heat effects involved in the kinetic analysis of a single (in principle isolated) nucleation event have been discussed as coupled fluxes (Chapter 5, Section 5).

The issue discussed in this section—the competition between different types of nucleation—is central to the control of microstructural development in materials. Many examples will be found throughout the book, notably in Chapter 14, Section 4 on engineering alloys.

6. SUMMARY

Nucleation that occurs naturally is almost always *heterogeneous* in that it occurs at nonuniformities in the system. We have seen that nucleation can be catalyzed in this way by *planar, linear,* or *point* heterogeneities. In the suppression of undesired types or the promotion of desired types, control of heterogeneous nucleation is at the heart of fundamental studies of transformations, or of their exploitation, whether in living systems or in industrial materials processing.

For nucleation on planar interfaces, the classical model is that of a spherical-cap-shaped embryo. We consider in the simplest case an original phase α and a β phase nucleating from it, where α and β have the same composition and density. The spherical-cap embryo is bounded by an α–β interface of uniform curvature analogous to that of a spherical embryo in a simple model for homogeneous nucleation. At the critical condition for nucleation, this interface has the same radius of curvature r^* in the two cases, given by

$$r^* = -\frac{2\sigma_{\alpha\beta}}{\Delta g},$$

where $\sigma_{\alpha\beta}$ is the energy per unit area of the α–β interface and Δg is the free-energy change per unit volume on transforming from α to β. However, the spherical cap has a smaller volume than the sphere, and its critical work of formation W^*_{het}, scaling directly with its volume, is given by

$$W^*_{\text{het}} = W^*_{\text{hom}} f(\phi),$$

where W^*_{hom} is the critical work of formation in the homogeneous case and $f(\phi)$ is a geometrical factor related to the contact angle ϕ:

$$f(\phi) = \frac{(2 - 3\cos\phi + \cos^3\phi)}{4}.$$

The contact angle ϕ is determined by the balance of interfacial energies at the interface with the nucleant substrate. Lower ϕ corresponds to a more potent nucleant, and when $\phi = 0$ the nucleation barrier is zero. The contact angle ϕ is the most widely used parameter to describe the catalytic effect of a nucleant substrate. The catalytic effect, however, is markedly reduced as the size of the substrate is reduced to approach that of the critical nucleus. Nucleation of spherical caps on a substrate is a *stochastic* process with overall kinetic behavior similar to that for homogeneous nucleation.

A contact angle can be defined, however, only for certain combinations of the interfacial energies. Otherwise, the substrate is "wetted" either by the original α phase in which case there is no catalytic effect on nucleation, or by the new β phase in which case there is no barrier to nucleation. In the latter case, there is still a barrier arising from the limited size of the nucleant substrates, and the nucleation of the transformation to β is *athermal*, occurring essentially *deterministically* as the driving force for transformation is increased. The work of formation as a function of embryo size (expressed as a linear dimension, or as the number of molecules within it) in this case has a minimum followed by a maximum, the energy difference between these decreasing as the driving force is increased.

The same form can be found for the work of nucleation on dislocations and on ions. Although these linear and point heterogeneities are of atomic dimensions, they can be effective catalysts for nucleation. The sharp structural discontinuity on the dislocation line and the concentration of charge on the ion stabilize small embryos and can lead to metastable configurations of β phase analogous to the wetting layer on a planar substrate.

The basis of preferred nucleation on a planar substrate is elimination of part of the interface between the substrate and the original α phase, lowering the work of embryo formation. However, this effect is augmented when a structure can act as a template for growth of the new phase. The structure may be that of the substrate itself, or it may that of an adsorbed or wetting layer on the substrate.

Typically there are varied sites for heterogeneous nucleation, and these are in competition with each other and with homogeneous nucleation. The critical work is lower for heterogeneous than for homogeneous nucleation, but this factor is offset by the lower number of molecules capable of participating in heterogeneous nucleation. The distribution of nucleation events depends on many factors and only rarely corresponds to the distribution of a single type of heterogeneity. Among other effects, the pattern of nucleation events may show the influence of *site saturation* or of *soft impingement* of solutal or thermal diffusion fields around transforming regions.

REFERENCES

[1] H. Nakanishi, M.E. Fisher, Multicriticality of wetting, prewetting and surface transitions, Phys. Rev. Lett. **49** (1982) 1565–1568.

[2] T. Young, An essay on the cohesion of fluids, Phil. Trans. Roy. Soc. Lond. **95** (1805) 65–87.

[3] J.W. Gibbs, *The Collected Works*, Vol. 1, Yale University Press, New Haven, CT (1948) p. 260.

[4] A.D. Alexandrov, B.V. Toshev, A.D. Scheludko, Nucleation from supersaturated water vapor on immiscible liquid substrates — Effect of the macroscopic geometry of the 3-phase system on the critical supersaturation and the line tension, Colloids Surf. A **79** (1993) 43–50.

[5] R. Aveyard, J.H. Clint, Liquid lenses at fluid/fluid interfaces, J. Chem. Soc. Faraday Trans. **93** (1997) 1397–1403.

[6] R.D. Gretz, Line-tension effect in a surface energy model of a cap-shaped condensed phase, J. Chem. Phys. **45** (1966) 3160–3161.

[7] R.D. Gretz, The line-tension effect in heterogeneous nucleation, Surf. Sci. **5** (1966) 239–251.

[8] J.Y. Wang, S. Betelu, B.M. Law, Line tension approaching a first-order wetting transition: Experimental results from contact angle measurements, Phys. Rev. E **63** (2001) 031601/1–11.

[9] B. Widom, Line tension and the shape of a sessile drop, J. Phys. Chem. **99** (1995) 2803–2806.

[10] A.I. Milchev, A.A. Milchev, Wetting behavior of nanodroplets: the limits of Young's rule validity, Europhys. Lett. **56** (2001) 695–701.

[11] J.W. Cahn, Critical point wetting, J. Chem. Phys. **66** (1977) 3667–3672.

[12] D. Bonn, D. Ross, Wetting transitions, Rep. Prog. Phys. **64** (2001) 1085–1163.

[13] E.M. Blokhuis, Nucleation of wetting layers, Phys. Rev. E **51** (1995) 4642–4654.

[14] M. Schick, P. Taborek, Anomalous nucleation at 1st-order wetting transitions, Phys. Rev. B **46** (1992) 7312–7314.

[15] J.O. Indekeu, Line tension at wetting, Int. J. Mod. Phys. B **8** (1994) 309–345.

[16] T. Pompe, S. Herminghaus, Three-phase contact line energetics from nanoscale liquid surface topographies, Phys. Rev. Lett. **85** (2000) 1930–1933.

[17] B.M. Law, Theory of nucleated wetting, Phys. Rev. Lett. **72** (1994) 1698–1701.

[18] R. Blossey, Nucleation at first-order wetting transitions, Int. J. Mod. Phys. B **9** (1995) 3489–3525.

[19] R.C. Tolman, The effect of droplet size on surface tension, J. Chem. Phys. **17** (1949) 333–337.

[20] G. Navascués, P. Tarazona, Contact-angle and line tension dependence on curvature, Chem. Phys. Lett. **82** (1981) 586–588.

[21] I.N. Stranski, R. Kaischew, Uber den Mechanismus des Gleichgewichtes kleiner Kriställchen, Z. Phys. Chem. B **26** (1934) 100–113.

[22] S. Auer, D. Frenkel, Line tension controls wall-induced crystal nucleation in hard-sphere colloids, Phys. Rev. Lett. **91** (2003) 015703/1–4.

[23] H. Vehkamäki, A. Määttänen, A. Lauri, I. Napari, M. Kulmala, Technical note: The heterogeneous Zeldovich factor, Atmos. Chem. Phys. Discuss. **6** (2006) 9069–9083.

[24] A.C. Zettlemoyer, *Nucleation Phenomena*, Elsevier, New York (1977).

[25] A.L. Greer, P.V. Evans, R.G. Hamerton, D.K. Shangguan, K.F. Kelton, Numerical modelling of crystal nucleation in glasses, J. Cryst. Growth **99** (1990) 38–45.

[26] G. Navascués, P. Tarazona, Line tension effects in heterogeneous nucleation theory, J. Chem. Phys. **75** (1981) 2441–2446.

[27] V. Talanquer, D.W. Oxtoby, Nucleation on a solid substrate: a density functional approach, J. Chem. Phys. **104** (1996) 1483–1492.

[28] J.J. Hoyt, M. Asta, A. Karma, Atomistic and continuum modeling of dendritic solidification, Mater. Sci. Eng. R **41** (2003) 121–163.
[29] K. Luo, M.-P. Kuittu, C. Tong, S. Majaniemi, T. Ala-Nissila, Phase-field modeling of wetting on structured surfaces, J. Chem. Phys. **123** (2005) 194702/1–12.
[30] J. Aizenberg, A.J. Black, G.M. Whitesides, Control of crystal nucleation by patterned self-assembled monolayers, Nature **398** (1999) 495–498.
[31] L. Gránásy, T. Pusztai, D. Saylor, J.A. Warren, Phase field theory of heterogeneous crystal nucleation, Phys. Rev. Lett. **98** (2007) 035703/1–4.
[32] N.H. Fletcher, Size effect in heterogeneous nucleation, J. Chem. Phys. **29** (1958) 572–576; Erratum. *ibid.* **31** (1959) 1136–1137.
[33] N.H. Fletcher, Nucleation by crystalline particles, J. Chem. Phys. **38** (1963) 237–240.
[34] D. Turnbull, Theory of catalysis of nucleation by surface patches, Acta Metall. **1** (1953) 8–14.
[35] M.J. Burke, S.E. Lindow, Surface properties and size of the ice nucleation site in ice nucleation active bacteria: theoretical considerations, Cryobiology **27** (1990) 80–84.
[36] D. Turnbull, Kinetics of heterogeneous nucleation, J. Chem. Phys. **18** (1950) 198–203.
[37] S.F. Jones, G.M. Evans, K.P. Galvin, Bubble nucleation from gas cavities — a review, Adv. Coll. Inter. Sci. **80** (1999) 27–50.
[38] R.P. Sear, Nucleation: Theory and applications to protein solutions and colloidal suspensions, J. Phys. Cond. Matter **19** (2007) 033101/1–28.
[39] J.C. Fisher, J.H. Hollomon, D. Turnbull, Nucleation, J. Appl. Phys. **19** (1948) 775–784.
[40] A. Ziabicki, Generalized theory of nucleation kinetics. II. Athermal nucleation involving spherical clusters, J. Chem. Phys. **48** (1968) 4374–4380.
[41] J.S. Im, V.V. Gupta, M.A. Crowder, On determining the relevance of athermal nucleation in rapidly quenched liquids, Appl. Phys. Lett. **72** (1998) 662–664.
[42] V.A. Shneidman, Analytical description of "athermal" nucleation and its relevance to rapidly quenched fluids, J. Appl. Phys. **85** (1999) 1981–1983.
[43] T.E. Quested, A.L. Greer, Athermal heterogeneous nucleation of solidification, Acta Mater. **53** (2005) 2683–2692.
[44] J. Feder, K.C. Russell, J. Lothe, G.M. Pound, Homogeneous nucleation and growth of droplets in vapours, Adv. Phys. **15** (1966) 111–178.
[45] S.A. Reavley, A.L. Greer, Athermal heterogeneous nucleation of freezing: numerical modeling for polygonal and polyhedral substrates, Philos. Mag. **88** (2008) 561–579.
[46] A.L. Greer, T.E. Quested, Heterogeneous grain initiation in solidification, Philos. Mag. **86** (2006) 3665–3680.
[47] A.D. Sutton, R.W. Balluffi, *Interfaces in Crystalline Materials*, OUP, Oxford (1995).
[48] H.I. Aaronson, Observations on interphase boundary structure, J. Microsc. **102** (1974) 275–300.
[49] D. Turnbull, Role of structural impurities in phase transformations, in: *Impurities and Imperfections*, Amer. Soc. Metals, Cleveland OH (1955), pp. 121–144.
[50] K.C. Russell, Nucleation in solids: the induction and steady state effects, Adv. Coll. Interface Sci. **13** (1980) 205–318.
[51] K.C. Russell, The role of excess vacancies in precipitation, Scripta Metall. **3** (1969) 313–316.
[52] C. Herring, Some theorems on the free energies of crystal surfaces, Phys. Rev. **82** (1951) 87–93.
[53] C. Herring, The use of classical macroscopic concepts in surface-energy problems, in: *Structure and Properties of Solid Surfaces*, Eds. R. Gomer, C.S. Smith, University of Chicago Press, Chicago (1953), pp. 5–81.
[54] W.W. Mullins, Solid surface morphologies governed by capillarity, in: *Metal Surfaces*, Amer. Soc. Metals, Metals Park, OH (1963), pp. 17–66.

[55] D.W. Hoffmann, J.W. Cahn, A vector thermodynamics for anisotropic surfaces: I. Fundamentals and applications to plane surface junctions, Surf. Sci. **31** (1972) 368–388.
[56] J.W. Cahn, D.W. Hoffmann, A vector thermodynamics for anisotropic surfaces — II. Curved and faceted surfaces, Acta Metall. **22** (1974) 1205–1214.
[57] J.K. Lee, H.I. Aaronson, Influence of faceting upon the equilibrium shape of nuclei at grain boundaries — I. Two-dimensions, Acta Metall. **23** (1975) 799–808.
[58] J.K. Lee, H.I. Aaronson, Influence of faceting upon the equilibrium shape of nuclei at grain boundaries — II. Three-dimensions, Acta Metall. **23** (1975) 809–820.
[59] H.I. Aaronson, M.R. Plichta, G.W. Franti, K.C. Russell, Precipitation at interphase boundaries, Metall. Trans. A **9A** (1978) 363–371.
[60] R.C. Cammarata, K. Sieradzki, Surface and interface stresses, Ann. Rev. Mater. Sci. **24** (1994) 215–234.
[61] R.C. Cammarata, R.K. Eby, Effects and measurement of internal surface stresses in materials with ultrafine microstructures, J. Mater. Res. **6** (1991) 888–890.
[62] A. Cacciuto, S. Auer, D. Frenkel, Breakdown of classical nucleation theory near isostructural phase transitions, Phys. Rev. Lett. **93** (2004) 166105/1–4.
[63] R.C. Cammarata, T.M. Trimble, D.J. Srolovitz, Surface stress model for intrinsic stresses in thin films, J. Mater. Res. **15** (2000) 2468–2474.
[64] H.B. Aaron, H.I. Aaronson, Comparison of relative interfacial energies of disordered interphase (α:θ) and grain (α:α) boundaries at grain boundary precipitates in Al–4% Cu during growth and at equilibrium, Acta Metall. **18** (1970) 699–711.
[65] K.C. Russell, Grain boundary nucleation kinetics, Acta Metall. **17** (1969) 1123–1131.
[66] W.C. Johnson, C.L. White, P.E. Marth, P.K. Ruf, S.M. Tuominen, K.D. Wade, K.C. Russell, H.I. Aaronson, Influence of crystallography on aspects of solid–solid nucleation theory, Metall. Trans. A **6A** (1975) 911–919.
[67] H.I. Aaronson, H.B. Aaron, The initial stages of the cellular reaction, Metall. Trans. **3** (1972) 2743–2756.
[68] K.C. Russell, H.I. Aaronson, Sequences of precipitate nucleation, J. Mater. Sci. **10** (1975) 1991–1999.
[69] P.E. Marth, H.I. Aaronson, T.L. Bartel, K.C. Russell, G.W. Lorimer, Application of heterogeneous nucleation theory to precipitate nucleation at GP zones, Metall. Trans. A **7A** (1976) 1519–1528.
[70] J.W. Palko, M. Durandurdu, J. Kieffer, Mechanically controlled, seeded formation of a nanoscale metastable phase in ionic compounds, Nano Lett. **4** (2004) 1769–1773.
[71] J.W. Cahn, The kinetics of grain boundary nucleated reactions, Acta Metall. **4** (1956) 449–459.
[72] J. Weertman, J.R. Weertman, *Elementary Dislocation Theory*, OUP, Oxford (1992).
[73] D. Hull, D.J. Bacon, *Introduction to Dislocations*, 4th edn, Elsevier, New York (2001).
[74] E. Hornbogen, Nucleation of precipitates in defect solid solutions, in: *Nucleation*, Ed. A.C. Zettlemoyer, Marcel Dekker, New York (1969), pp. 309–378.
[75] J.W. Christian, *The Theory of Transformations in Metals and Alloys, Parts I and II*, 3rd edn, Elsevier, Oxford (2002).
[76] D. Ma, Y. Li, Heterogeneous nucleation catastrophe on dislocations in superheated crystals, J. Phys. Cond. Matter **12** (2000) 9123–9128.
[77] J.W. Cahn, Nucleation on dislocations, Acta Metall. **5** (1957) 169–172.
[78] R. Gomez-Ramirez, G.M. Pound, Nucleation of a second solid phase along dislocations, Metall. Trans. A **4A** (1973) 1563–1570.
[79] U. Dehlinger, Präformierte Keime bei Umwandlungen und Ausscheidungen, Z. Metallk. **51** (1960) 353–356.
[80] C.C. Dollins, Nucleation on dislocations, Acta Metall. **18** (1970) 1209–1215.
[81] D.M. Barnett, On nucleation of coherent precipitates near edge dislocations, Scripta Metall. **5** (1971) 261–266.

[82] B.Ya. Lyubov, V.A. Solov'Yev, Possibility of stable segregations of solute atoms and nucleation of coherent centres of a new phase in the elastic stress field of an edge dislocation, Phys. Metals Metallogr. **19** (3), (1965) 13–22; Fiz. Metal. Metalloved. **19** (3) (1956) 333–345.
[83] F. Larché, Nucleation and precipitation on dislocations, in: Dislocations Ed. F.R.N. Nabarro, Vol. 4, North-Holland, Amsterdam (1979), pp. 135–153.
[84] J.J. Thomson, *Conduction of Electricity through Gases*, 2nd edn, CUP, Cambridge (1906), pp. 163–187.
[85] C.T.R. Wilson, Condensation of water vapour in the presence of dust-free air and other gases, Phil. Trans. Roy. Soc. Lond. A **189** (1897) 265–307.
[86] C. Henderson, *Cloud and Bubble Chambers*, Methuen, London (1970).
[87] M. Volmer, *Kinetik der Phasenbildung*, Theodor Steinkopf Verlag, Dresden (1939).
[88] K.C. Russell, Nucleation on gaseous ions, J. Chem. Phys. **50** (1969) 1809–1816.
[89] D.A. Glaser, Some effects of ionizing radiation on the formation of bubbles in liquids, Phys. Rev. **87** (1952) 665.
[90] J.O. Isard, P.F. James, A.H. Ramsden, An investigation of the possibility that electric fields could affect the nucleation of glass ceramics, Phys. Chem. Glasses **19** (1978) 9–13.
[91] V.B. Warshavsky, A.K. Shchekin, The effects of external electric field in thermodynamics of formation of dielectric droplet, Colloids Surf. A **148** (1999) 283–290.
[92] M. Shablakh, L.A. Dissado, R.M. Hill, Influence of electrical field on the nucleation of crystal growth in some cycloalcohols, J. Chem. Soc. Faraday Trans. II **79** (1983) 1443–1453.
[93] W.S. Tong, J.M. Rickman, K. Barmak, Quantitative analysis of spatial distribution of nucleation sites: Microstructural implications, Acta Mater. **47** (1999) 435–445.

PART II:
Experimental Measurements

CHAPTER 7

Crystallization in Liquids and Colloidal Suspensions

Contents

1. INTRODUCTION

The study of the nucleation of freezing in liquids has a long history, starting with Fahrenheit's systematic investigations of the crystallization of water (Chapter 1). While water is an exceptional liquid, this ability to be supercooled without crystallization is universal, common to all liquids.

Liquids and glasses (Chapter 8) are examples of *amorphous* phases, from the Greek $\alpha\mu o\rho\phi o\sigma$, meaning shapeless, without form.[1] However, this classification can be misleading, since it is often taken to mean without atomic structure or order. Liquids and glasses contain significant local order that can be directly inferred from X-ray diffraction studies. In Figure 1a, for example, the measured structure factor, $S(q)$, for liquid titanium is shown as a function of the scattering vector q ($= 4\pi \sin\theta_B/\lambda$), where θ_B is the Bragg angle and λ the X-ray wavelength). If there is no order at any length scale, such as in a dilute gas, the structure factor has a

[1] Oxford English Dictionary.

Pergamon Materials Series, Volume 15
ISSN 1470-1804, DOI 10.1016/S1470-1804(09)01507-7

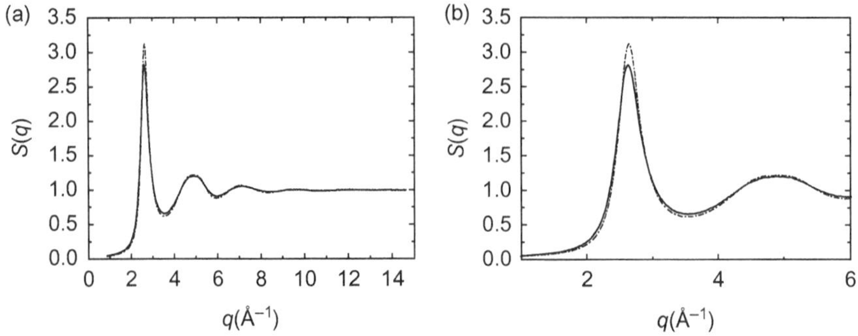

Fig. 1 (a) Structure factor, $S(q)$ as a function of scattering vector, q, for liquid titanium at 1700°C (solid line) and 1375°C (dash-dot line). (b) Enlarged view showing the sharpening of $S(q)$ with decreasing temperature [1].

constant value. Sharp peaks are observed in the structure factor for a well-ordered crystal phase. The character of $S(q)$ for the liquid phase falls between these extremes. The presence of peaks in $S(q)$ demonstrates that there is order in the liquid. That the peaks are broad indicates that this order extends only to nearest or next-nearest neighbors. The first and second peaks in $S(q)$ generally sharpen with decreasing temperature (seen more clearly in Figure 1b), showing that the liquid is becoming better ordered.

In this chapter, we review techniques for the measurement and analysis of crystal nucleation in supercooled liquids, focusing on metallic liquids. Representative maximum-supercooling and nucleation data are presented and compared with predictions from the nucleation theories discussed in earlier chapters. As will be discussed in Section 2.4 of this chapter, the local order in the liquid is directly coupled to the nucleation barrier; such coupling of nucleation processes, often noted in this book, is increasingly recognized as important.

2. MAXIMUM-SUPERCOOLING STUDIES

The homogeneous steady-state crystal nucleation rate in liquids rises rapidly with decreasing temperature, T, after a sharp onset. Figure 2, for example, shows the measured and calculated rates as a function of supercooling for liquid mercury.

For many liquids, particularly liquid metals, the crystal growth velocities are sufficiently high in the region that is accessible by supercooling that the time scale for crystallization is dominated by the time required to form nuclei. When nucleation does occur, the release of the heat of fusion with growth causes the temperature to rise

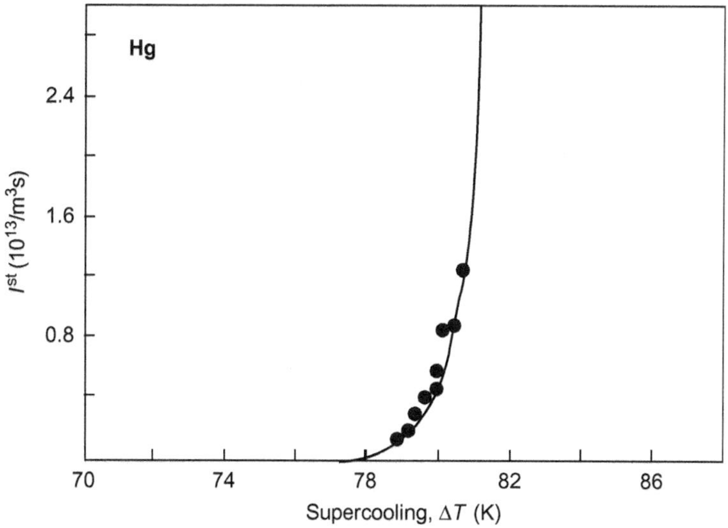

Fig. 2 The steady-state homogeneous nucleation rate as a function of supercooling for the solidification of liquid mercury, demonstrating the existence of a sharp onset. The points are experimental data [2]; the solid line is a fit to the classical theory of nucleation. (Reprinted from Ref. [3], copyright (1991), with permission from Elsevier.)

(a phenomenon known as *recalescence*). The minimum temperature T_{min} reached before recalescence is lower than, but can be taken to be a good approximation of, the onset temperature for nucleation. Corresponding to T_{min}, there is the maximum supercooling, ΔT_{max} ($= T_m - T_{min}$), where T_m is the equilibrium melting temperature (liquidus temperature, T_{liq}, for an alloy or other multicomponent liquids). Often the maximum supercooling is reported as the dimensionless *reduced supercooling* ΔT_r ($= (T_m - T_{min})/T_m$). The sharp rise in the nucleation rate on cooling, and the high growth rate, allow ΔT_{max} to be used as a probe of the nucleation rate in the liquids. Several reviews of supercooling studies exist (see Ref. [3] and the references therein).

For several reasons, the liquid-supercooling studies discussed in this chapter are predominantly for metals. Simple models can often be applied to interpret experimental results, and most industrial solidification processes apply to metals and alloys. An improved understanding of the phases that form, and how specific heterogeneous catalysts can be used to alter and control the nucleation rate, is of central importance. Finally, the supercooling behavior of covalently bonded liquids and polymeric liquids can be understood, to some level at least, within the phenomenology developed for liquid metals. Some supercooling data are provided in this chapter to demonstrate this. The extensions of nucleation theories to handle the distinctive topological and bonding properties of polymers are discussed in Chapter 11.

2.1 Experimental techniques

Typically, liquids are supercooled until the onset of crystallization is detected through a change in some macroscopic property, such as volume (by dilatometry), enthalpy (by calorimetry), or electrical resistivity. More microscopic probes, such as nuclear magnetic resonance (NMR), have also been used. Most frequently, ΔT_{max} is limited by heterogeneous nucleation on container walls, or on structural impurities within the liquid or at its surface. Such effects must be eliminated if the goal is to study homogeneous nucleation. There have been several attempts to supercool bulk samples of liquids by using clean crucibles made of materials that act only weakly as catalytic sites for heterogeneous nucleation. These are placed in vacuum or a reducing atmosphere to eliminate heterogeneous nucleation on surface oxide particles. With few exceptions, however, the supercooling obtained is smaller than expected for homogeneous nucleation. Consequently, more sophisticated methods have evolved to better eliminate the heterogeneous sites; the most common of these are illustrated in Figure 3.

2.1.1 Impurity isolation

The emulsion and substrate techniques pioneered by Vonnegut [4] and Turnbull [2, 5, 6] compartmentalize the heterogeneous particles in small volumes of the sample. In the *droplet emulsion method* (Figure 3a), the liquid

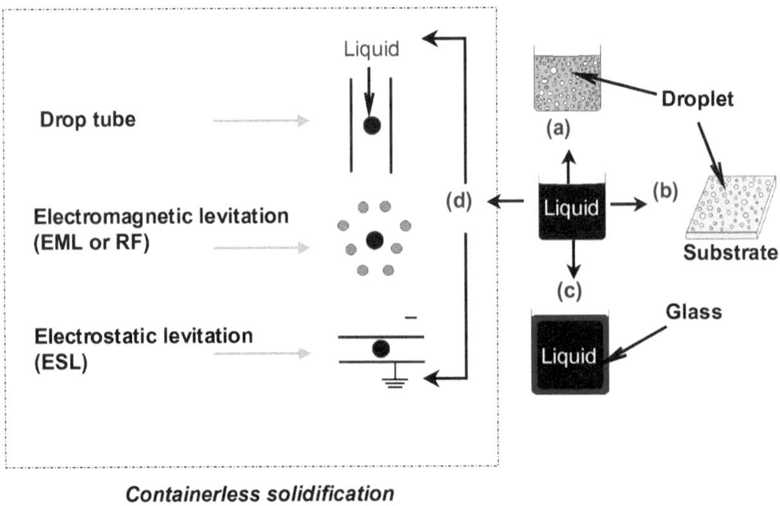

Fig. 3 Some of the most common and effective methods for removing heterogeneities from the melt to attain a supercooling of the liquid limited by homogeneous nucleation of the crystal: (a) and (b) impurity-isolation methods; (c) fluxing; (d) containerless solidification.

is dispersed into a large number of small droplets in an appropriate medium. If the dispersion is sufficiently fine (droplet diameter 100 μm), a significant number of droplets contain no heterogeneous sites. Assuming that the droplet surface does not catalyze nucleation, the homogeneously limited supercooling is the maximum observed in the droplet ensemble. To suppress droplet coalescence in the emulsion, and interactions with the container, the droplets are frequently coated with organic or inorganic surfactants for low-melting-temperature systems ($T_m < 500°C$). Droplets are often dispersed in a molten-salt mixture for studies of higher-melting-temperature systems. Dilatometry and, more recently, differential thermal analysis (DTA) and differential scanning calorimetry (DSC) have been used to determine the maximum supercooling [7, 8]. Calorimetric techniques are less sensitive to the initial stages of nucleation and they show larger reduced supercoolings (ΔT_r as large as 0.5, compared with 0.18 from dilatometric measurements). In the substrate technique, small droplets of liquid (10 to 100 μm in diameter) are placed on a heated substrate, typically common glass (Figure 3b). If the droplets are sufficiently small, are not contaminated during fabrication, and do not interact with the substrate, heterogeneous sites in the liquid will again be isolated [6]. To minimize surface oxidation, experiments are often performed in vacuum or a reducing atmosphere. Higher-melting-temperature liquids are more easily studied by this method than by the droplet-emulsion method. The maximum supercooling is typically measured by cooling the substrate and using optical microscopy to observe the temperature at which the particle surface roughens, indicating solidification. The maximum supercooling obtained by the substrate technique is typically smaller than from the emulsion methods, presumably reflecting the catalytic activity of the substrates.

The droplet diameters in the emulsion and substrate techniques are small, on the order of a few micrometers. For fluxing methods, the liquid is isolated from the container walls by a coating of a material that also dissolves the impurity particles or changes their structure to render them less active as heterogeneous sites (Figure 3c). The *flux* (or *slag* [9]) is a liquid or glassy phase, unlikely to serve as an active nucleation site for the crystalline phase of interest. Fluxing is a particularly useful technique for supercooling large samples and is often considered to have led to the production of the first "bulk" metallic glass [10].

2.1.2 Containerless processing

Containerless-solidification techniques allow studies of larger samples with diameters in the millimeter-to-centimeter range. The drop-tube and drop-tower are the most straightforward approaches and offer the chance to investigate solidification under microgravity conditions. In a typical drop-tube experiment, the sample is melted and then allowed to solidify while falling down a tube that is evacuated or filled with inert gas. In shorter, laboratory-scale facilities, the liquid is dispersed into droplets with a range of diameters. The solidified droplets are collected and their microstructures are examined as a function of the cooling rate, which is a function of the diameter. No direct record of the thermal history during solidification is obtained in such a post-mortem study, however. In taller drop tubes, the thermal history of a single falling droplet can be followed. When nucleation occurs, crystal growth is rapid, and the resulting recalescence event is detected by photosensors. Knowing the initial temperature and measuring the time elapsed during free fall before the recalescence event, ΔT_{max} can be estimated. Drop-tower experiments are made in a free-falling pod that contains instrumentation to measure some physical properties of the liquid during the drop. These experiments are not as acceleration-free as the drop-tube experiments, however, largely due to the small thrusters necessary to maintain the correct trajectory during the drop.

Neither drop-tube nor drop-tower methods are well suited for precision studies of supercooling, nor can they be used to measure the thermophysical and structural properties of the supercooled liquids. They have largely been replaced by levitation techniques. Depending on the levitation method used, the supercooling behavior of bulk samples of metallic, semiconductor, or insulating materials can be studied. Further, sample position and temperature can be accurately monitored and controlled. The most popular techniques are based on aerodynamic, acoustic, electrostatic, or electromagnetic levitation.

Aerodynamic and acoustic levitation. Aerodynamic levitation is achieved by controlled gas flow through an array of nozzles that are designed for the size and density of the samples of interest. The levitation force arises from the momentum transferred from the gas as it flows around the sample. Laser heating is used to melt the samples, allowing temperatures as high as 2700 K to be reached. In principle, this method enables studies of metallic, semiconducting, or nonconducting materials. The flowing gas makes temperature and positioning control difficult, however, and gas impurities may lead to an enhanced heterogeneous nucleation rate. This is a particularly severe problem for studies of highly reactive metals, which can quickly form thin surface films of oxide.

While acoustic-based levitation has been demonstrated, the large forces needed under terrestrial gravity tend to cause fragmentation of the samples when melted. An aero-acoustic variation that combines aerodynamic and acoustic forces has proven more successful (Figure 4). The sample-positioning control and containment are better than for aerodynamic or acoustic levitation alone. When heated, the gas flow provides a stable, quiescent boundary layer around the sample, suppressing the oscillations induced by the acoustic forces, which control the shape and rotation of the molten sample. This hybrid technique has

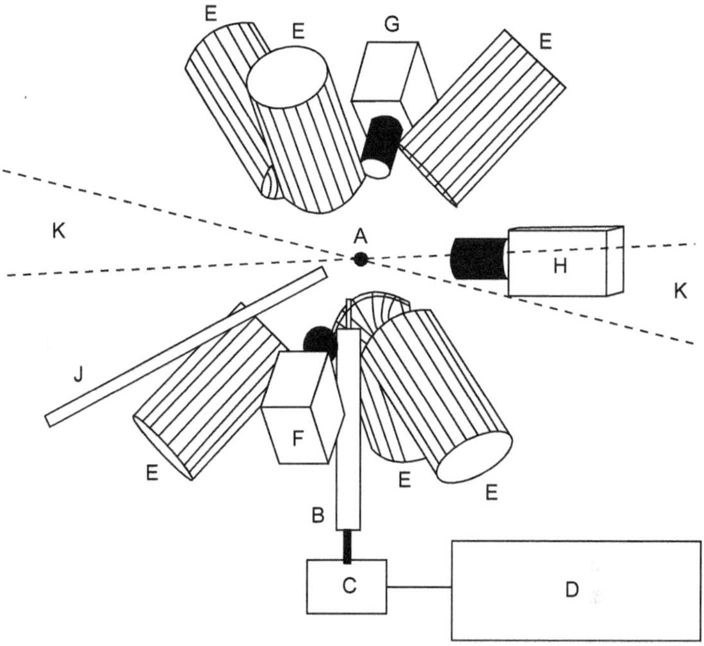

Fig. 4 An aero-acoustic levitation facility. A, levitated sample; B, gas-flow tube and heater; C, translation stage; D, flow-control system; E, acoustic transducers (triple axis); F, diode-laser specimen illuminator (triple axis); G, specimen position detector (triple axis); H, video camera; J, vacuum chuck; K, laser heating. (Reprinted with permission from Ref. [11], copyright (1994), American Institute of Physics.)

been used to process a variety of nonvolatile materials. It is particularly useful for processing nonconducting oxides and ceramics, which are best processed under an oxygen atmosphere.

RF levitation. In electromagnetic levitation (EML), a high-frequency electromagnetic field induces eddy currents within an electrically conducting sample [12]. A schematic of a precision EML facility is shown in Figure 5. In accord with Lenz's law, the induced eddy currents create a magnetic field that opposes the external field, leading to a levitation force. Because the current flow is confined near the sample surface, the effective electrical resistance is large and the eddy currents heat the sample. Using radio-frequency (rf) coils with a suitable geometry, samples can be heated to high temperatures and cooled, while maintaining levitation. For some terrestrial supercooling experiments, the samples have such a high density and low melting temperature that it is not possible to lower the temperature below T_m (or T_{liq}) while maintaining levitation. A cooling gas must then be introduced, raising the possibility of oxidation and heterogeneous nucleation. An EML space-based facility, TEMPUS, successfully

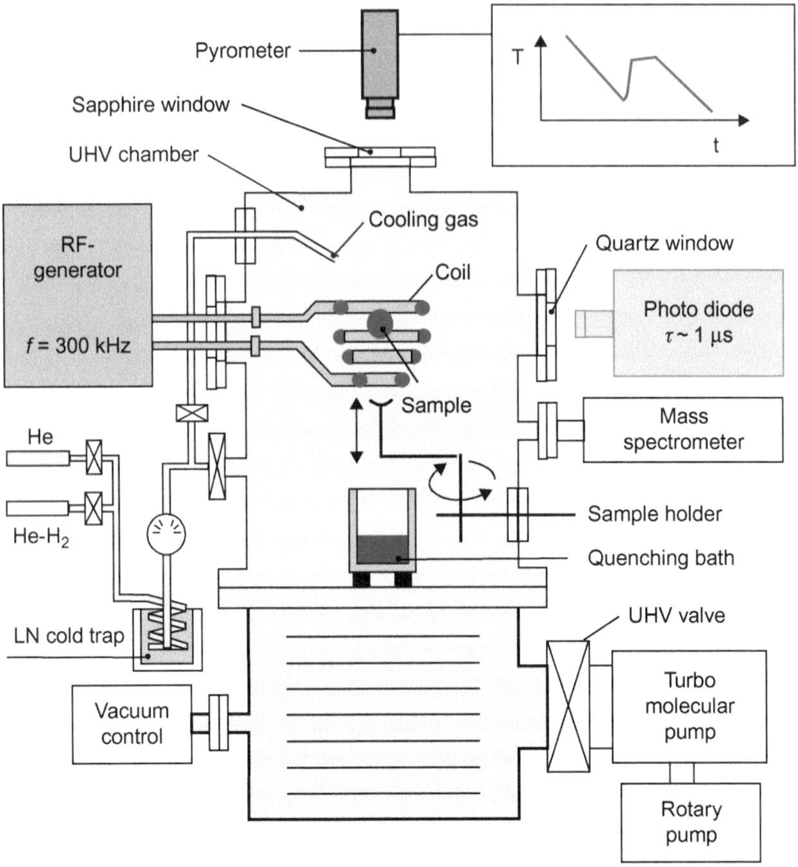

Fig. 5 Electromagnetic levitation (EML) facility for studies of supercooled liquid metallic samples. (Reprinted with permission from Ref. [12], copyright (1993), Maney Publishing.)

decoupled heating and positioning, allowing more precise supercooling studies in the microgravity conditions on board the space shuttle in earth orbit [13]. Additionally, precision monitoring of the rf-power and the known coil-sample coupling allowed accurate measurements of the thermophysical properties; the electrical resistivity of the liquid sample could also be measured [13]. Unfortunately, only metallic materials, or semiconducting materials that become metallic in the liquid phase (e.g. Si), can be studied by EML.

Electrostatic levitation. In electrostatic levitation (ESL), samples are levitated by Coulomb forces acting on a charged sample using an electrostatic field [14, 15]. The ESL method has several advantages; sample heating is decoupled from levitation, supercooling studies are made in an ultrahigh vacuum, and, like aerodynamic and aero-acoustic levitation, a variety of materials (metallic, semiconductor, or nonconducting) can be processed. Because there is no minimum in the electrostatic potential, sample position can be maintained only by using active computer feedback. Figure 6 shows a schematic diagram of the NASA ESL facility located at Marshall Space Flight Center, Alabama. Three sets of electrodes are used to position the samples during processing. Two dual-axis position-sensitive detectors are used to locate the sample and to provide input for a proportional-integral-differential-control-loop computer. The sample position is maintained by adjusting the voltages applied to the electrodes (see Ref. [16] for further details). Samples are charged by the photoelectric effect, using an ultraviolet source. The samples are heated with a high-intensity laser (such as continuous-wave 50 W Nd:YAG, 50 W CO_2, or more recently 50 W or higher diode lasers). Levitation experiments are typically performed at a pressure of approximately 10^{-8} torr. Figure 7 shows photographs of the ESL chamber and of a levitated sphere of solid and liquid zirconium.

In most of the current ESLs, samples are heated from one side, sometimes leading to large temperature gradients across the sample, which can cause large convective and surface-driven flow. The

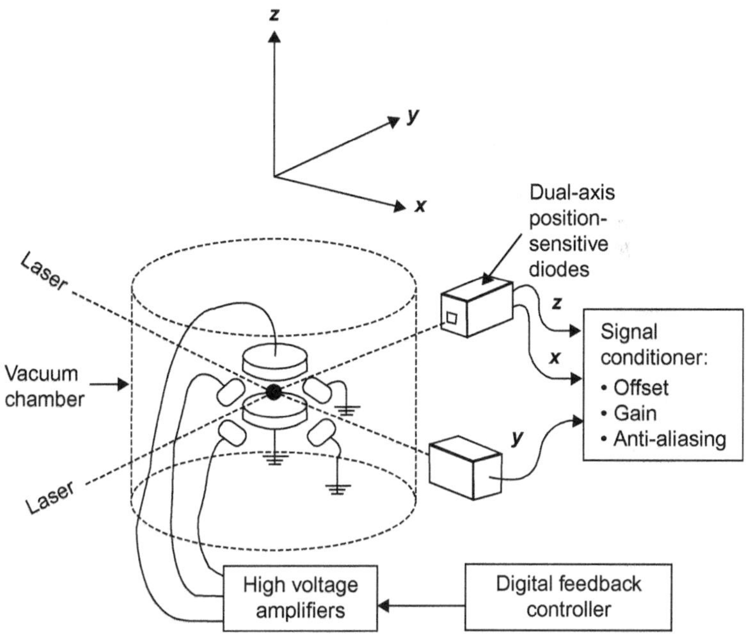

Fig. 6 The chamber in the NASA electrostatic levitation (ESL) facility located at Marshall Space Flight Center, Huntsville, AL. (Reprinted with permission from Ref. [16], copyright (1997), American Institute of Physics.)

(a) (b) (c)

Fig. 7 (a) The vacuum chamber and levitation electrodes of the ESL facility at NASA Marshall Space Flight Center; (b) a solid $Ti_{39.5}Zr_{39.5}Ni_{21}$ sphere suspended between the top and bottom electrostatic electrodes; (c) liquid $Ti_{39.5}Zr_{39.5}Ni_{21}$ after laser melting. (Courtesy Jan Rogers, NASA Marshall Space Flight Center, Huntsville, AL.)

Table 1 Comparison of levitation techniques for studies of supercooled liquid droplets

	Aerodynamic	EML	ESL
Levitation technique	Aerodynamic	Electromagnetic	Electrostatic
Atmosphere	Inert gas	Inert gas or vacuum	Vacuum (10^{-7} torr)
Materials that can be processed	All	Metals	All
Measurable properties of supercooled liquid:			
Specific heat capacity	No	Microgravity	Yes
Viscosity	No	Microgravity	Yes
Surface tension	No	Microgravity	Yes
Structure	Yes	Yes	Yes
Electrical resistivity	Yes	Yes	No
Density and thermal expansion	No	Yes	Yes

temperature gradients can be significantly reduced by using a highly symmetric arrangement of precision-power-controlled lasers for heating [17].

The relative merits and current measurement capabilities of containerless-levitation techniques are listed in Table 1.

2.2 Congruent freezing

Most maximum-supercooling studies have been made on liquids of a metallic element or an alloy liquid that freezes congruently, i.e. to a solid of identical composition without any partitioning of solute. Selected data in such systems obtained with the techniques described in Section 2.1 are presented here.

2.2.1 Experimental results

The ability of liquid metals to be deeply supercooled was dramatically illustrated by Turnbull's dilatometric measurements (Chapter 1, Figure 2) on liquid and solid mercury emulsions [2]. The liquid was cooled to ~ 60 K below the equilibrium melting temperature, a maximum supercooling of nearly one-third of the melting temperature. Representative data presented in Tables 2–6 (taken from Ref. [3]) demonstrate the universality of supercooling for all liquids. Some of the maximum-supercooling values obtained from the emulsion, substrate, and fluxing methods are listed in Table 2 for a wide variety of liquid metallic elements; typical sample volumes V and cooling rates ϑ are also listed. The maximum values obtained from containerless studies are listed in Table 3. The supercooling values for all metals are large, 10 to 30% of the melting temperature, indicating a large barrier to the formation of the new phase, consistent with the proposed noncrystallographic local order in the liquids. The supercooling is significantly greater for the containerless methods than for the container-based methods, likely signaling a failure to eliminate heterogeneous nucleation in the latter studies. It is interesting that the supercooling obtained by ESL is the least. The higher supercooling for drop-tube methods likely reflects a faster cooling rate; the differences between EML and ESL, however, are more difficult to understand.

For comparison, representative supercooling data on nonmetallic systems are listed in Tables 4 and 5. The range of ΔT_r is similar to that for liquid metals.

The values of the maximum observed supercooling ΔT_{max} for metals are collected in Table 6, together with the melting temperature T_m and, when available, the value of V/ϑ from Tables 2 and 3. The interfacial free energy and number of molecules in the critical nucleus (n^*), listed in the table, are derived using the classical theory of steady-state nucleation (Chapter 2); the technique for extracting a nucleation rate from the maximum-supercooling data is discussed next.

2.2.2 Work of cluster formation and interfacial free energy

Sample purity, droplet size, cooling rate, environment, and melt superheat have a significant influence on the maximum attainable supercooling and must be included in the analysis of supercooling data.

Table 2 Maximum supercoolings for solidification of liquid metallic elements, obtained using substrate, emulsion or fluxing techniques*

Metal	Emulsion			Substrate			Fluxed			Flux
	ΔT (K)	V (mol)	ϑ (K s^{-1})	ΔT (K)	V (mol)	ϑ (K s^{-1})	ΔT (K)	V (mol)	ϑ (K s^{-1})	
Ag	—	—	—	—	—	—	250 [18,19]	4.63	—	Soda-lime glass
	—	—	—	—	—	—	227 [6]	$\sim 10^{-9}$	~ 1	Pyrex
Al	175 [8]	$\sim 10^{-10}$	500	48 [6]	$\sim 10^{-8}$	~ 1	130 [6]	$\sim 10^{-8}$	~ 1	NaOH
Au	—	—	—	230 [6]	$\sim 10^{-9}$	~ 1	221 [6]	$\sim 10^{-9}$	~ 1	Pyrex
Bi	227 [8]	$\sim 10^{-10}$	0.3	90 [6]	$\sim 10^{-10}$	~ 1	—	—	—	—
Cd	110 [8]	$\sim 10^{-10}$	0.3	—	—	—	—	—	—	—
Co	—	—	—	330 [6]	$\sim 10^{-9}$	~ 1	310 [20]	0.3	—	Jena glass
Cu	—	—	—	236 [6]	$\sim 10^{-9}$	~ 1	180 [20]	0.3	—	Jena glass
Fe	—	—	—	295 [6]	$\sim 10^{-8}$	~ 1	250 [21]	2.7	—	Slag
Ga	174 [8]	$\sim 10^{-10}$	0.3	—	—	—	—	—	—	—
Ge	—	—	—	253 [6]	$\sim 10^{-10}$	~ 1	415 [22]	4×10^{-5}	~ 4	B_2O_3
Hg	88 [8]	$\sim 10^{-10}$	0.08	—	—	—	—	—	—	—
In	110 [8]	$\sim 10^{-10}$	0.3	—	—	—	—	—	—	—
Mn	—	—	—	—	—	—	308 [6]	$\sim 10^{-9}$	~ 1	Pyrex
Ni	—	—	—	319 [23]	$\sim 10^{-8}$	~ 1	319 [6]	$\sim 10^{-8}$	~ 1	Pyrex
Pb	153 [8]	$\sim 10^{-10}$	0.3	240 [24]	—	—	69 [6]	$\sim 10^{-10}$	~ 1	NaOH +KOH

Pd	—	—	—	332 [6]	$\sim 10^{-8}$	~ 1	310 [6]	~ 0.2	—	Jena glass
Pt	—	—	—	400 [25]	$\sim 10^{-8}$	—	—	—	—	
Rh	—	—	—	426 [25]	$\sim 10^{-8}$	—	—	—	—	
Sb	210 [8]	$\sim 10^{-10}$	0.3	135 [6]	$\sim 10^{-9}$	~ 1	—	—	—	—
Se	—	—	—	25 [6]	$\sim 10^{-8}$	~ 1	—	—	—	
Sn	191 [8]	$\sim 10^{-10}$	0.3	76.7 [26]	—	—	—	—	—	
Te	236 [8]	$\sim 10^{-10}$	0.3	—	—	—	—	—	—	

Note that sample 'volumes' V are specified in mole; ϑ is the cooling rate.
*(Reprinted from Ref. [3], copyright (1991), with permission from Elsevier.)

Table 3 Maximum supercoolings for the solidification of liquid metallic elements, obtained using containerless techniques

Metal	Drop tube	EML	ESL
		ΔT_{max} (K)	
Co		350 [27]	
Cu		266 [28]	
Fe		324 [29]‡	
		420 [30]	
		538 [27]	
Hf	530 [31]†		
Ir	430 [31]†		
Mo	520 [31]†		
Ni		341 [29]‡	239 [32]‡
		452 [27]	
		480 [30]	
Nb	525 [33]†		
Pt	380 [34]†		
Re	975 [35]‡		
Rh	450 [34]†		
Ru	330 [34]†		
Ta	650 [34]†		
Ti	350 [34]†		281 [32]‡
W	530 [35]‡		
Zr	430 [34]†		291 [32]‡

†High vacuum (HV) (0.5 mPa).
‡Ultra-high vacuum (UHV) (<10 nPa).

With the impurity isolation, container-based methods, the maximum supercooling is found in those droplets that are most resistant to nucleation. In an ensemble of many droplets of uniform volume, V_d, the probability of having no nuclei in a given droplet, given χ_V nuclei per unit volume (or equivalently the volume fraction of droplets containing no nuclei), is

$$P(0) = \exp(-V_d \chi_V). \tag{1}$$

If the cooling rate, $\vartheta = |dT/dt|$, is uniform from the melting temperature T_m to the minimum (i.e. maximum-supercooling) temperature T_{min},

$$P(0) = \exp\left(-\frac{V_d}{\vartheta}\int_{T_{min}}^{T_m} I^{st}(T)\mathrm{d}T\right). \tag{2}$$

Table 4 Maximum supercoolings for solidification of selected alkali halide melts, obtained by various methods*

Salt	ΔT_{max} (K)	T_m (K)	ΔT_r
CsBr	162	909.2	0.18
CsCl	152	919.2	0.17
CsF	153	955.2	0.16
CsI	206	894.2	0.23
KBr	162	1003.2	0.16
KCl	169	1049.2	0.16
KI	155	959.2	0.16
LiBr	94	820.2	0.11
LiCl	186	887.2	0.21
LiF	232	1115.2	0.21
NaBr	163	1028.2	0.16
NaCl	168	1074.2	0.16
NaF	281	1261.2	0.22
RbCl	163	988.2	0.16

*Data taken from Ref. [96]. (Reprinted from Ref. [3], copyright (1991), with permission from Elsevier.)

Table 5 Maximum supercoolings for solidification of selected organic liquids, obtained by various methods*

Molecule	ΔT_{max} (K)	T_m (K)	ΔT_r
$Br{\bullet}[CH_2]_2{\bullet}Br$	66.5	238.0	0.24
CBr_4	82^a	362.8	0.23
	82^b	362.8	0.23
CCl_4	50.4^a	249.5	0.20
	50.4^b	249.5	0.20
C_6H_6	70.2	278.6	0.25
$C_6H_5{\bullet}CO_2H$	120	394.7	0.30
diphenyl	86	344	0.25
H_2O	40.5	273.2	0.15
MeCl	55.6	174.0	0.32
naphthalene	94.4	353.6	0.27
NH_3	40.3	195.6	0.21

Superscript a and b represent two possible solid phases.
*Data taken from Ref. [96]. Reprinted from Ref. [3], copyright (1991), with permission from Elsevier.

If different techniques are equally sensitive to $P(0)$ and if heterogeneous nucleation is not a factor, the maximum observed supercooling varies with the ratio of the sample volume to the cooling rate, V_d/ϑ.

Table 6 Maximum supercoolings observed for solidification of liquid metallic elements, obtained by various methods*

Metal	ΔT_{max} (K)	V/ϑ ($m^3 s K^{-1}$)	T_m (K)	ΔT_r	σ_{ls} ($J m^{-2}$)	σ_{ls}^M (kJ mol−1)†	ΔH_f ($kJ mol^{-1}$)	$n*$	$r*$ (Å)
Ag	227	10^{-9}	1234.0	0.18	0.128	5.11	11.4	481	12.5
Al	175	2×10^{-13}	933.3	0.19	0.108	4.25	10.5	340	11.0
Au	230	10^{-9}	1336.2	0.17	0.137	5.45	12.8	505	12.7
Bi	227	3×10^{-10}	544.2	0.42	0.088	5.74	10.9	68	8.3
Cd	110	3×10^{-10}	594.1	0.19	0.058	2.71	6.4	398	12.7
Co	330	10^{-9}	1765.2	0.19	0.238	7.14	15.5	501	11.0
Cu	236	10^{-9}	1356.2	0.17	0.178	5.58	13.0	500	11.2
Fe	420	~0.003	1809.2	0.23	0.277	8.65	15.2	494	11.2
Ga	174	3×10^{-10}	302.9	0.57	0.077	3.35	5.59	38	5.6
Ge	415	10^{-5}	1210.2	0.34	0.300	14.4	32.2	74	7.3
Hg	88	10^{-9}	234.3	0.38	0.031	1.48	2.32	164	9.7
Hf	450	3×10^{-5}	2500.2	0.18	0.221	10.6	24.1	484	13.7
In	110	3×10^{-10}	429.6	0.26	0.036	1.86	3.27	370	13.2
Ir⁺	340	10^{-5}	2727.2	0.12	—	—	—	—	—
Mn	308	10^{-9}	1517.2	0.20	0.216	6.90	14.7	414	10.7
Mo⁺	520	4×10^{-5}	2888.2	0.18	—	—	—	—	—
Nb⁺	525	10^{-5}	2740.2	0.19	—	—	—	—	—
Ni	480	~0.003	1728.2	0.28	0.3	10.40	17.7	317	9.4
Pb	240	$\sim 10^{-8}$	600.6	0.40	0.06	3.51	4.98	185	11.0
Pd	332	10^{-8}	1825.2	0.18	0.207	7.49	16.7	503	12.1
Pt	380	8×10^{-6}	2042.2	0.19	0.239	8.80	19.7	464	11.9
Rh	450	7×10^{-6}	2239.2	0.20	0.301	10.4	22.6	403	11.0
Ru	330	10^{-5}	2583.2	0.13	0.221	7.58	19.7	916	14.4
Sb	210	3×10^{-10}	903.7	0.23	0.130	7.58	19.9	148	10.2
Se	25	10^{-8}	490.2	0.05	0.021	1.13	6.28	1503	21.4
Sn	191	3×10^{-10}	505.1	0.38	0.0705	4.10	7.08	120	9.2
Ta	650	4×10^{-6}	3253.2	0.20	0.301	12.5	24.7	540	13.2
Te	236	3×10^{-10}	723.2	0.33	0.125	7.89	17.6	87	8.9
Ti	350	4×10^{-6}	1940.2	0.18	0.202	8.23	18.8	480	12.6
Zr	430	4×10^{-5}	2125.2	0.20	0.193	9.49	20.1	426	13.3

⁺Heat of fusion not available.
† σ_{ls}^M is the molar interfacial free energy, $\sigma_{ls}^M = \sigma_{ls} V_M^{2/3} N_A^{1/3}$.
*Reprinted from Ref. [3], copyright (1991), with permission from Elsevier.

Assuming that the specific heats of the liquid and solid are the same, it follows that the steady-state nucleation rate I^{st} is (Chapter 2, Eqs. (14) and (55)):

$$I^{st} = A^* \exp\left(-\frac{16\pi}{3k_B T}\frac{\sigma_{ls}^3}{\Delta g^2}\right) = A^* \exp\left(-\frac{16\pi}{3k_B T}\frac{\sigma_{ls}^3}{\Delta h^2 \Delta T_r^2}\right), \quad (3)$$

where A^* is the dynamical prefactor, k_B is the Boltzmann constant and Δg is the volumetric Gibbs free energy difference per unit volume ($\Delta g = \Delta h \Delta T_r$);

Δh is the heat of fusion per unit volume and ΔT_r is the reduced supercooling. The data show a broad range of reduced supercooling values, between 0.1 and 0.5. It has been assumed that the temperature of maximum supercooling is approximated by the temperature at which the homogeneous nucleation rate shows its maximum, but classical theory predicts that the nucleation maximum is at $\Delta T_r \approx 0.56$–0.67, far in excess of the measured values. In fact, the temperature of the nucleation maximum is likely to be a poor approximation of the supercooling limit. A better comparison might be made between values of W^*, the work required to form the critical cluster. Using values for the interfacial energy determined later in this section, assuming Eq. (3) for the steady-state rate (computing values for A^* when data are available), and correcting for droplet sizes and cooling rates, we find $W^* = (60 \pm 2)k_B T$, in the range expected for the metastability limit from the classical theory of homogeneous nucleation.

The liquid–solid interfacial free energy, σ_{ls}, can be estimated from supercooling data, assuming homogeneous nucleation. The prefactor A^* in Eq. (3) contains the interfacial mobility, which is assumed to scale inversely with the viscosity. Although measured values for viscosity in the supercooled regime are generally not available, reasonable estimates for σ_{ls} can still be obtained because (1) existing measurements of viscosity show that it is approximately independent of temperature in the range sampled by the maximum supercooling measurements and (2) the value of σ_{ls} is relatively independent of the exact value for the prefactor, since it is raised to the third power in the exponential. Based on extrapolations of the measured viscosities for liquids at their melting temperature, $A^* \approx 10^{34 \pm 1}\,\mathrm{mol}^{-1}\,\mathrm{s}^{-1}$. The crystal growth velocity in liquid metals is sufficiently large in the supercooled regime that one nucleation event is sufficient to crystallize the droplet. Taking a reasonably small value (0.05) for $P(0)$ in Eq. (2), and assuming that the driving free energy is proportional to the supercooling, the interfacial free energies can be estimated; these are listed in Table 6.

It is well known that the crystal–vacuum interfacial energy is related to the heat of sublimation in molecular crystals with van der Waals bonding and to the crystal lattice energy in ionic crystals. It is reasonable, then, to expect that the interfacial free energies listed in Table 6 should be related to other properties that also depend on the bond energy. The heat of fusion ΔH_f is the obvious choice. It is a molar quantity, however, and σ_{ls} has units of energy per area. Comparisons, then, must be made with the molar interfacial free energy, defined as the free energy of a monolayer interface that contains the Avogadro number N_A of atoms,

$$\sigma_{ls}^{M} = \sigma_{ls} V_M^{2/3} N_A^{1/3}, \tag{4}$$

where V_M is the molar volume. The calculated values for σ_{ls}^M are also listed in Table 6 and are plotted as a function of heat of fusion ΔH_f in Figure 8.

Clearly, σ_{ls}^M scales with the heat of fusion; $\sigma_{ls}^M \approx 0.44(\Delta H_f)$. Values for σ_{ls}^M estimated from the supercooling data, but using different approximations for the Gibbs free energy difference between the liquid and the solid (see Appendix A in Ref. [3]), give similar results, with the multiplier for ΔH_f varying between 0.4 to 0.5. Such clear scaling behavior, for all metals (which have quite different crystal structures) indicates that whatever is the precise meaning of the interfacial free energy (see the problems discussed in Chapter 4), it is a fundamental parameter that determines the nucleation behavior. From rather simple experiments, σ_{ls}^M can be determined for each element and, as we will see later, for alloys and compounds.

Is this the most meaningful way to analyze the data, however? The structures and interatomic potentials of liquid metals are often assumed to be consistent with hard-sphere systems. From molecular dynamics studies of these models, σ_{ls}^M should scale with the melting temperature [36],

$$\sigma_{ls}^M = 0.0027T_m. \tag{5}$$

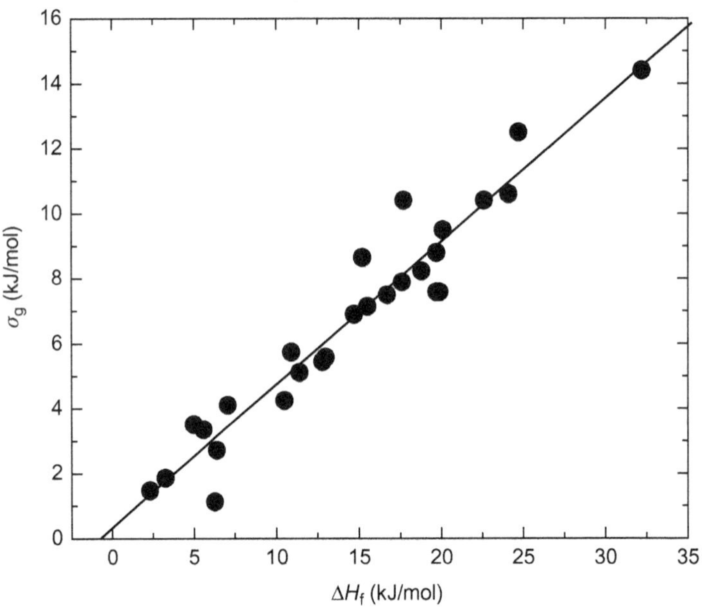

Fig. 8 The molar interfacial free energy, derived from maximum-supercooling experiments on elemental liquids, as a function of the molar heat of fusion. Because of uncertainties in V/ϑ, the data for iron, nickel and lead were excluded when determining the fit to the data. (Reprinted from Ref. [3], copyright (1991), with permission from Elsevier.)

As shown in Figure 9, the scaling with the melting temperature is dramatically worse than with the heat of fusion. However, the average slope is only slightly larger than expected from Eq. (5), equal to 0.0032. It is important to recognize that this scaling depends strongly on the type of bonding. The largest deviation from the linear correlation is for germanium.

2.2.3 Heterogeneous nucleation

Despite the efforts taken to minimize the catalytic sites for nucleation, there always remains the question of whether the nucleation measured is homogeneous or heterogeneous. As discussed in Chapter 6, the form of the nucleation rate is the same for both processes,

$$I^{\mathrm{st}} = A^{*} \exp\left(-\frac{W^{*}}{k_{\mathrm{B}}T}\right), \tag{6}$$

making it difficult to unambiguously identify the mechanism from supercooling data alone. Fundamental studies of heterogeneous nucleation are further complicated by uncertainties in the number of heterogeneous sites and their catalytic efficiencies. Nevertheless, it is possible in some situations to analyze heterogeneous nucleation processes in supercooled liquids. Two examples are provided here.

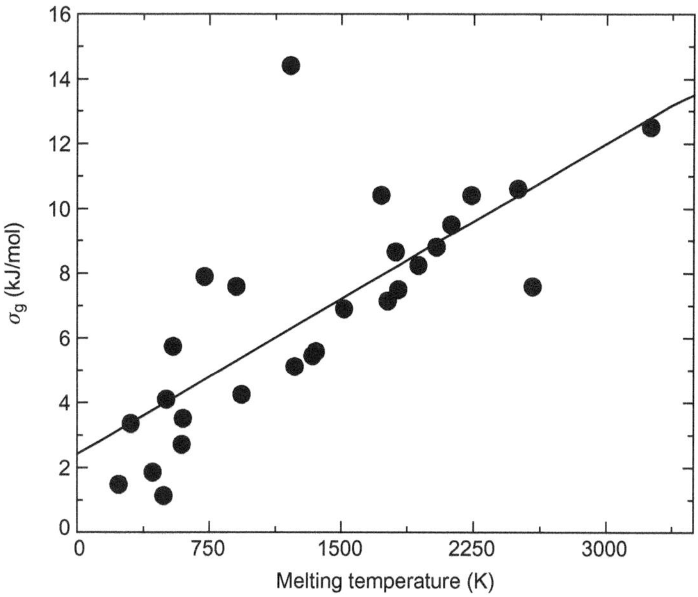

Fig. 9 The molar interfacial free energy, derived from maximum-supercooling experiments on liquid metallic elements as a function of the melting temperature; again data for iron, nickel and palladium were omitted when determining the fit.

Statistical analysis of supercooling data. While the forms of the steady-state nucleation rates are similar, A^* is typically smaller, by 8 to 10 orders of magnitude for heterogeneous nucleation than for homogeneous nucleation. It is then possible, at least in principle, to identify the type of nucleation by a careful statistical analysis of supercooling data taken from a single droplet.

Due to the fast crystal growth in liquid metals, it is assumed that transformation is completed after one nucleation event, and the droplet solidification is essentially adiabatic [37, 38]. The probability of the occurrence of χ nuclei appearing in a time interval, $0 \leq t \leq \tau$, within a volume V given a steady-state nucleation rate per unit volume, $I^{st}(T)$, is

$$P(\chi; 0, \tau) = \frac{I^{st}(T)V\tau \exp\left(-I^{st}(T)V\tau\right)}{\chi!}. \tag{7}$$

The probability that a single nucleus appears in the time interval $(0 \leq 0 \leq \tau)$ is obtained by setting χ equal to one in Eq. (7). The probability that a nucleus will appear in the time interval between τ and $\tau+\delta\tau$ is given by $P(1;\ \tau,\ \tau + \delta\tau) = P(1;\ 0,\ \tau + \delta\tau) - P(1;\ 0,\ \tau)$, or

$$\begin{aligned} \delta P(1; \tau, \tau + \delta\tau) = {} & I^{st}(T)V(\tau + \delta\tau)\exp\left(-I^{st}(T)V(\tau + d\tau)\right) \\ & - I^{st}(T)V\tau \exp(-I^{st}(T)V\tau). \end{aligned} \tag{8}$$

On expanding the exponential as a Taylor series, in the limit $\delta\tau \rightarrow 0$, Eq. (8) may be written as

$$\delta P(1; \tau, \tau + \delta\tau) = I^{st}(T)V\delta\tau \exp\left(-I^{st}(T)V\tau\right). \tag{9}$$

In a supercooling experiment, the temperature changes continuously. As for Eq. (2), we assume a constant cooling rate ϑ from the melting temperature T_m to the minimum temperature T_{min} giving

$$\delta P(1; \tau, \tau + \delta\tau) = \frac{I^{st}(T)V}{\vartheta} \delta T \exp\left(-\frac{V}{\vartheta} \int_{T_{min}}^{T_m} I^{st}(T)\, dT \right), \tag{10}$$

where δT corresponds to the temperature change in the time interval $\delta\tau$. Dividing by δT gives a probability density as a function of temperature,

$$P(1; T) = \frac{I^{st}(T)V}{\vartheta} \exp\left(-\frac{V}{\vartheta} \int_{T_{min}}^{T_m} I^{st}(T)\, dT \right). \tag{11}$$

This has a similar form to Eq. (2), as expected. By fitting Eq. (11) to the distribution of supercooling values, the magnitude of the prefactor can be estimated.

The shape of the distribution is sensitive to the temperature dependence of the nucleation prefactor and nucleation barrier. In particular, the distribution becomes narrower and less skewed toward lower temperatures (smaller third moment) as the magnitude and activation barrier of the prefactor increase. As an example, Figure 10 shows the measured frequency of occurrence of maximum-supercooling values for zirconium samples of three different purities. All droplets were approximately the same diameter (1.5–1.65 mm) and processed by ESL in a vacuum of 10^{-7} to 10^{-8} torr. Zirconium was chosen, since as for titanium, at high temperature oxygen tends to dissolve in the metal rather than forming a surface oxide that could catalyze nucleation. The same is true for other potential surface contaminants such as nitrogen and carbon, which can give heterogeneous nucleation sites in levitated single-droplet studies. The solid lines are fits to Eq. (11), assuming the steady-state rate in Eq. (3).

The peak of the distribution sharpens and shifts to a larger maximum supercooling with increasing sample purity, suggesting that nucleation in the less pure samples is predominantly heterogeneous. This is in agreement with the fit values for the prefactors, with A^* ranging from

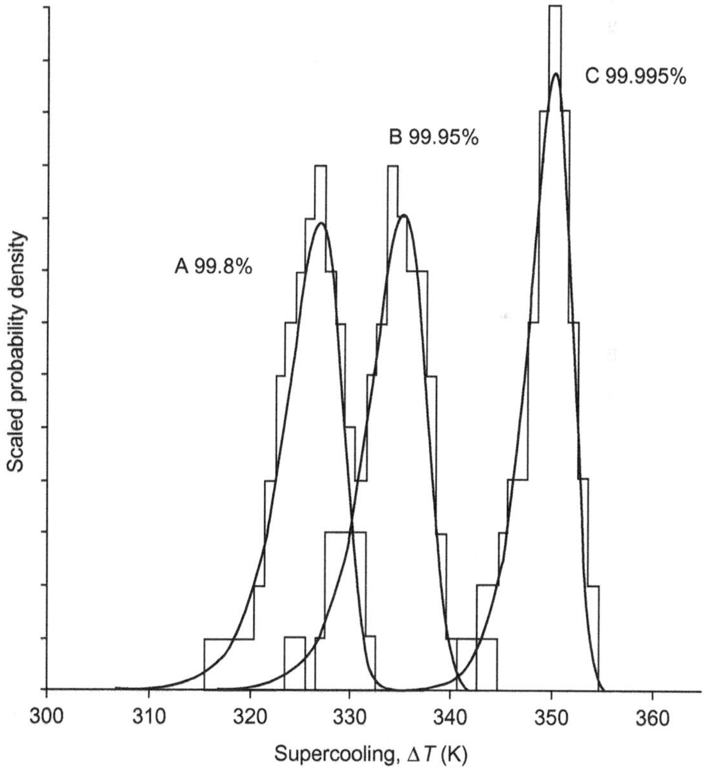

Fig. 10 Histograms of frequencies of the maximum supercooling in liquid zirconium samples with different impurity levels. The curves are fits to Eq. (11). (Reprinted from Ref. [37], copyright (1998), with permission from Elsevier.)

$10^{38}\,m^{-3}\,s^{-1}$ for the low-purity sample (A) to $10^{49}\,m^{-3}\,s^{-1}$ for the most pure sample (C). Similarly the nucleation barrier increases from 64 k_BT (A) to 88 k_BT (C). The most likely catalytic sites are oxides; the oxygen concentration is 1140 ppm in sample A, 540 ppm in B, and 50 ppm in C. It should be noted that the values for A^* and W^* from the purest samples are much larger than expected from the classical theory of nucleation ($A^* \approx 10^{39}\,m^{-3}\,s^{-1}$ and $W^* \approx 60\,k_BT$). These values were obtained assuming a temperature-independent interfacial free energy. If $\sigma_{ls} \propto T$, as expected for a diffuse interface (Chapter 4), the measured values ($A^* \approx 10^{40}\,m^{-3}\,s^{-1}$ and $W^* \approx 66\,k_BT$) are in much better agreement with the classical predictions [37, 38].

Such studies have been made on only a very limited number of liquids. It is, therefore, unknown to what extent heterogeneous nucleation plays a role in the maximum-supercooling data presented in Tables 2–6. However, the excellent scaling behavior with the heat of fusion obtained by assuming homogeneous nucleation suggests that the role of heterogeneous nucleation is small.

Supercooling measurements on entrained droplets. Entrained-droplet or "mush-quenching" techniques have been widely used for fundamental investigations of heterogeneous nucleation. In these studies, the melt solidifies partially on cooling to give a dispersion (volume fraction $<10\%$) of liquid droplets in a crystalline solid–solution matrix. The nucleation of solidification in the entrapped liquid droplets has been studied in a variety of systems (see for example [39–43]). Nucleation is probably heterogeneous, but some solid matrices may act as only weak catalytic sites. The technique has the advantage that intimate contact between the liquid and the solid matrix provides a good sink for heat evolved during crystallization. Uncertainties arise, however, from pressure effects, and difficulties in estimating the true supercooling due to changes in the liquidus temperature with changes in liquid composition.

Ultrafine microstructures are frequently obtained in rapidly quenched alloys. When these samples are reheated into the mushy zone, a particularly fine dispersion of liquid droplets can be obtained, with typical droplet diameters on the order of 20 nm and a population of more than 10^{15} droplets per unit volume. The ultrafine dispersion makes the solidification behavior very sensitive to nucleation processes. Furthermore, the well-controlled size distribution allows quantitative calorimetric studies to be made of the nucleation kinetics, typically using DSC [44]. With transmission electron microscopy (TEM), possible orientation relationships between the phases can be studied, important for gaining an understanding of the structural and chemical interactions underlying the mechanism of heterogeneous nucleation [45, 46].

Of particular relevance here, entrained-droplet techniques have permitted a quantitative analysis of heterogeneous-nucleation kinetics.

When large supercoolings ($\Delta T > 50\,\text{K}$) are required for heterogeneous nucleation, the spherical-cap model for nucleation (Chapter 6, Section 2.2) provides a good fit to the kinetic data [44, 47]. However, if the supercooling is very small, indicating potent nucleation and a small contact angle for a solid nucleus forming in the droplet on the matrix, the classical nucleation theory appears to fail [48, 49]. Attempts to fit the data from DSC scans to the classical model give unphysical values for the contact angle and for the number of heterogeneous sites. This is illustrated by the results collected in Table 7. For supercoolings smaller than $\sim 40\,\text{K}$, the derived values are not physically reasonable. The cause of the breakdown is unclear. The classical theory can include effects of substrate curvature (Chapter 6, Section 2.4), and such effects have not been taken into account. More fundamentally, it has been suggested that in this range of potent nucleation, an adsorption model may be more appropriate than a fluctuation-based nucleation approach [43, 50]. Beyond a critical supercooling limit, it becomes thermodynamically favorable to develop an adsorbed layer of a crystal phase, which becomes the basis for further growth. The failure of the classical theory may have a simpler origin, however. For such fine liquid droplets, the critical size is several times the size of the droplet itself; given the discussion in Chapter 4 of the diffuseness of the crystal–liquid interface, it is unreasonable to expect the classical theory to remain valid. The actual nucleation process in this extreme limit is unclear.

2.3 Noncongruent freezing

Many liquid–solid transformations of technological importance involve a change in composition, with solute being partitioned across the liquid–solid interface during freezing or melting; solute partitioning must be

Table 7 Solidification results for entrained droplets in mush-quenching studies*

Droplets	Matrix	Supercooling (K)	Contact angle (in degrees)	Number of sites per droplet
Al	Al_3Zr	0.2	—	—
Al	Al_3Ni	3.3	—	—
Pb	Cu	0.5	4	10^{-12}
Pb	Al	22	21	10^{-6}
Pb	Zn	30	23	10^{-8}
In	Al	13	27	10^{-7}
Cd	Al	56	42	30
Sn	Al	104	59	100

*Data taken from Ref. [51].

considered for a proper analysis of the nucleation kinetics (Chapter 5). Solute diffusion in the parent phase is critical for the nucleation step in solid-state precipitation and must be coupled with the interfacial processes (Chapter 9). However, for crystal nucleation in a liquid (in the presence of convection), and for many cases of crystal nucleation in a glass, transport at the interface remains dominant. The steady-state nucleation rate of a binary A–B phase from a parent phase of different composition then has a similar form to that for polymorphic nucleation (Chapter 5, Section 4),

$$I^{st} = A^* \exp\left(-\frac{W(a^*, b^*)}{k_B T}\right), \qquad (12)$$

where $W(a^*, b^*)$ is the reversible work of formation of an embryo of critical size n^*, composed of a^* atoms of species A and b^* atoms of B. Assuming, for simplicity, a sharp interface between the embryo and the original phase, the work of formation is approximately

$$W(a^*, b^*) = a^* \Delta\mu_A + b^* \Delta\mu_B + \zeta \bar{\sigma}_{ls} (a^* \bar{v}_A + b^* \bar{v}_B)^{2/3}. \qquad (13)$$

Here $\Delta\mu_A$ and $\Delta\mu_B$ are the differences between the Gibbs free energies of the original and new phases per molecule for A and B, ζ is a geometric factor, which is $(36\pi)^{1/3}$ for spherical clusters, $\bar{v}_A$ and $\bar{v}_B$ are the molecular volumes for A and B species respectively, and $\bar{\sigma}_{ls}$ is an effective interfacial free energy. As discussed in Chapter 5, A^* is a complex function of the composition of the liquid, the atomic mobilities of A and B at the interface, and the two mutually orthogonal curvatures at the saddle point in the surface of the work of cluster formation as a function of a and b. Depending on the relative attachment rates of the different species, the composition of the nucleating cluster might be very different from that expected from free-energy considerations alone (Chapter 5, Section 4.3). The simplifying assumption, that the compositions of the clusters remain constant throughout the nucleation step, allows Eqs. (12) and (13) to be used, but the values of $\Delta\mu_{A,B}$ and $\bar{\sigma}_{sl}$ are difficult to determine.

There have been only a few systematic studies of the maximum supercooling as a function of liquid composition. In metallic alloys, most studies have focused on simple eutectic or miscible alloys. Figure 11 shows supercooling data for the eutectic Sn–Bi alloy. The supercooling values parallel the liquidus curves, indicating that the composition-dependence of the nucleation rate is dictated primarily by changes in free energies [52]. Similar effects are found in other liquid alloys [16] and for nucleation in some silicate glasses (Chapter 8).

These experimental data indicate that nucleation can still be treated within the framework of the classical theory for alloy melts, as an interface-limited process. However, it is possible that if convection in the liquid could be reduced to a sufficient level (say in a microgravity

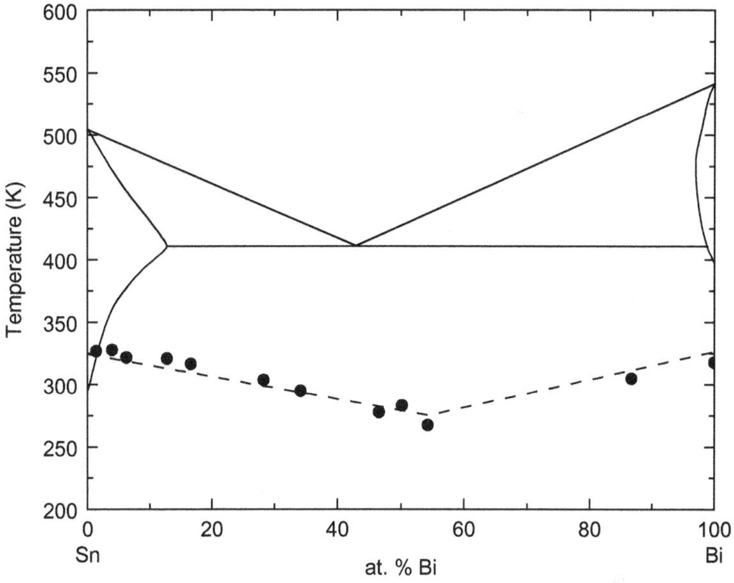

Fig. 11 Maximum-supercooling data for liquid Sn–Bi, superimposed on the equilibrium phase diagram. (Reprinted from Ref. [55], copyright (1983), with permission from Elsevier.)

environment), diffusion effects would become rate-limiting, rendering the classical theory inapplicable. The importance of stirring has been confirmed in a microgravity supercooling experiment (MSL-1) on the space shuttle [53], showing the influence on the secondary nucleation of the stable phase in ternary steels. Even if interfacial kinetic processes are rate-limiting, the nucleation may still be complicated. Desré et al., for example, have argued that in some cases nucleation may proceed by a composition fluctuation in the original phase, followed by structural or topological ordering within the phase-separated regions [54]. This is one example of a coupled phase transition that is not properly treated within the classical theory (see Section 5 and Chapter 10, Section 6 for further discussion of this point).

2.4 Nucleation of icosahedral quasicrystals

In 1952, Frank proposed a structural origin of the barrier for crystal nucleation that explained the large supercoolings observed in liquid metals. The crystal structures of many metals are described by one of two close-packed configurations of hard spheres, cubic close-packed (ccp) and hexagonal close-packed (hcp). Frank pointed out that another close-packed configuration, icosahedral packing, was slightly denser and energetically preferred. While incompatible with the long-range periodicity of crystal phases, icosahedral short-range order could be favored in metallic liquids, where long-range translational order is absent. For the liquid to solidify, this locally stable configuration must be replaced by that of the crystallizing phases, forcing a transition through higher-energy configurations and giving rise to a nucleation barrier.

Recently, more direct structural studies of liquids have become possible using the high X-ray intensities now available with synchrotrons, coupled with an advanced ability to confine reactive liquids without using physical containers. The discovery in 1984 of a new phase of condensed matter, the icosahedral quasicrystal (i-phase), which has the long-range order typical of a crystal phase but with an icosahedral rotational point-group symmetry that is forbidden for a periodic structure, opened the door for further novel investigations of liquid and crystal structures.

As will be discussed in Chapter 8, Holzer and Kelton [56] first demonstrated that the interfacial free energy between the icosahedral phase and a metallic glass of the same composition is extremely small, suggesting that the local atomic structures of the two phases are similar. Indications that this was likely came from earlier studies of the nucleation of quasicrystals from the liquid. Bendersky and Ridder [57], for example, found a very high density of quasicrystal grains ($\sim 10^{24}\,\mathrm{m}^{-3}$) in the smallest droplets solidified under vacuum in free fall in electrohydrodynamic atomization.

The first quantitative supercooling measurements in liquids that form quasicrystals were made by Holland-Moritz and coworkers [58, 59] on electromagnetically levitated droplets of Al–Cu–Fe and Al–Cu–Co liquids; these studies were later extended to other Al-based alloys [60]. The first measurements of the relative supercooling of Ti,Zr-based quasicrystals and crystal approximants were made recently for electrostatically levitated Ti–Zr–Ni liquids [61]. Selected data from these studies are collected in Table 8.

In both alloy classes, those solidifying to icosahedral quasicrystals show the smallest supercooling, consistent with a low barrier that is related to local icosahedral order in the liquid. Those solidifying to crystal approximant phases (λ-phase, μ-phase and C14 Laves phase) show a larger supercooling, reflecting greater differences between their tetrahedral structures and the local liquid structures. Simple crystal phases, such as β(Ti–Zr) and β(Al–Cu–Co), show the greatest supercooling due to the incompatibility of the local structures in the crystal and liquid. Studies of Al-based decagonal quasicrystal phases, which are quasiperiodic in two dimensions but periodic in the third, have $0.14 \leq \Delta T_r \leq 0.16$, corresponding to $\sigma_{ls}(T_{min}) = 0.11$–$0.16\,\mathrm{J\,m}^{-2}$ [60]. These studies support, but do not prove, the notion that an increased degree of polytetrahedral order lowers the interfacial free energy between the

Table 8 Reduced supercooling ($\Delta T_r = \Delta T_{max}/T_{liq}$) and calculated interfacial free energy for solidification of Al-transition metal alloy liquids to quasicrystals and crystal approximants*

Quasicrystal				Crystal Approximant			
Alloy	phase	ΔT_r	$\sigma_{ls}(T_{min})$ (J m^{-2})	Alloy	phase	ΔT_r	$\sigma_{ls}(T_{min})$ (J m^{-2})
$Al_{58}Cu_{34}Fe_8$	i-phase	0.09	0.09±0.01	$Al_{13}Fe_4$	λ-phase	0.12	0.16±0.01
$Al_{30}Cu_{34}Fe_6$	i-phase	0.09	0.09±0.01	$Al_{62}Cu_{25.5}Fe_{12.5}$	λ-phase	0.14	0.15±0.01
$Al_{72}Pd_{21}Mn_{17}$	i-phase	0.11	0.10±0.01	Al_5Fe_2	μ-phase	0.14	0.18±0.01
$Ti_{37}Zr_{42}Ni_{21}$	i-phase	0.09	0.06±0.01	$Ti_{37}Zr_{38}Ni_{25}$	C-14 Laves	0.14	0.10±0.01

*Data taken from Refs. [60] and [61].

crystallizing phase and the supercooled liquid, increasing the nucleation rate.

While in-situ X-ray diffraction studies clearly show that the Ti–Zr–Ni i-phase is the primary nucleating phase in $Ti_{39.5}Zr_{39.5}Ni_{21}$ alloys, cooling curves (Figure 12) indicate that it is metastable. The first recalescence event, indicating formation of the i-phase, is followed within a few seconds by a second recalescence as it transforms to the C14 Laves phase, which is stable at these elevated temperatures. As is clear from Table 8, the C14 Laves phase is also easy to nucleate, as expected from its polytetrahedral local structure. That a metastable i-phase forms before the C14 phase, which has a greater driving free energy for formation, indicates that the nucleation barrier is still smaller — this is, of course, reflected in the smaller value derived for the interfacial free energy.

Following the discussion in Chapter 4, more can be learned by examining the local order of the two phases (here the liquid and the i-phase or C14 phases), in addition to their interfacial free energies with respect to each other. Figure 13 shows the measured X-ray structure factor for the $Ti_{39.5}Zr_{39.5}Ni_{21}$ liquid as a function of supercooling temperature. The most interesting feature of the data is the enhancement of a shoulder on the high-q (momentum transfer) side of the second peak in $S(q)$ with increasing supercooling, in the temperature range where nucleation of the i-phase becomes favorable. This shoulder is consistent with local icosahedral order in the liquid. The enhancement of the shoulder with decreasing temperature demonstrates that the short-range order in the liquid becomes better defined and more pronounced, confirming predictions made earlier by Steinhardt et al. [62]. These conclusions of a growing icosahedral order in transition-metal liquids

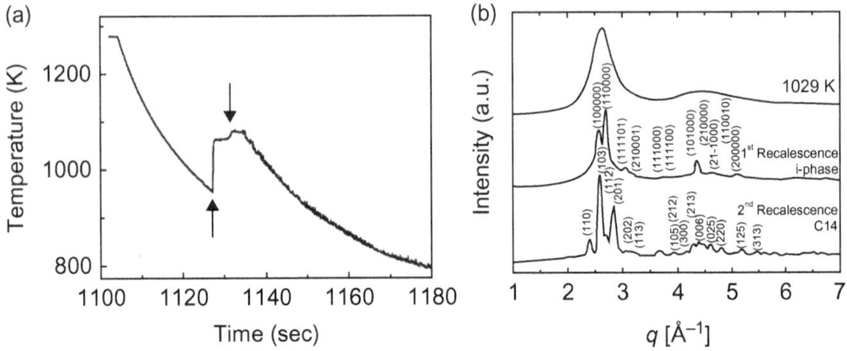

Fig. 12 (a) Cooling curve for an electrostatically levitated 2.5 mm droplet of $Ti_{39.5}Zr_{39.5}Ni_{21}$ (at.%) showing recalescence events (indicated by arrows); (b) X-ray diffraction pattern for supercooled liquid of Ti–Zr–Ni alloy at 1029 K, during the first recalescence to the i-phase, and during the second recalescence to the C14 phase. (Reprinted from Ref. [61], copyright (2003), American Physical Society.)

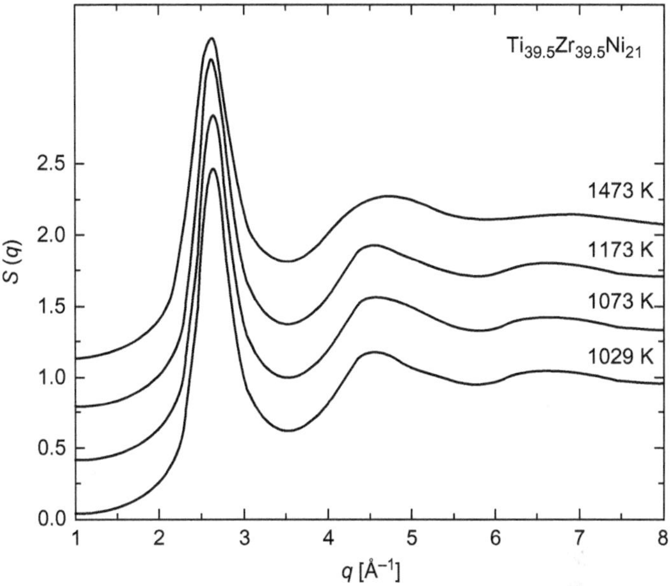

Fig. 13 $S(q)$ from a $Ti_{39.5}Zr_{39.5}Ni_{21}$ (at.%) liquid as a function of temperature. An increase in the intensity of the shoulder on the second peak is observed as the temperature is lowered below the liquidus (1083 K). (Reprinted from Ref. [61], copyright (2003), American Physical Society.)

have been further supported by a more detailed analysis of the scattering data [63, 64] and by ab-initio calculations [65, 66].

These data confirm that the metastable i-phase nucleates first because it has a short-range order similar to that of the liquid. The correspondence between local order in the liquid and the interfacial energy gives quantitative agreement with experimental supercooling data and confirms Frank's hypothesis. Close inspection shows that the asymmetry in the second peak in $S(q)$ is present even above the liquidus temperature, indicating that some icosahedral order already exists in the equilibrium liquid. Whether the critical cluster diameter for nucleation of the new phase is shorter or longer than the coherence length of the order in the liquid must be important. At equilibrium or near equilibrium (when liquid $Ti_{39.5}Zr_{39.5}Ni_{21}$ is held isothermally at $\Delta T \leq 30\,K$), the stable C14 phase is the first to nucleate, after a long time, due to the small driving free energy and the nucleation barrier arising from the icosahedral short-range order in the liquid. As the temperature of the liquid is decreased, the extent of the icosahedral order grows, and the nucleation of the metastable i-phase becomes possible below its metastable liquidus temperature. The critical size of an i-phase nucleus also decreases with decreasing temperature (due to an increasing driving free energy), eventually approaching the size of the regions of icosahedral short-range order, decreasing the nucleation barrier and increasing the probability that a stable critical cluster will form. Fluctuations in the liquid, then, act as a type of template for the formation of this ordered phase. It is not surprising that the local structures of network-forming "strong" [67] liquids or polymers influence the phases that form, but it is notable that the same is true in more "fragile" metallic liquids, despite weaker and more isotropic atomic interactions. The influence of preexisting local order in the liquid, then, is important in the liquid-to-crystal phase transition.

3. MEASUREMENTS OF THE NUCLEATION RATE

It is difficult to test quantitatively nucleation theories from supercooling experiments alone. As already indicated, the effects of heterogeneous sites are difficult to assess. An arbitrary value of the nucleation rate is assumed in order to analyze the supercooled limit, growth kinetics are neglected, and the statistics of nucleation in small volumes and at high cooling rates are often ignored. For a more critical evaluation of nucleation theories, it is necessary to obtain measurements of the magnitude and temperature dependence of the nucleation rate. Typically, these measurements are made on ensembles of droplets, in a manner similar to that in supercooling experiments. Due to the difficulties of these experiments, quantitative measurements have been made on only a limited number of metallic and organic liquids. In many cases, the data show a discrepancy between the measured nucleation prefactor A^* and the value predicted from the classical theory.

As for the supercooling studies, it is first important to establish that the nucleation rate is likely homogeneous and not heterogeneous. Turnbull determined the nucleation rate in coated mercury droplets using dilatometry to measure the fraction of droplets solidified as a function of time. Those data, reduced to a master isotherm by scaling the time with temperature, are compared with predictions for different nucleation modes in Figure 14. It is evident that nucleation is sensitive to the surface coating. Figure 14a shows the fraction solidified as a function of time for liquid droplets of mercury coated with mercury laurate. Figure 14b presents similar data for droplets coated with mercury acetate. The curves are from calculations for droplets of a given diameter or with a range of diameters, exhibiting nucleation throughout their volume or only at their surface. The samples coated with mercury laurate show results consistent with nucleation throughout their volume, while those coated with mercury acetate crystallized by nucleation at the surface only. Assuming that the volume nucleation is homogeneous, these data may be analyzed within the classical theory.

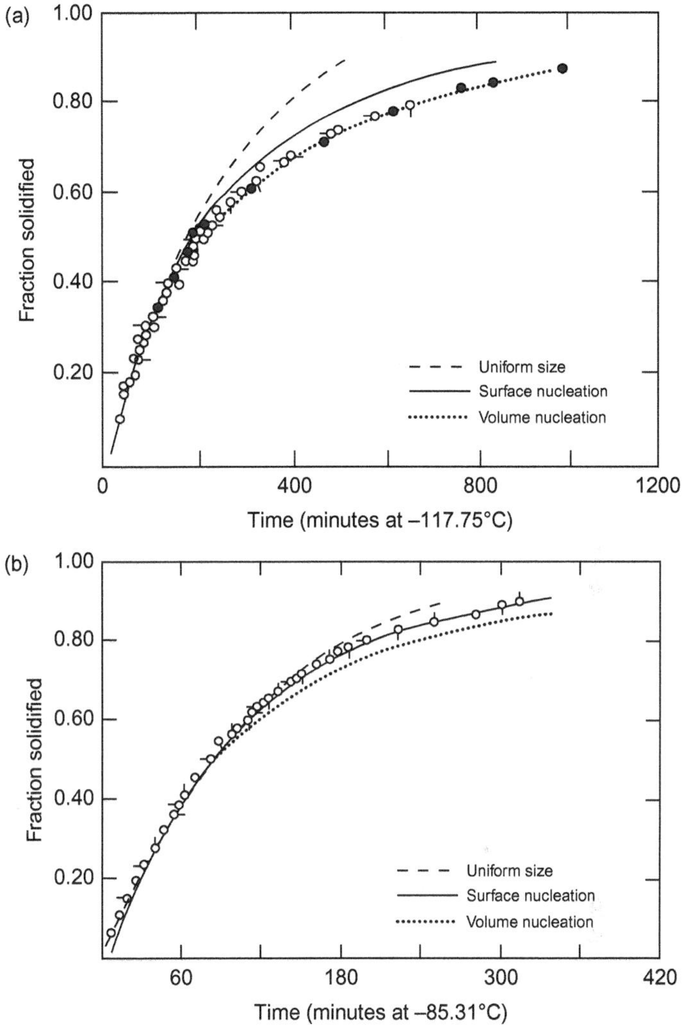

Fig. 14 (a) For mercury droplets coated with mercury laurate, the fraction solidified as a function of time at −117.75 °C, demonstrating nucleation throughout the droplet volume. (b) For mercury droplets coated with mercury acetate, the fraction solidified as a function of time at −85.31 °C, demonstrating nucleation at the droplet surfaces. The "uniform size" curves were calculated assuming a uniform distribution of droplet sizes; the other two curves assumed a distribution of droplet sizes. (Reprinted with permission from Ref. [2], copyright (1952), American Institute of Physics.)

3.1 Analysis within the classical theory

From Eq. (3), a plot of $\log I^{st}$ vs $(T\Delta T^2)^{-1}$ (where $\Delta T = T_m - T$) should give a straight line with slope proportional to σ_{ls}^3 and an intercept equal to

Table 9 Parameters derived from measurements of the steady-state nucleation rate for solidification $I^{st}(T)$ of selected systems*

System	σ_{ls} (J m^{-2})	log $A*$ (m^{-3}s^{-1})
Hg[a]	0.031	48.1
$C_{17}H_{36}$[b]	0.0072	36.5±2
$C_{18}H_{38}$[b]	0.0096	37.3±2
$C_{24}H_{50}$[b]	0.0082	30±4
H_2O[c]	0.032	52.5

[a]Ref. [2]
[b]Ref. [97]
[c]Ref. [98]
*Reprinted from Ref. [3], copyright (1991), with permission from Elsevier.

log A^*. The experimental data show this behavior; derived parameters for selected systems are listed in Table 9. Except for the linear alkane polymers, however, the measured values for A^* are orders of magnitude larger than predicted from the classical theory (for mercury, $A^*_{\text{theoretical}} \approx 10^{41}\,\text{m}^{-3}\,\text{s}^{-1}$). Turnbull first noted that the experimental values for A^* agree with theoretical predictions if σ_{ls} increases linearly with temperature, suggesting a sizable entropy deficit (*negentropy*) near the interface arising from ordering in the liquid (Chapter 4). That the data for the linear polymers give values for A^* that are in better agreement with classical theory predictions suggests that little additional ordering occurs there, implying that the local order (chains aligned parallel to each other) is similar in the solid and liquid phases.

The calculated interfacial free energies for the *n*-alkane liquids are also an order of magnitude smaller than for mercury. Turnbull and Spaepen attribute these two effects to the nonrandom probability for occurrence of linear conformations in the melt [68]. A small temperature dependence of the interfacial energy results because the localization of a small number of CH_2 groups forces the localization of a larger number of CH_2 groups in the same molecule without additional loss of entropy.

3.2 Fits to diffuse-interface models

An increasing order in the liquid as it approaches the interface with the cluster provides an entropic contribution to the crystal–liquid interfacial energy. Within a classical approach, this may be taken into account phenomenologically by assuming that the crystal–liquid interfacial energy has a linear dependence on temperature. This approach breaks down, i.e. the classical theory fails, when the radius of the cluster approaches the interfacial width. It is then necessary to take a more rigorous approach, comparing the nucleation data with predictions from theories that are not based on a Gibbsian sharp-interface analysis of the work of cluster formation (see Chapter 4). While the fundamental parameters needed to make such a direct comparison are often lacking, it is possible to make a valid assessment by using the different predictions for the work of cluster formation made by the thermodynamic models for nucleation that were discussed in Chapters 2 and 4.

Letting ϖ be the cube-root of the ratio of W^*_{NC} (the work to form a critical cluster in the nonclassical model of interest) to W^*_{CNT} (that computed from the classical theory) (i.e. $\varpi(T) = (W^*_{NC}/W^*_{CNT})^{1/3}$) the steady-state nucleation rate can be written

$$I^{st} = A^* \exp\left(-\varpi(T)^3 \frac{W^*_{CNT}}{k_B T}\right), \tag{14}$$

yielding (using $W^*{}_{CNT}$ from Chapter 2, Eq. (14))

$$\ln\left(\frac{I^{st}}{A^*}\right) = -\varpi(T)^3 \frac{16\pi}{3k_B T} \frac{\sigma^3_{CNT}}{\Delta g^2}, \tag{15}$$

where σ_{CNT} is the classical theory value for the solid-liquid interfacial free energy and Δg is the Gibbs free energy difference per unit volume. The nucleation data may now be analyzed in a similar way to that followed for the classical theory. Using the expected value for A^* for the classical theory, and plotting $\ln(I^{st}/A^*)$ as a function of $X \equiv \varpi(T)^3 \Delta g^{-2} T^{-1}$, a straight line should be obtained, with a slope that is proportional to σ^3_{CNT} and an intercept at the origin. Since the correct value for σ_{CNT} is unknown, the value of the slope does not allow a discrimination between the different models. But, those models that do not give a line passing near the origin can be eliminated.

The models that were fit to the experimental data were discussed in Chapter 4; for convenience, they are collected again in the Appendix and the expressions for the work of cluster formation are listed there. For the fits, the driving free energy was computed from the measured heat of fusion, Δh, and the specific heats of the two phases (Eq. (14) in Chapter 8); the interfacial mobility was computed from the viscosity using the Stokes–Einstein relation. (A more detailed discussion of the analysis is given in Chapter 8, Section 6.2.4) Figures 15a and 15b show $\ln(I^{st}/A^*)$ as a function of X for mercury and two of the n-alkane liquids, assuming the classical theory of nucleation with a temperature-independent interfacial free energy. While linear fits are obtained, the intercepts are incorrect (see Table 10). For example, the intercept for mercury is 14.3, or over 14 orders of magnitude larger than expected from the classical theory. This is worse agreement than obtained from the fits reported in Section 3.1 (Table 9), reflecting the error in the simple expression used in those fits for the driving free energy, $\Delta g \approx \Delta h \Delta T_r$. The intercepts of the n-alkane liquids are again close to zero, consistent with the results of Section 3.1. All of the models in Table 10 give similarly good linear fits to these data. Figures

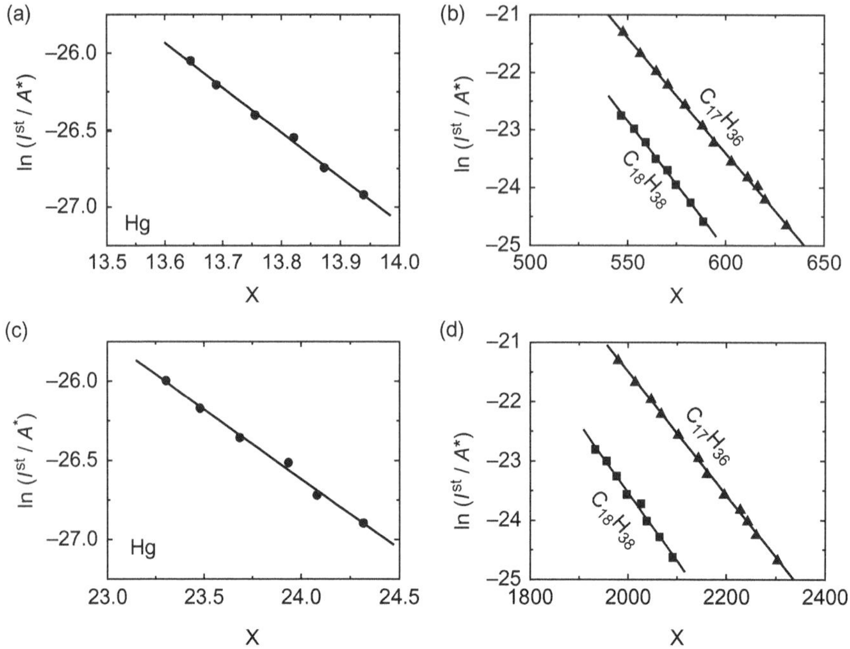

Fig. 15 Fits of models listed in Table 10 (see also Table 1 of the Appendix) to nucleation data for liquid mercury, $C_{17}H_{36}$ and $C_{18}H_{38}$ for classical nucleation theory (CNT, a and b) and the phenomenological diffuse-interface theory (DIT, c and d); $X = \varpi(T)^3 \Delta g^{-2} T^{-1}$. (Reprinted with permission from Ref. [69], copyright (1997), American Institute of Physics.)

15c and 15d show this for the phenomenological diffuse-interface theory (DIT).

Based on the data in Table 10, only two models correctly predict intercepts that lie near the origin, (DIT, Chapter 4, Section 3) and the semiempirical density-functional approximation (SDFA, Chapter 4, Section 4.3); all others give intercepts that are several orders of

Table 10 Intercept values and errors obtained by fitting experimental nucleation data for solidification to different models (see Appendix)

System	CNT	SCCT	DIT	SDFA	PDFA	MWDA
$C_{17}H_{36}$	0.6 ± 0.8	-1.5 ± 0.7	-0.7 ± 0.8	-0.5 ± 0.8	2.0 ± 0.8	0.5 ± 0.8
$C_{18}H_{38}$	1.5 ± 1.6	-1.3 ± 1.4	-0.2 ± 1.5	0.6 ± 1.5	2.9 ± 1.7	1.4 ± 1.6
Hg	14.3 ± 4.0	11.3 ± 3.7	5.7 ± 2.1	-8.9 ± 1.7	17.0 ± 4.2	20.1 ± 4.5

CNT, classical nucleation theory; SCCT, self-consistent CNT; DIT, diffuse-interface theory; SDFA, semiempirical density-functional approximation (DFA); PDFA, perturbative DFA; MWDA, modified weighted DFA (See Appendix). (Taken with permission from Ref. [69], copyright (1997), American Institute of physics.)

magnitude too large. In a way, it is not surprising that the DIT gives good agreement since, as we have seen, the CNT also gives good agreement if a temperature-dependent interfacial free energy is chosen. The DIT model provides a more systematic way of understanding this; it is a consequence of the diffuseness of the interface between small clusters and the original phase. The more sophisticated density-functional approaches confirm this.

4. CRYSTALLIZATION IN COLLOIDAL SUSPENSIONS

Soft-matter systems have at least one mesoscopic (nanometer to micrometer) length scale. Examples include polymers, biological macromolecules and colloids. *Colloids* (or *colloidal dispersions*) are homogeneous mixtures in which the unit size of the dispersed solute phase is much larger than that for the solvent [70]. The dispersed phase can be gas, liquid, or solid. Common examples include liquid or solid *aerosols*, *foams* such as whipped cream, *emulsions* such as mayonnaise, *gels* such as cheese or jelly, and *sols* such as paint or milk.

Colloidal suspensions, composed of solid particles uniformly dispersed in a liquid, can be used as model systems for the study of phase equilibria and transitions in atomic and molecular materials. The large sizes of the dispersed solid particles and the resulting longer transformation times allow accurate measurements of the crystallization kinetics. The particles in a colloid can be charged or uncharged. The ability to adjust the interaction potential of the physical system under study allows more accurate matching to the assumptions of computer simulations or theoretical models, leading to more quantitative comparisons. The interaction potentials can be tuned by adjusting the particle density and charge and by changing the concentration of the screening electrolyte. They can range from screened Coulomb to dipole–dipole (van der Waals) to hard-core potentials. Crystal nucleation and growth occur at lower packing fractions in weakly attractive charged colloids [71–73], than in hard-sphere systems. Computer calculations (Chapter 10) indicate that the soft repulsion for charged spheres favors an inhomogeneous density and lowers the crystal–liquid interfacial free energy [74].

Colloidal dispersions are not without complicating factors, however. The colloidal particles are never completely uniform in diameter. The variation in diameter can have a significant impact on the nucleation and growth rates. Experiments on hard-sphere colloids, for example, indicate that crystallization is suppressed when the range of the colloidal particle diameter exceeds approximately 12% [75].

Depending on the value of the particle volume fraction ς (which plays the role of a concentration) [76] and the electrolyte concentration, colloidal systems form either crystalline ccp or bcc phases, or liquid phases [77]. For a hard-sphere colloidal dispersion, the critical volume fraction for freezing, corresponding to the liquidus, is $\varsigma_l = 0.494$ [78]. For larger volume fractions, a liquid and solid phase coexist, with the system becoming completely solid at the solidus, $\varsigma_s = 0.545$. With increasing shear rates, colloidal systems undergo structural transformations. For a sufficiently high shear rate, colloidal crystals experience *shear melting*, a first-order structural transition to a colloidal liquid [79]. For a particle volume fraction of the shear-molten fluid ς_{fl}, the supersaturation that is driving crystallization is $(\varsigma_{fl}-\varsigma_l)/(\varsigma_s-\varsigma_l)$.

The solid circles in Figure 16 show measurements obtained by Sinn et al. [78] of the steady-state crystal nucleation rate per unit volume, I^{st}, as a function of volume fraction for a shear-melted "liquid" composed of hard spheres composed of a poly(methyl methacrylate) core with small hairs of poly(hydroxyl stearic acid) grafted onto the surface. Those data were obtained from light-scattering measurements (small-angle and large-angle coherent, i.e. Bragg). While these data show approximately the same dependence on ς_{fl} as the data obtained by others, the values are approximately two orders of magnitude higher than some.

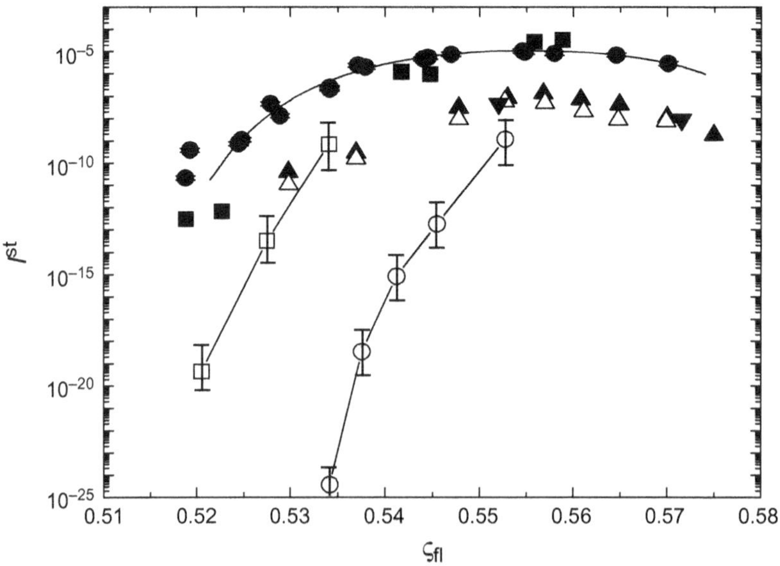

Fig. 16 The steady-state crystal nucleation rate, I^{st}, as a function of sample density, ς_{fl}, for a shear-melted "liquid" of hard spheres composed of a poly(methyl methacrylate) core with small hairs of poly(hydroxyl stearic acid) grafted onto the surface (●) [78]. The solid line is calculated from the classical theory of nucleation for an interfacial free energy, $\hat{\sigma}$, of 0.51 and a kinetic prefactor, A_c, equal to 0.013. Other data sets shown: ■ [84]; ▲ [80] (average); △ [80] (maximum); and ▼ [85]. Computer simulation predictions [82, 83]: □ (monodisperse); (○) (5% polydispersity.)

From the classical theory (Chapter 2), the steady-state nucleation rate per unit volume for crystallization in this colloidal dispersion can be written as [78]

$$I^{st} = A_c \frac{D\varsigma_{fl}^{5/3}}{r^5} \exp\left(-\frac{4\pi^3\hat{\sigma}^3}{27\varsigma_{cl}^2(\Delta\hat{\mu})^2}\right), \tag{16}$$

where D is the particle diffusion coefficient, r is the colloidal particle radius, $A_c = 6\sqrt{\hat{\sigma}}/\pi\varsigma_{cl}^{1/3}$, $\Delta\hat{\mu} = \Delta\mu/k_BT$ ($\Delta\mu$ is the chemical potential of the new phase less that of the original phase), $\hat{\sigma} = 4r^2\sigma_{ls}/k_BT$, and ς_{cl} is the cluster particle volume fraction (the average volume per particle in the crystal cluster is $\bar{v}_{cl} = 4\pi r^3/3\varsigma_{cl}$). Equation (16) fits the Sinn data well (solid line) for $\hat{\sigma} = 0.51$ and $A_c = 0.013$, suggesting that the classical theory is valid for these systems. Further, the fit value for $\hat{\sigma}$ is in good agreement with fits to other experimental data [80] and is consistent with results from hard-sphere calculations ($\hat{\sigma} = 0.55$ from Ref. [81]). However, the fit value for A_c is two orders of magnitude lower than expected from the classical theory for the measured value of $\hat{\sigma}$ and the appropriate colloidal-system length scales ($A_c \approx 1.6$). All of the data are in poor agreement with the nucleation rates obtained from computer simulations for a monodisperse colloidal system [82, 83]. As expected, the computed rates are lower when 5% polydispersity is introduced, making the disagreement with the data even more striking. The reasons for this discrepancy are unclear, although it has been suggested that the problem is a result of a misinterpretation of the experiments [82].

One of the most useful features of colloidal systems is that the cluster shapes, interfaces and internal structure can be observed directly. The concept of a critical fluctuation is common to all viable nucleation theories. Within the language of the classical theory (Chapter 2), this consists of a critical cluster containing n^* fundamental units, or "molecules." Cluster growth is stochastic, governed by the competition between the positive surface free energy and the negative volume driving free energy for the phase transformation. Clusters smaller than n^* tend to dissolve, those larger than n^* tend to grow, and those with n^* units are in unstable equilibrium. This behavior was first observed and measured in real-space imaging studies of the crystallization of nearly hard-sphere colloidal suspensions [72]. Figure 17 shows the size dependence of the difference in probability for a crystal cluster to grow or shrink as a function of the number of colloidal particles it contains. The critical cluster size is given by the point at which this difference changes from negative to positive. These data show that n^* lies between 50 and 160, in good agreement with predictions from computer simulations [86].

The structures of crystal colloidal clusters that are larger than the critical size are shown in Figure 18(a–c). In agreement with the fits to the experimental data discussed in this chapter, and with predictions from

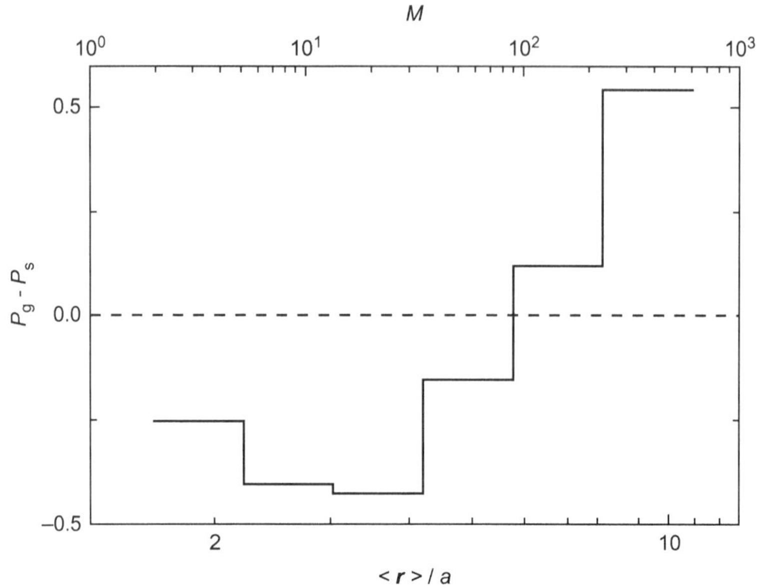

Fig. 17 The difference between the probability that a crystal is growing, P_g, or shrinking, P_s, as a function of the normalized crystal radius, $\langle \mathbf{r} \rangle / a$ (bottom axis) and the number of particles, M, in the crystal (top axis) for a particle volume fraction of, $\varsigma = 0.47$. The radius of the poly(methyl-methacrylate) spheres used in the study is represented by a; r is the radius of the crystal. The critical size lies in the range where the probability difference changes from positive to negative. (From Ref. [72], Reprinted with permission from AAAS.)

density-functional formulations of nucleation theory (Chapter 4), the interface between the crystal and liquid phases is broad. Further, the clusters are not compact, differing from the assumptions typically made in the classical theory (Chapter 2). Slices taken through the centers of the crystal clusters (Figures 18a and 18b) show stacked hexagonal layers (Figure 18d), in agreement with a bond-orientational analysis (see [72] for more detail).

The study of colloidal systems is an active and broad field, bringing fundamental new insights into nucleation processes. In this section, we have considered only some highlights; extensive reviews are available elsewhere [83, 87, 88].

5. NUCLEATION NEAR A MAGNETIC PHASE TRANSITION

Coupling between first-order phase transitions is rather common. For example, glass ceramics can be prepared by purposefully initiating

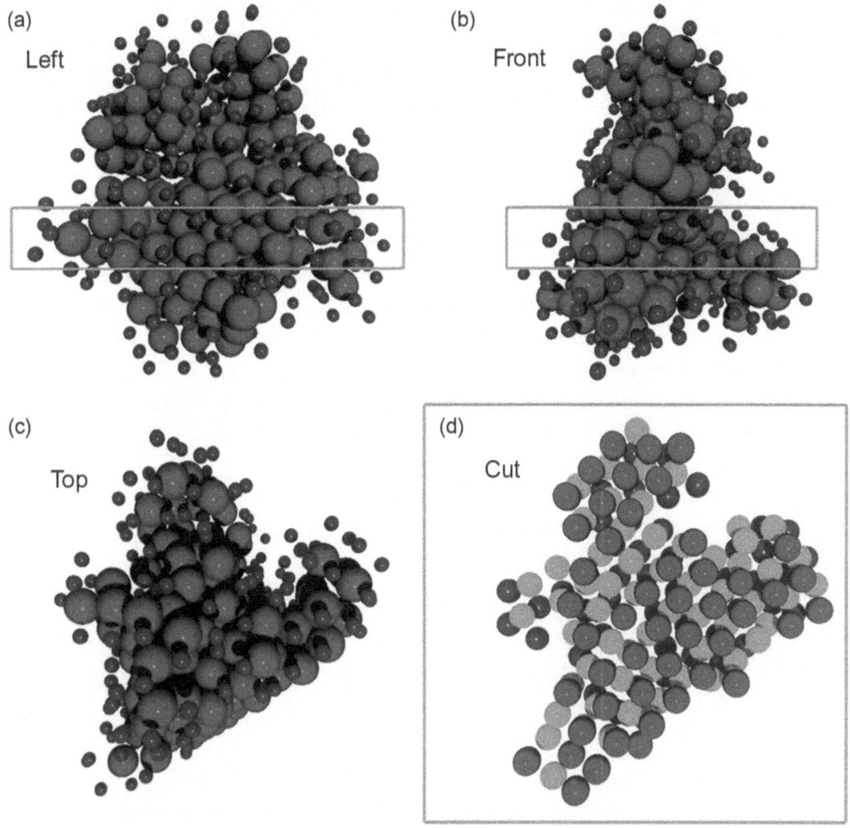

Fig. 18 (a) to (c) Crystal colloidal clusters that are larger than the critical size (observed by confocal microscopy). The larger lighter gray spheres represent colloidal particles that have crystal-type bonds and are part of crystal clusters; the smaller darker gray spheres represent those particles with at least one crystal-like bond with a larger gray particle. (d) A slice through the center of a cluster showing three stacked hexagonal layers (the shading of the spheres represents their layer location, i.e. dark gray spheres on one layer, medium gray spheres on the next and light gray spheres on the final layer, in order from front to rear). (From Ref. [72], Reprinted with permission from AAAS.)

processes such as precipitation and phase separation prior to devitrification (Chapter 14, Section 2). The influence of magnetic ordering on crystal nucleation in supercooled cobalt-based liquids demonstrates that higher-order phase transitions can also couple dramatically to the nucleation barrier.

Magnetic ordering in an electromagnetically (EML)-levitated cobalt–palladium liquid was first indicated by Faraday-balance measurements [89] and confirmed by muon-spin rotation experiments [90]. Figure 19 directly illustrates this ordering. A cobalt–samarium magnet encapsulated in a copper cylinder located near an EML-levitated sample of liquid

Fig. 19 Left: EML levitated $Co_{80}Pd_{20}$ sample at an supercooling of 100 K, far above the Curie temperature, T_C. The center of the sample coincides with the symmetry axis of the levitation coil, shown by the vertical line. Right: The sample at a deeper supercooling near T_C. The liquid has become ferromagnetic and is attracted by the cobalt–samarium magnet, moving the sample center to the left of the coil's symmetry axis. (Reprinted with permission from Ref. [91], copyright (2003), The Royal Society.)

$Co_{80}Pd_{20}$ produces no significant magnetic force on the sample for a supercooling of 100 K and the sample center remains on the symmetry axis of the levitation coil (Figure 19, left). When supercooled by 300 K, however, the sample moves away from the symmetry axis (Figure 19, right), indicating a strong attraction to the magnet and the onset of magnetic ordering.

It is difficult to supercool the liquid below the Curie temperature, suggesting an influence of this second-order phase transition from paramagnetic to ferromagnetic ordering in the sample (the Curie transition) on crystal nucleation from the supercooled liquid. This connection is most clearly observed from an examination of the measured maximum-supercooling temperatures as a function of cobalt concentration in cobalt–palladium alloys (Figure 20). For concentrations of cobalt lower than 70%, the classical theory of nucleation (shown by the dotted line) fits the data well. For alloys with a greater concentration, however, its predictions fall below the measured data. Measurements show that the deviation from the classical-theory predictions occurs when the maximum-supercooling temperature is near the Curie temperature.

As discussed in Section 2.2.3, an analysis of the distribution of maximum-supercooling temperatures (ΔT_{max}) can provide additional insight. As shown in Figure 21, the distribution of reduced supercoolings ($\Delta T_{max}/T_{liq}$, where T_{liq} is the liquidus temperature) in alloys containing greater than 70% cobalt are broad, indicating heterogeneous nucleation; fits to the distribution suggest a catalytic potency factor $f(\phi)$ (see Chapter 6) of 0.35. The distribution abruptly sharpens for higher cobalt concentrations, indicating a larger nucleation prefactor, interpretable as a higher potency. There is no reason to expect such a sudden change in potency, however, and experimental studies in similar, nonmagnetic alloys indicate that the potency remains essentially unchanged over a wide compositional range.

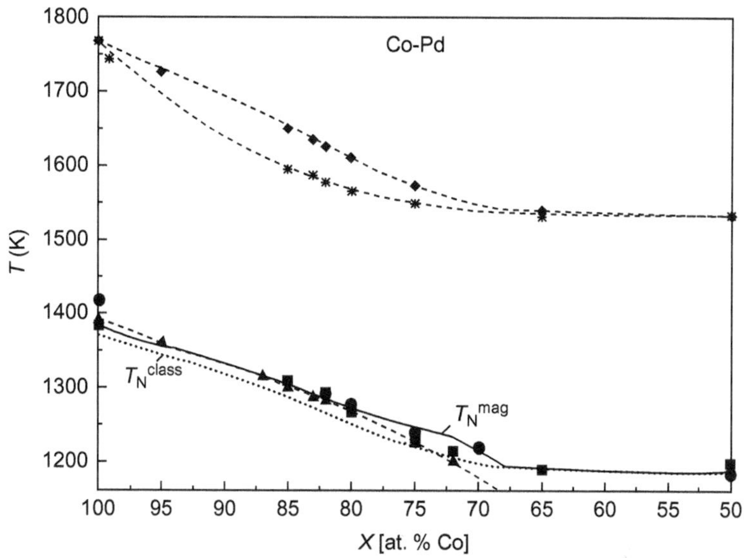

Fig. 20 Phase diagram showing the solidus and liquidus curves for cobalt–palladium alloys, and the measured maximum-supercooling values for solidification, as a function of cobalt concentration. Predictions from the classical theory of nucleation (dotted line) fit the data well for cobalt concentrations lower than 70 at.%, but fall below the data for higher concentrations. The incorporation of magnetic-ordering contributions to the driving free energy gives a reasonable fit to the data over the entire compositional range (solid line). (From Ref. [92], with permission; Taylor & Francis Ltd. http://www.informaworld.com.)

As shown in Figure 22, this abrupt change in the magnitude of the prefactor occurs when the temperature for the onset of nucleation, T_{min} approaches the Curie temperature, T_C. For temperatures near T_C, it is several orders of magnitude larger than would be predicted assuming no magnetic contribution to the nucleation barrier.

The change in nucleation rate can be understood by examining the change in driving free energy due to the onset of ferromagnetic ordering within the liquid below the Curie temperature [92]. For paramagnetic ordering, the partition function, Q, for a spin with total angular momentum, J, in an applied magnetic field, H_{appl}, is (see Ref. [95] for example)

$$\begin{aligned} Q &= \sum_{m_J=-J}^{J} \exp(-\mu_B g_L H_{appl} m_J/k_B T) \\ &= \frac{\exp\left(+\mu_B g_L H_{appl}(J+1/2)/k_B T\right) - \exp\left(-\mu_B g_L H_{appl}(J+1/2)/k_B T\right)}{\exp\left(+\mu_B g_L H_{appl}/2k_B T\right) - \exp\left(-\mu_B g_L H_{appl}/2k_B T\right)} \\ &= \frac{\sinh\left(\mu_B g_L H_{appl}(J+1/2)/k_B T\right)}{\sinh\left(\mu_B g_L H_{appl}/2k_B T\right)}, \end{aligned} \tag{17}$$

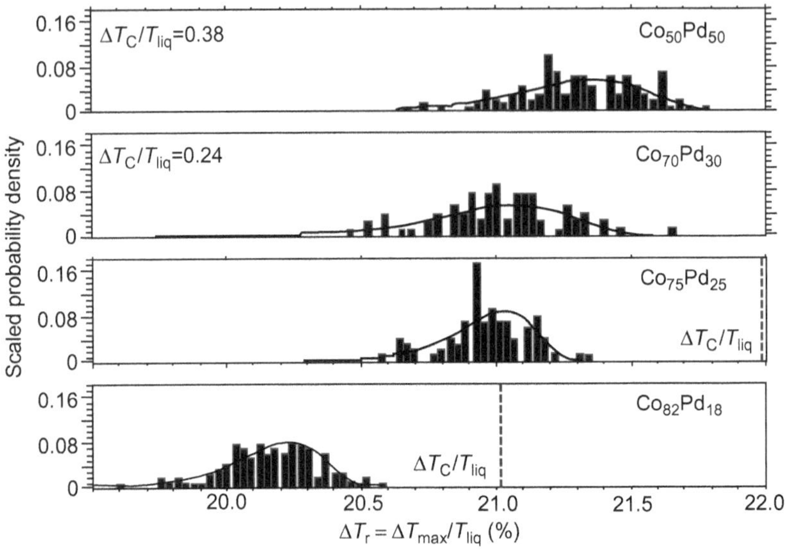

Fig. 21 Scaled probability density as a function of reduced supercooling for solidification of Co-Pd liquids (the values of ΔT_C $(=T_{liq}-T_C)$ are either listed or shown by a vertical dashed line); the solid line is a fit to Eq. (11). (Reprinted with permission from Ref. [91], copyright (2003), The Royal Society.)

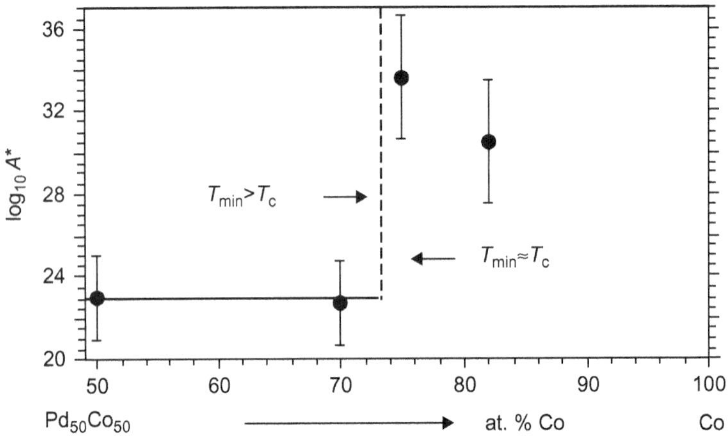

Fig. 22 Nucleation prefactor calculated from the classical theory of nucleation as a function of cobalt concentration, assuming that the difference in free energy between the liquid and solid phases, ΔG_{ls}, can be obtained from the measured heat of fusion and the specific-heat difference between the liquid and the solid (see Ref. [93]), and that the interfacial free energy scales with the entropy of fusion as $\sigma_{ls}(T) = \alpha \Delta S_f T (N_A V_M^2)^{-1/3}$, where α is a numerical factor that is a function of the structure of the crystal nucleus [94]. The nucleation (i.e. maximum-supercooling) temperature T_{min} and Curie temperature T_C are indicated. (Reprinted with permission from Ref. [91], Copyright (2003), The Royal Society.)

where m_J is the projection of J onto the direction of H_{appl}, taking on integer-spaced values between $-J$ and J. Here, $\mu_B = e\hbar/2m_e$ is the Bohr magneton (e is the charge of the electron, m_e is its mass, $\hbar$ is the Planck constant divided by 2π), and g_L is the Lande g-factor

$$g_L = 1 + \frac{J(J+1) + S(S+1) - L(L+1)}{2J(J+1)}, \tag{18}$$

where S and L are the total spin and orbital angular momentum respectively. The Helmholtz free energy for an ensemble of $\aleph$ spins per unit volume due to the magnetic contribution, F_M ($\approx G_M$ for a condensed phase), is

$$F_M = -\aleph k_B T \ln Q, \tag{19}$$

and the total magnetization M is

$$M = \aleph k_B T \left(\frac{\partial \ln Q}{\partial H_{appl}}\right)_T = \aleph g_L \mu_B J B_J \left(\frac{\mu_B g_L H_{appl}}{k_B T}\right), \tag{20}$$

where B_J is the Brillouin function,

$$B_J(x) = \frac{1}{J}\left((J+1/2)\coth\left((J+1/2)x\right) - 1/2\coth(x/2)\right). \tag{21}$$

Ferromagnetism is a more complex phenomenon, but a simple mean-field treatment is sufficient for a qualitative treatment of the coupling with the nucleation step of the first-order crystallization process. The total magnetic field at all points in a ferromagnetic condensed phase, H_T, is assumed to be proportional to the applied field, H_{appl} plus the magnetization due to the paramagnetic ordering, i.e. $H_T = H_{appl} + \lambda M$, where λ is an undetermined constant. If H_{appl} is set to zero, the field experienced by a given spin is due to the effective field of all other spins. Inserting this into Eq. (20), we obtain an expression for the magnetization as a function of temperature. For the case $J = ½$ so $g_L = 2$,

$$M = \aleph \mu_B B_{1/2}\left(\frac{2\lambda \mu_B M}{k_B T}\right) = \aleph \mu_B \tanh\left(\frac{\lambda \mu_B M}{k_B T}\right). \tag{22}$$

A self consistent-solution of the analogous equation to Eq. (22) for general values of J gives the Curie temperature for the ferromagnetic transition,

$$T_C = \frac{\aleph g_L^2 \mu_B^2}{3k_B} J(J+1)\lambda, \tag{23}$$

which scales with the unknown parameter, λ. Assuming the saturation magnetization for ccp cobalt to determine λ and J, the best fit is for $J = S = 1/2$, supporting our assumption used to obtain Eq. (22). For $J = 1/2$ the Helmholtz free energy per atom due to magnetic ordering is

$$F_M^{at} = -k_B T \ln\left(\frac{\sinh(2\mu_B \lambda M/k_B T)}{\sinh(\mu_B \lambda M/k_B T)}\right). \tag{24}$$

The entropy per atom associated with magnetic ordering is also readily computed

$$S_{\mathrm{M}}^{\mathrm{at}} = k_{\mathrm{B}}(\ln Q + T \ln Q)$$
$$= \aleph k_{\mathrm{B}}\left(\ln\left(\frac{\sinh(2\mu_{\mathrm{B}}\lambda M/k_{\mathrm{B}}T)}{\sinh(\mu_{\mathrm{B}}\lambda M/k_{\mathrm{B}}T)}\right) + T\ln\left(\frac{\sinh(2\mu_{\mathrm{B}}\lambda M/k_{\mathrm{B}}T)}{\sinh(\mu_{\mathrm{B}}\lambda M/k_{\mathrm{B}}T)}\right)\right), \quad (25)$$

and the enthalpy and Gibbs free energy follow directly from Eqs. (24) and (25) ($H_{\mathrm{M}}^{\mathrm{at}} \approx F_{\mathrm{M}}^{\mathrm{at}} + TS_{\mathrm{M}}^{\mathrm{at}}$ and $G_{\mathrm{M}}^{\mathrm{at}} \approx F_{\mathrm{M}}^{\mathrm{at}}$, ignoring the pressure-volume work).

The magnetic contributions for the cobalt–palladium alloys are assumed to be similar to those for pure cobalt. Based on Faraday-balance measurements, the Curie transition for liquid cobalt is 6 K lower than for the solid phase, producing a difference in the computed thermodynamic potentials for the two phases that should be added to the driving free energy for crystallization. The magnetic-ordering contributions $\Delta G_{\mathrm{M}}^{\mathrm{at}}$ $(= G_{\mathrm{M,l}}^{\mathrm{at}} - G_{\mathrm{M,s}}^{\mathrm{at}})$, $\Delta H_{\mathrm{M}}^{\mathrm{at}}$ and $\Delta S_{\mathrm{M}}^{\mathrm{at}}$, to the thermodynamic differences between the liquid and crystal phases are shown as a function of temperature in Figure 23. The magnetic contributions rise steeply as the temperature is decreased below the Curie temperature for the solid phase ($T_{\mathrm{C}}^{\mathrm{s}}$). Below the Curie temperature for the liquid phase

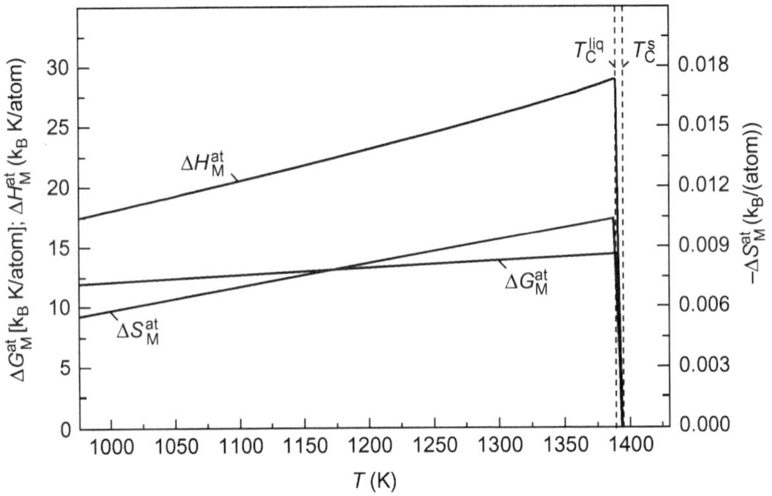

Fig. 23 Calculated contributions of magnetic ordering to the differences in Gibbs free energy $\Delta G_{\mathrm{M}}^{\mathrm{at}}$, enthalpy $\Delta H_{\mathrm{M}}^{\mathrm{at}}$, and entropy $\Delta S_{\mathrm{M}}^{\mathrm{at}}$, for the liquid and solid phases of cobalt. The Curie temperatures for the solid and liquid phases are indicated. (From Ref. [92], with permission; Taylor & Francis Ltd. http://www.informaworld.com.)

(T_C^{liq}), they decrease slowly. While an increase in the interfacial free energy is also expected, calculations show that it is negligible. The abrupt increase in the driving free energy due to ordering between T_C^s and T_C^{liq} should, therefore, dramatically lower the nucleation barrier, explaining the difficulty in supercooling below the region of the Curie transition.

The inclusion of the magnetic ordering contribution to the driving free energy gives much better agreement with the experimental data (Figure 20). This has been based on a mean-field treatment of the magnetic transition, however, and does not include the fluctuations that are inherent in second-order phase transitions, which could enhance the coupling. It is likely that other examples of coupling between higher-order phase transitions exist, and may have a profound influence on microstructural development.

6. SUMMARY

In this chapter, nucleation experiments in supercooled liquids were reviewed. Most attention was paid to studies of liquid metals, with a brief discussion of the covalently bonded polymeric and colloidal liquids. Key points include:

- To observe homogeneous nucleation effects, suppression of heterogeneous nucleation is essential. Early studies attempted to isolate the heterogeneous sites by dividing the liquids into large ensembles of essentially isolated droplets. This allowed studies of the ultimate supercooling to be made, probing the homogeneous nucleation limit.
- Recently, containerless techniques have been developed that allow supercooling studies on bulk samples. While probing the supercooling limit, these new techniques also allow measurements of the thermophysical properties and even the structures of the supercooled liquids, allowing more quantitative nucleation studies to be made.
- Most liquids can be supercooled from 10 to 30% of their melting temperatures before crystallizing, indicating a significant nucleation barrier. This is surprising for metals, since the densities and coordination numbers of the liquid and solid are similar. To explain this, it was hypothesized that the liquid develops icosahedral short-range order with supercooling. This connection between the nucleation barrier and the developing structure of the liquid was recently proven by correlating the preferential nucleation of a metastable icosahedral quasicrystal with icosahedral short-range order in the liquid.

- An analysis of the maximum-supercooling data shows that the interfacial free energy scales linearly with the heat of fusion.
- Quantitative fits to the classical theory of nucleation give experimental prefactors that are too large. This can be explained by assuming that the liquid near the cluster interface is ordered, indicating that the interface is not sharp, as assumed in the classical theory.
- The diffuse-interface model and the semiempirical density-functional approximation (Chapter 4) give better agreement with the experimental data than does the classical theory, consistent with the interface being diffuse.
- The classical theory fits nucleation data for uncharged colloidal systems well.
- As observed in colloidal systems, the interface between the nucleating cluster and the liquid is rough as well as diffuse.
- Ferromagnetic ordering in the liquid decreases the nucleation barrier near the Curie temperature. A mean-field calculation predicts that magnetic ordering introduces an additional contribution to the driving free energy for crystallization.

REFERENCES

[1] K. Sahu and K. F. Kelton, private communication, 2008.
[2] D. Turnbull, Kinetics of solidification of supercooled liquid mercury droplets, J. Chem. Phys. **20** (1952) 411–424.
[3] K.F. Kelton, Crystal nucleation in liquids and glasses, in: *Solid State Physics*, Eds. H. Ehrenreich, D. Turnbull, Academic Press, Boston, MA (1991), pp. 75–178 .
[4] B. Vonnegut, Variation with temperature of the nucleation rate of supercooled liquid tin and water drops, J. Coll. Sci. **3** (1948) 563–569.
[5] D. Turnbull, The subcooling of liquid metals, J. Appl. Phys. **20** (1949) 817.
[6] D. Turnbull, R.E. Cech, Microscopic observation of the solidification of small metal droplets, J. Appl. Phys. **21** (1950) 804–810.
[7] J.H. Perepezko, J.S. Paik, Thermodynamic properties of undercooled liquid metals, J. Non-Cryst. Solids **61&62** (1984) 113–118.
[8] J.H. Perepezko, B.A. Mueller, K. Ohsaka, Solidification of undercooled alloys, in: *Undercooled Alloy Phases*, Eds. E.W. Collings, C.C. Koch, The Metall. Soc. AIME, Warrendale, PA (1987), pp. 289–320 .
[9] T. Kattamis, M.C. Flemings, Dendrite structure and grain size of under-cooled melts, Trans. Metall. Soc. AIME **236** (1966) 1523–1532.
[10] H.W. Kui, A.L. Greer, D. Turnbull, Formation of bulk metallic glass by fluxing, Appl. Phys. Lett. **45** (1984) 615–616.
[11] J.K.R. Weber, D.S. Hampton, D.R. Merkley, C.A. Rey, M.M. Zatarski, P.C. Nordine, Aero-acoustic levitation: a method for containerless liquid-phase processing at high temperatures, Rev. Sci. Instrum. **65** (1994) 456–465.
[12] D.M. Herlach, R.F. Cochrane, I. Egry, H.J. Fecht, A.L. Greer, Containerless processing in the study of metallic melts and their solidification, Int. Mater. Rev. **38** (1993) 273–347.
[13] D.M. Herlach, D. Holland-Moritz, P. Galenko, *Metastable Solids from Undercooled Melts*, Elsevier, Oxford (2007).

[14] W.K. Rhim, M. Collender, M.T. Hyson, W.T. Simms, D.D. Elleman, Development of an electrostatic positioner for space material processing, Rev. Sci. Instrum. **56** (1985) 307–317.
[15] W.K. Rhim, S.K. Chung, D. Barber, K.F. Man, G. Gutt, A.J. Rulison, R.E. Spjut, An electrostatic levitator for high temperature containerless materials processing in 1-g, Rev. Sci. Instrum. **64** (1993) 2961–2970.
[16] A.J. Rulison, J.L. Watkins, B. Zambrano, Electrostatic containerless processing system, Rev. Sci. Instrum. **68** (1997) 2856–2863.
[17] S. Bossuyt, J. Schroers, W.K. Rhim, W.L. Johnson, Minimizing convection effects to measure diffusion in liquid droplets during high-temperature electrostatic levitation, Rev. Sci. Instrum. **76** (2005) 033909/1–9.
[18] G.L.F. Powell, The undercooling of silver, J. Aust. Inst. Metals. **10** (1965) 223–230.
[19] G.L.F. Powell, Heterogeneous nucleation of undercooled silver, Trans. Metall. Soc. AIME **239** (1967) 1244–1247.
[20] J. Fehling, E. Scheil, Supercooling of molten metals, Z. Metall. **53** (1962) 593–600.
[21] P. Bardenheuer, R. Bleckmann, Zur Frage der Primärkristallisation des Stahles: Unterkühlbarkeit und Keimbildung im flüssigen Zustand, Stahl. Eisen. **61** (1941) 49–53.
[22] G. Devaud, D. Turnbull, Undercooling of molten silicon, Appl. Phys. Lett. **46** (1985) 844–845.
[23] R.E. Cech, D. Turnbull, Microscopic observation of the solidification of Cu–Ni alloy droplets, J. Metals **3** (1951) 242–243.
[24] M.J. Stowell, T.J. Law, J. Smart, The growth, structure, melting and solidification of lead deposits on molybdenite and carbon substrates, Proc. Roy. Soc. Lond. A **318** (1970) 231–241.
[25] C.E. Mendenhall, L.R. Ingersoll, Movement of particles on a Nernst glower, Philos. Mag. **15** (1908) 205–214.
[26] T. Takahashi, W.A. Tiller, The supercooling dependence for nucleation of some metals on the liquid drop size, Acta Metall. **17** (1969) 643–650.
[27] J. Schade, A. McLean, W.A. Miller, Surface tension measurements on oscillating droplets of undercooled liquid metals and alloys, in: *Undercooled Alloy Phases*, Eds. E.W. Collings, C.C. Koch, The Metall. Soc. AIME, Warrendale, PA (1987), pp. 233–248 .
[28] R. Willnecker, D.M. Herlach, B. Feuerbacher, Nucleation in bulk undercooled nickel-base alloys, Mater. Sci. Eng. **98** (1988) 85–88.
[29] R. Willnecker, D.M. Herlach, B. Feuerbacher, Containerless undercooling of bulk Fe–Ni melts, Appl. Phys. Lett. **49** (1986) 1339–1341.
[30] D.W. Gomersall, S.Y. Shiraishi, R.G. Ward, Undercooling phenomena in liquid metal droplets, J. Aust. Inst. Metals. **10** (1965) 220–224.
[31] R.J. Bayuzick, W.H. Hofmeister, M.B. Robinson, Review on drop towers and long drop tubes, in: *Undercooled Alloy Phases*, Eds. E.W. Collings, C.C. Koch, The Metall. Soc. AIME, Warrendale, PA (1987), pp. 207–231 .
[32] G.W. Lee, A.K. Gangopadhyay, T.J. Rathz, J.R. Rogers, K.F. Kelton, private communication, 2002.
[33] L.L. Lacy, M.B. Robinson, T.J. Rathz, Containerless undercooling and solidification in drop tubes, J. Cryst. Growth **51** (1981) 47–60.
[34] W.H. Hofmeister, M.B. Robinson, R.J. Bayuzick, Undercooling of pure metals in a containerless, microgravity environment, Appl. Phys. Lett. **49** (1986) 1342–1344.
[35] B. Vinet, L. Cortella, J.J. Favier, P. Desré, Highly undercooled W and Re drops in an ultrahigh-vacuum drop tube, Appl. Phys. Lett. **58** (1991) 97–99.
[36] B.B. Laird, The solid-liquid interfacial free energy of close-packed metals: hard-spheres and the Turnbull coefficient, J. Chem. Phys. **115** (2001) 2887–2888.
[37] C.W. Morton, W.H. Hofmeister, R.J. Bayuzick, A.J. Rulison, J.L. Watkins, The kinetics of solid nucleation in zirconium, Acta Mater **46** (1998) 6033–6039.

[38] W.H. Hofmeister, C.W. Morton, R.J. Bayuzick, Monte Carlo testing of the statistical analysis of nucleation data, Acta Mater. **46** (1998) 1903–1908.

[39] C.C. Wang, C.S. Smith, Undercooling of minor liquid in binary alloys, Trans. AIME **188** (1950) 136–138.

[40] R.T. Southin, G.A. Chadwick, Heterogeneous nucleation in solidifying metals, Acta Metall. **26** (1978) 223–231.

[41] P.G. Boswell, G.A. Chadwick, Heterogeneous nucleation in entrained Sn droplets, Acta Metall. **28** (1980) 209–221.

[42] P. Ramachandrarao, K. Lal, Quenching from the mushy state — a new technique for the production of metastable phases, Mater. Sci. Eng. **41** (1979) 259–264.

[43] B. Cantor, Embedded droplet measurements and an adsorption model of the heterogeneous nucleation of solidification, Mater. Sci. Eng. A **178** (1994) 225–231.

[44] W.T. Kim, D.L. Zhang, B. Cantor, Nucleation of solidification in liquid droplets, Metall. Trans. A **22A** (1991) 2487–2501.

[45] C.R. Ho, B. Cantor, Effect of Ge on the heterogeneous nucleation of Cd solidification by Al, Philos. Mag. A **66** (1992) 141–149.

[46] D.L. Zhang, B. Cantor, The effect of dopants on the heterogeneous nucleation of solidification of Cd and Pb particles embedded in an Al matrix, Acta Metall. Mater. **40** (1992) 2951–2960.

[47] W.T. Kim, B. Cantor, Solidification of tin droplets embedded in an aluminium matrix, J. Mater. Sci. **26** (1991) 2868–2878.

[48] W.T. Kim, B. Cantor, Solidification behaviour of Pb droplets embedded in a Cu matrix, Acta Metall. **40** (1992) 3339–3347.

[49] W.T. Kim, B. Cantor, Heterogeneous nucleation of Al_2Cu in Al–Cu eutectic liquid droplets, Acta Metall. Mater. **42** (1994) 3045–3053.

[50] W.T. Kim, B. Cantor, An adsorption model of the heterogeneous nucleation of solidification, Acta Metall. Mater. **42** (1994) 3115–3127.

[51] B. Cantor, K.A.Q. O'Reilly, The nucleation of solidification in liquid droplets embedded in a solid matrix, Curr. Opin. Solid State Mater. Sci. **2** (1997) 318–323.

[52] J.H. Perepezko, J.S. Smith, Glass formation and crystallization in highly undercooled Te–Cu alloys, J. Non-Cryst. Solids **44** (1981) 65–83.

[53] M.C. Flemings, D.M. Matson, Alloy undercooling experiments in a microgravity environment, in: *Microgravity Science Laboratory (MSL-1) Final Report*, Ed. M.B. Robinson, NASA Technical Information Service, Springfield, VA (1998), pp. 282–290 .

[54] P.J. Desré, E. Cini, B. Vinet, Homophase-fluctuation-mediated mechanism of nucleation in multicomponent liquid alloys and glass-forming ability, J. Non-Cryst. Solids **288** (2001) 210–217.

[55] C.V. Thompson, F. Spaepen, Homogeneous crystal nucleation in binary metallic melts, Acta Metall. **31** (1983) 2021–2027.

[56] J.C. Holzer, K.F. Kelton, Kinetics of the amorphous to icosahedral phase transformation in Al–Cu–V Alloys, Acta Metall. Mater. **39** (1991) 1833–1843.

[57] L.A. Bendersky, S.D. Ridder, Nucleation behavior of Al–Mn icosahedral phase, J. Mater. Res. **1** (1986) 405–414.

[58] D. Holland-Moritz, D.M. Herlach, K. Urban, Observation of the undercoolability of quasicrystal-forming alloys by electromagnetic levitation, Phys. Rev. Lett. **71** (1993) 1196–1199.

[59] K. Urban, D. Holland-Moritz, D.M. Herlach, B. Grushko, The undercooling and phase formation behaviour of melts of quasicrystal–forming alloys, Mater. Sci. Eng. A **178** (1994) 293–298.

[60] D. Holland-Moritz, Short-range order and solid-liquid interfaces in undercooled melts, Int. J. Non-Equil. Proc. **11** (1998) 169–199.

[61] K.F. Kelton, G.W. Lee, A.K. Gangopadhyay, R.W. Hyers, T.J. Rathz, J.R. Rogers, M.B. Robinson, D.S. Robinson, First X-ray scattering studies on electrostatically-levitated metallic liquids – demonstrated influence of local icosahedral order on the nucleation barrier, Phys. Rev. Lett. **90** (2003) 195504/1–4.
[62] P.J. Steinhardt, D.R. Nelson, M. Ronchetti, Bond-orientational order in liquids and glasses, Phys. Rev. B **28** (1983) 784–805.
[63] T.H. Kim, K.F. Kelton, Structural study of supercooled liquid transition metals, J. Chem. Phys. **126** (2007) 054513/1–7.
[64] T.H. Kim, G.W. Lee, A.K. Gangopadhyay, R.W. Hyers, J.R. Rogers, A.I. Goldman, K.F. Kelton, Structural studies of a Ti–Zr–Ni quasicrystal-forming liquid, J. Phys. Cond. Matt. **19** (2007) 455212/1–10.
[65] N. Jakse, A. Pasturel, Local order of liquid and supercooled zirconium by ab-initio molecular dynamics, Phys. Rev. Lett. **91** (2003) 195501/1–4.
[66] N. Jakse, A. Pasturel, Local order of liquid and undercooled transition metal based systems: Ab-initio molecular dynamics study, Mod. Phys. Lett. B **20** (2006) 655–674.
[67] C.A. Angell, Formation of glasses from liquids and biopolymers, Science **267** (1995) 1924–1935.
[68] F. Spaepen, D. Turnbull, Crystal nucleation and crystal-melt interfacial tension in linear hydrocarbons, J. Polym. Sci. **63** (1978) 237–243.
[69] L. Gránásy, F. Igioi, Comparison of experiments and modern theories of crystal nucleation, J. Chem. Phys. **107** (1997) 3634–3644.
[70] R.J. Hunter, *Foundations of Colloid Science* 2nd edn, Oxford Univ. Press, Oxford (2001).
[71] J.K.S. Dhont, C. Smits, H.N.W. Lekkerkerker, A time-resolved static light scattering study on nucleation and crystallization in a colloid system, J. Coll. Inter. Sci. **152** (1992) 386–401.
[72] U. Gasser, E.R. Weeks, A. Schofield, P.N. Pusey, D.A. Weitz, Real-space imaging of nucleation and growth in colloidal crystallization, Science **292** (2001) 258–262.
[73] H.J. Schöpe, T. Palberg, A study on the homogeneous nucleation kinetics of model charged sphere suspensions, J. Phys. Cond. Matt. **4** (2002) 11573–11587.
[74] R.P. Sear, D. Frenkel, Continuous freezing in three dimensions, Phys. Rev. Lett. **90** (2003) 195701/1–4.
[75] P.N. Pusey, Colloidal suspensions, in: *Liquids, Freezing and Glass Transition*, Eds. J.P. Hansen, D. Levesque, J. Zinn-Justin, North-Holland, Amsterdam (1991), pp. 763–931 .
[76] W.G. Hoover, F.H. Ree, Melting transition and communal entropy for hard spheres, J. Chem. Phys. **49** (1968) 3609–3617.
[77] M.O. Robbins, K. Kremer, G.S. Grest, Phase diagrams and dynamics of Yukawa systems, J. Chem. Phys. **88** (1988) 3286–3312.
[78] C. Sinn, A. Heymann, A. Stipp, T. Palberg, Solidification kinetics of hard-sphere colloidal suspensions, Prog. Coll. Polym. Sci. **118** (2001) 266–275.
[79] B.J. Ackerson, N.A. Clark, Shear-induced melting, Phys. Rev. Lett. **46** (1981) 123–126.
[80] J.L. Harland, W. van Megen, Crystallization kinetics of suspensions of hard colloidal systems, Phys. Rev. E **55** (1997) 3054–3067.
[81] D.W. Marr, A.P. Gast, Interfacial free energy between hard sphere solids and fluids, Langmuir **10** (1994) 1348–1350.
[82] S. Auer, D. Frenkel, Prediction of absolute crystal-nucleation rate in hard-sphere colloids, Nature **409** (2001) 1020–1023.
[83] S. Auer, D. Frenkel, Quantitative prediction of crystal-nucleation rates for spherical colloids: a computational approach, Ann. Rev. Phys. Chem. **55** (2004) 333–361.
[84] K. Schätzel, B.J. Ackerson, Density fluctuations during crystallization of colloids, Phys. Rev. E **48** (1993) 3766–3777.
[85] Z. Cheng, *Colloidal Hard Sphere Crystallization and Glass Transition*, Princeton Univ. Press, Princeton, NJ (1998).

[86] S. Auer, D. Frenkel, Suppression of crystal nucleation in polydisperse colloids due to increase of the surface free energy, Nature **413** (2001) 711–713.
[87] H. Löwen, Colloidal soft matter under external control, J. Phys. Cond. Matt. **13** (2001) R415–R432.
[88] R.P. Sear, Nucleation: theory and applications to protein solutions and colloidal suspensions, J. Phys. Cond. Matt. **19** (2007) 033101/1–28.
[89] J. Reske, D.M. Herlach, F. Keuser, K. Maier, D. Platzek, Evidence for the existence of long-range magnetic ordering in a liquid undercooled metal, Phys. Rev. Lett. **75** (1995) 737–739.
[90] D. Herlach, C. Buhrer, D.M. Herlach, K. Maier, C. Notthoff, D. Platzek, J. Reske, Magnetic ordering in a supercooled Co–Pd melt studied by muon-spin rotation, Europhys. Lett. **44** (1998) 98–102.
[91] D.M. Herlach, D. Holland-Moritz, R. Willnecker, D. Herlach, K. Maier, Magnetic ordering and crystal nucleation in undercooled Co-based melts, Phil. Trans. Roy. Soc. Lond. A **361** (2003) 497–515.
[92] D. Holland-Moritz, F. Spaepen, The magnetic contribution to the driving force for crystal nucleation in undercooled Co–Pd melts, Philos. Mag. **84** (2004) 957–966.
[93] K.S. Dubey, P. Ramachandrarao, On the energy change accompanying crystallisation of undercooled melts, Acta Metall. **32** (1984) 91–96.
[94] D.R. Nelson, F. Spaepen, Polytetrahedral order in condensed matter, in: *Solid State Physics*, Eds. H. Ehrenreich, D. Turnbull, Academic Press, Boston, MA (1989), pp. 1–90 .
[95] A.H. Carter, *Classical and Statistical Thermodynamics*, Prentice Hall, Upper Saddle River, NJ (2001).
[96] A.G. Walton, Nucleation in liquids and solutions, in: *Nucleation*, Ed. A.C. Zettlemoyer, Marcel Dekker, New York (1969), pp. 225–307.
[97] D. Turnbull and R.L. Cormia, Kinetics of crystal nucleation in some normal alkane liquids, J. Chem. Phys. **34** (1961) 820–831.
[98] G.R. Wood and A.G Walton, Homogeneous nucleation kinetics of ice from water, J. Appl. Phys. **41** (1970) 3027–3036.

CHAPTER 8

Crystallization in Glasses

Contents

1. INTRODUCTION

As discussed in Chapter 7, liquids can solidify discontinuously through crystallization, initiated by the nucleation of ordered phases. However, if they are cooled fast enough, there is insufficient time for complete crystallization, and the remaining liquid then solidifies by a continuous process to become a *glass*, a configurationally frozen liquid. The role of

Pergamon Materials Series, Volume 15
ISSN 1470-1804, DOI 10.1016/S1470-1804(09)01508-9

nucleation in the formation and crystallization (*devitrification*) of glasses is the subject of this chapter.

There are several prominent families of glasses. The most familiar are the silicate glasses, used in a wide range of products from windows to glass vessels. Others include the phosphate glasses used in biomedical applications, the semiconductor chalcogenide glasses used as optical recording media, ionic glasses, polymeric glasses, and metallic glasses.

While not strictly glasses, noncrystalline, *amorphous* solids can be produced by means other than cooling the liquid. They can be formed by thin-film deposition from the vapor [1], by electrodeposition [2], by electroless deposition [3], by chemical precipitation [4], and by mechanical alloying and solid-state reactions [5, 6]. Where compared, the structures and properties of these amorphous solids can be indistinguishable from true glasses, produced by cooling the melt. In this chapter, the terms *amorphous phase* and *glass* are used interchangeably.

Although glasses themselves have many technological applications, often it is the crystallized, or *devitrified*, glass that is of most interest. A detailed knowledge of the crystallization processes and how to control them is, therefore, of practical interest, as will be discussed in Chapter 14, Section 2. Since glasses transform to ordered phases on a time-scale that is longer than that for the crystallization of supercooled liquids, nucleation and growth rates can be more directly and quantitatively measured than in liquids. These data allow more stringent tests of the nucleation theories discussed in Section I of this book. For example, since the glass is more deeply supercooled than is possible for liquids, nucleation can be studied at large departures from equilibrium, where the classical theory might be expected to fail. Time-dependent nucleation is studied much more easily in glasses than in liquids because of the slower kinetics. Measurements of the time-dependent nucleation rate as a function of temperature and sample thermal history provide a new probe of the cluster dynamics underlying nucleation phenomena, and as we shall see in Section 7.2, confirm the kinetic model of the classical theory of nucleation for polymorphic crystallization. While glasses do not have the long-range order that is characteristic of crystal or quasicrystal phases, they do show significant short-range order, evidenced by the broad peaks in their measured pair correlation functions (see [7] for example). Further, it is becoming apparent that some glasses, particularly metallic glasses, also have significant medium-range order [8], that is, ordering beyond the nearest neighbors. As we will see in this chapter, this order can have significant consequences for both the formation and devitrification of glasses. Since the most quantitative studies of devitrification have been made in the silicate and metallic glasses, they are the primary focus in this chapter; other glasses are briefly discussed in Chapter 14, Section 2.

2. THE GLASSY STATE

Some glasses, generally silicates, form naturally. For example, glassy tubes, *fulgurites*, can result when lightning strikes sand of suitable composition. During some meteor strikes on earth, small volumes of molten rock are formed and cooled quickly to form irregular glassy spheres, *tektites*. The best known example of a naturally occurring glass, however, is *obsidian* (volcanic glass), important in the Stone Age because it can be worked to form sharp implements such as knives and spear- and arrowheads. It was still widely used in the Americas at the time of the arrival of the Europeans in the fifteenth and sixteenth centuries.

Human glassmaking began some 5000 years ago in the Bronze Age, probably an accidental discovery from the earlier art of pottery glazing.[1] The oldest known piece of glass is a rod from 2600 BCE, found at Eshnunna in Babylonia (modern Iraq) [9]. Glass vases and utensils are found in the tombs of the Pharaohs and were widely used by the Romans. That these have survived so long demonstrates the stability of the glassy state. These "traditional glasses" are based on silica (SiO_2), usually in combination with lime (CaO) and soda (Na_2O). It was long believed that glass formation was rare, that metal alloys, for example, could not solidify as glasses. In the 1960s, however, it was discovered that if the cooling rates are sufficiently high, of order 10^6–10^{12} K s^{-1}, metallic glasses (*amorphous alloys*) can be formed [10]. Several quenching techniques for metallic-glass production were pioneered in the decade following their discovery [10] (see [11, 12] for a review of production techniques). In the 1990s, new classes of metallic glasses were discovered that could be produced at much lower quenching rates, approaching those used for the traditional silicate glasses [13]. These *bulk metallic glasses* (BMGs) can be produced in thicker cross-sections and can be molded and shaped like traditional glasses, leading to interesting applications and commercial products.

A characteristic of many glasses is their ability to continuously soften and reharden by heating and cooling. The nature of this glass transition, connecting the supercooled liquid and the glass, remains a major unsolved problem in condensed-matter physics. Unlike crystallization, the glass transition has no associated discontinuities in the volume, entropy, or enthalpy. It is dominated by kinetic effects, but it may also have an underlying thermodynamic character (see Refs. [14, 15] and the references therein for further discussion on this point).

[1] The earliest record of glass production comes from clay tablets discovered in Mesopotamia, one dating back to the XVII century BCE (evidence outlined by A.R. Lisella, "Glass Core Pendants from the Museum of the Studium Biblicum Francisanum - Jerusalem" Liber Annuus, Annual of the Studium Biblicum Franciscanum Jerusalem, **55** (2005) 435–456).

If crystallization of the liquid does not intervene, the volume, viscosity, and Gibbs free energy change smoothly on cooling into the supercooled regime (Figure 1). X-ray and neutron-diffraction studies show that the structures of liquids evolve continuously with decreasing temperature (Chapter 7). As the glass-transition temperature, T_g, is approached upon cooling, all the properties show a marked deviation from the trends extrapolated from higher temperature. At and below T_g, the evolution of the liquid structure is no longer kinetically possible within the imposed time scale of the cooling, and the liquid enters the configurationally frozen state of the glass. Unlike liquids, glasses are not *ergodic* (in *internal thermodynamic equilibrium*). The loss in entropy as T_g is approached on cooling causes a sharp increase in the specific heat that is reminiscent of the Schottky anomaly [16] (see example in Figure 1). Since the structures of glasses are less dependent on temperature, their properties are more independent of temperature than are those of liquids. The distinction between a glass and a liquid is most dramatic for the viscosity, due to its strong dependence on density. This is often used to give an operational definition of the glass transition, generally as the point at which the viscosity exceeds about 10^{13} poise (10^{12} Pa s). The structure of the configurationally frozen state of the glass is sensitive to the cooling rate. For example, in Figure 1 the cooling rate that is required to achieve a glass of state "1" is greater than that to achieve one of state "2." When annealed below T_g, glasses evolve toward a fully relaxed, but not unique, glassy structure. Upon reheating the glass through T_g, the system re-enters the supercooled-liquid region, and again becomes ergodic within reasonable experimental time scales. Above T_g, the viscosity becomes low enough that the material begins to flow under moderate stress, giving rise to a softening point that can be exploited in the fabrication of intricate glass shapes.

3. NUCLEATION IN GLASS FORMATION

A glass cannot form unless the liquid is cooled fast enough to avoid significant nucleation and growth of the more stable ordered phases before reaching the glass-transition temperature T_g. This led Turnbull to propose the first nucleation-based criterion for glass formation [18]. After a quench, the expected number of nuclei per unit volume, χ, is given by

$$\chi = \int_0^t I\,\mathrm{d}t \rightarrow \vartheta^{-1} \int_{T_1}^{T_2} I(T)\,\mathrm{d}T, \tag{1}$$

where I is the nucleation rate per unit volume, ϑ is the quenching rate, and T_1 and T_2 are the upper and lower temperatures for the range of significant nucleation. Turnbull argued that χ should be zero for glass

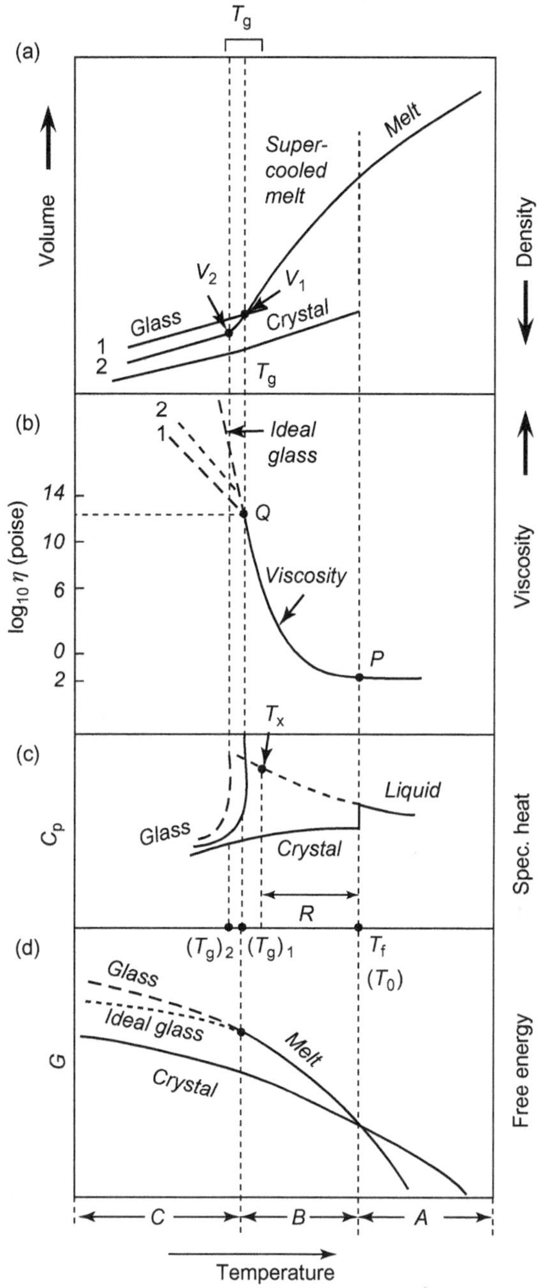

Fig. 1 A schematic diagram showing the variation, in the equilibrium-liquid, supercooled-liquid, and glassy states, of (from the top to bottom) specific volume, viscosity, specific heat at constant pressure, and Gibbs free energy. (Reprinted from Ref. [17], copyright (1996), with permission from Elsevier)

formation, but this is not realizable in practice. For example, for $Au_{81}Si_{19}$ (as is usual, metallic-glass compositions are quoted in at.%), one of the earliest known metallic glass formers, steady-state nucleation rates can be obtained from crystallization studies, and from these rates it is calculated that 10^{16}–10^{18} nuclei mol^{-1} form at the experimentally determined critical cooling rate for glass formation [19], clearly violating the Turnbull criterion for glass formation. Theoretical and experimental [20] considerations show that in many metallic glasses, there are significant populations of quenched-in nuclei that can have a profound impact on stability against crystallization when the glasses are heated to temperatures where the growth velocity is significant.

Thus, a nucleation criterion alone is insufficient to predict glass formation. A glass can be obtained by suppressing either the production of nuclei or their growth. A relevant criterion is, then, an operational one: a glass is produced if the degree of crystallization is less than some value, typically the lower limit for which it can be readily observed (i.e., a critical volume fraction, transformed, x_c). This is generally taken to be a volume fraction of one part per million (10^{-6}) [21]. Often calculations of glass-forming ability assume steady-state nucleation rates and macroscopic growth velocities. However, it is very unlikely that steady-state nucleation can be maintained during a rapid quench. The nucleation rate can be much lower than expected in steady-state, making glass formation easier and increasing the stability of the glass against crystallization [19].

As will be discussed in Section 7.2 of this chapter, the kinetic model underlying the classical theory of nucleation appears to describe quantitatively the evolution of the small clusters responsible for nucleation. The numerical method introduced in Chapter 3 to model time-dependent nucleation within the classical theory can then be readily extended to nucleation and growth under the nonisothermal conditions of a quench [19]. For a cluster of sufficiently large radius, r, the average cluster growth rate is [22]

$$\frac{\mathrm{d}r}{\mathrm{d}t} = \mathrm{C}\frac{16D}{\lambda^2}\left(\frac{3\bar{v}}{4\pi}\right)^{1/3} \sinh\left(\frac{\bar{v}}{2k_\mathrm{B}T}\left(\Delta g - \frac{2\sigma}{r}\right)\right), \tag{2}$$

where D is the molecular diffusion coefficient, λ is a jump distance (a length of order the molecular diameter), $\bar{v}$ is the molecular volume, σ is the crystal-glass interfacial free energy per unit area, Δg is the free energy decrease per unit volume for crystallization, k_B is the Boltzmann constant, T is the temperature in absolute units, and C is a constant that depends on the mechanism of growth [23]. Equation (2) applies only when nucleation and growth are rate-limited by interfacial processes and not by long-range diffusion. By using Eq. (2), a consistent treatment can be made of the evolution of clusters of all sizes. To model a quench, the continuous temperature change is divided into a large number of isothermal intervals of time duration $\delta t = \delta T/\vartheta$. At the end of each successive interval, the total number of nuclei generated during that interval is computed and stored, and the sizes of crystals growing on nuclei generated in previous intervals are evolved using Eq. (2). Assuming that the transformation is predominantly by nucleation and growth dispersed throughout the volume of the material, Johnson–Mehl–Avrami–Kolmogorov (JMAK) statistics may be assumed, giving the volume fraction transformed, $x(t, T)$ [24]

$$x(t,T) = 1 - \exp\left(-\sum_{i=1}^{m} \frac{4\pi}{3} \chi_i \left(r_i(t)\right)^3\right); \tag{3}$$

$r_i(t)$ is the time-dependent radius of the nuclei formed in the ith interval, χ_i is the population density per unit volume of new clusters produced in that interval and m is the number of intervals.

For illustration, Figure 2 shows the nucleation rate computed as a function of time for various quench rates using parameters appropriate for lithium-disilicate glass. As expected, the nucleation rate decreases markedly with increasing quench rate.

The lower nucleation rate gives fewer nuclei generated and a smaller volume fraction transformed during the quench than expected in steady state. These effects are most pronounced in more marginal glass formers, such as early metallic-glass compositions (Figure 3), for which the absolute quench rates of interest are higher. The number of quenched-in nuclei decreases much more strongly with increasing quench rate than

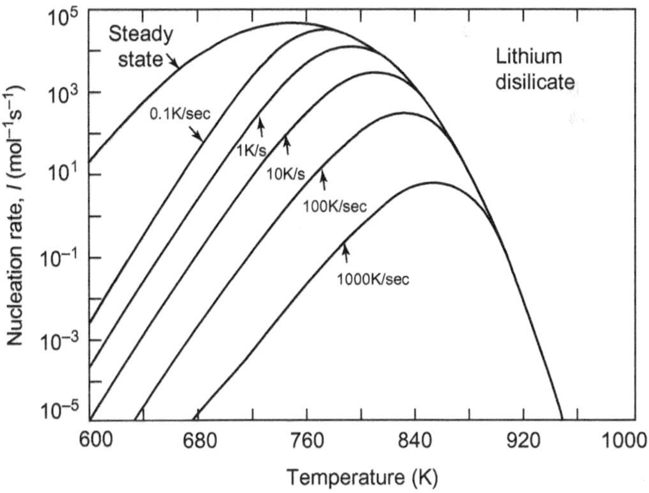

Fig. 2 Computed nucleation rate as a function of temperature for various quenching rates in lithium-disilicate liquid. (Reprinted from Ref. [19], copyright (1986), with permission from Elsevier.)

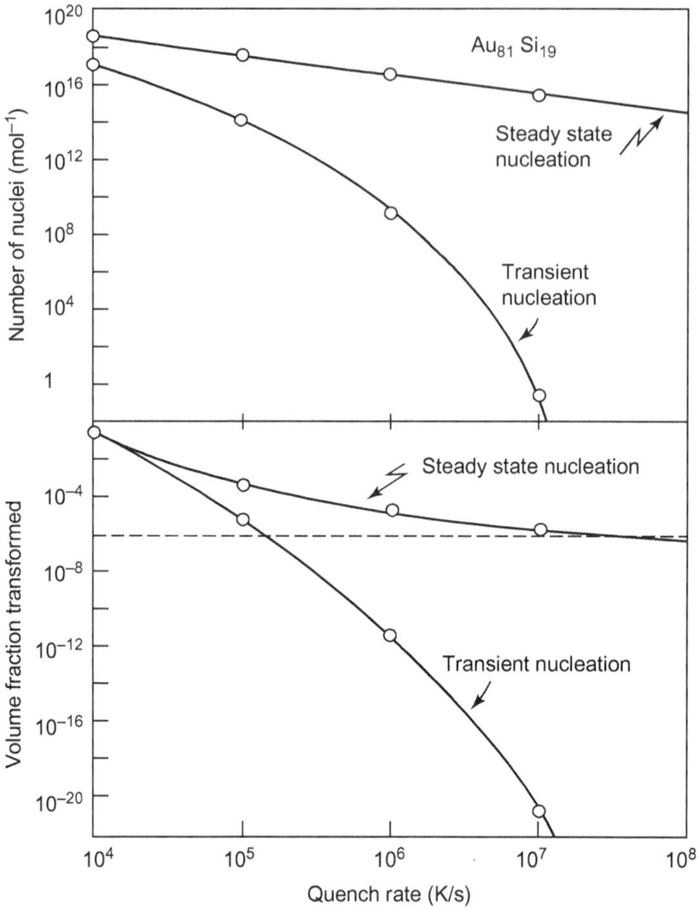

Fig. 3 Computed number of nuclei and volume fraction transformed in glassy $Au_{81}Si_{19}$ (at.%) as a function of liquid quench rate assuming steady-state or transient crystal nucleation. (Reprinted from Ref. [19], copyright (1986), with permission from Elsevier.)

predicted for steady-state nucleation. This was first observed experimentally in Fe–B glasses [20]. Assuming no solute partitioning during nucleation, and estimating the quenching rate in metallic-glass ribbons of different thickness, Greer demonstrated that the number of quenched-in nuclei scaled as $\vartheta^{-(2 \text{ to } 4)}$ [20]. Such transient nucleation effects can be a determining factor in whether it is possible to make a glass. In $Au_{81}Si_{19}$, for example, calculations based on steady-state nucleation predict a critical quenching rate ϑ_c of 10^7 to 10^8 K s^{-1} (for $x_c < 10^{-6}$). This is beyond the range of melt-spinning techniques used to obtain that glass, indicating, contrary to experiment, that it should not be possible to form this glass by that method. When transient nucleation effects are considered, the estimated ϑ_c drops to approximately 10^5 K s^{-1}, readily accessible by melt-spinning.

Similar behavior is observed in silicate glasses, although at the lower quench rates of interest transient effects are often too weak to influence glass formation [19]. In silicates, as in many BMGs, growth kinetics plays the key role in determining the volume fraction crystallized during the quench. However, the cluster distribution and population of quenched-in nuclei have strong implications on subsequent stability of the glass against crystallization [25, 26].

4. DEVITRIFICATION MECHANISMS

Given sufficient time, all glasses transform to one or more ordered phases; in most cases these are periodic crystal phases, but they can be more complicated such as the quasiperiodic icosahedral phases [27]. Since almost all glasses are multicomponent, the devitrification sequence is often complicated, making quantitative modeling of the nucleation difficult. In a few cases, the new phase has the same composition as the glass, but mostly a mixture of phases of different composition results. Devitrification may proceed directly from the initial glass, or it may be preceded by phase separation into two or more amorphous phases (Chapter 14, Section 2).

Three devitrification modes have been identified [28, 29]: *polymorphic*, *primary*, and *eutectic*, which are generally distinguished by the different resulting microstructures. In *polymorphic* crystallization, the glass transforms to a phase of the same composition, which may be a solid-solution phase of extended solubility or a metastable or stable compound. This is the easiest devitrification mode to model, since at constant temperature the growth rate is interface-controlled and the classical theory can be used to describe the nucleation step. In *primary* crystallization, a phase with a different composition from that of the parent glass forms, causing a compositional gradient and accompanying diffusion field to develop around the growing crystals. While the influence of diffusion on growth has been well studied, only recently has the influence on crystal nucleation been examined [30–32]. Primary crystallization does not go to completion, and is often followed by a different devitrification mode in the residual glassy matrix. In *eutectic* crystallization, a cooperative transformation occurs to two intimately interleaved phases having a mean composition that is the same as that of the original glass. Solute partitioning occurs along the interface between the two-phase crystalline regions and the glass, leaving the glass composition ahead of that interface unaffected. The effective growth rate of the two phases (at constant temperature) is, therefore, constant, allowing the crystallization kinetics to be modeled as for polymorphic crystallization.

As already discussed for liquids (Chapter 7, Section 5) and as will be discussed further (Chapter 10, Section 6), other phase transitions, even high-order ones, can couple to the nucleation barrier for crystallization. Phase separation, for example, is common in silicate glasses [33, 34]; some evidence indicates that it also occurs in metallic glasses [35]. If nucleation was difficult in the initial uniform glass, it might occur more readily, either within the phase-separated regions or at the interface between them (Chapter 14, Section 2.2).

5. MEASURING THE NUCLEATION RATE

For liquids, nucleation-rate measurements are generally restricted to a narrow temperature range near the liquidus. Only steady-state rates are investigated; the kinetic processes are too rapid to obtain accurate estimates of time-dependent effects. The low temperatures for devitrification, and the consequent slow atomic motion, allow both restrictions to be circumvented to some extent. Furthermore, nucleation rates are more easily measured directly, often by counting the number of nuclei that appear as a function of annealing time — a technique foreign to the study of supercooled liquids. Below T_g, the nucleation and growth rates are so low that accurate measurements cannot be made in reasonable times. In most cases, therefore, quantitative nucleation and crystallization studies are made at temperatures near or above T_g. The data, then, are actually from deeply supercooled liquids and not glasses. This is preferred for quantitative studies of nucleation, since a changing atomic mobility with structural relaxation that would accompany devitrification at lower temperatures would complicate the analysis. In practical applications (Chapter 14, Section 2), however, these complications may be unavoidable, particularly for primary crystallization, where the changing composition of the glass will likely affect both the thermodynamic and kinetic parameters during devitrification. Care has been taken in choosing the systems for discussion in this chapter to attempt to minimize these complications. Before looking at the nucleation data that can be obtained and how they are compared with theoretical predictions, it is useful first to examine the most common methods for measuring nucleation rates.

5.1 Modeling isothermal crystallization kinetics

Kinetic data, obtained by monitoring some physical parameters that change upon crystallization, are often used to extract information about nucleation and growth. For example, electrical resistivity measurements are commonly made because they can be very sensitive to the transformation microstructure [36]. The interpretation of such data, however, is often unclear. The rate of latent-heat evolution during crystallization scales linearly with the rate of transformation, making calorimetric measurements, such as differential scanning calorimetry (DSC) and differential thermal analysis (DTA), attractive. These techniques are sensitive, straightforward, and require only small quantities of sample. Isothermal measurements are typically analyzed using the JMAK equation [37–39], mentioned earlier (Eq. (3)) and discussed further here.

In the very earliest stages of polymorphic devitrification, impingement of the different crystallizing regions can be ignored, and the volume fraction of the glass that has transformed at time t (assuming an isotropic growth velocity and a unit volume of sample) is

$$x_{\mathrm{ext}} = \frac{4\pi}{3}\int_0^t I(\tau)\left(\int_\tau^t u(t')\,\mathrm{d}t'\right)^3 \mathrm{d}\tau, \tag{4}$$

where $I(\tau)$ is the time-dependent nucleation rate per unit volume and $u(t')$ is the time-dependent growth rate. As the crystallized regions grow, they eventually impinge and Eq. (4) no longer applies. If the sample size is much greater than any of the individual transformed regions and if growth proceeds uniformly throughout the sample, with spatially random nucleation, then the effects of impingement can be taken into account by assuming Poisson statistics, relating the true volume fraction, x, to the *extended volume fraction*, x_{ext} given by Eq. (4). The relationship is the JMAK equation $x = 1 - \exp(-x_{\mathrm{ext}}(t))$ (see the papers cited earlier, or Ref. [24]). For the interface-controlled polymorphic crystallization and constant nucleation and growth rates,

$$x = 1 - \exp\left(-\frac{\pi I u^3 t^4}{3}\right). \tag{5}$$

For interface-controlled growth on χ_0 quenched-in nuclei per unit volume with zero nucleation rate during the annealing treatment,

$$x = 1 - \exp\left(-\frac{4\pi\chi_0 u^3 t^3}{3}\right). \tag{6}$$

Equations (5) and (6) have the form

$$x = 1 - \exp\left(-(\kappa t)^m\right). \tag{7}$$

The kinetic parameters such as nucleation and growth rates are now contained within an effective kinetic constant κ, while the *Avrami exponent* m provides information on whether the transformation is one-, two-, or three-dimensional. It is often assumed that it is possible to infer whether the transformation is interface- or diffusion-controlled (polymorphic or primary), but Eq. (7) is not strictly valid for diffusion-controlled transformations (i.e., those rate-limited by long-range diffusion of solute, rather than by local interfacial kinetics). The conditions under which the JMAK approach may be used are limited and strictly defined [26, 40–42], and often inconsistent with the conditions under which it is applied. Nonetheless, it has yielded useful information on nucleation kinetics for polymorphic transformations under isothermal annealing. Any JMAK analysis must be supported by microscopy studies to determine whether its application is valid. Without such support and validation, *blind application of the JMAK analysis can yield misinterpretations.*

5.2 Nonisothermal crystallization kinetics

Much of the data in the literature are based on nonisothermal measurements of crystallization kinetics. If the transformation microstructure is known, computer modeling can be used to extract quantitative information from nonisothermal kinetic data [26, 41]. However, in most cases, data are analyzed using the Kissinger method [43, 44] to derive an effective activation energy for the transformation. The Ozawa method [45] is used to estimate the Avrami coefficient m to provide information about the transformation mechanism. In addition to all the problems for isothermal-analysis methods, the data from these nonisothermal techniques, particularly the Kissinger method, are strongly affected by sample thermal history and quenching rate, crystallization microstructure, and sample size. If the nucleation and growth stages are well separated, the activation energy extracted from a Kissinger analysis corresponds roughly to the average value of the activation energy for growth over the temperature range of significant heat evolution in the DSC or DTA. If the nucleation and growth stages overlap, the results from a Kissinger analysis are frequently misleading. In either case, the Kissinger analysis does not give meaningful information on nucleation processes or rates and should be avoided.

Other, more reliable, nonisothermal methods have been proposed for making quick estimates of nucleation rates. When the nucleation and growth stages are well separated, these methods are useful for locating the regime of significant nucleation rate [46]. For glass processing, this is often sufficient. While these methods themselves do not yield quantitative data on nucleation rates, they are useful for identifying the temperature range relevant for more detailed nucleation measurements.

5.3 Measurements of crystal population density

The accuracy of the nucleation rates obtained from devitrification kinetics is limited by the degree of validity of the transformation model. As noted earlier, the devitrification mechanism is often difficult to determine securely. A more direct method for estimating the nucleation rate is to count the number of crystals that form during annealing. Two methods are the most common.

5.3.1 Single anneal

For temperatures well above the maximum in the nucleation rate, nuclei can be formed and developed by annealing at a single temperature. The crystallites are counted directly in prepared sections of the sample examined using optical microscopy, or scanning (SEM) or transmission (TEM) electron microscopy. Though TEM hot-stage experiments can also be used for *in situ* studies of nucleation and growth, these transformations are generally dominated by surface effects in the thin foil. More reliable data are obtained from studies of samples annealed *ex situ* before preparation for TEM observations. Qualitative information on whether nucleation is steady-state or time-dependent, homogeneous or heterogeneous, can frequently be obtained from a simple analysis of the crystal size distribution. Figure 4(a–e), for example, shows distributions calculated for several different nucleation modes in polymorphic or eutectic crystallization. It has been assumed that there has not yet been any significant impingement of the crystallites; the growth velocity depends on the temperature only; and the nuclei grow isotropically. Since the largest crystallites nucleated first, increasing time points in the direction of decreasing crystal size. If transient nucleation effects are important (Figure 4b and d), the population density rises with decreasing crystal size, reflecting the acceleration of nucleation with increasing annealing time. The decreased population density for small crystals in Figure 4c and d reflects site saturation for heterogeneous nucleation. Even for homogeneous nucleation, however, the population density must eventually decrease due to the decrease in the volume of the original phase available for fresh nucleation (Figure 4a and b). A uniform size reflects growth only, indicating a population of quenched-in nuclei and negligible further nucleation during annealing (Figure 4e).

Interpretations become more difficult for primary crystallization. The influence of crystallites on each other can be significant, even when the mean size is small, due to impingement of their long-range diffusion fields. Further, a decreasing supersaturation for precipitation processes, for example, can decrease the nucleation rate in the later stages, mimicking the effects of heterogeneous nucleation in a polymorphic system (Figure 4f).

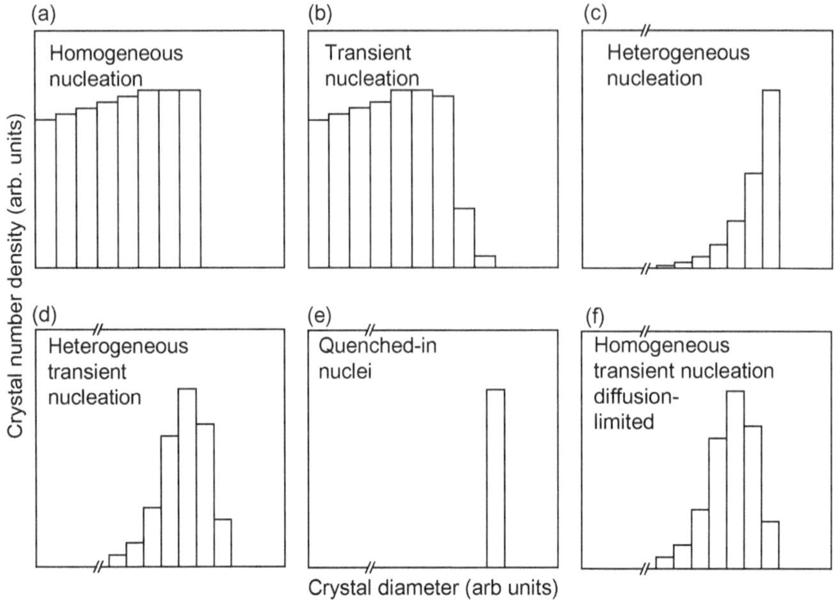

Fig. 4 Schematic histograms of the crystallite number density (vertical axes) as a function of crystal diameter (horizontal axes) for several nucleation mechanisms in polymorphic or eutectic (i.e., constant-growth-rate) devitrification. (Reprinted from Ref. [52], copyright (1991), with permission from Elsevier; reprinted from Ref. [53], copyright (1987) with permission from the Materials Research Society.)

More detailed analyses involve fitting the size distribution to plausible models for nucleation and growth, raising similar concerns about validity and uniqueness as for fitting the kinetic data. Since the distributions are obtained for small volume fractions transformed, however, crystallite overlap and impingement can often be neglected, considerably simplifying the modeling. When applying these techniques, it is also critical that the particle size distribution be derived while accounting for sectioning statistics. Even an ensemble of identical spheres appears with a distribution of sizes when a cross-section is examined. The techniques for dealing with this are outside the scope of this book; treatments can be found elsewhere [47–51].

5.3.2 Two-step anneal

For many silicate glasses and some of the recently discovered easily formed metallic glasses, the peak in the steady-state nucleation rate occurs at a temperature sufficiently low for growth from the nuclei to be slow. Nucleation rates can then be determined using a two-stage anneal. The sample is first annealed at a temperature T_N, where the nucleation rate is high and the growth velocity is small, to develop a population of

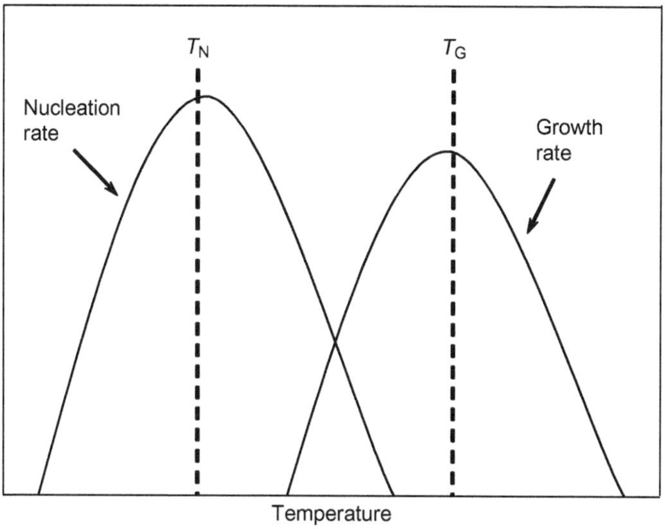

Fig. 5 Schematic illustration of the two-step annealing technique for the measurement of crystal nucleation rate in glasses. T_N is the temperature for the nucleation treatment; T_G is the temperature for the growth treatment.

nuclei (Figure 5). Those nuclei are then grown at a higher temperature T_G, where the growth velocity is large, but the nucleation rate is low enough to introduce no new nuclei. Since the number of molecules in the critical cluster, n, increases with temperature, those nuclei between $n^*(T_N)$ and $n^*(T_G)$ redissolve during the anneal at T_G, while those larger than $n^*(T_N)$ grow to observable size. The number of nuclei between the critical sizes is small compared with those above $n^*(T_G)$ and does not significantly perturb the results [22, 54–56]. In steady state, the nucleation rate is independent of the cluster size at which it is measured. The calculation of the nucleation rate at the critical size for the nucleating temperature should then give an accurate description of the measured rate.

6. HOMOGENEOUS NUCLEATION OF POLYMORPHIC CRYSTALLIZATION

When nucleation is homogeneous and the crystalline phase has the same composition as the parent glass, there is the greatest chance that the nucleation rate can be compared readily with predictions from theoretical models. The best data for this purpose come from silicate glasses, often from two-step anneals. Silicate glasses can be extremely resistant to devitrification. Most crystallize from the surface unless catalyzed by nucleating agents. There are a few cases, however, where crystal nucleation is homogeneous and dispersed through the sample volume. Such cases provide the best opportunities for quantitative studies of time-dependent and steady-state nucleation over a wide temperature range.

The reduced glass-transition temperature ($T_{rg} = T_g/T_{liq}$, where T_g is the glass-transition temperature and T_{liq} is the liquidus temperature) is a good predictor of whether devitrification is dominated by crystallization

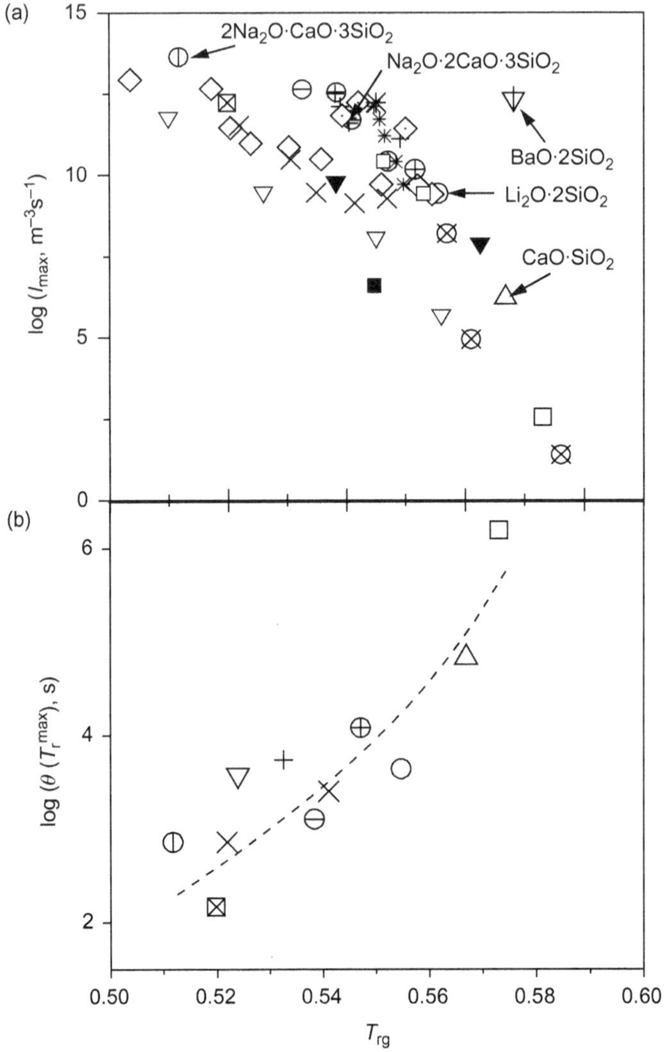

Fig. 6 (top) Maximum nucleation rate, I_{max}, as a function of the reduced glass-transition temperature, T_{rg}, for different silicate glasses; (bottom) induction time at the temperature of I_{max} as a function of T_{rg} (the dashed line is a guide to the eye). (Reprinted with permission from Ref. [57], copyright (2003), The Royal Society.)

at the surface or throughout the volume [34, 58, 59]. Figure 6a shows the maximum nucleation rate I_{max} in the volume and the corresponding T_{rg} for 51 glass compositions (both matching and nonmatching to stoichiometric crystal compositions) that are members of eight silicate glass families. The maximum rate drops off precipitously with increasing T_{rg}, becoming virtually undetectable for $T_{rg} > 0.58$. The induction time for nucleation measured at the temperature of I_{max} increases with T_{rg} (Figure 6b). These features are a consequence of the competition between the thermodynamic and kinetic aspects of nucleation. The liquid viscosity η abruptly increases as T_g is approached on cooling (Figure 7), radically slowing the rates of nucleation and growth of crystalline clusters. If T_g is much lower than T_{liq} (small T_{rg}), the driving free energy for crystallization is large at the onset of kinetic freezing, and the steady-state nucleation rate is high. Moving T_g toward T_{liq} (i.e., increasing T_{rg}) decreases the driving free energy at the onset, giving a lower nucleation rate. The dramatic change in I_{max} over a small range of T_{rg} is a consequence of the exponential dependence of the nucleation rate on the driving free energy (Chapter 2, Eq. (54)). The strong temperature dependence of the induction time θ is at first surprising given the expected weak dependence on driving free energy and the similarity of the viscosities at T_g for the various glasses. It is observed that T_{max}, the temperature of the maximum nucleation rate (I_{max}), increases with increasing T_{rg}. T_{max} is larger than T_g for low values of T_{rg} and eventually

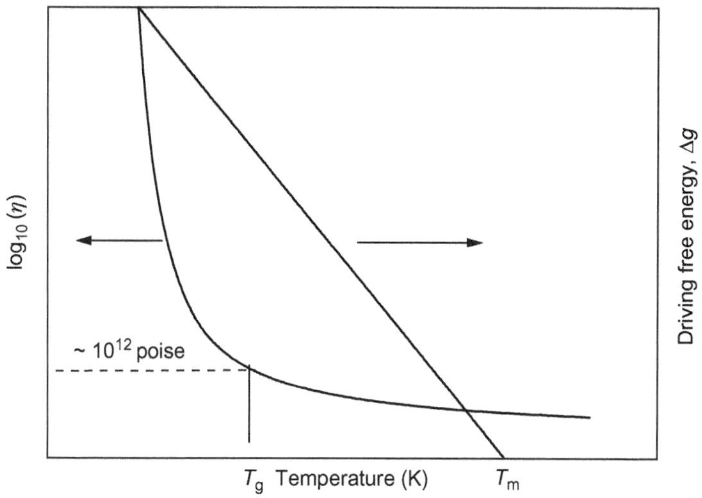

Fig. 7 Schematic illustration of the temperature dependence of the viscosity and the driving free energy that underlie the results shown in Fig. 6.

becomes smaller than T_g as T_{rg} increases. The induction time at T_{max} reflects this behavior since it scales with the viscosity.

The surface of a glass can be a potent catalyst for nucleation. Surface crystallization is dominant if $T_{rg} > 0.58$, even in the presence of internal nucleating agents. Supporting data and possible surface nucleation mechanisms are discussed in Section 10.1.

6.1 Steady-state nucleation in silicate glasses — available data

Measurements of the temperature-dependent steady-state nucleation rates for several well studied silicate glasses that crystallize polymorphically are shown in Figure 8. Relevant thermodynamic and kinetic data are summarized in Table 1. The nucleation data for these glasses show the temperature dependence predicted by the classical theory. For $Li_2O \cdot 2SiO_2$, nucleation rates can be measured from 698 to 803 K, with the peak rate occurring near 727 K. The magnitude of the peak rate

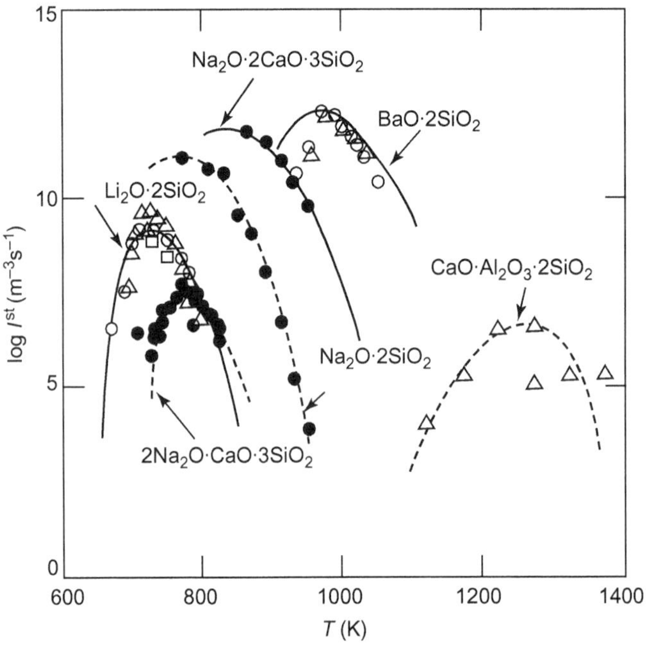

Fig. 8 The measured steady-state nucleation rate as a function of temperature for homogeneously nucleating silicate glasses: $Li_2O \cdot 2SiO_2$ (○, △, □); $Na_2O \cdot 2CaO \cdot 3SiO_2$ (●); $2Na_2O \cdot CaO \cdot 3SiO_2$ (●); $Na_2O \cdot 2SiO_2$ (●); $BaO \cdot 2SiO_2$ (○, △); $CaO \cdot Al_2O_3 \cdot 2SiO_2$ (△). The solid lines are fits to classical nucleation theory for those glasses where sufficient kinetic and thermodynamic data are available (the dashed lines are guides for the eye). (Reprinted from Ref. [52], copyright (1991), with permission from Elsevier.)

Table 1 Kinetic and thermodynamic parameters for selected silicate glasses

Glass	V_M (10^{-6} m^3 mol^{-1})	η (Vogel–Fulcher)	ΔH_f (kJ mol^{-1})	T_{liq} (K)	T_g (K)	Remarks
$BaO \cdot 2SiO_2$	73.34	$\log_{10}\eta = 1.83 + \frac{1701.9}{T - 795.6}$	37.5	1693	961	Congruent melting
$Na_2O \cdot 2CaO \cdot 3SiO_2$	126.6	$\log_{10}\eta = -3.86 + \frac{4893}{T - 547}$	91.3	1562	852	Congruent melting
$Li_2O \cdot 2SiO_2$	61.2	$\log_{10}\eta = -1.44 + \frac{3370}{T - 460}$	57.4	1306	683	Incongruent melting within 1 K of liquidus
		$\log_{10}\eta = 1.81 + \frac{1346.6}{T - 594.8}$				
$2Na_2O \cdot 3CaO \cdot 3SiO_2$	129.7	$\eta(T_{max}) = 10^{10}$ Pa s	–	–	743	Incongruent melting at 1422 K, liquidus 1473 K
$Na_2O \cdot SiO_2$	46.2	$\eta(T_{max}) = 10^{10}$ Pa s	51.8	1362	683	Congruent melting
$3BaO \cdot 5SiO_2$	–	–	26.1	–	–	Incongruent melting at 1696 K, liquidus at 1716 K
$CaO \cdot SiO_2$	–	–	56	1817	1030	Congruent melting
$CaO \cdot Al_2O_3 \cdot 2SiO_2$	100.8	–	167	1823	–	–

V_M is the molar volume, ΔH_f is the enthalpy of fusion, and η is the viscosity; T_{liq} is the liquidus temperature and T_g is the glass transition temperature, both measured in absolute units. (Reprinted from Ref. [52], Copyright (1991), with permission from Elsevier.)

depends on the source of the data, ranging from $\sim 7.8 \times 10^8$ to $4.25 \times 10^9 \mathrm{m}^{-3}\mathrm{s}^{-1}$. Since certain metallic impurities are known to catalyze nucleation in silicate glasses [60–62], variations in sample purity could explain this scatter. In $Li_2O \cdot 2SiO_2$, however, most impurities have little effect [63]; the nucleation rate is markedly enhanced, however, by the presence of water, probably due to increased atomic mobility since the viscosity is decreased [64].

6.2 Analysis of steady-state nucleation in silicate glasses

We discuss the information that can be obtained from studies of silicate and metallic glasses by first focusing on the best studied glass, lithium disilicate ($Li_2O \cdot 2SiO_2$). This would appear to be an ideal system, since crystallization is homogeneously nucleated throughout the volume, and there are measured data for the driving free energy and the viscosity as a function of temperature. Even for this glass, however, there are complications that make an unambiguous analysis difficult. These complications are also present in other glasses, so they are discussed at length here to draw attention to the problems that arise in the data analysis.

6.2.1 Classical theory

Assuming that the atomic or molecular diffusion coefficient D can be related to the viscosity η by the Stokes-Einstein relation

$$D = \frac{k_B T}{3\pi\lambda\eta}, \tag{8}$$

where λ is a length of order the atomic diameter, the classical expression for the steady-state nucleation rate I^{st} (Chapter 2, Eqs. (14) and (55)) can be written as

$$I^{st} = \frac{A}{\eta}\exp\left(-\frac{16\pi}{3k_B T}\frac{\sigma^3}{\Delta g^2}\right), \tag{9}$$

where $A = A^*\eta$ (A is the dynamical preterm in the nucleation rate, Chapter 2, Eq. (55)), σ is the crystal-glass interfacial free energy per unit area, and Δg is the volumetric driving free energy for crystallization. A plot of ln $I^{st}\eta$ versus $1/T\Delta g^2$ should give a straight line with slope proportional to σ^3 and intercept equal to ln A.

$$\ln(I^{st}\eta) = \ln A - \frac{16\pi}{3k_B T}\frac{\sigma^3}{\Delta g^2}. \tag{10}$$

The nucleation data for the polymorphic crystallization of $Li_2O \cdot 2SiO_2$ glass, shown in Figure 8, are used to illustrate this approach. Although $Li_2O \cdot 2SiO_2$ melts incongruently, the composition is fully liquid only 1 K above the incongruent melting temperature, and the congruent melting temperature can be approximated by the liquidus temperature. Curve *a* in Figure 9 is a plot of Eq. (10) using the viscosity data of Matusita and Tashiro [65, 66] and measured data for Δg (Table 1). The $\ln(I^{st}\gamma)$ data deviate significantly from linearity at low temperatures. It is possible that the measured nucleation is not in steady state there, since the transient times are large. Another possibility is that the viscosity data at low temperatures are incorrect. A much better fit is found, for example, if different viscosity data [64, 67] are used (curve *b*). Despite this, the Matusita and Tashiro viscosity data give better agreement with the measured induction times for transient nucleation (as will be shown in Figure 15b). Since the induction time is most sensitive to the atomic mobility, this suggests that the deviation at low temperatures is due to other factors.

A more significant problem is the value of the calculated prefactor A. Both sets of viscosity data give a value for A that is larger *by 20 to 25 orders of magnitude* than is theoretically predicted from Chapter 2, Eqs. (53) and (54) (Table 2). These problems are not unique to lithium disilicate; similar results are obtained for the other silicate glasses (Table 2). There are

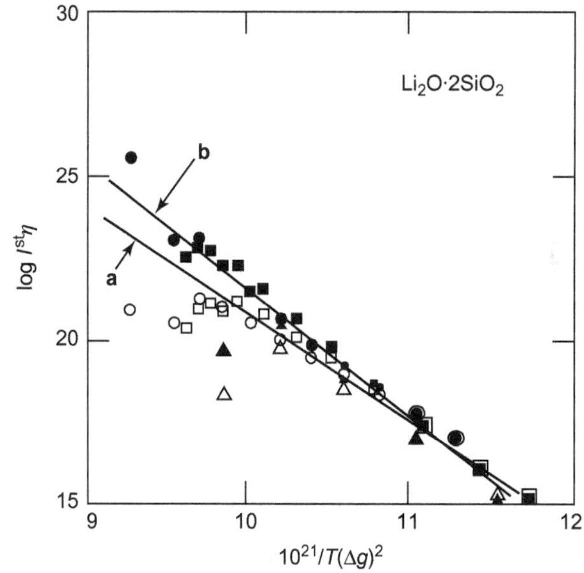

Fig. 9 The product of the steady-state nucleation rate I^{st} ($m^{-3}\ s^{-1}$) and the viscosity η (Pa s) as a function of the volumetric driving free energy for crystallization, Δg ($J\ m^{-3}$), for $Li_2O \cdot 2SiO_2$ glass. Curve a is based on the viscosity data of [65, 66]; curve b is based on the viscosity data of [64, 67]. Nucleation data are taken from Refs. [65, 66, 68]. (Reprinted from Ref. [52], copyright (1991), with permission from Elsevier.)

Table 2 Values of σ and A for silicate glasses that show homogeneous nucleation and polymorphic crystallization

Theoretical		Calculated					
		Constant σ		$\sigma = \sigma_0 + \sigma_1(T)$		$\sigma(T_{max})$ (J m^{-2})	
	A	A	σ	σ_0	σ_1	Kelton	James
Glass	(Pa m^{-3})	(Pa m^{-3})	(J m^{-2})	(J m^{-2})	(J m^{-2}K^{-1})	[52]	[69]
$Li_2O \cdot 2SiO_2$	$10^{33.0}$	$10^{53.2}$[a]	0.139	0.138	2.1×10^{-5}	0.153	0.143
		$10^{60.1}$[b]	0.147	0.125	3.7×10^{-5}	0.152	0.147
$Na_2O \cdot 2CaO \cdot 3SiO_2$	$10^{32.6}$	$10^{60.8}$	0.131	0.103	3.1×10^{-5}	0.130	0.108
$BaO \cdot 2SiO_2$	$10^{32.9}$	$10^{55.3}$	0.100	0.077	2.8×10^{-5}	0.104	0.101

Used viscosity data from:
[a][70] or
[b][67] and [64]. (Reprinted from Ref. [52], Copyright (1991), with permission from Elsevier.)

several possibilities to explain these discrepancies. Perhaps it is not the stable phase that is nucleating, but a metastable phase, in which case the free energy used in the analysis is incorrect. Perhaps the interfacial mobility does not scale inversely with the bulk viscosity. Perhaps, as we found in liquids, the interfacial free energy increases with increasing temperature. Most intriguing, however, is the suggestion that this indicates a fundamental breakdown of the classical theory of nucleation. We examine each possibility in turn.

6.2.2 Metastable phase nucleation

The composition of the lithium-disilicate glass is critical in determining which phase crystallizes. If the glass is lithia-rich ($\geq$35.5 mol% Li_2O), some lithium metasilicate crystals appear. If it is lithia-poor ($\leq$32.0 mol % Li_2O), a metastable miscibility gap develops in the supercooled liquid, leading to liquid-liquid phase separation. There is a continuing debate on whether even in the stoichiometric glass, however, the metasilicate phase might nucleate first if the interfacial energy were lower than for the stable phase. If true, the metasilicate could then serve as a heterogeneous nucleation site for the disilicate phase. Also, as will be discussed in Section 9, the nucleation rate of silicate glasses can be very sensitive to composition. If local fluctuations in composition are present, either through poor mixing, or due to incipient phase separation, these could alter the nucleation kinetics, possibly favoring the nucleation of a metastable phase or of a solid solution with composition different from the glass. Studies of homogeneous nucleation in stoichiometric $Na_2O \cdot 2CaO \cdot 3SiO_2$ glass, for example, suggest that the phase that nucleates is enriched in soda relative to the glass [71]. In any case, if it is not lithium disilicate that nucleates, the free energy used in the analysis of the nucleation-rate data presented here is incorrect, possibly leading to the observed discrepancies with the classical theory.

Despite many studies using a range of experimental techniques, including small-angle X-ray scattering (SAXS), dielectric relaxation, Raman spectroscopy, and X-ray photoelectron spectroscopy (XPS) and TEM, it remains unclear which phase nucleates first in $Li_2O \cdot 2SiO_2$ glass (see reviews by Zanotto [72] and Burgner et al. [73], and [57]). It seems, however, that even if some metasilicate does nucleate, it can be ignored for the present discussion, since it is not an effective heterogeneous nucleating agent for the disilicate and it disappears on further annealing [74, 75]). The question of whether the nuclei have a composition different from that of the final crystal is a more difficult one. If so, an analysis of the nucleation kinetics would require an approach of the kind outlined in Chapter 5 for multicomponent systems. Few measurements exist that can guide the choice of parameters needed for a proper analysis, however, so the issue has yet be resolved.

6.2.3 Incorrect interfacial mobility

The use of Eq. (9) assumes that the interfacial mobility relevant for crystal nucleation and growth is the same as the atomic mobility that underlies the viscous flow and molecular diffusion in the bulk glass. This may not be true, however, given the significant differences in the processes. The nucleation induction time is a more direct indication of the interfacial mobility relevant for nucleation. Assuming the Kashchiev expression (Chapter 3, Eq. (36)) for the transient time τ, the relation to the induction time $\theta = \pi^2\tau/6$, and using Eqs. (14) and (53) in Chapter 2, the steady-state nucleation rate (per mole) can be written as

$$I^{st} = \frac{2}{3\pi}\frac{N_A}{\theta}\left(\frac{|\Delta\mu|}{6\pi k_B T n^*}\right)^{-1/2} \exp\left(-\frac{16\pi}{3k_B T}\frac{\sigma^3}{\Delta g^2}\right). \tag{11}$$

where N_A is the Avogadro number and $\Delta\mu$ is the change in chemical potential for a single molecule, moving from the original phase to the new phase (the driving free energy per molecule). While this is a better approach, it does not reduce the disagreement with the preterm over that found in Figure 9 or Table 2. As we shall see next, the problem is that a central assumption in the classical theory is incorrect; the interface is not sharp.

6.2.4 Diffuse-interface and temperature-dependent interfacial free energy

Fits of Eq. (9) to the experimental nucleation rates give more reasonable values for A if the interfacial free energy is assumed to increase linearly with increasing temperature, that is, $\sigma(T) = \sigma_0 + \sigma_1 T$. A similar conclusion was reached when fitting nucleation data for supercooled liquids (Chapter 7). The values for σ_0 and σ_1 are obtained by assuming the theoretical values for A (Table 2) and fitting to the steady-state data. A measure of the relative entropic contribution to $\sigma(T)$ is given by the ratio $\sigma_1/\sigma(T)$; $\sigma_0/\sigma(T)$ gives the fraction of the enthalpic contribution. For metallic liquids, σ_0 is near-zero, indicating that the predominant contribution to the interfacial free energy is entropic. For the silicate glasses, however, σ_0 is large (Table 2), comparable in magnitude to $\sigma_1(T)$ at the temperature of the maximum nucleation rate. There is, then, a sizable enthalpic contribution to the interfacial free energy, which is not unexpected given the strong covalent bonding in the silicate glasses. The bond strength is sensitive to the directionality of the local bonding, which can be different in the crystal and glass phases.

Although the assumption of a temperature-dependent interfacial energy brings the values of A into line with what is expected theoretically, the critical radii at the large supercoolings where the nucleation rates are measured often are smaller than the unit cell of the growing crystal, possibly signaling a breakdown of the classical theory [76]. It is unclear that the critical nucleus need be as large as a unit cell, however, since the local order parameter differentiates whether an atom is in a local configuration more similar to that of the glass or crystal, defining to which phase it belongs. Further, as discussed in Chapter 4, the critical size computed by the classical theory is likely smaller than the true value. However, attempts to fit multiple-step annealing experiments also suggest a possible failure of the classical theory at large supercoolings (Section 7.2).

The temperature dependence of σ at a crystal–liquid interface reflects an ordering in the liquid near the interface (Chapter 4, Section 3 and Chapter 7, Section 3), the deficit in entropy (*negentropy*) giving a positive contribution to the free energy. Fundamentally, however, a temperature-dependent interfacial energy reflects the diffuseness of the interface, in conflict with the sharp interface picture assumed in the classical theory (Chapter 4, Sections 3 and 4). In Chapter 4, we saw that different approaches to this problem predicted different values for the work of cluster formation. Given the large amount of data available for the silicate glasses, it should be possible to sort out which of these approaches is most appropriate. Unfortunately, the fundamental parameters needed for such a direct comparison, such as the measured interfacial free energy, the partial structure factor of the liquid, etc., are often lacking.

As shown in Chapter 7 for the analysis of the liquid supercooling data (Section 3.2), it is possible to assess the approaches discussed in Chapter 4 by focusing on predictions for the work of cluster formation. As discussed there, let ϖ be the cube-root of the ratio of W^*_{NC}, the work to form a critical cluster in the nonclassical model of interest, to W^*_{CNT}, that computed from the classical theory $\varpi(T) = (W^*_{NC}/W^*_{CNT})^{1/3}$. The steady-state nucleation rate can then be written as

$$I^{st} = A^* \exp\left(-\varpi(T)^3 \frac{W^*_{CNT}}{k_B T}\right), \tag{12}$$

yielding

$$\ln\left(\frac{I^{st}}{A^*}\right) = -\varpi(T)^3 \frac{16\pi}{3k_B T} \frac{\sigma^3_{CNT}}{\Delta g^2}, \tag{13}$$

where σ_{CNT} is the classical theory value for the interfacial free energy between the ordered cluster and the glass. Using the value of A expected for the classical theory, and plotting $\ln(I^{st}/A^*)$, where I^{st} is the measured steady-state nucleation rate, as a function of $\varpi(T)^3 \Delta g^{-2}\, T^{-1}$, we should get a straight line through the origin with a gradient that is proportional to σ_{CNT}^3. As discussed in Chapter 7, Section 3.2, however, we do not know the correct value for σ_{CNT}, so the gradient does not allow us to evaluate the different models. But as we did there, we can eliminate those models that do not give a line passing near the origin. Gránásy and James applied this analysis to all silicate glasses that are presumed to crystallize polymorphically by homogeneous nucleation, and for which data exist [77]. The models that they considered have been discussed already in Chapter 4; for convenience, they are briefly reviewed in the Appendix and the expressions for the work of cluster formation are listed there.

To analyze the experimental nucleation data using these models, it is essential to know the driving free energy, since many of the approximate forms can lead to errors of up to tens of orders of magnitude [52]. For the fits, Δg was computed from measured thermodynamic data,

$$\Delta g = \left(\Delta h_{\rm f} - \int_T^{T_{\rm liq}} \Delta c_{\rm p}\, {\rm d}T'\right) - T\left(\frac{\Delta h_{\rm f}}{T_{\rm liq}} - \int_T^{T_{\rm liq}} \frac{\Delta c_{\rm p}}{T}\, {\rm d}T'\right), \tag{14}$$

where $\Delta h_{\rm f}$ is the enthalpy per unit volume from the liquid, $T_{\rm liq}$ is the liquidus temperature, and $\Delta c_{\rm p}$ is the specific heat difference between the liquid and solid phases (normalized to be per unit volume). The Stokes-Einstein relation was used to estimate the atomic mobility from the measured viscosity (Eq. (8)). The thermodynamic data used for these fits are listed in reference [77].

Figure 10 shows a graph of $\ln(I^{\rm st}/A)$ as a function of X (see caption of Figure 10) for nucleation in a $Li_2O \cdot 2SiO_2$ glass. From Table 3, only two models give intercepts close to the origin, the phenomenological diffuse-interface theory (DIT) and the semiempirical density-functional approximation (SDFA); all the others give intercepts that are many orders of magnitude too large. Even for the DIT and SDFA models, the intercept is not precisely at zero. That could be a result of measurement errors, in the nucleation data themselves, or in the values used for the input parameters (heat of fusion, specific heats of the glass and crystal, and the glass viscosity). The deviation may also signal heterogeneous nucleation, since in this case, the steady-state rate would be multiplied by a factor that accounts for the fraction of sites that can participate in the nucleation process, $x_{\rm N}$,

$$I^{\rm st} = A^* x_{\rm N} \exp\left(-\varpi(T)^3 \frac{W_{\rm CNT}^*}{k_{\rm B}T}\right), \text{ yielding}$$

$$\ln\left(\frac{I^{\rm st}}{A^*}\right) = \ln x_{\rm N} - \varpi(T)^3 \frac{16\pi}{3k_{\rm B}T} \frac{\sigma_{\rm CNT}^3}{\Delta g^2}. \tag{15}$$

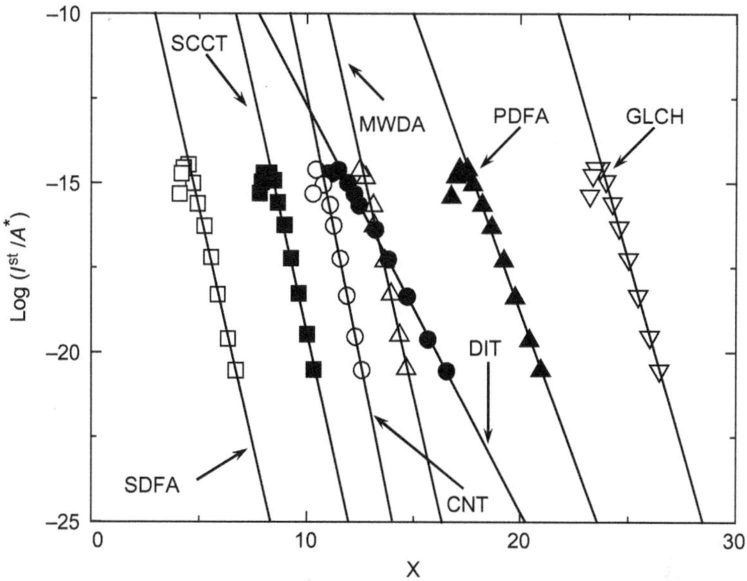

Fig. 10 Fits of models listed in Table 1 of the Appendix to $Li_2O \cdot 2SiO_2$ nucleation data. The expression for X is slightly different for the different models. It is expressed in units of $(\Delta h_f^2 T_{liq})^{-1}$ for the CNT, SCCT, SDFA, PDFA, MWDA, and GLCH models ($X \equiv \varpi(T)^3 \Delta g^{-2} T^{-1}/(\Delta h_f^{-2}\, T_{liq}^{-1})$) and in units of $\Delta h_f\, T_{liq}^{-1}$ for the DIT model ($X \equiv -\Delta g_{ct}\, \psi T^{-1}/\Delta h_f T_{liq}^{-1}$, where $\psi = 2(1+q)\xi^{-3} - (3+2q)\xi^{-2} + \xi^{-1}$, $q = \sqrt{1-\xi}$ and $\xi = \Delta g_{ct}/\Delta h_{ct}$). The subscript "ct" on the free energy and enthalpy differences between the glass and the solid in the expression for the DIT model signals that these quantities are evaluated at the center of the cluster. CNT — classical nucleation theory; SCCT — self-consistent CNT; SDFA — semiempirical density-functional approximation (DFA); PDFA — perturbative DFA; MWDA — modified weighted DFA; GLCH — Cahn-Hilliard/Ginzburg-Landau; DIT — diffuse-interface theory (Reprinted with permission from Ref. [77], copyright (1998), The Royal Society.)

Table 3 Intercept values and errors obtained by fitting experimental nucleation data

Glass	CNT	SCCT	DIT	SDFA	PDFA	MWDA	GLCH
LS_2	19±4	10±4	−0.5±2.0	−1.5±1.9	15±4	24±5	40±8
BS_2	24±5	18±5	−0.7±2.3	−4.9±1.9	22±5	30±7	49±10
N_2CS_3	125±66	92±52	−0.5±3.1	0.8±2.6	71±27	180±92	−410±328
NC_2S_3	56±18	40±19	−2.9±5.5	−9.2±1.8	49±15	72±25	156±94
CS	178±64	157±59	−3.6±4.8	−13±2	165±57	263±123	1837
NS	257	194±165	−10±3	−13±2	92±67	138±104	−108±65
LB_2	581	472	−13±3	−17±2	−336	202	−96±81
CAS_2	−15±4	−18±3	−17±2	−17±2	15±4	−14±4	−13±6

Glasses: $Li_2O \cdot 2SiO_2$ (LS_2), $BaO \cdot 2SiO_2$ (BS_2), $2Na_2O \cdot CaO \cdot 3SiO_2$ (N_2CS_3), $Na_2O \cdot 2CaO \cdot 3SiO_2$ (NC_2S_3), $CaO \cdot SiO_2$ (CS), $Na_2O \cdot SiO_2$ (NS), $Li_2O \cdot 2B_2O_3$ (LB_2) and $CaO \cdot Al_2O_3 \cdot 2SiO_2$ (CAS_2). (Reprinted with permission from Ref. [77], copyright (1998), The Royal Society.)

Since x_N is less than 1, a negative intercept is predicted. As seen from the data in Table 3, for $Li_2O \cdot 2SiO_2$ the errors are sufficiently small that the DIF and SDFA fits do intersect the abscissa at the origin, in agreement with extensive studies that indicate that nucleation in this glass is homogeneous.

In Figure 10, $Li_2O \cdot 2SiO_2$ glass was chosen for illustration. Similar results are obtained for other silicate glasses that nucleate by a homogeneous polymorphic mechanism (Table 3). Only the temperature dependences of the interfacial free energies that are predicted by the semiempirical density-functional approach and the phenomenological DIT describe the data. A similar conclusion was reached in Chapter 7 for nucleation in supercooled liquids. As we have seen, predictions from the CNT are in good agreement with nucleation data if a temperature-dependent interfacial free energy is used, reflecting ordering in the glass or liquid near the crystal nucleus. The DIT provides a more systematic way of accounting for this ordering, although it is also a phenomenological treatment. While more difficult to apply, the density-functional model provides a justification of this approach.

All of the fits for $Li_2O \cdot 2SiO_2$ in Figure 10 deviate significantly from linearity for low values of X, consistent with the trend that we saw in Figure 9. While not shown here, a similar deviation was observed in the analysis of the data for $CaO \cdot SiO_2$ [77]. Since these data are below the glass-transition temperature, this deviation may indicate that the nucleation rates were not steady-state; other suggestions include a reversal in the size dependence of the interfacial free energy for small cluster sizes [78] and a bimodal distribution of relaxation times, expected in glasses [79].

6.3 Polymorphic crystallization in metallic glasses

The crystallization kinetics of metallic glasses have been studied extensively, but unlike the case for silicate glasses there are few direct measurements of the nucleation rates, hindering quantitative comparisons with nucleation theories. Like the silicate glasses, crystallization often occurs from the surface, although there are as yet no studies of the conditions favoring surface crystallization that are as systematic as those for the silicates shown in Figure 6. Quenched-in nuclei can play an important role in metallic-glass crystallization, which is not usually the case in silicate glasses.

Nucleation studies in $Fe_{40}Ni_{40}P_{14}B_6$ (at.%) glass [80–82], which shows eutectic crystallization, and $Co_{33}Zr_{67}$, $Fe_{65}Ni_{10}B_{25}$ and $Ti_{67}Ni_{33}$ glasses [51], which crystallize polymorphically, provide good examples. In all these cases, the nucleation behavior was determined from the crystal size distributions, measured after annealing at a single temperature. Examples of heterogeneous and homogeneous nucleation are shown in Figure 11. Based on the arguments given in Figure 4, the increasing

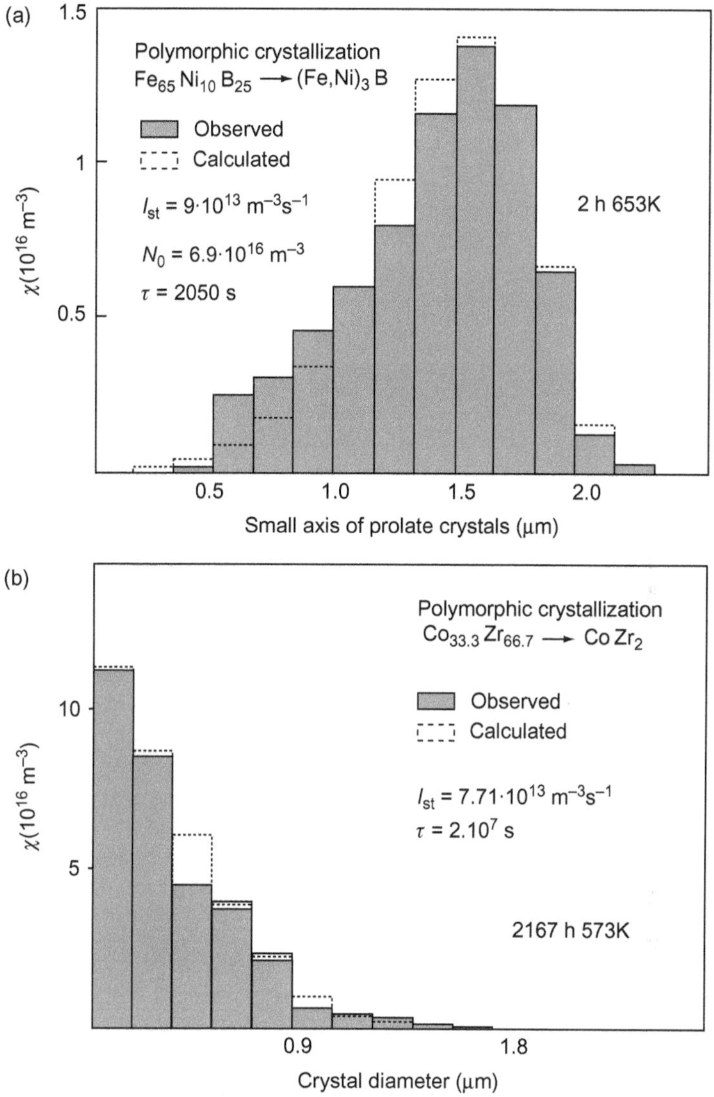

Fig. 11 Comparisons between measured and calculated size distributions of crystallites in polymorphically crystallizing glasses: (a) for $Fe_{65}Ni_{10}B_{25}$ glass, assuming transient heterogeneous nucleation; (b) for $Co_{33}Zr_{67}$ glass, assuming transient homogeneous nucleation. (Reprinted from Ref. [53], copyright (1987) with permission from the Materials Research Society.)

number of nuclei of decreasing size, which formed earlier and had a longer time to grow, and the site saturation at smaller size suggest that devitrification of $Fe_{65}Ni_{10}B_{25}$ is dominated by transient heterogeneous nucleation (Fig. 11a). By a similar analysis, $Co_{33}Zr_{67}$ devitrifies by transient homogeneous nucleation (Fig. 11b), with an increasing nucleation rate with decreasing size (later times) but no site saturation.

Most estimates of the steady-state nucleation rates in metallic glasses come from analyses of crystallization kinetics. Unfortunately, the required parameters, particularly the driving free energy, are often not available. This and the lack of quantitative nucleation data make it difficult to compare these results with theoretical predictions (see Ref. [52] for example). Recent time-dependent nucleation measurements (Section 7) have provided a more quantitative assessment of the steady-state nucleation rates [83].

7. TIME-DEPENDENT NUCLEATION

As mentioned in Section 3, since glasses are formed by cooling the melt on a time scale that is short compared with the time required for atomic rearrangements, the final cluster population depends on the details of the quench. Subsequent annealing treatments, therefore, reveal a time-dependent, or transient, nucleation behavior, as the cluster distribution evolves toward steady state at the annealing temperature (see Chapter 3 for a full treatment of transient nucleation within the classical theory). Gutzow et al. presented the first evidence for transient nucleation in glass-forming melts [84, 85]. It has been observed subsequently in a wide variety of systems, including amorphous silicon films [86] and metallic glasses [87, 88].

Transient nucleation has been studied most extensively in the silicate glasses. Figure 12 presents representative data for several glasses; in all

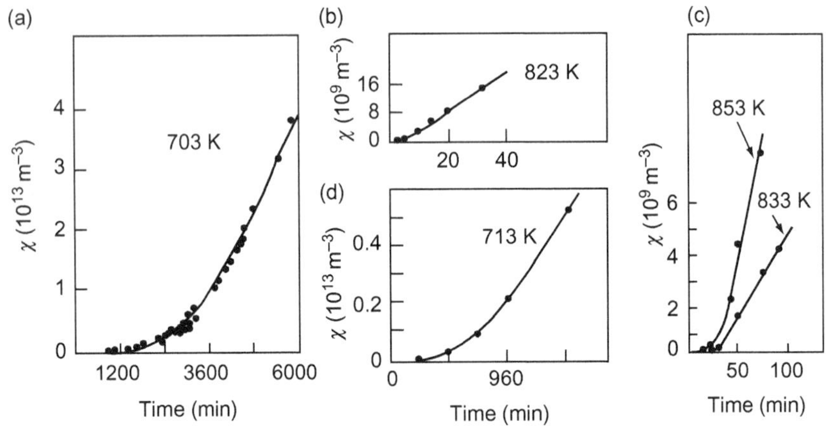

Fig. 12 The number of crystal nuclei measured as a function of time for several silicate glasses annealed at the indicated temperatures: (a) $Li_2O \cdot 2SiO_2$; (b) $Na_2O \cdot BaO \cdot SiO_2$; (c) $Na_2O \cdot CaO \cdot SiO_2$; (d) $Li_2O \cdot 2SiO_2$. (Reprinted from Ref. [89], copyright (1985), with permission from Elsevier.)

cases, the data follow the predictions from the classical theory discussed in Chapter 3. The rate of production of nuclei is low initially and rises to a constant steady-state rate with increased annealing time.

The data in Figure 12c demonstrate the influence of increasing temperature; the higher mobility gives a higher steady-state nucleation rate, that is, steeper gradient, and a shorter induction time, θ (see Chapter 3, Figure 1 and accompanying discussion for how θ and I^{st} are determined from these data). With increasing temperature above that of the maximum nucleation rate, the rate decreases sharply due to the decreasing driving free energy; the induction time decrease is more gradual, however, since it is primarily determined by the interfacial mobility (Chapter 3, Eq. (36)).

No evidence of transient nucleation was reported for the $Fe_{40}Ni_{40}P_{14}B_6$ glass mentioned in Section 6.3, but significant transient behavior was observed in the $Ti_{67}Ni_{33}$ glass (Figure 13a). Assuming an Arrhenius form for the induction time, the activation energy is 233 kJ mol^{-1}, similar to that for growth, 229 kJ mol^{-1}, as expected for a polymorphic transformation [51]. Unlike silicate glasses, the regions of significant nucleation and growth generally overlap in metallic glasses, hindering the use of two-step annealing techniques, which give more accurate nucleation data. The first two-step measurements of time-dependent nucleation were reported in a $Zr_{65}Al_{7.5}Ni_{10}Cu_{17.5}$ glass for a single nucleation temperature [90]; the high density and difficulties with TEM preparation technique, however, gave a significant measurement error. The first quantitative studies of time-dependent nucleation rates over a range of temperatures were recently reported for the nucleation of an icosahedral quasicrystal phase in a $Zr_{59}Ti_3Cu_{20}Ni_8Al_{10}$ bulk metallic glass [83] (Figure 13b).

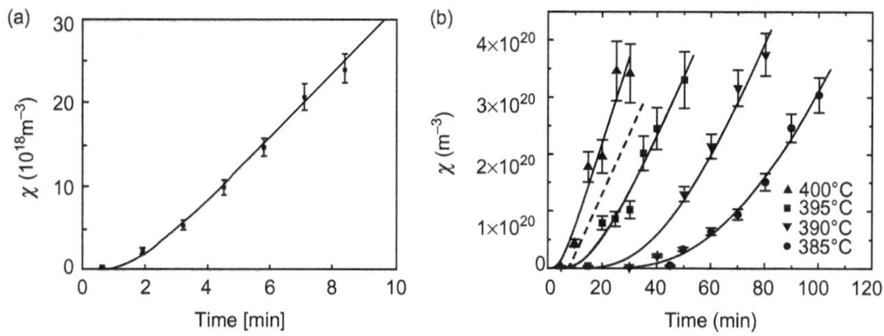

Fig. 13 Crystallite population densities as a function of time (a) in $Ti_{67}Ni_{33}$ glass after a single-step anneal for different times at 415°C (Reprinted from Ref. [51], copyright (1993), with permission from Elsevier.); (b) in $Zr_{59}Ti_3Cu_{20}Ni_8Al_{10}$ glass after a two-step anneal at the nucleation temperature 380°C. (Reprinted from Ref. [83], copyright (2009), American Physical Society.)

7.1 Simple analytical model for the induction time

Isothermal time-dependent nucleation data can be most simply analyzed using the Kashchiev expression (Chapter 3, Eq. (35)). Scaling the number of nuclei by the steady-state nucleation rate and the transient time, τ, at the annealing temperature, should give a master curve

$$\frac{\chi(t,\tau)}{I^{st}\tau} = \frac{t}{\tau} - 2\sum_{m=1}^{\infty}\frac{(-1)^n}{m^2}\exp\left(-\frac{m^2 t}{\tau}\right) - \frac{\pi^2}{6}. \tag{16}$$

Using this expression, nucleation data from $Li_2O \cdot 2SiO_2$ and $2Na_2O \cdot CaO \cdot 3SiO_2$ (Figure 14a) can be fit by a single curve that is independent of the glass composition and annealing conditions. The time-dependent nucleation data in the $Zr_{59}Ti_3Cu_{20}Ni_8Al_{10}$ bulk metallic glass taken at different temperatures follow the same scaling relation (Figure 14b). Since Eq. (16) was derived within the classical theory of nucleation, these results strongly support the use of that model.

Measured values for the induction time, θ (defined in Chapter 3, Figure 1), in $Li_2O \cdot 2SiO_2$ and $2Na_2O \cdot CaO \cdot 3SiO_2$ are shown in Figure 15a; the temperature dependence is clearly Arrhenian. In these two glasses, the activation energy for θ is close to that for viscous flow in the same temperature range [54]. This appears to support the use of the bulk mobility to describe the interfacial dynamics, but as we discussed in Section 6.2.3, this may not be strictly true. The calculated values for θ in $Li_2O \cdot 2SiO_2$ are compared with the measured data in Figure 15b. The solid and dashed curves are computed from a numerical solution of the coupled-rate equations of the classical theory using the viscosity data of

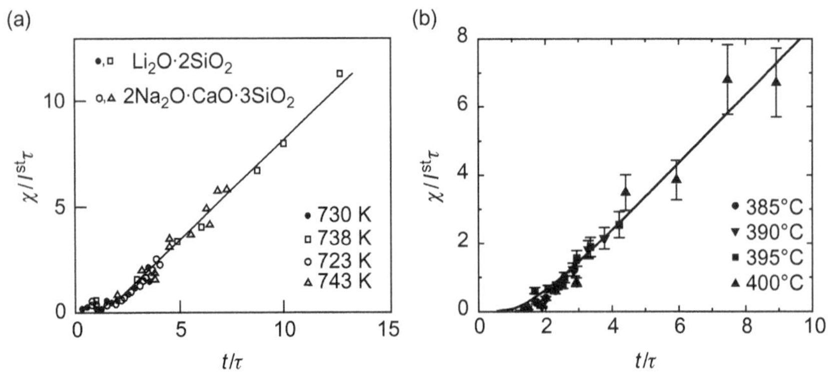

Fig. 14 The number of nuclei as a function of the time scaled by the product of the steady-state nucleation rate and the transient time for (a) $Li_2O \cdot 2SiO_2$ and $2Na_2O \cdot CaO \cdot 3SiO_2$ and (b) $Zr_{59}Ti_3Cu_{20}Ni_8Al_{10}$ glasses, demonstrating the fit of the Kashchiev expression (Eq. (35), Chapter 3) to the data. ((a) Reprinted from Ref. [91], copyright (1981), with permission from Elsevier.) and (b) Reprinted from Ref. [83], copyright (2009), American Physical Society.

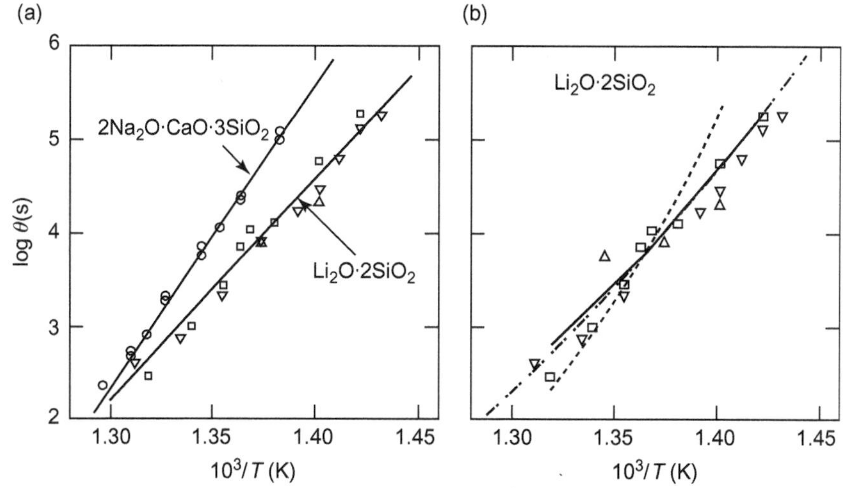

Fig. 15 (a) The induction time θ measured as a function of temperature for $Li_2O \cdot 2SiO_2$ glass (data from [54,91,93]) and $2Na_2O \cdot CaO \cdot 3SiO_2$ glass (data from Ref. [94]). The straight lines are Arrhenius fits to the data. (b) The results of a numerical calculation of θ for $Li_2O \cdot 2SiO_2$ using viscosity data from Ref. [66] (solid line) and Ref. [64] (dashed line). The dot-dash line is a prediction from the Kashchiev expression (Chapter 3, Eq. (36)) using the viscosity data from Ref. [70] and using Eq. (17) to compute the time to grow from n at the nucleating temperature to n at the growth temperature. (Reprinted from Ref. [52], copyright (1991), with permission from Elsevier.)

Matusita and Tashiro [66] and Gonzalez-Oliver et al. [64], respectively. A critical size at the growth temperature, 899 K, of $n^*(T_G) = 65$ is assumed. The data are better fit using the Matusita and Tashiro values. Using that viscosity data, and assuming that the time required to grow the nuclei from $n(T_N)$ to $n(T_G)$, is given by [55]

$$t_G = \frac{r^*(T_G) - r^*(T_N)}{u}, \tag{17}$$

where u is an effective growth rate for the nuclei and $r(T)$ is the radius of the critical nucleus at temperature T, the Kashchiev expression (Chapter 3, Eq. (36)) also fits the data well (dash-dot curve). A simpler expression for the induction time due to Hillig [92], $\tau = \pi\lambda^2/D$, where λ is the molecular diameter and D is the diffusion coefficient, is indistinguishable from the Kashchiev fit. The agreement between these two expressions reflects the long time required to grow from $n(T_N)$ to $n(T_G)$; the values for τ given by the two expressions are very different.

It may be, however, that the good agreement between the measured and computed induction times for nucleation is fortuitous. Other studies on a wide range of glasses have shown generally poor agreement, which has been argued to indicate a failure of the classical theory of nucleation

[95]. As is clear from Figure 15, however, the calculated values are sensitive to the assumed viscosity. Clearly, this is an important area for further research.

7.2 Test of the kinetic model

We have already seen that although measurements of the nucleation rates in liquids and glasses generally agree with predictions of the classical theory, there are clear indications that some aspects of that theory are flawed. Even when homogeneous nucleation rates are available, a detailed analysis is hampered by the lack of good data for Δg and η. Further, as discussed in Chapter 4, it is even questionable how well such macroscopic thermodynamic parameters describe the work of formation of clusters of a few atoms. For amorphous/crystal transitions, the interfacial energy σ is known only from fitted nucleation data, making independent comparisons impossible. As is clear from the discussion in Chapter 1, most theoretical attention has focused on these thermodynamic issues. The kinetic model, used in the classical theory and its many variants, has received much less theoretical and, until recently, virtually no experimental attention.

Multistep annealing treatments of silicate glasses can lead to a complicated time dependence of the nucleation rate, which sometimes reaches a maximum before settling down to the steady-state value (Figure 16). The intensity and position of the nucleation-rate maximum depend critically on the annealing temperatures and times. Within the classical theory, this behavior must result from the evolution of the cluster size distribution and should be predictable given the correct ensemble of kinetic coefficients and the equilibrium population density. Using measured data for Δg in $Li_2O \cdot SiO_2$ glass, and fitting the magnitude and temperature dependences of the steady-state nucleation rates and transient times to obtain the interfacial free energy and mobility as functions of temperature, the annealing treatments can be simulated numerically.

In Figure 17, the calculated number of nuclei is compared with experimental data obtained by nucleating at 758 K and growing at 899 K, following a preanneal at 713, 724, or 756 K. The apparent disagreement with the 713 K data is due to an error in the steady-state value originally reported at that temperature; shifting the data to take that into account produces good agreement. Such good agreement between theory and experiment is a vindication of the kinetic model underlying the classical theory of nucleation, indicating that nucleation proceeds as if it is governed by the evolution of a cluster distribution that is dictated by the bimolecular kinetic model of the classical theory. At the very least, it establishes a condition that must be satisfied by the kinetics of alternative Fokker-Planck formalisms (see Chapter 2).

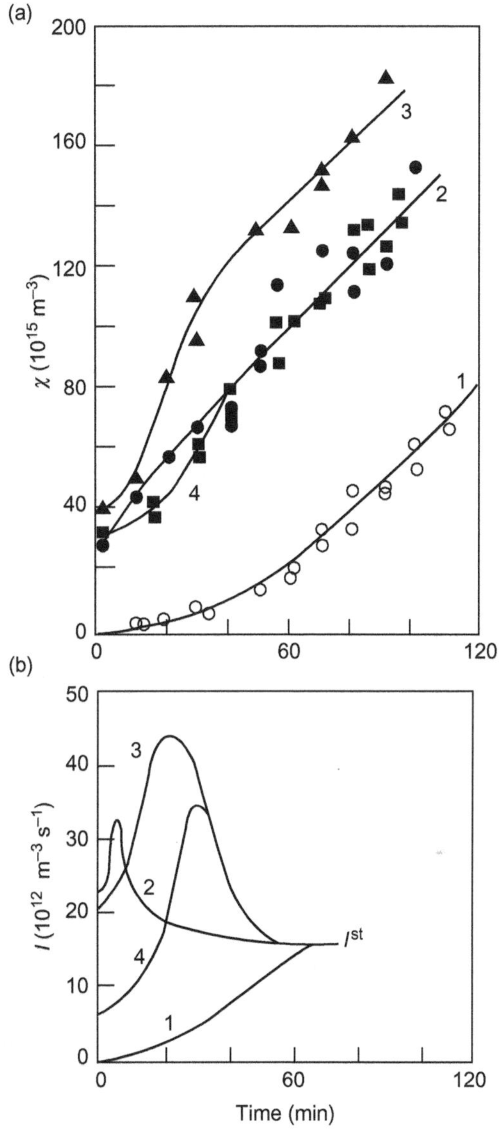

Fig. 16 (a) The number of nuclei as a function of annealing time at 753 K in $2Na_2O \cdot CaO \cdot 3SiO_2$ glass, (1) with no preannealing treatment (2) after a 4-h anneal at 743 K, (3) after a 3-h anneal at 733 K, (4) after a 65-h anneal at 723 K. (b) The nucleation rate derived from the local slope of the data in (a). (Reprinted from Ref. [94], copyright (1980), with permission from Elsevier.)

The cluster distributions calculated as a function of time reveal why the nucleation rate can reach a maximum under particular conditions (Figure 18). Here, the cluster density following a 65-h anneal at 703 K is

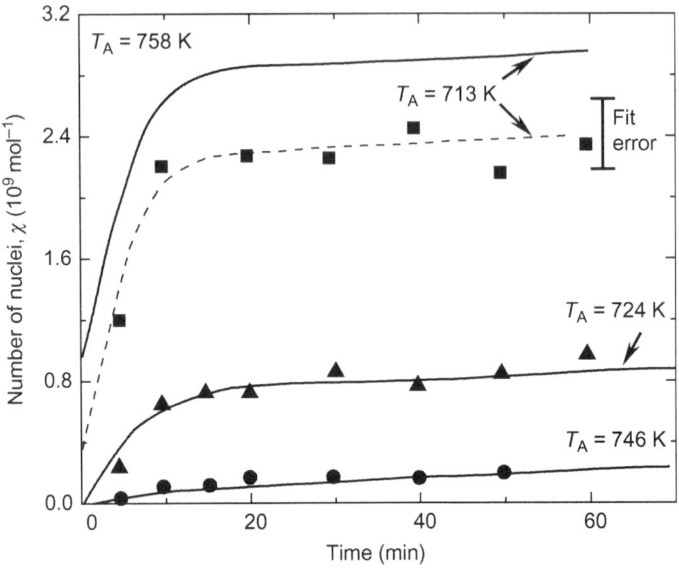

Fig. 17 Calculations (solid lines) and measurements (symbols) of the number of crystal nuclei χ in lithium-disilicate glass as a function of time at 758 K, after annealing 713 K for 18 h, 724 K for 4.5 h, or 746 K for 45 min. The 758 K anneal was followed by a growth treatment at 899 K. The dashed line is the simulation result for the 713 K anneal after shifting to match the initial number of nuclei. The maximum estimated error of the simulation, based on measurement uncertainties, is indicated. (Reprinted from Ref. [96], copyright (1988), American Physical Society.)

significantly below the steady-state population density at that temperature (curve a). On annealing at 738 K, the cluster distribution evolves toward the appropriate steady-state distribution (b–d). At larger sizes, however, the cluster population density increases and subsequently decreases, giving rise to a "pulse" of clusters that sweep through the distribution to $n(T_G)$.

Some recent three-step annealing studies of the nucleation rates in $Na_2O \cdot 2CaO \cdot 3SiO_2$ glasses with different concentrations of SiO_2 reveal difficulties when using the classical theory to describe primary crystallization [98] (see Section 9). Several different studies have also shown that the rates that are estimated from the induction time for nucleation do not reproduce the magnitude of the growth rates [23, 99]. These observations could indicate that a key feature of the kinetic model for nucleation, that is, single-molecule attachment/detachment at the interface, is not completely correct. Further studies are needed to clarify this point.

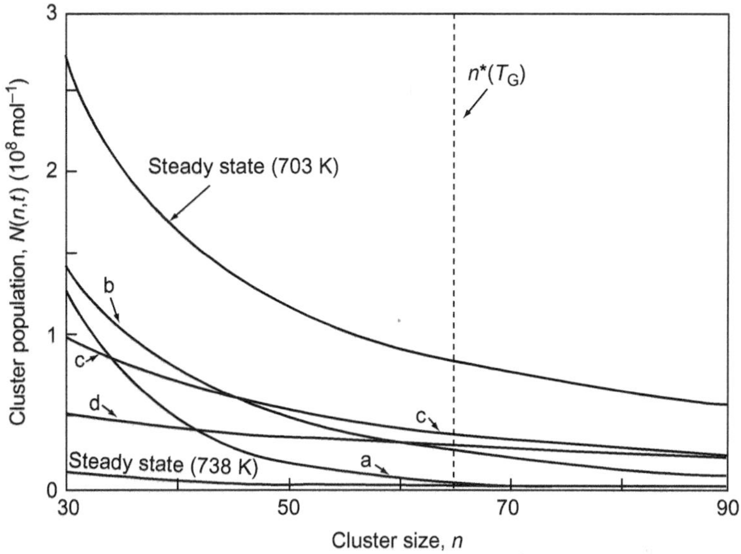

Fig. 18 The cluster size distribution in lithium-disilicate glass computed from a numerical solution to the classical-theory equations (see Chapter 2, Section 6) showing its evolution with annealing. The steady-state distributions at 703 and 738 K are shown. Curve a shows a distribution resulting from a 65-h anneal at 703 K. With annealing time at 738 K, the distribution evolves as shown by curves b (1000 s), c (2000 s), and d (3000 s). (Reprinted/Adapted from [97], copyright (1991), with permission from Wiley-Blackwell.)

8. CRYSTALLIZATION TO QUASICRYSTALS — A LOW NUCLEATION BARRIER

For a growing number of alloys that can be prepared as glasses, an icosahedral quasicrystal (i-phase) is the primary devitrification product, particularly in Zr- and Hf-based BMGs [100, 101] (see Ref. [102] for a more complete list). Often the devitrified glass is nanostructured, with grains as small as 5 nm in diameter, indicating a large nucleation rate and a slow growth rate. Atom-probe studies of Zr-based glasses suggest that liquid and glass stabilization and icosahedral short-range order (ISRO) are interlinked; elemental additions that stabilize the glass also stabilize icosahedral clusters [103]. The role of ISRO in the liquid phase on the crystal nucleation barrier was a key point discussed in Chapter 7 (Section 2.4). ISRO also plays an important role in glass formation and crystallization.

A low barrier for the nucleation of the i-phase from an amorphous phase was first demonstrated in a study of the polymorphic crystallization of $Al_{75}Cu_{15}V_{10}$ glass [104]. Interfacial attachment rates were derived from the measured growth rates, and Δg was calculated as a function of temperature from measured specific heats of the metallic glass and quasicrystal and the enthalpy of transformation. A fit to the nucleation data then gave an upper bound on the fit i-phase/amorphous interfacial free energy, $\sigma_{\text{i-a}}$, of 0.015 J m^{-2}. Recent measurements of the time-dependent nucleation rates of an i-phase in a $Zr_{59}Ti_3Cu_{20}Ni_8Al_{10}$ bulk metallic glass [83] allowed tighter bounds to be established, that is, $\sigma_{\text{i-a}} = 0.01 \pm 0.004$ J m^{-2}, in good agreement with the earlier measurement. Such a small value for $\sigma_{\text{i-a}}$ reflects a large amount ISRO in the glass. Further, $\sigma_{\text{i-a}}$ is almost an order of magnitude smaller than the interfacial free energies obtained for the nucleation of quasicrystals from supercooled liquids [105, 106]. The ISRO increases dramatically, then, with supercooling to the glass transition, implying a connection between the ISRO and the frustration underlying the glass transition in these and similar metallic glasses (see Ref. [107] for a discussion of the geometrical frustration basis for the glass transition).

9. PRIMARY CRYSTALLIZATION

While a limited number of transformations are approximately polymorphic (i.e., with the original and new phases of the same composition) most practical cases involve a change in composition. As discussed in Chapter 5, the required extensions to nucleation theories to address these problems depend on whether the difference in composition is manifested mainly as a change in the work of cluster formation, or requires a new kinetic model. Case studies of both extremes are given here.

Most studies of silicate glasses have focused on steady-state nucleation rates [108–111]. However, the example discussed here shows that time-dependent rates can often better reveal the true nucleation behavior. As for polymorphic crystallization, time-dependent nucleation rates have been most accurately measured in silicate glasses. The best studied case is $Na_2O \cdot 2CaO \cdot 3SiO_2$ glass. The steady-state nucleation rates, I^{st}, and induction times, θ, for as-quenched glasses of $Na_2O \cdot 2CaO \cdot 3SiO_2$ with different SiO_2 concentrations are shown in Figure 19. To indicate the amount of silica in the glass, the compositions are written as $(Na_2O \cdot 2CaO)_{1-X}(3SiO_2)_X$. In this notation, $Na_2O \cdot 2CaO$ and SiO_2 are treated as separate units; $X = 0.5$ represents the stoichiometric glass. The induction time for nucleation is defined by the intercept of the extrapolated linear region on the time axis. It provides a measure of the relaxation rate of the cluster distribution toward steady state. The steady-state nucleation rates decrease with increasing silica content; the temperatures of the peak nucleation rates are approximately independent of the SiO_2 concentration. The induction times increase with decreasing nucleation rate. Within the classical theory of nucleation, the

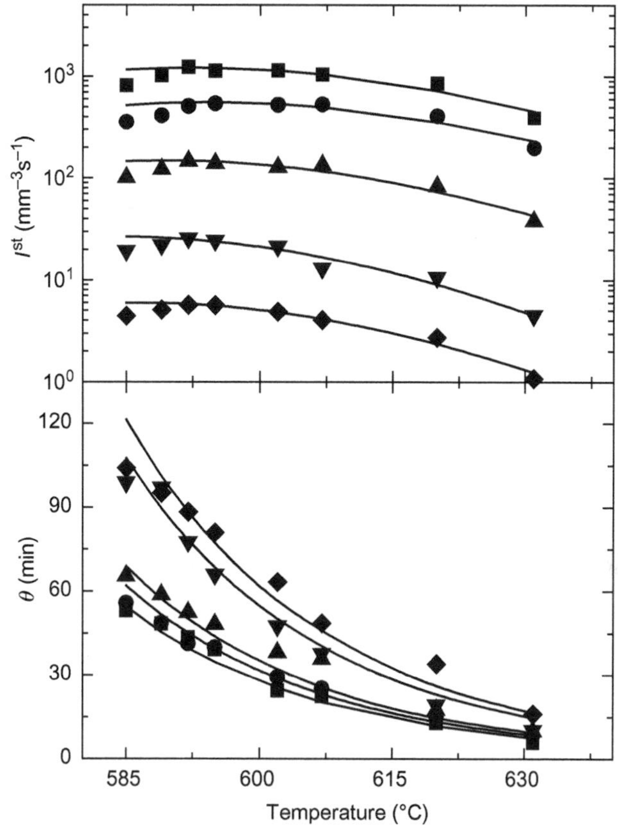

Fig. 19 Steady-state crystal nucleation rates (top) and induction times (bottom) as a function of temperature for $(Na_2O \cdot 2CaO)_{1-x}(3SiO_2)_x$ glasses of different composition: $X = 0.494$ (■); $X = 0.5$ (stoichiometric,); $X = 0.506$ (▲); $X = 0.52$ (▼); $X = 0.53$ (◆). Measurement uncertainties in I^{st} are comparable to the symbol sizes. A growth anneal of 5 min at 700°C was used. The solid lines through the points are a fit to the steady-state nucleation rates assuming a composition-dependent interfacial energy. (Reprinted from Ref. [112], copyright (1997), with permission from Elsevier.)

steady-state nucleation rate in a binary mixture is expected to have the form [52] (see Eqs. (12) and (13), Chapter 7)

$$I^{st} = A^* \exp\left(-\frac{W(a^*, b^*)}{k_B T}\right)$$
$$= A^* \exp\left(-\frac{a^* \Delta\mu_A + b^* \Delta\mu_B + \zeta\bar{\sigma}(a^*\bar{v}_A + b^*\bar{v}_B)^{2/3}}{k_B T}\right), \qquad (18)$$

where A^* is a dynamical prefactor that is linearly proportional to the atomic mobility, $\Delta\mu_A$ and $\Delta\mu_B$ are the differences between the chemical potentials of the original and new phases for A and B, γ is a geometric factor, which is $(36\pi)^{1/3}$ for spherical clusters, $\bar{v}_A$ and $\bar{v}_B$ are the molecular volumes for A and B species respectively, and $\bar{\sigma}$ is an effective interfacial free energy. The exact form of the induction time is less certain, but it should be inversely proportional to the atomic mobility and only weakly dependent on the driving free energy and the interfacial free energy [52].

From Figure 19, the change in the nucleation rate with composition is almost two orders of magnitude greater than the corresponding change in the induction time. As expected, changes in the growth velocity scale inversely with changes in the induction time. The large changes in the nucleation rate with silica content, therefore, are not explained by changes in the atomic mobility alone. Measurements of the liquid temperature show that the observed changes in I^{st} also cannot arise from changes in the driving free energy. If the classical theory applies, these changes must originate from the composition dependence of σ.

10. HETEROGENEOUS NUCLEATION

As already mentioned most silicate glasses nucleate from the surface or on heterogeneous sites in the volume of the glass; surface nucleation is also common in metallic glasses (Figure 20). It is generally favored over bulk nucleation because of a decreased interface energy penalty and higher diffusion coefficients. Since there is a finite set of sites, either on the surface or in the volume, where nucleation is favored, the number of nuclei in both cases saturates with time. Letting $\tilde{\chi}_S$ be the number of sites per surface area for surface nucleation, and $\tilde{\chi}_V$ the number per volume for heterogeneous nucleation throughout the volume, the rate of production of nuclei is given by

$$\frac{d\chi}{dt} = (\tilde{\chi}_{S\text{ or }V} - \chi)\, I^{\text{hetero}}_{S\text{ or }V}(t)$$
$$\Rightarrow \chi(t) = \tilde{\chi}_{S\text{ or }V}\left(1 - \exp\left(-\int_0^t I^{\text{hetero}}_{S\text{ or }V}(t')\,dt'\right)\right), \tag{19}$$

where $I^{\text{hetero}}_{S\text{ or }V}(t)$ is the time-dependent surface or volume heterogeneous nucleation rate.

10.1 Surface nucleation

As for homogeneous nucleation, surface nucleation rates can be steady-state or time-dependent [113, 114]. Due to the often rapid saturation of sites where surface nucleation can occur, even before crystallites become visible, rates are generally deduced from the crystallite size distributions obtained from single-step annealing treatments. The data obtained are

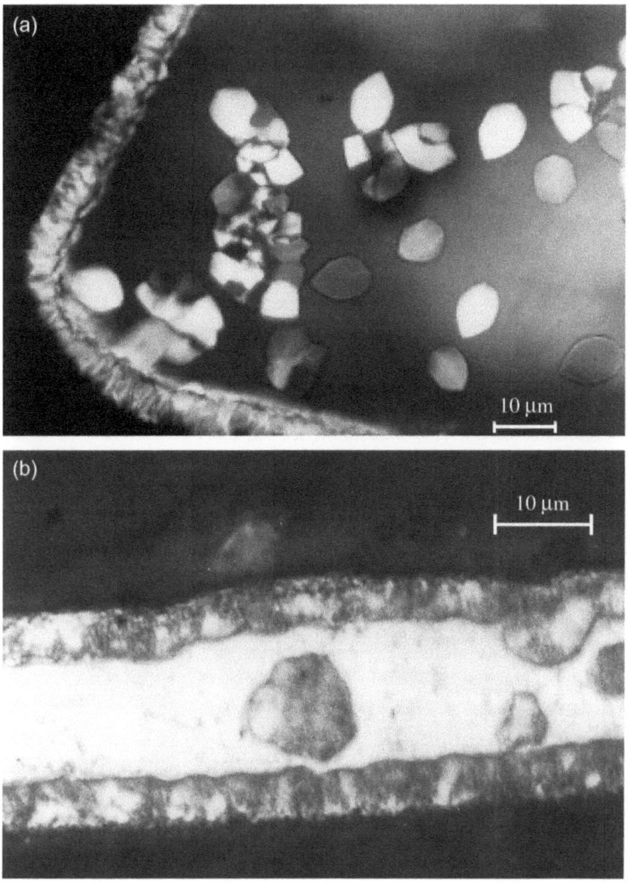

Fig. 20 (a) Transmission optical micrograph under slightly crossed nicols of μ-cordierite crystals nucleated at a fractured glass surface after annealing for 30 min at 960°C (Reprinted from Ref. [113], copyright (2000), with permission from Elsevier.) (b) Reflection optical micrograph of partially crystallized amorphous $Pd_{82}Si_{18}$ that has been partially crystallized in a nonisothermal differential scanning calorimetry scan, showing surface crystallization. (Reprinted from Ref. [36], copyright (1985), with permission from Elsevier.)

mostly qualitative, but nevertheless elucidate key features of surface nucleation.

10.1.1 Mechanical damage

Mechanical damage generally promotes surface nucleation, while polishing (e.g., in HF acid for silicates) impedes it. The data in Table 4 show the population density of nuclei increasing with increasing damage.

Table 4 Surface nucleation densities, N_S, for glasses with different degrees of mechanical damage

Surface Processing	Glass	Crystal	N_S (μm^{-2})	Reference
Powder	$2MgO \cdot 2Al_2O_3 \cdot 5SiO_2$	μ-Cordierite	10^{-3} to 10^{-1}	[123]
Ground	$2MgO \cdot 2Al_2O_3 \cdot 5SiO_2$	μ-Cordierite	$(0.2$ to $7) \times 10^{-2}$	[122]
	$CaO \cdot ZrO_2 \cdot 2Al_2O_3 \cdot SiO_2$	Willemite	3×10^{-2}	[115]
Mechanically Polished	$2MgO \cdot 2Al_2O_3 \cdot 5SiO_2$	μ-Cordierite	$(1$ to $10) \times 10^{-4}$	[122]
			2×10^{-4}	[116]
	$CaO \cdot ZrO_2 \cdot 2Al_2O_3 \cdot SiO_2$	Willemite	$(1$ to $3) \times 10^{-3}$	[115]
	$MgO \cdot CaO \cdot 2SiO_2$	Diopside	$(6$ to $10) \times 10^{-2}$	[117]
	$CaO \cdot Al_2O_3 \cdot 2SiO_2$	Anorthite	$(1$ to $4) \times 10^{-3}$	[119]
	$Na_2O \cdot 3CaO \cdot 6SiO_2$	Devitrite	1×10^{-1}	[118]
	Non-stoichiometric	β-Wollastonite	$(5$ to $10) \times 10^{-5}$	[118]
	Devitrite	Tridymite	$(5$ to $10) \times 10^{-5}$	[118]
	Devitrite	Devitrite	$(1.5$ to $7) \times 10^{-4}$	[118]
	Microscope slide	Devitrite	$(1$ to $6) \times 10^{-4}$	[118]
As-received	Float glass	Devitrite	$(3$ to $30) \times 10^{-4}$	[118]
	Float glass	Diopside	5×10^{-5}	[118]
	Float glass	Tridymite	1×10^{-5}	[118]
	Microscope slide	Devitrite	$(8$ to $30) \times 10^{-4}$	[118]
Fractured	$2MgO \cdot 2Al_2O_3 \cdot 5SiO_2$	μ-Cordierite	10^{-3} to 10^{-8}	[120]
			10^{-5} to 10^{-6}	[116]
Fire-polished	Microscope slide	Devitrite	"0"	[118]
	$2MgO \cdot 2Al_2O_3 \cdot 5SiO_2$	μ-Cordierite	"lowest"	[121]

Reprinted from Ref. [113], copyright (2000), with permission from Elsevier.

10.1.2 Nucleation at cracks and edges

Cracks and sharp edges are favored nucleation sites, partially explaining the sensitivity of nucleation to surface treatment. This may be due to elastic strain effects. The density difference between the glass and the crystal phase leads to strain fields and associated stresses that oppose nucleation and growth. This is generally assumed to be a stronger factor for growth; stress relaxation is argued to be faster than the slow stochastic development of nuclei, making it less important for nucleation. If some residual stress were present, however, it would presumably relax faster near sharp edges.

10.1.3 Solid particles in contact with the surface

It has long been known that particles in contact with smooth glass surfaces can trigger crystal nucleation (see for example, [124]). Examples include particles of precrystallized glass [125, 126], dust from furnace materials [127], and oxide particles. As for any heterogeneous nucleation, the catalytic activity of the articles depends on their structure, chemical composition, and on the thermodynamic stability of the glass and particle when in contact with the surface [128, 129]. Unstable particles react chemically with the glass surface, changing the surface composition; for example, B_4C can react to form B_2O_3. Most studies of the effects of dust remain qualitative (see [57] for a review of this subject).

10.1.4 Surface composition

As just mentioned, the surface composition can play a key role in the crystallization behavior. For example, preferential oxidation at the surface, leading to a composition change, is frequently the cause of surface crystallization in metallic glasses. This can be controlled partially by various surface treatments, such as the formation of a thin layer of NiO by a mild oxidation of $Pd_{40}Ni_{40}P_{20}$ to prevent the loss of phosphorus, inhibiting crystallization [130], or ion milling followed by coating with a suitable material [131].

10.2 Nucleation catalysis — Pt particles in silicate glasses

It is well known that particles introduced into glasses can accelerate crystallization. Slow crystal growth in glasses has allowed quantitative investigations of both the nucleation mechanism and the time-dependent nucleation rates. Heterogeneous nucleation is difficult to study, however, due to insufficient information about the number, size, and catalytic efficiency of the heterogeneous particles. In Chapter 13, a discussion of the importance of the surface structure and chemistry of heterogeneous inoculants is presented, focusing on solidification. Here we present an example of a quantitative study of heterogeneous nucleation kinetics. A second example illustrates how easily the situation can become complicated, showing the formation of small precipitates, presumably by homogeneous nucleation, which then act as heterogeneous nucleation sites leading to devitrification of the glass.

It is well known that the addition of noble metals in controlled quantities promotes crystal nucleation in silicate glasses. Platinum particles in particular have received much attention [132, 133], partially because glasses are frequently prepared in Pt crucibles. Turning again to $Li_2O \cdot 2SiO_2$, the model glass used to illustrate time-dependent and steady-state homogeneous nucleation behavior, Figure 21 illustrates the sensitivity of the steady-state nucleation rate to the Pt concentration.

Only a few measurements of time-dependent heterogeneous nucleation exist. As expected, the results look similar to those obtained for homogeneous nucleation. The introduction of nucleating particles generally has little effect on the induction time. For Pt in $Li_2O \cdot 2SiO_2$, the contact angles for crystal nuclei are 120–130° for all concentrations

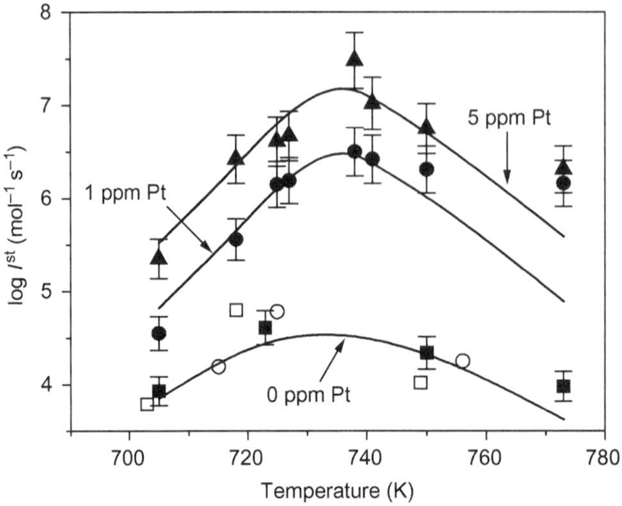

Fig. 21 Steady-state nucleation rates per mole as a function of temperature for undoped $Li_2O \cdot 2SiO_2$ glass (■), doped with 1 ppm Pt (●), and doped with 5 ppm Pt (▲). Homogenous nucleation rates from [54] (□) and [134] (○) are shown for comparison. (Reprinted from Ref. [135], copyright (1996), with permission from Elsevier.)

studied and over the temperature range measured. This agrees with the arguments presented in Chapter 6, Section 2.2; little effect is expected unless the contact angle is significantly below 100°.

11. SUMMARY

In this chapter, measurements of nucleation in glasses were discussed, focusing on the silicate and metallic glasses that have been most extensively studied. They crystallize much more slowly than supercooled liquids, and consequently allow more quantitative measurements of nucleation rates and more thorough kinetic studies than are possible in liquids. Some key points are summarized as follows.

- Crystal nucleation is significant in the analysis of glass formation; all glasses contain a population of quenched-in nuclei that have not had an opportunity to grow. When rapid quenching is necessary to produce a glass, time-dependent nucleation effects are important, and can be critical in restricting crystallization, thereby permitting a glass to form.
- During the quench of a liquid, time-dependent effects cause the crystal nucleation rate to fall below steady-state values, the cluster distribution

being unable to evolve fast enough to remain in steady state. Such effects reduce the population of quenched-in nuclei.

- Three devitrification modes, *polymorphic, primary,* and *eutectic,* can generally be identified from the devitrification microstructure.
- Slow crystallization in glasses permits the rate of nuclei production to be directly counted. Such experiments yield more reliable data than do calorimetric studies of overall crystallization kinetics.
- In silicate glasses, and recently in metallic glasses, two-step anneals have been used to quantify steady-state and time-dependent nucleation. Such experiments are not possible in supercooled liquids.
- In silicate glasses, steady-state nucleation rates as a function of temperature show the maximum expected from the classical theory of nucleation. The interfacial free energy, estimated from a classical analysis of these data, increases with increasing temperature, indicating that the crystal-glass interface, like the crystal–liquid interface (Chapter 4), is diffuse.
- As for liquids, nucleation rates in glasses are better fit by the semiempirical density-functional and the diffuse-interface theories than by the classical theory, consistent with the interface between the crystal and glass being diffuse.
- Fits to transient nucleation rates in silicate and metallic glasses are in good agreement with the analysis of Kashchiev.
- Based on a quantitative test using multistep annealing treatments in silicate glasses, the kinetic model for cluster evolution underlying the classical theory of nucleation appears quantitatively correct, at least for polymorphic crystallization of glasses.
- The nucleation barrier for the icosahedral quasicrystal phase is small, consistent with the existence of significant icosahedral short-range order in the glass.

REFERENCES

[1] W. Buckel, Elektronenbeugungs-Aufnahmen von dünnen Metallschichten bei tiefen Temperaturen, Z. Phys. **138** (1954) 136–150.

[2] A. Brenner, D.E. Couch, E.K. Williams, Electrodeposition of alloys of phosphorus with nickel or cobalt, J. Res. Nat. Bur. Stand. **44** (1950) 109–122.

[3] T. Watanabe, T. Tanabe, Formation and morphology of Ni–B amorphous alloy deposited by electroless plating, Mater. Sci. Eng. **23** (1976) 97–100.

[4] S.W. Charles, S. Wells, A. Meagher, S. Mørup, J. van Wonterghem, Properties of amorphous FeCoB alloy particles, J. Appl. Phys. **64** (1988) 5508.

[5] R.B. Schwarz, C.C. Koch, Formation of amorphous alloys by the mechanical alloying of crystalline powders of pure metals and powders of intermetallics, Appl. Phys. Lett. **49** (1986) 146–148

[6] L. Schultz, Formation of amorphous metals by solid-state reactions, Philos. Mag. B **61** (1990) 453–471

[7] R. Zallen, *The Physics of Amorphous Solids*, Wiley, New York (1983).

[8] H.W. Sheng, W.K. Luo, F.M. Alamgir, J.M. Bai, E. Ma, Atomic packing and short-to-medium-range order in metallic glasses, Nature **439** (2006) 419–425.

[9] "Glass," in *Encylopaedia Britannica*, Vol. **8**, pp. 181-196; Chicago: Encylopedia Britannica, Inc. (1977).

[10] W. Klement, P. Duwez, R.H. Willens, Non-crystalline structure in solidified gold–silicon alloys, Nature **187** (1960) 869–870.

[11] R.W. Cahn, Alloys rapidly quenched from the melt, in: *Physical Metallurgy*, Eds. R.W. Cahn, P. Haasen, North-Holland, Amsterdam (1983), pp. 1779–1852.

[12] C. Suryanarayana, F.H. Froes, R.G. Rowe, Rapid solidification processing of titanium alloys, Int. Mater. Rev. **36** (1991) 85–123.

[13] A. Inoue, A. Takeuchi, Recent progress in bulk glassy, nanoquasicrystalline and nanocrystalline alloys, Mater. Sci. Eng. A **375** (2004) 16–30.

[14] J. Jäckle, Models of the glass transition, Rep. Prog. Phys. **49** (1986) 171–231.

[15] C.A. Angell, The old problems of glass and the glass transition, and the many new twists, Proc. Natl. Acad. Sci. USA **92** (1995) 6675–6682.

[16] T. Odagaki, T. Yoshidome, T. Tao, A. Yoshimori, Specific heat anomaly at the glass transition, J. Chem. Phys. **117** (2002) 10151–10155.

[17] R.W. Cahn, A.L. Greer, Metastable States of Alloys, in: *Physical Metallurgy*, Eds. R.W. Cahn, P. Haasen, North Holland, Amsterdam (1996), pp. 1724–1830.

[18] D. Turnbull, Under what conditions can a glass be formed?, Contemp. Phys. **10** (1969) 473–488.

[19] K.F. Kelton, A.L. Greer, Transient nucleation effects in glass formation, J. Non-Cryst. Solids **79** (1986) 295–309.

[20] A.L. Greer, Crystallisation kinetics of $Fe_{80}B_{20}$ glass, Acta Metall. **30** (1982) 171–192.

[21] D.R. Uhlmann, A kinetic treatment of glass formation, J. Non-Cryst. Solids **7** (1972) 337–348.

[22] K.F. Kelton, A.L. Greer, Interpretation of experimental measurements of transient nucleation, in: *Rapidly Quenched Metals*, Eds. S. Steeb, H. Warlimont, North Holland, Amsterdam (1985), pp. 223–226.

[23] K.F. Kelton, M.C. Weinberg, Calculation of macroscopic growth rates from nucleation data, J. Non-Cryst. Solids **180** (1994) 17–24.

[24] J.W. Christian, *The Theory of Transformations in Metals and Alloys, Parts I and II* 3rd edn, Elsevier, Oxford (2002).

[25] K.F. Kelton, Numerical model for isothermal and nonisothermal crystallization of liquids and glasses, J. Non-Cryst. Solids **163** (1993) 283–296.

[26] K.F. Kelton, K. Lakshmi Narayan, L.E. Levine, T.C. Cull, C.S. Ray, Computer modeling of nonisothermal crystallization, J. Non-Cryst. Solids **204** (1996) 13–31.

[27] K.F. Kelton, Icosahedral and quasi-periodic structures, in: *Intermetallic Compounds: Principles and Practice*, Eds. J.H. Westbrook, R.L. Fleischer, Wiley, New York (1995), pp. 453–491.

[28] U. Herold, U. Köster, The influence of metal or metalloid exchange on crystallization of amorphous iron–boron alloys, *Rapidly Quenched Met., Proc. 3rd Int. Conf.*, Ed. B. Cantor, (1978), pp. 281–290.

[29] A.L. Greer, Amorphous phases, in: *Intermetallic Compounds: Principles and Practice*, Eds. J.H. Westbrook, R.L. Fleischer, Wiley, New York (1995), pp. 731–754.

[30] K.F. Kelton, A new model for nucleation in bulk metallic glasses, Philos. Mag. Lett. **77** (1998) 337–343.

[31] P.J. Desré, E. Cini, B. Vinet, Homophase-fluctuation-mediated mechanism of nucleation in multicomponent liquid alloys and glass-forming ability, J. Non-Cryst. Solids **288** (2001) 210–217.

[32] K.F. Kelton, Time-dependent nucleation in partitioning transformations, Acta Mater. **48** (2000) 1967–1980.

[33] E.D. Zanotto, P.F. James, A.F. Craivich, The effects of amorphous phase separation on crystal nucleation kinetics in barium oxide-silica glasses. Part 3. Isothermal treatments at 718 to 760°C; small-angle X-ray scattering results, J. Mater. Sci. **21** (1986) 3050–3064.
[34] P.F. James, Volume nucleation in silicate glasses, in: *Glasses and Glass Ceramics*, Ed. M.H. Lewis, Chapman and Hall, London (1989), pp. 59–105.
[35] A.A. Kündig, M. Ohnuma, T. Ohkubo, K. Hono, Early crystallization stages in a Zr–Cu–Ni–Al–Ti metallic glass, Acta Mater. **53** (2005) 2091–2099.
[36] K.F. Kelton, F. Spaepen, A Study of the devitrification of $Pd_{82}Si_{18}$ over a wide temperature range, Acta Metall. **33** (1985) 455–464.
[37] W.A. Johnson, R. Mehl, Reaction kinetics in processes of nucleation and growth, Trans. AIME **135** (1939) 416–442.
[38] M. Avrami, Kinetics of phase change 1 – general theory, J. Chem. Phys. **7** (1939) 1103–1112.
[39] A.N. Kolmogorov, On the statistical theory of metal crystallisation, Izv. Acad Nauk SSSR, Ser Fiz. **3** (1937) 355–359.
[40] J.W. Cahn, The time cone method for nucleation and growth kinetics on a finite domain, Mater. Res. Soc. Symp. Proc. **398** (1996) 425–437.
[41] K.F. Kelton, Analysis of crystallization kinetics, Mater. Sci. Eng. A **226** (1997) 142–150.
[42] L.E. Levine, K.L. Narayan, K.F. Kelton, Finite size corrections for the Johnson–Mehl–Avrami–Kolmogorov equation, J. Mater. Res. **12** (1997) 124–132.
[43] H.E. Kissinger, Reaction kinetics in differential thermal analysis, Anal. Chem. **29** (1957) 1702–1706.
[44] D.W. Henderson, Thermal analysis of nonisothermal crystallization kinetics in glass forming liquids, J. Non-Cryst. Solids **30** (1979) 301–315.
[45] T. Ozawa, Kinetics of non-isothermal crystallization, Polymer **12** (1971) 150–158.
[46] K.F. Kelton, On estimating the nucleation rate by differential scanning calorimetry, J. Am. Ceram. Soc. **75** (1992) 2449–2452.
[47] R.T. DeHoff, F.N. Rhines, N. Frederick, Determination of number of particles per unit volume from measurements made on random plane sections: the general cylinder and the ellipsoid, Trans. Met. Soc. AIME **221** (1961) 975–982.
[48] E.E. Underwood, *Quantitative Stereology*, Addison-Wesley, Reading, MA (1970).
[49] M. Barthel, D. Baither, A modified Saltykov technique for stereological grain and subgrain size measurement from TEM micrographs, Praktische Metallographie **24** (1987) 391–397.
[50] M. Barthel, D. Baither, Stereological determination of grain and subgrain size from TEM micrographs, Praktische Metallographie **26** (1989) 261–271.
[51] M. Buchwitz, R. Adlwarth-Dieball, P.L. Ryder, Kinetics of the crystallization of amorphous Ti_2Ni, Acta Metall. Mater. **41** (1993) 1885–1892.
[52] K.F. Kelton, Crystal nucleation in liquids and glasses, in: *Solid State Physics*, Eds. H. Ehrenreich, D. Turnbull, Academic Press, Boston (1991), pp. 75–178.
[53] U. Köster, M. Blank-Bewersdorff, Transient nucleation in Co-Zr metallic glasses, Mater. Res. Soc. Symp. Proc. **57** (1987) 115–127.
[54] P.F. James, Kinetics of crystal nucleation in lithium silicate glasses, Phys. Chem. Glasses **15** (1974) 95–105.
[55] A.M. Kalinina, V.M. Fokin, V.N. Filipovich, Method for determining the parameters characterizing the formation of crystals in glasses, Fiz. Khim Stekla **2** (1976) 298–305.
[56] A.M. Kalinina, V.M. Fokin, V.N. Filipovich, Induction period for crystal formation in lithium oxide–silica ($Li_2O.2SiO_2$) glass and its temperature dependence, Fiz. Khim. Stekla **3** (1977) 122–129.
[57] E.D. Zanotto, V.M. Fokin, Recent studies of internal and surface nucleation in silicate glasses, Phil. Trans. Roy. Soc. Lond. **361** (2003) 591–613.
[58] E.D. Zanotto, Isothermal and adiabatic nucleation in glass, J. Non-Cryst. Solids **89** (1987) 361–370.

[59] E.D. Zanotto, M.C. Weinberg, Trends in homogeneous crystal nucleation in oxide glasses, Phys. Chem. Glasses **30** (1989) 186–192.
[60] R.D. Maurer, Effect of catalyst size in heterogeneous nucleation, J. Chem. Phys. **31** (1959) 444–448.
[61] C.J.R. Gonzalez-Oliver, P.F. James, Internal and platinum initiated crystallization in $Na_2O \cdot 2CaO \cdot 3SiO_2$ glasses: early stages of growth, J. Microsc. **119** (1980) 73–80.
[62] C.S. Ray, W. Huang, D.E. Day, Crystallization kinetics of lithia-silica glasses: effect of composition and nucleating agent, J. Am. Ceram. Soc. **70** (1987) 599–603.
[63] P.F. James, B. Scott, P. Armstrong, Kinetics of crystal nucleation in lithium disilicate glass. A comparison between melts prepared in platinum and silica crucibles and between melts prepared from ordinary and high purity starting materials, Phys. Chem. Glasses **19** (1978) 24–27.
[64] C.J.R. Gonzalez-Oliver, P.S. Johnson, P.F. James, Influence of water content on the rates of crystal nucleation and growth in lithia-silica and soda-lime-silica glasses, J. Mater. Sci. **14** (1979) 1159–1169.
[65] K. Matusita, M. Tashiro, Effect of added oxides on the crystallisation of $Li_2O.2SiO_2$ glasses, Phys. Chem. Glasses **14** (1973) 77–80.
[66] K. Matusita, M. Tashiro, Rate of homogeneous nucleation in alkali disilicate glasses, J. Non-Cryst. Solids **11** (1973) 471–484.
[67] E.D. Zanotto, P.F. James, Experimental tests of the classical nucleation theory for glasses, J. Non-Cryst. Solids **74** (1985) 373–394.
[68] V.N. Filipovich, A.M. Kalinina, Relation of the temperature of the maximum crystal formation rate in glasses with the glass point, Izv. Akad. Nauk. SSSR, Neorganicheskie Materialy **7** (1971) 1844–1848.
[69] P.F. James, Kinetics of crystal nucleation in silicate glasses, J. Non-Cryst. Solids **73** (1985) 517–540.
[70] K. Matusita, M. Tashiro, Rate of crystal growth in $Li_2O \cdot 2SiO_2$ glass, Jap. J. Ceram. Assoc. (Yogyo-Kyokai-Shi) **81** (1973) 500–506.
[71] V.M. Fokin, O.V. Potapov, C.R. Chinaglia, E.D. Zanotto, The effect of pre-existing crystals on the crystallization kinetics of a soda-lime-silica glass. The courtyard phenomenon, J. Non-Cryst. Solids **258** (1999) 180–186.
[72] E.D. Zanotto, Metastable phases in lithium disilicate glasses, J. Non-Cryst. Solids **219** (1997) 42–48.
[73] L.L. Burgner, P. Lucas, M.C. Weinberg, P.C. Soares, Jr., E.D. Zanotto, On the persistence of metastable crystal phases in lithium disilicate glass, J. Non-Cryst. Solids **274** (2000) 188–194.
[74] E.D. Zanotto, M.L.C. Leite, The nucleation mechanism of lithium disilicate glass revisited, J. Non-Cryst. Solids **202** (1996) 145–152.
[75] P.C. Soares Jr., *Early states of crystallization of LS_2 glasses,* MSc dissertation, Universidade Federal de São Carlos, Brazil (1997).
[76] A. Hishinuma, D.R. Uhlmann, Nucleation kinetics in some silicate glass-forming melts, J. Non-Cryst. Solids **95&96** (1987) 449–456.
[77] L. Gránásy, P.F. James, Nucleation in oxide glasses: comparison of theory and experiment, Proc. Roy. Soc. Lond. A **454** (1998) 1745–1766.
[78] L. Gránásy, Diffuse interface theory for homogeneous vapor condensation, J. Chem. Phys. **104** (1996) 5188–5198.
[79] P. Harrowell, D.W. Oxtoby, The effect of a distribution of relaxation times on crystal nucleation in glass, in: Ceramic Transactions Ed. M.C. Weinberg, Vol. 30, Westerville, OH (1993), pp. 35–44.
[80] D.G. Morris, Crystallisation of the Metglas 2826 amorphous alloy, Acta Metall. **29** (1981) 1213–1220.
[81] R.S. Tiwari, S. Ranganathan, M. von Heimendahl, TEM of the kinetics of crystallization of Metglas 2826, Z. Metallk. **72** (1981) 563–568.

[82] R.S. Tiwari, Analysis of steady state crystal nucleation in Metglas 2826, J. Non-Cryst. Solids **83** (1986) 126–133.
[83] Y.T. Shen, T.H. Kim, A.K. Gangopadhyay, K.F. Kelton, Icosahedral order, frustration and the glass transition: evidence from time-dependent nucleation and supercooled liquid structure studies, Phys. Rev. Lett. **102** (2009) 057801/1–4.
[84] I. Gutzow, E. Popov, S. Toschev, M. Marinov, Rost Kristallov, in: *Proc. VII. Int. Congress on Crystallization*, Moscow, Vol. 8, p. 95 (1966).
[85] I. Gutzow, S. Toschev, Non-steady state nucleation in the formation of isotropic and anisotropic phases, Kristall. Tech. **3** (1968) 485–498.
[86] U. Köster, Crystallization of amorphous silicon films, Phys. Stat. Solidi A **48** (1978) 313–321.
[87] C.V. Thompson, A.L. Greer, F. Spaepen, Crystal nucleation in amorphous $(Au_{100-y}Cu_y)_{77}Si_9Ge_{14}$alloys, Acta Metall. **31** (1983) 1883–1894.
[88] U. Köster, M. Blank-Bewersdorff, Transient nucleation in zirconium-based metallic glasses, Mater. Sci. Eng. **97** (1987) 313–316.
[89] I. Gutzow, D. Kashchiev, I. Avramov, Nucleation and crystallization in glass-forming melts old problems and new questions, J. Non-Cryst. Solids **73** (1985) 477–499.
[90] T.K. Croat, Time dependent nucleation in a bulk metallic glass forming alloy, Mater. Res. Soc. Symp. Proc. **481** (1998) 137–142.
[91] V.M. Fokin, A.M. Kalinina, V.N. Filipovich, Nucleation in silicate glasses and effect of preliminary heat treatment on it, J. Cryst. Growth **52** (1981) 115–121.
[92] W.B. Hillig, Theoretical and experimental investigation of nucleation leading to uniform crystallization of glass, in: *Symposium on Nucleation and Crystallization in Glasses and Melts*, Columbus OH: American Ceramic Society (1962), pp. 77–89.
[93] M.F. Barker, T.H. Wang, P.F. James, Nucleation and growth kinetics of lithium disilicate and lithium metasilicate in lithia-silica glasses, Phys. Chem. Glasses **29** (1988) 240–248.
[94] A.M. Kalinina, V.N. Filipovich, V.M. Fokin, Stationary and nonstationary crystal nucleation rate in a glass of $2Na_2O.CaO.3SiO_2$ stoichiometric composition, J. Non-Cryst. Solids **38&39** (1980) 723–728.
[95] L. Gránásy, P.F. James, Transient nucleation in oxide glasses: the effect of interface dynamics and subcritical cluster population, J. Chem. Phys. **111** (1999) 737–749.
[96] K.F. Kelton, A.L. Greer, Test of classical nucleation theory in a condensed system, Phys. Rev. B **38** (1988) 10089–10092.
[97] A.L. Greer, K.F. Kelton, Nucleation in lithium disilicate glass: a test of classical theory by quantitative modeling, J. Am. Ceram. Soc. **74** (1991) 1015–1022.
[98] K.L. Narayan, K.F. Kelton, Effects of multi-step annealing treatments on crystal nucleation in $Na_2O.(2CaO)_{1-x}(SiO_2)_x$ glasses, Acta Mater. **46** (1998) 3159–3164.
[99] L. Gránásy, P.F. James, Nucleation and growth in cluster dynamics: a quantitative test of the classical kinetic approach, J. Chem. Phys. **113** (2000) 9810–9821.
[100] L.Q. Xing, T.C. Hufnagel, J. Eckert, W. Löser, L. Schultz, Relation between short-range order and crystallization behavior in Zr-based amorphous alloys, Appl. Phys. Lett. **77** (2000) 1970–1972.
[101] J. Saida, M. Matsushita, A. Inoue, Nanoicosahedral quasicrystalline phase in Zr–Pd and Zr–Pt binary alloys, J. Appl. Phys. **90** (2001) 4717–4724.
[102] K.F. Kelton, Crystallization of liquids and glasses to quasicrystals, J. Non-Cryst. Solids **334&335** (2004) 253–258.
[103] B.S. Murty, K. Hono, Nanoquasicrystallization of Zr-based metallic glasses, Mater. Sci Eng. A **312** (2001) 253–261.
[104] J.C. Holzer, K.F. Kelton, Kinetics of the amorphous to icosahedral phase transformation in Al–Cu–V alloys, Acta Metall. Mater. **39** (1991) 1833–1843.
[105] D. Holland-Moritz, Short-range order and solid–liquid interfaces in undercooled melts, Int. J. Non-Equil. Proc. **11** (1998) 169–199.

[106] G.W. Lee, A.K. Gangopadhyay, T.K. Croat, T.J. Rathz, R.W. Hyers, J.R. Rogers, K.F. Kelton, Link between liquid structure and the nucleation barrier for icosahedral quasicrystal, polytetrahedral, and simple crystalline phases in Ti–Zr–Ni alloys: verification of Frank's hypothesis, Phys. Rev. B **72** (2005) 174107/1–10.
[107] G. Tarjus, S.A. Kivelson, Z. Nussinov, P. Viot, The frustration based approach of supercooled liquids and the glass transition: a review and critical assessment, J. Phys. Cond. Matt. **17** (2005) R1143–R1182.
[108] D.G. Burnett, R.W. Douglas, Nucleation and crystallization in the soda–baria–silica system, Phys. Chem. Glasses **12** (1971) 117–124.
[109] P.F. James, E.G. Rowlands, Kinetics of crystal nucleation and growth in barium disilicate glass, *in Phase Transformations*, Vol. 2, The Institution of Metallurgists, London (1979) p. III/27–III/29
[110] Z. Strnad, R.W. Douglas, Nucleation and crystallization in soda–lime–silica system, Phys. Chem. Glasses **14** (1973) 33–36.
[111] P.F. James, Nucleation in glass-forming systems – a review, in: Advances in Ceramics Eds. J.H. Simmons, D.R. Uhlmann, G.H. Beall, Vol. 4, American Ceramic Society, Columbus, OH (1982), pp. 1–48.
[112] K.L. Narayan, K.F. Kelton, First measurements of time-dependent nucleation as a function of composition in $Na_2O.2CaO.3SiO_2$ glasses, J. Non-Cryst. Solids **220** (1997) 222–230.
[113] R. Müller, E.D. Zanotto, V.M. Fokin, Surface crystallization of silicate glasses: nucleation sites and kinetics, J. Non-Cryst. Solids **274** (2000) 208–231.
[114] U. Köster, B. Punge-Witteler, G. Steinbrink, Surface crystallization of metal–metalloid–glasses, Key Eng. Mater. **40&41** (1990) 53–62.
[115] P.W. McMillan, The crystallisation of glasses, J. Non-Cryst. Solids **52** (1982) 67–76.
[116] N.S. Yuritsyn, V.M. Fokin, A.M. Kalinina, V.N. Filipovich, Crystal nucleation and growth in the surface crystallization of cordierite glass, Glass Phys. Chem. **20** (1994) 116–124.
[117] E.D. Zanotto, Surface nucleation in a diopside glass, J. Non-Cryst. Solids **130** (1991) 217–219.
[118] E.D. Zanotto, Surface crystallization kinetics in soda–lime–silica glasses, J. Non-Cryst. Solids **129** (1991) 183–190.
[119] E. Wittman, E.D. Zanotto, Surface nucleation and growth in anorthite glass, J. Non-Cryst. Solids **271** (2000) 94–99.
[120] S. Reinsch, R. Müller, W. Pannhorst, Active nucleation sites at cordierite glass surfaces, Glastech. Ber. Glass Sci. Technol. **67C** (1994) 432–435.
[121] M. Yamane, W. Park, Effect of surface condition on nucleation and crystallization of a cordierite glass, Presentation at TC 7 Meeting, International Commission on Glasses, Jena, Germany (1990).
[122] R. Müller, D. Thamm, Surface-induced nucleation of cordierite glass, in: *Proc. 4th Int. Otto-Schott-Coll.* Friedrich-Schiller-Universität, Jena: Freidrich-Schiller-Universität (1990), p. 86.
[123] R. Müller, S. Reinsch, R. Sojref, M. Gemeinert, Nucleation at cordierite glass powder surface, in: *Proc. XVII Int. Congr. on Glass, Beijing, People's Republic of China*, Beijing Inernational Academic Publishers (1995), pp. 564–569.
[124] R.A.F. de Réaumur, Art de faire nouvelle espèce de porcelaine, Mém. Acad. R. Sci., Paris (1739) 370–388.
[125] E.M. Rabinovich, Crystallization and thermal expansion of solder glasses in the $PbO–B_2O_3–ZnO$ system, Ceram. Bull. **58** (1979) 595–605.
[126] Y. Ding, A. Osaka, Y. Miura, Enhanced surface crystallization of ß-barium borate on glass due to ultrasonic treatment, J. Am. Ceram. Soc. **77** (1994) 749–752.
[127] R. Müller, Surface nucleation in cordierite glass, J. Non-Cryst. Solids **219** (1997) 110–118.

[128] A. Dobreva, I. Gutzow, Activity of substrates in the catalyzed nucleation of glass-forming melts. I. Theory, J. Non-Cryst. Solids **162** (1993) 1–12.
[129] A. Dobreva, I. Gutzow, Activity of substrates in the catalyzed nucleation of glass-forming melts. II. Experimental evidence, J. Non-Cryst. Solids **162** (1993) 13–25.
[130] A.G. Escorial, A.L. Greer, Surface crystallization of melt-spun $Pd_{40}Ni_{40}P_{20}$ glass, J. Mater. Sci. **22** (1987) 4388–4394.
[131] U. Herold, Ph.D. thesis, Dept. Mech. Eng., Ruhr-Universität Bochum, 1982.
[132] G. Rindone, Influence of platinum nucleation on crystallization of a lithium silicate glass, J. Am. Ceram. Soc. **41** (1958) 41–42.
[133] D. Cronin, L.D. Pye, Platinum catalyzed crystallization of Li_2O–$2SiO_2$ glass, J. Non-Cryst. Solids **84** (1986) 196–205.
[134] J. Deubener, R. Brückner, M. Sternitzke, Induction time analysis of nuleation and crystal-growth in disilicate and metasilicate glasses, J. Non-Cryst. Solids **163** (1993) 1–12.
[135] K.L. Narayan, K.F. Kelton, C.S. Ray, Effect of Pt doping on nucleation and crystallization in $Li_2O.2SiO_2$ glass: experimental measurements and computer modeling, J. Non-Cryst. Solids **195** (1996) 148–157.

CHAPTER 9

Precipitation in Crystalline Solids

Contents

1. PHASE TRANSFORMATIONS IN THE SOLID STATE

In previous chapters, the nucleation of ordered phases, typically crystals, from amorphous phases such as liquids and glasses, was examined. Many of the effects that were ignored there become dominant in solid-state transformations such as precipitation, void nucleation in radiation-damaged materials, clustering in ion-implanted solids, nucleation on dislocations and grain boundaries, recrystallization, etc. These include:

- stress effects due to a difference in density between the initial and final phase;
- influence of interfacial coherence;
- anisotropy of the parent phase;

Pergamon Materials Series, Volume 15
ISSN 1470-1804, DOI 10.1016/S1470-1804(09)01509-0

- possible roles of heterogeneities such as dislocations, vacancy clusters, grain boundaries, twin boundaries, stacking faults, etc.;
- dominance of long-range diffusion over interfacial attachment as the rate-limiting step in nucleation.

These effects complicate the comparison between nucleation data and predictions from nucleation theories. Since they are of practical concern, they are discussed in Section III of this book. For the discussion in this chapter, examples are chosen where the influence of heterogeneous nucleation, partial interfacial coherence, and anisotropy in the parent phase are minimal, allowing a focus on the influence of interfacial coherency strain and long-range diffusion on nucleation. As will be shown, however, the analysis can be complicated by other factors such as a nonequilibrium vacancy concentration at the precipitation annealing temperature (often called the reaction temperature in the literature), possible precursor stages of decomposition prior to precipitation, and coarsening that may occur simultaneously with precipitate nucleation and growth.

Two case studies are discussed. The first is cobalt precipitation from solid-solution Cu–Co alloys, in which the cobalt has substituted for the copper on the lattice. The second is oxygen precipitation from a solid solution where the oxygen resides in interstitial sites in silicon. Both are examples where long-range diffusion plays an important role during nucleation. In the second case, quantitative nucleation data allow tests of the kinetic models for classical and coupled-flux nucleation.

New techniques are emerging, such as three-dimensional atom-probe field-ion microscopy (3DAP), which enable structural studies at atomic length scales and provide new insights into the mechanisms of solid-state transformations. These have the potential to radically deepen our understanding of solid-state nucleation processes; this is illustrated briefly in the discussion of cobalt precipitation.

2. PRECIPITATION IN Cu–Co

When the temperature of a homogeneous solid solution is lowered, the phase often becomes supersaturated in one or more components and precipitation can occur. Studies of the precipitation rates for a wide range of metallic alloys have been made to examine homogeneous nucleation theories (see Aaronson and LeGoues [1] for a critical review). The most complete tests are for cobalt precipitation in a Cu–Co alloy. Like the elemental solid phases of Cu and Co, the Cu–Co solid-solution (matrix) phase is ccp. The γ-Co precipitate is almost pure cobalt, ccp above 390°C, and fully coherent (Chapter 6, Section 2.6) with the matrix. The lattice mismatch between γ-Co and the matrix phase is small, allowing the precipitates to form coherently, at least initially, giving little strain contribution to the driving free energy.

2.1 Measurements of nucleation rate

Nucleation rates for precipitation have been determined by a variety of methods, including changes in the electrical resistivity [2, 3], the rate of precipitate appearance as a function of annealing time using transmission electron microscopy (TEM) [1, 4], and studies of the cluster population and composition fluctuations in the solid-solution phase using 3DAP methods [3, 5].

The electrical resistivity is extremely sensitive to the solute concentration in the matrix phase [6], and is easily measured. Assume that the resistance, R, of a sample that was heat-treated to take all the cobalt into solution (i.e., a fully *solution-treated* sample), R_S, is sufficiently close to the resistance of a sample where precipitation has gone to completion, R_T [7]. The fraction transformed (fraction of the total amount that could be precipitated were the reaction to go to completion) x as a function of annealing time, t, may be computed directly from the measured values of $R(t)$.

$$x(t) = \frac{R_S - R(t)}{R_S - R_T}. \tag{1}$$

As discussed in Chapter 8, Section 5.1, $x(t)$ approximately follows the relation, $x(t) = 1 - \exp(-(\kappa t)^m)$, where κ is the effective rate constant for precipitation and m is the Avrami exponent, which provides information about the precipitation mechanism.

As shown in Figure 1, the nucleation rate is a very strong function of the supersaturation. Nucleation is, therefore, virtually completed in the earliest stage of the precipitation process due to rapid solute depletion, and the population (number per unit volume) χ_p of β precipitates in the original solid-solution phase (α) is approximately constant for a given annealing temperature and alloy composition (ignoring any influence of coarsening). Under such conditions, $m = 3/2$; precipitation proceeds essentially by diffusion-limited growth on a fixed ensemble of centers [8], as is experimentally observed [2]. In this case, χ_p is given by [9]:

$$\chi_p = \left(\frac{3\pi}{4}\right) 2^{-3/2} \left(\frac{X_\beta - X_\alpha^i}{X_\alpha^0 - X_\alpha^i}\right) \left(\frac{\kappa}{D}\right)^{3/2}, \tag{2}$$

where X_β is the mole fraction of solute in the precipitate phase, X_α^i is the mole fraction of solute in the α phase next to the growing precipitate, and X_α^0 is the mole fraction of solute in the α phase before precipitation began, assumed to be uniform throughout. (Note that throughout this discussion

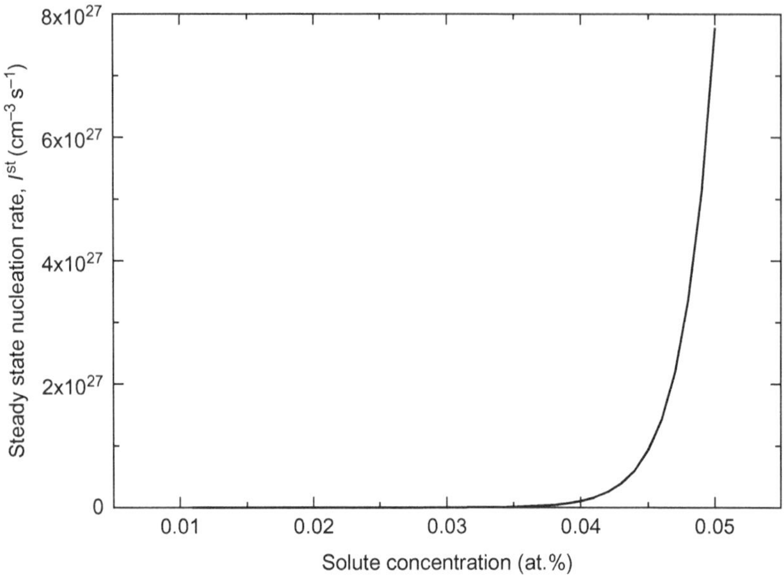

Fig. 1 Nucleation rate as a function of supersaturation for a given temperature (computed using parameters derived later in this section from Cu–Co precipitation data).

of Co precipitation, concentrations are expressed as atomic fraction). Since at any temperature $\chi_p = \int_0^\infty I^{st}\, dt$, it may be assumed that [2]

$$\chi_p \propto I^{st}. \tag{3}$$

Although this assumption can certainly be questioned, an analysis of the Cu–Co measurements using Eqs. (1)–(3) gives essentially the same results as a more sophisticated treatment (see Ref. [2]). The nucleation rates derived from electrical-resistivity studies for alloys of different solute concentration are shown by the open symbols in Figure 2.

While the electrical-resistivity measurements were accurate and the quenching rates from the solution temperature above the solvus were rapid (avoiding further precipitation during cooling), it is unclear how accurate is the estimate of the number of precipitates obtained by such an indirect method. Questions have been raised, for example, over the validity of the resistivity method to accurately evaluate the number when the precipitates are very small. However, the agreement with estimates of the nucleation rates obtained from direct measurements of the precipitate population [1, 4] is surprisingly good.

Figure 3 shows the induction times for nucleation, measured by extrapolating the number of nuclei produced (determined from TEM investigations) as a function of annealing time to the time axis. This is

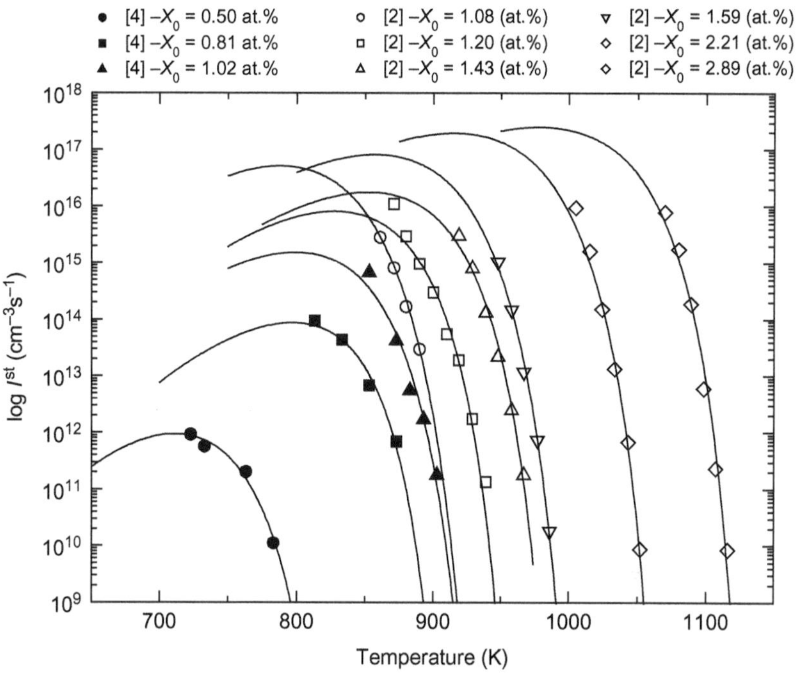

Fig. 2 Measured nucleation rates for cobalt precipitation in a Cu–Co solid-solution: open symbols from resistivity measurements [2]; closed symbols from TEM studies [1, 4]. The lines are fit to the classical theory of nucleation, assuming no strain contributions.

similar to the methods used to study transient nucleation in glasses (see Chapter 8, Sections 6 and 7). However, a significant difference is that in the glass studies, the population density of nuclei was obtained following a two-step annealing treatment, whereas here only one annealing treatment is given. The cluster size at which the induction times are measured is, therefore, unclear. Since in most cases the dominant contribution to the induction time is the time required to establish the steady-state distribution at the critical size, the data shown here should be correct, at least to within an order of magnitude.

2.2 Fits to steady-state classical theory

As for gases, liquids, and glasses, the nucleation data are first analyzed using the classical theory. From Chapter 2, Eqs. (14) and (53),

$$I^{st} = A^* \exp\left(-\frac{16\pi}{3k_B}\frac{\sigma^3}{T\Delta g^2}\right) = AD\exp\left(-\frac{16\pi}{3k_B}\frac{\sigma^3}{T\Delta g^2}\right), \tag{4}$$

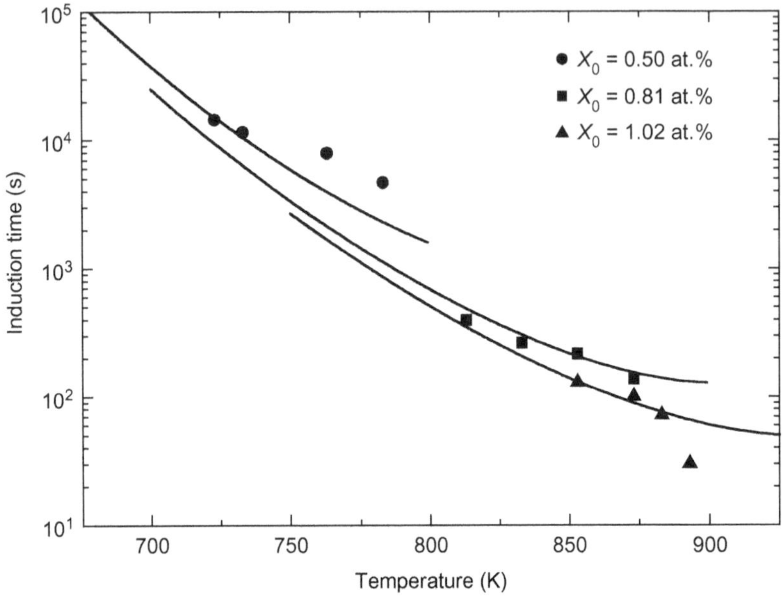

Fig. 3 Measured induction times for nucleation of cobalt precipitates in a Cu–Co solid-solution from TEM studies [1, 4]. The lines are computed from the classical theory of nucleation, assuming no strain contribution. The calculated induction times are much smaller than the measured values; they have been multiplied by 10^4 so that the temperature dependences can be compared.

where A^* is dynamical preterm for the nucleation rate, D is the solute diffusion coefficient, σ is the interfacial free energy between the precipitate and the solid-solution phase, and A is the portion of A^* that is independent of the atomic mobility

$$A = \frac{24n^{*2/3}N_0}{\lambda^2}\left(\frac{|\Delta\mu|}{6\pi k_B T n^*}\right)^{1/2}. \tag{5}$$

The driving free energy per unit volume, Δg, is $\Delta g = \Delta\mu/\bar{v}$, where $\Delta\mu$ is the difference in chemical potential for the precipitating species in the original and new phase and $\bar{v}$ is the solute atomic volume; N_0 is the number of atoms of the precipitating species in the original phase (in the same units as the nucleation rate, i.e., per mole or per unit volume), λ is the atomic jump distance, n^* is the number of atoms in the critical nucleus, k_B is the Boltzmann constant, and T is the temperature in absolute units. Initially ignoring strain contributions to the driving free energy, $\Delta\mu$ is simply a function of the supersaturation, s,

$$\Delta\mu = -k_B T\ \ln\left(\frac{X_\alpha^0}{X_\alpha^{eq}}\right) = -k_B T \ln s, \tag{6}$$

where X_α^{eq} is the equilibrium solute concentration. Recasting Eq. (4), then

$$\ln\left(\frac{I^{st}}{D}\right) = \ln A - \frac{16\pi\bar{v}^2}{3k_B^3}\sigma^3\frac{1}{T^3(\ln s)^2}. \tag{7}$$

The parameters used in the fits are listed in Table 1. Since diffusion is substitutional, the vacancy concentration is assumed to be in equilibrium at the reaction temperature. Note that per mole, $N_0 = X_\alpha^0 N_A$, where N_A is the Avogadro number, as confirmed recently by fits to the measured precipitate populations [10].

As observed in Figure 2, the steady-state nucleation data are fit well by Eq. (7) for all concentrations; the derived values for log A are listed in Table 2. The interfacial free energies obtained are in reasonable agreement with those expected from a nearest-neighbor bond model [2]. For nucleation in liquids and glasses, the measured values for A were larger than the theoretical one by several orders of magnitude, reflecting the diffusiveness of the interface between the clusters of the new phase and the parent phase. Here, however, the measured values are lower by four to six orders of magnitude than the theoretical values, clearly requiring a different explanation.

Up to now, possible strain contributions to the driving free energy have been ignored. Small precipitates are coherent with the matrix phase [12] with a small lattice mismatch [12, 13]. The resulting strain energy is positive and scales with the volume of the precipitate cluster, effectively decreasing the driving free energy. Including dilatational strain the driving free energy for cluster formation is

$$\Delta\mu = -k_B T\ \ln s + \frac{E}{9(1-\nu_P)}\delta^2\bar{v}_{Co}, \tag{8}$$

where E is the Young modulus, ν_P is the Poisson ratio of the matrix phase and $\delta \equiv (\bar{v}_{Co} - \bar{v}_{CuCo})/\bar{v}_{CuCo}$. The computed effect on the nucleation rate is small, less than an order of magnitude, which is insufficient to explain the discrepancies between the computed and fit values for the prefactor, listed in Table 2.

Table 1 Parameters used to fit the nucleation rates for γ-Co precipitation in Cu–Co alloys

Equilibrium solubility (from [2])[a]	Diffusion coefficient (from [11])	Miscellaneous
$X_\alpha^{eq} = C_1\exp(-\Delta H_S/k_BT)$	$D = D_0\exp(-\Delta H/k_BT)$	$\bar{v} = 1.099 \times 10^{-29}\,m^3$
$C_1 = 7.658$ (at.%)	$D_0 = 8.4 \times 10^{-5}\,m^2\,s^{-1}$	$\lambda = 2.2 \times 10^{-10}\,m$
$\Delta H_S = 0.570$ ev	$\Delta H_D = 2.26$ ev	

[a]Quoted as wt.% in [2]; converted here to at.%.

Table 2 Values for log A and σ from nucleation rates for γ-Co precipitation in Cu–Co alloys

X_{Co} in at.%	log $A_{measured}$	log $A_{theoretical}$	σ (J m^{-2})
0.5[a]	30.5	35.4 ± 0.04	0.22
0.81[a]	30.0	35.8 ± 0.03	0.18
1.02[a]	31.5	35.9 ± 0.03	0.20
1.08[b]	33.5	36.0 ± 0.03	0.23
1.2[b]	31.5	36.0 ± 0.03	0.19
1.43[b]	31.4	36.2 ± 0.03	0.19
1.594[b]	32.1	36.2 ± 0.02	0.20
2.21[b]	31.7	36.4 ± 0.02	0.20
2.894[b]	30.5	36.6 ± 0.02	0.18

[a] [1,4].
[b] [2].

The validity of the classical theory of nucleation can be investigated further by examining the agreement between the measured and calculated critical cluster radii, r^*. For small supersaturations (from Eq. (6)),

$$\Delta\mu = -k_B T \ \ln\left(\frac{X_\alpha^0}{X_\alpha^{eq}}\right) = -k_B T \ \ln\left(1 + \frac{X_\alpha^0 - X_\alpha^{eq}}{X_\alpha^{eq}}\right) \approx -\frac{k_B T}{X_\alpha^{eq}}(X_\alpha^0 - X_\alpha^{eq}), \quad (9)$$

and the radius of the critical cluster is given by (from Chapter 2, Eq. (19)):

$$r^* = \frac{2\sigma}{|\Delta g|} \approx \frac{2\sigma\, \bar{v} X_\alpha^{eq}}{k_B T(X_\alpha^0 - X_\alpha^{eq})}. \quad (10)$$

The high resolution possible with 3DAP techniques allows accurate measurements of the populations of small clusters. It is commonly assumed that the radii of the smallest precipitates that are measured using these techniques correspond to r^*. In fact, a continuous distribution of cluster sizes exists below r^*, with the cluster population increasing dramatically with decreasing size (Chapters 2 and 3), raising questions about this approach. However, the values for r^* computed from Eq. (10) are in good agreement with the data (Figure 4a), both in magnitude and in the dependence on solute concentration, supporting this approach and the validity of the classical theory of nucleation. More recent data show that the agreement is worse for lower concentrations and lower annealing temperatures (Figure 4b). From Chapter 2, Eqs. (14) and (19), the work of critical cluster formation can be written as $W^* = (4\pi/3)\sigma r^{*2}$. A larger than predicted critical size should, therefore, give a nucleation rate that is lower than that calculated from the classical theory, which is in conflict with the experimental results and the fits to the classical theory discussed earlier in this section.

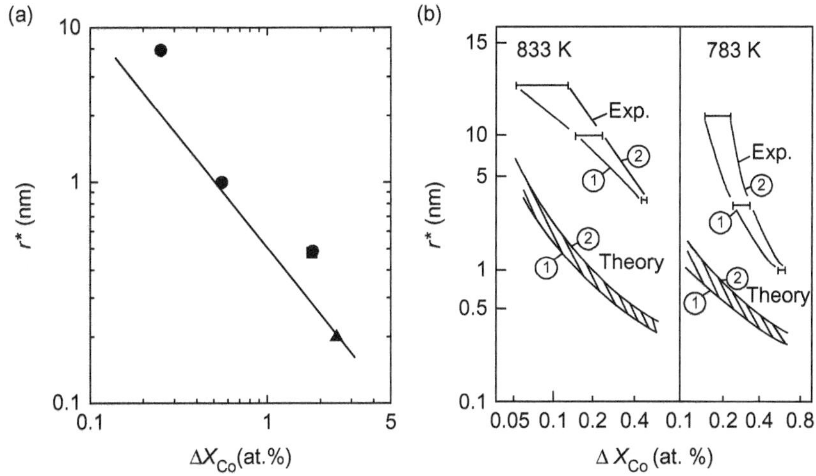

Fig. 4 (a) Radii of the smallest cobalt precipitates measured using FIM (field-ion microscopy) after annealing at 833 K (■) and 783 K (●) as a function of ΔX_{Co} ($= X_\alpha^0 - X_\alpha^{eq}$, where X_α^0 is the initial concentration of cobalt in the Cu–Co alloy and X_α^{eq} is the equilibrium concentration) (Reprinted from Ref. [14], copyright (1986), with permission from IOP Publishing Limited.). The critical size calculated from the classical theory of nucleation assuming a regular solution at one composition [15] is designated by (▲) and the straight line shows the $(\Delta X_{Co})^{-1}$ dependence predicted by Eq. (10). (b) Comparison between measured values of r^* from FIM and those calculated from the classical theory using two different measurements of the solvus curve (line 1 using data from [16] and line 2 using data from [17]). (Reprinted from Ref. [29], copyright (1990), with permission from Elsevier.)

2.3 Induction time for nucleation

The induction time for nucleation, θ, may be fit using one of the various analytical approximations discussed in Chapter 3; the fits in Figure 3 were made using the expression due to Kashchiev (Chapter 3, Eq. (36)). While the predicted temperature dependence of θ is in reasonable agreement with the data, θ itself is four orders of magnitude smaller than the measured value. This is little affected by including strain effects in the driving free energy since the induction time essentially scales inversely with the driving free energy (Chapter 3, Eq. (36)).

It is useful to examine the product $I^{st}\theta$, which is a function of only the thermodynamic factors. From the classical theory, this is (Chapter 2, Eq. (50) and Chapter 3, Eq. (36))

$$I^{st}\theta = Zk^+(n^*)N^{eq}(n^*)\left(\frac{\pi^2}{6}\frac{4}{\pi^3 k^+(n^*)Z^2}\right) = \frac{2}{3\pi Z}N^{eq}(n^*)$$
$$= \frac{2}{3\pi Z}N_0 \exp\left(-\frac{16\pi}{3k_B}\frac{\sigma^3}{T\Delta g^2}\right) \tag{11}$$

Table 3 Values determined from fits to $I^{st}\theta$ for γ-Co precipitation in Cu–Co alloys

X_{Co} in at.%	$\ln\left(\frac{2N_0}{3\pi Z}\right)_{meas}$	$\ln\left(\frac{2N_0}{3\pi Z}\right)_{calc}$	σ (J m^{-2})
0.5	42.1 ± 1.4	47.9 ± 0.5	0.20 ± 0.02
0.81	42.0 ± 0.8	49.4 ± 0.5	0.17 ± 0.01
1.02	48.2 ± 1.2	49.5 ± 0.5	0.21 ± 0.02

where Z is the Zeldovich factor $k^+(n^*)$ is the forward molecular attachment rate for a critical cluster, and $N^{eq}(n^*)$ is the equilibrium number of clusters of size n^*. (An identical product is obtained when the coupled-flux nucleation rate for a dilute solution is used instead of the classical theory expression (Chapter 5)). Following a similar analysis to that used to interpret the steady-state nucleation rate data (Eq. (7)), and using Eq. ((6)) to compute the driving free energy,

$$\ln(I^{st}\theta) = \ln\left(\frac{2}{3\pi Z}N_0\right) - \frac{16\pi\bar{v}^2}{3k_B^3}\sigma^3\frac{1}{T^3(\ln s)^2}. \tag{12}$$

Since it is unclear how to interpret the induction times obtained from the resistivity measurements discussed earlier, only the TEM data were used in this comparison. As expected, $\ln(I^{st}\theta)$ scales linearly with $(T^3(\ln s)^2)^{-1}$ (Figure 5). The line intercept gives a quantity that is dependent only on the number of sites and the Zeldovich factor; the calculated and measured values are in disagreement, with the measured values being approximately two orders of magnitude too small for the 0.5 and 0.81 at.% alloys. The derived values for the interfacial free energy are similar to those obtained by fitting the nucleation rates directly (Table 3).

2.4 Beyond the classical theory

The small critical sizes in Figure 4 indicate that nucleation is occurring in a system that is far from equilibrium, where the classical theory would not be expected to remain valid (see Chapter 4). LeGoues and Aaronson [4] have investigated this, also fitting the nucleation data to more sophisticated Cahn–Hilliard [18–20], continuum nonclassical [21], and Cook–deFontaine discrete lattice [21–25] models.

The Cahn–Hilliard model is a density-functional theory (DFT) discussed in Chapter 4, Section 4. Assuming the square density approximation, the work of formation of a critical nucleus can be written as (following Refs. [1] and [21]),

$$W^* = \int_V (\Delta\mu + \kappa(\nabla X)^2)d\mathbf{r}, \text{ where } \Delta\mu = \mu(X_\alpha^0) - \mu(\bar{X}_\alpha) - (X_\alpha^0 - \bar{X}_\alpha)\left(\frac{\partial\mu}{\partial X}\right)_{\bar{X}_\alpha}, \tag{13}$$

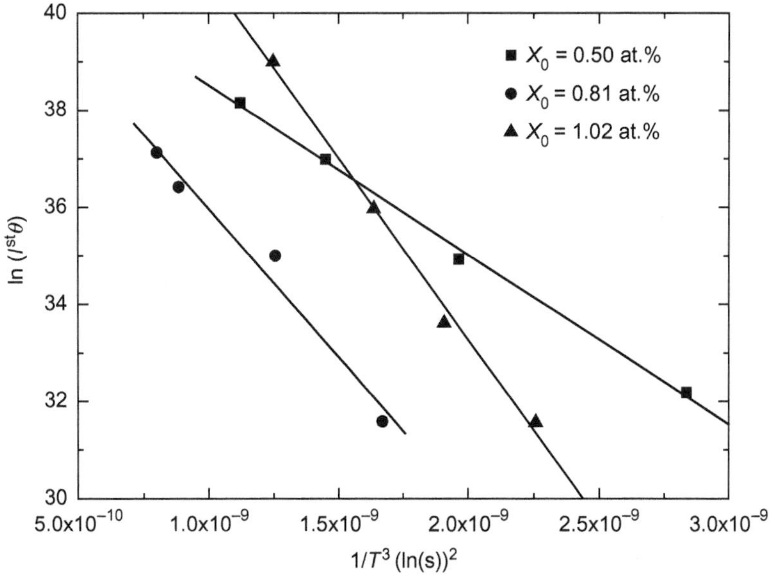

Fig. 5 Measured products of induction times and steady-state nucleation rates for cobalt precipitation in a Cu–Co solid-solution from TEM studies [1, 4]. The lines are fits to Eq. (11). The fit parameters are compiled in Table 3.

where V is the cluster volume, κ is the gradient-energy coefficient (see Chapter 4, Eq. (48)), and $\mu(X_\alpha^0)$ and $\mu(\bar{X}_\alpha)$ are the chemical potentials of the solute atoms in homogeneous α solid solutions of initial concentration X_α^0 and average concentration $\bar{X}_\alpha$. Cook and deFontaine developed a discretized version of the Cahn–Hilliard DFT for a nonuniform system composed of two types of atoms by expressing the concentration gradients as a sum over p lattice sites of the concentration differences between nearest-neighbor sites, giving for the work of cluster formation

$$W = \frac{1}{N_V}\sum_p \left(\Delta\mu + \frac{\kappa}{2a^2}\sum_r (X(p+r) - X(p))^2\right), \tag{14}$$

where N_V is the number of atoms per unit volume, a is the cubic lattice parameter, and r is the vector from a site to its nearest-neighbor sites; $\Delta\mu$ is a function of the homogeneous concentration surrounding the site p, $X(p)$. Assuming a regular solution, the expression for the work of cluster formation is [21]:

$$W = \frac{1}{N_V}\sum_p \left(\begin{array}{r} k_B T\left(X(p)\ln\dfrac{X(p)}{\bar{X}_\alpha} + (1 - X(p))\ln\dfrac{1 - X(p)}{1 - \bar{X}_\alpha}\right) - 2k_B T_c (X(p) - \bar{X}_\alpha)^2 \\ + \dfrac{k_B T_c}{a^2 z}\sum_r (X(p+r) - X(p))^2 \end{array} \right), \tag{15}$$

where z is the lattice coordination number (12 for the ccp lattice considered here) and T_c is a critical temperature for the miscibility gap. Following the usual procedure, W^* is determined by the condition $\partial W/\partial X(p) = 0$ for all p, yielding

$$k_B T \ln\frac{X(p)(1-\bar{X}_\alpha)}{\bar{X}_\alpha(1-X(p))} - 4k_B T_c(X(p)-\bar{X}_\alpha) + \frac{2k_B T_c}{a^2 z}(X(p+\boldsymbol{r}) - 2X(p) + X(p-\boldsymbol{r})) = 0. \quad (16)$$

As for the classical theory, as shown in Eq. (8), a term should be added to account for elastic strain [22, 23, 25]

$$W_{\text{total}} = W + \sum_{p'}(X(p)-\bar{X}_\alpha)(X(p')-\bar{X}_\alpha)\left(\sum_{h\neq 0}\Omega(h)\exp(i\boldsymbol{k}(h)\cdot\boldsymbol{r}_{p',p})\right), \quad (17)$$

where W is the work of cluster formation given in Eq. (14). $\Omega(h)$ is a function of the Fourier transforms of elastic-coupling parameters between solute pairs, solvent pairs, and solute–solvent pairs; h is an element of the triplets, h_j, given by $h_j = m_j/L$, where L is the edge length of the first Brillouin zone for the allowed sites in reciprocal space, and m_j is an integer. The distance between sites p' and p is $\boldsymbol{r}_{p',p} \equiv \boldsymbol{r}_{p'} - \boldsymbol{r}_p$ and $\boldsymbol{k}(h) = 2\pi h_j \boldsymbol{q}_j$, where $\boldsymbol{q}_j$ are the reciprocal lattice translation vectors. The summation over h in Eq. (17) is over the allowed sites in the first Brillouin zone. As before, the critical work is obtained from the condition $\partial W_{\text{total}}/X(p) = 0$, giving

$$k_B T \ln\frac{X(p)(1-\bar{X}_\alpha)}{\bar{X}_\alpha(1-X(p))} - 4k_B T_c(X(p)-\bar{X}_\alpha) + \frac{2}{a^2}\left(\frac{k_B T_c}{z} - K\right)(X(p+\boldsymbol{r}) - 2X(p) + X(p-\boldsymbol{r})) + \sum_{p'}(X(p')-\bar{X}_\alpha)\sum_{h\neq 0}\Omega(h)\exp(i\boldsymbol{k}(h)\cdot\boldsymbol{r}_{p',p}) = 0. \quad (18)$$

The critical temperature for the miscibility gap shifts from that in Eq. (16) when strain effects are included.

Assuming an equilibrium vacancy population at the reaction temperature, the predicted nucleation rates from this more sophisticated treatment are several orders of magnitude below the data (Figure 6). If it is assumed that the vacancy concentration is greater, however, equal to the equilibrium concentration at the higher solution-annealing temperature, excellent agreement is obtained with the experimental data. However, even with the arguments of Aaronson and LeGoues [1], it is difficult to understand how the excess quenched-in vacancy concentration could survive long enough at the reaction temperature [15, 26].

It should be pointed out that all of the fits to the precipitation data that have been discussed in this chapter are based on the kinetic model of the classical theory of nucleation, despite the central importance of long-range diffusion. As discussed in Chapter 5, for this case, a theory that

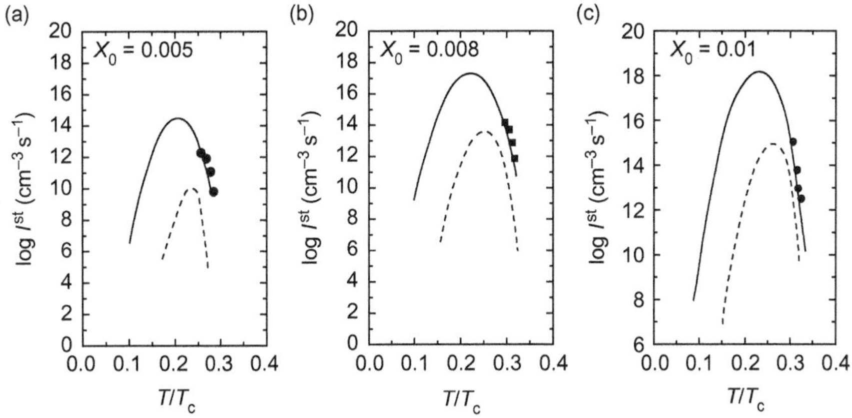

Fig. 6 Comparison between measured nucleation rates (•) of cobalt precipitates in Cu–Co solid-solutions of three different compositions and those calculated from the Cook–deFontaine model assuming (i) that the diffusion coefficient is defined by a quenched-in vacancy concentration that corresponds to the equilibrium concentration at the solution-annealing temperature (solid curve) and an equilibrium vacancy concentration at the precipitation (or reaction) temperature (dashed curve). (Adapted from Ref. [1], copyright (1992), with kind permission of Springer Science and Business Media.)

takes account of the coupling between the stochastic diffusion and interfacial attachment fluxes, such as the coupled-flux model (Chapter 5, Section 5), should be used instead of the classical theory. The smaller nucleation pre-factors measured for cobalt precipitation (Table 2) and the correspondingly longer induction times are consistent with expectation from that model (Chapter 5, Figure 8). However, based on Chapter 5, Figure 12 only about one order of magnitude decrease below the steady-state nucleation rate would be expected, not the four orders of magnitude observed, indicating that there must be other important factors.

2.5 Spinodal mechanism, precursor ordering or coarsening?

In addition to the question of which nucleation theory should be used and the usual uncertainty about the value and meaning of the interfacial free energy, there are apparent discrepancies for expected scaling between the steady-state nucleation rate and the nucleation induction times, and uncertainties about the correct vacancy concentration and atomic mobility. Further, as has been mentioned several times, the possibility of heterogeneous nucleation must also be considered. Phases of the same or other types might form during quenching, giving rise to preexisting growth centers or catalytic sites for nucleation. Defects such as voids and dislocations might also form due to the precipitation of excess vacancies, again forming sites for heterogeneous nucleation. Only in the case of the 1.02 at.% alloy are the measured and calculated preterms in reasonable agreement, leading some to argue that it is only at this composition that nucleation was homogeneous [10]. The smaller intercepts of the other two alloys reflect the smaller population of sites where nucleation can occur. However, it seems that there would also be a significantly different effective interfacial free energy value for the 1.02 at.% alloy, which is not observed.

As already mentioned, these data have been measured far from equilibrium, raising the possibility that precipitation might be better described by a spinodal mechanism rather than nucleation and growth. This has been investigated using scattering techniques such as small-angle X-ray (SAXS) and neutron (SANS) scattering techniques as well as the direct imaging techniques such as 3DAP and high-resolution electron microscopy (HREM). Some evidence for a spinodal-type process has been obtained from TEM and SANS studies of more concentrated (>3 at.% Co) alloys [27]. However, bright-field-zone-axis incidence (BFZA) TEM studies, which allow precise measurements of population and radius for precipitates as small as 1 nm diameter in the electron-transparent region of the specimen, showed no evidence for the expected regularity in particle spacing that would be characteristic of a spinodal mechanism.

SANS studies have suggested precursors to precipitation [14, 26, 28, 29] that were characterized by small-amplitude composition fluctuations with a large spatial extent, consistent with spinodal decomposition. A precursor stage was also inferred from 3DAP studies [30], suggesting that it is characterized by fractal-like aggregates of cobalt [31]. More recent SANS studies, however, using polarized neutrons, are in contradiction with this interpretation, showing good agreement with the cluster size distributions expected from standard nucleation processes [32] and showing no evidence for precursor stages. The possibility that chemical ordering precedes precipitation is reminiscent of the coupling between icosahedral or magnetic ordering and the nucleation barrier discussed in Chapter 7 for liquid crystallization.

The precursor stages could also be in agreement with a key prediction of the coupled-flux model for nucleation such that the solute concentration around subcritical clusters is enhanced rather than depleted (Chapter 5, Section 5 and [33–35]). Figure 7a–c shows the development of cobalt precipitates in a Cu–1 at.% Co alloy on annealing at 723 K, determined from a 3DAP study [5]. After aging at 30 min (Figure 7a), regions of increased cobalt concentration are observed, but no precipitates are visible. Precipitates are observed after annealing for 2 h (Figure 7b) and the material has the highest precipitate population of any of the annealing treatments. The particles are only a few nanometers in diameter, consistent with earlier measurements (Figure 4). The interfacial width is large, as expected from the discussion in Chapter 4, and incompatible with classical-theory assumptions. With

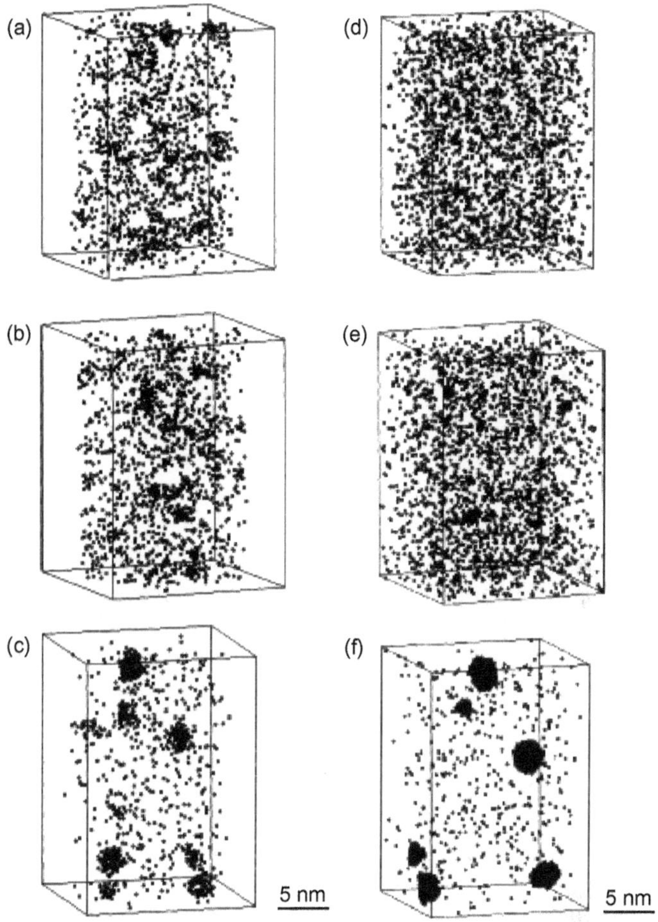

Fig. 7 Distribution of cobalt during precipitation from Cu–1 at.% Co alloy measured using 3DAP for annealing times of (a) 0.5 h, (b) 2 h, and (c) 24 h. Predictions from a lattice model calculation assuming regular solution after (d) 600 Monte-Carlo steps (MCS), (e) 1200 MCS, and (f) 29000 MCS. (Reprinted with permission from Ref. [5], copyright (2003), The Royal Society.)

further aging, the precipitates become larger but their population has decreased, consistent with coarsening. As will be discussed in Chapter 10, realistic computer modeling studies of nucleation processes are now possible. This is illustrated here, where the results of the 3DAP studies are in reasonable agreement with the results of dynamical Ising Monte-Carlo calculations shown in Figure 7c–e [5]. The peak precipitate population predicted by the Ising calculation ($(2.4 \pm 0.2) \times 10^{24}\,\mathrm{m}^{-3}$) is in good agreement with the measured peak population [$(1.0 \pm 0.4) \times 10^{25}\,\mathrm{m}^{-3}$]. These calculations were made for the simple case of a regular solution and diffusion by direct exchange of atoms on an fcc lattice; more realistic assumptions would likely lead to even better agreement.

Coarsening occurring simultaneously with nucleation has been argued previously [1, 36]. The high supersaturation leading to a burst of production of small clusters would shut down nucleation quickly due to the rapid depletion of solute in solution and the overlapping diffusion fields, a process sometimes called catastrophic nucleation [37]. Coarsening at long times and time-dependent nucleation (Chapter 3) at short times restrict the available window for measurements of steady-state nucleation rates (see [37] and [1]) and could partially explain the difficulties in fitting the existing data to nucleation theories. Clearly, although well studied, precipitation in Cu–Co alloys remains a complex and interesting subject.

3. OXYGEN PRECIPITATION IN SILICON

Cobalt precipitation in Cu–Co alloys, discussed in the previous section, is an example of nucleation that is governed by substitutional diffusion with strain effects that are small due to the close lattice match between the precipitate and solid-solution phases. In this section, a case is examined where the lattice strain is large, the atomic mobility is governed by interstitial diffusion, and defects such as vacancies and interstitials play important roles. The precipitation process also has significant technological importance.

Silicon single crystals grown by the Czochralski method, in which a rotating seed crystal placed in molten silicon is slowly withdrawn, contain a concentration of approximately $10^{24}\,m^{-3}$ ($\approx 10^{-4}\%$) of dissolved oxygen atoms, due to contamination from the SiO_2 crucible containing the melt [38]. These oxygen atoms are electrically neutral and are located in interstitial sites (Figure 8) that are slightly displaced away from the $\langle 111 \rangle$ bonds connecting the silicon atoms [39]. At the lower temperatures subsequently used for device fabrication, the high oxygen supersaturation frequently causes extensive precipitation ($Si+2O \rightarrow SiO_2$) during very large scale/ultra large scale integration (VLSI/ULSI) fabrication. These precipitates (hereafter called oxygen precipitates) can be beneficial, gettering transition-metal impurities [40, 41], and increasing the high-temperature strength of the wafer [42, 43], but if they form in active regions they can drastically impair the device performance [44]. Currently, industrial control of oxygen precipitation relies on a series of preannealing treatments for different times and temperatures, established empirically for a particular device fabrication process. An ability to quantitatively predict the precipitation population for arbitrary isothermal and nonisothermal annealing sequences would have

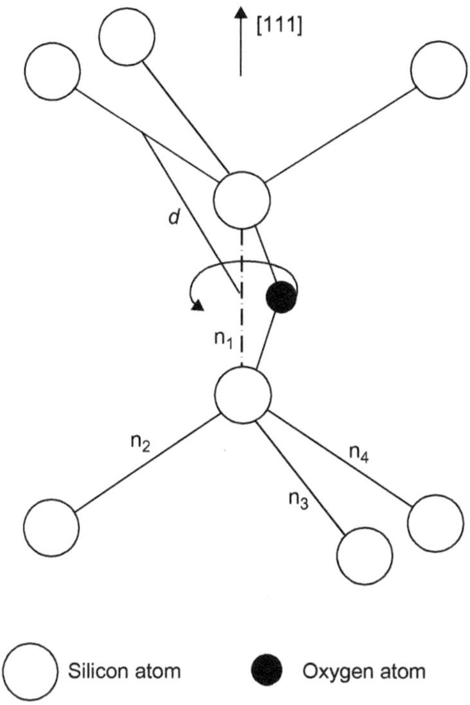

Fig. 8 Location of off-axis interstitial oxygen in silicon. The mean diffusion jump distance *d* is shown and the Si and O bonds are indicated. Oxygen atoms may be located along any of the four nearest-neighbor silicon atom bonds, n_1-n_4. (Reprinted from Ref. [45], copyright (1994), with permission from Elsevier.)

significant practical benefit; consequently, the nucleation kinetics have been studied extensively. These data coupled with the well-known structural and thermodynamic properties of silicon are also of fundamental interest for testing kinetic models of nucleation in solids. Further, as will be discussed in Section 3.5, electrically active defects, called thermal donors (TDs), may in the future give an unprecedented opportunity to study the development of extremely small oxygen clusters, allowing fundamental examinations of the dynamics of nucleation theories.

3.1 Nucleation, diffusion, and solubility data

While precipitation in silicon has been much studied, mostly the focus has been on the as-grown material. The nucleation behavior is, however, a strong function of the thermal history [46, 47], which is dependent on the local growth conditions, making quantitative modeling difficult. Falster et al. have studied the precipitation kinetics as a function of initial oxygen concentration in silicon wafers; three different oxygen concentrations were used, $6.1310 \times 10^{23}\,m^{-3}$, $7.1310 \times 10^{23}\,m^{-3}$, or $8.031 \times 10^{23}\,m^{-3}$, designated here as "low," "med," or "high". All wafers were first annealed at 1000°C for 15 min to erase any differences in thermal history during growth (see [48]). These standardized wafers were then annealed at different times for temperatures between 400°C and 750°C, to produce a population of nuclei, that were subsequently grown to visible size by a two-step treatment at temperatures where additional nucleation was negligible. Annealing at 800°C was required to promote those clusters larger than the critical size at the nucleation temperature to a size that exceeds the critical size at the final growth temperature (1000°C). The final precipitate population was measured by the "cleave-and-etch" method [49]; the counting detection limit was approximately $8.3 \times 10^{12}\,m^{-3}$. These data are used for the discussion presented here.

Figure 9 shows the precipitate population as a function of annealing time at temperatures between 400°C and 750°C in samples of two different oxygen concentrations. All samples were subsequently annealed for 4 h at 800°C and 16 h at 1000°C to grow the nuclei to observable size. The measured population of oxygen precipitates depends strongly on the annealing time and on the initial oxygen concentration in the silicon. The concentration dependence is largely due to the increased driving free energy for precipitation, but it also reflects a smaller concentration dependence of the diffusivity [50]. Two peaks are observed, one near 500°C and the other near 650°C. With increased annealing time, the double peak vanishes, however, producing a broad peak near 600°C, consistent with the previous identification of a peak in the steady-state nucleation rate near this temperature [51].

Carbon is known to catalyze nucleation [52]. However, the concentration of carbon in the wafers studied was below the detection limit, and the results are reproducible taking the concentration of oxygen as the only variable. Based on these and other considerations, it is reasonable to assume that SiO_2 precipitation is initiated by homogeneous nucleation.

As is clear from the cases examined in this and previous chapters, quantitative modeling requires knowledge of the atomic mobility and the driving free energy (a function of the supersaturation for precipitation processes).

The expression for the diffusion coefficient for oxygen in silicon, D_O, is in good agreement with measurements for $T > 650°C$ and $T < 450°C$. Due to experimental difficulties, however, few direct measurements are available between 450°C and 650°C — the temperature range where the nucleation rate of oxygen precipitates is significant. Oxygen-loss measurements and thermal-donor formation suggest that the diffusion rate can be higher than expected, by several orders of magnitude.

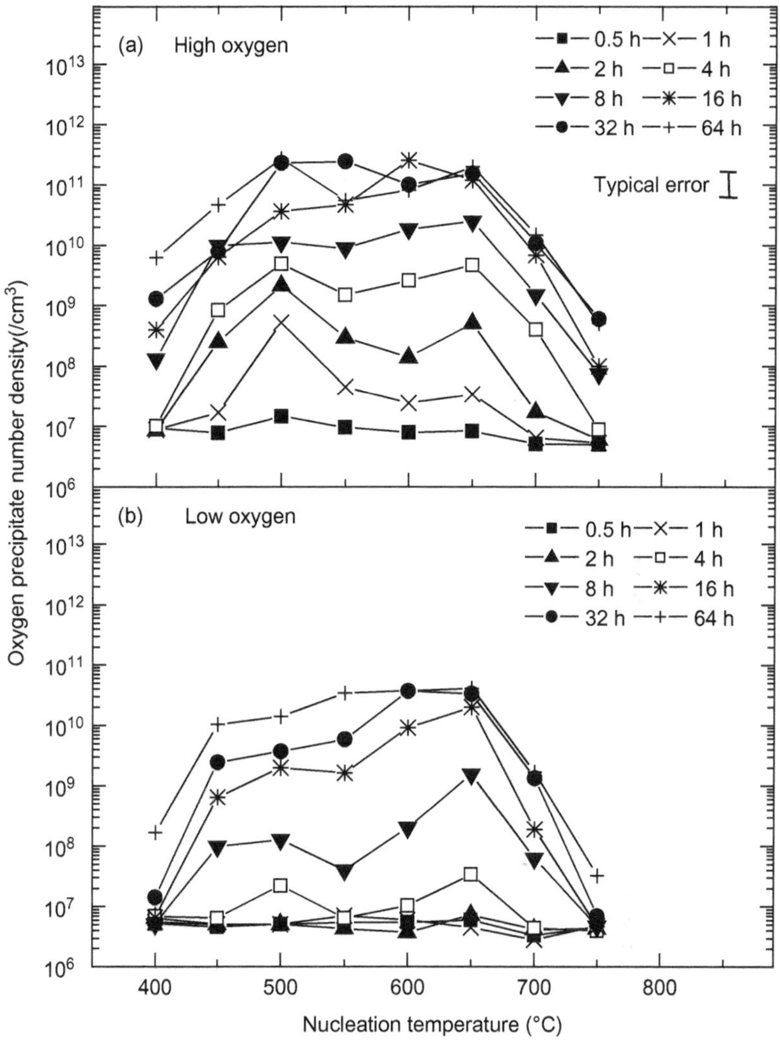

Fig. 9 The measured oxygen-precipitate population as a function of time between 500 and 750°C. These nucleation treatments were followed by anneals of 4 h at 800°C and 16 h at 1000°C for (a) high-oxygen ($8.031 \times 10^{23}\,m^{-3}$) and (b) low-oxygen ($6.131 \times 10^{23}\,m^{-3}$) samples. A typical measurement error is indicated. (Reprinted with permission from Ref. [55], copyright (2003), The Royal Society.)

Recently, D_O in this temperature range was deduced from dislocation-unlocking experiments [50]. The enhancement was much less than predicted by previous studies. These data, and the "normal" diffusion behavior, are shown in Figure 10.

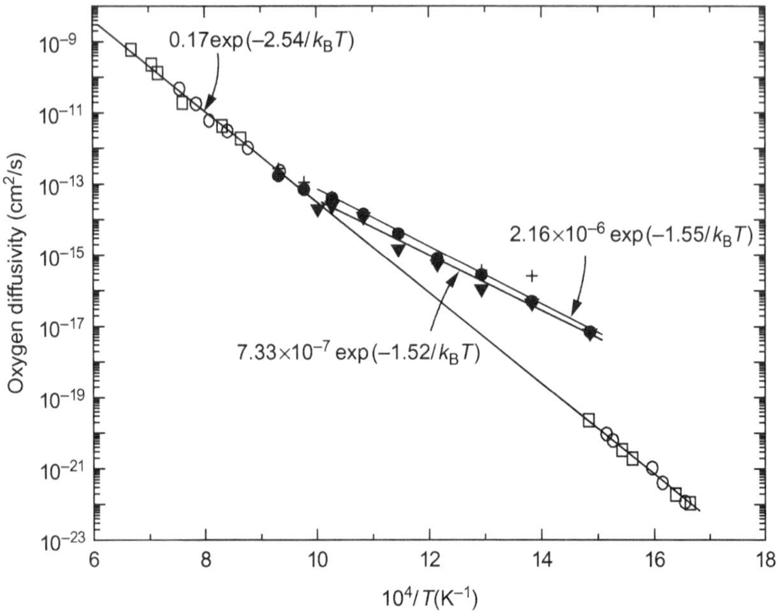

Fig. 10 Measured values of normal oxygen diffusion (open symbols) and a fit to the data (Eq. (2)), shown by the closed line. Also shown are the values obtained from dislocation-unlocking experiments for low-oxygen (▼) and high-oxygen (•) samples with lines indicating fits to the data. The values deduced from fits of the precipitates calculated from the coupled-flux nucleation model are also shown (+). (Reprinted with permission from [50], copyright (2001), American Institute of Physics.)

3.2 Fits to the classical theory of nucleation

First, assume the classical theory of nucleation and only supersaturation contributions to the driving free energy. Equation (6) is used to compute $\Delta\mu$ the chemical potential difference between oxygen in an SiO_2 precipitate and dissolved in the silicon crystal. In some cases, $\Delta\mu$ can be time-dependent if the depletion of the dissolved oxygen is sufficiently large with precipitate formation and growth. Let C_α^0 and $C_\alpha(t)$ be the initial and time-dependent oxygen concentration of oxygen dissolved in Si, respectively, and let C_α^{eq} be the equilibrium dissolved oxygen concentration; values of C_α^{eq} and D_O were computed from the expressions given in Table 4. The evolution of the cluster distribution for the experimental annealing treatments was computed numerically, using the Runge–Kutta method to solve the coupled rate equations of the classical theory (Chapter 2), and assuming diffusion-controlled rates. The interfacial free energy, σ, was determined by fitting to the precipitate population at 650°C. For comparison with experimental data, the time-dependent nucleation rates of the oxide precipitates were

Table 4 Parameters used to fit the nucleation rates for oxygen precipitates in Czochralski silicon

Oxygen equilibrium solubility (from Ref. [53])	Oxygen interstitial diffusion coefficient (from Ref. [39])	Miscellaneous
$C_O^{eq} = C_1 \exp(-\Delta H/k_B T)$	$D_O = D_O \exp(-\Delta H/k_B T)$	$\bar{v} = 3.45 \times 10^{-29} m^3$
$C_1 = 2.21 \times 10^{27} (m^{-3})$	$D_O = 1.3 \times 10^{-5} m^2 s^{-1}$	$\lambda = 2.15 \times 10^{-10} m$
$\Delta H_S = 1.03$ ev	$\Delta H_D = 2.53$ ev	

computed at the critical size for a final growth temperature of 1000°C. An initial steady-state distribution at 1000°C was assumed, consistent with the preannealing treatment for 15 min at 1000°C for all samples. As for the experimental data, the effects of nucleation anneals of duration 0.5, 1, 2, 4, 8, 16 and 32 h were calculated. Only those clusters that made it past n^*_{1000}, the critical size at 1000°C, were counted, consistent with the growth treatments after the nucleation anneals.

A detailed discussion of the comparison between the calculated and experimental data is provided in Ref. [48]. Figure 11 illustrates the results for a sample containing an initial oxygen concentration equal to $8.031 \times 10^{23} m^{-3}$. The computed populations are in good agreement with the experimental data for temperatures above 600°C, although above 650°C, the calculated precipitate population rises more rapidly than the measured population. Further, the classical theory of nucleation fails to reproduce the composition dependence of the precipitate population. A dramatic disagreement is observed for nucleation temperatures lower than 600°C where the populations are underestimated by several orders of magnitude.

Increasing the diffusion coefficient to agree with the results of dislocation-unlocking experiments (Figure 10) dramatically improves the agreement between the calculated and measured data (Figure 12b). However, the calculated precipitate population is too large at the lowest annealing temperatures. This reflects the inability of the common formulation of the classical theory to correctly handle long-range diffusion fluxes. The data in Figure 12c were computed using the coupled-flux model for nucleation discussed in Chapter 5, which links the interfacial and long-range diffusion fluxes; this is discussed in the next section.

3.3 Fits to the coupled-flux model for nucleation with no account of strain

The assumption of a constant composition in the parent phase adjacent to the developing clusters of the new phase is incorrect for solid-state

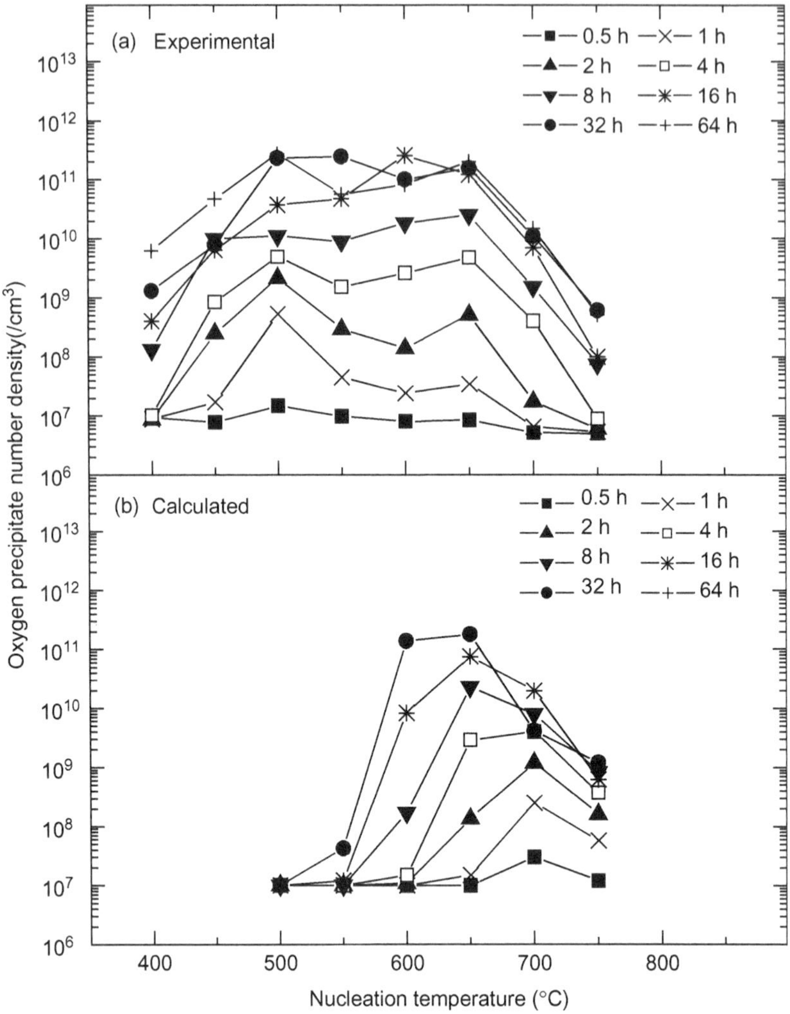

Fig. 11 The oxygen precipitate population as a function of the nucleation time at temperatures between 500 and 750°C, followed by a 4-h anneal at 800°C and 16 h at 1000°C in high-oxygen ($8.031 \times 10^{23}\,m^{-3}$) samples: (a) measurements; (b) calculated values. (Reprinted with permission from [48], copyright (1999), American Institute of Physics.)

precipitation processes where the diffusion rates are much slower than interfacial attachment processes. This can have profound consequences on the nucleation. If, for example, an embryo becomes critical in unstable equilibrium with the parent phase of average composition, the depletion of solute in the parent phase immediately surrounding the cluster will locally decrease the driving free energy and increase the critical size. For subcritical clusters, on the other hand, the dissolution rate depends on

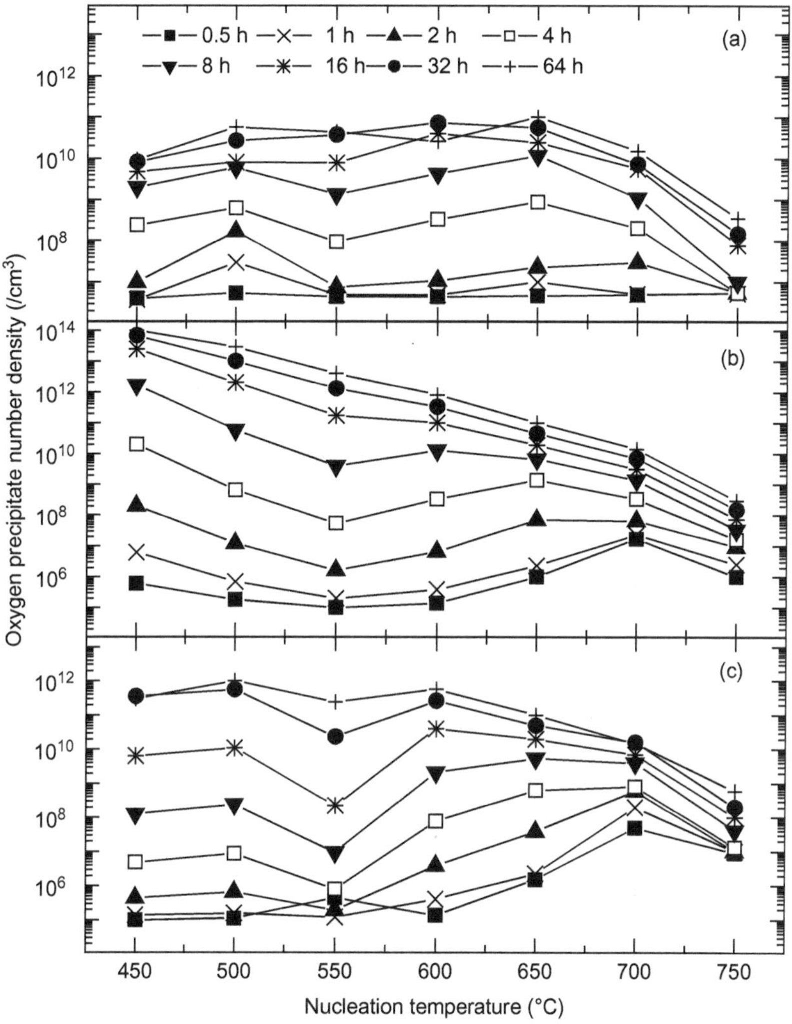

Fig. 12 The oxygen precipitate population as a function of the nucleation time at temperatures between 450 and 750°C, followed by a 4-h anneal at 800°C and 16 h anneal at 1000°C (medium-oxygen ($7.131 \times 10^{23}\,m^{-3}$) sample): (a) measurements; (b) classical theory calculation assuming enhanced diffusion; (c) coupled-flux calculation. (Reprinted with permission from [54], copyright (2000), American Institute of Physics.)

size, leading to the solute concentration in the matrix near a cluster being itself dependent on size. To take account of these and similar effects, a variable describing the changing composition of the parent phase as a function of cluster growth must be introduced. The correct treatment, where the composition of both the cluster and the parent phase dynamically change, is at present intractable. A simpler approach,

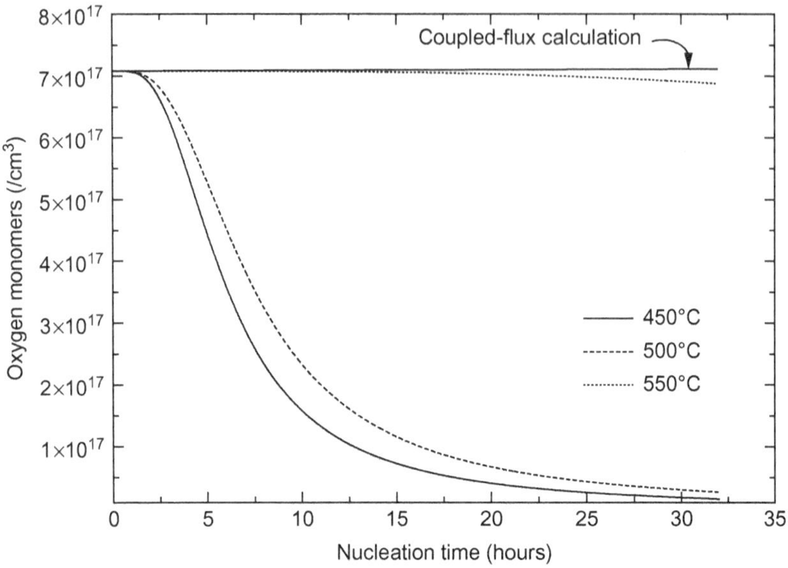

Fig. 13 The calculated oxygen single-atom depletion as a function of annealing time at the nucleation temperatures indicated. Results from the coupled-flux calculation are also shown.

considering a constant composition cluster, was introduced in Chapter 5. The oxygen precipitate population calculated with the coupled-flux model (Figure 12c) is in much better agreement with the experimental data (Figure 12a) over the complete range of nucleation temperatures, even matching the observed "double peak."

The predicted oxygen single-atom depletion during precipitation is also different with the classical theory and the coupled-flux model. The experimental data show no measurable oxygen depletion during the nucleation and 800°C growth treatments. In contrast, calculations from the classical theory predict that the single-atom depletion is extremely large at the nucleation temperatures (Figure 13). In agreement with the experimental data, the coupled-flux nucleation model predicts no significant depletion. The incorrect prediction of the classical theory is a consequence of the incorrect kinetic model, leading to rates that are much larger than expected for diffusion-limited precipitation.

3.4 Fits to the coupled-flux model for nucleation accounting for strain and enhanced diffusion

Although the agreement between the experimental data and the coupled-flux calculations is much better than with the predictions from the classical theory, there remain problems. For example, for nucleation times greater than 16 h, more nuclei are predicted than are observed, particularly at the lower annealing temperatures (Figure 12). Also, quantitative agreement is poor for all nucleation times at the lower nucleation temperatures. These points are demonstrated more clearly by comparing the computed and measured populations as a function of time at 450°C (Figure 14a). The computed data underestimate the precipitate population for smaller annealing times, suggesting that the diffusion rate is enhanced even above the values used, estimated from the dislocation-unlocking experiments. The computed population is over an order of magnitude high, however, for long annealing times, indicating that the nucleation rate decreases rapidly; since single-atom depletion is negligible, some other mechanism must be responsible for this decrease.

The volume of a SiO_2 precipitate is approximately twice that of the matrix with the same number of silicon atoms, leading to a compressive stress in the precipitate, and a positive contribution to the work of cluster formation (see discussion in Section 2). Calculations show that unless this stress is relieved, the nucleation of the oxygen precipitates is shut down. Various stress-relief mechanisms have been proposed [44, 56, 57]; the emission of silicon self-interstitials is commonly accepted [52, 58]. Many of these diffuse rapidly to the wafer surface, voids, or stacking faults,

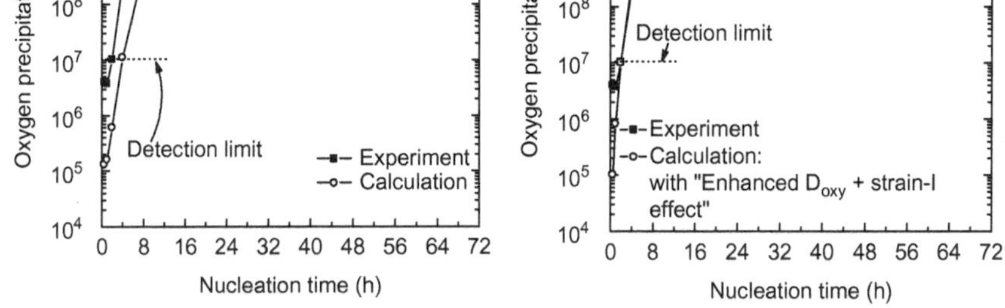

Fig. 14 (a) The best agreement between the measured population as a function of nucleation time at 450°C and that computed from the coupled-flux model using the diffusion rates estimated from dislocation-unlocking experiments, but with no consideration for accumulated strain and strain relief (medium-oxygen, $7.131 \times 10^{23}\,m^{-3}$). (b) The agreement when interstitial compensation for strain is included. (Reprinted with permission from Ref. [55], copyright (2003), The Royal Society.)

where they are annihilated. Depending on the emission rate, however, some remain in the matrix, causing a supersaturation of interstitials. The work to form a cluster containing n oxygen atoms is

$$W(n,t) = -nk_{\mathrm{B}}T \ \ln\frac{C(t)}{C_{\alpha}^{\mathrm{eq}}} + k_{\mathrm{B}}T \ \ln\frac{C_{\mathrm{I}}(t)}{C_{\mathrm{I}}^{\mathrm{eq}}} + 4\pi r^2\sigma + \Delta G_{\varepsilon}, \tag{19}$$

where $C_{\mathrm{I}}(t)$ and $C_{\mathrm{I}}^{\mathrm{eq}}$ are the time-dependent and equilibrium interstitial concentration, respectively, σ is the interfacial free energy between the precipitate and the solid-solution phase, r is the cluster radius, and ΔG_{ε} is the residual strain energy. (Note that the time-dependent concentrations give rise to a time-dependent work of cluster formation in this case). Tan and Taylor [52] and others [59] have concluded that silicon ejection is a very efficient mechanism for relieving strain, allowing the last term in Eq. (18) to be ignored.

The number of silicon interstitials should be half the number of oxygen interstitial atoms consumed in the precipitation process, $\Delta O_{\mathrm{I}}(t)$. This proportionality factor (i.e., 1/2) is decreased, however, since, as already mentioned, most of the interstitials diffuse to sinks, introducing a new proportionality factor, β, that must be fit, $C_{\mathrm{I}} = \beta\Delta O_{\mathrm{I}}(t)$. With the introduction of interstitial compensation for strain, the agreement between the experimental and calculated data becomes extremely good (Figure 14). Further, this good agreement is obtained for all nucleation temperatures and oxygen concentrations. The growing supersaturation of interstitials decreases the nucleation rate at low temperatures and long annealing times, in agreement with observation. However, the calculations show that a diffusion rate that is enhanced even above that measured from the dislocation-unlocking experiments is required.

3.5 Thermal donors

It is clear that taking account of the diffusion flux is critical for this and any other precipitation problem. But, is the agreement shown in Figures 12 and 14 sufficient to argue that the theory is valid? Likely not. The critical sizes at the lower nucleation temperatures calculated using the classical theory are of order only a few atoms, making the assumed work of cluster formation (Eq. (19)) clearly wrong. As discussed several times up to now, it is essential that the diffuse nature of the interface be considered in this limit. While that is readily done within the kinetic formalism of the coupled-flux model, it has not yet been attempted. The interfacial free energies, and now the proportionality factor, α, which accounts for the annihilation of the self-interstitials, are only fitting parameters. More microscopic information is required to go beyond these arguments, but a final population of nuclei is not sufficient. The 3DAP studies discussed in Chapter 9, Section 6 are clearly one approach although, for ensemble evolution, it is difficult to measure the kinetics that is needed to compare with nucleation theories. This may be possible from studies of oxygen precipitation in silicon, however.

Electrically active centers form upon annealing oxygen-rich silicon at 300–550°C [60–62]. These TDs are generally assumed to be small clusters of oxygen, containing only a few atoms. Early studies of TD kinetics used electrical resistivity to compute the concentration; they concluded that all TDs were of one specific type, identified as an SiO_4 donor complex. It is now known that donor centers of several types evolve sequentially with annealing time [63]. The measured TD populations show maxima with sufficiently long annealing time. The rate of TD generation and the maximum TD population, along with the accompanying depletion in free interstitial oxygen, O_i, are strong functions of the initial oxygen concentration and the annealing temperature. Initial experiments at 450°C indicated that the formation rate of thermal donors scales with the fourth power of the oxygen concentration, while the TD population scales with the third power [61]. Recent studies confirm this, but also indicate that the concentration dependence of the rate of TD formation is a strong function of the temperature, scaling with the oxygen concentration squared for $T < 400°C$, but increasing to the eighth power of the oxygen concentration at 500°C.

It is tempting to identify TD formation as an early stage of cluster formation during oxygen precipitation. If true, the formation kinetics and concentration dependences should be predicted by a quantitative model for time-dependent nucleation. Initial calculations made using parameters determined from the oxygen-precipitation studies support this [55]. Studies of TD formation, then, may provide an unprecedented opportunity to refine nucleation models, providing new experimental data for the evolution of small clusters.

4. SUMMARY

In this chapter, we have briefly examined two cases of nucleation in crystalline solids. There are some similarities with nucleation processes in glasses, including the possible importance of transient nucleation. But, there are also important differences. Key points include:

- Long-range diffusion is critically important in precipitation processes where the new phase has a different chemical composition from the parent phase.
- In glasses, stress due to the difference in density between the initial and final phases is typically assumed to relax on the same time scale as

cluster growth, making it of little consequence. Stress effects can be critical in nucleation processes in the crystalline state, however.

- Defects such as vacancies and interstitial atoms are important in solid-state crystallization processes, coupling to nucleation in complex ways.
- Nucleation data for cobalt precipitation in Cu–Co alloys are reasonably well fit by the classical theory of nucleation. The measured nucleation prefactor, however, is four to six orders of magnitude lower than predicted. The reasons for this are as yet unclear.
- Coarsening simultaneous with nucleation and growth can be significant in solid-state precipitation.
- The nucleation kinetics of oxygen precipitates in Czochralski silicon cannot be fit to the classical theory of nucleation. This is an example of a coupling between long-range diffusion and interfacial attachment. The nucleation kinetics is fit well by the coupled-flux model (Chapter 5) when fluxes of vacancies and interstitial silicon atoms are considered.
- Thermal donors in silicon are believed to be small oxygen clusters. The kinetics of their evolution, then, may provide direct information on the behavior of the smallest clusters, allowing a more quantitative study of the evolution of the cluster size distribution.

The two examples of solid-state precipitation for this chapter were chosen to be sufficiently simple that they could be analyzed within the analytical theories developed in Chapters 2–5. The increasing use of atomistic probes such as 3DAP, however, has indicated that the nucleation process may be much more complex than previously thought, with possible precursor clusters and simultaneous nucleation, growth, and coarsening processes. The analysis becomes even more difficult for decomposition processes in multicomponent alloys, where, for instance, the diffusion mechanism and correlation effects and the range of the vacancy–solute interactions become critical [64]. An analytical approach is incapable of handling the complexity of such processes; understanding is increasingly based on computer simulations [5, 64–66], as considered in the next chapter.

REFERENCES

[1] H.I. Aaronson, F.K. LeGoues, An assessment of studies on homogeneous diffusional nucleation kinetics in binary metallic alloys, Metall. Trans. A **23A** (1992) 1915–1945.

[2] I.S. Servi, D. Turnbull, Thermodynamics and kinetics of precipitation in the copper–cobalt system, Acta Metall. **14** (1966) 161–169.

[3] S. Esmaeili, D. Vaumousse, M.W. Zandbergen, W.J. Poole, A. Cerezo, D.J. Lloyd, A study on the early-stage decomposition in the Al–Mg–Si–Cu alloy AA6111 by electrical resistivity and three-dimensional atom probe, Philos. Mag. **87** (2007) 3797–3816.

[4] F.K. LeGoues, H.I. Aaronson, Influence of crystallography upon critical nucleus shapes and kinetics of homogeneous FCC–FCC nucleation. IV. Comparisons between theory and experiment in Cu–Co alloys, Acta Metall. **32** (1984) 1855–1864.

[5] A. Cerezo, S. Hirosawa, I. Rozdilsky, G.D.W. Smith, Combined atomic-scale modelling and experimental studies of nucleation in the solid state, Phil. Trans. Roy. Soc. Lond. **361** (2003) 463–477.
[6] N.F. Mott, H. Jones, *The Theory of the Properties of Metals and Alloys*, Dover Publications, Inc., New York (1936) pp. 286–296
[7] A. Davidson, M. Tinkham, Phenomenological equations for the electrical conductivity of microscopically inhomogeneous materials, Phys. Rev. B **13** (1976) 3261–3267.
[8] J.W. Christian, *The Theory of Transformations in Metals and Alloys*, Pergamon Press, Oxford (1975) pp. 534–542
[9] H.K. Hardy, T.J. Heal, Report on precipitation, Prog. Metal. Phys. **5** (1954) 143–278.
[10] M.J. Stowell, Precipitate nucleation: does capillarity theory work?, Mater. Sci. Technol. **18** (2002) 139–144.
[11] R. Döhl, M.-P. Macht, V. Naundorf, Measurement of the diffusion coefficient of cobalt in copper, Phys. Stat. Sol. A **86** (1984) 603–612.
[12] H.P. Degischer, Diffraction contrast from coherent precipitates in elastically-anisotropic materials, Philos. Mag. **26** (1972) 1137–1151.
[13] C.G. Richards, W.M. Stobbs, Determination of the in situ misfits of coherent particles, J. Appl. Cryst. **8** (1975) 226–228.
[14] W. Wagner, The size of critical nuclei in Cu-rich CuCo, J. Phys. F: Metal Phys. **16** (1986) L239–L243.
[15] H. Wendt, P. Haasen, Atom probe field ion microscopy of the decomposition of Cu–2.7at%Co, Scripta Metall. **19** (1985) 1053–1058.
[16] B. Frauhauf, W. Gust, R. Kampmann, B. Predel, E. Wachtel, R. Wagner, (unpublished research (1990)), see refs. [1] and [29].
[17] R. Bormann, (unpublished research (1990)), see refs. [1] and [29].
[18] J.W. Cahn, J.E. Hilliard, Free energy of a nonuniform system I. Interfacial free energy, J. Chem. Phys. **28** (1958) 258–267.
[19] J.W. Cahn, Free energy of a nonuniform system. II. Thermodynamic basis, J. Chem. Phys. **30** (1959) 1121–1124.
[20] J.W. Cahn, J.E. Hilliard, Free energy of a nonuniform system. III. Nucleation in a two-component incompressible fluid, J. Chem. Phys. **31** (1959) 688–699.
[21] F.K. LeGoues, Y.W. Lee, H.I. Aaronson, Influence of crystallography upon critical nucleus shapes and kinetics of homogeneous FCC–FCC nucleation. II. The non-classical regime, Acta Metall. **32** (1984) 1837–1843.
[22] H.E. Cook, D. deFontaine, On the elastic free energy of solid solutions—I microscopic theory, Acta Metall. **17** (1969) 915–924.
[23] H.E. Cook, D. deFontaine, On the elastic free energy of solid solutions—II. Influence of the effective modulus on precipitation from solution and the order–disorder reaction, Acta Metall. **19** (1971) 607–616.
[24] H.E. Cook, D. deFontaine, J.E. Hilliard, A model for diffusion on cubic lattices and its application to the early stages of ordering, Acta Metall. **17** (1969) 765–773.
[25] F.K. LeGoues, H.I. Aaronson, Y.W. Lee, Influence of crystallography upon critical nucleus shapes and kinetics of homogeneous FCC–FCC nucleation—III. The influence of elastic strain energy, Acta Metall. **32** (1984) 1845–1853.
[26] W. Wagner, Early stages of precipitation in dilute CuCo alloys, Z. Metallk. **80** (1989) 873–882.
[27] M.F. Chisholm, *Spinodal Nucleation in Binary Alloys*, Dissertation, Carnegie Mellon Univ., Pittsburgh (1986). Available from Diss. Abstr., Int. B **47**(10) (1987) 4268.
[28] W. Wagner, J. Piller, H.-P. Degischer, H. Wollenger, Comparative FIM, SANS and TEM study of the Cu–2 at.%Co decomposition, Z. Metallk. **76** (1985) 693–700.
[29] W. Wagner, The influence of precursor fluctuations on the kinetics of α-Co precipitation in dilute CuCo alloys, Acta Metall. Mater. **38** (1990) 2711–2719.

[30] R.P. Setna, J.M. Hyde, A. Cerezo, G.D.W. Smith, M.F. Chisholm, Position sensitive atom probe study of the decomposition of a Cu–2.6 at.% Co alloy, Appl. Surf. Sci. **67** (1993) 368–379.
[31] X. Jiang, W. Wagner, H. Wollenberger, FIM-AP investigation of the early stage decomposition in a Cu–0.8 at.% Co alloy, Z. Metallk. **82** (1991) 192–197.
[32] T. Ebel, R. Kampmann, W. Wagner, Investigation of the nucleation stage in a Cu–0.8 at. % Co alloy by means of polarized neutron scattering, Phys. B **180–181** (1992) 357–358.
[33] K.C. Russell, Linked flux analysis of nucleation in condensed phases, Acta Metall. **16** (1968) 761–769.
[34] K.F. Kelton, Kinetic model for nucleation in partitioning systems, J. Non-Cryst. Sol. **274** (2000) 147–154.
[35] K.F. Kelton, Time-dependent nucleation in partitioning transformations, Acta Mater. **48** (2000) 1967–1980.
[36] R. Hattenhauer, P. Haasen, The decomposition kinetics of Cu–1 at.% Co at 823 K, studied by bright-field-zone-axis-incidence transmission electron microscopy, Philos. Mag. A **68** (1993) 1195–1213.
[37] K.C. Russell, Nucleation in solids: the induction and steady state effects, Adv. Colloid Inter. Sci. **13** (1980) 205–318.
[38] W. Lin, The incorporation of oxygen into silicon crystals, in: *Oxygen in Silicon*, Ed. F. Shimura, Academic Press, Boston (1994), pp. 9–52 .
[39] R.C. Newman, Oxygen diffusion and precipitation in Czochralski silicon, J. Phys.: Condens. Matter **12** (2000) R335–R365.
[40] T.Y. Tan, E.E. Gardner, W.K. Tice, Intrinsic gettering by oxide precipitate induced dislocations in Czochralski Si, Appl. Phys. Lett. **30** (1977) 175–176.
[41] W.J. Patrick, S.M. Hu, W.A. Westforp, The effect of SiO_2 precipitation in Si on generation currents in MOS capacitors, J. Appl. Phys. **50** (1979) 1399–1402.
[42] S.M. Hu, W.J. Patrick, Effect of oxygen on dislocation movement in silicon, J. Appl. Phys. **46** (1975) 1869–1874.
[43] S.M. Hu, Dislocation pinning effect of oxygen atoms in silicon, Appl. Phys. Lett. **31** (1977) 53–55.
[44] A. Borghesi, B. Pival, A. Sassella, A. Stella, Oxygen precipitation in silicon, J. Appl. Phys. **77** (1995) 4169–4244.
[45] R.C. Newman, Diffusion of oxygen in silicon, in: *Oxygen in Silicon*, Ed. F. Shimura, Academic Press, Boston (1994), pp. 290–352.
[46] Y. Shimanuki, H. Furuya, I. Suzuki, K. Maurai, Effects of thermal history on microdefect formation in Czochralski silicon crystals, Jap. J. Appl. Phys. **24** (1985) 1594–1599.
[47] G. Fraundorf, P. Fraundorf, A. Craven, A. Frederick, J. Moody, R.W. Shaw, The effects of thermal history during growth on O precipitation in Czochralski silicon, J. Electrochem. Soc. **132** (1985) 1701–1704.
[48] K.F. Kelton, R. Falster, D. Gambaro, M. Olma, M. Cornara, P.F. Wei, Oxygen precipitation in silicon—Experimental studies and theoretical investigations within the classical theory of nucleation, J. Appl. Phys. **85** (1999) 8097–8111.
[49] D.G. Schimmel, A comparison of chemical etches for revealing <100> silicon crystal defects, J. Electrochem. Soc. **123** (1976) 734–741.
[50] S. Senkader, P.R. Wilshaw, R.J. Falster, Oxygen-dislocation interactions in silicon at temperatures below 700°C: dislocation locking and oxygen diffusion, J. Appl. Phys. **89** (2001) 4803–4808.
[51] N. Inoue, K. Watanabe, K. Wada, J. Osaka, Time lag in nucleation of oxide precipitates in silicon due to high temperature preannealing, J. Cryst. Growth **84** (1987) 21–35.
[52] T.Y. Tan, W.J. Taylor, Mechanisms of oxygen precipitation: some quantitative aspects, in: *Oxygen in Silicon*, Ed. F. Shimura, Academic Press, Boston (1994), pp. 353–390.

[53] R.A. Craven, Oxygen precipitation in Czochralski silicon, in: *Semiconductor Silicon*, Ed. H.R. Huff, Proceedings Electrochem. Soc., New York, **81–85** (1981) p. 254–271.
[54] P.F. Wei, K.F. Kelton, R. Falster, Coupled-flux nucleation modeling of oxygen precipitation in silicon, J. Appl. Phys. **88** (2000) 5062–5070.
[55] K.F. Kelton, Diffusion-influenced nucleation: a case study of oxygen precipitation in silicon, Phil. Trans. Roy. Soc. Lond. **361** (2003) 429–446.
[56] J. Vanhellemont, C. Claeys, A theoretical study of the critical radius of precipitates and its application to silicon oxide in silicon, J. Appl. Phys. **62** (1987) 3960–3967.
[57] J. Vanhellemont, C. Claeys, Erratum: A theoretical study on the critical radius of precipitates and its application to silicon oxide in silicon, J. Appl. Phys. **71** (1992) 1073.
[58] T.Y. Tan, U. Gösele, Point defects, diffusion processes, and swirl defect formation in silicon, Appl. Phys. A **A37** (1985) 1–17.
[59] K.H. Yang, H.F. Kappert, G.H. Schwuttke, Minority carrier lifetime in annealed silicon crystals containing oxygen, Phys. Stat. Sol. A **50** (1978) 221–235.
[60] C.S. Fuller, J.A. Ditzenberger, N.B. Hannay, E. Buehler, Diffusivity and solubility of copper in germanium, Phys. Rev. **93** (1954) 1182–1189.
[61] W. Kaiser, H.L. Frisch, H. Reiss, Mechanism of the formation of donor states in heat-treated silicon, Phys. Rev. **112** (1958) 1546–1554.
[62] J. Michel, L.C. Kimerling, Electrical properties of oxygen in silicon, in: *Oxygen in Silicon*, Ed. F. Shimura, Academic Press, Boston (1994), pp. 251–2289.
[63] M. Claybourn, R.C. Newman, Activation energy for thermal donor formation in silicon, Appl. Phys. Lett. **51** (1987) 2197–2199.
[64] Z.G. Mao, C.K. Sudbrack, K.E. Yoon, G. Martin, D.N. Seidman, The mechanism of morphogenesis in a phase-separating concentrated multicomponent alloy, Nat. Mater **6** (2007) 210–216.
[65] F. Soisson, G. Martin, Monte Carlo simulations of the decomposition of metastable solid solutions: Transient and steady-state nucleation kinetics, Phys. Rev. B **62** (2000) 203–214.
[66] C.K. Sudbrack, R.D. Noebe, D.N. Seidman, Compositional pathways and capillary effects during isothermal precipitation in a nondilute Ni–Al–Cr alloy, Acta Mater. **55** (2007) 119–130.

CHAPTER 10

Computer Models

Contents

1. INTRODUCTION

As is evident in Chapters 7–9, experiments that can unambiguously test nucleation theories are generally difficult to design and to perform. Even when data are available, their interpretation is complicated by an incomplete knowledge of the atomic interactions and mobilities for the material studied. As discussed in Chapter 4, even the structures of the cluster and of the interface between the cluster and the parent phase can be uncertain. Computer simulations, based on known atomic interactions and dynamics, can be used to "experimentally" test nucleation theories and to identify relevant parameters that must be incorporated into revised theories. Only in the early 1970s did available computing power become adequate to begin to realistically treat phase transitions. However, even with the calculation capabilities currently available, the ensembles that can be considered remain relatively small and the potential interactions are often idealized, frequently yielding only qualitative conclusions. Collectively, these results demonstrate unambiguously that under appropriate

Pergamon Materials Series, Volume 15
ISSN 1470-1804, DOI 10.1016/S1470-1804(09)01510-7

conditions, supercooled liquids and supersaturated vapors change phase by a fluctuation-based mechanism of nucleation and growth, verifying a fundamental assumption of all realistic nucleation theories. As suggested by the density-functional calculations in Chapters 4, Section 4, computer studies demonstrate that the assumption of a sharp interface between the cluster and the original phase is incorrect. They also support many other conclusions from those treatments, in particular showing that the structures of the first small clusters to form often are different from that of the final phase. In this chapter, some salient features of these calculations are discussed briefly.

2. OVERVIEW OF COMPUTER METHODS

The time evolution of an initial system configuration is generally simulated using one of two methods: molecular dynamics (MD) or Monte Carlo. Both are in principle easy to understand and straightforward to apply, but limited computing resources have required the development of specialized techniques to allow a comparison of the computed results with experimental data. A discussion of these techniques is beyond the scope of this chapter, which provides a very brief view of how MD and Monte Carlo calculations have led to further insight into nucleation processes. More information on the methods can be found in Refs. [1–3].

2.1 Molecular dynamics

The concept of MD is clear. The time evolution of an ensemble of interacting point particles is obtained by numerically solving Newton's equations of motion. The evolution of the ensemble is described as a trajectory in phase space, characterized by a set of M independent generalized coordinates, q_i, and corresponding velocities, $\dot{q}_i \equiv \mathrm{d}q_i/\mathrm{d}t$, that describe the state of the system. For a conservative system, like this one, the forces are given by the derivatives of the potential energy. Hamilton's variational principle states that the trajectory will be the one for which $\int \mathscr{L}(\{q_i\}, \{\dot{q}_i\}, t)\,\mathrm{d}t$ is an extremum, where $\mathscr{L}$, the Lagrangian, (equal to the kinetic energy minus the potential energy) is the solution of Lagrange's equations of motion,

$$\frac{\mathrm{d}}{\mathrm{d}t}\left(\frac{\partial \mathscr{L}}{\partial \dot{q}_i}\right) - \frac{\partial \mathscr{L}}{\partial q_i} = 0 \quad \text{for } i = 1, \ldots, M. \tag{1}$$

The Hamiltonian, $\mathscr{H}$, is defined in terms of the generalized coordinates, q_i, and the generalized momenta, $p_i = \partial \mathscr{L}/\partial \dot{q}_i$,

$$\mathscr{H} = \sum_i \dot{q}_i p_i - \mathscr{L}. \tag{2}$$

It can be shown that the equations of motion are

$$\dot{q}_i = \frac{\partial \mathcal{H}}{\partial p_i} \quad \text{and} \quad \dot{p}_i = -\frac{\partial \mathcal{H}}{\partial q_i}, \tag{3}$$

and that $\mathcal{H}$ is a conserved quantity (the total energy) if it does not contain an explicit time dependence [4]. By numerically integrating these equations of motion, MD methods allow the trajectory through phase space to be observed directly and all of the thermally accessible configurations of the ensemble to be explored.

While MD simulations can model phase transitions closely, they require large computer resources to simulate macroscopic systems. The computational requirements typically limit the ensemble sizes to 10^4–10^6 molecules and the small time steps required (10^{-15}–10^{-14} s) limit the total simulation times to of order 10^{-8} s, making it difficult to have direct comparisons with experimentally measured nucleation rates. Further, as in experimental investigations, heterogeneous nucleation on container walls can be significant, although this can be avoided by assuming periodic boundary conditions. Finally, the short simulation times generally ensure that the nucleation rates obtained from MD calculations are time-dependent, as discussed in more detail in Section 4.

2.2 Monte Carlo

The random nature of the Monte Carlo technique differs from the deterministic approach of MD. It is based on a stochastic process that produces a Markovian chain of configurations that are Boltzmann-weighted. The average potential energy can be defined as

$$\langle U \rangle = \int U(\xi) P(\xi)\, \mathrm{d}\xi, \tag{4}$$

where $U(\xi)$ is a function of the possible particle configurations of the system, ξ, each occurring with probability, $P(\xi)$,

$$P(\xi) = \frac{\exp(-U(\xi)/k_\mathrm{B}T)}{\int \exp(-U(\xi)/k_\mathrm{B}T)\,\mathrm{d}\xi} = \frac{\exp(-U(\xi)/k_\mathrm{B}T)}{Q}, \tag{5}$$

where Q is the canonical partition function, and k_B and T have their usual meaning. Equation (4) can be solved directly by randomly selecting all possible configurations, evaluating $P(\xi)$ and $U(\xi)$ for each, and summing their products. The enormous number of possible configurations, however, makes this straightforward approach impractical. A faster calculation method, proposed by Metropolis et al., is to select configurations with a frequency determined by $P(\xi)$ and compute $\langle U \rangle$ by adding the energy from each selected configuration without weighting [5].

A calculation of the probability for a given configuration requires that Q be known (Eq. (5)). However, since Monte Carlo models generate new state configurations from previous ones, only the energy differences between the original and new states are needed to compute the transition probabilities. The relative probability for generating the mth state from the nth state is

$$\frac{P_n(t)}{P_m(t)} = \frac{e^{-E_n/k_B T}/Q}{e^{-E_m/k_B T}/Q} = e^{-\Delta E/k_B T}, \tag{6}$$

where E_n and E_m are the energies of the nth and mth states respectively and $\Delta E = E_n - E_m$.

Monte Carlo methods are widely used to study phase transitions in lattice models such as the *Ising model*. In the simplest Ising model, spins sit on a square or triangular lattice and have only nearest-neighbor interactions. The spins can have only two states, $s_i = \pm 1$, analogous to the spins pointing up or down. The energies of the system are computed from the interactions of the spins with each other and with an applied field, which breaks the symmetry. The field can be varied to change the thermodynamic properties of the system, analogous to changing the driving free energy. The transition rates between states must be chosen so that they satisfy detailed balance, but there is considerable latitude in choosing their exact form. For example, for the transition rate from state n to state m, $W_{n \to m}$, Metropolis et al. choose

$$\begin{aligned} W_{n\to m} &= \frac{e^{-\Delta E/k_B T}}{\tau_0} \qquad \Delta E > 0 \\ &= \frac{1}{\tau_0} \qquad \Delta E < 0. \end{aligned} \tag{7}$$

where τ_0 is the characteristic time required to attempt a spin flip on a lattice. In two or more dimensions, the Ising model undergoes a phase transition between a disordered phase at high temperature and an ordered one at low temperature, providing, for example, a simplified model of ferromagnetism. Other phase transitions can also be mapped onto lattice models. In particular, Ising models have been used to probe many of the basic assumptions of nucleation theories and have guided the development of new models [6–9].

The calculation steps for an Ising-model calculation can be written in a straightforward way, allowing them to be easily implemented in a computer program. An initial state for the system is chosen, corresponding to some distribution of up and down spins on a lattice of a given type. A systematic sweep is made through the lattice, flipping the spin on each lattice site in turn. The decision whether to keep the new or old spin configuration is based on the change in energy of the system (Eq. (6)).

For illustration, the steps in a Metropolis-based Ising algorithm are given below (from Ref. [2]):

1. Choose a lattice type, size, and initial distribution of spins.
2. Choose each lattice site in turn.
3. Calculate the change in energy ΔE, if the sign of the spin on that site is reversed.
4. Generate a random number $0 < R_n < 1$.
5. If $R_n < \exp(-\Delta E/k_B T)$, accept the spin reversal; if not, retain the original spin orientation.
6. Move to the next site and perform steps 3–4, continuing until the sweep through the lattice sites is complete.

In the following sections, a few key results for nucleation, obtained from computer models, are presented.

3. STEADY-STATE NUCLEATION

Monte Carlo, Ising, and molecular-dynamics simulations have demonstrated that nucleation processes are indeed stochastic, have illuminated the limits of the classical theory of nucleation, and have provided valuable information about cluster structure. Of the many simulation investigations that have been performed, only a few are selected here to illustrate points widely supported by other studies. Most assume a Lennard-Jones (LJ) potential (see Chapter 4, Eq. (93)), frequently choosing parameters appropriate for argon. Although such a central potential is well suited only for studies of nucleation in inert gases, it serves to illustrate universal properties of nucleation without the complexities involved in the use of more realistic interatomic potentials. Most of the conclusions discussed in this chapter are drawn from simulations using LJ potentials or spin/magnetic field interactions.

3.1 The validity of the stochastic model

As discussed in Chapters 1 and 2, since the time of Gibbs it has been generally believed that nucleation is inherently a stochastic process; this can now be directly demonstrated by simulations. A key concern in any computer calculation is the unambiguous identification of a cluster of the new phase. In Ising model calculations, the "phase" can be defined in terms of the neighboring spins. The type of information that can be obtained from an Ising model with an applied field is illustrated in Figure 1. To aid the visualization, only the spins of the new phase that are aligned with the applied field B are shown. Figure 1a shows a 3D Ising lattice before nucleation has occurred; almost all of the spins are isolated (corresponding to isolated atoms or molecules). Figure 1b shows the development of small clusters of aligned spins, indicating the beginning of nucleation. In the last figure in this sequence (Figure 1c), these clusters have grown to large size.

Figure 2 shows the size of the largest cluster as a function of simulation time. Clusters at different points in the lattice begin to grow rapidly at time steps near 10 and 40. Both subsequently shrink in size,

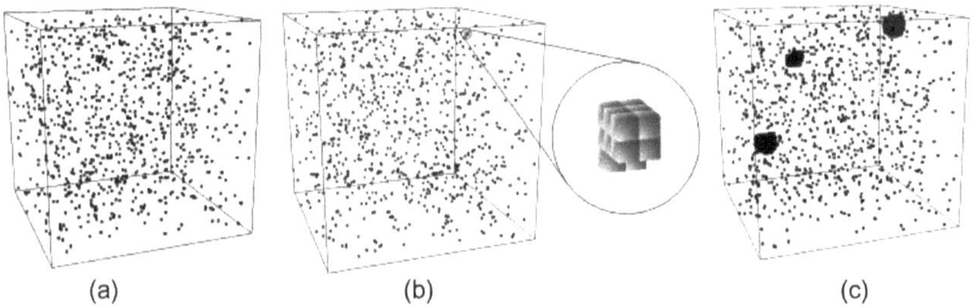

Fig. 1 Three dimensional visualization of nucleation in an Ising model containing 112 × 112 × 112 spins in a normalized applied field of $h = 3/2$, assuming cubic clusters ($h = 2B/k_BT$, see Section 3.2): (a) 11 time steps; (b) 127 time steps; (c) 160 time steps. (Reprinted with permission from Ref. [10], copyright (2000), American Institute of Physics.)

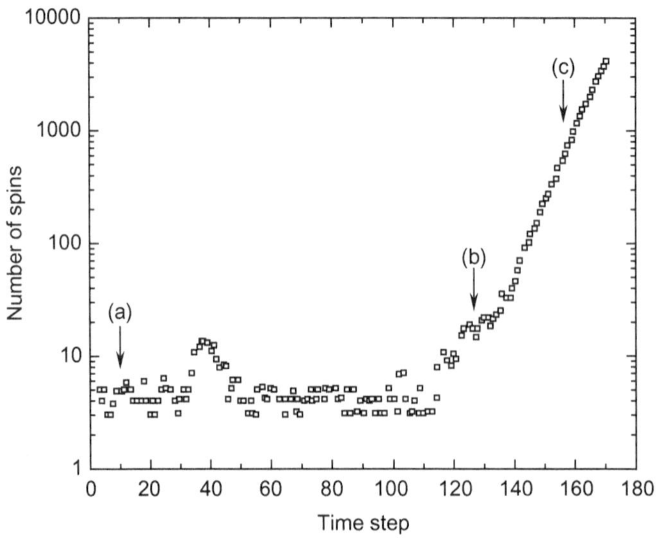

Fig. 2 The size of the largest cluster as a function of time in an Ising-model calculation for a temperature of $T/T_c = 0.25$ and a normalized applied field of $h = 3/2$, assuming cubic clusters ($h = 2B/k_BT$, see Section 3.2). The points (a), (b) and (c) correspond to the three states of the system shown in Figure 1. (Reprinted with permission from Ref. [10], copyright (2000), American Institute of Physics.)

however, indicating that they were not successful in overcoming the nucleation barrier. Another cluster begins to grow after about 120 time steps. Growth is slow at first, but becomes faster after approximately 140 time steps. Such behavior is consistent with the kinetic picture from the classical theory introduced in Chapter 2. The initially slow growth is due to the diffusive nature of cluster evolution for sizes near the critical size, transitioning to a constant growth velocity as the clusters become larger than the critical size and the volume contribution to the work of cluster formation becomes dominant. These calculations demonstrate that clusters fluctuate in size, eventually dissolving unless the fluctuation produces a cluster larger than the critical size, which is biased (although not guaranteed) to grow. Three attempts lead to only one successful case of cluster growth. This behavior demonstrates the inherent stochastic nature of nucleation, the fundamental assumption of all realistic nucleation theories.

The stochastic process of nucleation and the existence of a nucleation barrier (corresponding to a critical size in the classical theory) are

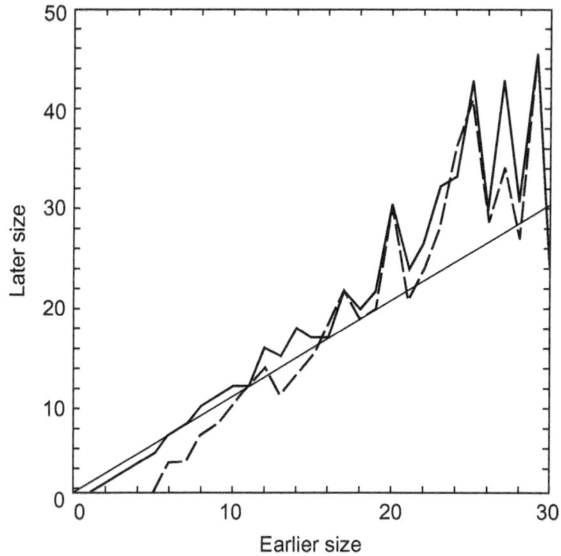

Fig. 3 MD calculation of the mean (solid curve) and median (dashed curve) number of molecules in solid-like clusters in an LJ liquid containing 10^6 molecules. The number of molecules in the cluster is plotted as a function of the number in the cluster at a previous time (5 time-steps earlier). For comparison, a solid line is shown to indicate the case where the number of molecules in the clusters remained constant. The cluster distributions for both the earlier and later times are normalized to the total numbers of clusters of size 1 to 20. (Reprinted figure from Ref. [11], copyright (1990), American Institute of Physics.)

demonstrated in Figure 3, showing results obtained from an MD calculation of a supercooled LJ liquid (an ensemble of 10^6 molecules) [11]. The number of molecules in solid-like clusters at a given time ("later") is plotted against that number five time steps earlier ("earlier"). The solid straight line represents the case where these numbers are the same. The simulation data (dashed curves) show that solid-like clusters containing fewer than ten molecules tend on average to shrink, with the numbers falling below the solid line. Clusters containing more than about 18 molecules tend to grow, with the numbers rising above the solid line. The accuracy of these results does not allow a precise identification of the critical size, but one clearly exists for a cluster containing between 10 and 18 molecules.

3.2 Dependence on driving free energy

The dependence of the nucleation rate on the driving free energy provides a good test of the thermodynamic basis of a nucleation theory. Figure 4 shows the nucleation rate as a function of a normalized field h ($h = 2B/k_BT$), where B is the applied magnetic field); h is the functional analog to $\ln s$ (where s is the supersaturation). The rates were taken to be

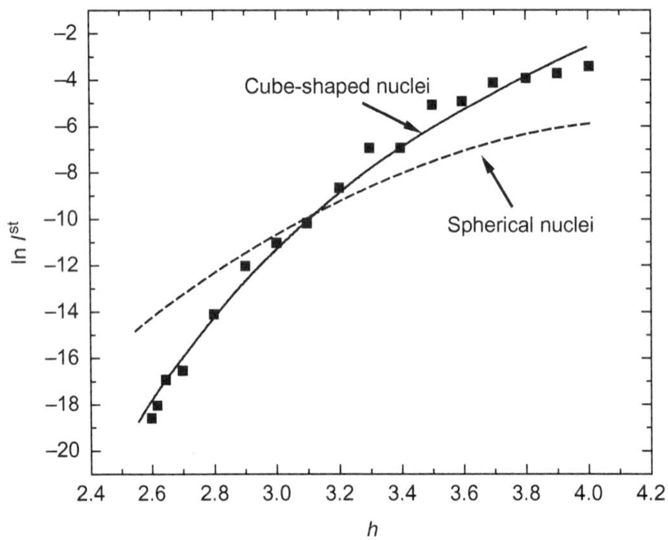

Fig. 4 The nucleation rate as a function of the normalized applied field, h ($= 2B/k_BT$). Fits are shown for spherical and cube-shaped critical clusters, with W^* calculated, respectively, from Eq. (18) in Chapter 2 and Eq. (9) in this chapter, at $T/T_c = 0.25$, where T_c is the critical temperature for the phase transition. The nucleation prefactor A^* was used as a fitting parameter. (Reprinted with permission from Ref. [10], copyright (2000), American Institute of Physics.)

the inverse nucleation time in the MD calculation that produced the data shown in Figures 1 and 2; this is valid as long as the nucleation rate is low. The fits are to the standard expression for the steady-state nucleation rate, I^{st}

$$I^{st} = A^* \exp\left(\frac{-W^*}{k_B T}\right), \tag{8}$$

where A^* is the dynamical prefactor and W^* is the reversible work of forming a critical cluster. As pointed out in the caption for Figure 1, cube-shaped clusters are formed; if the work for spherical clusters is used (Eq. 18, Chapter 2), the agreement between the predictions from Eq. (8) and the simulation results is poor. However, excellent agreement is obtained using the work of formation for cube-shaped clusters,

$$\frac{W^*}{k_B T} = -l^{*3}h + \frac{\sigma O^*}{k_B T} = \frac{2\sigma l^{*2}}{k_B T}, \tag{9}$$

where l^* is a dimensionless edge-length of the critical cluster, σ is the interfacial free energy, and O^* is the total surface area.

Such good agreement is surprising, given the arguments made in previous chapters about fundamental flaws in the thermodynamic model for the work of cluster formation that is used in the classical theory, particularly the assumption of capillarity (i.e. that the cluster of the new phase can be viewed as having a sharp surface). As we shall see later in this chapter, MD calculations support arguments made in previous chapters for a diffuse cluster interface. A close inspection of Figure 4 hints at a deviation of the data from the classical theory prediction at the largest values of h. The critical cluster size decreases and the cluster interface becomes more diffuse with increasing h.

If d ln $A^*/\mathrm{d}h$ is negligible, the number of atoms or molecules in the critical cluster, n^*, can be estimated as a function of h from the nucleation rate data in the simulation by using the nucleation theorem discussed in Chapter 2, Section 11:

$$n^* \approx \frac{\mathrm{d} \ln I}{\mathrm{d}h}. \tag{10}$$

As shown in Figure 5, there is poor agreement between these data and classical-theory predictions (assuming spherical clusters, Chapter 2, Eq. (15)). Much better agreement is obtained for cube-shaped clusters:

$$n^* = l^{*3} = \frac{2W^*}{k_B T h}. \tag{11}$$

This worsens with increasing h (Figure 5), again indicating a failure of the assumption of capillarity for small clusters (here containing fewer than 10 molecules).

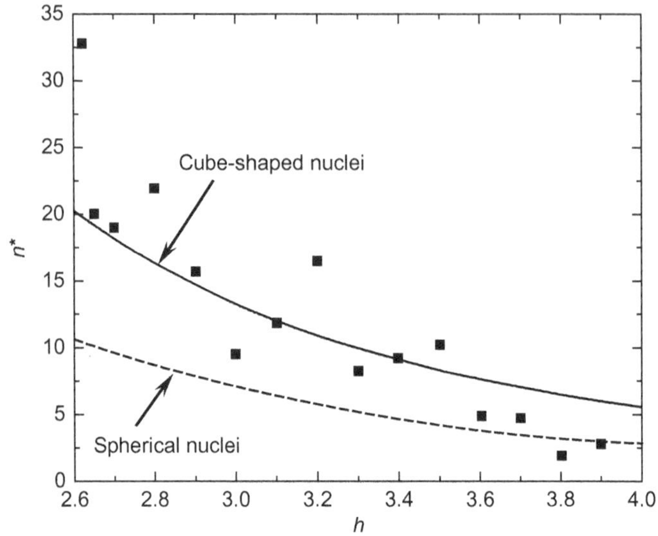

Fig. 5 The critical cluster size (■), as a function of normalized applied field h ($= 2B/k_BT$), extracted from the nucleation rates calculated in an Ising-model simulation (shown in Figure 3). The solid line is a prediction of the classical theory for cube-shaped nuclei and the dashed line is for spherical nuclei. (Reprinted with permission from Ref. [10], copyright (2000), American Institute of Physics.)

3.3 Equilibrium cluster distribution

As discussed in Chapter 2, if the nucleation rate is not high enough to lead to significant depletion of the single-molecule population during the nucleation stage of the phase transition, the cluster size distribution should become time-invariant. This is supported by computer calculations. Figure 6, for example, shows the sizes of solid-like clusters obtained in an MD calculation for an ensemble containing 10^6 molecules for calculations from 100 to 150 time steps [11]. The distribution is nearly independent of time, consistent with it being in steady-state. More surprisingly, however, it approximately follows the exponential distribution expected for clusters significantly smaller than the critical size, i.e. $N(n) \approx N_0 \exp(-W(n)/k_BT)$ for $n \leq n^*$, where $N(n)$ is the number of clusters containing n molecules, $W(n)$ is the corresponding work of cluster formation and N_0 is the number of single molecules in the original phase (10^6, here) (see Chapter 2, Eq. (12)). Fits to this expression give a critical size between 11 and 18 molecules, depending on which cluster size is taken as the lower limit of the distribution; this is consistent with the size range expected from Figure 3. Fits could not be made for clusters containing fewer than about three molecules, consistent with the discussion in Chapter 4 regarding the invalidity of the capillarity

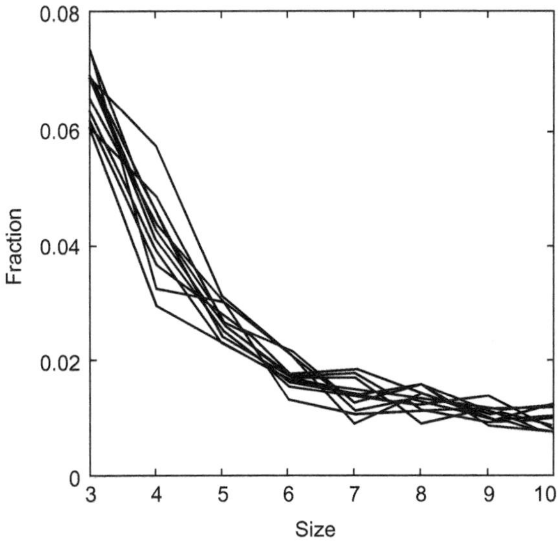

Fig. 6 Solid-like cluster size distribution for times ranging from 100 to 150 time-steps for an MD calculation in an ensemble of 10^6 molecules. The weak time dependence is consistent with a steady-state distribution. The cluster distributions for both the earlier and later times are normalized to the total numbers of clusters of size 1 to 20. (Reprinted figure from Ref. [11], copyright (1990), American Physical Society.)

expression for such small sizes, i.e. the assumption that the interface between a cluster of the new phase and the original phase is sharp. However, it is remarkable that capillarity works as well as it does down to such small cluster sizes.

3.4 Nucleation prefactor

Much of the discussion in this book has centered on validating the thermodynamic model of nucleation and formulating more accurate expressions for the work of critical cluster formation W^*. Little has been said about the prefactor A^*, which cannot be determined from macroscopic considerations [12]. This approach is reasonable since the nucleation prefactor is a weak function of the nucleation barrier. Furthermore, even an order-of-magnitude error in A^* would be difficult to discern from experimental data, which generally have uncertainties of this magnitude. The inadequate knowledge of A^* does, however, lead to significant difficulties in deciding between competing nucleation theories. The uncertainty in the value of the work of cluster formation has led to assertions regarding the prefactor that have been difficult to evaluate. A dramatic example is the Lothe–Pound correction [13–15], where it was argued that, when applied to vapor condensation, the kinetic formulation of the classical theory ignored important contributions to the free energy from cluster translations and rotation. If these were included, the correct prefactor was argued to be of order 10^{17} larger than the classical one. It is now generally accepted that this view is incorrect.

Ising models have been particularly useful for studies of A^* (e.g. [8, 16, 17]). They show, for example, that the prefactor is not independent of cluster size [7], but for a 2D Ising model near the critical temperature (T_c) $A^* \propto n^{-v}$, where the exponent $v \approx 2.1$ [6]. The value of v is smaller at lower temperatures [7,18,19]. As discussed in Chapter 4, Section 5.2, based on a field-theoretic treatment, Langer also suggested a nonclassical prefactor.

The effects of cluster size can be observed from a recent 2D square-lattice Ising-model calculation [20]. The nucleation prefactor as a function of cluster size, $A^*(n)$, obtained by fitting the simulated cluster distributions,

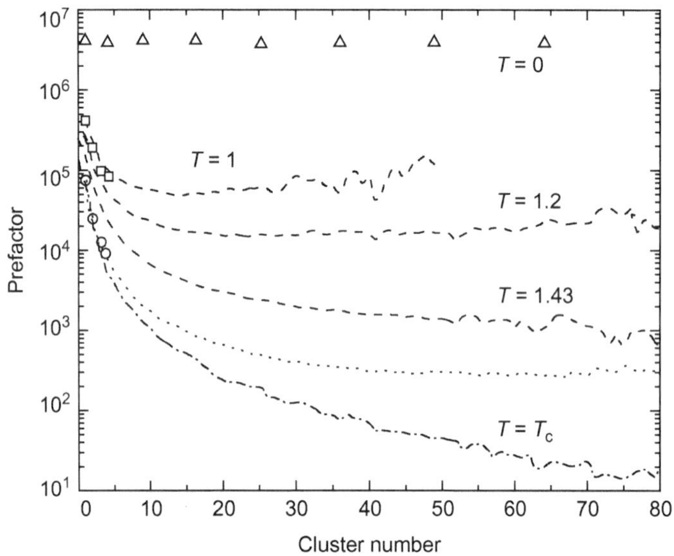

Fig. 7 Nucleation prefactors for a 2D Ising-model calculated from Eq. (30) of Ref. [20] (lines). Ising-model simulation results for $T = 1$ (□) and $T = T_c$ (○) for $n \leq 4$ (T_c is the critical temperature for the phase transition). The classical value (△) for $T = 0$ is independent of cluster size. The dotted line is the pre-exponential for $T = T_c$ for a *reduced supercooling of* −0.1 (see Eq. (5) of Ref. [20]). The deviation from the lower curve (no *reduced supercooling*) is due to coagulation of the clusters on the lattice. (Reprinted with permission from Ref. [20], copyright (1999), American Institute of Physics.)

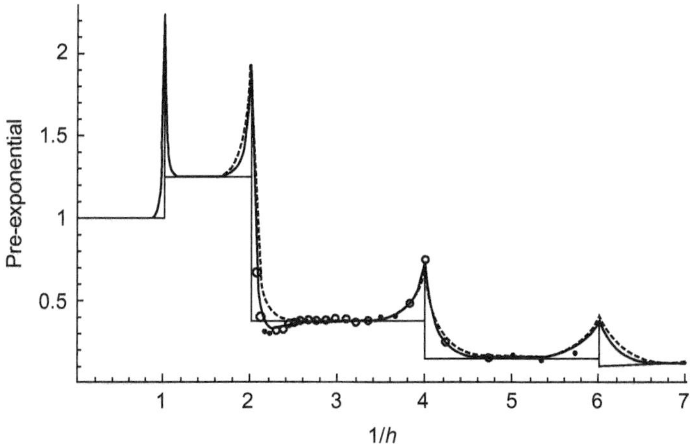

Fig. 8 Preexponential at $T = 0$ (linear segments) and at $T = 0.2J/k_B$, where J is the interaction energy between two neighboring spins on a two-dimensional spin lattice (dashed curve and solid lines correspond to different energy approximations). (Reprinted figure from Ref. [21], copyright (2003), American Physical Society.)

differs from that expected from the classical theory, falling far below the expected value and showing a significant cluster-size dependence (Figure 7). The deviation from the classical theory values increases as the temperature is increased toward the critical temperature. The data at T_c are contaminated by cluster coagulation (since the critical size is now so large); a corrected curve (dotted line) is shown. Clearly, these data indicate that the prefactor predicted by the classical theory is in error.

A very interesting feature has emerged from studies at high driving free energies (corresponding to strong applied fields). Calculations indicate the presence of sharp peaks in the low-temperature prefactor as a function of the field (Figure 8). So far, this has been identified only in Ising calculations; it is doubtful whether it could be observed in experimental data. However, since this is a recent result, further studies are needed.

4. TIME-DEPENDENT NUCLEATION RATE

The nucleation rates of the crystal phase computed for a molecular-dynamics simulation of an ensemble of 10^4 molecules interacting through a LJ potential are shown as a function of driving free energy in Figure 9 [22]. In these studies, the ensemble was first equilibrated in the stable liquid region of the phase diagram and then cooled below the melting temperature. The nucleation rates were deduced in a manner analogous

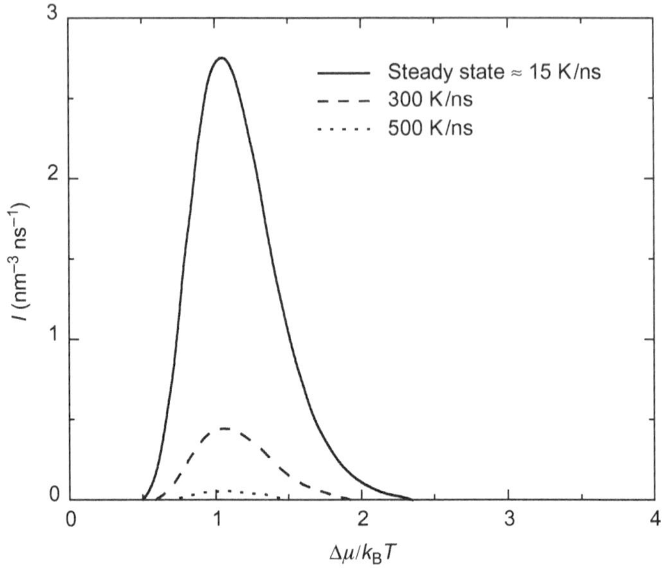

Fig. 9 Time-dependent and steady-state nucleation rates as a function of chemical potential difference at a pressure of 3000 atm. assuming a LJ potential for three different cooling rates. (Reprinted from Ref. [22], copyright (2000), American Physical Society.)

to that used in the study of supercooled liquids (Chapter 7), derived from the probability of finding at least one nucleus at a given time $P(t)$,

$$P(t) = 1 - P(0, t) = 1 - \exp\left(-V \int_0^t I(t')\,\mathrm{d}t'\right), \tag{12}$$

where $P(0,t)$ is the probability that there are no nuclei, $I(t)$ is the time-dependent nucleation rate, and V is the sample volume.

For comparison, the steady-state nucleation rates were also predicted from the classical theory using driving free energies as a function of temperature that were derived from the measured chemical potentials, and rates of single-molecule addition to the critical cluster that were derived from measurements of the crystal growth rate. The interfacial free energy was assumed to scale with the enthalpy of fusion. The agreement between these rates and those obtained from the MD simulation is poor, unless the cooling rate in the simulation is sufficiently low. This is very similar to the effect of the quenching rate on nucleation in glasses (Chapter 8). It again emphasizes the occurrence of a nonsteady-state or time-dependent nucleation rate when the driving free energy for nucleation is changed so rapidly that the cluster size distribution cannot relax fast enough to attain the steady-state form for the actual temperature. Such transient effects are clearly manifest in the nucleation rates derived from the MD simulation as a function of cooling rate. As shown in Figure 9, the nucleation rate is approximately equal to the steady-state rate for a cooling rate of 15 K/ns. The measured nucleation rate falls further and further below the calculated steady-state value as the cooling rates increase (300 and 500 K/ns).

Ising model calculations support these conclusions [23]. Figure 10, for example, shows the number of postcritical clusters appearing as a function of time. The size and time dependencies are identical to those found for classical-theory calculations (Chapter 3, Figs. 3 and 4) and in experimental data obtained from nucleation experiments in glasses (Chapter 3 Fig. 1 and Chapter 8 Figs. 12–14). The rate of cluster production is initially low, indicating that the nucleation rate is time-dependent. The number of clusters eventually increases linearly with time, consistent with steady-state nucleation. The fits are to an asymptotic solution to the Zeldovich–Frenkel equation in the growth regime (see Chapter 2, Section 9). The decrease in the number of clusters with long annealing times is due to cluster coagulation and cannot be described within the classical theory of nucleation. This decrease is reminiscent of that observed in the experimental data for

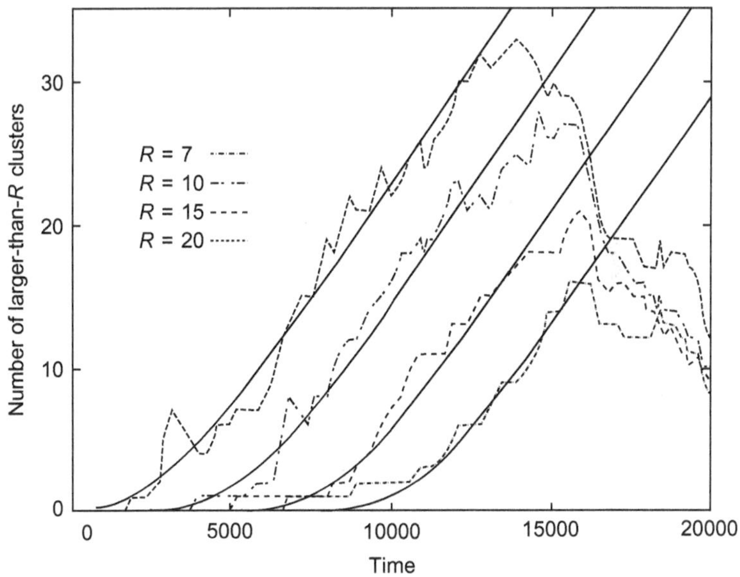

Fig. 10 The number of clusters exceeding the critical size ($r^* = 5$ for the conditions used in the Ising-model calculations). The solid lines are a fit to an asymptotic solution to the Zeldovich–Frenkel equation in the growth regime. The dashed lines are the simulation data. Time is measured in the number of Monte Carlo steps. (Reprinted from Ref. [23], copyright (1999), American Physical Society.)

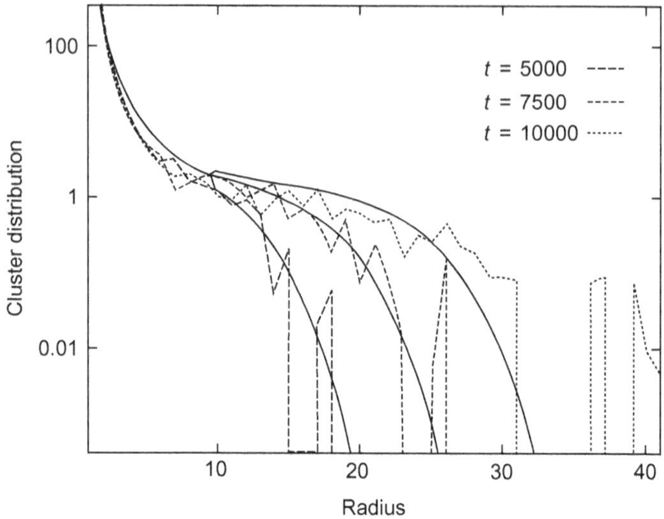

Fig. 11 Population of clusters with radius larger than the critical radius ($r^* \approx 5$) as a function of time. (Radius measured in lattice spacings). The solid lines are fits to asymptotic solutions obtained for classical theory (Chapter 3, Section 4.3); the dashed lines are the Ising-model calculation results. The decrease in cluster population with long annealing time is due to cluster coagulation. (Reprinted from Ref. [23], copyright (1999), American Physical Society.)

precipitation in Cu–Co alloys (Chapter 9, Section 2); there, however, it is due to coarsening of the precipitates.

The time-dependent nucleation rate reflects the time evolution of the cluster distribution. The distribution eventually approaches a time-invariant steady-state value, first for small clusters and extending to larger clusters with longer annealing times (Figure 11). As described for the classical theory in Chapter 3, the distribution grows like a front, at a rate determined by the effective diffusion coefficient in cluster-size space.

As already discussed, experimental measurements of the time-dependent nucleation rates in silicate and metallic glasses (Chapter 8) are in quantitative agreement with predictions from the classical theory, even for complicated preannealing treatments. The good agreement found here with rates from MD simulations further demonstrates that the kinetic approach of the classical theory captures the essential features of the evolution of the cluster size distribution.

5. CLUSTER PROPERTIES

By directly following the fluctuations in the local order parameter within regions of the original phase, computer calculations allow detailed studies of cluster evolution, cluster structure, and the nature of the interface between the original phase and small regions of the new phase. These support many of the arguments made in Chapter 4, elucidating areas where the classical theory of nucleation is inadequate.

One example is illustrated in Figure 12 showing the typical configurations as a function of temperature for a 100-atom cluster during a Monte Carlo simulation of clusters appearing in the vapor

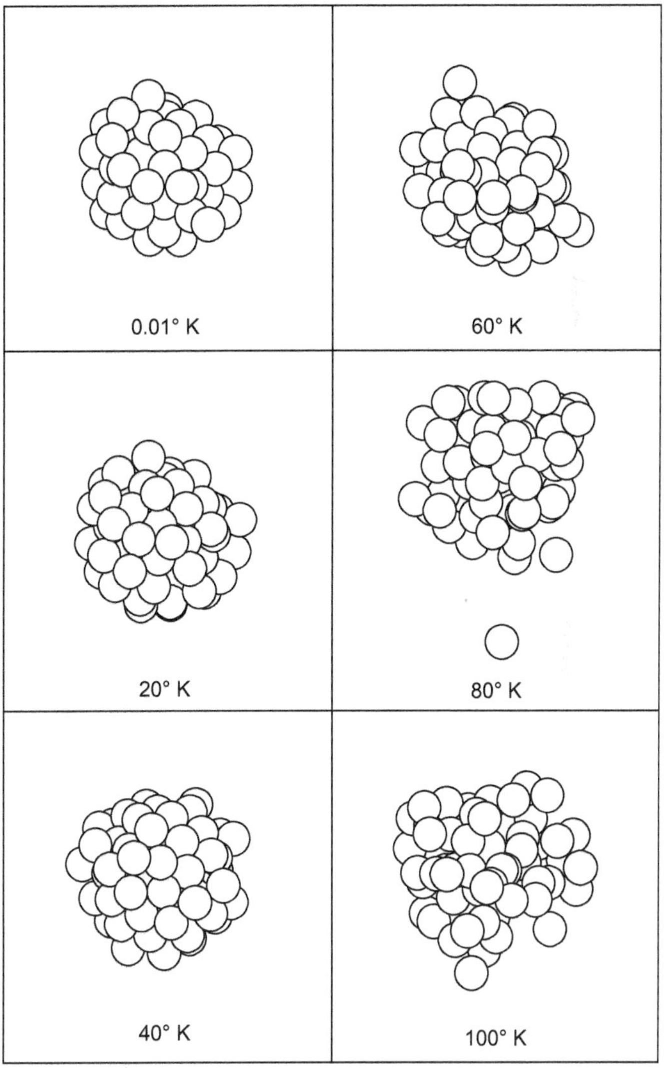

Fig. 12 Computed images of a 100-atom cluster during a Monte Carlo simulation. (Reprinted with permission from Ref. [24], copyright (1973), American Institute of Physics.)

phase of argon, assuming a LJ potential. Below 40 K, the clusters become more crystal-like; at higher temperatures (particularly above the triple point, 84 K), they are more liquid-like. Only at the lowest temperature (nearly absolute zero), is the cluster compact. The diffuseness of the interface increases significantly with increasing temperatures. Computed radial density distributions for these argon clusters also show that the density at the center of small clusters is less than that of the bulk liquid.

At the end of the nineteenth century, Ostwald proposed a "step-rule" (*Stufenregel*) to describe phase evolution during a first-order transformation, arguing that the phase that forms first is not necessarily the stable phase, but the phase that is closest in free energy to the original phase [25]. Over 30 years later, Stranski and Totomanow revised this rule, arguing that the phase that forms first is the one that has the lowest barrier to formation [26], relating phase selection to the work of cluster formation. We have discussed one example of this in Chapter 7, Section 2.4, where the nucleation of a metastable Ti–Zr–Ni icosahedral phase is preferred to the nucleation of the stable C14 Laves phase, because the *i*-phase is structurally more similar to the supercooled liquid, giving it a smaller nucleation barrier.

As expected for a central potential, the stable crystal phase in LJ systems is a ccp crystal. Theoretical calculations, however, have suggested that a metastable bcc phase is the one that should nucleate [27, 28], and this has been observed in experiments on rapidly cooled melts [29]. To observe nucleation in reasonable times, early computer simulations were made at very large supercoolings, of order half the melting temperature [11, 30–35]. These showed the nucleation of both bcc and ccp crystals. This is not surprising since the driving free energies for all possible ordered phases are so large that all nucleation barriers are close to zero. Calculations made closer to equilibrium, using a bond-orientational order parameter analysis, show that the cluster changes character with size. Clusters smaller than the critical size have a bcc-like character, in agreement with theoretical predictions, while clusters equal to or larger than the critical size are more ccp-like [36]. Both the bcc and ccp clusters show significant liquid-like disorder, consistent with the diffuse interface for small clusters discussed in Chapter 4. Interestingly, the type of structural order varies within a given cluster. As shown in Figure 13, ccp order is dominant in the center of the cluster. While bcc order is low in the cluster center, it increases with radial distance and peaks at the cluster interface, at a value higher than the ccp order. Upon moving into the liquid phase, both the ccp and bcc order decrease to zero, as expected. This surface wetting of the ccp nucleus by bcc order explains the bcc character of

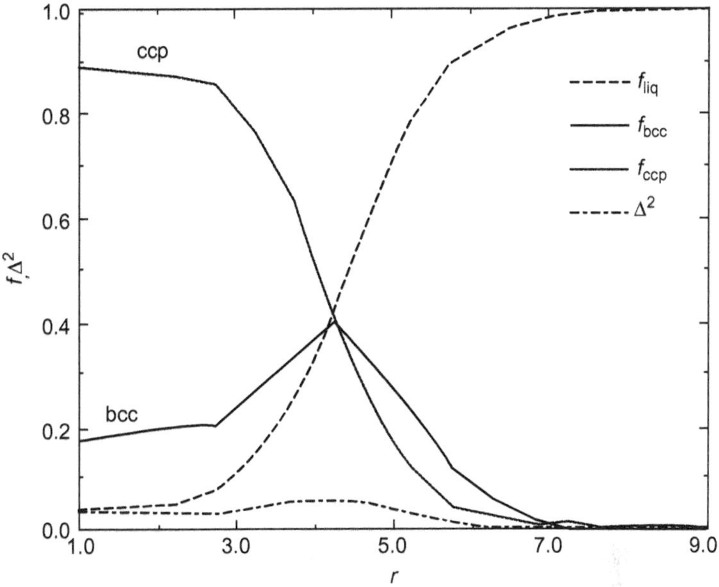

Fig. 13 Parameters defining bcc (f_{bcc}), ccp (f_{ccp}) and liquid (f_{liq}) order with radial distance from the center of the critical cluster. Δ^2 is defined from the distributions of the number of connections per particle, determined from a dot product of bond-orientational parameters. The distribution of the number of connections for different structures are represented as *n*-dimensional vectors. $\Delta^2 = (\mathbf{v}_{cl} - (f_{liq}\mathbf{v}_{liq} + f_{bcc}\mathbf{v}_{bcc} + f_{ccp}\mathbf{v}_{ccp}))$, where $\mathbf{v}_{cl}$, $\mathbf{v}_{liq}$, $\mathbf{v}_{bcc}$ and $\mathbf{v}_{ccp}$ are the characteristic vectors for the cluster, equilibrated liquid, bcc and ccp structures respectively. (Reprinted figure from Ref. [36], copyright (1995), American Physical Society.)

the subcritical clusters, which are dominated by the presence of the interface.

6. COUPLED PHASE TRANSITIONS

Computer studies have shown that in some cases fluctuations from nearby critical points can dramatically lower the nucleation barrier for a first-order phase transition. It is known that the range of attraction between spherical particles in a colloidal suspension, for example, dramatically influences the phase diagram [37, 38]. For an interaction distance that is large compared with the size of the dispersed particles, the phase diagram resembles that of a normal atomic system such as argon, showing the expected vapor, liquid and solid phases and coexistence between them. However, as the interaction distance shortens, a fluid–fluid critical point moves toward the triple point. As shown in

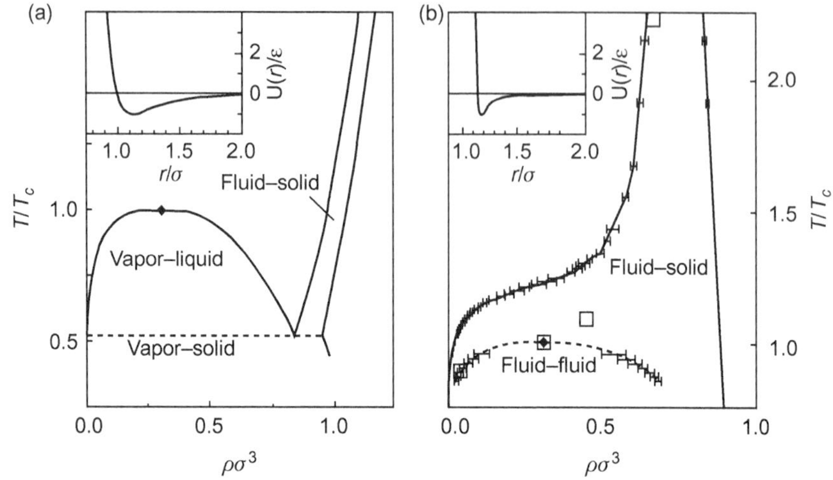

Fig. 14 Phase diagram for a Lennard-Jones system for a long-range attractive interaction (a) and a short-range range attractive interaction (b). The shapes of the potentials are shown in the insets; ρ is the number density in units of σ^{-3}. (From Ref. [39]. Reprinted with permission from AAAS.)

Figure 14, this behavior can be reproduced using a LJ potential with a variable range of interaction [39]. With decreasing interaction distance the vapor phase vanishes and only the liquid and solid are left.

It is of interest to examine the possible influence of the metastable critical point on nucleation in the region of the critical temperature, T_c. The free-energy landscape can be expressed in terms of the number of particles belonging to a crystal nucleus, n_{crys} and the number of particles in a high-density cluster that could be either liquid-like or solid-like, n_{clus}. For nucleation far away from T_c, n_{clus} is proportional to n_{crys}, indicating that the only high-density regions in the liquid are those containing the crystal phase (Figure 15a). Near T_c, however, the behavior is very different, with n_{clus} first increasing while n_{crys} remains essentially zero. This indicates that the first step toward the formation of a critical nucleus is the creation of a dense liquid drop. Subsequent crystal nucleation occurs within the liquid drop (Fig. 15b).

What is most surprising is the catalytic influence of the dense-fluid fluctuations on the nucleation barrier (Fig. 15c). Near T_c, the barrier is reduced by almost 50 k_BT over that on either side! The mechanism of coupling between the critical point and the nucleation barrier depends on the precise nature of the two phase transitions, but it is clear that proximity of a critical point can profoundly influence the nucleation rate and, therefore, the resulting microstructure. The nature of this coupling and its formal incorporation into nucleation theories are of significant interest [40–42].

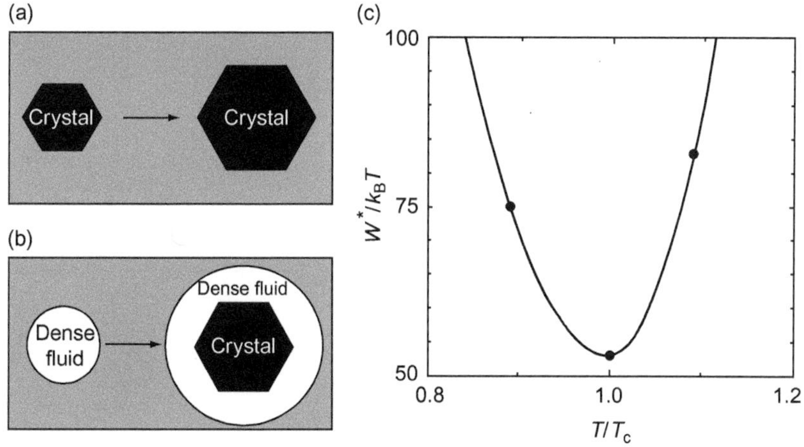

Fig. 15 (a) Schematic illustration of a nucleation process far from T_c, with the crystal nucleus forming directly from the homogeneous liquid; (b) nucleation near T_c with the initial formation of regions of dense fluid, followed by preferential nucleation of the crystal phase inside them; (c) the influence of the critical point on the nucleation barrier. (From Ref. [39]. Reprinted with permission from AAAS.)

7. IMPACT OF DIFFUSION ON NUCLEATION

As discussed earlier, most theories focus on nucleation processes that involve interface-limited kinetics, appropriate for first-order phase transitions where the chemical compositions of the original and new phases are the same. For many phase transitions, however, this is not the case and the nucleation processes are influenced by long-range solute diffusion. The Lifshitz–Slyozov–Wagner (LSW) type theories discussed in Chapter 4 are appropriate for diffusion-limited kinetics. The coupled-flux models discussed in Chapter 5 have argued in favor of both interface- and diffusion-limited kinetics, providing a way to model nucleation processes that fall between the limits of the classical and LSW theories.

Recently, several key aspects of the coupled-flux approach were verified from kinetic Monte Carlo simulations of a cubic-lattice-gas model [43]. The Arrhenius rate processes considered were free diffusion on the lattice (D) and interfacial attachment (IA), characterized by the rate constants k_D and k_{IA}, respectively. The interfacial detachment rate was determined by assuming detailed balance. To study sufficiently large systems, the system was "seeded" with spherical clusters of uniform size, rather than generating clusters by quenching from high-temperature configurations. Results for each condition of seeded cluster size and level of supersaturation were obtained by averaging over 64 independent simulations, each containing 125 clusters. Calculations were made as a function of initial mole fraction X_0 of solute, which represents the fractional population of lattice sites.

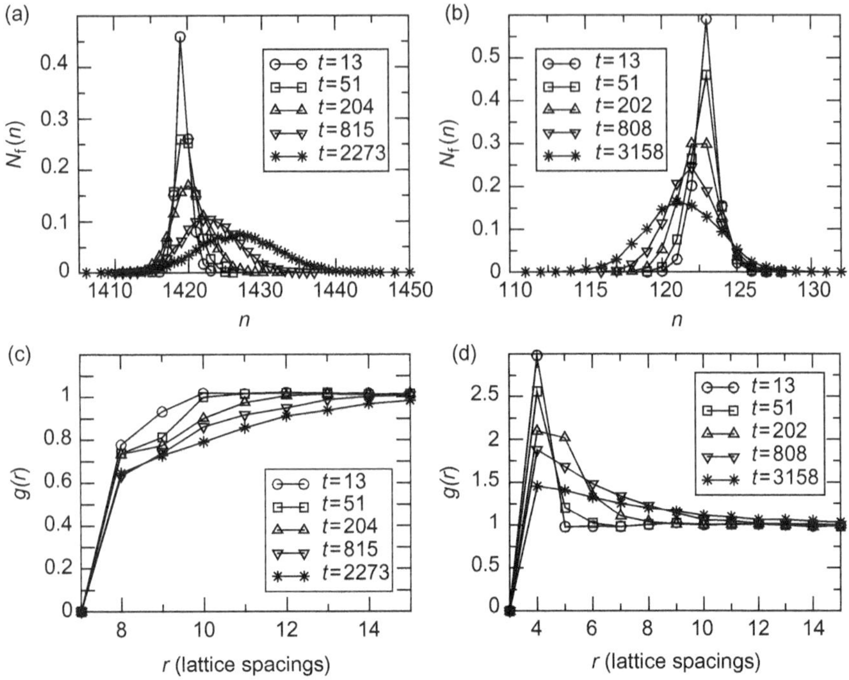

Fig. 16 (a) Size distribution of clusters, $N_f(n)$ as a function of calculation time (measured in units of the inverse of the interfacial attachment rate, k_{IA}). The initial radius of the seed cluster was seven lattice spacings (1419 particles), which is above the critical size for the initial mole fraction, ($X_0 = 0.1\%$). The ratio of rate constants for diffusion and for interfacial attachment, k_D/k_{IA}, is set equal to one. (b) $N_f(n)$ as a function of time for seed clusters of initial radius 3 lattice spacings (123 particles), which is below the critical size; $k_D/k_{IA} = 0.105$. (c) The normalized radial distribution function $g(r)$ of the particle mole fraction (fraction of occupied lattice sites) around the cluster for the conditions of (a). (d) The normalized $g(r)$ corresponding to (b). (Reprinted from Ref. [43], copyright (2008), with permission from Elsevier.)

As shown in Figure 16, starting with seeded clusters that are larger than the critical size (Figure 16a), the average size (the peak of the distribution) grows with time, and the single-molecule population density lies below its average value in regions far from growing clusters (Figure 16b), which is the expected behavior for cluster growth. In contrast, if the seeded clusters are smaller than the critical size, the average cluster shrinks with time (Figure 16c) and the regions near the clusters become enriched, i.e. the single-molecule population density exceeds its average value (Figure 16d). An enhanced solute level near subcritical clusters is one of the key predictions of the coupled-flux model (see Chapter 5, Section 5.3).

For high diffusion rates ($k_D/k_{IA} \gg 1$), little enrichment or depletion of solute is observed, and nucleation can be described by interface-limited theories. The significance of the enrichment/depletion increases with decreasing k_D/k_{IA}. For subcritical clusters, which are on average shrinking, the solute-enriched neighborhood should cause them to shrink more slowly than predicted from interface-limited models. Likewise, the depletion of solute around postcritical clusters should retard their growth compared to predictions. The Monte Carlo simulations confirm these effects and demonstrate that cluster growth rate is determined by the local solute level around the cluster (Figure 17, top). The average cluster growth rate, $d\langle n\rangle/dt$, for clusters of initial radius $r_0 = 3$ (i.e. 3 lattice spacings) is shown as a function of solute mole fraction.

The bulk solute content (○) corresponds to cases where $k_D/k_{IA} = 42.5$. A range of initial bulk solute contents (far from the growing cluster) is considered. In this case, the high diffusion rate (compared with the interfacial attachment rate) ensures that the neighborhood composition stays close to the bulk composition. The cluster growth is, then, essentially interface-controlled and can be described, for example, by the classical theory of nucleation. As shown by the dotted line, $r_0 = 3$ is the critical cluster size for a solute mole fraction of 0.162%; for lower solute contents the clusters shrink, and for higher they grow. For lower diffusion rates, the composition near the clusters is different from the bulk composition. This is seen for clusters of initial radius $r_0 = 3$ and bulk solute mole fractions of 0.1% (i.e. shrinking, ▽) or 0.2% (growing, △). These data show that the resulting composition near the cluster depends on the value of k_D/k_{IA} (shown for the corresponding data point). But, given the local composition, clusters grow at approximately the rate expected for an interface-limited case (fast diffusion) with a bulk solute content of the same value.

Predictions from the coupled-flux model (Figure 17, bottom) can be compared to those already seen from Monte Carlo simulations (Figure 17, top) for clusters of the same initial radius. The circles correspond to the case of fast diffusion relative to the IA rate (i.e. $\xi D/6D_i \gg 1$, where D_i and D are the effective diffusion constants governing interfacial attachment and volume diffusion (i.e. on the lattice) and ξ is a composition-dependent correction factor to ensure the correct limit of diffusion-limited growth for large clusters (Chapter 5, Section 5.2). Like the simulation results, the local composition and the bulk composition are essentially the same, indicating that the process is interface-limited. Coupled-flux calculations are shown for initial bulk mole fractions of 0.1% (i.e. shrinking, ▽) or 0.2% (growing, △). As for the simulation, the resulting composition near the cluster depends on the value of $\xi D/6D_i$ (shown for the corresponding data point). While showing the correct trends, the agreement remains semiquantitative. The divergence near the critical concentration is likely due to assumptions made in the derivation

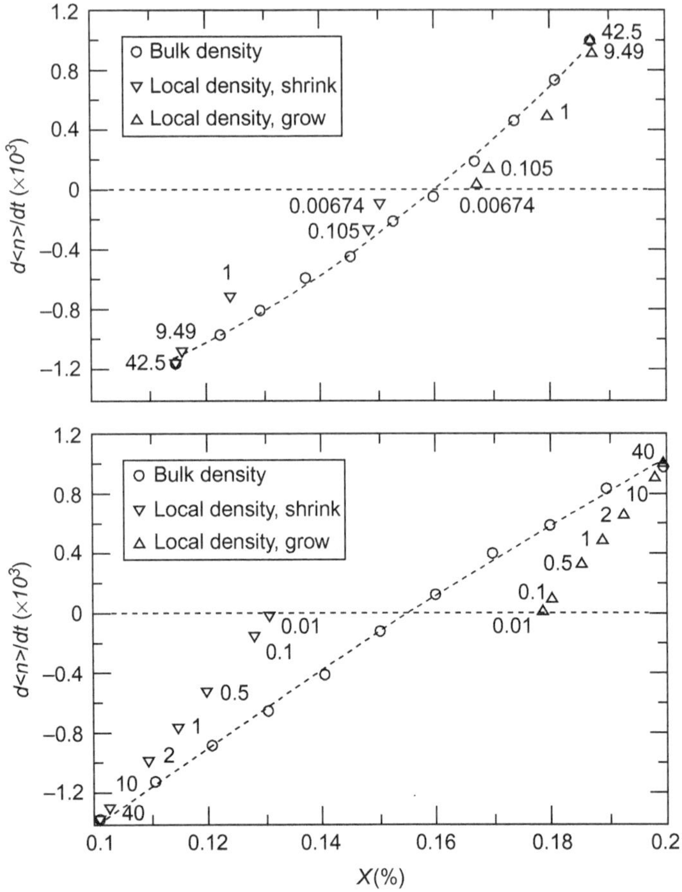

Fig. 17 (a) The growth rate, $d\langle n\rangle/dt$, for seeded clusters of radius 3 lattice spacings (○) as a function of the bulk lattice concentration for interface-limited conditions, $k_D/k_{IA} = 42.5$. The growth rates for clusters under diffusion-limited conditions (△, with adjacent numbers corresponding to k_D/k_{IA}). Negative growth rates (shrinking clusters) are observed for simulations where the initial atom fraction is $X_0 = 0.1\%$. Positive growth rates (growing clusters) are observed for an initial atom fraction of $X_0 = 0.2\%$. (b) The calculated growth rate (○) from the coupled-flux model for seeded clusters of radius 3 as a function of the bulk concentration at high diffusion rate ($\xi D/6D_i \gg 1$). The growth rates (△, with adjacent numbers corresponding to k_D/k_{IA}) from the Monte Carlo simulation for seed clusters of radius 3 for an initial mole fraction of 0.1% (shrinking) and 0.2% (growing). The growth rates are plotted as a function of the average shell composition. (Reprinted from Ref. [43], copyright (2008), with permission from Elsevier.)

of the coupled-flux model in Chapter 5, in particular the use of a classical-type model for the work of cluster formation and the neglect of enthalpy differences in the matrix near and far away from the cluster. However, it is important to note that the predictions from both interface-limited theories, such as the classical theory, and diffusion-limited LSW-type models fail to even qualitatively fit the Monte Carlo results.

8. SUMMARY

Computer models provide a valuable new type of "experimental data" in nucleation studies, since all input parameters are directly known. This chapter briefly reviewed of some of the results that elucidate fundamental aspects of the nucleation process. Key points are listed below.

- There are two common methods for simulating the time evolution of a system configuration:
  - The Monte Carlo technique is based on a stochastic process that produces a Markovian chain of configurations that are Boltzmann-weighted. Monte Carlo methods are widely used to study phase transitions in lattice models, such as the Ising model.
  - Molecular-dynamics (MD) techniques follow the time evolution of a system by numerically solving a set of dynamical equations. However, they require enormous computer resources to model macroscopic systems.
- Monte Carlo calculations with Ising models demonstrate that:
  - nucleation is stochastic, as argued by Gibbs and assumed in almost all models for nucleation;
  - nucleation can be described reasonably well by the classical theory when the driving free energy is small, i.e. close to equilibrium, but becomes less accurate with increasing driving free energy;
  - the prefactor in the classical theory is in error, at least when applied to a square Ising lattice.
- Molecular-dynamics simulations demonstrate that:
  - the basic kinetic approach of the classical theory captures the essential features of the evolution of the cluster distribution underlying nucleation processes, in agreement with experimental studies in silicate and metallic glasses;
  - the cluster interface is diffuse and for small clusters the order parameter is not uniform, even at their center, in agreement with the density-functional calculations discussed in Chapter 4.
- Subcritical clusters do not necessarily have the structure corresponding to that of the new phase, and may have a complex mixture of structures.

- Nearby high-order phase transitions may couple to the nucleation process and dramatically lower the nucleation barrier.
- The coupled-flux model provides a reasonably good description of nucleation processes involving both long-range diffusion and interfacial attachment when the two rates are similar. Neither the classical theory nor the diffusion-limited LSW theories are appropriate for such cases.

REFERENCES

[1] D.C. Rapaport, *The Art of Molecular Dynamics*, Cambridge University Press, Cambridge, UK (1995).

[2] D.P. Landau, K. Binder, *A Guide to Monte Carlo Simulations in Statistical Physics*, Cambridge University Press, Cambridge, UK (2000).

[3] D. Frenkel, B. Smit, *Understanding Molecular Simulation*, Vol. 1, Academic Press, San Diego (2002).

[4] H. Goldstein, *Classical Mechanics*, Addison-Wesley, Reading, MA (1950).

[5] N. Metropolis, A.W. Rosenbluth, M.N. Rosenbluth, A.H. Teller, E. Teller, Equation of state calculations by fast computing machines, J. Chem. Phys. **21** (1953) 1087–1092.

[6] M.E. Fisher, The theory of equilibrium critical phenomena, Rep. Prog. Phys. **30** (1967) 615–730.

[7] K. Binder, D. Stauffer, Statistical theory of nucleation, condensation and coagulation, Adv. Phys. **25** (1976) 343–396.

[8] D. Stauffer, A. Coniglio, D.W. Heermann, Monte Carlo experiment for nucleation rate in the three-dimensional Ising model, Phys. Rev. Lett. **49** (1982) 1299–1302.

[9] D.W. Heermann, W. Klein, Percolation and droplets in a medium-range three-dimensional Ising model, Phys. Rev. B **27** (1983) 1732–1735.

[10] S. Wonczak, R. Strey, D. Stauffer, Confirmation of classical nucleation theory by Monte Carlo simulations in the 3-dimensional Ising model at low temperature, J. Chem. Phys. **113** (2000) 1976–1980.

[11] W.C. Swope, H.C. Andersen, 106 particle molecular-dynamics study of homogeneous nucleation of crystals in a supercooled atomic liquid, Phys. Rev. B **41** (1990) 7042–7054.

[12] E.M. Lifshitz, L.P. Pitaevskii, *Physical Kinetics*, Vol. 10, Pergamon, Oxford (1981).

[13] J. Lothe, G.M. Pound, On the statistical mechanics of nucleation theory, J. Chem. Phys. **45** (1966) 630–634.

[14] J. Lothe, G.M. Pound, Concentration of clusters in nucleation and the classical phase integral, J. Chem. Phys. **48** (1968) 1849–1852.

[15] K. Nishioka, G.M. Pound, Theory of the replacement partition function for crystals in homogeneous nucleation, Acta Metall. **22** (1977) 1015–1021.

[16] D. Stauffer, Ising droplets, nucleation and stretched exponential relaxation, Int. J. Mod. Phys. **3** (1992) 1059–1070.

[17] P.A. Rikvold, H. Tomita, S. Miyashita, S.W. Sides, Metastable lifetimes in a kinetic Ising-model — dependence on field and system size, Phys. Rev. E **49** (1994) 5080–5090.

[18] E. Stoll, K. Binder, T. Schneider, Evidence for Fisher's droplet model in simulated two-dimensional cluster distributions, Phys. Rev. B **6** (1972) 2777–2780.

[19] V.A. Shneidman, Dynamics of an Ising ferromagnet at $T \leq T_c$ from the droplet model approach, Physica A **190** (1992) 145–160.

[20] V.A. Shneidman, K.A. Jackson, K.M. Beatty, On the applicability of the classical nucleation theory in an Ising system, J. Chem. Phys. **111** (1999) 6932–6941.

[21] V.A. Shneidman, G.M. Nita, Nucleation preexponential in dynamic Ising models at moderately strong fields, Phys. Rev. E **68** (2003) 021605/1–11.
[22] H.E.A. Huitema, J.P. van der Eerden, J.J.M. Janssen, H. Human, Thermodynamics and kinetics of homogeneous crystal nucleation studied by computer simulation, Phys. Rev. B **62** (2000) 14690–14702.
[23] V.A. Shneidman, K.A. Jackson, K.M. Beatty, Nucleation and growth of a stable phase in an Ising-type system, Phys. Rev. B **59** (1999) 3579–3589.
[24] J.K. Lee, J.A. Barker, F.F. Abraham, Theory and Monte Carlo simulation of physical clusters in the imperfect vapor, J. Chem. Phys. **58** (1973) 3166–3180.
[25] W. Ostwald, Studien über die Bildung und Umwandlung fester Körper: 1 Abhandlung Ubersättigung and Uberkaltung, Z. Phys. Chem. **22** (1897) 289–330.
[26] I.N. Stranski, D. Totomanow, Keimbildungsgeschwindigkeit und Ostwaldsche Stufenregel, Z. Phys. Chem. **A 163** (1933) 399–408.
[27] S. Alexander, J.P. McTague, Should all crystals be bcc? Landau theory of solidification and crystal nucleation, Phys. Rev. Lett. **41** (1978) 702–705.
[28] W. Klein, F. Leyvraz, Crystalline nucleation in deeply quenched liquids, Phys. Rev. Lett. **57** (1986) 2845–2848.
[29] W. Löser, T. Volkmann, D.M. Herlach, Nucleation and metastable phase formation in undercooled Fe–Cr–Ni melts, Mater. Sci. Eng. A **173** (1994) 163–166.
[30] J. Yang, H. Gould, W. Klein, Molecular-dynamics investigation of deeply quenched liquids, Phys. Rev. Lett. **60** (1988) 2665–2668.
[31] S. Nosé, F. Yonezawa, Isothermal–isobaric computer simulations of melting and crystallization of a Lennard-Jones system, J. Chem. Phys. **84** (1986) 1803–1814.
[32] R.D. Mountain, A.C. Brown, Molecular dynamics investigation of homogeneous nucleation for inverse power potential liquids and for a modified Lennard-Jones liquid, J. Chem. Phys. **80** (1984) 2730–2734.
[33] C.S. Hsu, A. Rahman, Interaction potentials and their effect on crystal nucleation and symmetry, J. Chem. Phys. **71** (1979) 4974–4986.
[34] M.J. Mandell, J.P. McTague, A. Rahman, Crystal nucleation in a three-dimensional Lennard-Jones system. II. Nucleation kinetics for 256 and 500 particles, J. Chem. Phys. **66** (1977) 3070–3075.
[35] M.J. Mandell, J.P. McTague, A. Rahman, Crystal nucleation in a three-dimensional Lennard-Jones system: a molecular dynamics study, J. Chem. Phys. **64** (1976) 3699–3702.
[36] P.R. ten Wolde, M.J. Ruiz-Montero, D. Frenkel, Numerical evidence for bcc ordering at the surface of a critical fcc nucleus, Phys. Rev. Lett. **75** (1995) 2714–2717.
[37] A.P. Gast, W.B. Russell, C.K. Hall, Polymer-induced phase separations in nonaqueous colloidal suspensions, J. Coll. Inter. Sci. **96** (1983) 251–267.
[38] A.P. Gast, W.B. Russell, C.K. Hall, An experimental and theoretical study of phase transitions in the polystyrene latex and hydroxyethylcellulose system, J. Coll. Inter. Sci. **109** (1986) 161–171.
[39] P.R. ten Wolde, D. Frenkel, Enhancement of protein crystal nucleation by critical density fluctuations, Science **277** (1997) 1975–1978.
[40] R.P. Sear, Scaling theory for the free-energy barrier to homogeneous nucleation of a noncritical phase near a critical point, J. Chem. Phys. **116** (2002) 2922–2927.
[41] V. Talanquer, D.W. Oxtoby, Crystal nucleation in the presence of a metastable critical point, J. Chem. Phys. **109** (1998) 223–227.
[42] D.W. Oxtoby, Crystal nucleation in simple and complex fluids, Phil. Trans. Roy. Soc. Lond. A **361** (2003) 419–428.
[43] J. Diao, R. Salazar, K.F. Kelton, L.D. Gelb, Impact of diffusion on concentration profiles around near-critical nuclei and implications for theories of nucleation and growth, Acta Mater **56** (2008) 2585–2591.

PART III:
Further Topics and Applications

CHAPTER 11

Crystallization in Polymeric and Related Systems

Contents

1. INTRODUCTION

In this third part of the book, we cover nucleation in a broad range of contexts. The theoretical treatments described in Part I are for systems in which the entities involved in nucleation are atoms or small molecules; these treatments are not always appropriate, as shown particularly by the systems considered in this chapter and the next. In the present case, the nature of polymer molecules can significantly modify nucleation kinetics, and also leads to some nucleation phenomena not seen in other systems. Polymer crystals can be formed from solution (the subject of most early studies [1]) and from the melt (the focus of more recent work [2]).

An obvious feature of polymers is their high molecular weight. Yet this is not the key feature distinguishing their crystal nucleation from that in small-molecule systems. The effects of molecular size (keeping a spheroidal shape) have been analyzed in general terms by Oxtoby [3]. For atomic systems, such as metals, the range of attractive forces between the species is similar to their diameter; for large molecules, such as some proteins, these forces operate over distances much shorter than the diameter. Oxtoby's density-functional calculations suggest that, in the

Pergamon Materials Series, Volume 15
ISSN 1470-1804, DOI 10.1016/S1470-1804(09)01511-9

latter case, density or concentration increases can precede crystalline ordering, a conclusion supported by computer simulation [4]. Extreme examples of large-species systems are provided by *colloids*. While colloids are in some ways quite distinct, in their dynamics of crystal nucleation they can serve as good models for atomic and small-molecule systems [5], as discussed in Chapter 7, Section 4. Overall, for spheroidal species, size does not much change the models that are the basis for analyzing nucleation kinetics. That polymers show different nucleation behavior is attributable not to their large molecular weight, but to their chain-like nonspheroidal shape.

Polymers are never 100% crystalline, and some cannot crystallize at all. To understand this, it is necessary to outline some aspects of the structure of polymer chains, even though a full review lies well beyond the scope of this book. Introductory treatments have been given, for example, by Windle and by Callister [6, 7]. The chains are formed by the joining of monomers. The simplest is *polyethylene*, consisting of a sequence of methylene ($-CH_2-$) units, giving a zig-zag carbon backbone with only hydrogen atoms attached. Although each carbon atom retains the tetrahedral arrangement of its four bonds, there is rotation about those bonds and the polyethylene chain is highly flexible. In view of its simplicity, polyethylene was used in early studies of crystallization [1], and flexible, linear polymers of this kind remain the focus of the work described in this chapter. There are many ways in which polymer chains can become more complex and more difficult to crystallize: instead of being just linear they may show branching; linear chains can be occasionally cross-linked and may even form network structures. All such features limit chain mobility and the ability to order into a crystalline structure. Even when only linear chains are involved, disorder in the chains themselves can inhibit crystallization. For monomers more complicated than ethylene, one or more hydrogen atoms are replaced by another atom or a side group R. When such monomers are joined, there is the possibility that joining may be head-to-tail (the usual case, due to repulsion of the side groups) or head-to-head; any irregularity in this pattern hinders crystallization. Even when the monomer joining is exclusively head-to-tail, isomerism within the monomers opens up further possibilities. One example of this is *stereoisomerism*: the side groups may be all on the same side of the chain (*isotactic* configuration), may alternate sides (*syndiotactic*), or may be randomly arranged (*atactic*) (Figure 1). These configurations cannot be interchanged by rotation about the C–C bonds. The randomness in the atactic configuration inhibits crystallization.

Flexible chains, of which polyethylene is the archetype, are found in solution or in the melt in a *random coil* configuration. Attachment of large

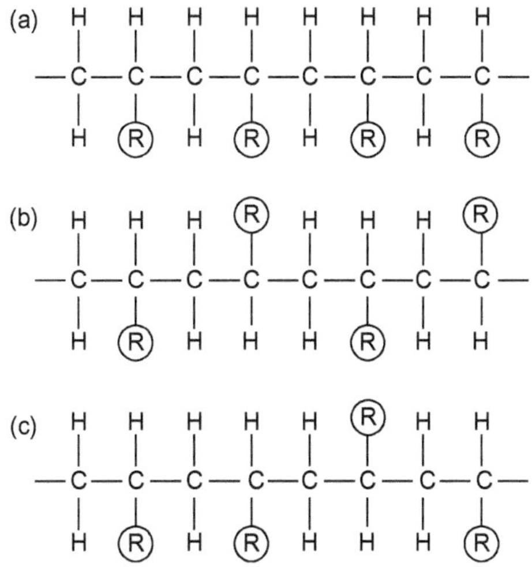

Fig. 1 A linear polymer with a side group R: (a) isotactic; (b) syndiotactic; (c) atactic. (From Ref. [7], copyright 2000, reprinted with permission of John Wiley & Sons, Inc.)

side groups reduces flexibility, and this is reduced further by the incorporation of inflexible units such as benzene rings into the polymer backbone. As described by Phillips [2], there is a wide range of semirigid and rigid polymers, which can show distinctive features in their crystallization. Often referred to as *liquid-crystalline polymers,* their rigidity can lead to states of partial order even if the structural sequence along a given chain is random. Such systems are among those considered briefly in Section 7.

So far, we have considered only *homopolymers,* which incorporate only one type of monomer, but a *copolymer* chain includes more than one type. For two monomer types, the sequence along the chain may alternate, or be random, or the monomers may be grouped into sequences of like kind to form a *block copolymer.* Polymer chains of different types may also be attached in branching configurations to form a *graft copolymer.* These possibilities introduce extra complexity in analyzing crystallization and nucleation.

Polymer chains can vary greatly in length and in a given sample there is usually a spread in length. As will be seen later (particularly in Section 6), the length or molecular weight of a polymer is relevant for its mobility and therefore for the kinetics of nucleation and growth of polymer crystals. In a condensed phase, polymer chains (like spaghetti) have a tendency to lie parallel to each other, giving local regions of crystallinity. This *fringed micelle* model for crystallinity is not valid for most crystals in polymers, but it may explain the partial crystallinity found in some polymers even after rapid quenching of the melt [8]. In most cases, however, a growing crystal must *reel in* polymer chains from the surrounding melt. That this is possible at all is due to *reptation* [9], in which a polymer chain can move locally parallel to itself without the neighboring chains having to move out of the way. In effect, the molecule migrates along a tube defined by its neighbors. Diffusional transport by such a process is quite different from the Brownian motion of atoms and small molecules, and leads to distinctive kinetics. Reptation is hindered by large side groups, cross-links and chain entanglements. It also provides a straightforward way of interpreting some of the kinetic effects of chain length (molecular weight) (Section 6 [2]).

Polymer single crystals grown from solution or the melt typically have the form of thin lamellae, 10–20 nm thick and with lateral dimensions up to 10 μm. These can grow as thin ribbons (*fibrils*) that can twist and branch, ultimately generating the radial structure seen within large crystalline regions called *spherulites*. As first clearly shown by Keller [1], the chains in a single crystal of a polymer such as polyethylene are folded, such that the chain axis is normal to the lamellar plane. There has been much controversy about the nature of the folding. Figure 2 shows a highly ordered structure with *adjacent re-entry* of the chains. This is a good approximation to the structure of solution-grown crystals [2]. For crystals grown from the melt, however, the behavior is more complex, including some adjacent re-entry, but with some nonadjacent re-entry, and even pseudorandom coils entering and leaving a crystal many times [2]. Between crystals the complex chain conformations preclude thin boundaries and account for the residual noncrystallinity found even in simple linear polymers such as polyethylene.

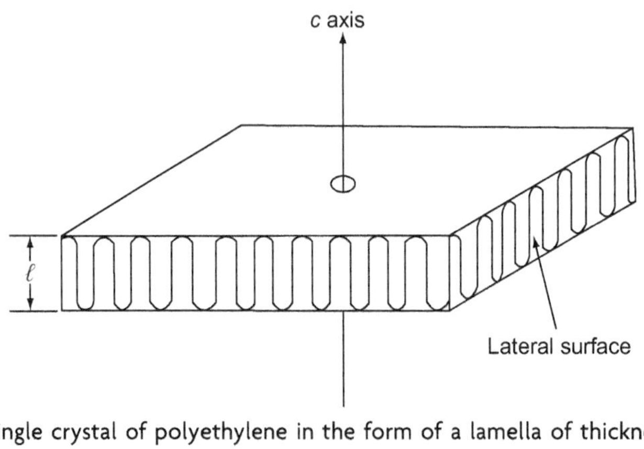

Fig. 2 A single crystal of polyethylene in the form of a lamella of thickness ℓ with a structure of folded chains. Growth occurs by addition of chains to the lateral surfaces. (Adapted from Ref. [74] courtesy of the National Institute of Standards and Technology, U.S. Department of Commerce.)

Because polymer crystals are so thin and their thickness is inversely proportional to the supersaturation at which they grew (Section 6), the melting point of a typical crystalline polymer is spread over a temperature range and dependent on the thermal history of the sample. For a given polymer, variations in molecular weight and in the degree of branching also contribute to this spread.

Even with a basic crystal structure of parallel chains, there are opportunities for variants of the packing to give different phases (*polymorphs*). *Polypropylene* is a simple vinyl polymer in which the side group R (Figure 1) is a methyl (–CH_3) group. Isotactic polypropylene is a common polymer. Its backbone is flexible, but the bulkiness of the side groups imposes a twist, giving left- and right-handed helical conformations with equal probability. In addition, the side groups are tilted so that they are not normal to the main chain axis; as a result, a chain of a given handedness has distinguishable "up" and "down" senses of the helix. When the chains are packed parallel to each other in a crystal structure, variants can arise according to how left-handed, right-handed, "up" and "down" chains are arranged as discussed, for example, by Cheng and Lotz [10].

The following sections explore various ways in which the long-chain nature of polymers affects crystal nucleation in such systems.

2. HOMOGENEOUS NUCLEATION

2.1 Experimental studies

Attempts to detect homogeneous nucleation of polymer crystals have focused on crystallization from the melt rather than from solution. Early studies used the emulsion technique described in Chapter 7, Section 2.1 to subdivide the melt and to reduce the effect of heterogeneous nucleants [11, 12]. In this way, it was easily shown that the usual crystallization of bulk melts must be heterogeneously nucleated and that much greater supercoolings of the melt can be achieved in emulsified samples. In the early work, the finest droplets of polymer melt had a diameter of ~3 μm. Supercoolings as large as 100 K (reduced supercooling $\Delta T_r = 0.22$) were reported for isotactic polypropylene [12], and isothermal nucleation kinetics were determined in droplet dispersions held at supercoolings just smaller than this value. Taking this nucleation to be homogeneous, the kinetics were interpreted in terms of classical nucleation theory, but the parameters used were dependent on rather arbitrary assumptions about the shape of the nuclei and the number of polymer chains involved. The number of chains was adjusted to achieve a fitting of the data giving a reasonable value of the kinetic prefactor (A^* in Chapter 2, Eq. (55)); however no attempt was made to consider whether A^* for long chains might have to be treated differently from that for atoms or small molecules [12]. See also the discussion of A^* in Chapter 7, Section 3.1.

The emulsion technique has continued to be developed in later work, mainly to widen the range of polymer melts and carrier liquids and to attain finer droplet dispersions with narrower size distribution. The methods of generating dispersions have broadened, from the original precipitation from solution and redispersion [11, 12], to include solid-state shear pulverization [13], polymerization in emulsion [14], and stirring and sonification [15]. By such methods droplets can be made with diameters in the range 30–300 nm [14, 15].

As a carrier liquid for a droplet emulsion, water is attractive because its polar nature, quite distinct from that of the dispersed polymer, should reduce any influence the carrier might have on the crystallization of the droplets; water-based dispersions can be made and do allow large supercoolings to be achieved. Uriguen et al. [16] have found supercoolings as large as 121 K for a modified polypropylene; this is larger than the maximum supercooling of 100 K for polypropylene dispersed in a surfactant, reported by Burns and Turnbull [12]. The comparison of these two supercoolings suggests that there may have been residual heterogeneous nucleation catalysis in the work of Burns and Turnbull. We can note that the observation of isothermal nucleation kinetics does not prove that the nucleation is homogeneous.

Among the possibilities for reducing the influence of heterogeneities, the substrate technique (Chapter 7, Section 2.1) has also been used successfully for polymers. Massa and Dalnoki-Veress [17] have studied the crystallization of droplets of poly(ethylene oxide) on a polystyrene substrate. The droplets are made by the break-up (*dewetting*) of a thin film of the polymer. The dewetting, specifically the parting of the original blanket film, is nucleated at defects that might also serve to nucleate crystallization. As the film forms into droplets, the droplets pull away from these defects; in this way, the dewetting process itself has a useful self-cleaning effect. The droplets formed are larger than those of interest in the emulsion technique; they have volumes as large as 500 μm^3. Figure 3 shows the nucleation kinetics within droplets of poly(ethylene oxide) held at −5°C (supercooling 69 K, $\Delta T_r = 0.20$). Within a given size range in which there are N droplets, the number of uncrystallized (still amorphous) droplets, N_{amor} decreases with time t according to the probability P per unit time of having a nucleation event in the droplet:

$$\frac{d}{dt}\left(\frac{N_{amor}}{N}\right) = -P\left(\frac{N_{amor}}{N}\right) = -\frac{1}{\tau}\left(\frac{N_{amor}}{N}\right). \tag{1}$$

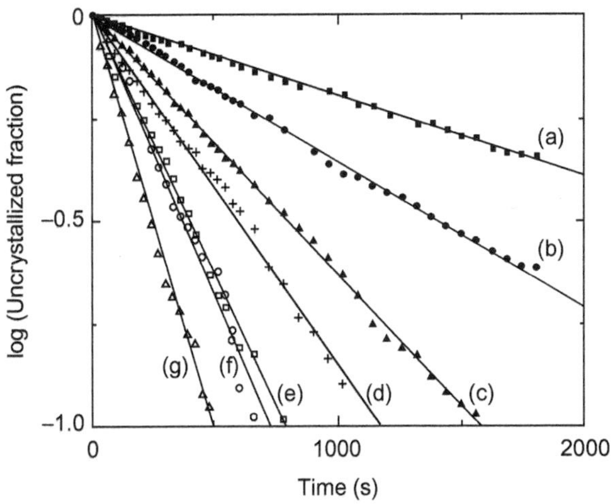

Fig. 3 Crystallization of dispersions of poly(ethylene oxide) droplets held at −5°C on a polystyrene substrate. The data are binned according to the base area of the droplets in μm^2: (a) 44–50; (b) 50–75; (c) 75–100; (d) 100–125; (e) 125–150; (f) 150–175; (g) 175–225. (Adapted with permission from Ref. [17], copyright (2004) by the American Physical Society.)

The probability is the reciprocal of the time constant τ characterizing the exponential decay in the population of uncrystallized droplets; from Eq. (1), we have

$$\frac{N_{\text{amor}}}{N} = \exp\left(\frac{-t}{\tau}\right). \qquad (2)$$

As can be seen in Figure 3, this time dependence is followed within each size range. Importantly, however, P is found to scale linearly with the *volume* V_d of the droplets. This rules out any nucleation catalysis at the droplet surfaces and is consistent with homogeneous nucleation.

Within a narrow temperature range, the nucleation rate can be measured. It is convenient to display the variation with temperature T in terms of the volume-normalized time constant τV ($\equiv V_\text{d}/P$) (Figure 4). As shown in the inset, $\ln(\tau V_\text{d})$ varies linearly with $1/T(T_\text{m}-T)^2$; this is exactly as expected from classical nucleation theory (Chapter 7, Eq. (3)), taking the kinetic pre-factor A^* and the liquid-solid interfacial energy σ_ls to be independent of T in this narrow range, and taking the free energy change on crystallization Δg to be linearly proportional to the supercooling below the equilibrium melting temperature T_m (Chapter 6, Eq. (24)).

Polymers offer possibilities for creating fine dispersions not available in other types of system. Block copolymers allow dispersions to be engineered. When molten, and if the repulsion between blocks is sufficiently large, they form very regular dispersions (*mesophases*) with a variety of morphologies [18]. Such materials can be used to obtain dispersions of spheroidal

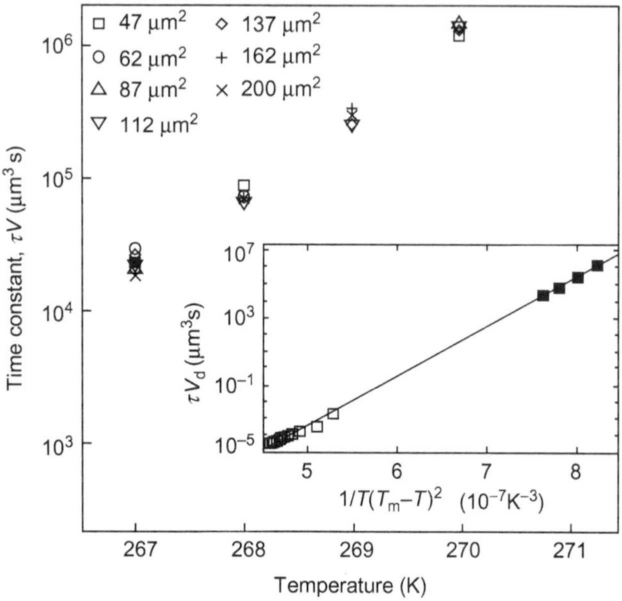

Fig. 4 The data collected in the main plot show the volume-normalized time constant τV_d as a function of temperature T for each of the size ranges indicated in Figure 3. The inset shows that the linear dependence of τV_d on $1/T(T_m - T)^2$, expected from classical nucleation theory, fits nucleation rates derived from the data in Figure 3 (solid symbols) and from data on 12-nm droplets in a block co-polymer (open symbols) [19]. (Adapted with permission from Ref. [17], copyright (2004) by the American Physical Society.)

microdomains, derived from polymer blocks. Such domains (for example the 25-nm diameter spheres of polyethylene studied by Loo et al. [18]) can crystallize without destroying the mesophase dispersion. That crystallization can occur independently in such microdomains has been established directly by atomic force microscopy [19]. In this latter work, Reiter et al. [19] studied 12-nm diameter spheres of poly(ethylene oxide) formed in a block copolymer with a hydrogenated polybutadiene. Although the polybutadiene remains amorphous, poly(ethylene oxide) can crystallize. The nucleation rates within the 12-nm diameter spheres are, remarkably, consistent with the data on the same polymer in much larger form presented in Figure 4. The single straight line in the inset links the two data sets and suggests that the nature of homogenous nucleation within isolated domains remains unaffected by domain size, even down to diameters around 10 nm [17].

2.2 Computer simulations

As reviewed, for example, by Paul and Smith [20], there have now been many computer simulations of the structure and dynamics of long-chain polymers. With modern computing power, it is possible to quantify the interatomic interactions sufficiently to model key phenomena such as chain folding without the calculation becoming intractable. It is particularly important to have good relative values for the stiffnesses of bond lengths, bond angles and torsional angles [21]. A common simplification is the *united atom* approach in which there is no explicit modeling of the H-atoms. Thus in polyethylene, each methylene unit ($-CH_2-$) is treated as a bead and the polymer is a chain of such beads connected by springs [21, 22]. The applications of molecular-dynamics simulations to polymer crystallization (briefly reviewed in [21, 23]) are relatively few, and have mostly focused on deposition of chains on an existing substrate (Section 5) or crystal (Section 6). Homogeneous nucleation from the melt has been treated for alkanes up to dodecane [22] and for isotactic polypropylene [23], but these studies did not focus on the chain conformations in the early stages. The details of homogeneous nucleation of long-chain systems appear to have been studied mainly for crystallization from solution.

As is widely recognized [24, 25], nucleation in long-chain polymers is of particular interest because the number of monomers in a critical crystal nucleus may be far fewer than in a single chain. Whereas in small-molecule systems, each molecule participates in only one nucleus at a time, a single chain can be part of several nuclei. For long chains in dilute solution, many of the phenomena of nucleation and crystallization can be understood by considering a single chain, and indeed crystallization of single chains has been experimentally observed [24]. In contrast, in small-molecule systems, crystallization must involve many molecules. The transition from one regime to the other can be followed as a function of chain length in polymers. As chains become shorter, it becomes thermodynamically possible for a single chain to crystallize, by folding, under the conditions for multichain crystallization. Our focus is on the case of long chains.

Hu et al. [24] have used Monte Carlo simulations and simple analytical approximations to derive the free energy of a single chain in terms of the fraction that is ordered or is in a random coil conformation. The energy of the chain favors order, the entropy favors disorder. Figure 5 shows the energy variations around the equilibrium melting temperature; these schematic curves show that a nucleation barrier is expected for crystallization of the random coil. The height of this barrier is not sensitive to the chain length.

More detail on chain conformations in the very early stages of single-chain nucleation is provided by the molecular-dynamics simulations of polyethylene by Muthukumar and coworkers [21, 25]. Figure 6 shows the evolution of a chain of 2000 backbone carbons. Initially *baby nuclei* are formed by local folding. Just as polymer chains are reeled into larger

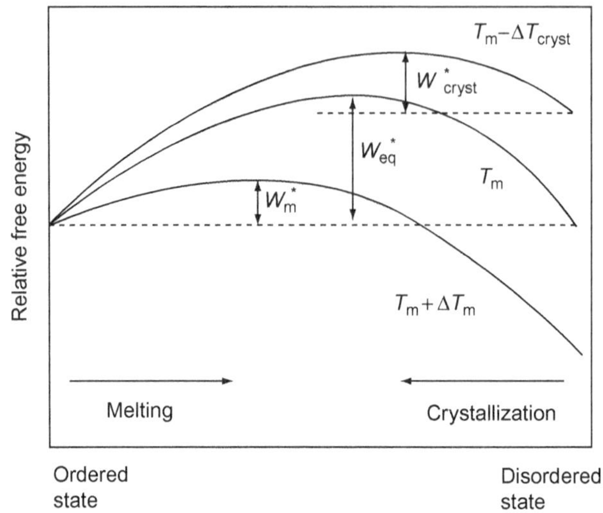

Fig. 5 Schematic diagram showing free-energy curves near the equilibrium melting temperature T_m of a single polymer chain. The energy barriers at equilibrium (W^*_{eq}), for crystallization (W^*_{cryst}) at a supercooling ΔT_{cryst} and for melting (W^*_m) at a superheating ΔT_m are marked. (Reprinted with permission from Ref. [24], copyright (2003) American Chemical Society.)

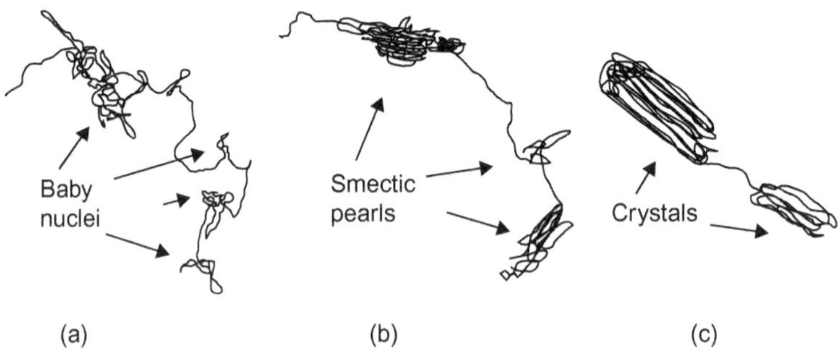

Fig. 6 Molecular-dynamics simulation of a single chain of polyethylene with 2000 monomers showing, at successive times, the development of crystals by reeling in the connecting chain. (From Ref. [25], with permission.)

crystals during growth, the single chain is likewise reeled into the baby nuclei. As the nuclei grow, and the orientational order within them increases, they can be termed *smectic pearls*, and ultimately identifiable crystals are formed. (The term *smectic* here refers to a parallel alignment of molecules in a well-defined layer, as found in smectic phases in liquid crystals [26].) The chain connecting the smectic pearls, or crystals, is first pulled taut without decrease of the distance between the original nuclei. Then, as the chain continues to be reeled in, the pearls or crystals do get closer and ultimately merge (Figure 7). As discussed by Muthukumar [25], the reeling of the chain into a crystal can be well modeled by the Fokker–Planck kinetics familiar from classical nucleation theory.

When baby nuclei have formed as in Figure 6a, the system has density fluctuations. Many small-angle X-ray scattering studies of polymer crystallization have found evidence for density fluctuations, and in some cases these have been taken to indicate that a liquid–liquid spinodal-type instability precedes crystal nucleation [27]. The work of Muthukumar [25], on the other hand, suggests that a structure of linked baby nuclei can explain the scattering observations; in particular, an analytical model for the entropy of connecting chains can be used to estimate the wave vector at which the scattered intensity is maximum. Dielectric relaxation measurements provide evidence for ordering in polymer melts (in particular the development of interfaces within them) before any detectable small-angle scattering [28]; this argues against the concept of a spinodal instability in the melt being a necessary precursor to crystallization.

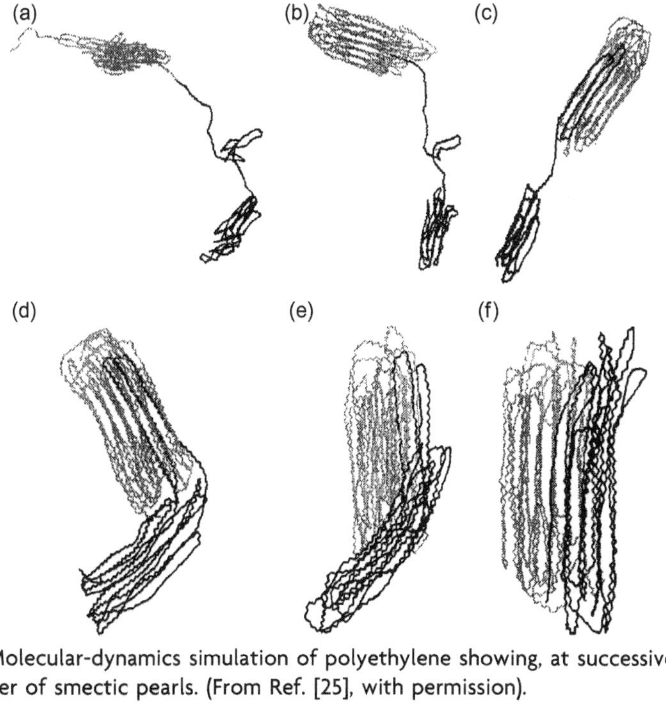

Fig. 7 Molecular-dynamics simulation of polyethylene showing, at successive times, the merger of smectic pearls. (From Ref. [25], with permission).

3. MEMORY EFFECTS

In Chapter 7, Section 2.4, it was noted for metallic systems that the nucleation of crystallization in a melt may be influenced by ordering in the melt. Such effects are much stronger for the crystallization of polymeric melts, because of the range of possible structures in the melt (for example, chain alignments, cross-links, and entanglements) and because such structures can have much longer lifetimes than would be usual in atomic or small-molecule systems. The crystallization of polymeric melts can be strongly influenced by the history of the melt, behavior commonly labeled as *memory effects*. In this section, we consider the influence of thermal history; in Section 4, the treatment is extended to the influence of stress and flow. Even when considering only thermal history, there is a wide range of memory effects.

The most obvious of these is *crystalline memory* in which, after melting, some crystals are retained. Given the temperature range over which the melting of most polymers occurs, it is easy to obtain partially melted samples. It has long been recognized that retained crystals acting as seeds on subsequent cooling (a phenomenon termed *self-nucleation*) could be useful in exercising microstructural control [29]. Nucleation on the retained crystals is athermal (Chapter 6, Section 2.5) and instantaneous [30]. While crystalline memory can be completely erased by heating the melt to a high enough temperature, it can be retained for limited times just above the melting temperature range. On holding the melt at a given temperature, the residual population of crystals decreases exponentially with time (Figure 8); there are fewer active seeds and consequently in the fully transformed material, the spherulite diameter is greater [31]. The decrease in the number of nucleation centers typically levels off at a nonzero value representing the population of infusible foreign particles (Section 5). The characteristic decay time decreases for higher holding temperature.

Memory effects in partially melted polymers extend beyond just the retention of some crystals. In the molten region around a residual crystal, the chains can retain some *orientation memory* of the previous spherulite. On cooling, many new, finer spherulites can form within the volume of the previous one. A memory effect is evident in that the pattern of the new spherulites mimics the previous structure, fitting within the outline of the original spherulite [29, 32]. It has also been suggested that the conformation of individual chains can be retained after the crystal has melted. For example, a retained helical conformation of isotactic polypropylene chains is a possible explanation of enhanced spherulitic growth rates [33]. As will be noted later (Section 5), for some polymers the selection of a particular crystalline polymorph can be important. This selection is nucleation-controlled and is itself subject to memory effects [34, 35].

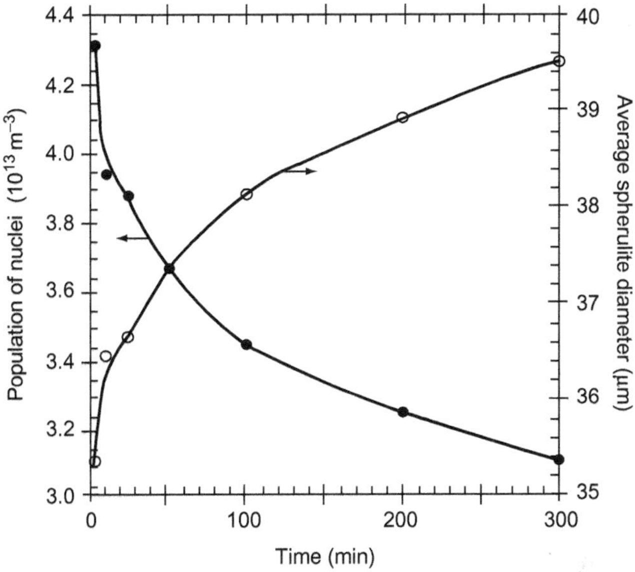

Fig. 8 Isothermal crystallization of syndiotactic polypropylene after partial or complete melting. The number of athermal nuclei decreases, and the resulting spherulite size increases, with the holding time at 180°C. The highest observed melting temperature for this polymer is 125°C. (From Ref. [31], copyright (2000), reprinted with permission of John Wiley & Sons, Inc.)

4. ORIENTATION-INDUCED NUCLEATION

Whether in solution or in the melt, imposing an extensional flow can cause polymer chains to align and straighten. This can increase the rates of nucleation and overall crystallization by many orders of magnitude, an effect quite distinct from anything known in atomic and small-molecule systems, and can induce crystallographic preferred orientation. The effects of flow on nucleation have been extensively reviewed [36, 37]. A particular reason for interest is the industrial importance of the production of oriented fibers. In contrast, shear-induced crystallization at the surface of extruded and injection-molded polymer artefacts is of concern as it can lead to whitening, distortion, and brittleness [38, 39]. Related effects are found in crystallization of prestressed polymer films [40].

The acceleration of nucleation under flow has thermodynamic and kinetic contributions that are difficult to separate, given that they become significant under the same conditions [37]. The thermodynamic effect is that aligned and oriented chains have a lower entropy than random coils; this makes the liquid state less stable and raises the melting temperature of the polymer. In this section, however, we focus on the kinetic effects of flow. With longer duration of flow, more and more nuclei are created [36], the effect being to accelerate the process seen in quiescent melts. Shear-induced smectic ordering in a melt of isotactic polypropylene [41] has been linked to the formation of smectic pearls shown in Figure 6.

Of most interest, however, is the completely distinct *row-nucleation* found beyond a critical flow rate (Figure 9); remarkable oriented aggregates (*shish-kebabs*) form both in solution and in melts [42]. The *shish* (skewer) is formed first as chains align with the extensional flow. The transition from random coil to the stretched conformation is discontinuous and apparently first-order [43]. While in general longer chains are more likely to be aligned, there is a strong dependence on the initial conformation of the chains and entanglements. Even if all chains were of the same length, stochastically there can develop a low population of stretched chains in a melt or solution mostly containing random coils. Stretched chains can develop regions of overlap (Figure 10); these regions are crystalline and once initiated spread rapidly to form a crystalline fiber of a few locally parallel chains. Once aligned chains are present in sufficient number, the nucleation of shishs is more deterministic than stochastic. Entanglements prevent perfect alignment of the chains, and the shish typically has dangling chains (*cilia*) emerging from it.

Next the *kebabs* nucleate and grow on the shish. While flow is essential for shish formation, it is not essential for kebabs; nevertheless

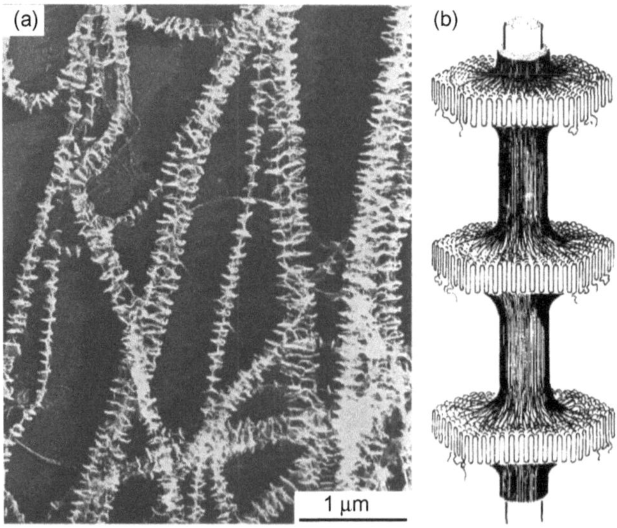

Fig. 9 Shish kebabs of polyethylene crystallizing from a stirred solution: (a) electron micrograph, (b) a possible model for their structure. (Adapted from Ref. [42], copyright (1970), with kind permission of Springer Science and Business Media.)

their formation is normally under continuing flow. It has long been known that the kebabs consist of folded chains parallel to the axis of the shish, giving an epitaxial match. Smaller kebabs are more stable, and may be associated with cilia attached to the shish [43]. Molecular-dynamics simulations, in which each monomer is represented by a bead, can illustrate features of both shish (Figure 10) and kebab (Figure 11) formation. Experimental observations are consistent with the mechanisms suggested by the simulation [44]. Chains near a shish can stretch out parallel to the shish or can fold into crystalline kebabs. The latter process is favored by faster crystallization and lower flow rate [43]. The kebabs appear to have a well-defined spacing [42, 44], but the origins of this have yet to be analyzed.

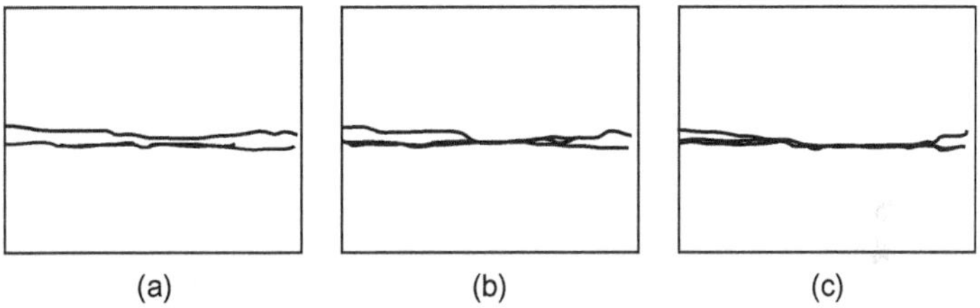

Fig. 10 Molecular-dynamics modeling of the development of a shish from 1000-monomer chains of polyethylene. (Reprinted with permission from Ref. [43], copyright (2003), American Institute of Physics.)

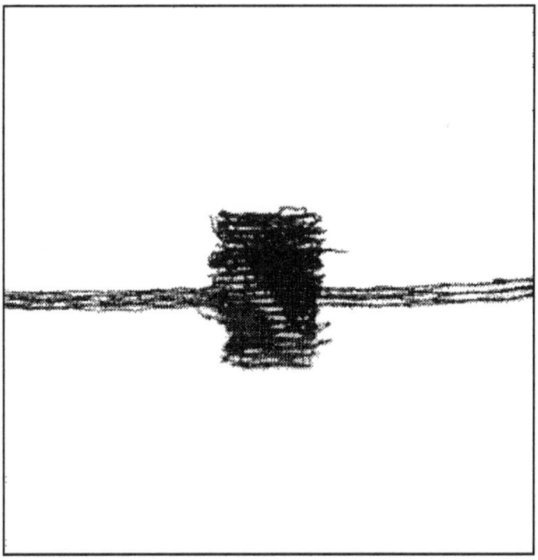

Fig. 11 Molecular-dynamics modeling of kebab formation on a stable shish in polyethylene. The shish is composed of 7 long (500 monomer) chains and the kebab of 44 short (180 monomer) chains. (Reprinted with permission from Ref. [43], copyright (2003), American Institute of Physics.)

Memory effects (Section 3) are evident also in polymeric samples oriented by flow or stress. Thus the crystallization of a polymer is influenced not only by flow during crystallization, but by the flow history of the sample [45]. Other effects can arise from particular crystallization morphologies: melting of kebabs previously formed under flow gives local volumes of melt that subsequently crystallize unusually fast. This is interpreted to be because of the low density of entanglement in the melt formed from the kebabs [46].

5. NUCLEATION ON FOREIGN PARTICLES

There is clear evidence for the nucleation, often athermal (Chapter 6, Section 2.5), of polymer crystallization on foreign particles. Possible unintended sources of particles able to act as substrates for nucleation include residues from the polymerization catalyst, and contamination from the container or from the surface during polymer processing in operations such as pelletizing. Nucleation can also occur on intended additives such as fillers [47, 48] and pigments [49, 50]. In analyzing the side effects (for example, nucleation catalysis) of such additives, it is important to note that they may be due to secondary additions such as the surfactant used to facilitate dispersion of the pigment [51]. There is much interest in polymer–matrix composites, and in the crystallization behavior of the matrix [52]. It is commonly found that the crystallization is accelerated in the presence of the reinforcing phase, and it is reasonable to suppose that heterogeneous nucleation can play a dominant role in this effect. Examples of reinforcements promoting crystallization include: carbon fibers [53]; single-walled carbon nanotubes [54]; AlN particles [55]; nanoparticles of silica [56], or zinc oxide [57]. Polymer–clay nanocomposites, in which the polymer intercalation leads to dispersion of the clay as platelets of atomic-scale thickness, have attractive properties, and in these it is clear that the clay platelets can be substrates for heterogeneous nucleation [58–60].

Of most direct interest, however, is the deliberate addition of *nucleating agents*, used commercially for two key reasons. Firstly, addition of these agents accelerates crystallization and thereby the rate of manufacture of molded products for which a semicrystalline condition is desired [61]; this is especially important for slow crystallizing polymers such as the widely used isotactic polypropylene [62]. Secondly, the agents are effective in controlling microstructure and achieving the small spherulite sizes [63, 64], which underlie improved mechanical properties such as increased ultimate tensile strength [65], Young modulus [65], and impact strength [66, 67]. Nucleating agents are also important as *clarifying agents* because the finer microstructures show reduced haze and turbidity and increased transparency [62].

Common nucleating agents contain benzene rings (i.e. they are *aromatic*) and are effective for a wide range of polymers. This lack of specificity in their action led to the suggestion that the nucleating action does not depend on structural matching (epitaxy) at the interface with the polymer [68]. Indeed, it is found that inorganic glass surfaces can stimulate nucleation [69]. However, in most cases epitaxy is important, as shown particularly in studies of nucleation on large single crystals of potent nucleating agents [70]. Commonly, these aromatic compounds can be prepared as thin plates with large {0 0 1} surfaces. Anthracene can be taken as typical of a wide range of fused-ring aromatic compounds, all isomorphous on their {0 0 1} surface. Polyethylene nucleates on this surface with its {1 0 0} planes parallel to them. Polyphenyls are another common family of nucleating agents: in this case, different periodicity in their {0 0 1} surface matches with the {1 1 0} planes of polyethylene. Such nucleating agents are effective also for polypropylene, aliphatic polyesters, and polyamides, as these have crystal periodicities very similar to those of polyethylene [70]. There have also been studies of epitaxy on prismatic, non-{0 0 1}, surfaces of nucleating agents [71], and often a given nucleating agent has at least two crystallographically distinct surfaces, permitting heteroepitaxial growth of a polymer crystal. A consequence of the importance of epitaxy is that it is possible to select nucleating agents to favor a particular crystallographic polymorph of a given polymer [67, 72]. A case of particular interest here is isotactic polypropylene, which shows α and β polymorphs. As noted above, nucleating agents are used in industrial processing of this polymer. The α-phase is thermodynamically more stable and is more common, but there are agents that promote formation of the β-phase. The β-phase is mechanically much tougher because the stress around cracks induces transformation to the α-phase [62]; this *transformation toughening* is analogous to that in some ceramics and metallic alloys.

Given the above, it is of interest to consider the design of nucleating agents. As an example, Blomenhofer et al. [62] have explored the nucleation potency, for the α- and β-phases of isotactic polypropylene, of a family of aromatic compounds, the 1,3,5-trisamides. The particular examples studied have a benzene ring as the core, amides capable of forming hydrogen bonds, and an apolar group used to adjust the crystallographic spacings of the additive. An example of the screening of different compounds in this work is given in Figure 12, where (a) shows the opacity induced by nucleation of the light-scattering β-phase and (b) shows a transparent ring where the additive level induces particularly copious nucleation of α-phase. The potency of the compounds for nucleation of particular polymorphs is attributed to epitaxial matching [62]. A suitable

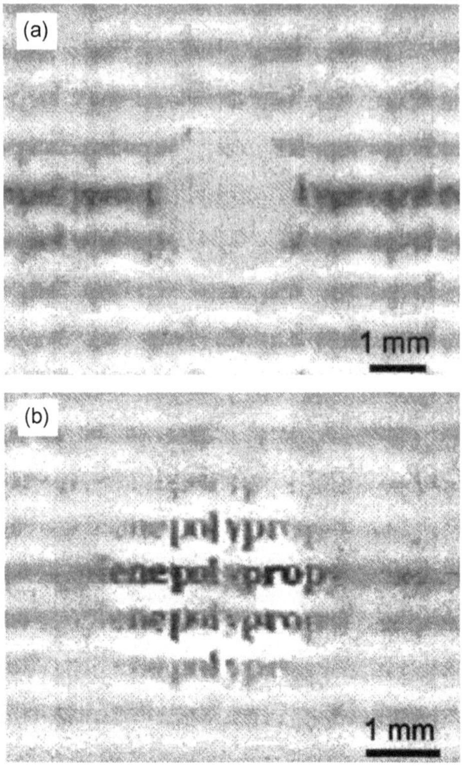

Fig. 12 Screening tests of isotactic polypropylene in which spots of different additives give a radial composition variation. In (a) the additive is an effective nucleant for the light-scattering β-phase; in (b) the particular addition level in a ring around the spot nucleates the α-phase in such a fine dispersion that transparency is enhanced. (Adapted with permission from Ref. [62], copyright (2005) American Chemical Society.)

organic nucleating compound may not be easy to disperse in the polymer, and it can then be of interest to use hybrid systems in which, for example, the compound is dispersed as a coating on inorganic nanoparticles [57].

Molecular-dynamics simulations have also proven useful in the screening of potential nucleating agents. Nagarajan and Myerson [23] have modeled the nucleation of isotactic polypropylene on a variety of derivatives of dibenzylidene sorbitol. The molecular-dynamics results were analyzed by counting the number of oriented backbone carbon atoms as a measure of the number of nucleated crystallizing sites. The orientation of importance is the adoption of a helical conformation of the kind found in the crystalline lattice. A typical simulation included four polypropylene chains, each with 90 backbone carbon atoms, and this was sufficient to show that benzene rings on either side of sorbitol in the nucleant molecule give it a stiffness favorable for nucleation. The differences in nucleation potency indicated by the molecular-dynamics are consistent with experimental observations. This encouraging result suggests prospects for molecular design of nucleating agents.

6. SECONDARY NUCLEATION: THE LAURITZEN–HOFFMAN THEORY

The analysis of nucleation in polymer crystallization is often in the context of growth. As noted in Section 1 polymer crystals, whether grown from solution or from the melt, take the form of faceted lamellae. As for the growth of faceted crystals in general (mechanisms are reviewed in Ref. [73]), it is expected that attachment of new molecules can occur most easily at surface ledges (commonly termed *niches* in the polymer literature). Attachment is much more difficult for polymer chains than for single atoms or small molecules. When a polymer chain starts to be attached to the crystal surface there is a large drop in its entropy; this unfavorable change constitutes an energy barrier. Crystal growth depends on addition of new chains. This is clearly the case for short chains, which can be reeled completely into the crystal. For long chains, complete reeling in is impossible and new chains must be attached for growth to continue. In atomic and small-molecule systems, dislocations reaching the surface of a faceted crystal can facilitate its growth by creating perpetual ledges. Growth at such ledges adopts a spiral form incompatible with the attachment of polymer chains to form an aligned folded array (Figure 2). Growth without such ledges requires ongoing nucleation of this attachment of new chains. This is *secondary nucleation* and is distinct from the primary nucleation considered so far in this chapter.

The classical model for secondary nucleation is that of Lauritzen and Hoffman [74, 75]. This has been steadily refined, and continues to be the basis of analytical treatments of polymer crystal growth [10] that are able to explain most phenomena. Modern versions of the Lauritzen–Hoffman (L-H) theory (for example, as reviewed in Ref. [76]) are detailed and complex; a full presentation of this theory for growth lies well outside the scope of this book. Even so, a brief outline of the theory is useful as it has relevance for the heterogeneous nucleation of polymers on substrates, and because it provides a quantitative basis for consideration of the distinctive polymer transport kinetics arising from reptation (Section 1).

The basis of the L-H theory is surface nucleation of an attached chain, followed by lateral growth of the next layer of crystal by continuing deposition of the chain in a series of folds. Though capable of extension to other cases, this theory was developed for the growth of polyethylene crystals from the melt. Figure 13a depicts a polyethylene crystal. The plane of folds parallel to the page is one face of the lamellar form. The chains are perpendicular to this face. The thickness of the lamella is the dimension ℓ. Whether growing from solution or the melt, ℓ is inversely proportional to the driving force for crystallization, a relationship successfully predicted by the L-H theory [10, 76]. The dimension L is the coherence length of the lattice. Defects in the structure prevent lateral growth from a nucleation site on the surface spreading over a span greater than L. The growth velocity of this crystal, represented by the vector v, is measured normal to the most closely packed plane in polyethylene, of the {1 1 0} type. A new polymer chain may originally

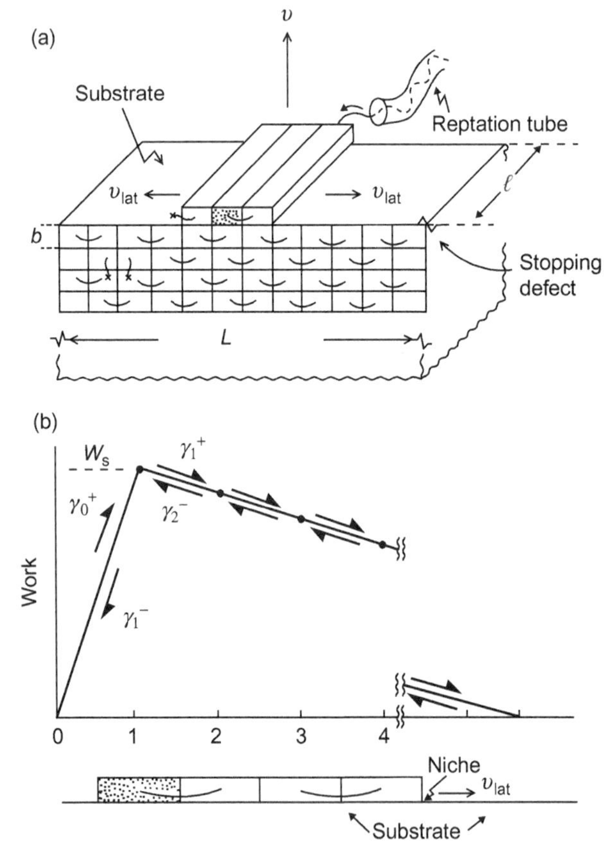

Fig. 13 The Lauritzen-Hoffman (L-H) theory for polymer crystal growth. (a) a folded-chain crystal can grow at a rate represented by the vector v as new stems are added on the top surface. (b) the work associated with stem addition, with rate constants. (Adapted from Ref. [76], copyright (1997), with permission from Elsevier.)

attach in a two-dimensional random coil on this surface [77], but then aligns with the existing chains to form a new *stem* (segment of length ℓ); the end of the first stem is shaded in Figure 13b. In the classical L-H theory the work W_s, of attaching this stem is calculated from the increased area of the interface forming the stem sides:

$$W_s = 2b\ell\sigma_{ls}, \tag{3}$$

where b is the thickness of a layer of chains. The main contribution to W_s, when it is evaluated more carefully, is the loss of entropy of the chain [10, 76, 78]. Once the first stem is in place, the chain can fold at the lamellar surface and lay down parallel stems along the niche, giving a lateral growth velocity described by the vector υ_{lat}. While a full treatment includes a specific fold energy [76], a simple picture of the work is shown in Figure 13b: the key barrier is the work to lay down the initial stem, as subsequent growth is favorable. The lamellar thickness ℓ maximizes the growth velocity υ, balancing the opposing effects of the nucleation barrier W_s and the driving force for crystallization, each of which increases with increasing ℓ [10]. The rate constant γ_0^+ (Figure 13b) is analogous to those in Chapter 2, Eq. (25), and is given by

$$\gamma_0^+ = \beta \exp\left(\frac{-W_s}{k_B T}\right), \tag{4}$$

where β is a prefactor governed by reptation and therefore different from that for atoms or small molecules. In long-chain systems, growth from the melt of large folded-chain lamellae would be virtually impossible without reptation — the forced diffusion of a chain inside the effective tube formed by neighboring molecules in the melt. The prefactor for a chain of n –CH_2– units is

$$\beta = \frac{\kappa}{n}\left(\frac{k_B T}{h}\right) \exp\left(\frac{-Q_{rept}}{RT}\right), \tag{5}$$

where k_B, T, and R ($= N_A k_B$) have their usual meanings, κ is a constant related to the frictional force on a monomer, and h is the Planck constant [76]. The activation energy Q_{rept} is that for reptation, essentially the same as that for diffusion, but distinct from that for viscous flow [76]. The greater friction on a longer chain leads to β being inversely proportional to n. The velocity at which a chain can be reeled into a crystal is inversely proportional to n, and the time for complete reeling in is proportional to n^2. Thus the crystal growth rate is lower for polymers of higher molecular weight and there is a segregation effect in which crystals are preferentially formed from shorter chains [10].

Three growth regimes can be distinguished (Figure 14). In regime I, the time for complete lateral growth is small compared to the interval between nucleation events; in this case, the liquid/solid interface remains

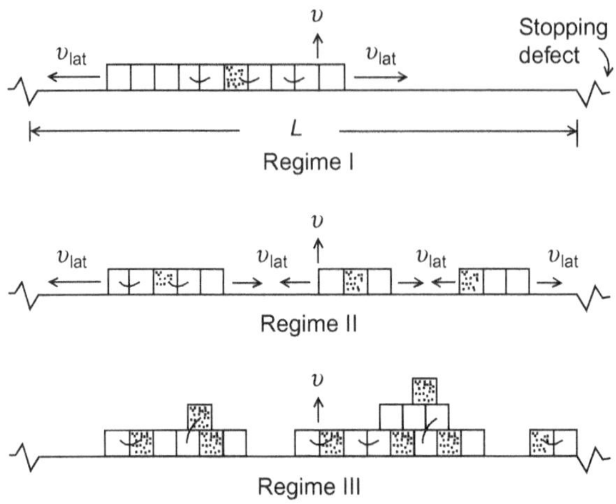

Fig. 14 The three regimes of crystal growth in the L-H theory: I, one nucleation event leads to complete coverage of the surface plane; II, there are multiple events; III, nucleation occurs even on partial layers. The shaded areas are the ends of the nucleating stems in each layer. (Adapted from Ref. [76], copyright (1997), with permission from Elsevier.)

flat. In regime III, the time for lateral growth is large compared to the nucleation interval; the interface is then rough. Regime II is an intermediate case. Each regime has distinctive growth kinetics, and clear evidence for them (Figure 15) supports the L-H theory [79].

7. RIGID MOLECULES: ISOTROPIC-TO-NEMATIC TRANSITION

In this section, we consider the nucleation of crystals from rigid rod-like molecules or particles, in solution or colloidal suspension. In addition to human-made rigid polymers [80] and carbon nanotubes, such particles are common, for example proteins [81], viruses [80, 82], and clay particles [83, 84], and their behavior attracts a wide range of interest. The main technological importance lies with liquid crystals [85] and liquid-crystalline polymers. The flexible-chain polymers considered so far in this chapter show crystallization mechanisms quite distinct from those of atoms or simple spheroidal molecules. Rigid molecules cannot adopt the folding mechanisms possible for flexible chains. Even when rigid molecules are rather short and deviate only slightly from sphericity, for example prolate ellipsoids with the major and minor axes in the ratio 1.25 [86], the nucleation is found to be quite different from that for spherical particles. While the nucleation of crystals from spherical molecules or particles, and from long flexible chains, is relatively well understood, that

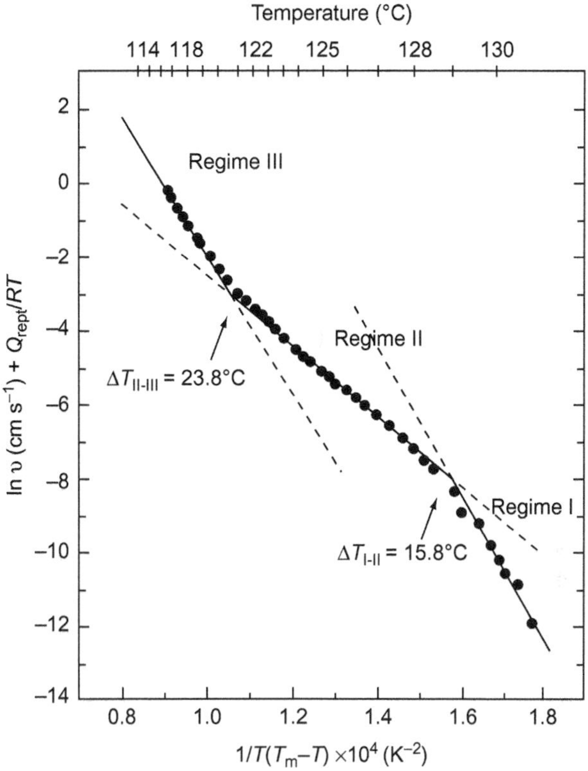

Fig. 15 The growth rate for polyethylene crystals forming in the melt, plotted according to the L-H theory against $1/T(T_m-T)$, where T is the temperature and T_m is the equilibrium melting temperature (417.9 K in this case). Q_{rept} is the activation energy for reptational diffusion (24 kJ mol^{-1} in this case). As the supercooling ΔT ($=T_m-T$) is increased, three growth regimes, I, II and III are found. (Adapted with permission from Ref. [79], copyright (2002) American Chemical Society.)

from rigid nonspherical molecules or particles remains the subject of active study so far without any straightforward analytical treatment.

The state in which neighboring molecules are aligned parallel to each other without any long-range positional order is termed *nematic*. Molecules without well-defined order along their chains cannot crystallize according to the mechanisms for flexible chains described earlier in this chapter, but if they are rigid, nematic alignment is common. The suspension of particles can also be isotropic. In the nematic phase, there is a high concentration of particles with high orientational order, in the isotropic phase a low concentration with no orientational order. As the overall concentration of particles is increased, the system can undergo spinodal decomposition into the high- and low-concentration phases.

Alternatively, at appropriate supersaturation, the high-concentration nematic phase can nucleate and grow to form crystals often called *tactoids*. There is plenty of experimental evidence that this isotropic-to-nematic phase transition is first-order. For example, experiments on a *nematogen* liquid crystal, in which supersaturation is induced by a sudden pressure increase, show the existence of a metastable supersaturated state with an induction time for nucleation of the nematic phase [85].

The very wide range of transitions found in liquid–crystal systems [86] lies well beyond our scope, but the relatively simple isotropic-to-nematic transition serves to illustrate some of the distinct problems with analysis of nucleation when the molecules or particles are nonspherical. Since the nematic crystal has aligned molecules, its interfacial energy with the surrounding isotropic phase and the interfacial attachment kinetics are both strongly anisotropic, and the crystal itself is of course not spherical. Furthermore, the existence of the crystal can induce orientation ordering of the molecules in the surrounding phase. In colloids with short rigid rods (length-to-diameter ratio of 2, for example), the rods around a crystal can align perpendicular to its surface thereby greatly impeding growth, an effect termed *self-poisoning* [86].

The isotropic-to-nematic transition in colloids has been widely studied in Monte Carlo simulations. Schilling and Frenkel found that for mildly nonspherical particles (rods with length-to-diameter ratio of 2 and prolate ellipsoids with axial ratio of 1.25) it was impossible to define the nucleation path and therefore to adopt standard analyses of barrier crossing. Although the classical nucleation theory predicts that critical work of nucleation and critical nucleation radius diverge to infinity for infinitely long rods, it may be that nucleation mechanisms become simpler for rods of intermediate length [87]. Cuetos and Dijkstra [87] applied Monte Carlo simulations to rigid rods with length-to-diameter ratios in the range 5 to 15. They found that for high enough length-to-diameter ratio, the isotropic-to-nematic transition is clearly first-order, and that at appropriate supersaturation the simulation shows classical nucleation kinetics. In the simulation, nematic clusters appear and disappear, and after a sufficient induction time a cluster may grow. In this case, there is a well-defined nucleation path, and the classical theory can be applied with thermodynamic input parameters from Onsager's analysis of oriented systems [88]. It is found that the classical theory gives reasonable predictions for the critical work of nucleation W^*, but overestimates the critical size (number of molecules) n^*. The complexity of the nucleation is illustrated in Figure 16, which shows contour plots for density (a, b) and for the degree of nematic alignment (c, d) for a subcritical embryo (a, c) with $n = 20$ and for a critical nucleus with $n^* = 95$ (b, d). In this case the length-to-diameter ratio of the

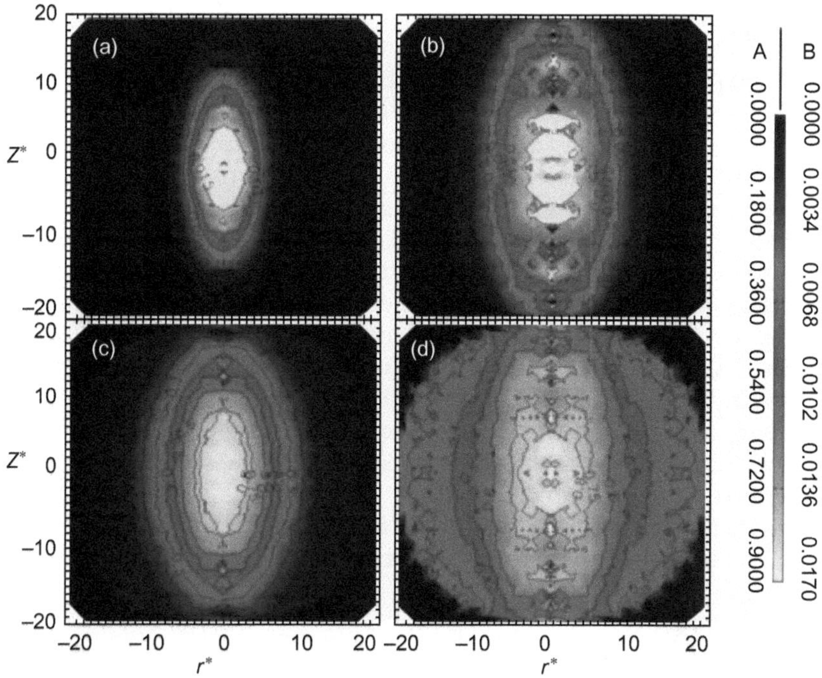

Fig. 16 Monte Carlo simulations of a nematic cluster of rigid rods (length-to-diameter ratio 15) forming in an isotropic phase. The contour plots show the concentration of rods (a, b, scale B) and the degree of their nematic alignment (c, d, scale A) for a subcritical embryo (a, c) with $n = 20$ and for a critical nucleus with $n^* = 95$ (b, d). (Reprinted with permission from [87], copyright (2007) by the American Physical Society.)

particles is 15. The nematic clusters are prolate ellipsoids with axial ratio ≈ 1.7, with smoothly varying density. It is striking that the degree of nematic alignment varies in a similar way, but extends further from the cluster center. Cuetos and Dijkstra note that there is a need for better nucleation theories for nonspherical particles [87].

8. SUMMARY

This chapter has focused on linear polymers with flexible chains, capable of attaining a high degree of crystallinity. In broad terms, the formation of polymer crystals, whether from solution or from the melt, shows familiar nucleation phenomena: isothermal and athermal nucleation kinetics, heterogeneous catalysis, and possible homogeneous nucleation. Yet the long-chain nature of polymers has profound effects. The motion of the chains by reptation is very different from small-molecule diffusion.

A chain can contain many more monomers than a crystal nucleus and a given chain can participate in many nuclei; this gives distinctive types of interaction between nuclei. The complexity of chain conformations and packing, and the relative sluggishness of chain motion, even in the melt, give strong thermal history or memory effects on crystal nucleation. Flow-induced alignment of the chains in solution or in the melt has a dramatic effect on nucleation, giving a new mode of *shish-kebab* crystallization not seen without flow, and quite unknown in nonpolymeric systems. Nucleating agents are widely used in polymer processing; there has been some progress in understanding their operation and in optimizing their performance. The long-chain nature of polymers also affects their crystal growth, and underlies the important growth mechanism of continuing secondary nucleation on crystal surfaces.

The formation of crystals from rigid nonspherical molecules is at yet poorly understood, yet is of importance in a very wide range of systems. Existing work has served mainly to illustrate the complexity of the processes involved, which include the influence of the crystalline cluster on the molecular or particle alignment in the surrounding phase.

REFERENCES

[1] A. Keller, A note on single crystals in polymers: evidence for a folded chain configuration, Philos. Mag. **2** (1957) 1171–1175; Pl. 44–47.
[2] P.J. Phillips, Polymer crystals, Rep. Prog. Phys. **53** (1990) 549–604.
[3] D.W. Oxtoby, Crystal nucleation in simple and complex fluids, Phil. Trans. Roy. Soc. Lond. A **361** (2003) 419–428.
[4] P.R. ten Wolde, D. Frenkel, Enhancement of protein crystal nucleation by critical density fluctuations, Science **277** (1977) 1975–1978.
[5] U. Gasser, E.R. Weeks, A. Schofield, P.N. Pusey, D.A. Weitz, Real-space imaging of nucleation and growth in colloidal crystallization, Science **292** (2002) 258–262.
[6] A.H. Windle, A Metallurgist's Guide to Polymers, in: *Physical Metallurgy*, 4th edn, Eds. R.W. Cahn, P. Haasen, Vol. III, North-Holland, Amsterdam (1996), pp. 2663–2740; Chapter 32.
[7] W.D. Callister, Jr., *Materials Science and Engineering. An Introduction*, 5th edn, New York, Wiley (2000), Chapter 15, pp. 446–470.
[8] A. Keller, Crystalline polymers: An introduction, Faraday Disc. **68** (1979) 145–166.
[9] P.G. De Gennes, *Scaling Concepts in Polymer Physics*, Cornell University Press, Ithaca, NY (1979).
[10] S.Z.D. Cheng, B. Lotz, Enthalpic and entropic origins of nucleation barriers during polymer crystallization: the Hoffman–Lauritzen theory and beyond, Polymer **46** (2005) 8662–8681.
[11] R.L. Cormia, F.P. Price, D. Turnbull, Kinetics of crystal nucleation in polyethylene, J. Chem. Phys. **37** (1962) 1333–1340.
[12] J.R. Burns, D. Turnbull, Kinetics of crystal nucleation in molten isotactic polypropylene, J. Appl. Phys. **37** (1966) 4021–4026.
[13] A.H. Lebovitz, K. Khait, J.M. Torkelson, Sub-micron dispersed-phase particle size in polymer blends: overcoming the Taylor limit via solid-state shear pulverization, Polymer **44** (2003) 199–206.

[14] F.M. Bauers, R. Thomann, S. Mecking, Submicron polyethylene particles from catalytic emulsion polymerization, J. Am. Chem. Soc. **125** (2003) 8838–8840.
[15] A. Taden, K. Landfester, Crystallization of poly(ethylene oxide) confined in miniemulsion, Macromolecules **36** (2003) 4037–4041.
[16] J.I. Uriguen, L. Bremer, V. Mathot, G. Groeninckx, Preparation of water-borne dispersions of polyolefins: new systems for the study of homogeneous nucleation of polymers, Polymer **45** (2004) 5961–5968.
[17] M.V. Massa, K. Dalnoki-Veress, Homogeneous crystallization of poly(ethylene oxide) confined to droplets: the dependence of the crystal nucleation rate on length scale and temperature, Phys. Rev. Lett. **92** (2004) 255509/1–4.
[18] Y.-L. Loo, R.A. Register, A.J. Ryan, Polymer crystallization in 25-nm spheres, Phys. Rev. Lett. **84** (2000) 4120–4123.
[19] G. Reiter, G. Castelein, J.-U. Sommer, Direct visualization of random crystallization and melting in arrays of nanometer-size polymer crystals, Phys. Rev. Lett. **87** (2001) 226101/1–4.
[20] W. Paul, G.D. Smith, Structure and dynamics of amorphous polymers: computer simulations compared to experiment and theory, Rep. Prog. Phys. **67** (2004) 1117–1185.
[21] C. Liu, M. Muthukumar, Langevin dynamics simulations of early-stage polymer nucleation and crystallization, J. Chem Phys. **109** (1998) 2536–2542.
[22] K. Esselink, P.A.J. Hilbers, B.W.H. van Beest, Molecular-dynamics study of nucleation and melting of *n*-alkanes, J. Chem. Phys. **101** (1994) 9033–9041.
[23] K. Nagarajan, A.S. Myerson, Molecular-dynamics of nucleation and crystallization of polymers, Cryst. Growth Design **1** (2001) 131–142.
[24] W. Hu, D. Frenkel, V.B.F. Mathot, Intramolecular nucleation model for polymer crystallization, Macromolecules **36** (2003) 8178–8183.
[25] M. Muthukumar, Molecular modelling of nucleation in polymers, Phil. Trans. Roy. Soc. Lond. A **361** (2003) 539–556.
[26] J.W.G. Goodby, G.W. Gray, *Structures and Textures of Liquid Crystals*, CRC Press, Boca Raton (2007).
[27] P.D. Olmsted, W.C.K. Poon, T.C.B. McLeish, N.J. Terrill, A.J. Ryan, Spinodal-assisted crystallization in polymer melts, Phys. Rev. Lett. **81** (1998) 373–376.
[28] A. Wurm, R. Soliman, J.P.G. Goossens, W. Bras, C. Schick, Evidence of pre-crystalline-order in super-cooled polymer melts revealed from simultaneous dielectric spectroscopy and SAXS, J. Non-Cryst. Solids **351** (2005) 2773–2779.
[29] W. Banks, M. Gordon, A. Sharples, The crystallization of polyethylene after partial melting, Polymer **4** (1963) 289–302.
[30] J.A. Martins, J.J.C.C. Pinto, Evaluation of the instantaneous nucleation density in the isothermal crystallization of polymers, Polymer **43** (2002) 3999–4010.
[31] P. Supaphol, J.E. Spruiell, Crystalline memory effects in isothermal crystallization of syndiotactic polypropylene, J. Appl. Polym. Sci. **75** (2000) 337–346.
[32] S. Schneider, X. Drujon, B. Lotz, J.C. Wittmann, Self-nucleation and enhanced nucleation of polyvinylidene fluoride (α-phase), Polymer **42** (2001) 8787–8798.
[33] X. Zhu, Y. Li, D. Yan, Y. Fang, Crystallization behavior of partially melting isotactic polypropylene, Polymer **42** (2001) 9217–9222.
[34] K. Cho, D.N. Saheb, J. Choi, H. Yang, Real time in situ X-ray diffraction studies on the melting memory effect in the crystallization of β-isotactic polypropylene, Polymer **43** (2002) 1407–1416.
[35] K. Cho, D.N. Saheb, H.C. Yang, B.I. Kang, J. Kim, S.S. Lee, Memory effect of locally ordered alpha-phase in the melting and phase transformation behavior of beta-isotactic polypropylene, Polymer **44** (2003) 4053–4059.
[36] H. Janeschitz-Kriegl, E. Ratajski, M. Stadlbauer, Flow as an effective promoter of nucleation in polymer melts: a quantitative evaluation, Rheol. Acta, **42** (2003) 355–364.

[37] J. van Meerveld, G.W.M. Peters, M. Hütter, Towards a rheological classification of flow induced crystallization experiments of polymer melts, Rheol. Acta **44** (2004) 119–134.
[38] G. Eder, H. Janeschitz-Kriegl, S. Liedauer, Crystallization processes in quiescent and moving polymer melts under heat-transfer conditions, Prog. Polym. Sci. **15** (1990) 629–714.
[39] S. Liedauer, G. Eder, H. Janeschitz-Kriegl, On the limitations of shear-induced crystallization in polypropylene melts, Int. Polym. Process. **10** (1995) 243–250.
[40] A. Keller, M.J. Machin, Oriented crystallization in polymers, J. Macromol. Sci. B Phys. B **1** (1967) 41–91.
[41] L. Li, W.H. de Jeu, Shear-induced smectic ordering in the melt of isotactic polypropylene, Phys. Rev. Lett. **92** (2004) 075506/1–3.
[42] A.J. Pennings, J.M.A.A. van der Mark, A.M. Kiel, Hydrodynamically induced crystallization of polymers from solution: III. Morphology, *Kolloid Z. Z. Polym.* **237** (1970) 336–358.
[43] I. Dukovski, M. Muthukumar, Langevin dynamics simulations of early stage shish-kebab crystallization of polymers in extensional flow, J. Chem. Phys. **118** (2003) 6648–6655.
[44] P. Zhu, J. Tung, G. Edward, Effects of sodium benzoate on distributions of lamellar structures of sheared isotactic polypropylene: a study using small-angle synchrotron X-ray scattering, Polymer **46** (2005) 10960–10969.
[45] S.P. Kim, S.C. Kim, Crystallization kinetics of poly(ethylene-terephthalate) — memory effect of shear history, Polym. Eng. Sci. **33** (1993) 83–91.
[46] S. Yamasaki, M. Hikosaka, A. Toda, I. Wataoka, K. Yamada, K. Tagashira, Nucleation and morphology of polyethylene under shear flow, J. Macromol. Sci., Phys. B **42** (2003) 499–514.
[47] D. Vesely, G. Ronca, Epitaxial nucleation of crystallization at polymer–filler interfaces, J. Microsc. **Pt 2**(201), (2000) 137–143.
[48] W. Zheng, X.H. Lu, S.C. Wong, Electrical and mechanical properties of expanded graphite-reinforced high-density polyethylene, J. Appl. Polym. Sci. **91** (2004) 2781–2788.
[49] P. Jacoby, B.H. Bersted, W.J. Kissel, C.E. Smith, Studies on the beta crystalline form of isotactic polypropylene, J. Polym. Sci. Pt B **24** (1986) 461–491.
[50] J. Broda, Nucleating activity of the quinacridone and phthalocyanine pigments in polypropylene crystallization, J. Appl. Polym. Sci. **90** (2003) 3957–3964.
[51] T. Nagy, J.L. White, The effect of colorants on the properties of rotomolded polyethylene parts, Polym. Eng. Sci. **36** (1996) 1010–1018.
[52] A. Maffezzoli, J.M. Kenny, L. Nicolais, A macrokinetic approach to crystallization modeling of semicrystalline thermoplastic matrices for advanced composites, J. Mater. Sci. **28** (1993) 4994–5001.
[53] Y. Lee, R.S. Porter, Crystallization of poly(etheretherketone) (PEEK) in carbon-fiber composites, Polym. Eng. Sci. **26** (1986) 633–639.
[54] S.Y. Yang, J.R. Castilleja, E.V. Barrera, K. Lozano, Thermal analysis of an acrylonitrile-butadiene-styrene/SWNT composite, Polym. Degradation Stability **83** (2004) 383–388.
[55] R.K. Goyal, Y.S. Negi, A.N. Tiwari, Preparation of high performance composites based on aluminum nitride/poly(ether–ether–ketone) and their properties, Eur. Polym. J. **41** (2005) 2034–2044.
[56] S. Jain, H. Goossens, M. van Duin, P. Lemstra, Effect of in situ prepared silica nano-particles on non-isothermal crystallization of polypropylene, Polymer **46** (2005) 8805–8818.
[57] J.G. Tang, Y. Wang, H.Y. Liu, L.A. Belfiore, Effects of organic nucleating agents and zinc oxide nanoparticles on isotactic polypropylene crystallization, Polymer **45** (2004) 2081–2091.

[58] E. Di Maio, S. Iannace, L. Sorrentino, L. Nicolais, Isothermal crystallization in PCL/clay nanocomposites investigated with thermal and rheometric methods, Polymer **45** (2004) 8893–8900.
[59] J.W. Huang, C.C. Kang, T.H. Chen, Isothermal crystallization of poly(ethylene-*co*-glycidyl methacrylate)/clay nanocomposites, J. Appl. Polym. Sci. **97** (2005) 1051–1064.
[60] Y. Li, J.H. Ma, Y.M. Wang, B.R. Liang, Synthesis and characterization of sulfonated poly (ethylene terephthalate)/montmorillonite nanocomposites, J. Appl. Polym. Sci. **98** (2005) 1150–1156.
[61] S.S. Song, J.L. White, M. Cakmak, Crystallization kinetics and nucleating agents for enhancing the crystallization of poly(paraphenylene sulfide), Polym. Eng. Sci. **30** (1990) 944–949.
[62] M. Blomenhofer, S. Ganzleben, D. Hanft, H.-W. Schmidt, M. Kristiansen, P. Smith, K. Stoll, D. Mäder, K. Hoffmann, "Designer" nucleating agents for polypropylene, Macromolecules **38** (2005) 3688–3695.
[63] K.A. Narh, J.A. Odell, A. Keller, G.V. Fraser, The effect of nucleation density on the kinetics of crystallization and on the resulting structure and thermal properties of polymers crystallized during cooling, J. Mater. Sci. **15** (1980) 2001–2009.
[64] J.O. Iroh, J.P. Berry, Heterogeneous nucleation of short glass fiber polypropylene composites, Polymer **34** (1993) 4747–4751.
[65] J.O. Iroh, J.P. Berry, Mechanical properties of nucleated polypropylene and short glass fiber-polypropylene composites, Eur. Polym. J. **32** (1996) 1425–1429.
[66] K. Trongtorsak, P. Supaphol, S. Tantayanon, Effect of calcium stearate and pimelic acid addition on mechanical properties of heterophasic isotactic polypropylene/ethylene–propylene rubber blend, Polym. Testing **23** (2004) 533–539.
[67] M. Obadal, R. Cermak, N. Baran, K. Stoklasa, J. Simonik, Impact strength of beta-nucleated polypropylene, Int. Polym. Process. **19** (2004) 35–39.
[68] F.L. Binsbergen, Heterogeneous nucleation in the crystallization of polyolefins: Part 1. Chemical and physical nature of nucleating agents, Polymer **11** (1970) 253–267.
[69] P.C. Guschi, J.U. Otaigbe, Crystallization kinetics of low-density polyethylene and polypropylene melt-blended with a low-T_g tin-based phosphate glass, J. Appl. Polym. Sci. **90** (2003) 3445–3456.
[70] J.C. Witmann, B. Lotz, Epitaxial crystallization of polymers on organic and polymeric substrates, Prog. Polym. Sci. **15** (1990) 909–948.
[71] D. Parikh, P.J. Phillips, Epitaxial crystallization of polyethylene on single crystals of acrylidine, J. Polym. Sci. C **28** (1990) 167–172.
[72] W.H. Kai, B. Zhu, Y. He, Y. Inoue, Crystallization of poly(butylene adipate) in the presence of nucleating agents, J. Polym. Sci, Pt B **43** (2005) 2340–2351.
[73] W.A. Tiller, *The Science of Crystallization: Microscopic Interfacial Phenomena*, CUP, Cambridge (1991) §2.2, pp 74–90.
[74] J.I. Lauritzen, Jr., J.D. Hoffman, Theory of formation of polymer crystals with folded chains in dilute solution, J. Res. Natl. Bur. Stand. A **64A** (1960) 73–102.
[75] J.D. Hoffman, J.I. Lauritzen, Jr., Crystallization of bulk polymers with chain folding: theory of growth of lamellar spherulites, J. Res. Nat. Bur. Stand. A **65A** (1961) 297–336.
[76] J.D. Hoffman, R.L. Miller, Kinetics of crystallization from the melt and chain folding in polyethylene fractions revisited: theory and experiment, Polymer **38** (1997) 3151–3212.
[77] J.I. Lauritzen, Jr., J.D. Hoffman, Extension of theory of growth of chain-folded polymer crystals to large undercoolings, J. Appl. Phys. **44** (1973) 4340–4352.
[78] D. Turnbull, F. Spaepen, Crystal nucleation and crystal-melt interfacial tension in linear hydrocarbons, J. Polym. Sci. C Polym. Symp. **63** (1977) 237–243.

[79] J.P. Armistead, J.D. Hoffman, Direct evidence of regimes I, II, and III in linear polyethylene fractions as revealed by spherulite growth rates, Macromolecules **35** (2002) 3895–3913.
[80] M.P. Lettinga, K. Kang, A. Imhof, D. Derks, J.K.G. Dhont, Kinetic pathways of the nematic-isotropic phase transition as studied by confocal microscopy on rod-like viruses, J. Phys. Cond. Matter **17** (2005) S3609–S3618.
[81] J. Viamontes, P.W. Oakes, J.X. Tang, Isotropic to nematic liquid crystalline phase transition of F-actin varies from continuous to first order, Phys. Rev. Lett. **97** (2006) 118103/1–4.
[82] Z. Dogic, Surface freezing and a two-step pathway of the isotropic-smectic phase transition in colloidal rods, Phys. Rev. Lett. **91** (2003) 165701/1–4.
[83] M.P.B. van Bruggen, J.K.G. Dhont, H.N.W. Lekkerkerker, Morphology and kinetics of the isotropic-nematic phase transition in dispersions of hard rods, Macromolecules **32** (1999) 2256–2264.
[84] Z.X. Zhang, J.S. van Duijneveldt, Isotropic-nematic phase transition of nonaqueous suspensions of natural clay rods, J. Chem. Phys. **124** (2006) 154910/1–7.
[85] N. Oranth, F. Strohbusch, Kinetics of the isotropic-nematic phase transition in liquid crystals. Nucleation from isotropic states of 4,4′-octylcyanobiphenyl supercooled by shock waves, J. Phys. Chem. **92** (1988) 1338–1341.
[86] T. Schilling, D. Frenkel, Nucleation in suspensions of anisotropic colloids, Comp. Phys. Commun. **169** (2005) 117–121.
[87] A. Cuetos, M. Dijkstra, Kinetic pathways for the isotropic-nematic phase transition in a system of colloidal hard rods: a simulation study, Phys. Rev. Lett. **98** (2007) 095701/1–4.
[88] L. Onsager, The effect of shape on the interaction of colloidal particles, Ann. NY Acad. Sci. **51** (1949) 627–659.

CHAPTER 12

Dislocation-Mediated Transformations

Contents

1. INTRODUCTION

Dislocations were introduced in Chapter 6, Section 3, where it was noted that, as static defects in a lattice, they can be favorable sites for nucleation of new phases. In this chapter, we consider nucleation phenomena dependent on the generation, motion, or rearrangement of dislocations. The importance of such phenomena is wide-ranging, from the brittleness or ductility of bulk crystalline materials, to the stress state in thin films, to the nucleation of broad classes of phase transformations. In all these cases, the phenomena are best analyzed in terms of the motion of dislocations, rather than of atoms or molecules, in contrast to all the types of nucleation considered so far in this book.

As noted in Chapter 6, Section 3, dislocations have long-range strain fields, and these are best treated at the continuum level. Classical analyses of dislocation behavior have been based at this level [1]. Equally,

Pergamon Materials Series, Volume 15
ISSN 1470-1804, DOI 10.1016/S1470-1804(09)01512-0

however, there are important mechanisms that can only be analyzed at the atomic level. In recent years, great advances have been made in developing modeling techniques to span these length scales, including large-scale molecular-dynamics simulations [2], phase-field [3] and density-functional [4] approaches, and finite-element calculations with constitutive relations derived from interatomic potentials [5]. Interestingly, atomistic simulations may be used to provide the physical basis for much improved continuum-level treatments [6], an approach that is important because it is computationally intractable to model relevantly large systems at an atomistic level. In parallel, progress has been made with experimental techniques, resolving effects attributable to individual dislocations (Section 2.3 of this chapter).

2. NUCLEATION OF DISLOCATIONS

2.1 Homogeneous nucleation of a dislocation loop

In plastic flow of crystals mediated by dislocation glide, dislocations are continually eliminated at free surfaces, grain boundaries and incoherent interphase interfaces, and they can also be rendered sessile in various ways. There is, therefore, a need for a continual supply of glissile dislocations. This does not normally involve nucleation, however, as dislocations multiply during plastic deformation by a variety of mechanisms, including Frank–Read sources at which a glissile segment of dislocation line can bow out to form a series of dislocation loops on a given glide plane [1]. Even in well-annealed metals, the dislocation density is high enough (of the order of $10^{10}\,m^{-2}$) to provide sufficient dislocation sources for all subsequent plasticity. However, there are situations in which dislocation nucleation is required: yielding of initially dislocation-free crystals (such as metallic whiskers [7], or at small surface indents [5, 8, 9], or around misfitting nanoparticles [10]); the generation of misfit dislocations in epitaxial thin films beyond a critical film thickness [11]; the brittle-to-ductile transition [12]; and the creep of superalloy single crystals [13]. In the following subsections, we examine some of these cases in more detail. First, we consider how a dislocation might nucleate within a defect-free crystal.

When dislocations are introduced into a crystalline solid, the increase in the internal energy is large enough, and the increase in entropy small enough that the free energy of the system is always raised: in thermal equilibrium, the population of dislocations (unlike vacancies) is never greater than zero. On the other hand, under an applied shear stress, there is a driving force for the formation of a dislocation loop (Figure 1), and homogeneous nucleation in this case has been analyzed, for example, by Hirth and Lothe [1]. When a loop is formed, the overall energy is raised by the line energy of the dislocation itself, Λ per unit length (Chapter 6, Eq. (43)). The resolved stress $\mathfrak{s}_{\mathrm{res}}$ on the system does work by moving the area within the loop (shaded in Figure 1) by the Burgers vector (assumed here to be parallel to the resolved traction on the glide plane) of length b. The work of formation W_{loop} of a loop of radius R is given by an expression of the form

$$W_{\mathrm{loop}} = 2\pi R\Lambda - \pi R^2(b\mathfrak{s}_{\mathrm{res}} \pm \Gamma_{\mathrm{SF}}), \tag{1}$$

where the term $\pm\Gamma_{\mathrm{SF}}$ is the energy (per unit area) of a stacking fault (see below) and is present only if the Burgers vector of the dislocation is not a lattice vector. This work has a maximum as a function of R, the corresponding value W^*_{loop} representing the nucleation barrier. The nucleation rate I is then given by

$$I = \gamma N_0 \exp\left(-\frac{W^*_{\mathrm{loop}}}{k_{\mathrm{B}}T}\right), \tag{2}$$

where N_0 is the number (per unit volume) of atomic sites, k_{B} and T have their usual meanings, and γ is the frequency of atomic motion enlarging the loop, approximated at the critical radius R^* by [1]

$$\gamma = \frac{8\pi R^* v}{b}, \tag{3}$$

where v is the Debye frequency. At lower temperatures, the loop cannot grow by single-atom movement, but must nucleate a double kink over

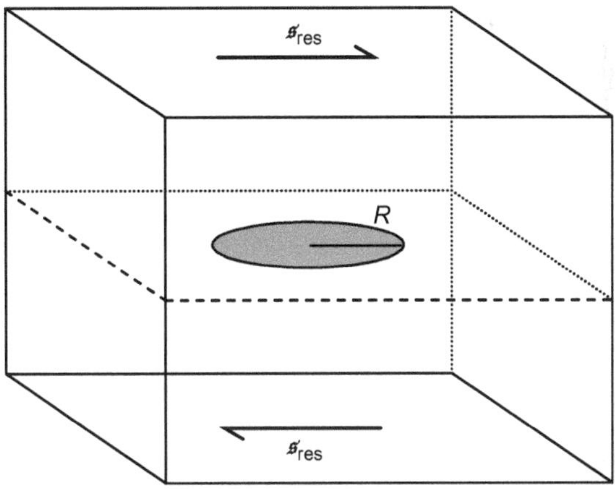

Fig. 1 A schematic illustration of a circular dislocation loop of radius R nucleating homogeneously within a single crystal subjected to an applied shear stress $\mathfrak{s}_{res}$. If the dislocation has a Burgers vector that is not a lattice vector, the shaded area is a stacking fault in the crystal.

the Peierls barrier; under these conditions, γ is much lower, as discussed by Hirth and Lothe [1]. Full expressions for the line energy Λ can be rather complex [1], but as noted in Chapter 6, Eq. (43), Λ is proportional to $\mu_s b^2$, where μ_s is the shear modulus of the material. Thus, by differentiating Eq. (1) and setting $\mathrm{d}W_{\mathrm{loop}}/\mathrm{d}R = 0$, we find that R^* is proportional to b and W^*_{loop} is proportional to b^3. Obviously short Burgers vectors are preferred, and therefore nucleation of *partial dislocations* can be expected. (Partial dislocations and the associated stacking faults are briefly introduced in Section 4.2 in this chapter) If the loop shown in Figure 1 is of a partial dislocation, a stacking fault (shaded) is created (or destroyed, if the nucleation occurs within an existing fault), and the term $\pm\Gamma_{\mathrm{SF}}$ is needed in Eq. (1). Substitution of values shows that homogeneous nucleation of a full loop in bulk material would occur only at stresses approaching the theoretical strength of a perfect crystal. However, the line energy Λ is dependent on the volume of material in the strain field of the dislocation. Consequently, nucleation is much easier near a free surface, particularly nucleation of a half-loop on a glide plane at a shallow inclination to the surface (see Figure 7, later in this chapter) for which image forces, attracting the dislocation toward the surface, are less important. At a surface, such a loop creates a step, or destroys an existing step, leading to another energy term so that the overall work of nucleation is of the form [11]

$$W_{\mathrm{loop}} = C_{\mathrm{loop}}\mu_s b^2 R - C_{\mathrm{WD}}\mathfrak{s}_{\mathrm{res}} b R^2 \pm C_{\mathrm{SF}}\Gamma_{\mathrm{SF}} R^2 \pm C_{\mathrm{step}}\mu_s b^2 R. \tag{4}$$

The constants (C_{loop}, C_{WD}, C_{SF}, and C_{step}, respectively, for the terms giving the contributions of the line energy, work done, stacking faults and steps) include geometrical factors. Considering only the first two contributions, it is readily seen that R^* and W^*_{loop} are inversely proportional to the resolved stress $\mathfrak{s}_{\mathrm{res}}$. The variation with R of the energy contributions in Eq. (4) is illustrated for a particular case in Section 2.4 (Figure 8).

This simple approach, based on macroscopic parameters, predicts that thermal activation energies are far too small to play a role in homogeneous nucleation of dislocations. Recently, as more accurate calculations have become possible, it has been of interest to revisit this question. The Peierls–Nabarro model of a dislocation treats the relative displacement between two atomic layers as a continuous distribution of infinitesimal dislocations. Extending this approach, Xu and Argon [14] have used a variational boundary-integral method to calculate the work of homogeneous nucleation. Although this method does not treat atomic-scale configurations, it is the favored way to calculate the work [5]. Xu and Argon show, for a variety of cubic-close packed (ccp) metals and for silicon, that even at resolved shear stresses as high as one half of the ideal shear strength of the crystal, thermal motion has no effect on the homogeneous nucleation of dislocation loops.

The dislocation loop considered in Figure 1 is formed on a single glide plane. Although not treated here, it should be noted that loops of other kinds (*prismatic* loops) can be formed when supersaturated vacancies condense into clusters. The nucleation kinetics of these clusters can be treated by classical theory [1] (Chapter 14, Section 5.2).

In a nanocrystalline material, the limited extent of the perfect crystal may permit special mechanisms to operate. For example, it is suggested that there can be homogeneous nucleation of loops of noncrystallographic partial dislocations. For these dislocations, the Burgers vector continuously grows in magnitude during nucleation [15].

2.2 Ductility and brittleness

In a cracked crystalline material, the competition between crack extension or crack blunting when the material is stressed determines whether the material is brittle or ductile [16]. The blunting of the crack tip occurs by dislocation emission, and the conditions for such emission have, therefore, been extensively studied both experimentally and theoretically. Experiments have often been based on cleavage cracks in dislocation-free single-crystal silicon [12, 17, 18]. Such experiments show the operation of slip systems that have the lowest energy barrier to dislocation emission from crack-front cleavage ledges. Relatively rare heterogeneous nucleation events are followed by dislocation multiplication by the usual mechanisms [17].

The most thorough early calculations were by Rice and Thomson [19], who derived the elastic strain field around an atomically sharp crack normal to the surface of the solid under a variety of loading conditions, and analyzed the interaction of the crack with a preexisting dislocation. This work was extended by Rice [20]; here we consider only his result for Mode-I loading of the crack, that is, tensile loading normal to the crack plane. The loading condition can be characterized by the applied strain-energy release rate $\mathfrak{G}$, for which there is a critical value $\mathfrak{G}_{\text{emit}}$ at which a fully formed straight dislocation line is emitted from the crack tip. Rice showed that for dislocation emission on a glide plane inclined at an angle ϕ to the crack plane

$$\mathfrak{G}_{\text{emit}} = \frac{8\Gamma_{\text{us}}}{(1+\cos\phi)\sin^2\phi}, \tag{5}$$

where Γ_{us} is the unstable stacking energy characterizing the barrier to glide. Similar analytical treatments based on continuum elasticity (e.g., [21]) capture the essentials of the phenomenon, but greater detail can now be provided by atomistic calculations. Such approaches, based on potentials derived using the embedded-atom method, have been used to examine dislocation nucleation at a straight crack edge [22, 23]. For example, Zhu et al. [23] have performed a calculation for copper taking both the crack and glide planes to be of {1 1 1} type, with $\phi = 70.53^\circ$. The unstable stacking energy $\Gamma_{us} = 158\,\mathrm{mJ\,m^{-2}}$. The three-dimensional atomistic simulation (total number of atoms 103,920) gives $\mathcal{G}_{emit} = 1.629\,\mathrm{J\,m^{-2}}$, significantly greater than the value $\mathcal{G}_{emit} = 1.067\,\mathrm{J\,m^{-2}}$ given by the continuum analytical approach (Eq. (5)). Zhu et al. noted that the latter approach ignores the contribution of crack-front surface production to the inhibition of dislocation emission. It has been suggested, however, that the continuum approach is better for Mode-II loading (i.e., shear on the plane of the crack) and that it can be improved by accounting for surface stress [22]. From the work of Zhu et al. [23], Figure 2 shows the critical nucleation condition for a dislocation loop at the crack front. On further loading, this loop would advance in both forward and lateral directions on the glide plane.

Dislocation nucleation at a straight crack front is often termed homogeneous. Zhu et al. applied their atomistic calculation to derive the work of nucleation, and they concluded that it is too high for thermally activated "homogeneous" nucleation to be feasible. As also noted in continuum analyses [21], any heterogeneity on the crack front can greatly reduce the work of nucleation, but the kinetics have not yet been derived from an atomistic calculation, even though this should now be feasible.

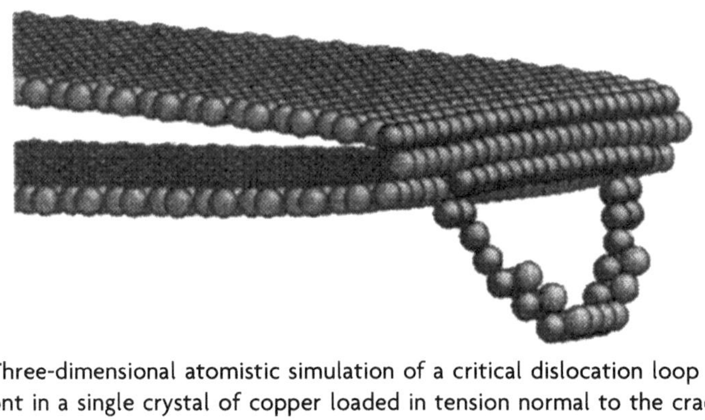

Fig. 2 Three-dimensional atomistic simulation of a critical dislocation loop at a crack front in a single crystal of copper loaded in tension normal to the crack plane. For clarity, all of the perfectly coordinated atoms are invisible, leaving only those on the crack surface and in the core of the dislocation. The crack plane and dislocation glide plane are both of {111} type. (Reprinted with permission from Ref. [23], copyright (2004) by the American Physical Society.)

2.3 Instrumented indentation

The advances in calculation methods just outlined have been matched by new measurement techniques for mechanical properties. Notable among these is instrumented indentation (often termed *nanoindentation*), which is capable of detecting the nucleation and motion of individual dislocations [24]. In this technique, an indenter tip is pressed into the surface of a sample, under load or displacement control. The technique is distinct from conventional indentation: the indent is small (typically <400 nm deep and <1 μm wide); and the load and displacement are monitored during indentation. Figure 3 shows typical load–displacement curves on indenting a single-crystalline ccp metal. As has been extensively analyzed, instrumented indentation can be used not only to study plasticity (on loading) but also to extract elastic constants (on unloading) [25]. Interest in loading curves of the type shown focuses on the sharp discontinuities in displacement, seen under load control as *pop-in* events. The displacement is fully elastic until the onset of plasticity indicated by the first pop-in. Analysis of the stress distribution under the indenter (for these studies, usually a spherical indenter, or the near-spherical tip of a Berkovich pyramidal indenter [24]) shows that, at the onset of plasticity, the maximum shear stress under the indenter is close

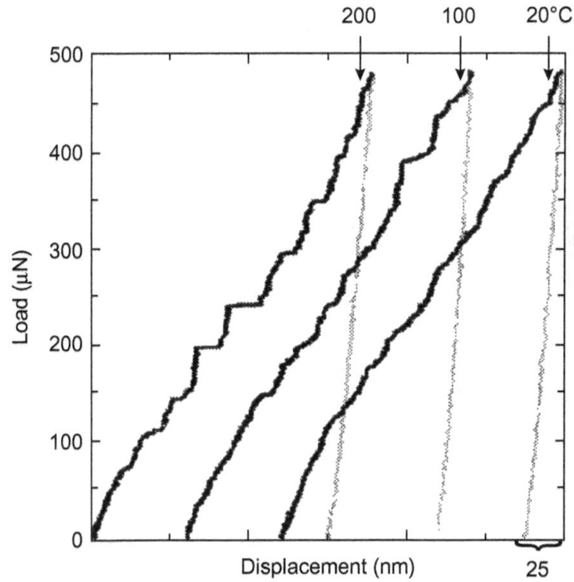

Fig. 3 Instrumented indentation of a (1 0 0) Pt single crystal under load control. The load–displacement curves (loading in black, unloading in gray) show pop-ins, marking bursts of plastic flow. (Reprinted with permission from Ref. [24], copyright (2004), American Institute of Physics.)

to the ideal shear strength of the crystal [26], and it has, therefore, been usual to conclude that the pop-in is a signature of homogeneous dislocation nucleation. It is plausible that nucleation is necessary, as the small area under the indent may be dislocation-free. Subsequent pop-ins are likely to be associated with the activation of existing dislocation sources.

Evidence for the initial homogeneous nucleation is provided by observations of bubble rafts that are two-dimensional analogs of atomic systems [8]. Under indentation, nucleation of dislocation pairs is observed to occur at well-defined locations within the crystal under the tip (Figure 4); the dislocations subsequently move in opposite directions, one making its way to the top surface of the crystal. The conditions for such nucleation in real three-dimensional crystals have been extensively analyzed. Approaches taken span the length scales from the continuum (finite-element modeling), through density-functional theory to the atomistic (molecular dynamics) [4, 26]. The most comprehensive modern work involves computationally efficient methods for continuum elastic calculations on relevant length and time scales, based on constitutive relations derived from interatomic potentials, and incorporating an instability criterion for dislocation nucleation verified at an atomistic level [5, 8]. Such approaches are capable of predicting the site of dislocation nucleation under an indenting tip, and the slip plane and Burgers vector for the

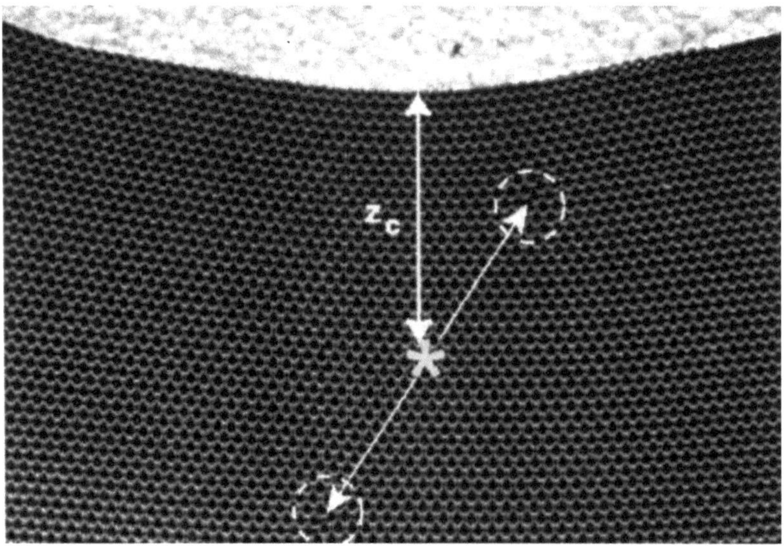

Fig. 4 Indentation of a soap-bubble raft causing nucleation (at $*$) of a pair of edge dislocations that then move apart. The distance z_c is 0.78 of the radius of the indenter tip. (Reprinted from Ref. [8], copyright (2004), with permission from Elsevier.)

dislocation thus formed [5,27]. As an example, Figure 5 shows a molecular-dynamics simulation of the homogeneous nucleation of dislocations in copper under a spherical indenter. Simulations of this type suggest that the first pop-in involves homogeneous nucleation of glissile dislocation loops, which then lead to dislocation multiplication through mechanisms similar to those of the Frank–Read source [27]. They also show the limitations of criteria for dislocation nucleation based on a critical resolved shear stress: the critical condition can depend on the sign of the shear, and is clearly affected by the local overall stress state [5]. While such studies provide useful insights, they are often conducted for 0 K, thus avoiding effects of thermal activation. Calculation of the work of nucleation can best be made using the variational boundary-integral method noted in Section 2.1 [14].

Simulations of the kind outlined earlier are for indentation of atomically flat surfaces. If the indented surface has a preexisting surface step, this can significantly reduce the load required for initiation of plasticity [28, 29]. The most recent instrumented-indentation studies have focused on the effects of temperature and loading rate on the first pop-in [9, 24]. As seen in Figure 3, increasing temperature lowers the load for the onset of plasticity, an effect also seen for lower loading rate. Such effects suggest a role for thermal activation of the onset of plasticity, in which the rate of pop-in events $\dot{N}$ (proportional to the probability of their occurrence) is expressible in the form

$$\dot{N} = \gamma_{\mathrm{V}} \exp\left(-\frac{W_\ell^* - \mathfrak{s} v^*}{k_{\mathrm{B}} T}\right), \tag{6}$$

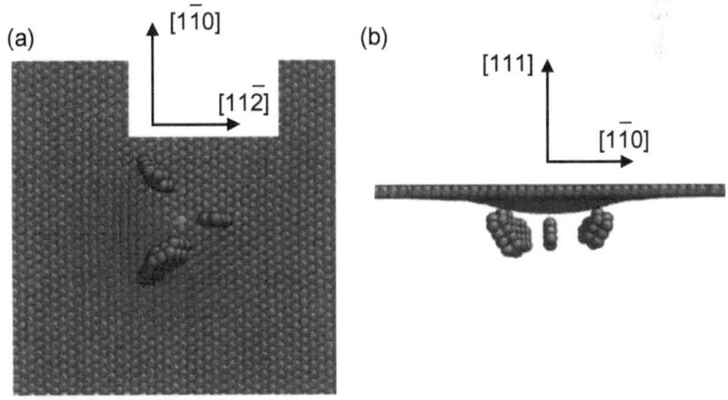

Fig. 5 Molecular-dynamics simulation of a (111) surface of single-crystal copper under a spherical indenter, showing three homogeneously nucleated dislocations: (a) view down the surface normal, (b) side view along [11$\bar{2}$]. For clarity, all the perfectly coordinated atoms are invisible, leaving only those on the surface and in the dislocation cores. (Reprinted from Ref. [5], copyright (2004), with permission from Elsevier.)

where γ_V is the attempt frequency per unit volume, W_ℓ^* is the work of dislocation nucleation, $\mathfrak{s}$ is the applied stress, and v^* is the activation volume. Indeed the critical conditions for the first pop-in show a significant variation, consistent with stochastic behavior of the kind implied in Eq. (6) [30]. Such behavior is susceptible to quantitative analysis [30]. Fitting data on the distribution of critical loads for the first pop-in for more than 2000 indentations on (1 1 0) platinum, Schuh et al. [30] have derived $\gamma_V = 1.8 \times 10^{-25}\,\mathrm{s}^{-1}\,\mathrm{m}^{-3}$, $W_\ell^* = 0.28$ eV, and $v^* = 1.02 \times 10^{-29}\,\mathrm{m}^3$. On this basis, Schuh et al. discount the possibility of homogeneous nucleation, for which $\gamma \approx 10^{41}\,\mathrm{s}^{-1}\,\mathrm{m}^{-3}$, $W_\ell^* \approx 1.3$ eV and $v^* = 10^{-27}\,\mathrm{m}^3$ [30]. In contrast, they suggest that dislocation nucleation must be heterogeneous. The most likely heterogeneity is considered to be a vacancy or a vacancy cluster, defects not considered in the analog shown in Figure 4, or the type of simulation illustrated in Figure 5. Despite the likely heterogeneous nature of the nucleation, the critical shear stress ($\sim$4.4 GPa for platinum) is still a large fraction of the ideal shear strength of the defect-free metal ($\sim$5.3 GPa) [30]. Atomistic simulations of the mechanical behavior under a nanoindenter may need to include distributions of point defects before they can provide realistic descriptions.

We note in passing that instrumented indentation has also been used to study the nucleation of plastic flow in glassy materials in which dislocation mechanisms cannot apply. In metallic glasses plastically deformed at ambient temperature, shear is sharply localized into bands 10–20 nm thick. Schuh et al. [31] have used pop-in events in loading curves (similar to those in Figure 3) to analyze the nucleation of these shear bands in Pd- and Mg-based bulk metallic glasses. They suggest that the nucleation is kinetically limited and that a typical shear-band nucleus (conceptually similar to the shaded area in Figure 1) attains a diameter of approximately 0.5 μm before the strain rate becomes sharply localized and rapid propagation of the band ensues.

2.4 Loss of coherency in thin films

There is much technological interest (for electronic and optoelectronic devices) in epitaxial thin films of semiconductors. Ideally these are dislocation-free, being deposited on dislocation-free, single-crystal substrates. There is in general a mismatch of lattice parameter between film and substrate, leading to large biaxial stresses in the film if lattice coherency is maintained. Thin heteroepitaxial films are initially coherent with the substrate, but later lose coherency when their thickness h exceeds a critical value h_c. At least in semiconductor devices, the introduction of dislocations is undesirable. The misfit dislocations are located at the interface with the substrate, which could be far from the relevant region of the device. However, the misfit dislocations mostly do not extend right across the substrate, but terminate at the top surface of the thin film; the *threading dislocations* linking the interface to the top surface cause problems. The loss of coherency may be undesirable also because the coherency strain itself may induce useful properties in the deposit.

The critical thickness h_c can be evaluated by considering an energy balance in equilibrium, as reviewed by Nix [32]. The elastic strain energy in the uniformly strained coherent film is proportional to the square of the misfit strain and increases linearly with film thickness h. The introduction of misfit dislocations at the film/substrate interface reduces uniform strain in the film, but adds the line energy of the dislocations. Consistent with the decaying strain field around a dislocation (Chapter 6, Eq. (42)), the line energy of the dislocations increases only as the logarithm of h. For this reason, it always becomes energetically favorable to introduce misfit dislocations as h is increased.

Another approach to estimate h_c is to assume that there are threading dislocations and balance the forces on them [32–34]. The coherency stress tends to distort the threading dislocations to become misfit dislocations along the interface, a distortion opposed by the line tension of the dislocations. As reviewed by Hull and Bean [11], the resulting estimates of h_c in equilibrium (i.e., without kinetic constraints) are successful in fitting data on some metallic systems in which there is a substantial population of threading dislocations. For semiconductors, however, there is evidence that h_c can be substantially exceeded (giving continued growth of a *metastable* strained layer), especially at low temperatures. It is natural to attribute this to a kinetic barrier, most likely the nucleation of dislocations.

The calculations of equilibrium h_c are complex and depend on parameters subject to some uncertainty. Thus the ideal h_c is not known with great precision. Furthermore, there are uncertainties in the experimental determination of h_c. Nevertheless, the discrepancy, particularly for fully covalent systems, is clear (Figure 6). When dislocations are already present, even in very low concentrations, there are many mechanisms for dislocation multiplication. A simple example is the regenerative Frank–Read source. One of many mechanisms observed specifically in thin films is that proposed by Hagen and Strunk [35], in which two intersecting misfit dislocations react to form two new mobile dislocations. However, even when dislocations are present, the sluggishness of the plastic response allows growth of rather thick overlayers that are highly strained and almost fully coherent with the substrate [36]. In many cases, it is doubtful that multiplication can account for all the misfit dislocations needed for strain relaxation. As outlined by Hull and Bean [11], a typical required density of misfit dislocations is higher than $10^{10}\,m^{-2}$. This can be compared with typical defect densities in silicon of

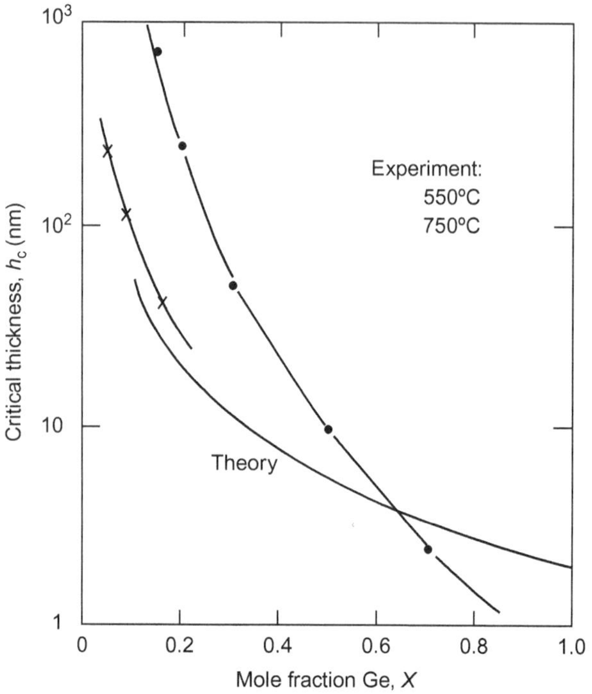

Fig. 6 Plots of the experimentally observed critical thickness in the Ge_XSi_{1-X}/Si (1 0 0) system, together with predictions of the Matthews–Blakeslee [33] theory. (Adapted from Ref. [11], with permission, Taylor & Francis Ltd. http://www.informaworld.com.)

10^5 to $10^6\,m^{-2}$ and in III–V semiconductors (GaAs and InP) of 10^7 to $10^8\,m^{-2}$. At least for large misfits, the generation of all the required misfit dislocations seems to require a contribution from true dislocation nucleation.

For the reasons outlined in Section 2.1, nucleation of dislocations is most likely to occur at the free surface of the film. It is known to occur both during growth and during subsequent annealing [37]. Nucleation of a half-loop at a surface defect is followed by the development of misfit and threading dislocations (Figure 7). Following Hull and Bean [11], we examine a particular example in some detail. For a diamond-structure overlayer (e.g. Si or Ge), the dislocations have Burgers vectors $(a/2)\langle 1\,1\,0\rangle$ and the glide planes are $\{1\,1\,1\}$. For epitaxy on a (1 0 0) substrate, the glide planes are inclined at 54.7° to the top surface. Of the three slip directions in a plane, one is parallel to the top surface and to the interface with the substrate. Interfacial dislocations with this Burgers vector would be of screw character and would not relax the misfit. The other two slip directions are at 60° to the dislocation line vector and dislocations of the

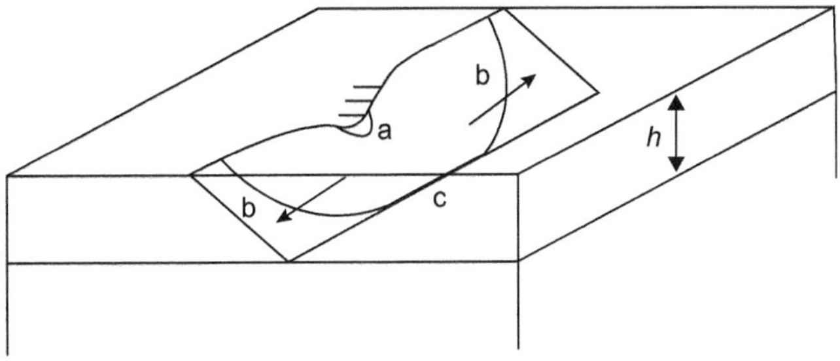

Fig. 7 Schematic illustration of the formation of misfit dislocations in an epitaxial deposited film of thickness h exceeding the critical value for loss of coherency. After nucleation of a dislocation half-loop at the top surface at (a), the loop expands on the inclined glide plane, ultimately giving threading dislocations (b) and a misfit dislocation at the film/substrate interface (c). (Adapted, with permission, from Ref. [37] by Annual Reviews www.annualreviews.org.)

resulting mixed edge/screw character are the ones of interest. An example of a calculation of the critical work of nucleation W_{ℓ}^{*} for such dislocations in $Ge_{0.4}Si_{0.6}$ on Si (1 0 0) is shown in Figure 8 for the case where the dislocation is eliminating a surface step. The contributions to the work of nucleation have already been considered in Eq. (4). The line energy of the half-loop E_{loop} is positive and increases with loop radius R. On the other hand, the formation of the loop decreases the uniform strain energy of the film E_{strain} and the energy of the surface step E_{step}. The total energy shows a maximum, defining the work of nucleation and the critical radius R^{*}.

Even given the large strains that can be present in heteroepitaxial thin films, the work of formation of a critical dislocation loop is often too large for thermal activation to be significant [38]. Calculations such as those underlying Figure 8 have a number of uncertainties, but it appears that the work of nucleation is unattainably high (>100 eV) for misfit strains below 1%. In such a case, nucleation requires a local stress concentration. For misfit strains of $>4\%$, however, the work of nucleation is low (<1 eV).

The strain in heteroepitaxial semiconductor thin films can induce surface roughening (Figure 9). There is no nucleation barrier for this morphological instability, driven by reduction in the overall strain energy. Either during deposition or on subsequent annealing, surface diffusion gives first a pattern of incipient islands, then two-dimensional ridges with deep cusp-like valleys between, and finally three-dimensional islands. Such morphological changes are of interest for many reasons, including the fabrication of quantum-dot devices [37]. The specific case of strained heteroepitaxial Si–Ge on Si (1 0 0) has been reviewed by Gao and Nix [37]. Studies on subcritical films (with $h<h_c$) show the development of surface roughening in the absence of any misfit dislocations. Similar studies on supercritical films ($h>h_c$) show the

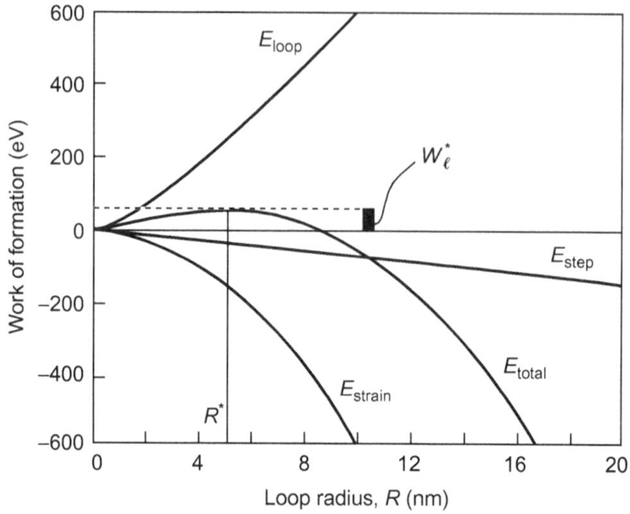

Fig. 8 Energy changes accompanying the formation of a 60° glide dislocation half-loop annihilating a step at the surface of a $Ge_{0.4}Si_{0.6}$ film grown on a Si(1 0 0) substrate. There are contributions from the dislocation line energy E_{loop}, from the uniform strain energy in the film E_{strain}, and from the surface step E_{step}. The work of nucleation W_ℓ^* and critical loop radius R^* follow from the form of the total energy E_{total}. (From Ref. [11], with permission, Taylor & Francis Ltd. http://www.informaworld.com.)

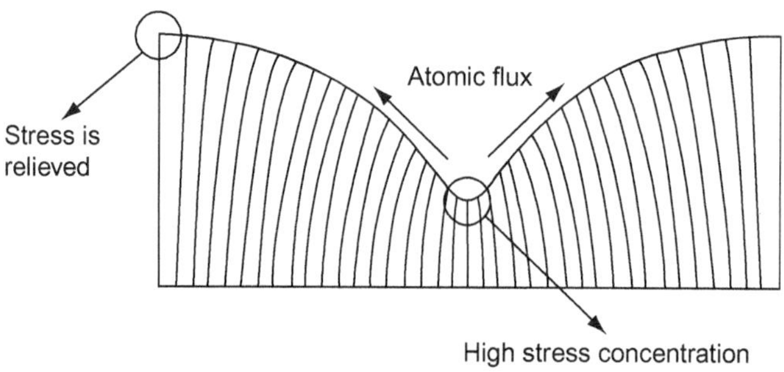

Fig. 9 A schematic illustration of lattice spacings showing the stress relaxation that can occur in a compressive thin film when there is surface roughening. There is a stress concentration at the cusp from which atoms are transported by surface diffusion. (Reprinted, with permission, from Ref. [37] by Annual Reviews www.annualreviews.org.)

development of two-dimensional ridges parallel to $\langle 100 \rangle$ before the generation of misfit dislocations. When the misfit dislocations do appear, they are aligned parallel to $\langle 110 \rangle$, and the resulting changed stress state induces realignment of the ridges parallel to $\langle 110 \rangle$. Given this complexity, it can be understood why there was uncertainty about whether misfit dislocations induced surface roughness or vice versa. It is now clear, however, that surface roughening *precedes* the nucleation of misfit dislocations, and that deep valleys (Figure 9) provide the surface sites necessary for nucleation. The stress concentration at these valleys is such that the work of nucleation of a dislocation half-loop is reduced effectively to zero [37]. The valleys are a likely type of defect to nucleate a dislocation as depicted at point (a) in Figure 7. Their existence must often invalidate approaches based on uniform film thickness, such as the calculations shown in Figure 8.

It should be noted that dislocations can arise during growth of thin films by other mechanisms also. Mass transport at the surface of a growing film naturally generates ledges. Subsequent addition and rearrangement of atoms can generate a sessile dislocation trapped by further deposition on top [37]. Such dislocations are quite distinct from the glissile dislocations that have been the focus of the above discussion, and that once nucleated at the surface move away from it under the influence of the misfit strain in the film (Figure 7).

So far, we have considered the loss of coherency in planar systems in which layer-by-layer growth is favored. In some cases, however, the deposited material forms islands. For semiconductors, a number of studies have shown that the islands are initially coherent, with apparently only a small energy barrier for their formation. For example, Eaglesham and Cerullo [39] showed that for growth of Ge on Si (1 0 0), the Ge islands are fully coherent up to a thickness (50 nm) some 50 times the equilibrium coherent layer thickness, indicating a considerable barrier for dislocation nucleation.

3. RECRYSTALLIZATION

The term *recrystallization* is sometimes used where *crystallization* would be more appropriate, for example, in the devitrification of glasses (Chapter 8) or precipitation from solution (as discussed in Ref. [40]). Here, we use *recrystallization* in its strict sense of denoting the formation of a new grain structure on annealing a metal that has been heavily deformed. It is one of several important processes occurring on annealing a polycrystalline metal and has been extensively reviewed [41]. Recrystallization takes place entirely in the solid state, and metallographic studies show that it proceeds by nucleation and growth. New grains, which are relatively dislocation-free, grow from a distribution of centers into the surrounding matrix, which has a high dislocation density. On completion of recrystallization, the stored energy of cold work [42] is reduced to zero. The reduction in stored energy during recrystallization is essentially from the reduction in dislocation density, and this constitutes the driving force for recrystallization, as the energy involved is large compared to the total grain-boundary energy of the system before or after recrystallization. As revealed by a variety of studies, the new grains are separated from the matrix by high-angle grain boundaries. The grain structure that is formed has no clear relationship with the grain structure before deformation, and the crystallographic texture (preferred orientation) produced by deformation can be changed by the recrystallization. The nucleation stage of recrystallization is important in determining the grain size after annealing. The density of nucleation sites is lower for lower deformation levels. In the *critical strain annealing* technique, this is exploited to produce single crystals: a sample is lightly deformed to give just one recrystallization nucleus and, therefore, forms a single crystal on annealing. There has also been much work on the potential role of nucleation in governing the changes in texture.

When a crystalline metal is deformed, the individual grains are also deformed, and the dislocation density in the material rises dramatically, typically from $10^{10}\,\mathrm{m}^{-2}$ before deformation to $10^{15}\,\mathrm{m}^{-2}$ after heavy deformation. There is also a rise in the concentration of point defects — vacancies and self-interstitials — arising from the dragging of sessile dislocation jogs. It is clear that the increased defect concentrations raise the free energy of the material and that there is an enhanced driving force for microstructural change. On annealing, that change takes place in two stages: recovery and recrystallization. In recovery, the excess point-defect concentrations anneal out and the dislocations rearrange; a few dislocations of opposite sign annihilate, while others of like sign align into subgrain boundaries. Recovery reduces the driving force for recrystallization (by up to 70%, depending on the metal, its purity, and the extent of prior deformation [42]), yet the processes of recovery turn out to be essential in providing the conditions for recrystallization to occur, through subgrain formation as illustrated in Figure 10.

In analyzing the nucleation stage of recrystallization, it is first of interest to determine whether homogenous nucleation might be possible. The free energy change on recrystallization arises mainly from the reduction in dislocation concentration; a typical value is $1\,\mathrm{MJ\,m^{-3}}$ [42], very low compared to those found for changes of phase. The interfacial free energy is that typical of high-angle grain boundaries, of order $1\,\mathrm{J\,m^{-2}}$. With these values, the radius and work of formation of the critical cluster, using Eqs. (14) and (19), Chapter 2, are of order $r^* = 1\,\mu\mathrm{m}$ and $W^* = 10^9 k_\mathrm{B}T$, respectively. Since the largest W^* for a reasonably observable nucleation rate is 60 $k_\mathrm{B}T$, it can clearly be seen that

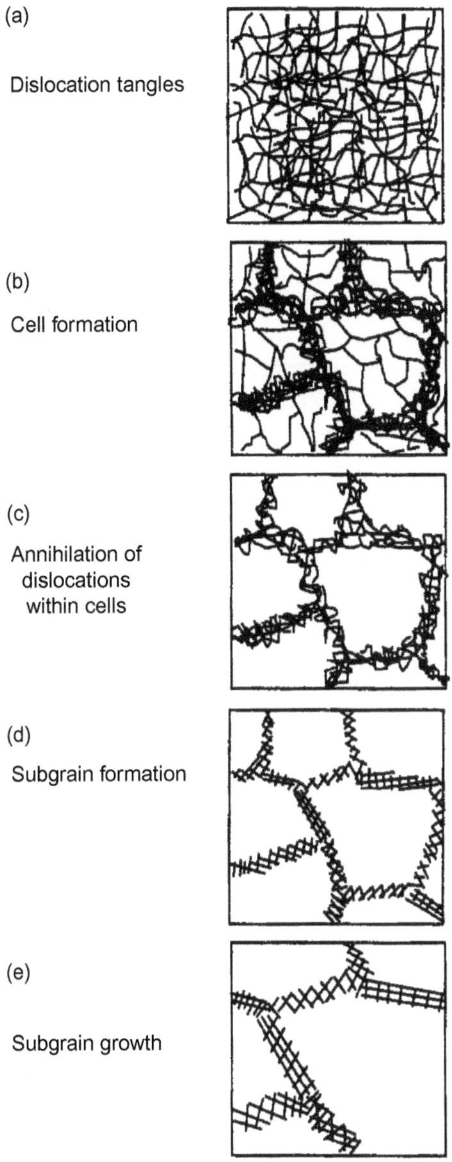

Fig. 10 Schematic illustration of the dislocation rearrangements in recovery of a plastically deformed metal. (Reprinted from Ref. [41], copyright (1995), with permission from Elsevier.)

homogenous nucleation of a new grain is not a realistic possibility. Indeed, even a reduction of W^* by heterogeneous nucleation on a substrate (Chapter 6, Section 2.2) seems unlikely to bring about nucleation at a measurable rate.

It appears instead that the recrystallization "nucleus" is an already existing region in which the dislocation density is low. Such a region can grow by the bulging of a preexisting high-angle grain boundary, and is likely to retain some dislocations from the prior structure (Figure 11a). This phenomenon of *strain-induced grain-boundary migration* is prevalent at low strain [41]. For bulging to occur, there must be a difference in dislocation density across the grain boundary. This may occur by coalescence of subgrains (Figure 11b).

Away from a high-angle grain boundary, recrystallization can still nucleate by growth of subgrains in the recovered microstructure. Whichever is the location of nucleation, there is no evidence for the creation of any new crystallographic orientation (except by twinning, as discussed further in connection with *dynamic recrystallization*). On annealing, there is a tendency for subgrains to grow, analogous to the coarsening seen in grain growth. In grain growth, the process can be *normal* in which the average grain size increases with time while maintaining a unimodal grain size distribution, or can be *abnormal* in which a few favored grains grow at the expense of their neighbors, giving a bimodal size distribution. The subgrain evolution leading to recrystallization must be analogous to abnormal grain growth. The

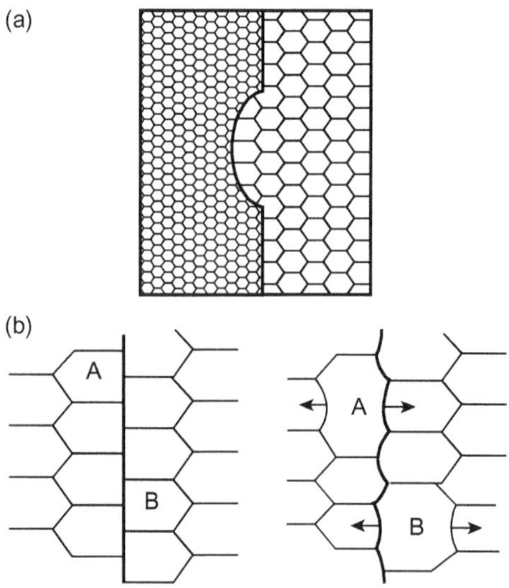

Fig. 11 (a) A boundary between two grains each containing dislocations arranged as subgrain boundaries. Strain-induced grain-boundary migration (SIBM) gives a bulge of material with low dislocation density. (b) Subgrains at A and B (left) can coalesce (right), giving the difference in dislocation density necessary to initiate SIBM. (Reprinted from Ref. [41], copyright (1995), with permission from Elsevier.)

similarity of the two processes is emphasized by the name *secondary recrystallization* sometimes given to abnormal grain growth.

The favored subgrains are those with boundaries that are more mobile, and that also have a driving force for motion. The boundary mobility is principally determined by the misorientation between the lattices on either side; high-angle boundaries can be over 10^3 times more mobile than low-angle boundaries, as shown in Figure 12 [43]. Thus the essence of the nucleation, in this "Cahn–Cottrell" mechanism [40], is that growth of the recrystallized grains proceeds from existing subgrains that are highly misoriented (by at least 15°) with respect to their surroundings. As such, the process is distinct from more conventional nucleation in which there is a clearly defined critical fluctuation. Nonetheless, nucleation of a recrystallized grain is still rare. Typically only a very small fraction, typically 10^{-6}, of the subgrains grow into recrystallized grains. The energetic factors favoring some subgrains could be a local variation in the stored energy of cold work, or a greater size of the subgrain itself. Recently, the grain-boundary bulging that constitutes nucleation of recrystallization has been modeled using simulations of grain and subgrain growth that allow the examination of the roles of boundary topology, and variations in subgrain size, energy, and

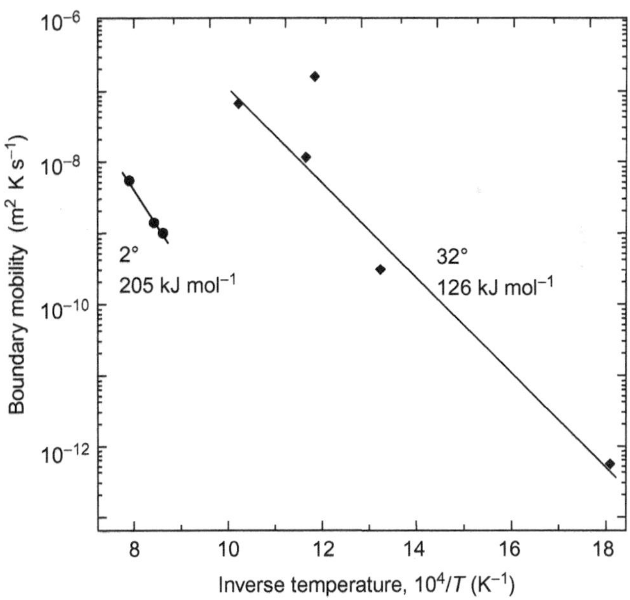

Fig. 12 Grain-boundary mobility as a function of temperature in high-purity copper bicrystals. The boundary misorientations and the corresponding activation energies of the mobility are marked. (Adapted from Ref. [43], copyright (1973), with permission from Elsevier.)

boundary mobility [44]. A typical subgrain evolution with marked bulging of the original grain boundaries is shown in Figure 13; in this case, the bulging arises only from the local topological heterogeneities.

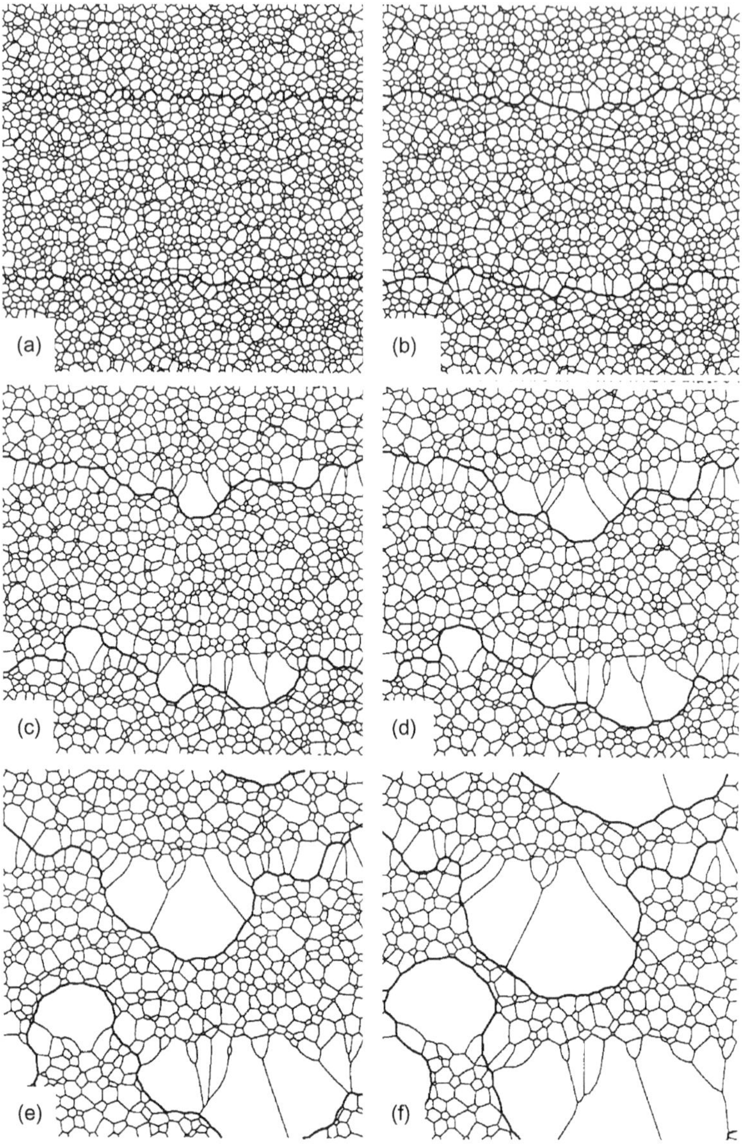

Fig. 13 A two-dimensional simulation of grain boundary (heavier lines) and subgrain boundary migration. (a) Initial configuration; (b)–(f) evolution sequence, showing the development of bulges (recrystallization nuclei) on the grain boundaries through subgrain growth. (From Ref. [44], with permission, Taylor & Francis Ltd. http://www.informaworld.com.)

It is concluded that it is difficult to predict the rate of nucleation or number of nuclei per unit area of grain boundary, even when the deterministic evolution of the subgrain structure is studied in detail [44].

The development of strong recrystallization texture in some rolled metals offers an example of the role of local variations in the microstructure. In ccp metals, for example, a strong "cube" texture (i.e. with the $\langle 1\,0\,0 \rangle$ axes of the cubic structure parallel to the rolling, traverse, and normal directions of the sheet) can develop on recrystallization. The deformed microstructure can be inhomogeneous, with bands of this texture separated by high-angle boundaries from the neighboring regions. Enhanced recovery in these "cube" regions gives larger subgrains and lower stored energy; the recrystallization nuclei then develop in these regions and grow to consume the rest of the material. It is now generally accepted that the development of the strong cube texture in recrystallization can be explained by such oriented nucleation, rather than by oriented growth [40].

Particularly when there are no significant local variations in microstructure, that is in a deformed material showing uniform texture and degree of recovery, it is of interest to identify specific events that give rise to nucleation. One clear possibility is that of subgrain coalescence, which would give larger subgrains, separated by a twist boundary (Figure 14). In the boundary, there are two sets of equally spaced screw dislocations; coalescence occurs if these can migrate to dislocation sinks such as high-angle boundaries. But the screw dislocations must be connected to edge segments on the other faces of the subgrain and migration then requires climb of these edge segments. On opposite sides of the subgrain, the segments are of opposite sign, so that climb in one direction occurs by complementary production and absorption of point defects. The transport of these, principally vacancies, can be by lattice diffusion across the subgrain at higher temperatures or by diffusion along the screw dislocations themselves at lower temperatures, giving distinctive kinetics, which can be tested against transmission electron microscopy observations [45]. Since the spacing of the two sets of screw dislocations must remain equal, there is a requirement for two dislocation sinks on adjoining edges of the subgrains. Experimentally it is found that nucleation indeed often occurs on prior high-angle grain boundaries in a grain in which there is an intersection with a misoriented band. A subgrain within this grain, when at the intersection, has the two dislocation sinks necessary to permit coalescence and the adjacent high-angle mobile boundaries to permit growth. Despite the understanding that has been developed of dislocation mechanisms such as that in Figure 14, quantitative predictions are hindered by the complexity and possible heterogeneities of the deformed microstructure, and by the effects of heterogeneities and solute content on the recovery processes.

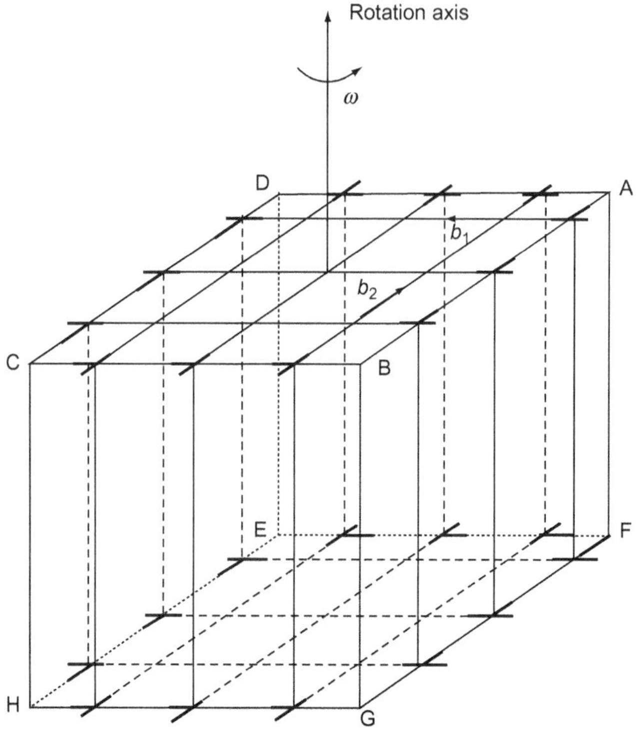

Fig. 14 An idealized subgrain, misoriented by twist angle ω with respect to its neighbor on the other side of the face ABCD. By elimination of the screw dislocations on this face, ω can be reduced to zero and the subgrains can coalesce. The rate of this is limited by the climb of the attached edge dislocations on the side faces BCHG, etc., of the subgrain, which are tilt boundaries. (Reprinted from Ref. [45], copyright (1984), with permission from Elsevier.)

It is certainly clear that greater deformation, leading to greater dislocation density, enhances the probability of forming a critical subgrain during recovery. For this reason, recrystallization is promoted by the presence of hard second-phase particles in the alloy, as deformation can be concentrated around these. Particle-stimulated recrystallization can be exploited in obtaining desired microstructures [46].

There has been extensive work on the modeling of recrystallization kinetics, including predictions of grain structure and texture, as reviewed by Rollett [47]. The modeling mostly deals with grain growth, and nucleation frequencies usually have to be arbitrarily specified.

The Cahn–Cottrell mechanism can generate grains only in the orientations of existing subgrains. The mechanism for the generation of new grain orientations appears to be twinning [48], as has been explored particularly for *dynamic recrystallization* [49]. This is recrystallization that occurs *during* deformation, and it can be nucleated at grain boundaries, precipitate particles, or deformation (shear) bands. In many cases, however, and particularly when grain-boundary sliding is a significant deformation mode, grain-boundary junctions (*triple junctions*) are the preferred nucleation sites. Sliding along the grain boundaries is blocked at the triple junctions, leading to stress concentrations and folding deformation in one grain at the junction. This deformation is favored when the plane of the sliding is parallel to a slip plane. It leads to the creation of a recrystallization nucleus by twinning, and this is accompanied by grain-boundary migration to accommodate the new grain. This type of nucleation is favored by high strain, low strain rate, high temperature, and large grain size — factors that promote sliding and stress concentration.

The nucleation of recrystallization (static or dynamic) does not involve stochastic behavior of the kind associated with surmounting an energy barrier. On the other hand, it is stochastic in the sense of being dependent on the chance appearance of favorable grain-boundary and dislocation configurations.

4. TWINNING AND MARTENSITIC TRANSFORMATIONS

4.1 Dislocation-mediated phase transformations

The phase transformations considered so far in this book have been *reconstructive* (also known as *diffusive*), the transfer of atoms or molecules from one phase to another involving diffusive-type jumps and being largely uncoordinated. In polymeric systems (Chapter 11), the nature of the molecules imposes some coordination on atomic motion, but even so, there is no one-to-one correspondence of atomic positions before and after transformation. *Displacive* transformations, in contrast, involve coordinated (sometimes called *military*) motion of atoms such that each atom moves less than one atomic spacing relative to its neighbors and reconstruction of the local topology is avoided. Such transformations, also called *shear* or *diffusionless*, are widespread; an archetype is the formation of *martensite*. Martensite is a hard phase found in quenched steels, but the type of transformation by which it is formed is also found in nonferrous alloys, ceramics, inorganic compounds, polymers, solidified gases, and even in biogenic proteins [50–52]. Martensitic transformations involve no diffusion; after they are nucleated, growth is typically fast (near the speed of sound) even at low temperatures approaching absolute zero. In many cases, they show *athermal* kinetics, in that the degree of transformation depends on the temperature to which the sample has been cooled but is independent of time at that temperature; *isothermal* transformation is also known. The initial and final phases have a crystallographic orientation relationship. The transformation from one lattice to the other involves a shear of the order of 20%, and there is therefore typically a large strain energy associated with the transformation. General characteristics of displacive, martensitic transformations have been extensively reviewed [50, 53].

The interphase interface in a martensitic transformation is a wall of glissile dislocations, the ease of movement of such a wall explaining the fast kinetics of the transformation. The motion of a wall of dislocations is also involved in *deformation twinning*, a closely related phenomenon that, however, does not involve a change of phase. The types of transformation and nucleation considered in this section are also relevant for such phenomena as the shape memory effect, superelasticity, ferroelasticity, and transformation-induced plasticity.

Normally, martensitic transformations occur when the driving force for transformation is too small for homogeneous nucleation to be possible [50, 54]. An early analysis by Hollomon and Turnbull [55] emphasizes this point: for the homogeneous nucleation rate to be detectable, the energy of the interface surrounding the nucleus would have to be $\leq 10\,\mathrm{mJ\,m^{-2}}$, whereas the actual energy of the interface is $\sim 200\,\mathrm{mJ\,m^{-2}}$. The driving forces for martensitic transformations are normally comparatively small for alloys, because their nondiffusive nature dictates that the initial and final phases have the same compositions.

Studies of martensite nucleation have often centered on the Fe–Ni system, which has a high-temperature ccp phase that transforms martensitically to a body-centered cubic (bcc) phase on cooling. (The ccp structure is obtained when a motif of one atom is combined with a face-centered cubic lattice, and is often termed fcc.) Reflecting the athermal kinetics noted earlier, on cooling the martensitic transformation starts at a characteristic *martensite-start temperature* (M_s). The system is suitable for detailed analysis because the nickel content can be used to adjust M_s.

Evidence for the heterogeneous nature of martensite nucleation is found mainly in small-particle experiments (reviewed briefly in [56, 57]). Early work by Cech and Turnbull [58] on 30–100 μm Fe–Ni particles solidified in a drop-tube found that the smaller particles did not undergo any martensitic transformation on cooling. Further analyses, for example Ref. [59], focused on the kinetics of the transformation and show that there must be a distribution of effectiveness of nucleation sites. The martensite-start temperature (M_s) depends on composition, but decreases strongly with particle diameter for a given composition. Plastic deformation of fine particles can induce the martensitic transformation, however. More recent extension of such studies to ultra-fine (20–200 nm) particles [60] has found that transformation can still occur, contrary to the expectation from extrapolating the Cech and Turnbull results; it was suggested that there may be a role for thermal activation.

Similar studies have focused on precipitate particles: in Cu–Fe alloys, iron can precipitate out within the ccp Cu-rich matrix as a coherent ccp phase, which would be expected to undergo a martensitic transformation to a bcc phase on cooling. For coherent dislocation-free particles (5–30 nm), this transformation is suppressed, but except in the finest particles it can be triggered by plastic deformation [61]. As noted earlier, any martensitic transformation is hindered by the elastic strain energy associated with the shear shape change. Nucleation of the transformation is, therefore, easier near a free surface or a medium that is more elastically compliant [62], as shown by the transformation of precipitate particles when the matrix is dissolved away [63]. For particles embedded in an amorphous matrix, direct nucleation of martensite is replaced by a new transformation path (via an intermediate phase) for particle diameters less than 100 nm [64]. Size effects are also found in polycrystalline materials, where M_s is clearly reduced as grain size is reduced [65, 66], and again new transformation paths can be taken [67].

It is commonly observed in martensitic transformations that the appearance of the first martensite particle triggers many others. This *autocatalytic* or *sympathetic* nucleation gives an avalanche or *burst* of transformation and greatly impedes interpretation of the nucleation kinetics [68]. For this reason, small-particle experiments of the kind just outlined have particular importance.

Although homogeneous nucleation of martensite must be rare, it can be observed if the driving force is sufficient. For example, Lin et al. [68] have shown that defect-free coherent ccp Fe–Co precipitates in a ccp Cu-rich matrix can transform to the bcc phase on cooling. In this case, the driving force for the transformation is approximately $10\,\mathrm{kJ\,mol^{-1}}$ — seven times higher than is needed for the onset of the same martensitic transformation in bulk alloys where it can be heterogeneously nucleated. It is of clear interest and importance to understand the nature of the heterogeneous nucleation that is normally found.

4.2 The Olson–Cohen nucleation mechanism

For transformations that can proceed by the glide of a wall of dislocations, it is not surprising that heterogeneous nucleation is also dislocation-mediated. The theory for this case has developed by a number of authors, including Ahlers [69]. We follow the treatment by Olson and Cohen [70–72], beginning with the simplest case of shear transformations in a cubic ccp metal. In the ccp structure, close-packed planes of atoms (Figure 15) are stacked, each atom sitting on top of three atoms in the layer below to form a tetrahedron. In this way, an 'A' layer is followed by a 'B' layer and a 'C' layer, a sequence that then repeats (Figure 16a). Other stacking sequences are possible, notably two-layer repeats (ABABAB, etc.), which give the hexagonal close-packed (hcp) structure. In the ccp structure, the close-packed planes are of the type {111} and are the slip planes on which the dislocations glide, giving plastic flow. On top of an A layer, the passage of a *perfect dislocation* (i.e., one for which the Burgers vector is a lattice vector) would move a B atom to a B position, the vector displacement being, for example, $(a/2)[1\bar{1}0]$ on (111). It can be seen in Figure 15, however, that the likely path from a B position to a B position is via a C position, involving two displacements $(a/6)[1\bar{2}1]$ and $(a/6)[2\bar{1}\bar{1}]$, which can be taken to be the Burgers vectors of two *partial dislocations*. The passage of both partial dislocations past a given spot on the glide plane restores the original structure, but the passage of only one leaves a changed structure (Figure 16c) in which there is a local region of hcp-type stacking. Between the partial dislocations, there is a *stacking fault*, a lattice defect having a positive energy. Stacking faults of various kinds are of importance in understanding the plastic flow of crystals, a topic that lies beyond our present scope (for details see standard texts, e.g., [1]). The two partial

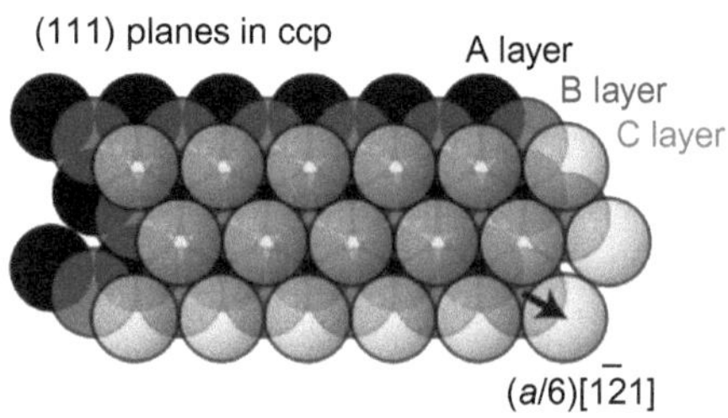

Fig. 15 Stacking of close-packed planes of atoms. This ABC stacking generates the ccp structure. The arrow is the Burgers vector of the partial dislocation involved in the glide illustrated in Figure 16.

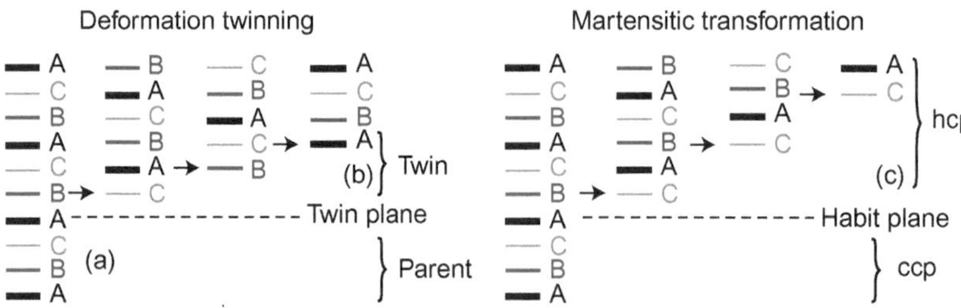

Fig. 16 The glide of partial dislocations can change an initial ccp structure (a) into a ccp twin (b), or into hcp (c).

dislocations repel each other, with a force per unit length inversely proportional to their separation. The positive energy of the stacking fault corresponds to an attractive force between the dislocations, independent of their separation. Thus there is an equilibrium separation of the partial dislocations, associated with the minimum energy on curve (a) in Figure 17.

The stacking fault just described can be thought of as an embryo of the hcp structure. Passage of further partial dislocations on every second plane generates a thicker embryo. If the dislocation glide is occurring under thermodynamic conditions such that, macroscopically, the ccp-to-hcp transformation is favorable, then there must be a critical size above which the metastable embryo can grow. As shown by Olson and Cohen [70], this can be analyzed in terms of the number of close-packed planes in the embryo. The energy per unit area, Γ_{SF}, of the stacking fault can be considered to have components from bulk and interfacial free energies:

$$\Gamma_{SF} = n\rho_A(\Delta G_{ch} + E_{ch}) + 2\Gamma_{ch}, \tag{7}$$

where n is the number of planes in the fault, ρ_A is the density of atoms in a close-packed plane (moles m^{-2}), ΔG_{ch} is the free energy change for the ccp-to-hcp transformation ($J\,mol^{-1}$), E_{ch} is the strain energy associated with the shear to form the new phase ($J\,mol^{-1}$), and Γ_{ch} is the free energy ($J\,m^{-2}$) of the interface (on the close-packed planes) between the ccp matrix and the hcp embryo. The terms ΔG_{ch} and E_{ch} are taken to have the

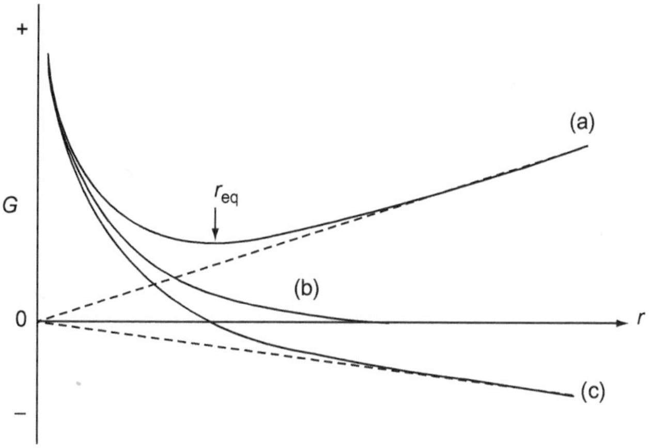

Fig. 17 Sketch of the variation of the total free energy G of a stacking fault as a function of the separation r of the partial dislocations bounding the faulted region for different cases of the stacking-fault energy: (a) $\Gamma_{SF} > 0$, (b) $\Gamma_{SF} = 0$, (c) $\Gamma_{SF} < 0$. As the driving force for transformation increases, the sequence (a) to (c) is followed, with the onset of athermal nucleation corresponding to (b). (From Ref. [70], copyright (1976), with kind permission of Springer Science and Business Media.)

values characteristic of bulk so that deviations from bulk values are encapsulated into Γ_{ch}, which must therefore vary somewhat with n [70]. The measurement of Γ_{ch} in representative alloys has been briefly reviewed by Olson and Cohen [70]; typical values are in the range 5–20 mJ m^{-2}. Under conditions in which ΔG_{ch} is negative, it is possible for Γ_{ch} to be zero or negative, if ΔG_{ch} or n is of sufficient magnitude. As shown in Figure 17, curves (b) and (c), once Γ_{ch} is zero or negative the stacking fault can spread without barrier on the close-packed planes. Thus $\Gamma_{ch} = 0$ can be taken to be the critical condition for nucleation of the hcp phase. At a given temperature (setting the value of ΔG_{ch}), the number of planes n must exceed a critical value for nucleation to occur. The existence of such a critical number is consistent with the absence, noted in Section 4.1, of martensitic transformations in small particles. For a given value of n, there is a critical value of ΔG_{ch} and, therefore, a critical supercooling (scaling inversely with n) for nucleation to occur on cooling. This onset of nucleation on cooling is represented schematically by the sequence of curves (a), (b), and (c) in Figure 17. This is athermal nucleation (analogous to that discussed in Chapter 6, Section 2.5). The concept of a spectrum of initial defects (dislocation configurations on n planes), giving a spectrum of critical supercoolings for nucleation of the hcp phase, fits well with the athermal kinetics often observed for martensitic and related transformations. In particular, the largest defects would correspond to the martensite-start temperature (M_s); Olson and Cohen [70] show that for the common Fe–Cr–Ni system, this critical value of n is up to 10. They suggest that before the critical supercooling is reached, the metastable embryo could (among other possibilities) be spheroidal, formed by a local dissociation of perfect dislocations between pinning points into partial dislocations (Figure 18). For a critical $n = 10$, there must be five dislocations of suitable Burgers vector in a sequence of every second plane. Such a configuration is highly improbable, which is consistent with the known sparseness of nucleation sites (typically 10^{11}–10^{13} m^{-3} [70]). However, there are microstructural features such as interfaces with inclusion particles, incoherent twin boundaries, and even some grain boundaries, which can contain the types of dislocation array required for nucleation [70]. As reviewed by Christian [54], in situ microscopy observations that the product phase forms first as a thin layer parallel to close-packed {111} planes are a strong evidence in support of the Olson–Cohen mechanism for nucleation of the ccp-to-hcp transformation.

4.3 Deformation twinning

If a partial dislocation passes on *every* close-packed plane (not every second plane as in the previous example), then the embryo formed is of a

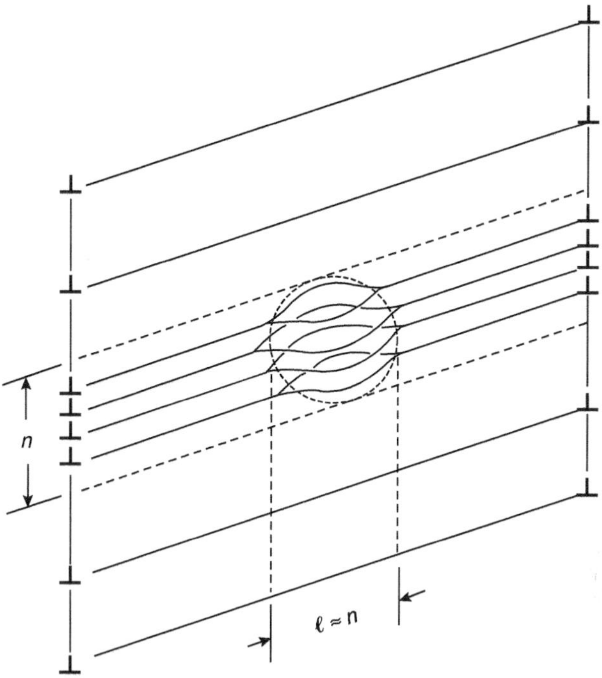

Fig. 18 Spheroidal critical nucleus for a martensitic or twinning transformation, forming on a segment of a tilt grain boundary by the bowing out of dislocations between pinning points. The distance ℓ between these points is similar to the height n of the boundary segment containing the required number of dislocations at the appropriate spacing. (From Ref. [70], copyright (1976), with kind permission of Springer Science and Business Media.)

ccp *twin* (i.e., with the ABC stacking sequence inverted; Figure 16b). Thus the same nucleation analysis, again relying on favorable local configurations of dislocations, can apply for twinning. In this case, the driving free energy for the change, ΔG (analogous to ΔG_{ch} in Eq. (7)) is not attributable to a change of phase, but rather to the work done by an externally applied stress when the sample shears through the twinning operation. Thus the analysis applies for *deformation twinning*, a form of plasticity likely to dominate over dislocation glide if there is a shortage of active slip systems or at high deformation rates. As reviewed by Christian and Mahajan [73], deformation twinning has a distinct nucleation stage. Studies have recently been extended to deformed nanocrystalline metals, in which deformation twinning is found to be nucleated (as expected from the earlier discussion of the Olson–Cohen mechanism) from grain-boundary dislocations [74]. Twinning is most readily nucleated in a particular range of grain diameter, for example, 5–7 nm for aluminum.

4.4 The austenite-to-martensite transformation

Variants of the Olson–Cohen mechanism have been applied to transformations from bcc to ccp, from bcc to hcp, from ccp to face-centered tetragonal, and to deformation twinning in bcc [71]. Metallurgically, however, most interest lies in the usual martensitic transformation in ferrous alloys, that is, from *austenite* (ccp) to *martensite* (essentially a distorted bcc *ferrite*). This transformation typically shows athermal kinetics and can occur at temperatures as low as 4.2 K [75]. It is significantly more complex than the ccp-to-hcp transformation considered in Section 4.2. The *lattice deformation* that effects the change of phase is accompanied by a lattice-invariant deformation (by slip or twinning) in martensite to give a strain-free interface with the austenite. In addition, there is plastic deformation of the austenite to accommodate the shear strain as the martensite is formed (typically in the shape of a lenticular lath spanning an austenite grain). While the nucleation stage of the transformation might not display all the complexities associated with later growth, the variation of the M_s temperature with applied stress does suggest that the total shape change is already established in the critical nucleus [54].

Olson and Cohen [71] have analyzed the ccp-to-bcc transformation in terms of the hard-sphere lattice-deformation model of Bogers and Burgers [76], in which the bcc structure is generated from the ccp structure by two invariant-plane strains. Passage of partial dislocations with Burgers vectors of the type $(a/18)\langle 112\rangle$ on every plane in a stack of type $\{111\}$ converts some $\{111\}$ planes of ccp into $\{110\}$ planes of bcc. These $\{110\}$ planes then need to be sheared to attain the correct stacking sequence for the bcc structure. The first shear can be modeled as the passage of normal partial dislocations (with Burgers vectors of the type $(a/6)\langle 112\rangle$ as considered earlier) on *every third* plane of a stack of close-packed planes in ccp. The subsequent shears, as detailed by Olson and Cohen [71], are rather complex to ensure that the plane of the stacking fault induced in the original ccp structure remains unrotated. Olson and Cohen take the free energy Γ_{am} of the austenite–martensite interface (parallel to the close-packed planes in austenite, and analogous to Γ_{ch} in Eq. (7)) to have the value 15 mJ m^{-2}, and thereby derive that the critical thickness of the stacking fault is 13.5 planes, thus corresponding to four or five properly spaced dislocations. Some observations do confirm the existence of just such configurations [77]. The most favorable nucleation sites are at twin and grain boundaries, especially on triple lines.

Using this model to fit the data of Cech and Turnbull on small droplets [58], it is possible to derive a size distribution for the faults in the austenite, of the form

$$N(n) = N_V \exp(-\kappa n), \qquad (8)$$

where $N(n)$ is the number of faults per unit volume of size n or greater, $N_V = 2 \times 10^{17}\ \mathrm{m}^{-3}$, and $\kappa = 0.84$. This exponential form is very similar to the relevant part of the particle size distribution found in some analyses of the heterogeneous nucleation of solidification (Chapter 13, Figure 12).

While athermal kinetics is common, martensite nucleation can also proceed isothermally. Olson and Cohen [72] have considered various possible rate-limiting steps for embryo growth, which might explain isothermal kinetics. Embryo thickening, increasing the number of planes in the original stacking fault, requires the generation of additional dislocations. As has generally been noted in this chapter, thermal activation of new dislocation loops is impossible, but new loops can be formed by extension of existing dislocations mediating the transformation and interaction of those dislocations with the natural population of dislocations not parallel to the fault plane. This *pole mechanism* is athermal and not capable of accounting for isothermal nucleation kinetics [72]. The lattice-invariant deformation accompanying the transformation requires the successive nucleation of dislocation loops, but a quantitative analysis shows that this is also not a likely rate-limiting step for martensite nucleation [72]. Finally, it is necessary to analyze growth of the martensite embryo within the fault plane. Reasonable resistance to dislocation motion in this plane does not invalidate the Olson–Cohen mechanism, but does have an effect on the kinetics. At high temperatures, the resistance is low and temperature-independent; the nucleation kinetics is then athermal. At lower temperatures, thermally activated motion of partial dislocations can underlie isothermal growth of martensite embryos and can explain isothermal nucleation kinetics. The stress $\mathfrak{s}$ acting on partial dislocations because of the stacking-fault energy Γ_{SF} is given by

$$\mathfrak{s} = \frac{-\Gamma_{SF}}{nb}, \tag{9}$$

where n is the number of planes in the fault and b is the length of the Burgers vector. The activation energy $Q_{\mathfrak{s}}$ for dislocation motion under an applied stress $\mathfrak{s}$ is given by

$$Q_{\mathfrak{s}} = Q_0 - (\mathfrak{s} - \mathfrak{s}_\mu)v^*, \tag{10}$$

where Q_0 is the activation energy without an applied stress, $\mathfrak{s}_\mu$ is the minimum, temperature-independent resistance to dislocation motion, seen at high temperature, and v^* is the activation volume defined by

$$v^* = -\frac{\partial Q_{\mathfrak{s}}}{\partial \mathfrak{s}}. \tag{11}$$

Combining Eqs. (7), (9), and (10), and rearranging, we have

$$Q_{\mathfrak{s}} = Q_0 + \left(\mathfrak{s}_\mu + \frac{\rho_A E}{b} + \frac{2\Gamma_{am}}{nb}\right)v^* + \left(\frac{\rho_A}{b}v^*\right)\Delta G_{am}. \tag{12}$$

It is reasonable to take the activation volume v^* to be independent of temperature over the range relevant for observations of martensite nucleation. The activation energy for isothermal nucleation is then approximately proportional to the chemical driving force ΔG_{am} for the austenite-to-martensite transformation, as is commonly observed. Furthermore Eq. (12) shows that at a given temperature (and therefore given ΔG_{am}), the activation energy should be inversely correlated with n. A distribution of embryos with varying n is then predicted to give a distribution of activation energies, as is observed experimentally [72].

While the basic features of the Olson–Cohen model appear to apply to the nucleation of martensite from austenite, there is a significant range of behavior. For example, as pointed out by Kajiwara [78], plastic deformation of the austenite can have an important effect on the transformation kinetics. When the shape change of a martensite lath is readily accommodated by plastic flow of the austenite, autocatalytic nucleation is suppressed. The resulting dislocation multiplication can, however, still accelerate martensite nucleation nearby.

Martensite nucleation has also been considered in terms of a lattice instability [79, 80]. In some systems, it has been observed that as the conditions for the onset of a martensitic transformation are approached, an elastic constant can decrease, perhaps even to zero. Such behavior is associated with marked elastic anisotropy in which soft phonon modes facilitate shear on particular crystallographic planes. In such a way, an initial phase could become unstable against transformation to a new phase. In ferrous alloys undergoing the martensite transformation, however, there are no anomalies in the elastic constants as M_s is approached. Any lattice instability conceivable under extreme driving force is obscured by the intervention of the conventional transformation heterogeneously nucleated at small driving force. Any reduction in elastic stiffness would assist martensite nucleation, but appears not to be necessary for it.

4.5 Nucleation of bainite and related phases

So far we have considered the nucleation of transformations that are entirely diffusionless. It is important to note, however, that a martensite-type nucleation mechanism can still apply in some transformations showing diffusion (partitioning of solute between original and new phases). A good example is the *bainite* transformation in which austenite transforms to sheaves of ferrite plates with particles of *cementite* (Fe_3C) embedded between or within the ferrite plates. Bainite is an important structure in steels, as has been reviewed by Bhadeshia [81]. Closely related is *acicular ferrite*, which can be considered to be bainite nucleated within the austenite grains rather than at grain boundaries. Although the overall bainite transformation undoubtedly involves the redistribution of carbon, it is now established that the precipitation of cementite follows the initial coherent growth of ferrite that has the same carbon content as the austenite and is, therefore, supersaturated. This lack of carbon partitioning between the ferrite and austenite impedes the transformation and leads to the *incomplete reaction phenomenon* [81]. It has been outlined earlier that for dislocation-mediated martensitic nucleation, the activation energy is proportional to the driving force for the transformation (Eq. 12). This is in marked contrast to a normal diffusional transformation for which the work of nucleation is inversely proportional to the square of the driving force (Chapter 2, Eq. (14)). Studies of bainite nucleation kinetics [82] show behavior characteristic of the martensitic mechanism, and consistent with the absence of carbon diffusion during nucleation. The martensitic mechanism appears to apply also to the nucleation of *Widmanstätten ferrite*, the initial growth of which involves partitioning of carbon (a fast, interstitial solute), but still no partitioning of substitutional solute elements [83], an intermediate case termed *paraequilibrium*. The decomposition of austenite is considered further in Chapter 14, Section 4.2.

5. SUMMARY

There is a wide range of nucleation phenomena for which the kinetics can best be analyzed in terms of the motion of dislocations rather than atoms or molecules. Dislocations are extended line defects with long-range strain fields; their energies are so high that dislocation-mediated transformations are rarely thermally activated. When a crystalline material is plastically deformed, large numbers of dislocations are generated, but by interactions of existing dislocations rather than nucleation. Nevertheless there are cases when, locally and in small volumes, dislocations must be generated in material initially without dislocations: then nucleation is required. Examples considered in this chapter have included blunting of crack tips by emission of dislocations, small-scale indentation, and the loss of coherency between thin films and the substrate onto which they are deposited. In each case, dislocation nucleation is driven by an applied shear stress, the critical nucleus size and work of nucleation decreasing as the stress is increased. The nucleation is rarely, if ever, homogeneous. Heterogeneities favoring nucleation include surface steps, ledges on crack fronts, vacancy clusters under indents, and strain-induced surface roughening in thin films. Classical analyses based on linear elasticity applied at a macroscopic level have been superseded by a wide range of calculational tools covering length scales down to the atomistic; the newer approaches have transformed the understanding of dislocation configurations and their associated energies. In parallel, new methods of mechanical testing, such as instrumented (nano) indentation, are becoming capable of resolving the nucleation and motion of individual dislocations. Despite greatly improved understanding, quantitative prediction of kinetics is often prevented by the complexity of the heterogeneities and their evolution leading to dislocation nucleation.

Recrystallization is the formation of a new grain structure in a heavily deformed metal, driven by a reduction in the dislocation density. Homogeneous nucleation is not observed, and classical nucleation theory is not applicable. On annealing, dislocations rearrange to form subgrains, and growth of favored subgrains constitutes nucleation. The selection of these subgrains is influenced by their size and stored energy (from the prior deformation), and by the topology and mobility of the grain and subgrain boundaries around them. Nucleation events are sparsely distributed; although not dependent on surmounting an energy barrier, the behavior is stochastic, relying on the chance appearance of favored subgrains. While individual mechanisms, for example, of subgrain coalescence, are well understood, the complexity of the microstructural changes leading to nucleation again precludes quantitative prediction of kinetics.

Martensitic transformations and deformation twinning occur at interfaces that are walls of glissile partial dislocations. Olson and Cohen have shown that local dissociation of dislocations can generate embryos bounded by such walls. The embryos become transformation nuclei when the driving force for transformation exceeds a critical value that scales inversely with the number of crystal planes within the embryo. This gives athermal nucleation kinetics of the kind commonly associated with martensitic transformations. The dislocation configurations necessary for nucleation are highly improbable, and their absence inhibits the transformation of small volumes. Nucleation of martensite can also proceed under isothermal conditions. The kinetics then show a very distinctive feature: the temperature dependence of the nucleation rate is described by an activation energy that is approximately proportional to the driving force for the transformation.

A universal feature of these dislocation-mediated transformations is that, with present analyses, the microstructural features necessary for nucleation involve levels of complexity that prevent quantitative prediction of nucleation kinetics. Nevertheless, as will be discussed in

Chapter 14, Section 4.2, the understanding of underlying mechanisms can be exploited to achieve practical control over the nucleation of dislocation-mediated transformations in industrial materials processing.

REFERENCES

[1] J.P. Hirth, J. Lothe, *Theory of Dislocations*, 2nd edn, Wiley, New York (1982).

[2] V. Yamakov, D. Wolf, S.R. Phillpot, A.K. Mukherjee, H. Gleiter, Dislocation processes in the deformation of nanocrystalline aluminium by molecular-dynamics simulation, Nat. Mater **1** (2002) 1–4.

[3] A. Minami, A. Onuki, Dislocation formation in two-phase alloys, Phys. Rev. B **70** (2004) 184114/1–10.

[4] M. Fago, R.L. Hayes, E.A. Carter, M. Ortiz, Density-functional-theory-based local quasicontinuum method: prediction of dislocation nucleation, Phys. Rev. B **70** (2004) 100102R/1–4.

[5] T. Zhu, J. Li, K.J. Van Vliet, S. Ogata, S. Yip, S. Suresh, Predictive modeling of nanoindentation-induced homogeneous dislocation nucleation in copper, J. Mech. Phys. Solids **52** (2004) 691–724.

[6] R.E. Miller, A. Acharya, A stress-gradient based criterion for dislocation nucleation in crystals, J. Mech. Phys. Solids **52** (2004) 1507–1525.

[7] G.L. Pearson, W.T. Read Jr, W.L. Feldmann, Deformation and fracture of small silicon crystals, Acta Metall. **5** (1957) 181–191.

[8] J. Li, T. Zhu, S. Yip, K.J. Van Vliet, S. Suresh, Elastic criterion for dislocation nucleation, Mater. Sci. Eng. A **365** (2004) 25–30.

[9] C.A. Schuh, A.C. Lund, Application of nucleation theory to the rate dependence of incipient plasticity during nanoindentation, J. Mater. Res. **19** (2004) 2152–2158.

[10] M.F. Ashby, L. Johnson, On the generation of dislocations at misfitting particles in a ductile matrix, Philos. Mag. **20** (1969) 1009–1022.

[11] R. Hull, J.C. Bean, Misfit dislocations in lattice-mismatched epitaxial films, Crit. Rev. Solid State Mater. Sci. **17** (1992) 507–546.

[12] C. St. John, The brittle-to-ductile transition in pre-cleaved silicon single crystals, Philos. Mag. **32** (1975) 1193–1212.

[13] T.M. Pollock, A.S. Argon, Creep resistance of CMSX-3 nickel-base superalloy single-crystals, Acta Metall. Mater. **40** (1992) 1–30.

[14] G. Xu, A.S. Argon, Homogeneous nucleation of dislocation loops under stress in perfect crystals, Philos. Mag. Lett. **80** (2000) 605–611.

[15] M.Yu. Gutkin, I.A. Ovidko, Special mechanism for dislocation nucleation in nanomaterials, Appl. Phys. Lett. **88** (2006) 211901/1–3.

[16] A. Kelly, W.R. Tyson, A.H. Cottrell, Ductile and brittle crystals, Philos. Mag. **15** (1967) 567–586.

[17] A.S. Argon, B.J. Gally, Selection of crack-tip slip systems in the thermal arrest of cleavage cracks in dislocation-free silicon single crystals, Scripta Mater **45** (2001) 1287–1294.

[18] B.J. Gally, A.S. Argon, Brittle-to-ductile transitions in the fracture of silicon crystals by dynamic crack arrest, Philos. Mag. A **81** (2001) 699–740.

[19] J.R. Rice, R. Thomson, Ductile versus brittle behaviour of crystals, Philos. Mag. **29** (1974) 73–97.

[20] J.R. Rice, Dislocation nucleation from a crack tip—an analysis based on the Peierls concept, J. Mech. Phys. Solids **40** (1992) 239–271.

[21] G. Xu, A.S. Argon, M. Ortiz, Critical configurations for dislocation nucleation from crack tips, Philos. Mag. A **75** (1997) 341–367.

[22] J. Knap, K. Sieradzki, Crack tip dislocation nucleation in FCC solids, Phys. Rev. Lett. **82** (1999) 1700–1703.
[23] T. Zhu, J. Li, S. Yip, Atomistic study of dislocation loop emission from a crack tip, Phys. Rev. Lett. **93** (2004) 025503/1–4.
[24] A.C. Lund, A.M. Hodge, C.A. Schuh, Incipient plasticity during nanoindentation at elevated temperatures, Appl. Phys. Lett. **85** (2004) 1362–1364.
[25] W.C. Oliver, G.M. Pharr, An improved technique for determining hardness and elastic modulus using load and displacement sensing indentation experiments, J. Mater. Res. **7** (1992) 1564–1583.
[26] K.J. Van Vliet, J. Li, T. Zhu, S. Yip, S. Suresh, Quantifying the early stages of plasticity through nanoscale experiments and simulations, Phys. Rev. B **67** (2003) 104105/1–15.
[27] J. Li, K.J. Van Vliet, T. Zhu, S. Yip, S. Suresh, Atomistic mechanisms governing elastic limit and incipient plasticity in crystals, Nature **418** (2002) 307–310.
[28] J.D. Kiely, J.E. Houston, Nanomechanical properties of Au (1 1 1), (0 0 1), and (1 1 0) surfaces, Phys. Rev. B **57** (1998) 12588–12594.
[29] J.A. Zimmerman, C.L. Kelchner, P.A. Klein, J.C. Hamilton, S.M. Foiles, Surface step effects on nanoindentation, Phys. Rev. Lett. **87** (2001) 165507/1–4.
[30] C.A. Schuh, J.K. Mason, A.C. Lund, Quantitative insight into dislocation nucleation from high-temperature nanoindentation experiments, Nat. Mater **4** (2005) 617–621.
[31] C.A. Schuh, A.C. Lund, T.G. Nieh, New regime of homogeneous flow in the deformation map of metallic glasses: elevated temperature nanoindentation experiments and mechanistic modeling, Acta Mater **52** (2004) 5879–5891.
[32] W.D. Nix, Mechanical properties of thin films, Metall. Trans. A **20A** (1989) 2217–2245.
[33] J.W. Matthews, A.E. Blakeslee, Defects in epitaxial multilayers: 3. Preparation of almost perfect multilayers, J. Cryst. Growth **32** (1976) 265–273.
[34] L.B. Freund, The stability of a dislocation threading a strained layer on a substrate, J. Appl. Mech. Trans. ASME **54** (1987) 553–557.
[35] W. Hagen, H. Strunk, New type of source generating misfit dislocations, Appl. Phys. **17** (1978) 85–87.
[36] B.W. Dodson, J.Y. Tsao, Relaxation of strained-layer semiconductor structures via plastic flow, Appl. Phys. Lett. **51** (1987) 1325–1327.
[37] H. Gao, W.D. Nix, Surface roughening of heteroepitaxial thin films, Ann. Rev. Mater. Sci. **29** (1999) 173–209.
[38] D.E. Jesson, S.J. Pennycook, J.M. Baribeau, D.C. Houghton, Direct imaging of surface cusp evolution during strained-layer epitaxy and implications for strain relaxation, Phys. Rev. Lett. **71** (1993) 1744–1747.
[39] D.J. Eaglesham, M. Cerullo, Dislocation-free Stranski-Krastanov growth of Ge on Si (1 0 0), Phys. Rev. Lett. **64** (1990) 1943–1946.
[40] R.D. Doherty, Recrystallisation and texture, Prog. Mater. Sci. **42** (1997) 39–58.
[41] F.J. Humphreys, M. Hatherly, *Recrystallization and Related Annealing Phenomena*, Pergamon, Oxford (1995).
[42] M.B. Bever, D.L. Holt, A.L. Titchener, Stored energy of cold work, Prog. Mater. Sci. **17** (1972) 5–177.
[43] R. Viswanathan, C.L. Bauer, Kinetics of grain boundary migration in copper bicrystals with [0 0 1] rotation axes, Acta Metall. **21** (1973) 1099–1109.
[44] D. Weygand, Y. Bréchet, J. Lépinoux, On the nucleation of recrystallization by a bulging mechanism: a two-dimensional vertex simulation, Philos. Mag. B **80** (2000) 1987–1996.
[45] R.D. Doherty, J.A. Szpunar, Kinetics of sub-grain coalescence—a reconsideration of the theory, Acta Metall. **32** (1984) 1789–1798.
[46] M.L. Bhatia, Recrystallisation—some applied aspects, Prog. Mater. Sci. **42** (1997) 59–77.

[47] A.D. Rollett, Overview of modeling and simulation of recrystallization, Prog. Mater. Sci. **42** (1997) 79–99.
[48] A. Berger, P.-J. Wilbrandt, F. Ernst, U. Klement, P. Haasen, On the generation of new orientations during recrystallization—recent results on the recrystallization of tensile-deformed fcc single crystals, Prog. Mater. Sci. **32** (1988) 1–95.
[49] H. Miura, T. Sakai, H. Hamaji, J.J. Jonas, Preferential nucleation of dynamic recrystallization at triple junctions, Scripta Mater **50** (2004) 65–69.
[50] C.M. Wayman, H.K.D.H. Bhadeshia, Phase transformations, nondiffusive, in: *Physical Metallurgy,* 4th edn, Eds. R.W. Cahn, P. Haasen, Vol. 2, North-Holland, Amsterdam (1996), pp. 1507–1554; Chap. 16.
[51] P.M. Kelly, L.R.F. Rose, The martensitic transformation in ceramics—its role in transformation toughening, Prog. Mater. Sci. **47** (2002) 463–557.
[52] R.C. Pond, Y.W. Chai, S. Celotto, Martensitic transformations in 'unfamiliar' systems, Mater. Sci. Eng. A **378** (2004) 47–51.
[53] J.W. Christian, *The Theory of Transformations in Metals and Alloys,* 3rd edn, Pergamon, Amsterdam (2002), Part II, Chapters 21–23, pp. 961–1075.
[54] J.W. Christian, *The Theory of Transformations in Metals and Alloys,* 3rd edn, Pergamon, Amsterdam (2002), Part II, pp. 1062–1069.
[55] J.H. Hollomon, D. Turnbull, Nucleation, Prog. Metal Phys. **4** (1953) 333–388.
[56] X.Q. Zhao, B.X. Liu, Is homogeneous nucleation of martensitic transformation in iron-base alloys possible?, Scripta Mater **38** (1998) 1137–1143.
[57] I.T. Walker, A.L. Greer, Displacive transformations in Fe–Ni nanophase alloys, Mater. Sci. Eng. A **304–306** (2001) 905–909.
[58] R.E. Cech, D. Turnbull, Heterogeneous nucleation of the martensite transformation, Trans. AIME **206** (1956) 124–128.
[59] C.L. Magee, The kinetics of martensite formation in small particles, Metall. Trans. **2** (1971) 2419–2430.
[60] S. Kajiwara, S. Ohno, K. Honma, Martensitic transformations in ultra-fine particles of metals and alloys, Philos. Mag. A **63** (1991) 625–644.
[61] K.E. Easterling, G.C. Weatherly, On the nucleation of martensite in iron precipitates, Acta Metall. **17** (1969) 845–852.
[62] H.Y. Yu, S.C. Sanday, B.B. Rath, On the heterogeneous nucleation of martensite, Mater. Sci. Eng. B **32** (1995) 153–158.
[63] K.E. Easterling, P.R. Swann, Nucleation of martensite in small particles, Acta Metall. **19** (1971) 117–121.
[64] T. Waitz, H.P. Karnthaler, Martensitic transformation of NiTi nanocrystals embedded in an amorphous matrix, Acta Mater **52** (2004) 5461–5469.
[65] P.E. Reyes-Morel, J.-S. Cherng, I.-W. Chen, Transformation plasticity of CeO_2-stabilized tetragonal zirconia polycrystals: II. Pseudoelasticity and shape memory effect, J. Am. Ceram. Soc. **71** (1988) 648–657.
[66] J. Tu, B. Jiang, T.Y. Hsu, J. Zhong, The size effect of the martensitic transformation in ZrO_2-containing ceramics, J. Mater. Sci. **29** (1994) 1662–1665.
[67] T. Waitz, D. Spišák, J. Hafner, H.P. Karnthaler, Size-dependent martensitic transformation path causing atomic-scale twinning of nanocrystalline NiTi shape memory alloys, Europhys. Lett. **71** (2005) 98–103.
[68] M. Lin, G.B. Olson, M. Cohen, Homogeneous martensitic nucleation in Fe–Co precipitates formed in a Cu matrix, Acta Metall. Mater. **44** (1993) 253–263.
[69] M. Ahlers, Martensitic transformation. Model, Z. Metallk. **65** (1974) 636–642.
[70] G.B. Olson, M. Cohen, Nucleation: Part I. General concepts and the FCC→HCP transformation, Metall. Trans. A **7A** (1976) 1897–1904.
[71] G.B. Olson, M. Cohen, A general mechanism of martensitic nucleation. Part II. FCC→BCC and other martensitic transformations, Metall. Trans. A **7A** (1976) 1905–1914.

[72] G.B. Olson, M. Cohen, A general mechanism of martensitic nucleation. Part III. Kinetics of martensitic nucleation, Metall. Trans. A **7A** (1976) 1915–1923.
[73] J.W. Christian, S. Mahajan, Deformation twinning, Prog. Mater. Sci. **39** (1995) 1–157.
[74] Y.T. Zhu, X.Z. Liao, S.G. Srinivasan, E.J. Lavernia, Nucleation of deformation twins in nanocrystalline face-centered-cubic metals processed by severe plastic deformation, J. Appl. Phys. **98** (2005) 034319/1–8.
[75] S.A. Kulin, M. Cohen, On the martensitic transformation at temperatures approaching absolute zero, Trans. AIME **188** (1950) 1139–1148.
[76] A.J. Bogers, W.G. Burgers, Partial dislocations on the {1 1 0} planes in the bcc lattice and the transition of the fcc into the bcc lattice, Acta Metall. **12** (1964) 255–261.
[77] X. Zhang, Y.Y. Li, On the nucleation and growth of ferrous martensites, in: Eds. C.M. Wayman, J. Perkins, *Proc. Int. Conf. Martensitic Transformations (ICOMAT-92)*, Monterey Inst. Adv. Studies, Monterey CA (1992), pp. 203–208.
[78] S. Kajiwara, Continuous observation of isothermal martensite formation in Fe–Ni–Mn alloys, Acta Metall. **32** (1991) 407–413.
[79] K.C. Russell, Nucleation in solids: the induction and steady state effects, Adv. Coll. Interface Sci. **13** (1980) 205–318.
[80] G.B. Olson, A.L. Roitburd, Martensitic nucleation, in: *Martensite*, Eds. G.B. Olson, W.S. Owen, MIT Press, Cambridge, MA (1992), pp. 149–174 .
[81] H.K.D.H. Bhadeshia, *Bainite in Steels*, Inst. of Materials, London (1992).
[82] H.K.D.H. Bhadeshia, A rationalization of shear transformations in steels, Acta Metall. Mater. **29** (1981) 1117–1130.
[83] A. Ali, H.K.D.H. Bhadeshia, Nucleation of Widmanstätten ferrite, Mater. Sci. Technol. **6** (1990) 781–784.

CHAPTER 13

Solidification

Contents

1. INTRODUCTION

While solidification is of practical concern in a very wide range of fields, from ice damage on crops (Chapter 16, Section 2.3) to the processing of chocolate (Chapter 17, Section 3), this chapter focuses on metals and alloys, for which solidification has a key role in large-scale industrial processing. Solidified metallic microstructures consist of crystalline grains of one or more phases. Recent years have seen great advances in understanding dendritic and eutectic growth in solidification [1]. When solidification kinetics is dominated by growth and not nucleation (in cases such as surface-melting treatments), good agreement can be obtained between observations and predictions for complex microstructures [2]. The variation of microstructure with processing conditions can be represented on *microstructure-selection maps* [3]. While some progress

Pergamon Materials Series, Volume 15
ISSN 1470-1804, DOI 10.1016/S1470-1804(09)01513-2

has been made in extending the use of such maps to cases, such as droplet solidification, in which the microstructure is influenced mainly by nucleation [4], quantitative prediction of nucleation kinetics remains elusive.

Particularly in industrial practice, solidification is initiated at small supercooling and the nucleation must be heterogeneous. The fundamentals of heterogeneous nucleation on interfaces have been covered in Chapter 6, Section 2, and form an important background for this chapter; of particular relevance are the treatments of the spherical-cap model (Chapter 6, Section 2.2), substrate size and shape, (Chapter 6, Section 2.4) and athermal nucleation (Chapter 6, Section 2.5).

Quantitative prediction of nucleation kinetics is hindered firstly by inadequate knowledge of the input parameters. With the spherical-cap geometry in the classical theory (Chapter 6, Section 2.2, Figure 3), the nucleation kinetics is very dependent on the contact angle ϕ, and cannot be well predicted as ϕ and the underlying interfacial energies are not accurately measured or calculated. Furthermore, heterogeneous nucleation must usually involve a variety of nucleant surfaces with ill-defined characteristics. Secondly, as discussed in Chapter 6, Section 2.1, the classical theory breaks down for potent nucleation catalysis (small ϕ), which is often the case of interest for solidification. Despite these difficulties, it will be shown (mainly using the industrially relevant example of grain refinement of aluminum alloys, Section 3) that quantitative predictions of grain size can be made in some cases.

Nucleation is important because the size and shape of the grains in metals and alloys can determine performance in structural and functional applications. It is established practice to promote or to inhibit nucleation to obtain desired microstructures. For ambient-temperature structural use, a fine grain size is generally desired, because it simultaneously gives greater strength and greater toughness. On the other hand, directional solidification, to give columnar grains with boundaries parallel to the main stress axis, or to give a single crystal, is beneficial for creep resistance at elevated temperatures, as required for example in turbine blades [5]. Control of solidification is also important in welding, where an equiaxed grain structure in the weld metal reduces susceptibility to center-line cracking and facilitates ultrasonic inspection of welds [6, 7]. Metallic artefacts are mostly made by casting into shaped molds or dies (*casting* alloys), or by solid-state deformation of basic castings in the form of sheet or billet (*wrought* alloys). In the latter case, the final grain structure is mainly the result of solid-state processes — the deformation itself, perhaps followed by recrystallization and grain growth. Even so, the final microstructure is likely to be significantly affected by the prior as-cast microstructure.

Conventional solidification processing (*casting*) is considered in Section 2, where it will be seen that the natural initiation of grains in the liquid is not always by nucleation and that the kinetics of grain initiation is far from the only factor influencing the solidified microstructure. Controlled initiation of grains by addition of nucleant particles is considered in Section 3. Based on the analysis of conventional processing, Section 4 examines the nucleation laws appropriate for quantitative modeling of grain-structure development. Unconventional processing, such as rapid solidification, is considered in Section 5; in this, large supercoolings can be achieved and can give extremely high nucleation populations similar to those found in the devitrification of glasses (Chapter 8).

Finally we consider a quite different kind of nucleation during solidification. It is common for there to be some porosity in castings, and the phenomenon has been widely studied because of its detrimental effects, notably on surface quality and on mechanical properties. The nucleation of the pores is analyzed in Section 6.

2. MICROSTRUCTURE OF CASTINGS

2.1 Mechanisms of grain initiation in ingots

When a liquid alloy solidifies in a mold, there can be up to three distinct zones in the microstructure (Figure 1) [8, 9]. In contact with the mold wall, the *chill zone* is formed, in which the grains are fine, equiaxed and have a largely random crystallographic orientation. The nucleation of grains in this zone may be on the wall itself or on heterogeneities in the melt, and is driven by the chilling that occurs when the liquid first makes contact with the cold wall. The zone can be eliminated by heating the mold. As solidification continues towards the center of the casting, the grains become elongated, giving the *columnar zone*. In the chill zone, some grains have the preferred growth direction (typically $\langle 1\,0\,0 \rangle$ for metals) more closely antiparallel to the heat flow; these grains grow at lesser supercooling and, by projecting ahead of other grains at the liquid–solid interface, come to dominate in the columnar zone, thereby giving it a preferred crystallographic orientation [10]. For pure metals, the columnar zone continues to the center of the casting (Figure 1a), but for impure metals and alloys there is often an *equiaxed zone* in the final stages of solidification (Figure 1b), in which a random crystallographic orientation is restored. In both wrought and casting alloys, nonuniform microstructures (mixed columnar and equiaxed) are undesirable, and (as outlined in Section 1) mostly an entirely equiaxed microstructure (Figure 1c) is strongly

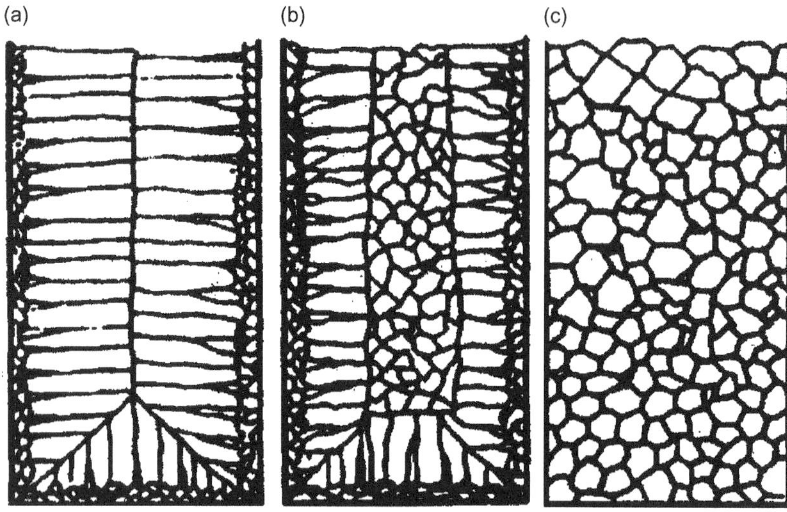

Fig. 1 Schematic grain structures in a simple ingot casting: (a) chill zone and then completely columnar growth; (b) as in the previous case except that there is a columnar-to-equiaxed transition in the final stages of solidification; (c) completely equiaxed. (Reprinted with permission from Ref. [8], copyright (1967), McGraw-Hill Companies.)

preferred. For this reason, the *columnar-to-equiaxed transition* (CET) has been the subject of much study [11–13]. Studies have mostly been for wrought alloys with low solute contents, which solidify predominantly to a single phase showing the CET clearly. In contrast, casting alloys are often near eutectic compositions, and for these the grains shown schematically in Figure 1 would be two-phase eutectic cells. As analyzed in Section 3, fully equiaxed structures can be obtained by adding nucleant particles to the melt. The present discussion, however, focuses on the origin of the equiaxed zone in the absence of added particles.

The equiaxed zone must involve grains dispersed and growing in the liquid ahead of the columnar grains. The dispersed grains may nucleate in the liquid ahead of the columnar grains, or may have been transported there having been formed earlier elsewhere. They can nucleate or survive only if the liquid is supercooled. This is possible because a significant supercooling ΔT_{col} (Figure 2 [14, 15]) is required to drive the columnar front at velocity v. This arises largely from solute partitioning that, by altering the liquid composition at the front, lowers the local liquidus temperature. The extent of the liquid ahead of the front in which there is such *constitutional supercooling* is dependent, among other things, on the temperature gradient $\mathfrak{G}$ in the solidifying region during casting. The competition between equiaxed and

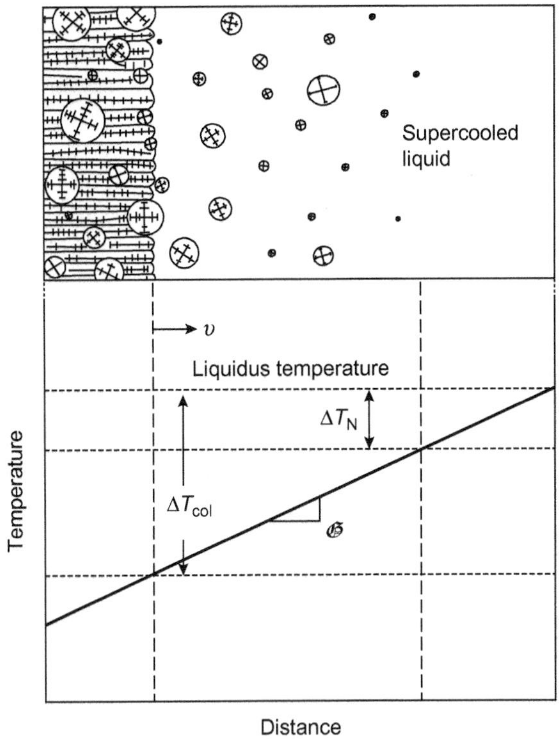

Fig. 2 A schematic view of the columnar-to-equiaxed transition ([14], adapted from Ref. [15]). Solidification proceeds from left to right. The columnar front moves with velocity v and the temperature gradient throughout the solidifying region is $\mathcal{G}$. ΔT_N is the supercooling for heterogeneous nucleation on substrates in the liquid, while ΔT_{col} is the supercooling (relative to the liquidus of the bulk liquid) for advance of the columnar front. (Adapted from Ref. [14], with permission, Taylor & Francis Ltd. http://www.informaworld.com.)

columnar solidification is thus affected by alloy composition and by solidification conditions.

A number of possible origins have been suggested for the dispersed grains leading to the equiaxed zone. These include heterogeneous nucleation on unspecified nucleants in the melt. This nucleation may occur within the supercooled liquid just ahead of the columnar front [16, 17]. It may also occur when the liquid is chilled on first contact with the mold. Grains not attached to the mold wall are swept into the bulk liquid. These grains survive only if the liquid superheat is not too great, and then in the so-called *big-bang* phenomenon a fine equiaxed structure can be obtained throughout the casting [18]. This is favored in small castings in which a significant supercooling can quickly be established throughout the liquid volume.

In larger castings there are cases where the equiaxed zone has its origin in grain fragments originating at the top and settling into the bulk [19]. Unless counteracting measures are taken, the top of the liquid in an ingot casting (such as in Figure 1) is cooler than the bulk and a solidified skin can form. A mechanism by which grain fragments might detach from solid was first observed in the solidification of cyclohexanol [20]. This organic liquid solidifies with morphologies similar to those of metals and alloys and, being transparent, allows the solidification to be readily observed. With suitable solute added, cyclohexanol solidifies dendritically. The main stems of the dendrites and the side arms have similar diameters, but the root of a side-arm (where it is attached to the main stem) tends to be narrower because solute partitioning lowers the liquidus temperature immediately around the main stem. It was observed that fluctuations in temperature and growth rate, caused by convective flow in the liquid, could cause the narrowed roots of the side-arms to remelt; the detached side-arms could then be transported into the bulk liquid. Similar detachment of grains at remelting roots has been observed in the early stages of formation of the chill zone [21]. In welding, a partially molten region can be formed in the heat-affected zone of the materials to be joined. It has been noted that grain detachment from this region can promote equiaxed solidification in the weld [22].

When the melt is inoculated (Section 3) with nucleant particles to give a fully equiaxed grain structure, the nucleation of grains on the added particles can completely swamp all other mechanisms. But in the absence of inoculation, a given casting is likely to have the grains in the equiaxed zone initiated by more than one mechanism. There is evidence in different cases for all of the above mechanisms of grain initiation. Particular mechanisms are more or less likely to be significant, depending on the alloy composition and processing conditions. For example, dendritic break-up or grain detachment is favored by alloying to give a large range of freezing temperature and by flow of the liquid, from the initial pouring or from convection. The exact mechanisms are beyond the scope of this chapter, but the phenomenon of break-up is significant; microstructures apparently indicating dispersed nucleation may have a quite different origin not susceptible to any nucleation analysis.

2.2 Columnar-to-equiaxed transition

In controlled directional solidification (for example Bridgman solidification along a fine capillary [18, 23]), liquid flow is limited; heterogeneous nucleation in the supercooled liquid just ahead of the columnar front is then the most likely origin of the CET, rather than break-up and transport of grains from elsewhere. This particular case has been much studied. Both experiment and microstructural modeling (for example Refs. [23, 24]) show that the CET is gradual. As equiaxed grains begin to nucleate ahead of the main solidification front, they are engulfed in the columnar grain structure. Eventually, fully equiaxed structures result when grains nucleated ahead of the columnar front completely block its advance. Hunt [15] presented an analytical model for the CET; as Hunt's model clearly demonstrates the role of nucleation, it is taken as the basis for the following discussion.

In this model, a constant temperature gradient $\mathfrak{G}$ is taken to apply throughout the solidifying region (Figure 2). Hunt presents a geometrical argument that blocking of columnar growth is complete (i.e. the final grain structure is fully equiaxed), when the volume fraction of equiaxed grains in the melt at the front is 0.49 (corresponding to an extended volume fraction of 0.66 when impingement is ignored) [15]. This critical value of the volume fraction has been confirmed by microstructural modeling [25]. The supercooling $\Delta T_{\rm col}$ (Figure 2) driving the advance of the columnar front with velocity υ is given by

$$\Delta T_{\rm col} = 2\left\{\frac{2Q\upsilon\sigma_{\rm ls}}{D\Delta s}\right\}^{1/2}, \tag{1}$$

where $\sigma_{\rm ls}$ is the liquid–solid interfacial energy, D is the diffusivity of the solute in the liquid, and Δs is the entropy of melting per unit volume [15]. The parameter Q is the *growth-restriction factor*, given (for an alloy with a single solute) by

$$Q = m(k-1)X_0, \tag{2}$$

where m is the liquidus slope of the alloy, k is the solute partition coefficient, and X_0 is the content of alloying element. As can be inferred from Eq. (1), at a given supercooling the solidification velocity is inversely proportional to Q.

Hunt assumes that there is classical heterogeneous nucleation on substrates in the melt, with a well defined onset supercooling $\Delta T_{\rm N}$, and that site saturation (i.e. one nucleation event per particle) occurs very early in solidification. The spatial extent of the supercooled zone (Figure 2) is greater for lesser $\mathfrak{G}$. Hunt [15] derives that fully equiaxed growth is expected for:

$$\mathfrak{G} < 0.617 N_0^{1/3}\left(1 - \frac{\Delta T_{\rm N}^3}{\Delta T_{\rm col}^3}\right)\Delta T_{\rm col}, \tag{3}$$

where N_0 is the number of nucleant particles per unit volume.

As the main processing variables $\mathfrak{G}$ and υ (which feature in Eqs. (1) and (3)) can be controlled independently in Bridgman solidification, they can be used as the basis of maps of the CET. Maps computed by Hunt for Al–Cu alloys are presented in Figure 3. These show the effects of $\Delta T_{\rm N}$, N_0 and X_0. For a given set of values of these variables, Figure 3a shows the boundaries of the $\mathfrak{G}$–υ regions in which the grain structure is predicted to be fully equiaxed, fully columnar or mixed. As $\mathfrak{G}$ is decreased there is a

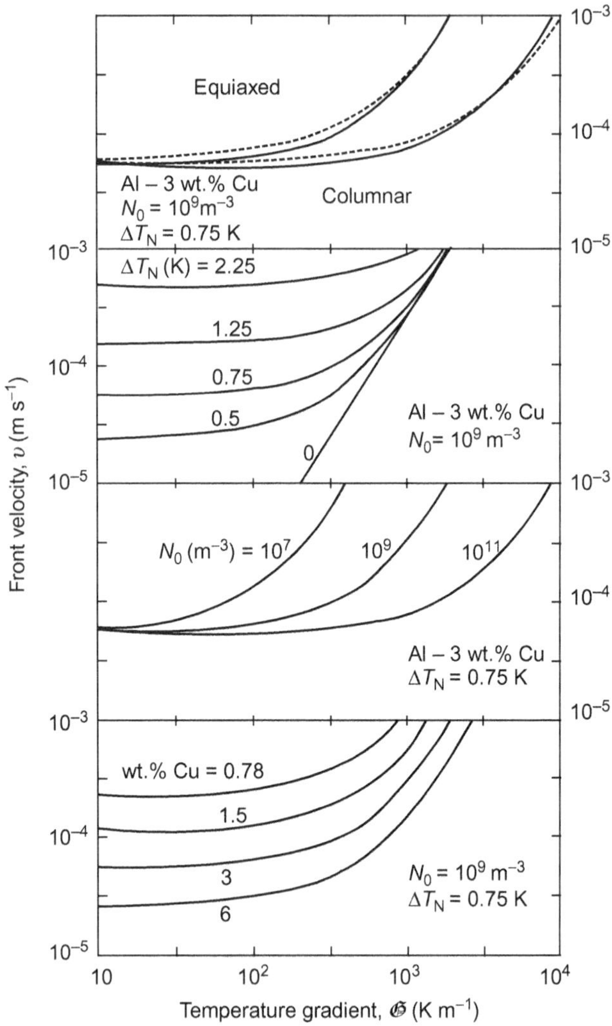

Fig. 3 Maps of growth velocity v (m s^{-1}) versus temperature gradient $\mathcal{G}$ (K m^{-1}) showing the columnar-to-equiaxed transition (CET) in Al–Cu alloys from the analytical treatment by Hunt [15]. In (a) the bounds of fully equiaxed and fully columnar grain structures show that the CET is gradual. (The solid lines are for the model, consistent with Eq. (3), that assumes nucleation is complete at ΔT_N and $\mathcal{G}$ is almost constant; the dashed lines are for a more complete analysis.) In the other parts of the figure, only the bound of fully equiaxed structures is used to show the effects of: (b) the onset supercooling for nucleation ΔT_N; (c) the population of nucleants N_0; (d) the copper content. (Adapted from Ref. [15], copyright (1984), with permission from Elsevier.)

gradual transition to equiaxed structures. As υ is decreased, $\mathfrak{G}$ has to be decreased further for the CET to occur. At the lowest values of υ there is a distinct regime in which columnar structures are obtained independent of $\mathfrak{G}$. This occurs when ΔT_{col} is smaller than ΔT_N, so that nucleation ahead of the columnar front is not possible. In the remaining parts of Figure 3, for clarity, only the boundary of the fully equiaxed region is shown. Figure 3b shows how this boundary is affected by the nucleation potency of the nucleants as expressed by ΔT_N. Greater potency (smaller ΔT_N) permits the CET to be observed at smaller υ. Increasing the population density of nucleant particles clearly also favors equiaxed structures, but its effect is on the critical $\mathfrak{G}$ value for the CET (Figure 3c). Increasing the solute content X_0 inhibits columnar growth and increases ΔT_{col}, affecting the critical values of both $\mathfrak{G}$ and υ (Figure 3d).

Approaches to more complete modeling of the CET have been reviewed by Spittle [13]. Realistic modeling of nonregular grain structures must include stochastic effects. In the work of Gandin and Rappaz [24], for example, a cellular-automaton (CA) grid is defined in which solidification takes place from cell to cell. Nucleation events are generated at random locations according to a Gaussian distribution of melt supercooling (as discussed in Section 4). The temperature distribution and heat flow are calculated using a finite-element (FE) algorithm, taking account of the release of the latent heat of solidification at the solid–liquid interface. The combined CA–FE model, in which the microstructural and thermal aspects are integrated, gives calculated grain structures in good agreement with experiment and consistent with the predictions of the Hunt model shown in Figure 3 [23, 24].

In solidifying melts, it is highly unlikely that there would be a single value of nucleation supercooling ΔT_N, as taken in the Hunt model. Nevertheless, this model is useful in highlighting the distinct roles of ΔT_N and of nucleant population density N_0 in promoting equiaxed solidification. Inoculation of melts, discussed in the next section, is used to overwhelm the natural CET and to obtain completely equiaxed solidification. The roles of ΔT_N and N_0 are important in understanding the minimum requirements for successful inoculation.

3. GRAIN REFINEMENT BY INOCULATION

3.1 Refinement of aluminum alloys: Phenomena and mechanisms

Inoculation is the use of additives to ensure that a material solidifies to a fine, equiaxed grain structure. While it is applied to a wide variety of alloy types from copper [26] to intermetallics [27], and to other materials such as polymers (Chapter 11, Section 5), the predominant use, by far, is in aluminum casting [28] (Figure 4 [30]). The great majority of aluminum alloys are cast with an additive (termed *inoculant* or *grain refiner*). Grain refinement by this method is used in the DC (direct chill) casting of billets of wrought alloys, reducing macrosegregation, avoiding center-line cracking and permitting higher casting speeds. Grain refinement is also important in shaped casting in foundries, giving improved feeding, reduced porosity, and reduced hot-tearing. In this case, there is an optimum degree of inoculation, with larger or smaller populations of nucleating particles degrading the castability [31]. There can also be many other benefits of added grain refiner, ranging from enhanced response to subsequent heat treatment to better surface quality and improved mechanical properties. Wrought alloys are, of course, subjected to heavy mechanical reduction after casting. This destroys the as-cast grain structure, but the effects of refinement are important not only in facilitating the casting itself and subsequent processing, but also in influencing aspects of the final microstructure, notably the selection of second-phase intermetallics [32, 33]. In shaped casting, there is use of eutectic and hypereutectic Al–Si alloys in which there are particular roles for melt additives to modify the eutectic microstructure [34] or to act as a nucleation catalyst for silicon as the primary solidification phase [35].

Fig. 4 Micrographs of cross-sections of TP-1 test [29] samples of superpurity aluminum. Three test series (different added refiners, different holding temperatures) show the effect holding time on grain-refiner performance. (Reprinted with permission from Ref. [30], copyright (2000), Maney Publishing.)

In the latter case, AlP is the most potent nucleant phase, formed in the melt itself or added directly in an inoculant. The present discussion, however, is focused on aluminum alloys with low solute content in which α-Al is the primary and predominant solidification phase.

Addition levels of refiner are quoted in kg $tonne^{-1}$, i.e. ppt (parts per thousand). Typical addition levels are in the range 0.25 to 3.0 ppt to achieve effective grain refinement, considered to be achieved when the grain size is $<200\,\mu m$. There are several commercially available inoculants; most are based on the system Al–Ti–B, and the most common composition is Al – 5 wt% Ti – 1 wt% B (henceforth Al–5Ti–1B). Refiners such as this give consistent performance, but problems motivating detailed study include the following:

- At best, only ~1% of the added nucleant particles successfully nucleate aluminum grains.
- The nucleant particles can agglomerate into larger clusters, giving particular quality problems in products such as thin foil and lithographic sheet.
- The presence of certain solutes, notably zirconium [36], chromium or silicon, can hinder the refining action of an inoculant.

This last effect is termed *poisoning* and an example is shown in Figure 5 [37]; in the presence of zirconium, the refiner action is severely degraded on holding in the melt. The mechanisms of poisoning are considered in Section 3.3. Refiners based on the Al–Ti–C system have also attracted much attention, primarily because of their resistance to agglomeration and to poisoning by some solutes, notably by zirconium. The following coverage is mostly of the predominant Al–Ti–B refiners, but Al–Ti–C refiners are also considered.

Most studies of grain refinement rely on tests using small melt volumes, under conditions designed to mimic full-scale casting. An example is the Alcan or TP-1 test [28, 29] using just $100\,cm^3$ of melt, for which the cooling rate is $\sim 3.5\,K\,s^{-1}$, roughly matching that in DC casting. The grain-size results quoted in this section, for example those in Figures 4 and 5, are all from TP-1 tests. The results in Figure 4 show the results of TP-1 tests on superpurity (99.999%) aluminum inoculated with different Al–Ti–C refiners [30].

Under typical conditions of holding the melt with added refiner, the grain size obtained on subsequent casting starts to increase after 20 to 30 min of holding time. As shown in Figure 5, this *fade* is largely avoided if the melt is stirred just before casting [37], implying that settling of the relatively dense refiner particles is the main cause. However, there may also be a contribution to fade from particle agglomeration.

In Al–Ti–B refiners, including the common Al–5Ti–1B, all the boron is combined in TiB_2 particles. The refiner consists of an α-Al matrix with

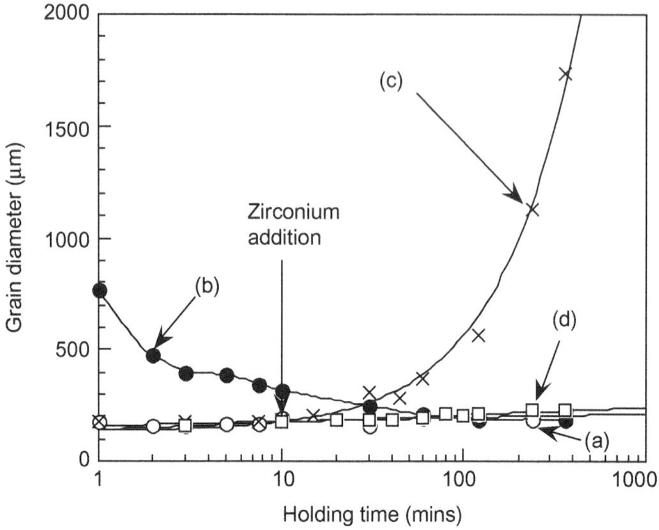

Fig. 5 The grain diameter of CP-Al refined with 1 ppt Al–5Ti–1B as a function of holding time before casting: (a) a standard refiner shows little fade with holding time at 800°C; (b) a poor refiner (held at 760°C) converges on the standard behavior; (c) poisoning of refiner action at 800°C after addition of 0.05 wt% Zr; and (d) a tantalum-modified refiner (held at 800°C) is resistant to the poisoning action of the same zirconium addition. In each case, the melt is stirred before the sample is taken for a TP-1 test. (Data from Ref. [37].)

embedded particles of TiB_2 and Al_3Ti. On addition to the melt, the α-Al matrix melts and dissolves and the TiB_2 particles survive. The dilution of overall titanium content is such that Al_3Ti particles also dissolve; this appears to be so fast as to preclude any direct influence of Al_3Ti particles in the grain refinement [38]. Thus it appears that the TiB_2 particles must be responsible for the refinement. However, they are not always good nucleants for α-Al [39], and appear to be fully effective only if there is some excess titanium in the melt (i.e., titanium beyond that in the TiB_2 phase) [40]. In contrast, if Al_3Ti particles are stable in the melt (at high enough titanium content), they are found to be extremely effective nucleants via a peritectic reaction. As reviewed by Schumacher et al. [41], it thus seems that effective grain refinement by Al–Ti–B refiners involves some type of combined action of Al_3Ti (which is a good nucleant but not stable in the melt) and TiB_2 (particles of which are stable in the melt, but not intrinsically good nucleants). The nature of this combined action was a matter of speculation; it was suggested that the TiB_2 could act to preserve the Al_3Ti locally, through: (i) a surrounding shell of borides [42, 43], (ii) survival in cavities (Chapter 6, Section 2.4, Figure 14; [44]), or (iii) a layer adsorbed on the borides (Chapter 6, Section 2.1; [45]).

As described in the next section, recent microscopical studies have substantially resolved this question.

3.2 Microscopical studies of nucleation

It would be good to obtain information on crystallographic matching and local composition at the interface between the inoculant particles and the nucleating α-Al by using a technique such as transmission electron microscopy (TEM). This is not possible during conventional solidification, and attention has therefore turned to cast structures. While some useful studies have been made of the microstructure of the grain refiner itself and of refined alloys [46, 47], there are difficulties. Given the low efficiency of refinement, inoculant particles in the center of grains that they have nucleated are of course rare compared to particles pushed into grain-boundary regions. Not only are nucleation sites rare in thin foils, but often the thinning procedure necessary for TEM observation can cause the nucleant particles to fall out during specimen preparation. Even if an embedded particle is found in the thin foil, it is difficult to be certain that the particle nucleated the grain rather than being engulfed by it. In the case of α-Al nucleating on TiB_2 particles, the potential role of an Al_3Ti coating may be obscured, because the Al_3Ti may be removed by peritectic reaction with the melt as solidification proceeds.

Some of these problems can be solved by stopping growth at an early stage; this can be achieved by embedding the nucleant particles in a metallic glass acting as an analog of the liquid, as demonstrated by Schumacher and co-workers [36, 41, 48, 49–51]. In the case taken in Figure 6, a glass-forming alloy melt (of composition $Al_{85}Y_8Ni_5Co_2$, at.%) containing TiB_2 particles from added Al–5Ti–1B refiner was rapidly quenched by melt spinning to obtain a glassy alloy ribbon with the embedded particles. The figure shows the type of TEM examination that can then be conducted on the particles. Figure 6a shows a single TiB_2 particle in the glassy matrix. The crystal structure of TiB_2 is hexagonal (C32 in the *Strukturbericht* classification), and the particle has the characteristic form of a hexagonal platelet bounded by prismatic $\{10\bar{1}0\}$ and basal $\{0001\}$ faces; in the image shown, the sixfold rotation axis lies in the plane of the thin foil and the electron beam is parallel to $\langle 11\bar{2}0 \rangle$ permitting the basal faces, and some prismatic faces, to be viewed edge-on. The other images in the figure are corresponding dark-field micrographs. Figure 6b is formed using a diffracted beam from TiB_2 and shows that the boride particle is a single crystal with no obvious defects other than surface ledges. Figure 6c shows α-Al crystallites that formed on the surface of the boride during the rapid quench. Their appearance shows that TiB_2 is a good nucleating substrate for α-Al, with the crystals presumably nucleating in the early stages of the quench.

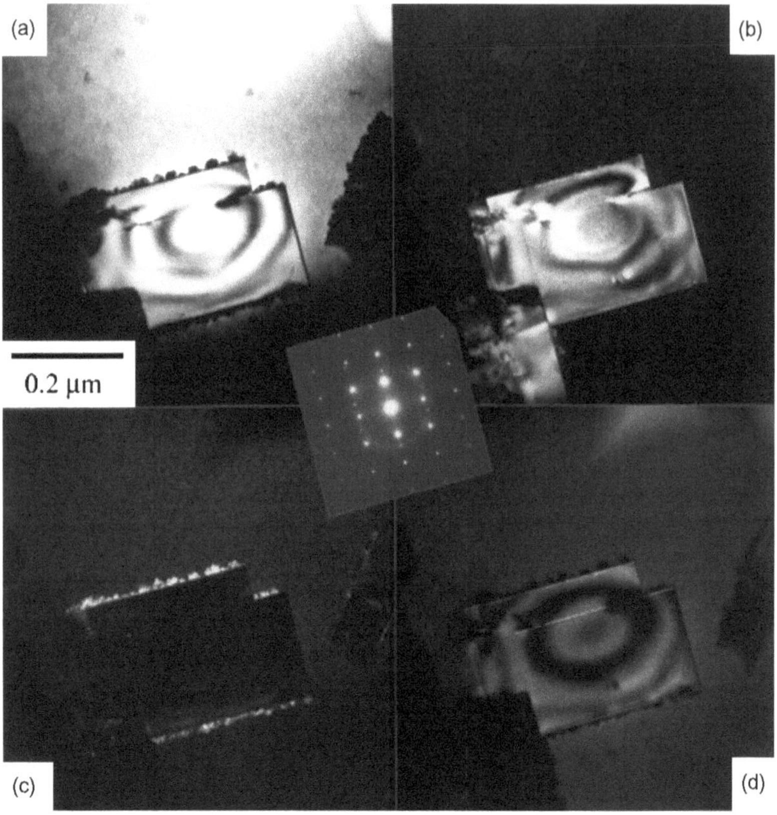

Fig. 6 Transmission electron microscopy of a TiB_2 particle in an Al-based glassy alloy matrix: (a) bright-field micrograph, showing side view of hexagonal TiB_2 platelet; dark-field micrographs showing (b) the single-crystalline TiB_2, (c) the α-Al crystallites forming in preferred orientation on the basal (0 0 0 1) faces of the TiB_2 particle and (d) the Al_3Ti layer coating the particle; and in the center the corresponding electron diffraction pattern. (Reprinted with permission from Ref. [41], copyright (1998), Maney Publishing.)

However, the rapid increase in viscosity of the glass-forming melt as it is cooled does effectively stifle crystal growth. This stifling of growth is the reason that many α-Al crystallites are observed. This aspect of the behavior in the glass-forming melt is in marked contrast to conventional grain refining in which it appears that each inoculant particle gives rise to at most one nucleation event. It is noticeable in Figure 6c that most of the α-Al crystallites are contributing to the image; this shows that they are not in random crystallographic orientations, but have formed in an epitaxial relationship with the substrate. In the center of Figure 6 is the electron diffraction pattern that shows spots attributable to TiB_2, to α-Al, and to Al_3Ti. The dark-field micrograph

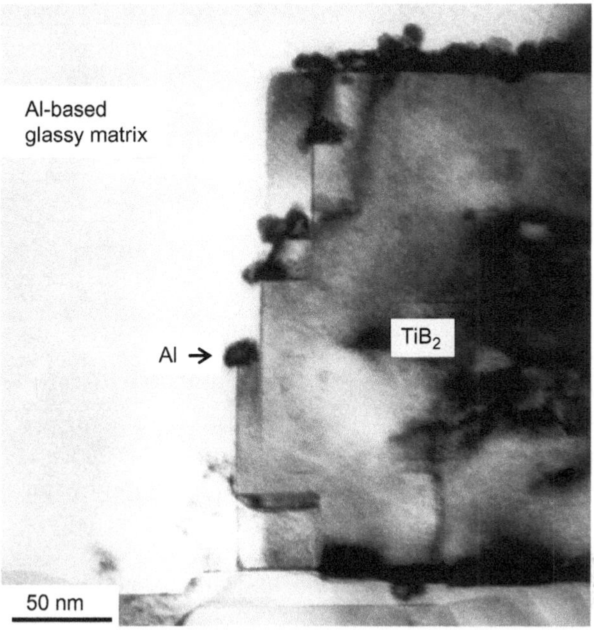

Fig. 7 TiB_2 particles are used to nucleate grains in casting of aluminum. The mechanism of nucleation can be studied by embedding the particles in an Al-based glassy analog of the liquid. This bright-field transmission electron micrograph [41] shows α-Al crystallites formed after annealing the sample. The labeled crystallite sits on a (0 0 0 1) orientation ledge on the $(10\bar{1}0)$ face of the particle, but avoids contact with the $(10\bar{1}0)$ face. (Reprinted with permission from Ref. [41], copyright (1998), Maney Publishing.)

Figure 6d shows that the Al_3Ti exists as a thin layer on the boride. In contrast, when a modified refiner without (significant) excess Ti is used, the metallic glass experiments show no Al_3Ti layer, and no α-Al crystallites forming on the TiB_2 particles [41].

It is evident in Figure 6 that the α-Al crystallites do not uniformly coat the boride, but form preferentially on its basal {0 0 0 1} faces. In the close-up of a similar sample in Figure 7, an α-Al crystallite has formed on a ledge. The ledge is in (0 0 0 1) orientation and is coated by the α-Al crystal, which avoids contact with the side face of the boride. The crystal therefore appears to have formed on the ledge only because the (0 0 0 1) orientation is highly favored and not because of low-energy contacts with more than one face in a re-entrant corner. From these observations comes the idea (exploited in the model in Section 3.4) that it is the large basal faces of the borides that are important for nucleation of the α-Al, and not any special surface sites (such as ledges).

From electron diffraction patterns, it is possible to extract the orientation relationships between the various phases. The crystal structure of Al_3Ti is tetragonal ($D0_{22}$ type), its unit cell being essentially a stack of two α-Al (cubic close-packed, A1 type) unit cells with chemical ordering on the sites. In this structure, the {1 1 2} planes are pseudo-close-packed. The orientation relationships are simple — the close-packed atomic planes in each phase are parallel

$$\{0001\}_{TiB_2}//\{112\}_{Al_3Ti}//\{111\}_{Al}$$

and the close-packed directions are parallel

$$\langle 11\bar{2}0\rangle_{TiB_2}//(\langle 201\rangle \text{ or } \langle 110\rangle)_{Al_3Ti}//\langle 110\rangle_{Al}.$$

As shown in Figure 8, the in-plane interatomic spacing in TiB_2 is about 5.9% larger than that of α-Al. On the other hand, the average spacing in Al_3Ti is about 3.7% smaller than that for α-Al [36].

High-resolution electron microscopy [48] of the surface of a boride (Figure 9) shows contrast consistent with a thin Al_3Ti layer coherent with the underlying TiB_2 particle. The in-plane lattice parameter of the Al_3Ti layer is therefore larger than its stress-free value. If varying degrees of coherency are possible, then the parameter of the Al_3Ti could vary along the dashed line indicated in Figure 8, and might even be a perfect match with α-Al. The combination of the TiB_2 and Al_3Ti phases may be significant also because the {1 1 2} planes of Al_3Ti providing good matching with α-Al are not the faces naturally bounding isolated Al_3Ti particles in Al–Ti alloy melts [52].

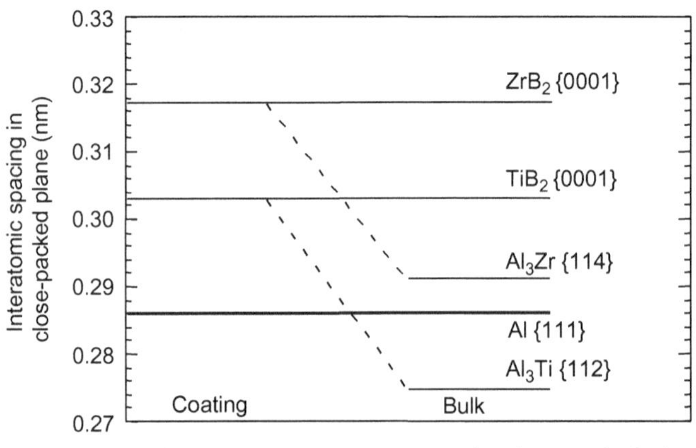

Fig. 8 Relative values of the interatomic spacings in the close-packed planes of the phases possibly involved in the nucleation of α-Al. The aluminide phase is a coating on the boride and, when thin enough, is coherent with the boride, adopting the same interatomic spacing. The diagram shows schematically how the spacing in the aluminides would decrease toward the value characteristic of the bulk phases as interfacial dislocations are introduced when the coatings are thicker. (Reprinted with permission from Ref. [36], copyright (1999), Maney Publishing.)

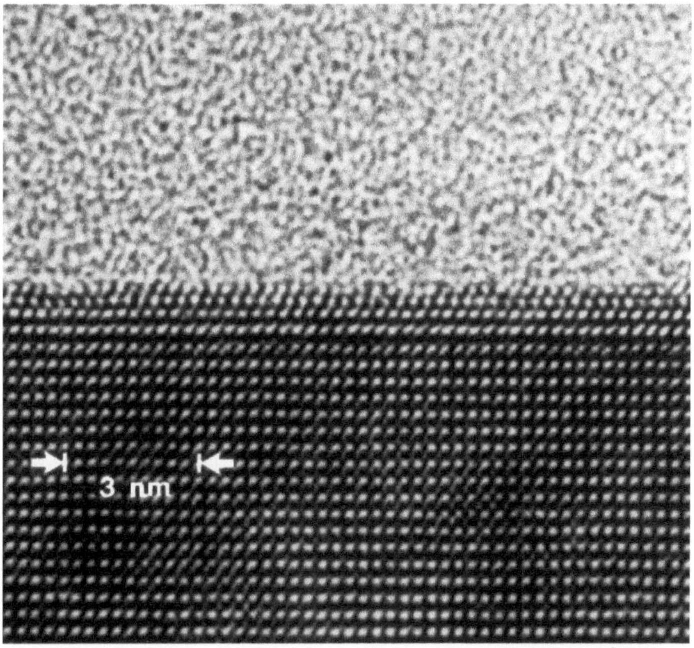

Fig. 9 Cross-section high-resolution transmission electron micrograph of the surface of a TiB_2 particle embedded in an Al-based glassy matrix (speckle contrast at top). The [0 0 0 1] axis of the hexagonal particle is vertical, in the plane of the micrograph. On the boride is a coherent surface layer, which has lattice spacings consistent with Al_3Ti. (Adapted from Ref. [48], with permission.)

An Al_3Ti coating on TiB_2 particles has been found in the refiner itself [47]. Observations of the kind shown in Figures 6 and 9 provide evidence that an Al_3Ti-like layer can survive, in melts in which bulk Al_3Ti would not be stable, by adsorption on the TiB_2 particles. Jones [45] has proposed a hypernucleation model for aluminum grain refinement, based on adsorption effects. It must be noted, however, that there is as yet no quantitative analysis of the adsorption energies that could stabilize the Al_3Ti phase in melts with low titanium contents. Nevertheless, the existence of an Al_3Ti coating would facilitate the interpretation of several effects. The development of the layer on holding in the melt may explain the improvement in performance during the *contact time*. It may also explain the larger changes seen when badly made refiners can dramatically improve their performance on longer holding in the melt [37] (Figure 5, curve b). Destruction of the layer in the presence of some solutes is a likely mechanism for at least some poisoning effects (Section 3.3). The layer may also act as a glue between TiB_2 particles, aiding their undesirable agglomeration [48].

The above discussion has been for Al–Ti–B inoculants. Al–Ti–C inoculants consist of Al_3Ti and TiC particles in an α-Al matrix. Their action is similar to that of Al–Ti–B in that only one of these phases, TiC, survives addition to aluminum melts to act as nucleant particles. TiC is cubic and the particles are predominantly octahedral, bounded by {111} faces; direct microstructural observations confirm that the TiC particles do act as nucleants [53]. The action of the TiC particles appears to be different from that of the TiB_2 particles in that there is no evidence of the existence of, or of the need for, an intervening Al_3Ti layer to facilitate nucleation. The absence of a layer may help to explain some of the resistance to poisoning. The most significant difference between the TiB_2-based refiners and the TiC-based refiners is in the stability of these phases themselves. TiB_2 appears to be very stable, even in melts with substantial superheat. On the other hand, TiC is unstable, transforming progressively to Al_4C_3 on holding in the melt [30]. This transformation is sufficiently sluggish (apparently limited by the slow nucleation and growth of the stable Al_4C_3 [54]) to pose no threat to grain refinement under normal conditions, but it does speed up markedly at higher temperature. The progressive dissolution of the TiC particles, and the lack of a layer on them, may account for the alleviation of agglomeration problems with Al–Ti–C refiners. The instability of TiC particles has also precluded microscopical studies of nucleation of the kind shown in Figure 6 for TiB_2 particles.

3.3 Poisoning

Figure 5 (curve c) gives an example of the poisoning of refinement that can occur in the presence of particular solutes. A possible mechanism is that reaction between solutes to give intermetallic particles can reduce overall solute levels in the melt. There is then a reduced level of growth restriction by solute (Eq. (2)), which as will be seen later (Figure 16) would impede refinement. There is little evidence for this mechanism, however. In most cases, poisoning appears to be a direct effect on the nucleation mechanism.

The example in Figure 5 is the poisoning of an Al–Ti–B refiner in the presence of zirconium. This effect has been studied using a Zr-containing, Al-based glassy alloy matrix (composition $Al_{87}Ni_8Zr_5$, at.%). TEM studies (analogous to those in Figure 6) show that on holding in the melt the TiB_2 particles are eventually converted to ZrB_2 [36]; on these there is no evidence of an aluminide coating, and no evidence of α-Al crystallites nucleating on their surfaces. Figure 8 shows how the lattice parameters of the boride and aluminide phases would be affected by Zr substitution. The lattice parameter of ZrB_2 is significantly larger than that of TiB_2 and is ~10.8% larger than the relevant in-plane parameter of α-Al; this seems certain to inhibit epitaxial nucleation of α-Al. Doping of the refiner with tantalum imparts resistance to poisoning (Figure 5, curve d). The phases TaB_2 and Al_3Ta are very stable and have atomic spacings only slightly greater than TiB_2 and Al_3Ti.

Silicon is a poisoning element of particular significance in casting alloys. The poisoning mechanism has been studied in Si-containing metallic glasses [51]. For silicon contents greater than 3 wt%, the TiB_2 particles are coated not with Al_3Ti, but with $TiSi_2$, which appears not to be a good nucleant for α-Al. The silicon content for the onset of poisoning is less for higher titanium contents, consistent with poisoning through $TiSi_2$ being the primary solidification phase [55].

Al–Ti–C refiners are resistant to zirconium poisoning, but are susceptible to silicon poisoning. Although TEM studies in metallic glass matrices have not been possible, microstructural studies of refined alloys have elucidated the mechanisms. Poisoning is severe and essentially instantaneous when silicon is added to the melt at a level greater than 3 wt%. With this silicon content, TiC particles are replaced by SiC particles that are not good nucleants. As SiC is isomorphous with TiC, it may form directly on, or by conversion of the surface of the TiC particles; in this way, the poisoning effect is rapid. At lower silicon contents, the fading of refiner action is apparently unrelated to silicon content, and instead reflects the replacement of TiC by Al_3C_4, which is more stable in typical melt compositions. As Al_3C_4 is hexagonal and without any good crystallographic match with cubic TiC, it appears to form independently of the TiC particles. The TiC particles dissolve slowly only as the carbon in the system is taken up by the nucleation and growth of the Al_3C_4; in this way, the fading of the refiner action is gradual [54].

3.4 Thermal modeling

The microscopical studies described in Sections 3.2 and 3.3 have been important in understanding the mechanisms of nucleation and poisoning. These studies, however, do not assist directly in understanding why grain refinement is so inefficient, with typically $\leqslant 1\%$ of the particles nucleating grains. The TEM observations of TiB_2 particles, for example, show nucleation of α-Al on all or none of the particles. Modeling to understand the inefficiency of grain refinement was first attempted by Maxwell and Hellawell [56]. Nucleation events are naturally spread in time, and the crystal growth from the nuclei formed early must act to stifle later events. In principle, the stifling might occur by (i) crystallization reducing the available volume of melt, (ii) a changing composition of the residual melt, i.e. soft impingement of the solute diffusion fields, or (iii) soft impingement of the thermal diffusion fields. Importantly, Maxwell and Hellawell noted that the dominant mechanism had to be (iii), since the volume fraction transformed when nucleation ceases is very small (typically 10^{-4}) and the thermal diffusivity is much greater than (by typically $10^4\times$) the solutal diffusivity. They considered the cooling of an aluminum melt containing a population of refining particles, and noted that growth of an increasing number of grains would give an accelerating rate of heat release, eventually surpassing the rate of external heat extraction and giving a temperature increase (Figure 10). This *recalescence* limits the number of nucleation events and therefore contributes to the inefficiency of grain refinement. A complete description of the temperature distribution during casting would be complex, taking into account the nature of the external heat extraction and the localized recalescence around each growing crystal. However, Maxwell and Hellawell recognized that, to a good approximation, the melt can be treated as spatially isothermal. This is because the thermal diffusion length greatly exceeds (by typically 10^2 to $10^3\times$) the separation between the nucleant particles.

Maxwell and Hellawell [56] computed the shape of the cooling curve (Figure 10) and the consequent number of grain nucleation events under various conditions, but did not make any direct comparison with

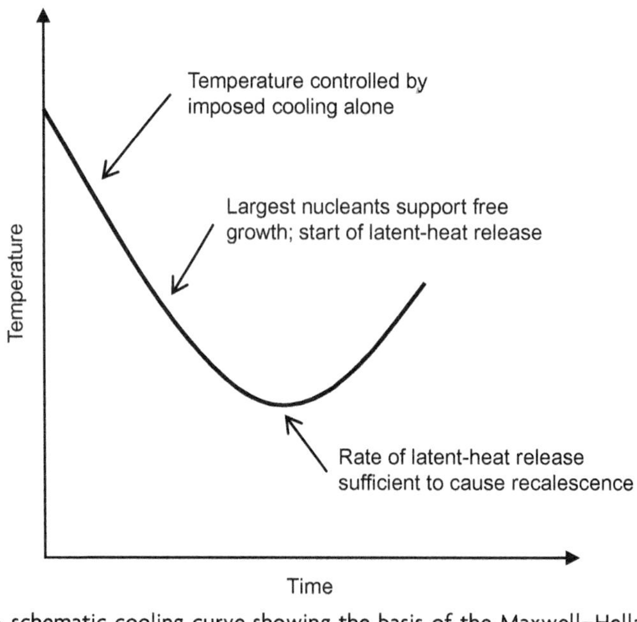

Fig. 10 A schematic cooling curve showing the basis of the Maxwell–Hellawell model [56]. The latent heat released from growing crystals is eventually sufficient to outweigh the external heat extraction, giving recalescence. After the minimum temperature has been passed, there is no further nucleation of new grains.

experiment. They did note, however, that the important factors controlling the grain size are: number and potency of nucleant particles, cooling rate, and the solute content in the melt. They used the spherical-cap model (Chapter 6, Section 2.2) to calculate the heterogeneous nucleation frequency on particles in the melt. For ease of computation, the particles were taken all to have the same size. The solute content is important because it restricts the rate of growth of the crystals. Maxwell and Hellawell took the crystals to be spherical; at the onset of recalescence the grains are still so small that they have not reached the onset of the morphological instability into dendritic growth [57]. Using the invariant-size approximation for the diffusion fluxes [58], the growth rate (dr/dt, where r is radius and t is time) at a given supercooling is proportional to the diffusivity of the solute in the liquid and inversely proportional to the growth-restriction factor Q defined in Eq. (2). (This matches the dependence of υ on Q noted in Eq. (1).) Values of Q are, to a good approximation, linearly additive for all the solutes in the melt, provided they do not interact to form complexes or precipitates that would effectively reduce the overall solute content [59]. Values of m, k and Q for common solutes in aluminum are given in Table 1 (using data from [60]), from which it can be seen that titanium is the most growth-restricting solute. Addition of Al–Ti–B or Al–Ti–C refiners at typical levels (1 to 3 ppt) increases the titanium content dissolved in the melt. This extra titanium must be taken into account in analyzing the growth restriction in the inoculated aluminum.

In effective grain refinement, nucleation on the inoculant particles dominates over nucleation elsewhere, and so must occur at small

Table 1 The values of m (liquidus slope) and k (equilibrium partition coefficient), calculated from parameters in [60], for common solutes in aluminum

Solute	m (K wt%$^{-1}$)	k	Q (K)
Cr	2.6	1.75	0.19
Cu	−2.5	0.145	0.21
Fe	−2.925	0.03	0.28
Mg	−5.84	0.48	0.30
Mn	−1.2	0.62	0.05
Ni	−3.5	0.004	0.35
Si	−6.62	0.12	0.58
Ti	25.63	7	15.38
Zn	−1.65	0.43	0.09

The solutes are compared in terms of the *growth-restriction parameter* Q (from Eq. (2)), for a solute content of 0.1 wt%.

supercoolings. This is likely to be a regime of *athermal nucleation* as analyzed in Chapter 6, Section 2.5. The α-Al initially formed on the {0001} face of a TiB_2 particle can be considered to be a spherical-cap nucleus with a very small contact angle ϕ. More likely, the α-Al is an adsorbed layer surviving even above the liquidus temperature, as suggested in the *hypernucleation* model [45] noted in Section 3.2. The nucleation is not then controlled by the initial formation of α-Al, but by the barrier to *free growth* from the inoculant particle. The hexagonal {0001} face of a TiB_2 particle can be approximated as a circle of radius R_{Nuc}, and the analysis in Chapter 6, Section 2.5 can then be applied directly. (The β coefficient given in Chapter 6, Table 1 can be used to evaluate more precisely the free-growth supercooling for hexagonal substrates [61].) As the α-Al formed on the {0001} face grows, the radius of curvature of its interface with the melt must at first decrease (inset in Figure 11). This growth can continue only until the radius of curvature of the liquid–solid interface reaches the critical value r^* for nucleation (which depends on the temperature at that instant). If r^* exceeds R_{Nuc},

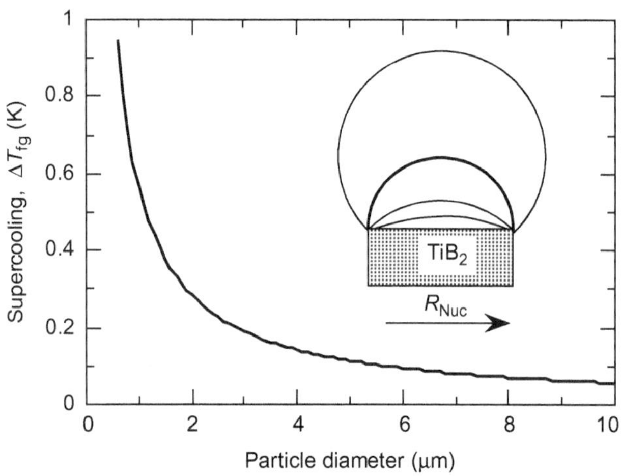

Fig. 11 The supercooling necessary to initiate free growth from a planar circular particle face of a given diameter. Such a face approximates the planar, hexagonal faces found on TiB_2 particles in Al–Ti–B refiners. This supercooling ΔT_{fg} arises from the Gibbs–Thomson shift of the solid–liquid equilibrium and is given by Eq. (4). The inset shows crystal growth following nucleation on one {0001} face of a boride particle. Thickening of the crystal reduces the radius of curvature of its interface with the liquid. As this radius cannot go below the critical nucleus radius r^*, there is a barrier to free growth if $R_{Nuc} < r^*$. Further growth past the critical hemispherical condition (for which the solid–liquid interface has minimum radius of curvature) is then possible only by increasing the supercooling to reduce r^*. (Adapted from Ref. [62], with permission.)

free growth of the crystal from the particle is not possible and the α-Al is a dormant nucleus. When the supercooling is increased, r^* is decreased, and when the condition $r^* \leq R_{Nuc}$ is reached, free growth of the crystal is possible through the minimum-radius hemispherical shape of the liquid–solid interface. (The shape of the nucleus at the onset of free growth from a hexagonal substrate is shown in Chapter 6, Figure 18a [61] and is closely approximated by a hemisphere.) Since the supercooling is small, the analysis in Chapter 6, Eq. 24 applies, and the supercooling for the onset of free growth ΔT_{fg} and R_{Nuc} are related by Eq. 25, Chapter 6, which can be written in more specific form as:

$$\Delta T_{fg} = -\frac{2\sigma_{ls}}{\Delta s R_{Nuc}}, \tag{4}$$

where σ_{ls} is the liquid–solid interfacial energy and Δs is the entropy of fusion. The variation of the free-growth supercooling with particle diameter is shown in Figure 11 for aluminum. Actual particle diameters in typical refiners vary over a wide range (Figure 12), but many are of the

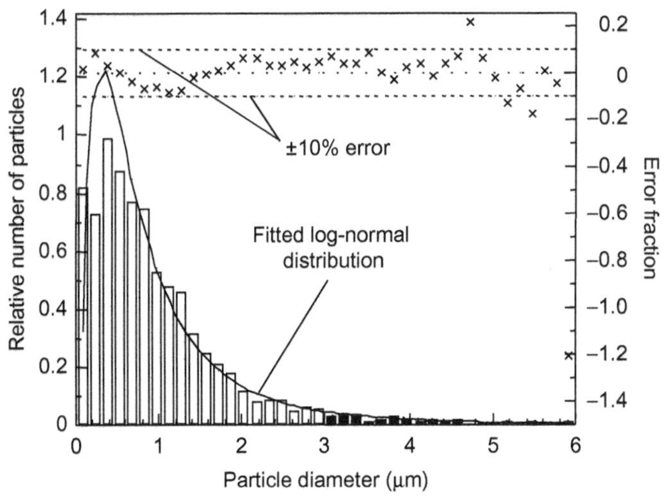

Fig. 12 The TiB_2 particles in a commercial Al–5Ti–1B (wt%) refiner are hexagonal platelets that can be approximated as disks. The measured diameter distribution of the disks is shown, fitted by a log-normal distribution with the indicated relative error. The relative populations (shown) are estimated from intersections in SEM micrographs of polished surfaces, corrected for sectioning effects. Absolute populations are then estimated from the known volume fraction of TiB_2 phase. The black area (in this case for diameters $> 3\,\mu m$) shows the fraction of the particles calculated by the free-growth model to be active for an addition level of 1 ppt in CP-Al cooled at $3.5\,K\,s^{-1}$. The refiner is very inefficient as under normal conditions only the largest particles initiate grains. (From Ref. [63], with permission.)

order of a few micrometers [57], consistent with the small observed supercoolings, certainly <1 K, and typically ~ 0.2 K. This agreement is evidence for the applicability of the free-growth model. The *nature* of the inoculant particles (including whether they are suitably coated, with Al_3Ti for example) determines the ease of initial formation of α-Al. But if the particles are potent nucleants, it is their *size* that determines the effective value of ΔT_N.

Chapter 6, Figure 15 shows the form of the energy barrier that must be surmounted for α-Al formed on a basal face of a TiB_2 particle to become a transformation nucleus. Using the analysis associated with Chapter 6, Figure 17, it was noted that the energy barrier is such that thermal activation of nucleation can occur on particles with radius of ~ 1 μm only when the supercooling is within less than 1 part in 10^7 of the value given by Eq. (4). Thus nucleation is completely *deterministic*, occurring at a supercooling ΔT_{fg} determined by the particle radius [64]. Turnbull [65] considered athermal nucleation of this type to analyze the solidification kinetics of liquid mercury droplets. He used the dependence of solidified population on supercooling to derive the size distribution of nucleating *surface patches*. This analysis is inverted in the present case of aluminum inoculation when a measured size distribution is used to predict the solidification kinetics.

In the work of Maxwell and Hellawell [56] and in later work on the free-growth model, cooling curves are calculated by dividing time into a series of isothermal steps; the number of grains initiated in each step is calculated, as is the total latent heat release from the incremental growth of all the grains initiated in earlier steps. This latent heat release, together with the imposed external heat extraction during the time increment, is used to calculate the cooling rate and thereby the temperature during the next increment. Once recalescence starts, there are no more initiation events and the number of grains per unit volume χ_V is determined.

The free-growth model has been applied mostly to analyze the results of TP-1 tests on the grain refinement of commercial-purity CP-Al using Al–5Ti–1B refiner [57]. The input parameters for the calculations are given in Table 2. The particle size distribution shown in Figure 12 was used [57, 63]. All these inputs come from the literature or from independent measurements; while there can be some selectivity in the values chosen, the input parameters are essentially not adjustable. The measured grain diameters in Figure 13 show that the inoculation becomes less efficient as the refiner addition level is raised. This behavior is matched well by the model predictions. The good agreement, both in the trend of the data and the absolute magnitude of the grain diameter, strongly supports the validity of the model, in particular the concept that the grain-refining efficiency is

Table 2 Parameters in the grain-size calculations used in the *free-growth model* [57]. The material parameters are mostly for pure aluminum.

Quantity	Symbol	Units	Value
Liquid–solid interfacial energy	σ_{ls}	$J\,m^{-2}$	0.158
Entropy of fusion per unit volume	Δs	$J\,K^{-1}\,m^{-3}$	1.112×10^6
Enthalpy of fusion per unit volume	Δh	$J\,m^{-3}$	9.5×10^8
Heat capacity of melt per unit volume	c_p	$J\,K^{-1}\,m^{-3}$	2.58×10^6
Diffusivity in melt (Ti in Al)	D	$m^2\,s^{-1}$	2.52×10^{-9}
Cooling rate in TP-1 test	ϑ	$K\,s^{-1}$	3.5

The sources of the data are cited in Ref. [57].
(Reprinted from Ref. [57], copyright (2000), with permission from Elsevier.)

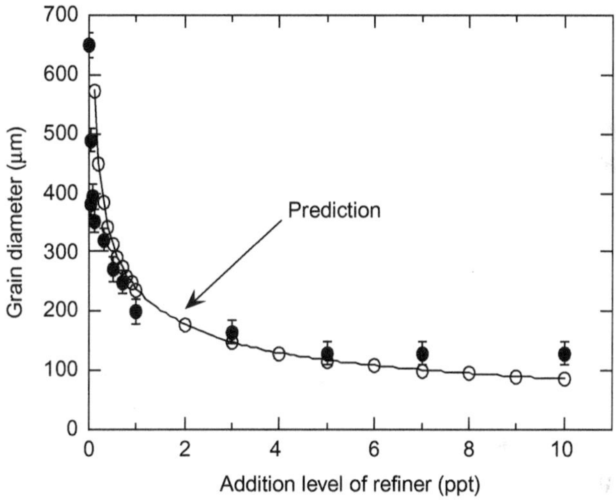

Fig. 13 Grain size (mean linear intercept) for CP-Al inoculated with Al–5Ti–1B at various levels. The grain diameters measured in TP-1 tests (•) are compared with the model predictions (○) assuming a cooling rate of 3.5 $K\,s^{-1}$. Good agreement is found, even though the model has no adjustable parameters. (Reprinted from Ref. [57], copyright (2000), with permission from Elsevier.)

limited by recalescence. The *efficiency* can be defined by

$$\text{Efficiency} = \frac{\text{number of grains per unit volume}}{\text{number of inoculant particles per unit volume}}. \tag{5}$$

The data in Figure 13 are replotted in Figure 14 to show the efficiency more clearly. Per unit volume, addition of 10^{12} particles gives only 10^{10} grains, while addition of 10^{15} particles gives 10^{12} grains — i.e. the efficiency decreases from 1% to 0.1% as the addition level is increased. Also shown in Figure 14 are the predictions of the original

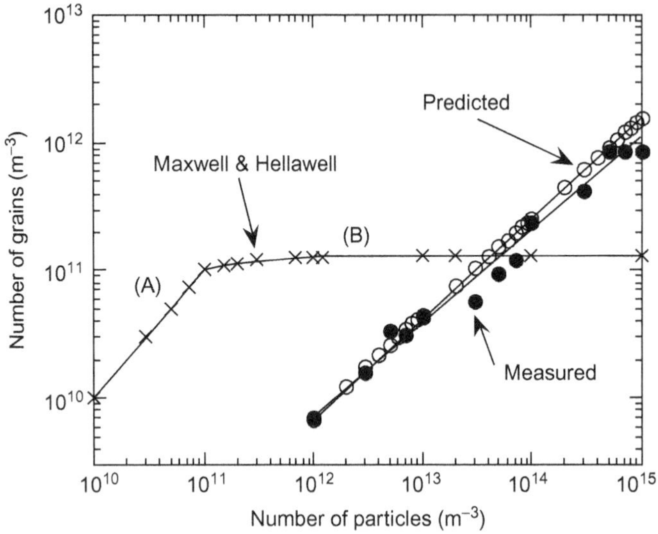

Fig. 14 The number of grains per unit volume as a function of the number of refiner particles per unit volume. The predictions of the Maxwell–Hellawell model [56] show two regimes: (A) in which there is 100% efficiency (1 grain per particle), and (B) in which the number of grains saturates. Also shown are data (•) calculated from the grain diameters measured in TP–1 tests and shown in Figure 13. These are compared with predictions (○) of the free-growth model. The free-growth predictions are qualitatively different from those of the Maxwell–Hellawell model, and are a much better fit to the data. (Reprinted from Ref. [57], copyright (2000), with permission from Elsevier.)

Maxwell–Hellawell [56] model (i.e. classical spherical-cap nucleation), showing two regimes. With a low population of particles ($<10^{11}\ m^{-3}$), there is time for a grain to nucleate on every particle, giving 100% efficiency. With larger populations, the number of grains nucleated does not continue to increase with the number of particles, but instead tends to saturate, giving decreasing efficiency. The measured data do not show these two regimes, and are much better matched by the free-growth model. The prediction of two distinct regimes arises in the Maxwell–Hellawell model because of the assumption of nucleation on particles of a uniform size. In this way, if a given temperature were maintained long enough, nucleation would eventually occur on all the particles and 100% efficiency would be achieved; this is regime A, favored by low particle populations. With the free-growth model, on the other hand, the distribution of particle diameter means that at any supercooling only a fraction of the particles could ever initiate grains. The shape of the measured diameter distribution (Figure 12) is such that a large fraction of the particles may never reach the supercooling at which they would become active; an efficiency of 100% is then impossible. Figure 12 shows,

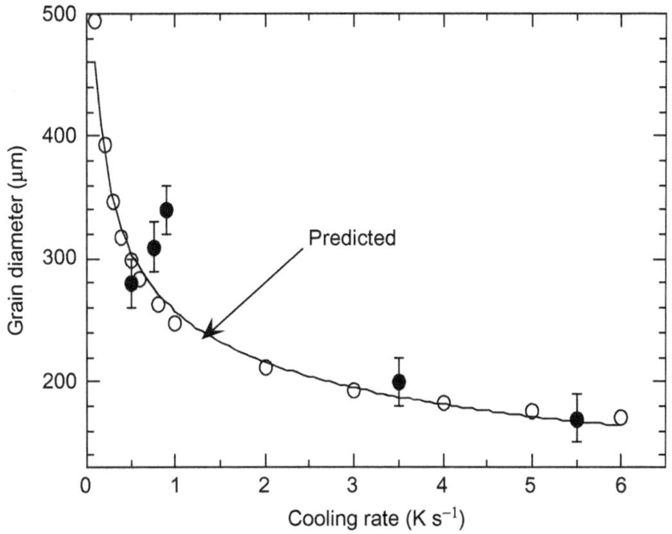

Fig. 15 The grain size of CP-Al with 5 ppt addition of Al–5Ti–1B as a function of cooling rate, measured (•) in modified TP-1 tests, and compared with predictions (○) of the free-growth model. (Reprinted from Ref. [57], copyright (2000), with permission from Elsevier.)

for a particular calculation (taking an addition level of 1 ppt in CP-Al, cooled at $3.5\,\mathrm{K\,s^{-1}}$), that nucleation of aluminum grains occurs only on particles down to a diameter of $3\,\mu\mathrm{m}$, corresponding well with the measured maximum supercooling of 0.2 K (Figure 11).

When the cooling rate is varied in the grain-refinement tests [57], the grain size varies as shown in Figure 15. Despite the limited number of measurements, it can be concluded that the data are consistent with the model predictions. This provides evidence in support of the thermal basis of the modeling.

Figure 16 shows the effects of solute content on grain size. The data are taken from the work of Spittle and Sadli [66] in which varying amounts of the solutes Cr, Cu, Fe, Mg, Mn, Si, Zn, or Zr were added to high-purity aluminum. For each case, Eq. (2) has been used with appropriate parameters to calculate the growth-restriction factor Q for the added solute and has been added to the Q for the extra titanium arising from the 2 ppt refiner addition used in the experiments. There is fair agreement between the measurements and the predictions. This agreement, and the clear trend to smaller grain size at larger Q, provides evidence for the relevance of growth restriction as included in the modeling. At low Q values, the measured grain sizes can considerably exceed the predicted values. When actual grain sizes are greater than

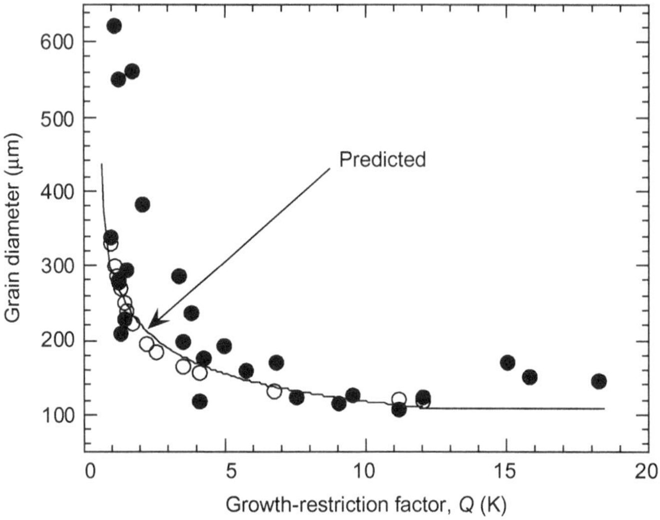

Fig. 16 Grain size as a function of growth-restriction parameter Q (Eq. (2)) for a standard TP-1 test with 2 ppt addition of Al–5Ti–1B refiner. The measured data (•) are for a wide range of solutes and addition levels [66], and are compared with predictions (○) from the free-growth model, taking parameters appropriate for each chosen solute addition. (Reprinted from Ref. [57], copyright (2000), with permission from Elsevier.)

~400 μm, however, the grain structure is columnar and therefore not treated by the modeling at all.

The modeling appears successful in predicting the grain size, but while the assumption of an isothermal melt may be reasonable for small volumes such as in the TP-1 test it cannot apply in larger castings, where there is no overall supercooling of the melt before grain initiation but there are clear temperature gradients and progressive solidification from the mold walls [67]. To deal with such a case it is appropriate to undertake more comprehensive modeling, of the kind considered in Section 4.

The above discussion has been based on Al–Ti–B refiners, in particular Al–5Ti–1B. In Al–Ti–C refiners, the TiC particles are octahedra rather than hexagonal platelets, and α-Al can form equally on each octahedral face. The onset of free growth from such particles can be treated using Chapter 6, Eq. 37, with the appropriate value of the β coefficient from Chapter 6, Table 1. The shape of the nucleus at the onset of free growth is shown in Chapter 6, Figure 19 [61]. The modeling described above for Al–Ti–B refiners [57] works also for Al–Ti–C refiners [68]. The smaller particle size of TiC compared to TiB_2 correlates with the greater nucleation supercooling that has been suggested for Al–Ti–C refiners [69].

The assumption of athermal nucleation, according to a free-growth model, also appears to be valid for analyzing the grain size in cast magnesium alloys [70, 71]. It may also be applicable in interpreting the supercoolings associated with the action of biogenic ice-nucleating agents [72] (Chapter 16, Section 2.3).

3.5 Design of grain refiners

Having established that a free-growth model with measured particle size distribution as input can make quantitative predictions of grain size, it is of interest to use the modeling to design refiners with optimized particle size distributions. Two idealized types of diameter distribution have been used: normal (i.e. Gaussian) and log-normal. The former is mathematically more straightforward, but has the disadvantage of giving small numbers of particles with unphysical negative diameter; the latter is a better match to typical size distributions in metallurgical microstructures and fits well in the present case (Figure 12). Normal [73] and log-normal [62, 74] distributions give predictions that are qualitatively in agreement for the effects of varying mean diameter and distribution width while holding constant the volume fraction of nucleant phase. Setting the width of the log-normal distribution to match that in a commercial Al–5Ti–1B refiner (geometric standard deviation = 0.88), the effects of a variation in the arithmetic mean diameter are shown in Figure 17. There is a clear minimum in grain size at intermediate values of mean particle diameter. Large particles (for fixed volume fraction) are necessarily few and give large grain sizes. On the other hand, small particles, though numerous, become active only at greater supercooling when the greater crystal growth rate impairs grain refinement.

The optimum refinement is at intermediate sizes. Interestingly, empirical development has brought the mean particle diameter in commercial refiners very close to this optimum. Figure 17 also confirms that at the optimum the *number efficiency* (Eq. (5)) is low. Alternatively, the efficiency can be evaluated in terms of the volume fraction of nucleant phase (i.e. TiB_2) initiating grains. As the active particles are the largest, this *volume efficiency* is substantial, approximately 80%. Thus the low number efficiencies usually quoted can be a misleading guide to refiner effectiveness.

It has been established that differences in the particle size distribution in commercial refiners do show some correlation with grain-refining performance [75]. So far, however, there has been little attempt to develop new, improved refiners. One exception is the work of Han et al. in which ultrasound has been applied during refiner production to obtain a narrower particle size distribution [76]. In agreement with

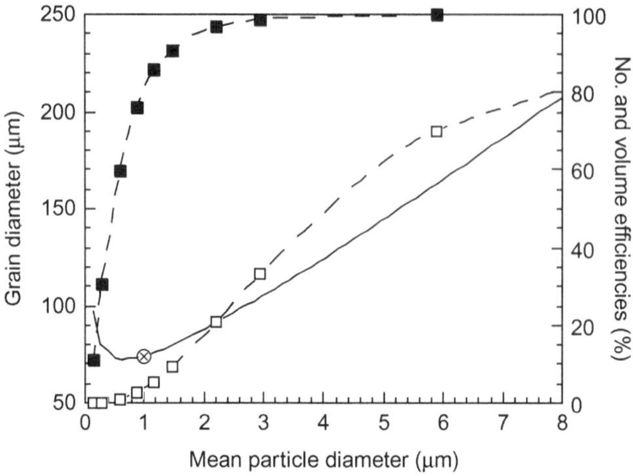

Fig. 17 Free-growth model predictions for CP-Al inoculated with a volume fraction of refiner particles corresponding to 2 ppt addition of Al–5Ti–1B and cooled at 3.5 K s^{-1}, showing: grain diameter (solid line), volume efficiency (■) and number efficiency (□) of refining (Eq. (5)) as a function of arithmetic mean particle diameter, for a fixed width of size distribution (geometric standard deviation = 0.88). The mean diameter in a commercial Al–5Ti–1B refiner (Figure 12) (⊗) is close to the minimum-grain-size optimum. (Adapted from Ref. [62], with permission.)

calculation [74], the narrower distribution does give a significantly finer grain size [76].

In practice, it may be more important to have a uniform grain size than a fine grain size. One reason why grain size might vary across a casting is nonuniformity in cooling rate, the effects of which are shown in Figure 18. These calculations are for idealized refiners with a Gaussian particle size distribution of fixed width (standard deviation) 0.5 μm. As the average particle diameter is increased (again holding constant the total volume fraction of nucleant phase), the efficiency (measured as number or volume fraction) increases and the grain size becomes increasingly insensitive to cooling rate.

The calculations summarized here have all assumed a nucleant particle shape characteristic of a commercial Al–5Ti–1B refiner. As seen in TEM, these particles are hexagonal prisms. In the modeling, these are approximated as disks with thickness 35% of their diameter. As the active surfaces are the hexagonal/circular (0 0 0 1) faces, the particles would be equally effective if thinner. The effectiveness of refiners, measured in terms of the refining achieved for a given volume fraction of nucleant phase, could be increased arbitrarily if the particles were thinner. For Al–Ti–B refiners, in which the nucleant phase is the hexagonal TiB_2, changing the aspect ratio of the particles is possible, at least in principle.

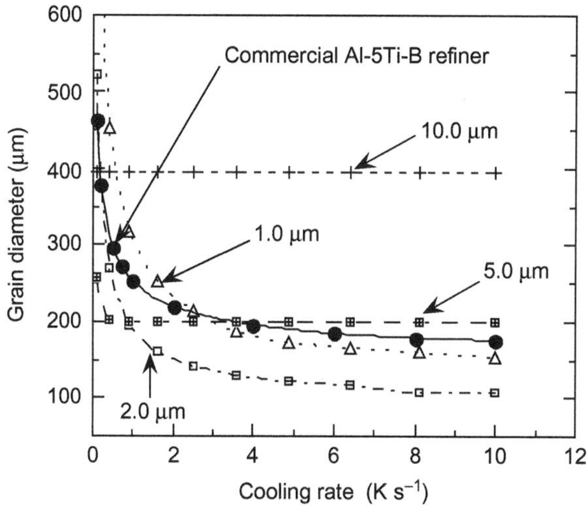

Fig. 18 Free-growth model predictions of the performance of idealized grain refiners in which the particle size distribution is Gaussian (with, in this case, a standard deviation of 0.5 μm). The dependence of grain size of CP-Al on cooling rate is compared for idealized refiners with different average particle diameters (labels on curves), and for 2 ppt addition of commercial Al–5Ti–1B refiner (with particle size distribution as in Figure 12). For the idealized refiners the volume fraction of inoculant particles is set to match that of the commercial refiner. (From Ref. [73], with permission.)

In contrast, for Al–Ti–C refiners in which the nucleant phase is cubic TiC with octahedral morphology, changing the particle aspect ratio is not a design option.

The most important reason to inoculate melts is to suppress columnar growth. As noted in Section 2.2, equiaxed growth is favored by a larger population of nucleant particles and particularly by a smaller supercooling to activate the particles. The free-growth model implies, therefore, that larger particles would be better for suppressing columnar growth. Preliminary studies of the effectiveness of different refiners in DC-cast ingots do appear to support this [75]. Of particular interest is the contrast between Al–Ti–B and Al–Ti–C refiners. For example, commercial refiners of composition Al–3Ti–0.15C (wt%) have active TiC inoculant particles with average diameters typically less than a quarter of those of active TiB_2 particles. Thus it is expected that the Al–Ti–C and Al–Ti–B refiners should show different performance. It is suggested from experiment that Al–Ti–C refiners are preferable to Al–Ti–B at higher addition levels, higher growth restriction, and lower cooling rate or temperature gradient [69, 77]. Many of these trends appear consistent with the modeling results in Figures 17 and 18.

4. NUCLEATION LAWS FOR SOLIDIFICATION MODELING

Comprehensive computer simulation and modeling of solidification must include, among many factors, macromodeling of fluid flow and heat transfer, and micromodeling of phase nucleation and growth [78]. As the laws of heat and mass transfer and of crystal growth are well established, it is likely that the derivation of the nucleation law is the greatest outstanding problem in developing a predictive model for solidification. The difficulties with quantitative prediction of nucleation kinetics were identified in Section 1. These difficulties were overcome in the modeling of grain refinement of aluminum described in Section 3.4. Quantitative prediction was possible in that particular case because:

- grain initiation is completely dominated by events on added nucleant particles;
- grain initiation on those particles is governed by the free-growth criterion and not by nucleation itself;
- the free-growth criterion depends on particle diameter, and not on unmeasured quantities such as the contact angle of spherical-cap nuclei; and
- the particle diameter distribution is readily measurable.

In that case, the nucleation law could be calculated from independently measured parameters and did not have to be deduced from the microstructure itself. In modeling other cases that are not so well characterized, the nucleation law involves adjustable parameters.

The early observations of Oldfield [79] showed that there can be a clear correlation between the population of grains and the maximum supercooling reached in the melt. In Oldfield's case, the population χ was proportional to the square of the supercooling ΔT. This corresponds to a nucleation law for cooling in which the nucleation rate $d\chi/d\Delta T$ is linearly proportional to ΔT. Such simple, empirical laws predict continuous rather than the effectively instantaneous nucleation given by a simple classical nucleation analysis [15]. They are easy to apply in solidification modeling and have been widely used.

Microstructural modeling yielding realistic nonregular grain structures is achievable for example using the *cellular automaton* [80] approach outlined in Section 2.2 of this chapter. Nucleation is assumed to be on sites that become active instantly at critical supercoolings. In the CA–FE modeling of Rappaz and Gandin [80] now used in commercial software, for example, $d\chi/d\Delta T$ has a Gaussian distribution about a maximum as a function of supercooling. As shown in Figure 19, it is common to superpose two Gaussians, one for nucleation in the bulk liquid with a maximum nucleation rate at an supercooling $\Delta T_{V,max}$ and one for nucleation on the mold wall with a maximum at $\Delta T_{S,max}$. For each

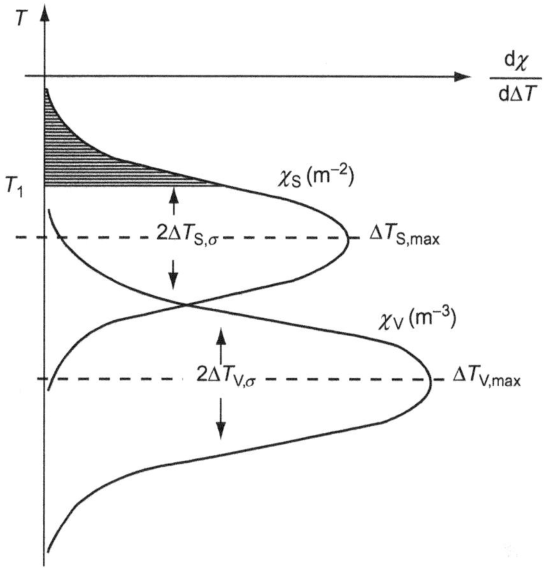

Fig. 19 Schematic illustration of the nucleation laws used in CA–FE modeling (for example, as in [24, 80]). There are separate Gaussian distributions for nucleation on the mold wall (with density of sites $\chi_S\,m^{-2}$) and in the bulk liquid (with density of sites $\chi_V\,m^{-3}$). The shaded area corresponds to the density of nucleation sites activated on the mold walls during supercooling to the temperature T_1. (Adapted from Ref. [80], copyright (1993), with permission from Elsevier.)

Gaussian, it is necessary to specify also the width of the distribution ($\Delta T_{V,\sigma}$, $\Delta T_{S,\sigma}$) and the total possible number of nucleation events ($\chi_V\,m^{-3}$, $\chi_S\,m^{-2}$). Such laws envisage a spectrum of nucleation sites, each site becoming active instantaneously when its critical supercooling is reached. This approach is completely compatible with, and is justified by, the free-growth model. With a measured particle size distribution as in Figure 12, the nucleation rate $d\chi/d\Delta T$ increases strongly, roughly exponentially, with ΔT in the range of interest of small ΔT; this is true also for the Gaussian distributions shown in Figure 19. CA–FE modeling has been applied to grain refinement in directional solidification, as noted in Section 2.2 [23].

5. GRAIN REFINEMENT WITHOUT INOCULATION

5.1 Spontaneous grain refinement in supercooled melts

As noted in Chapter 7, techniques such as electromagnetic and electrostatic levitation permit containerless processing of metallic melts and the attainment of very large supercoolings (Chapter 7, Table 3).

When the supercooled liquid alloy is clean and has no added inoculants, the solidification is expected to be columnar. Triggered directional solidification has been used as the basis for measuring growth velocity as a function of supercooling. Unexpectedly, melts with solidification triggered in particular ranges of supercooling can show fine equiaxed grains. Figure 20 [81] shows the marked refinement obtained at small and large supercoolings in Cu–Ni alloys. This *spontaneous* grain refinement is analyzed in this section; it provides a cautionary example of grain initiation by processes other than nucleation.

It has already been noted in Section 2.1 that dendrite break-up during the pouring of a casting can lead to a fine, equiaxed grain structure. Although the melts in experiments of the kind giving the data in Figure 20 are quiescent, the grain refinement observed at small supercooling appears also to involve dendrite break-up. Dendrites have a large surface area, and coarsening to reduce the area can naturally lead to localized remelting and break-up [20, 82], as observed directly at the surfaces of solidifying samples [19]. Grain refinement at small supercooling is observed only for alloys and not for pure melts, but the required solute content can be very low. For example, 0.01 at.% oxygen in nickel [83] or 0.018 at.% sulfur in copper [84] is sufficient to give some grain refinement at small supercooling. The solute may contribute to the effect by promoting the formation of fine dendrites and by facilitating subsequent local remelting. The phenomenon may also involve some fluid flow. Even without the agitation of a poured melt, there is flow induced by the

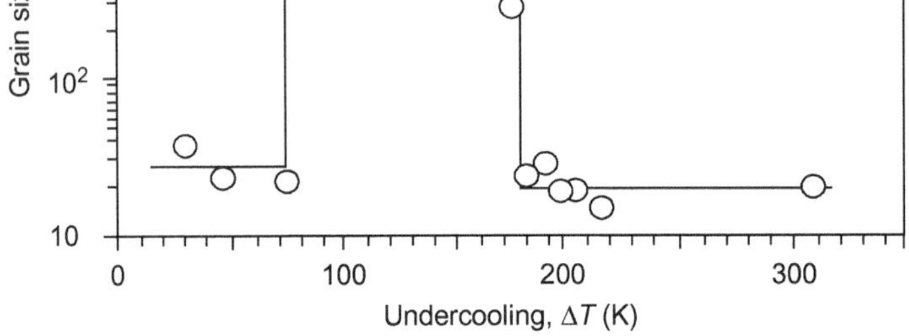

Fig. 20 The grain size in $Cu_{30}Ni_{70}$ (at.%) electromagnetically levitated droplets, with solidification triggered at selected supercoolings. There are clear transitions in grain structure, from refined equiaxed to coarse columnar at ΔT_1^* and back to refined equiaxed at ΔT_2^*. (Reprinted with permission from Ref. [81], Copyright (1994) by the American Physial Society.)

solidification shrinkage. Also, many of the experiments (as for Figure 20) have been conducted on electromagnetically levitated droplets in which the liquid is stirred by the levitation forces.

As shown in Figure 20, grain refinement is also found at large supercooling. Unlike the effect at small supercooling, this is found for pure metals as well as alloys. It was first reported for nickel melts supercooled more than 140 to 150 K [85]. The equiaxed grain diameter in samples solidified at these large supercoolings was less than one tenth of the diameter of the coarse columnar grains obtained at smaller supercoolings. Similar phenomena have now been demonstrated in other pure metals [86, 87] and in several alloys [88–90]. Observation of the temperature distribution in solidifying levitated drops shows that a thermal front, associated with the solidification front, crosses the drop from the point at which nucleation was triggered [91]. Despite this directionality of solidification, the grain structure formed behind the front is equiaxed.

Walker [85] suggested that a fine equiaxed structure could arise from copious nucleation triggered by a pressure pulse. Solidification shrinkage could indeed generate the conditions for cavitation ahead of the solid, while collapse of a cavity would generate a significant pressure pulse [92]. Alternatively, copious nucleation might naturally result from the rapid increase in nucleation frequency as the supercooling is increased. But the grain refinement is observed even when the supercooling is held fixed — in drops equilibrated at a given temperature and then touched with an external needle to trigger solidification. In addition, even if copious nucleation did arise throughout a supercooled drop, the subsequent recalescence (reheating) would most likely cause remelting of many of the small nuclei; without rapid heat extraction, this would greatly limit the refinement achievable [93]. And in any case, the progression of solidification across the drop rules out the possibility that the grain refinement is the result of copious nucleation throughout the melt at a critical temperature. It can be concluded that the grain refinement observed at large supercooling is not attributable to nucleation.

At large supercooling, solidification is rapid, and the solid that forms can be highly stressed. There is therefore a driving force for recrystallization, which could give grain refinement. Whether the grain structure has recrystallized or not is often discernible from the microstructure, because the composition variations corresponding to the dendritic microsegregation within the original grains are unaffected by the subsequent recrystallization. Recent studies of Cu–Sn offer an example where the grain refinement at large supercooling is clearly attributable to recrystallization [94]. In metallographic sections, the boundaries of the fine equiaxed grains are superposed on, but show no correspondence with, the segregation pattern of dendrites, which are on a larger scale. On the other hand, the Cu–O system shows a fine equiaxed grain structure exactly corresponding to the segregation pattern; each grain shows coring with a spheroidal rather than dendritic shape [95]. In this case, the grain refinement occurred during the solidification itself. In many cases, and the Cu–Ni system illustrated in Figure 20 is an example, grain refinement results from both solidification and subsequent recrystallization. It has been shown that in such a case, recrystallization can be prevented by rapid cooling in the solid state, but that grain refinement still occurs, at a critical supercooling set by the solidification process [96]. Thus recrystallization may have a role in the grain refinement observed at large supercooling, but there is nevertheless a clear refinement phenomenon in the solidification itself.

Just as for the refinement seen at small supercooling, the effect is attributable to dendrite break-up rather than to copious independent nucleation. There is some evidence for break-up from the preferred crystallographic orientation of grains in equiaxed structures that have been rapidly cooled to limit the time for reorientation of fragments [97]. The mechanisms of break-up remain controversial and lie largely beyond the scope of this book. They have been studied mainly for electromagnetically levitated drops in which solidification is triggered at selected supercoolings. For this case, Karma [81, 82] has presented a simple model to account for the grain refinement both at small and at large supercooling. When solidification is triggered, there is rapid growth of dendrites across the drop, accompanied by recalescence. At the end of this first stage, the drop is partly solidified, with a dendritic network throughout and a temperature set by the equilibrium between that network and the interdendritic liquid. This gives a thermal plateau that has a duration Δt_{pt} depending mainly on the rate of heat extraction. During the thermal plateau, the dendrites are subject to coarsening and in a time Δt_{bu} may break up into fragments; Δt_{bu} is determined largely by the thickness of the dendrite trunks and side-arms, related to the initial supercooling that set the growth velocity. In Karma's approach, the geometry of coarsening is simplified and is modeled as a Rayleigh instability of cylindrical dendrite trunks. It is found that for intermediate supercoolings $\Delta t_{\mathrm{bu}} > \Delta t_{\mathrm{pt}}$, so that break-up does not occur and columnar grain structures are found. At small and large supercoolings, however, $\Delta t_{\mathrm{pt}} > \Delta t_{\mathrm{bu}}$, so that break-up does occur. In this way, as the supercooling is increased, two microstructural transitions are predicted: from grain-refined to columnar at ΔT_1^* and from columnar to grain-refined at ΔT_2^*, as indicated by the data in Figure 20. The model is far from quantitative and does not take account of the stabilizing effect of continuing solidification during the thermal plateau [98]. It also does not treat the related phenomenon that at ΔT_2^* there is a change in the dependence of the velocity v of the solidification front on the supercooling ΔT. For $\Delta T < \Delta T_2^*$,

υ is as expected from dendrite growth theory and increases with ΔT approximately according to $\upsilon \propto \Delta T^{2.5}$; for $\Delta T > \Delta T_2^*$, υ varies linearly with ΔT [99]. Mullis and Cochrane [100–102] have shown that this transition may be associated with a dendrite instability in which their tips split at high growth velocities, and that this may contribute to the sharp onset of grain refinement for $\Delta T > \Delta T_2^*$.

5.2 Grain structures on the margin of glass formation

For conventional solidification of alloys, occurring close to the liquidus, the importance of latent heat release has been emphasized; indeed recalescence limits the attainable degree of grain refinement (Section 3.4). When crystallization is from the glassy state, however, the crystal growth rate is so low that recalescence is negligible (Chapter 8). In this section, an attempt is made to survey crystallization kinetics over the temperature range from the liquidus down to the glass transition temperature. This is done using a simple approach first adopted for analyzing the attainment of ultrafine grain sizes by rapid quenching of metallic melts [93]. The focus is on homogeneous nucleation as this may give grain sizes smaller than any conventional processing.

For simplicity only polymorphic solidification (i.e., without solute partitioning) is analyzed. The atomic rearrangement processes at the solid–liquid interface may be *diffusion-limited* or *collision-limited*. The former case refers to the need for diffusive-type jumps as the interface moves and does not imply any need for long-range partitioning. In this diffusion-limited case, the maximum interface speed υ is the diffusive speed D/λ , where D is the atomic diffusivity in the liquid and λ is the jump distance or atomic diameter. This case applies to, for example, the growth of intermetallic compounds and is consistent with glass formation. In the collision-limited case, on the other hand, the interfacial processes are not diffusive. The maximum speed of the solid–liquid interface is then approximately the speed of sound, precluding glass formation. Collision-limited kinetics was analyzed first for pure metals [103], but applies also for random solid solutions and permits solute trapping [104]. These two cases thus serve to illustrate a range of glass-forming ability even though only polymorphic solidification is considered.

Figure 21 [93] shows homogeneous nucleation rates and growth rates for the diffusion-limited and collision-limited cases, calculated with parameters typical for metallic (e.g. nickel-based) alloys. The calculations are for isothermal transformations with steady-state nucleation, even though (as will be emphasized below) such conditions cannot always be maintained during solidification. (The transient effects favoring glass formation on cooling have been considered in Chapter 8, Section 3.) For the diffusion-limited case, a glass-forming alloy is considered with a glass-transition

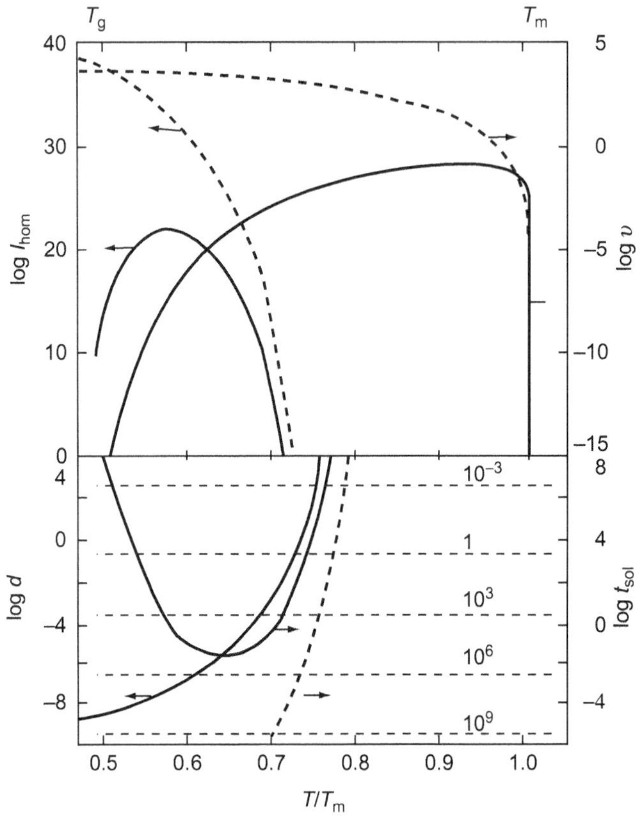

Fig. 21 Calculated solidification kinetics for typical alloys showing diffusion-limited (solid lines) and collision-limited (dashed lines) interfacial kinetics. Shown are the $\log_{10}$ of: I_{hom} (homogeneous nucleation rate, $m^{-3}\,s^{-1}$), υ (crystal growth rate, $m\,s^{-1}$), d (grain size, $(\upsilon/I_{hom})^{1/4}$, m), and t_{sol} (solidification time, $d/2\upsilon$, s). The grain size is the same for both types of kinetics. The horizontal lines are labeled with cooling rates in $K\,s^{-1}$. If the solidification time falls below these lines, recalescence will preclude grain refinement and the calculated value of d will not apply. (Adapted from Ref. [93], copyright (1991), with permission from Elsevier.)

temperature $T_g = 0.5T_m$ (the equilibrium melting temperature), and a Vogel–Fulcher–Tammann temperature dependence of viscosity from which the interfacial kinetics is derived [105]. The crystal growth rate is assumed to be limited by the interfacial kinetics. In the collision-limited case a similar assumption would be unrealistic, as the growth is limited by the thermal and curvature supercoolings. Since planar growth under these conditions is not stable, a simple dendrite growth law $\upsilon \propto \Delta T^3$ (found for pure nickel [106]) is used, modified at high ΔT to limit υ to the speed of sound. In each case, classical homogeneous nucleation theory was used [105].

Figure 21 shows that substantial supercooling ΔT is required for the onset of homogeneous nucleation. At smaller ΔT, heterogeneous nucleation will dominate. If latent heat release during solidification is ignored, isothermal transformation is possible, and the resulting grain size d is related to the growth rate υ and the bulk nucleation rate I_{hom} by [107]

$$d = \left(\frac{\upsilon}{I_{\text{hom}}}\right)^{1/4}. \tag{6}$$

The grain size calculated from Eq. (6) is shown in Figure 21 and is the same for both diffusion-limited and collision-limited kinetics. The curve for d demonstrates grain sizes finer than obtainable conventionally ($d < 100\,\mu\text{m}$) for large supercooling ($T/T_{\text{m}} < 0.68$). However, this refinement can be observed only if recalescence during growth does not stifle further nucleation.

A simple analytical approach permits estimation of the latent heat release rate under isothermal conditions. As the crystals grow, even at a constant rate υ, the rate of release of latent heat accelerates as the surface area of the crystals increases. If, within the solidification time (i.e. before the grains impinge), the latent heat release rate becomes greater than the external heat extraction rate (characterized by the cooling rate ϑ that would be found in the absence of the heat release), then recalescence is significant. As discussed by Greer [93], this is the case when the solidification time ($t_{\text{sol}} = d/2\upsilon$, where d is the grain diameter and υ is the solidification velocity) falls below a critical value given by

$$t_{\text{sol}} = \frac{3\Delta h}{c_{\text{p}}\vartheta}. \tag{7}$$

where c_{p} and Δh are the liquid heat capacity and latent heat per unit volume. Contours of $(3\Delta h/c_{\text{p}}\vartheta)$ are plotted on Figure 21 for comparison with t_{sol}.

At small and moderate supercoolings ($T/T_{\text{m}} = 0.8$ to 1.0), Figure 21 shows that recalescence is not important when the nucleation can only be homogeneous. For collision-limited growth, the solidification time t_{sol} (dashed line) becomes so small that recalescence would dominate well before the large supercoolings where grain refinement by copious homogeneous nucleation might occur. The nucleation is stifled in this case, even for imposed cooling rates greatly in excess of the maximum ($\sim 10^6\,\text{K}\,\text{s}^{-1}$) achievable in typical rapid solidification (by melt spinning). In the diffusion-limited case, however, cooling rates $> \sim 10^6\,\text{K}\,\text{s}^{-1}$ are sufficient to render the latent heat release unimportant. In that case, grain refinement due to copious nucleation should become observable just as glass formation becomes possible. Figure 21 is based on isothermal kinetics. Although this may be useful in considering whether or not recalescence is likely to be important, and in demonstrating the refinement achievable at large ΔT, it cannot be used to obtain quantitative estimates of grain size.

The prediction arising from Figure 21, that ultrafine microstructures can be attained under conditions on the margin of glass formation, has been amply verified by experiment. When glass-forming alloy compositions are melt-spun, they can show a transition from a glassy to a fine-grained structure through the thickness, which appears to arise from copious homogeneous nucleation in the bulk [108]. The combination of slow growth rate and rapid external heat extraction ensures that a high supercooling (and therefore rapid nucleation) can be maintained even as solidification progresses. Recently, the interest in such systems has focused on Al–TM–Ln (TM ... transition metal; Ln ... lanthanide) alloys, which can give nanophase composites in which α-Al crystallites are dispersed in an Al-based glassy matrix [109, 110]. These microstructures can be obtained directly by rapid quenching on the margin of full glass formation, or by subsequent annealing and partial devitrification of fully glassy precursors obtained by more rapid quenching (Chapter 14, Section 2.2). The crystallites are 3 to 10 nm in diameter and occupy up to 25% volume fraction. The partially devitrified alloys are up to 50% harder than the fully glassy alloys, while avoiding the embrittlement that often accompanies annealing of metallic glasses. The similarity of structure and properties of material made by different routes confirms that the microstructural refinement achieved by devitrification of a glassy alloy is very closely related to that achieved in melts undergoing rapid quenching.

Apart from the Al-based nanophase composites of interest for their mechanical properties, similar nanophase composite microstructures in iron-based systems are of interest for their magnetic properties. Commercially important microcrystalline hard-magnetic materials based on Fe–Nd–B alloys can be made (apart from a powder metallurgical route) equally by controlled quenching directly to the microcrystalline structure or by devitrification of faster-quenched glassy material [111]. Also of interest are soft-magnetic materials based on the formation of α-Fe nanocrystallites in compositions based on Fe–Si–B [112]. In this case, a typical composition, $Fe_{73.5}Cu_1Nb_3Si_{13.5}B_9$, has added copper to stimulate crystallite nucleation [113] and niobium to inhibit growth of the α-Fe crystallites (see Chapter 14, Section 2.2). These additions, of course, are similar in intent to those involved in conventional grain refinement. While the copper is evidently successful in giving a fine dispersion of nucleant particles, such additions are not always necessary; nanometer-scale dispersions of α-Fe crystallites giving good soft magnetic properties can also be obtained by partial devitrification of amorphous alloys such as $Fe_{91}Zr_7B_2$ [114].

The Al-based and Fe-based glasses just quoted do not show polymorphic crystallization (the types of crystallization have been presented in Chapter 8, Section 3). The solute partitioning accompanying the growth of the α-Al or α-Fe crystallites plays an important role in the microstructural refinement by impeding growth. The compositions in these systems must be chosen to avoid polymorphic crystallization to make glass formation possible in the first place. Nevertheless, the basic forms of behavior shown by the curves for the diffusion-limited case in Figure 21 still apply.

The slow crystal growth in devitrification greatly facilitates quantitative analysis of solidification kinetics. There is good evidence in a number of glasses that the crystal nucleation during crystallization is homogeneous [115]. In others, growth occurs on nuclei that appear to have been formed homogeneously during the quench; the population of these nuclei is strongly dependent on quench rate [115]. In either case there is the prospect, as indicated in Figure 21, of obtaining very fine grain sizes. The heat release that could stifle such an effect is not significant for the glass-forming system until near the minimum in the solidification–time curve shown in Figure 21. At lower temperatures, a metallic glass can remain closely isothermal as crystallization proceeds. However, near the minimum external heat extraction may be necessary to prevent the heat of crystallization, leading to accelerating nucleation and growth rates and a runaway, or *explosive,* reaction. In explosive crystallization, much of the grain refinement effect would be lost.

6. POROSITY

In most cases, solidification involves shrinkage, and it is consequently necessary to feed liquid toward the solidifying region. If the feeding is inadequate, hydrostatic tension in the remaining liquid leads to *shrinkage porosity.* An extreme case is the large-scale *macroporosity* that develops when a local hot spot in a casting remains part-liquid, surrounded by material that has completely solidified. Even if such hot spots are avoided, castings can suffer from a distribution of many small pores. One origin of such *microporosity* is the significant *solidification range* between the liquidus and solidus temperatures found in many alloys. Within this range there exists a part-liquid, part-solid *mushy zone.* As the solidified fraction increases, and particularly for an interconnected network of dendrites, the liquid flow through this zone, necessary to feed the solidification shrinkage, becomes more difficult and negative pressure may arise. The other origin of microporosity is dissolved gas in the liquid metal. On cooling, gas solubility in the liquid increases, yet *gas porosity* may form because partitioning of the dissolved gas from the solid as it grows raises the dissolved gas concentration in the liquid. The development of microporosity in castings and how it may be modeled have been extensively reviewed [116, 117]. It is important to have a good treatment of pore nucleation in the modeling because alloy properties are affected not only by the overall percentage porosity but also by the size distribution of the pores. The largest pores, for example, have the strongest detrimental effect on a cast component's fatigue resistance [118].

The fracture of liquids by spontaneous nucleation of vapor bubbles was analyzed by Fisher [119]. He considered the work of formation of a spherical bubble of radius r in a liquid of pressure p_l. There is a positive contribution to the work, $4\pi r^2\sigma$, where σ is the liquid surface energy (i.e. surface tension). The emptying of the volume of the pore requires a work that is simply the pressure–volume product $(4/3)\pi r^3 p_l$. The liquid at the given temperature has a vapor pressure p_v. Filling the pore with vapor at this pressure reduces its work of formation by $(4/3)\pi r^3 p_v$. To these terms considered by Fisher, we can add the effect of dissolved gas. Corresponding to the concentration of dissolved gas there is, in equilibrium, an external gas pressure p_g, and this pressure in the pore reduces its work of formation by $(4/3)\pi r^3 p_g$. Overall, then, the reversible work of pore formation, W_{pore}, is given by

$$W_{\text{pore}} = 4\pi r^2\sigma + \frac{4}{3}\pi r^3(p_l - p_v - p_g), \tag{8}$$

from which by differentiation the critical radius for pore nucleation, r^*, is

$$r^* = \frac{2\sigma}{(p_v + p_g - p_l)}. \tag{9}$$

In the case considered by Fisher [119], fracture of liquids under hydrostatic tension, there is no dissolved gas ($p_g = 0$), and the vapor pressure is negligible ($p_v = 0$); pore nucleation can then occur only if p_l is sufficiently negative. This is of practical interest in connection with, for example, the cavitation in liquids that leads to damage of marine propellers. Although not of interest in alloy solidification, for completeness we note that, as the temperature of a pure ($p_g = 0$) liquid is raised, nucleation of a pore (bubble) can occur when the vapor pressure exceeds the ambient pressure ($p_v > p_l$); this triggers boiling. Finally, in the case relevant for gas porosity in metals, even at temperatures at which the vapor pressure is negligible, a pore can nucleate when the gas pressure corresponding to the dissolved concentration exceeds the ambient pressure ($p_g > p_l$). Whittenberger and Rhines [120] showed that for typical microporosity in castings, there are significant local contributions from both an increasing p_g during solidification and a decreasing (or increasingly negative) p_l.

We next consider how to relate p_g to the concentration of dissolved gas. Because of its prevalence and practical importance, microporosity has been particularly widely studied for aluminum castings, in which hydrogen is the only dissolved gas of importance. The dissociation of the hydrogen molecule on solution,

$$H_2(gas) \leftrightarrow 2H(in liquid Al), \tag{10}$$

means that the concentration of dissolved atomic hydrogen C_H is related to the equilibrium pressure by Sievert's law:

$$C_H = S(p_g)^{1/2}, \tag{11}$$

where S is an empirically determined constant. The *supersaturation ratio* α_s is the factor by which C_H exceeds the concentration C_0 that would be in equilibrium with gas at the actual pressure of the liquid p_l:

$$\alpha_s = \frac{C_H}{C_0} = \left(\frac{p_g}{p_l}\right)^{1/2}, \tag{12}$$

and the *supersaturation*, *s*, is $(\alpha_s - 1)$. Lee and Hunt [121] used X-ray radiography of samples solidifying in a temperature-gradient stage to determine the temperatures of pore nucleation in Al–Cu alloys. The high diffusivity of hydrogen permits the increase in C_H in the liquid during solidification to be estimated simply from the solidified volume fraction using the lever rule (the dissolved gas concentration in the solid phase can be taken to be negligible). Lee and Hunt found a Gaussian distribution of nucleation temperatures of the kind shown in Figure 19, corresponding to mean supersaturation ratio $\alpha_s = 2.0$ with a standard deviation of 0.5. They noted that this magnitude of α_s is far too low for homogeneous nucleation. They concluded that hydrogen-induced pores must nucleate heterogeneously, perhaps on oxides, or that the pores grow from entrapped nonsoluble gas pockets in crevices in oxides or other inclusions. Nucleation from such pockets was noted by Fisher in analyzing the fracture of liquids under hydrostatic tension [119]. In the case of aluminum alloys, the effects of bilayered oxide films have been widely studied [122], and gas trapped within the bilayers could play a role in pore nucleation. If so, nucleation would correspond to the onset of free growth as already analyzed extensively for crystal nucleation in inoculated liquids (Section 3.4), and would be dependent above all on the size of the preexisting defects. Related examples of the nucleation of bubbles in liquids are considered in Chapter 16, Section 3 and Chapter 17, Section 4. Lee and Hunt [121] studied the effects of added grain refiner on pore nucleation in aluminum alloys. They concluded that grain-refiner particles do not have a direct nucleating effect. There is an increase in the average α_s necessary for pore nucleation, an effect attributed to inhibited hydrogen diffusivity in the finer-grained mushy zone.

7. SUMMARY

The focus of this chapter has been the industrial processing of metals, for which the control of grain initiation in casting is important. While directional solidification tends to give a columnar grain structure, there can be a natural transition to an equiaxed grain structure. Suppression of this transition is used to obtain columnar or single-crystal grain structures, but in most cases, an equiaxed grain structure is desired. Equiaxed structures can be promoted under normal solidification conditions by the addition of an inoculant, but can also arise spontaneously in highly supercooled liquids. Although heterogeneous nucleation must be involved, in the cases just cited it is rarely, if ever, the controlling factor in grain initiation. In the natural CET, and in spontaneous grain refinement at high supercooling, grain initiation can involve the break-up and dispersal of preexisting crystals. In inoculation, the criterion limiting grain initiation is that for free growth from the surface of the added particles. In all these cases, the balance between latent heat release and external heat extraction has a dominant influence on the solidified grain structure.

Nucleation itself, homogeneous or heterogeneous, does become dominant in the initiation of new grains at high supercooling if recalescence can be suppressed. This is the case for melts on the margin of glass formation, and is very closely related to devitrification. In this case, exceptionally fine grain sizes can be achieved.

In well-inoculated alloys, grain initiation is controlled by the condition for free growth from the added particles. This depends on the size of the nucleant particles rather than their nature, facilitating modeling by eliminating the dependence on parameters with unknown values. If the size distribution of the nucleant particles is measured, quantitative predictions can be made for grain size as a function of alloy composition and processing conditions. This provides a basis for general nucleation laws to be applied in more comprehensive modeling of solidification, including microstructural development. It also permits the design of particle size distributions to optimize grain-refining performance.

Porosity can form in solidifying alloys through restricted flow of liquid to feed solidification shrinkage and through dissolved gas becoming supersaturated in the liquid as a result of partitioning out of the growing solid. The supersaturations necessary for pore nucleation are so small that the process must be heterogeneous. It is likely to be analyzable in terms of a free-growth model analogous to that used for grain initiation in inoculated melts, but the nucleation sites have yet to be characterized.

REFERENCES

[1] W.J. Boettinger, S.R. Coriell, A.L. Greer, A. Karma, W. Kurz, M. Rappaz, R. Trivedi, Solidification microstructures: recent developments, future directions, Acta Mater. **48** (2000) 43–70.

[2] W. Kurz, Solidification microstructure processing maps: theory and application, Adv. Eng. Mater. **3** (2001) 443–452.

[3] S.C. Gill, W. Kurz, Rapidly solidified Al–Cu alloys 1: experimental determination of the microstructure-selection map, Acta Metall. Mater. **41** (1993) 3563–3573.

[4] A.F. Norman, K. Eckler, A. Zambon, F. Gärtner, S.A. Moir, E. Ramous, D.M. Herlach, A. L. Greer, Application of microstructure-selection maps to droplet solidification: a case study of the Ni–Cu system, Acta Mater. **46** (1998) 3355–3370.

[5] F.L. Versnyder, M.E. Shank, The development of columnar grain and single crystal high temperature materials through directional solidification, Mater. Sci. Eng. **6** (1970) 213–247.

[6] H.W. Kerr, Solidification and grain structure in welds, In: *International Trends in Welding Science and Technology*, Eds. S.A. David, J.M. Vitek, ASM International, Materials Park, OH (1993), pp. 157–166.

[7] N.H. Tyas, Grain Refinement of Austenitic Stainless Steels to Facilitate Ultrasonic Inspection, *Ph.D. Thesis*, University of Cambridge (2000).

[8] R.W. Heine, C.R. Loper, Jr., P.C. Rosenthal, *Principles of Metal Casting*, McGraw Hill, New York (1967) p. 694

[9] T.F. Bower, M.C. Flemings, Formation of the chill zone in ingot solidification, Trans. Met. Soc. AIME **239** (1967) 216–219.

[10] J.A. Spittle, Preferred orientations in cast metals, Mater. Sci. Technol. **21** (2005) 546–550.

[11] S.C. Flood, J.D. Hunt, Columnar to equiaxed transition, in: *Metals Handbook*, 4th edn., Vol. 15, *Casting*, ASM, Metals Park OH (1988), pp. 130–136.

[12] J.E.C. Hutt, D.H. St John, The origins of the equiaxed zone — review of theoretical and experimental work, Int. J. Cast Metals Res. **11** (1998) 13–22.

[13] J.A. Spittle, Columnar to equiaxed grain transition in as solidified alloys, Int. Mater. Rev. **51** (2006) 247–269.

[14] A.L. Greer, Control of grain size in solidification, in *Solidification and Casting*, Eds. B. Cantor, K.A.Q. O'Reilly, Institute of Physics, Bristol, UK (2003), pp. 199–247 (Chapter 13).

[15] J.D. Hunt, Steady state columnar and equiaxed growth of dendrites and eutectic, Mater. Sci. Eng. **65** (1984) 75–83.

[16] L. Northcott, The influence of alloying elements on the crystallization of copper. Part II. — Large additions and the part played by the constitution, J. Inst. Metals, **65** (1939) 173–204.

[17] W.C. Winegard, B. Chalmers, Supercooling and dendritic freezing in alloys, Trans. ASM **216** (1954) 1214–1224.

[18] B. Chalmers, *Principles of Solidification*, Wiley, New York (1964).

[19] R.T. Southin, Nucleation of the equiaxed zone in cast metals, Trans. Met. Soc. AIME **239** (1967) 220–225.

[20] K.A. Jackson, J.D. Hunt, D.R. Uhlmann, T.P. Seward, III, On the origin of the equiaxed zone in castings, Trans. Met. Soc. AIME **236** (1966) 149–158.

[21] A. Ohno, T. Motegi, H. Soda, Origin of equiaxed crystals in castings, Trans. ISIJ **11** (1971) 18–23.

[22] B.P. Pearce, H.W. Kerr, Grain-refinement in magnetically stirred GTA welds of aluminum alloys, Metall. Trans. B **12B** (1981) 479–486.

[23] M. Vandyoussefi, A.L. Greer, Application of cellular automaton—finite element model to the grain refinement of directionally solidified Al–4.15 wt% Mg alloys, Acta Mater. **50** (2002) 1693–1705.

[24] C.-A. Gandin, M. Rappaz, A coupled finite-element cellular-automaton model for the prediction of dendritic grain structures in solidification processes, Acta Metall. Mater. **42** (1994) 2233–2246.
[25] S.G.R. Brown, J.A. Spittle, Computer simulation of grain growth and macrostructure development during solidification, Mater. Sci. Technol. **5** (1989) 362–368.
[26] O.L. Bustos, W. Reif, Grain refinement of superpure copper by additions 2, Metallurgy **48** (1994) 123–127.
[27] T.T. Cheng, The mechanism of grain refinement in TiAl alloys by boron addition — an alternative hypothesis, Intermetallics **8** (2000) 29–37.
[28] D.G. McCartney, Grain refining of aluminium and its alloys using inoculants, Int. Mater. Rev. **34** (1989) 247–260.
[29] *Standard Test Procedure for Aluminum Alloy Grain Refiners: TP-1, 1987*, The Aluminum Association, Washington, DC (1987).
[30] M. Vandyoussefi, J. Worth, A.L. Greer, Effect of instability of TiC particles on grain refinement of Al and Al–Mg alloys by addition of Al–Ti–C inoculants, Mater. Sci. Technol. **16** (2000) 1121–1128.
[31] J.A. Spittle, Grain refinement in shape casting of aluminium alloys, Int. J. Cast Metals Res. **19** (2006) 210–222.
[32] M.W. Meredith, A.L. Greer, P.V. Evans, The effect of grain-refining additions on intermetallic selection in dilute aluminium alloys, In: *Light Metals 1998*, Ed. B. Welch, TMS, Warrendale, PA (1998), pp. 977–982.
[33] C.M. Allen, K.A.Q. O'Reilly, P.V. Evans, B. Cantor, The effect of vanadium and grain refiner additions on the nucleation of secondary phases in 1xxx Al alloys, Acta Mater. **47** (1999) 4387–4403.
[34] S.C. Flood, J.D. Hunt, Modification of Al–Si eutectic alloys with Na, Metal. Sci. **15** (1981) 287–294.
[35] C.R. Ho, B. Cantor, Heterogeneous nucleation of solidification of Si in Al–Si and Al–Si–P alloys, Acta Metall. Mater. **43** (1995) 3231–3246.
[36] A.M. Bunn, P. Schumacher, M.A. Kearns, C.B. Boothroyd, A.L. Greer, Grain refinement by Al–Ti–B refiners in aluminum melts: a study of the mechanisms of poisoning by zirconium, Mater. Sci. Technol. **15** (1999) 1115–1123.
[37] A.M. Bunn, A.L. Greer, A.H. Green, M.A. Kearns, Effects of transition metals on the potency of Al–Ti–B grain refiners, in *Proc. 4th Dec. Int. Conf. on Solidification Processing (SP'97)*, Eds. J. Beech, H. Jones, University of Sheffield, Sheffield (1997), pp. 264–267.
[38] T.W. Clyne, M.H. Robert, Stability of intermetallic aluminides in liquid aluminium and implications for grain refinement, Met. Technol. **7** (1980) 177–185.
[39] P.S. Mohanty, J.E. Gruzleski, Mechanism of grain refinement in aluminum, Acta Metall. Mater. **43** (1995) 2001–2012.
[40] J. Pearson, M.E.J. Birch, D. Hadlet, Recent advances in aluminium grain refinement, in *Solidification Technology in the Foundry and Casthouse*. Metals Society, London (1983), pp. 74–78.
[41] P. Schumacher, A.L. Greer, J. Worth, P.V. Evans, M.A. Kearns, P. Fisher, A.H. Green, New studies of nucleation mechanisms in Al-alloys: implications for grain-refinement practice, Mater. Sci. Technol. **14** (1998) 394–404.
[42] M. Johnsson, L. Bäckerud, Nucleants in grain refined aluminum after addition of Ti-containing and B-containing master alloys, Z. Metallkd. **83** (1992) 774–780.
[43] M. Vader, J. Noordegraaf, The effectiveness of a grain refiner reinforced by a built-in energy content, In: *Light Metals 1990*, Ed. C.M. Bickert, TMS, Warrendale, PA (1990), pp. 851–857.
[44] D. Turnbull, Kinetics of heterogeneous nucleation, J. Chem. Phys. **18** (1950) 198–203.
[45] G.P. Jones, Grain refinement of castings using inoculants for nucleation above liquidus, In: *Solidification Processing 1987*, Eds. J. Beech, H. Jones, Institute of Metals, London (1988), pp. 496–497.

[46] W.T. Kim, B. Cantor, W.D. Griffith, M.R. Jolly, TEM characterisation of melt-spun Al–3Ti–1B and Al–5Ti–1B alloys, Int. J. Rapid Solid. **7** (1992) 245–254.
[47] B.J. McKay, P. Cizek, P. Schumacher, K.A.Q. O'Reilly, Heterogeneous nucleation of α-Al in Al–Ni–Si alloys, In: *Light Metals 2000*, Ed. R.D. Peterson, TMS, Warrendale, PA (2000), pp. 833–838.
[48] P. Schumacher, A.L. Greer, High-resolution transmission electron microscopy of grain-refining particles in amorphous aluminium alloys, In: *Light Metals 1996*, Ed. W. Hale, TMS, Warrendale, PA (1996), pp. 745–753.
[49] B.J. McKay, P. Cizek, P. Schumacher, K.A.Q. O'Reilly, Heterogeneous nucleation in an Al–Ni–Si alloy studied using a metallic glass technique, Mater. Sci. Eng. A **304** (2001) 240–244.
[50] P. Schumacher, P. Cizek, Heterogeneous nucleation mechanism in Al–Fe–Si amorphous alloys, Mater. Sci. Eng. A **304** (2001) 215–219.
[51] P. Schumacher, B.J. McKay, TEM investigation of heterogeneous nucleation mechanisms in Al–Si alloys, J. Non-Cryst. Solids **317** (2003) 123–128.
[52] L. Arnberg, L. Bäckerud, H. Klang, Grain refinement of aluminum. 2. Intermetallic particles in Al–Ti–B-type master alloys for grain refinement of aluminium, Met. Technol. **9** (1982) 7–13.
[53] A. Tronche, A.L. Greer, Electron-backscatter diffraction study of inoculation of Al, Philos. Mag. Lett. **81** (2001) 321–328.
[54] A. Tronche, M. Vandyoussefi, A.L. Greer, Instability of TiC particles in Al melts inoculated with an Al–Ti–C grain refiner, Mater. Sci. Technol. **18** (2002) 1072–1078.
[55] T.E. Quested, A.T. Dinsdale, A.L. Greer, Thermodynamic evidence for a poisoning mechanism in the Al–Si–Ti system, Mater. Sci. Technol. **22** (2006) 1126–1134.
[56] I. Maxwell, A. Hellawell, A simple model for grain refinement during solidification, Acta Metall. **23** (1975) 229–237.
[57] A.L. Greer, A.M. Bunn, A. Tronche, P.V. Evans, D.J. Bristow, Modelling of inoculation of metallic melts: application to grain refinement of aluminium by Al–Ti–B, Acta Mater. **48** (2000) 2823–2835.
[58] H.B. Aaron, D. Fainstein, G.R. Kotler, Diffusion-limited phase transformations: a comparison and critical evaluation of the mathematical approximations, J. Appl. Phys. **41** (1970) 4404–4410.
[59] T.E. Quested, A.T. Dinsdale, A.L. Greer, Thermodynamic modelling of growth-restriction effects in aluminium alloys, Acta Mater. **53** (2005) 1323–1334.
[60] 2nd ed., T.B. Massalski (ed.), *Binary Alloy Phase Diagrams*, Vol. 1, ASM International, Metals Park, OH (1990).
[61] S.A. Reavley, A.L. Greer, Athermal heterogeneous nucleation of freezing: numerical modeling for polygonal and polyhedral substrates, Philos. Mag. **88** (2008) 561–579.
[62] T.E. Quested, A.L. Greer, P.S. Cooper, The variable potency of TiB_2 nucleant particles in the grain refinement of aluminium by Al–Ti–B additions, Mater. Sci. Forum **396–402** (2002) 53–58.
[63] A.L. Greer, Grain refinement of alloys by inoculation of melts, Phil. Trans. Roy. Soc. Lond. A **361** (2003) 479–495.
[64] T.E. Quested, A.L. Greer, Athermal heterogeneous nucleation of solidification, Acta Mater. **53** (2005) 2683–2692.
[65] D. Turnbull, Theory of catalysis of nucleation by surface patches, Acta Metall. **1** (1953) 8–14.
[66] J.A. Spittle, S.B. Sadli, Effect of alloy variables on grain refinement of binary aluminium alloys with Al–Ti–B, Mater. Sci. Technol. **11** (1995) 533–537.
[67] T.E. Quested, A.L. Greer, Grain refinement of Al alloys: mechanisms determining as-cast grain size in directional solidification, Acta Mater. **53** (2005) 4643–4653.

[68] A. Tronche, A.L. Greer, Effect of solute elements on the grain structures of Al–Ti–B and Al–Ti–C grain-refined Al alloys, In: *Continuous Casting*, Eds. K. Ehrke, W. Schneider, DGM and Wiley-VCH, Weinheim (2000), pp. 218–223.
[69] W. Schneider, M.A. Kearns, M.J. McGarry, A.J. Whitehead, A comparison of the behaviour of AlTiB and AlTiC grain refiners, In: *Light Metals 1998*, Ed. B. Welch, TMS, Warrendale, PA (1998), pp. 953–961.
[70] R. Günther, C.H. Hartig, R. Bormann, Grain refinement of AZ31 by (SiC)p: theoretical calculation and experiment, Acta Mater. **54** (2006) 5591–5597.
[71] M. Qian, Heterogeneous nucleation on potent spherical substrates during solidification, Acta Mater. **55** (2007) 943–953.
[72] A.L. Greer, T.E. Quested, Heterogeneous grain initiation in solidification, Philos. Mag. **86** (2006) 3665–3680.
[73] A. Tronche, A.L. Greer, Design of grain refiners for aluminium alloys, In: *Light Metals 2000*, Ed. R.D. Peterson, TMS, Warrendale, PA (2000), pp. 827–832.
[74] T.E. Quested, A.L. Greer, The effect of the size distribution of inoculant particles on as-cast grain size in aluminium alloys, Acta Mater. **52** (2004) 3859–3869.
[75] W. Schneider, T.E. Quested, A.L. Greer, P.S. Cooper, A comparison of the family of Al–Ti–B refiners and their ability to achieve a fully equiaxed grain structure in DC casting, In: *Light Metals 2003*, Ed. P.N. Crepeau, TMS, Warrendale, PA (2003), pp. 953–959.
[76] Y.F. Han, D. Shu, J. Wang, B.O. Sun, Microstructure and grain refining performance of Al–5Ti–1B master alloy prepared under high-intensity ultrasound, Mater. Sci. Eng. **430** (2006) 326–331.
[77] P.C. Van Wiggen, J.K. Belgraver, AlTiC grain refiners with nucleation power, In: *Light Metals 1999*, Ed. C.E. Eckert, TMS, Warrendale, PA (1999), pp. 779–785.
[78] D.M. Stefanescu, G. Upadhya, D. Bandyopadhyay, Heat transfer — Solidification kinetics modeling of solidification of castings, Metall. Trans. A **21A** (1990) 997–1005.
[79] W. Oldfield, A quantitative approach to casting solidification: freezing of cast iron, Trans. ASM **59** (1966) 945–961.
[80] M. Rappaz, Ch.-A. Gandin, Probabilisitic modelling of microstructure formation in solidification processes, Acta Metall. Mater. **41** (1993) 345–360.
[81] M. Schwarz, A. Karma, K. Eckler, D.M. Herlach, Physical mechanisms of grain refinement in solidification of undercooled melts, Phys. Rev. Lett. **73** (1994) 1380–1383.
[82] A. Karma, Model of grain refinement in solidification of undercooled melts, Int. J. Non-Equil. Proc. **11** (1998) 201–233.
[83] B.L. Jones, G.M. Weston, The structural features of undercooled nickel and nickel–oxygen alloys, J. Aust. Inst. Metal. **15** (1970) 189–194.
[84] R.T. Southin, G.M. Weston, The influence of alloy additions on the structure of undercooled copper ingots, J. Aust. Inst. Metal. **18** (1973) 74–81.
[85] J.L. Walker, The influence of large amounts of undercooling on the grain size of nickel, In: *The Physical Chemistry of Process Metallurgy, Part 2*, Ed. G.R. St. Pierre, Interscience, New York (1959), pp. 845–853.
[86] G.L.F. Powell, The undercooling of silver, J. Aust. Inst. Metal. **10** (1965) 223–230.
[87] M. Costa Agra Mello, C.S. Kiminami, Undercoolability of copper bulk samples, J. Mater. Sci. Lett. **8** (1989) 1416–1417.
[88] T.Z. Kattamis, M.C. Flemings, Dendrite structure and grain size of undercooled melts, Trans. AIME **236** (1966) 1523–1532.
[89] L.A. Tarshis, J.L. Walker, J.W. Rutter, Experiments on the solidification structure of alloy castings, Metall. Trans. **2** (1971) 2589–2597.
[90] K.F. Kobayashi, P.H. Shingu, The solidification process of highly undercooled bulk Cu–O melts, J. Mater. Sci. **23** (1988) 2157–2166.
[91] D.M. Herlach, R.F. Cochrane, I. Egry, H.J. Fecht, A.L. Greer, Containerless processing in the study of metallic melts and their solidification, Int. Mater. Rev. **38** (1993) 273–347.

[92] G. Horvay, The tension field created by a spherical nucleus freezing into its less dense undercooled melt, Int. J. Heat Mass Transf. **8** (1965) 195–243.
[93] A.L. Greer, Grain refinement in rapidly solidified alloys, Mater. Sci. Eng. A **A133** (1991) 16–21.
[94] R.F. Cochrane, S.E. Battersby, A.M. Mullis, Dendrite growth velocity — undercooling relationships and microstructural evolution in highly undercooled dilute Cu–Sn alloys, In: *Solidification 1998*, Eds. S.P. Marsh, J.A. Dantzig, R. Trivedi, W. Hofmeister, M.G. Chu, E.J. Lavernia, J.H. Chun, TMS, Warrendale, PA (1998), pp. 245–254.
[95] R.F. Cochrane, S.E. Battersby, A.M. Mullis, Determination of the solidification velocity in the Cu and Cu–O systems by the melt encasement technique, In: *Solidification 1998*, Eds. S.P. Marsh, J.A. Dantzig, R. Trivedi, W. Hofmeister, M.G. Chu, E.J. Lavernia, J.H. Chun, TMS, Warrendale, PA (1998), pp. 223–232.
[96] R.F. Cochrane, D.M. Herlach, Grain refinement produced by solidification into undercooled melts, in *Proc. 7th Eur. Symp. Materials and Fluid Sciences in Microgravity*, ESA (SP-295), Paris (1989), pp. 147–154.
[97] R.F. Cochrane, D.M. Herlach, B. Feuerbacher, Grain refinement in drop-tube-processed nickel-based alloys, Mater. Sci. Eng. A **133** (1991) 706–710.
[98] A.M. Mullis, R.F. Cochrane, On the Karma model for spontaneous grain refinement at high solid fractions, Int. J. Non-Equil. Proc. **11** (2000) 283–297.
[99] R. Willnecker, D.M. Herlach, B. Feuerbacher, Evidence of nonequilibrium processes in rapid solidification of undercooled metals, Phys. Rev. Lett. **62** (1989) 2707–2710.
[100] A.M. Mullis, A model for spontaneous grain refinement in alloy systems at low undercooling, Mater. Sci. Eng. A **226–228** (1997) 804–808.
[101] A.M. Mullis, R.F. Cochrane, Grain refinement and the stability of dendrites growing into undercooled pure metals and alloys, J. Appl. Phys. **82** (1997) 3783–3790.
[102] A.M. Mullis, R.F. Cochrane, Grain refinement and growth instability in undercooled alloys at low undercooling, J. Appl. Phys. **84** (1998) 4905–4910.
[103] D. Turnbull, Under what conditions can a glass be formed?, Contemp. Phys. **10** (1969) 473–488.
[104] M.J. Aziz, Model for solute redistribution during rapid solidification, J. Appl. Phys. **53** (1982) 1158–1168.
[105] C.V. Thompson, A.L. Greer, F. Spaepen, Crystal nucleation in amorphous $(Au_{100-y}Cu_y)_{77}Si_9Ge_{14}$ alloys, Acta Metall. **31** (1983) 1883–1894.
[106] E. Schleip, R. Willnecker, D.M. Herlach, G.P. Görler, Measurements of ultrarapid solidification rates in greatly undercooled bulk melts with a high speed photosensing device, Mater. Sci. Eng. **98** (1988) 39–42.
[107] W.F. Johnson, R.F. Mehl, Reaction kinetics in processes of nucleation and growth, Trans. AIME **135** (1939) 416–458.
[108] W.J. Boettinger, Microstructural variations in rapidly solidified alloys, Mater. Sci. Eng. **98** (1988) 123–130.
[109] Y.H. Kim, A. Inoue, T. Masumoto, Ultrahigh tensile strengths of $Al_{88}Y_2Ni_9Mn_1$ or $Al_{88}Y_2Ni_9Fe_1$ amorphous alloys containing finely dispersed fcc Al particles, Mater. Trans. JIM **31** (1990) 747–749.
[110] H. Chen, Y. He, G.J. Shiflet, S.J. Poon, Mechanical properties of partially crystallized aluminum-based metallic glasses, Scripta Metall. Mater. **25** (1991) 1421–1424.
[111] R. Coehoorn, J. Duchateau, Preferential crystallite orientation in melt spun Nd–Fe–B permanent magnet materials, Mater. Sci. Eng. **99** (1988) 131–135.
[112] Y. Yoshizawa, S. Oguma, K. Yamauchi, New Fe-based soft magnetic alloys composed of ultrafine grain structure, J. Appl. Phys. **64** (1988) 6044–6046.
[113] M. Ohnuma, K. Hono, S. Linderoth, J.S. Pedersen, Y. Yoshizawa, H. Onodera, Small-angle neutron scattering and differential scanning calorimetry studies on the copper clustering stage of Fe–Si–B–Nb–Cu nanocrystalline alloys, Acta Mater. **48** (2000) 4783–4790.

[114] K. Suzuki, N. Kataoka, A. Inoue, T. Masumoto, High saturation magnetization and soft magnetic properties of bcc Fe–Zr–B and Fe–Zr–B–M (M = transition metal) alloys with nanoscale grain size, Mater. Trans. JIM **32** (1991) 93–102.
[115] A.L. Greer, Crystal nucleation in glass-forming alloys, Mater. Sci. Eng. **97** (1998) 285–290.
[116] P.D. Lee, A. Chirazi, D. See, Modeling microporosity in aluminum–silicon alloys: a review, J. Light Metals **1** (2001) 15–30.
[117] R.C. Atwood, S. Sridhar, P.D. Lee, Equations for nucleation of hydrogen gas pores during solidification of aluminium seven weight percent silicon alloy, Scripta Mater. **41** (1999) 1255–1259.
[118] J.F. Major, Porosity control and fatigue behavior in A356–T61 aluminum alloy, AFS Trans. **105** (1998) 901–906.
[119] J.C. Fisher, The fracture of liquids, J. Appl. Phys. **19** (1948) 1062–1067.
[120] E.J. Whittenberger, F.N. Rhines, Origin of porosity in castings of magnesium–aluminum and other alloys, J. Metals **4** (1952) 409–420.
[121] P.D. Lee, J.D. Hunt, Measuring the nucleation of hydrogen porosity during the solidification of aluminium–copper alloys, Scripta Mater. **36** (1997) 399–404.
[122] J. Campbell, An overview of the effects of bifilms on the structure and properties of cast alloys, Metall. Mater. Trans. B **37B** (2006) 857–863.

CHAPTER 14

Transformations in the Solid Phase

Contents

1. INTRODUCTION

In the conventional processing of engineering materials, a common objective is to obtain a fine, uniform dispersion of phases in the microstructure. This can be beneficial for optimizing properties in systems as diverse as glass-ceramics and high-strength steels. Focusing on nucleation, Section 2 examines how microstructural control can be exercised in devitrification, not only in oxide glasses to yield conventional glass-ceramics, but also in other glass types, notably metallic. Before examining nucleation within a crystalline phase, the role of nucleation in crystal melting is discussed in Section 3. The possibilities for microstructural control through heterogeneous nucleation are vastly

Pergamon Materials Series, Volume 15
ISSN 1470-1804, DOI 10.1016/S1470-1804(09)01514-4

greater in crystalline solids than in glasses. Heterogeneities in the solids include grain boundaries, grain-boundary junctions and triple points, interphase boundaries, dislocations, vacancy clusters, and irradiation-induced defects. All of these defects can be exploited to catalyze nucleation. The principles of such catalysis have been largely covered already; some examples from the wide range of metallurgical practice are explored in Section 4. The extreme case of instability under irradiation highlights some interesting nucleation phenomena not met so far in the book; these are treated in Section 5.

2. DEVITRIFICATION

Corning Glass Works ushered in the field of glass-ceramics in 1957 [1] with the introduction of Corningware®, a thermal and shock-resistant cookware. The superior properties of this glass-based material were the result of the devitrified microstructure, a glassy/polycrystal composite containing micro and nanosized crystals. Glass-ceramics are produced by the controlled crystallization of glasses, normally by techniques developed to induce high rates of internal crystal nucleation. These methods allow glasses to be fabricated into intricate shapes using conventional forming techniques and subsequently crystallized. Bulk metallic glasses can also devitrify to glassy/nanocrystal composites, some with exceptional physical properties of technological interest. Guided by the previous work in silicate glass-ceramics, efforts are underway to develop techniques to control the microstructural evolution in these glasses as well. A different level of control is required for the use of chalcogenide glasses in memory applications, requiring that the material undergo many amorphization/devitrification steps. These three cases are discussed briefly in this section to illustrate some of the methods of nucleation control.

2.1 Glass-ceramics

Since their introduction as a novel material for cookware, glass-ceramics have found many important applications, including high-strength hermetic seals, connectors, high-vacuum feed-throughs, tiles for floors and internal and external wall cladding, infrared-transmitting cook-tops (hobs), regenerators in turbine engines, biomaterials, and many others. Some glass-ceramics, such as Macor®, also can be precision-machined with conventional metal-working tools, allowing the production of intricate parts. Most glass-ceramics of commercial importance are based on silica-containing glasses, divided into the three groups of silicates, aluminosilicates and fluorosilicates. Reviews of the applications for each of these groups can be found elsewhere [2, 3].

The key to producing glass-ceramics tailored for particular applications is an ability to control their devitrification microstructure. Glasses can be stabilized against crystallization by a high barrier either to the nucleation of the crystal phase or to the growth of nuclei. Since the driving free energy for crystallization is essentially set by the glass composition, controlled devitrification to a nanostructured glassy composite requires either (1) lowering the interfacial free energy between the glass and the crystallizing phase and/or (2) inhibiting growth. As we illustrate with two examples, the enhanced nucleation rates can be obtained by precipitating heterogeneous nucleant particles or by inducing glassy phase separation prior to devitrification. Soft-impingement of diffusion fields during growth limited by solute partitioning can give rise to small crystal sizes.

2.1.1 Precipitation of heterogeneous nucleation sites

The techniques for producing the first glass-ceramics grew out of earlier work by Stookey [4] and others on photosensitive glasses, where exposure to UV or X-rays and subsequent annealing treatments below the softening point (near T_g) caused dissolved gold, silver, or copper to precipitate colloidal metal crystals within the glass. These precipitates served as potent heterogeneous nucleation sites for the subsequent crystallization of the glass, lowering the nucleation barrier. Stookey recognized that this could readily be extended to specific impurities dissolved in the glass that would precipitate readily and provide suitable catalytic sites for glass devitrification and developed the following criteria:

(i) The components of the catalyst particle should remain soluble at the melting and glass-forming temperatures, but have limited solubility in the glass at low temperatures.
(ii) The nucleation barrier for precipitation of the catalyst should be small, requiring that the original phase be highly supersaturated (increasing the driving free energy for transformation) and it has a reasonably low interfacial free energy with the glass.
(iii) The precipitating species should diffuse more rapidly at low temperatures than the constituents of the glass itself.
(iv) There should be reasonable matching of the crystal structures and lattice parameters of the precipitate crystal and the devitrifying phase to lower the interfacial free energy between the two, increasing the catalytic efficiency (decreasing the contact angle) for heterogeneous nucleation (Chapter 6, Section 2).

Several different nucleants have been explored. Metals such as gold, silver, and platinum, are known to promote nucleation in silicate glasses and were widely studied in the 1960s and 1970s for the production of glass-ceramics. Now, there are more effective nucleants, including P_2O_5 and photo-produced colloidal gold and silver particles in lithium-disilicate glasses [2], and there is widespread use of particular metal oxides [5]. Additions of TiO_2, ZrO_2, and Ta_2O_5, for example, lead to the formation of 100-nm-diameter, and even smaller, crystals of β-quartz in a lithium-aluminosilicate glass, SiO_2–Al_2O_3–Li_2O–MgO–ZnO; such fine dispersions are of interest, for example, in promoting transparency [5, 6]. Adding ~2 mol.% each of TiO_2 and ZrO_2 gives a very fine devitrification microstructure with crystallites less than 50 nm in diameter and a population density approaching $10^{23}\,m^{-3}$ (Figure 1a). Crystallization of the glass by continuous heating (approximating commercial heat treatments) leads to a mixture of β-quartz crystallites of similar size and much smaller crystallites of $ZrTiO_4$ (Figure 1b). Transmission electron microscopy (TEM) studies of the earliest stages of crystallization made by annealing first in the temperature range where nucleation of the β-quartz is appreciable, followed by an anneal at a temperature where the growth rate is high, show fewer, but larger grains. Significantly, each grain has a small $ZrTiO_4$ crystal at its center [6].

2.1.2 Phase separation

Beginning in the 1950s, primarily from electron microscopy studies, it became obvious that phase separation occurs in many glass-forming systems. It has been most studied in the silicate glasses; the many

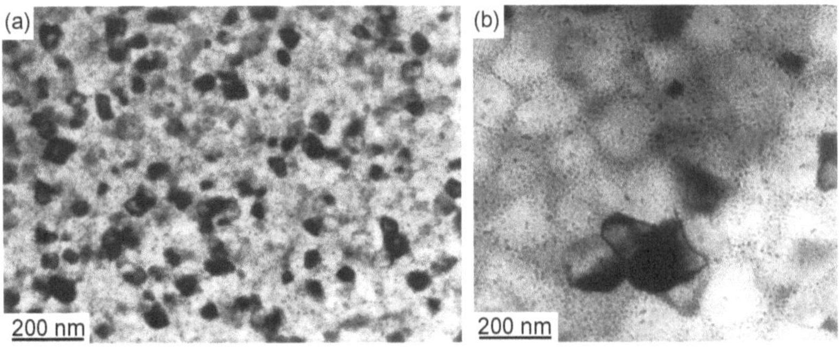

Fig. 1 (a) Microstructure of devitrified SiO_2–Al_2O_3–Li_2O–MgO–ZnO glass, showing a nano-sized grain mixture of metastable 50-nm grains of β-quartz and still smaller grains of $ZrTiO_4$. (b) TEM micrograph of the glass annealed in a temperature range where the nucleation rate is high but the growth rate of β-quartz is low, followed by an anneal where the growth rate is high, showing impinging grains of β-quartz, each containing a crystallite of $ZrTiO_4$ at its center. Note also the precipitation of $ZrTiO_4$ in the residual amorphous phase located at the crystal grain boundaries. (Reprinted, with permission, from Ref. [2] by Annual Reviews www.annualreviews.org.)

examples include SiO_2–Al_2O_3–Li_2O, SiO_2–Al_2O_3–MgO, and SiO_2–Al_2O_3–K_2O (see [3] and [7]). It was recognized that phase separation could play an important role in glass stability and devitrification, and could be exploited for developing new glasses and for improved microstructural control [3]. Nucleation in such cases involves coupled processes (Chapter 5, Section 5; Chapter 7, Section 5 and Chapter 10, Section 6), and it is important to analyze both the kinetics of the phase separation itself and its subsequent effect on devitrification. As discussed in Chapter 1, contrasting phase transformations in metastable and unstable regimes, phase separation can occur by either a localized large fluctuation (nucleation) or a continuous increase in the amplitude of long-wavelength composition fluctuations (spinodal transformation). The length scale is often (although not always) sufficiently coarse in the silicate and phosphate glasses that the phase-separated regions can be readily identified in optical micrographs, but these can be misleading. Phase separation is best identified from small-angle X-ray scattering (SAXS) [8–14], and 3D atom-probe field-ion microscopy (3DAP) [15–18] studies. In the cases where it has been studied, the phase separation appears to be well described as a spinodal transformation, fitting the predictions of Cahn-Hilliard theory [19, 20]. There can be preferential nucleation of crystal phases near phase-separated regions (Figure 2; [21]); the possible reasons for this include the following [7]:

(i) The homogeneous nucleation of phases might be more thermodynamically favored in regions of the phase-separated glass than in the uniform glass.
(ii) Since the viscosity is likely a strong function of glass composition, the mobility might be much higher in the phase-separated regions, accelerating the nucleation kinetics.
(iii) Finally, the interfaces between the phase-separated regions could be catalytic sites for nucleation.

A different microstructure is found in a devitrified $(Al_2O_3)_5(SiO_2)_{95}$ glass, which shows a phase-separated morphology upon quenching (Figure 3a; [22]). Here, primary crystallization to mullite ($3Al_2O_3 \cdot 2SiO_2$) gives a morphology that reflects the microstructure of the glassy phase-separated regions, rather than tracing the boundaries (Figure 3b). The atomic mobility is higher in the alumina-rich regions of the phase-separated mixture. The lower mobility in the glassy alumina-poor regions arrests crystal growth, causing the crystallites to reflect the original phase-separation morphology.

2.1.3 Other proposed mechanisms

The two mechanisms just discussed have received the most attention. They represent the simplest ways to increase the internal nucleation

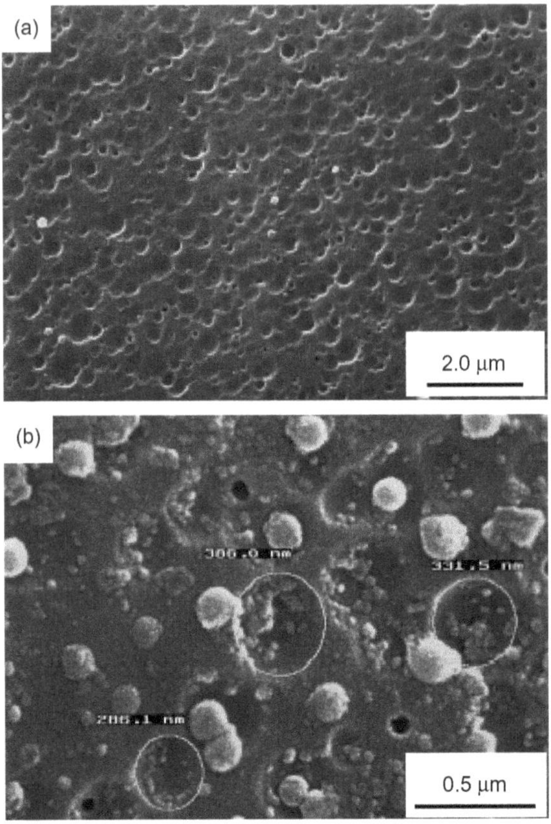

Fig. 2 (a) SEM micrograph of $(SiO_2)_{61.5}(Al_2O_3)_{9.4}(Na_2O)_{9.2}(K_2O)_{7.7}(CaO)_6(ZrO_2)_{0.5}(TiO_2)_{0.2}(P_2O_5)_{1.9}(CeO_2)_{0.3}(Li_2O)_{0.5}(B_2O_3)_{0.3}F_{2.5}$ glass showing phase separation of CaO-P_2O_5-rich droplet regions, revealed by etching for 10 s in a 3% HF solution; (b) preferential crystallization near the phase-separated regions. (Reprinted from Ref. [21], Copyright (1999), with permission from Elsevier.)

rate during devitrification, that is, binary phase separation, the dominant nucleation of a particular phase, or a regular sequence of phase nucleation (such as precipitation followed by nucleation). There are glasses, however, where internal nucleation cannot be catalyzed, requiring control of surface nucleation to devitrify the glass, by using fine powder of the same glass or fine particles of a heterogeneous catalyst such as Al_2O_3 [3]. Kokubo, for example, used the approach of surface nucleation to develop a bioactive glass-ceramic for bone replacement [23]. There are also more complex processes, however, such as the simultaneous nucleation of multiple phases, phase separation into multiple glassy phases with different composition, and combinations of mechanisms. For example, SiO_2–Al_2O_3–MgO–Na_2O–K_2O–P_2O_5–F glasses

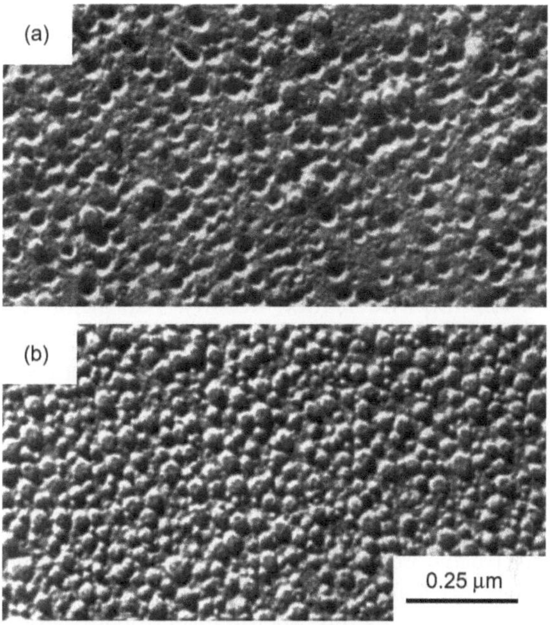

Fig. 3 Replica electron micrographs demonstrating the influence of annealing on an $(Al_2O_3)_5(SiO_2)_{95}$ phase-separating glass (a) as quenched, showing a phase-separated glass with the two regions of the glass having different etching characteristics; (b) after annealing at 950°C for 10 h, showing the formation of crystals of mullite in the alumina-rich regions of the phase separated glass. (Reproduced in part from Ref. [22], copyright (1969), with permission from Wiley-Blackwell.)

separate into three phases; annealing leads to the heterogeneous nucleation of mica and the homogeneous nucleation of calcium apatite [24]. These mechanisms are reviewed in more detail in Ref. [3].

2.2 Nanocrystallization in metallic glasses and coupled phase transformations

The nanostructures found in the crystallized Zr-based glasses discussed in Chapter 8, Section 8, reflect extremely high nucleation rates for the icosahedral quasicrystal phase (i-phase), argued in that section to be due, at least in part, to strong icosahedral short-range order (ISRO) in the glass. Nanostructure formation in devitrified metallic glasses is more common than this, however, also being found in glasses that do not form the i-phase with its low nucleation barrier. The cause of such high nucleation rates is currently debated. Suggested mechanisms are similar to those outlined for the silicate glasses in Section 2.1 in this chapter, including quenched-in nuclei [25], heterogeneous precipitate formation

[26], and phase separation [16]. At least three additional factors favoring nanocrystallization have been identified: fluctuations in local composition [27, 28], diffusion-controlled nucleation [29], and effects of local stress [30].

All metallic glasses that form devitrification nanostructures have a composition different from that of the primary crystallizing phase, providing a clue to the mechanism. The Al-3d transition-metal rare-earth (Al-TM-RE) glasses are an example. In many cases, the primary crystallizing nanophase is α-Al (Figure 4a), ruling out ISRO as an explanation for the high nucleation density. The population density of crystallites is often very large; a 10-min anneal of an $Al_{88}Y_7Fe_5$ glass at 245°C, for example, produces a density of the order of $10^{22}\,m^{-3}$ [31]. Comparing the measured size distribution of crystallites (Figure 4b) with those in Figure 4 in Chapter 8, it might be concluded that nucleation is time-dependent and heterogeneous, since the population density decreases sharply on the small-diameter side of the maximum. Instead, the shape of the size distribution probably reflects soft-impingement of the long-range diffusion fields during both crystal nucleation and growth. The high nucleation rate suggests that nucleation is coupled with other processes, such as the precipitation of nanosized catalytic sites or nanoscale phase separation.

2.2.1 Precipitation of heterogeneous nucleation sites

As in the silicate glasses, a large population density of grains of the primary crystallizing phase would occur if the metallic glass were supersaturated in elements that when precipitated would serve as potent heterogeneous nucleation sites. Such appears to be the case in the well-studied Fe–B–Si glasses when small amounts of insoluble elements such as copper are added [15, 32–36]. Elements such as niobium are also added to restrict crystal growth (according to the principles discussed in Chapter 13, Sections 2.2 and 3.4). Primary crystallization in glasses with these additions produces a microstructure of 10-nm-diameter grains of the bcc α-Fe(Si) phase embedded in the residual glass. This nanocomposite (Finemet®) has attracted attention due to its excellent magnetic properties, high permeability, and high saturation magnetization [26, 37].

Shown in Figure 5 are the results of 3DAP measurements on a $Fe_{73.5}Si_{13.5}B_9Nb_3Cu_1$ metallic glass [38]. The copper is homogeneously distributed in the as-quenched alloy (Figure 5a), but precipitates on annealing (Figures 5b and c) [38].

The crystallization sequence for the alloy is illustrated in Figure 6. Annealing the as-quenched glass (1) causes the precipitation of the supersaturated copper as a high population of small clusters (2). When the clusters are sufficiently large to be effective heterogeneous nucleation sites, primary crystallization of α-Fe(Si) starts. The rejection of the solutes

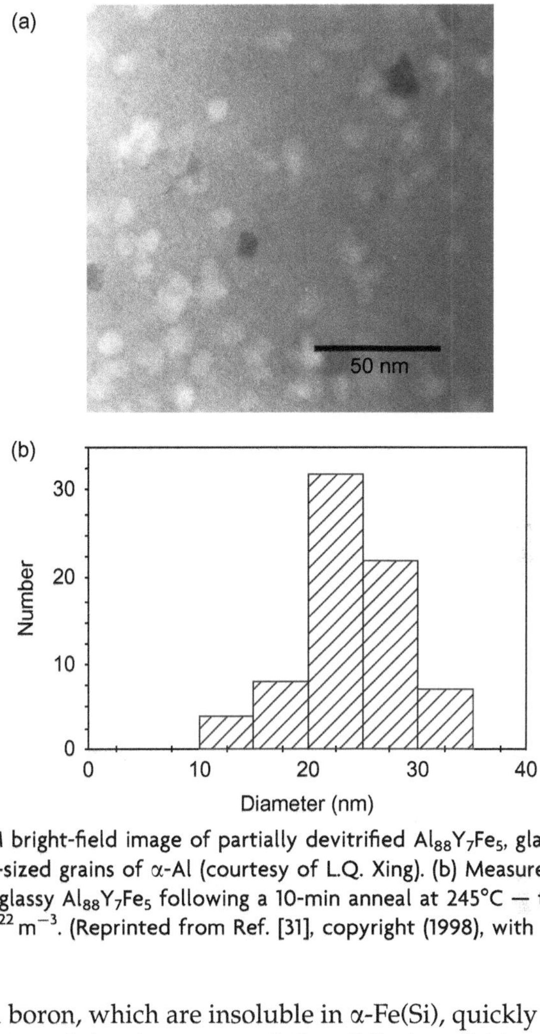

Fig. 4 (a) TEM bright-field image of partially devitrified $Al_{88}Y_7Fe_5$, glass, showing dendritic nano-sized grains of α-Al (courtesy of L.Q. Xing). (b) Measured particle size distribution in glassy $Al_{88}Y_7Fe_5$ following a 10-min anneal at 245°C — the total grain density is $\sim 10^{22}\,m^{-3}$. (Reprinted from Ref. [31], copyright (1998), with permission from Elsevier.)

niobium and boron, which are insoluble in α-Fe(Si), quickly arrests phase growth due to soft-impingement of the diffusion fields of the nearby crystallites [15].

2.2.2 Phase separation

As we have seen in the silicate glasses, phase-separation-produced compositional fluctuations can also serve as sites for heterogeneous nucleation. Field-ion micrographs of a slowly cooled $Zr_{41.2}Ti_{13.8}Cu_{12.5}$-$Ni_{10.0}Be_{22.5}$ metallic glass first indicated phase separation, on a length scale of order 20 nm, into regions with differing zirconium and beryllium concentrations [16]. Simultaneous wide- and small-angle X-ray scattering measurements [39] are consistent with this interpretation. Based on fits to the crystallization kinetics and energy-filtered TEM studies,

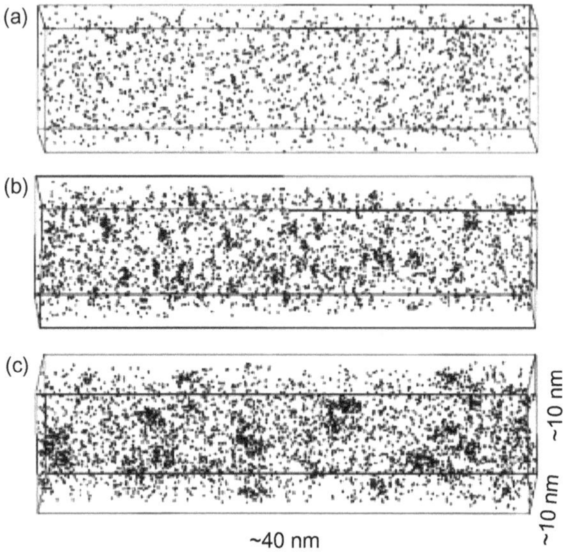

Fig. 5 A 3DAP map of copper in an $Fe_{73.5}Si_{13.5}B_9Nb_3Cu_1$ metallic glass: (a) as-quenched; after annealing at 400°C for (b) 5 min and (c) 60 min. (Reprinted from Ref. [38], copyright (1999), with permission from Elsevier.)

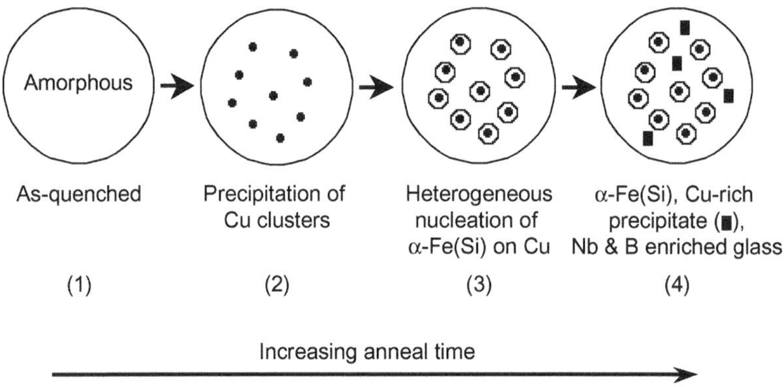

Fig. 6 Schematic diagram of the crystallization of Finemet® glass. (Adapted from Ref. [15], copyright (1995), with permission from Elsevier.)

low-temperature annealing of $Al_{88}La_2Gd_6Ni_4$ metallic glasses is argued to give nanoscale phase separation [40, 41], which couples to crystal nucleation. SAXS measurements suggest phase separation in $Al_{92}Sm_8$ glasses [42]. 3DAP studies give particularly convincing evidence for nanoscale phase separation and its role in nanoscale crystallization [43, 44]. For example, the compositional fluctuations on the nanoscale in an as-quenched $Al_{89}Ni_6La_5$ metallic glass (Figure 7) would explain the formation of the high population density of α-Al crystals during devitrification. Recently, however, much of the evidence for phase separation in metallic glasses has been questioned [45]. Time will eventually resolve this controversy, but, given the widely accepted reports of phase separation in silicate glasses, it would be unusual if phase separation were not also active in metallic glasses.

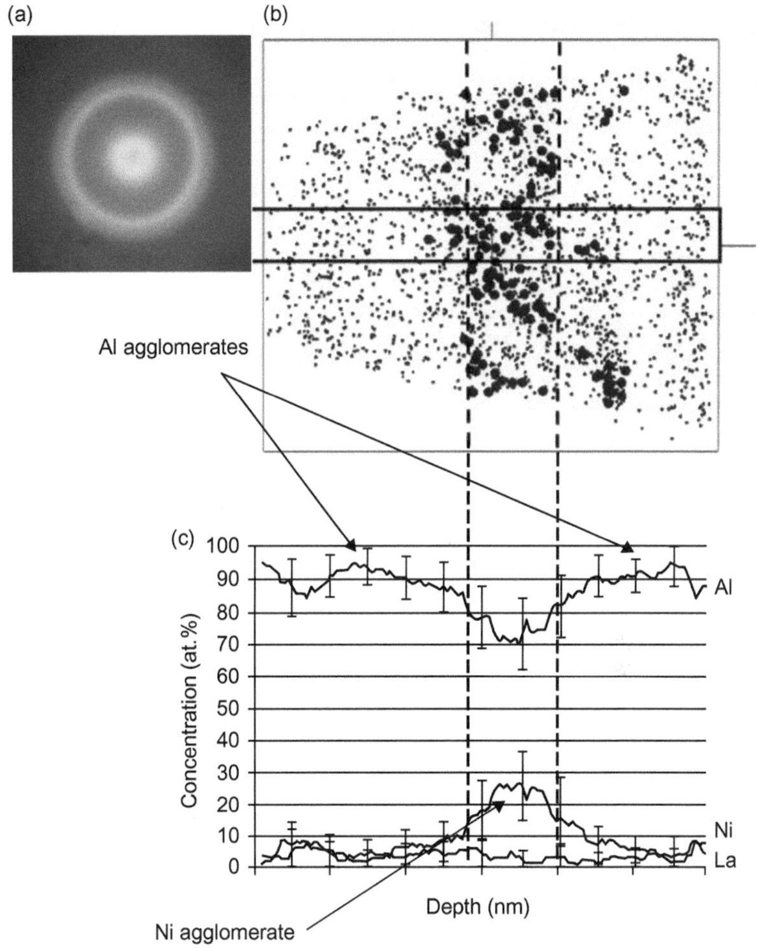

Fig. 7 (a) TEM selected-area diffraction pattern from an as-quenched $Al_{89}Ni_6La_5$ metallic glass; (b) 3DAP of the Ni atom distribution within an 11 × 11 × 12 nm volume — the bold dots show regions where the Ni concentration is greater than 10 at.%; (c) concentration profiles for the Al, Ni, and La atoms measured by 3DAP along the axis of the cylindrical region shown in (b). The arrows indicate the regions where the Al content exceeds the average of 89 at.%. (Reprinted with permission from Ref. [43], copyright (2008), American Institute of Physics.)

2.2.3 Mechanisms based on solute partitioning

As for the silicate glasses, other mechanisms than these have been proposed to explain the nanoscale devitrification microstructure. A point that has received particular attention is the role of composition on glass formation and crystallization. As already noted, glasses that undergo nano-devitrification typically do so to phases different in composition from the host glass. Further, the devitrification generally proceeds by stages, with primary crystallization to a low volume fraction (of order 10%) of a nanocrystal phase, followed by the more complete crystallization of the glass at higher temperatures. Three approaches to this are briefly discussed here; one is based on thermodynamic considerations and the others focus on kinetic processes. They carry a step further the notion of coupling, discussed so far for first-order phase transitions (e.g., precipitation and phase separation). At this time, there is only indirect evidence that these mechanisms are active in nano-devitrification. That they are theoretically possible, however, makes them of considerable interest for future investigations of nucleation processes.

Composition fluctuations

In the classical theory of nucleation, fluctuations in structure, density, and chemistry occur together. These fluctuations can sometimes couple to have unexpected and significant consequences on the nucleation barrier, and hence the nucleation rate. Desré et al. first proposed a two-stage nucleation mechanism by which fluctuations in the local chemical order can catalyze the structural transition to the final phase [27, 28]. If the fluctuations were to cause the composition in a small region to approach that of the nucleating phase, the nucleation barrier inside the spatial extent of the fluctuation, W^{poly} (referring to polymorphic, i.e., partitionless, transformation), would be smaller than for nucleation in the uniform material (Figure 8). Correspondingly, the critical size for nucleation (n^*_{cf}) would be smaller than the critical size calculated for nucleation from the uniform glass composition (n^*). The influence on the nucleation rate, then, is similar to that of a heterogeneous catalyst. There are differences, however. The lifetime of the fluctuation must be long compared with the time required for the polymorphic nuclei to develop, roughly the transient time for nucleation. Also, the spatial extent of the chemical fluctuation must be large enough to accommodate a nucleus size n^*, or the cluster cannot grow sufficiently to avoid dissolution after the lifetime of the fluctuation.

Coupled kinetic fluxes

The large sizes, and consequent relative immobility, of the RE and Zr atoms relative to the other elements in glasses such as $Al_{88}Y_7Fe_5$ and $Zr_{59}Ti_3Cu_{20}Ni_8Al_{10}$ make long-range diffusion an important limiting

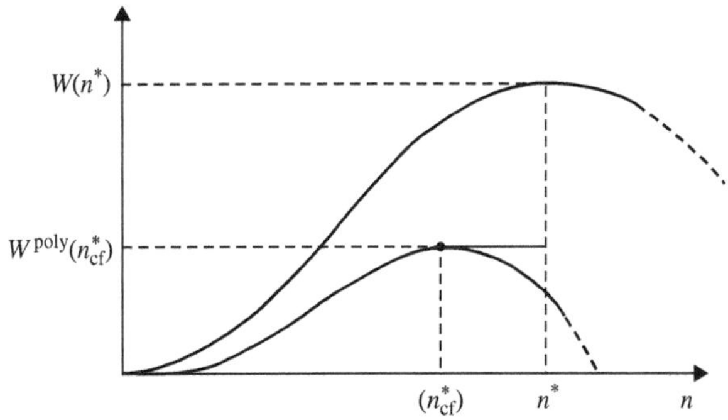

Fig. 8 Schematic illustration of the work of critical cluster formation in the homogeneous medium, $W(n^*)$, and inside the fluctuated region where essentially polymorphic nucleation can occur, $W^{\text{poly}}(n^*_{\text{cf}})$; n^* is the critical size in the homogeneous material and n^*_{cf} is the critical size in the region of chemical fluctuation. (Adapted from Ref. [28], copyright (2001), with permission from Elsevier.)

factor for crystal nucleation and growth. As discussed in Chapter 5, a correct treatment of nucleation in such cases must take account of the coupling between the stochastic long-range diffusion and interfacial attachment fluxes. A key prediction from the coupled-flux model that was discussed there and was confirmed by computer simulations in Chapter 10 (Section 7) is that the composition of the matrix near subcritical clusters is intermediate between that of the crystallizing phase and the initial amorphous phase. For the precipitation of α-Al in the Al-RE-TM glasses, for example, this would imply that the amorphous phase adjacent to clusters is *enriched* in aluminum, rather than depleted as for a growing postcritical cluster. A significant population of small clusters exists in quenched glasses [46, 47]. Even clusters larger than the critical size at the annealing temperature should be surrounded by a matrix rich in solute, due to the larger critical sizes at the higher temperatures where they were formed. When the glasses are annealed, these clusters should grow quickly until the nearby excess solute is consumed, producing the large population of small nanocrystals. The nucleation and growth rates subsequently slow down due to the consumption of the excess aluminum near the clusters, the overlapping diffusion fields from the growing crystallites and overall decreasing supersaturation [29]. While these ideas are supported by computer calculations, they have yet to be demonstrated experimentally.

Diffusional asymmetry

A different type of coupling is exemplified in the much-studied family of metallic glasses based on a combination of one or more early transition metals (ETM: Ti, Zr, Hf) with one or more late transition metals (LTM: Fe, Co, Ni, Cu). Glasses of this type show comparatively wide composition ranges for glass formation and form the basis for a number of bulk metallic glasses, notably those based on zirconium [30, 48]. Glass formation in metallic systems is favored by atomic size differences [49], and the LTM-ETM systems illustrate this well. In the archetypal Ni–Zr system, for example, the atomic volume of the zirconium is more than twice that of the nickel [50, 51]. Our interest is in the consequences of such differences for interdiffusion and for crystal nucleation and growth. The following discussion focuses on amorphous Ni–Zr, but the effects can in principle be found for any system with components of significantly different mobilities; indeed the kinetics of interdiffusion in such a case was first analyzed for silicate glasses [52].

Amorphous nickel–zirconium has been chosen for the present discussion because its interdiffusion kinetics has been extensively studied, particularly in compositionally modulated multilayers deposited as thin films [51]. In such multilayers, the period of modulation can be as short as a few nanometers, and in this way quantitative diffusion studies can be made on length scales relevant for nucleation or the early stages of growth of crystals. The key point of interest is that at typical temperatures of measurement, the smaller nickel atoms, essentially behaving as interstitial diffusers, have a diffusivity D_{Ni} as much as $10^5 \times$ that of the zirconium atoms D_{Zr} [51, 53]. In any diffusion couple, then, the interdiffusion (mixing) fluxes within the amorphous phase are very asymmetric. The flux of Ni atoms into the Zr-rich side of a diffusion couple is much greater than vice versa, in effect giving a net transport of volume in which the Zr-rich side of the couple swells relative to the Ni-rich side, generating stresses. In this subsection, we briefly describe how these stresses can profoundly affect interdiffusion kinetics in a way that is relevant for primary crystallization.

When interdiffusion is slow, the stresses generated by the diffusional asymmetry can be relaxed by viscous flow and their effects are insignificant. This is the regime in macroscopic samples in which the interdiffusivity $\tilde{D}$ is given by the classical Darken relation [53]:

$$\tilde{D} = X_{\mathrm{Zr}} D_{\mathrm{Ni}} + X_{\mathrm{Ni}} D_{\mathrm{Zr}}, \tag{1}$$

where X_{Ni} and X_{Zr} are the mole fractions of the two species and D_{Ni} and D_{Zr} are their diffusivities. The interdiffusivity is a weighted average of the diffusivities of the two species. In our case, when $D_{\mathrm{Ni}} \gg D_{\mathrm{Zr}}$, the value of $\tilde{D}$ is dominated by the fast component (Ni). At short diffusion distances, when interdiffusional mixing is relatively much faster, significant stresses can develop, and in that case the interdiffusivity is given by the Nernst–Planck relation [53]:

$$\frac{1}{\tilde{D}} = \frac{X_{\mathrm{Zr}} \bar{V}_{\mathrm{Zr}}^2 \rho^2}{D_{\mathrm{Ni}}} + \frac{X_{\mathrm{Ni}} \bar{V}_{\mathrm{Ni}}^2 \rho^2}{D_{\mathrm{Zr}}}, \tag{2}$$

where $\bar{V}_{Ni}$ and $\bar{V}_{Zr}$ are the partial molar volumes of the two species and ρ is the density (mol m^{-3}) of the average composition. Notwithstanding the difference between $\bar{V}_{Ni}$ and $\bar{V}_{Zr}$, the product $\bar{V}\rho$ is of order one for both species, and thus Eq. (2) sets out an averaging of the inverse diffusivities. Equations (1) and (2) can be considered to represent the models for, respectively, parallel and series conduction. In the latter case, the value of $\tilde{D}$ is dominated by the slow component (Zr). Between these two regimes, the rate-limiting step for interdiffusion is viscous flow and $\tilde{D}$ is a function of the wavelength of the composition profile. The stresses that lead to deviation from the Darken regime take time to develop. Thus the early stages of interdiffusion, independent of length scale, are fast and governed by Eq. (1). Overall, the result is that the interdiffusion of two species such as Ni and Zr in amorphous Ni–Zr is complex. Even at a fixed overall composition and fixed temperature, $\tilde{D}$ varies over orders of magnitude with time and diffusion distance. The representation of this in Figure 9 is based on measurements of $\tilde{D}$ quantitatively fitted to a model treating the transition between the Darken and Nernst–Planck regimes [30, 52].

It has already been noted that in devitrification, the need for solute partitioning favors finer-scale crystallites. In cases of widely differing diffusivities for the components in metallic glasses, crystal growth rates have been interpreted as limited by the slowest component [54] and the fastest component [55]. Both are correct: the early stages of growth

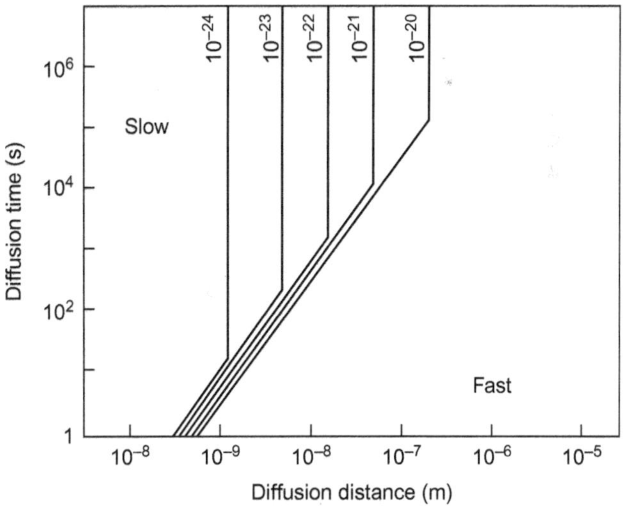

Fig. 9 The Ni-Zr interdiffusivity ($\tilde{D}$, contours labeled in m^2 s^{-1}) as a function of diffusion time and distance in amorphous $Ni_{55}Zr_{45}$ (at.%) at 529 K. Because $D_{Ni} \gg D_{Zr}$, the $\tilde{D}$ can vary by orders of magnitude. At long times, $\tilde{D}$ shows a transition from the slow Nernst-Planck regime to the fast Darken regime as the diffusion distance is increased to macroscopic values. (From Ref. [30], with permission.)

are in the Nernst–Planck (slow) regime and the effective value of $\tilde{D}$ increases as the length scale of interdiffusion (in the composition profile associated with solute partitioning) increases on subsequent growth (Figure 9). Nucleation necessarily involves short length scales, but also relatively short times in the sense that it precedes growth. In an LTM–ETM system such as amorphous Ni–Zr, the rate of crystal nucleation may be limited by the fast D_{Ni}, while the rate of the immediately following growth may be limited by D_{Zr}. Since the differing atomic sizes give $D_{Ni} \gg D_{Zr}$, they favor nucleation relative to growth and thus promote crystallization on finer scales.

2.3 Switching between glassy and crystalline states

The crystallization of glasses has so far been considered in connection with their stability and with the production of fine and uniform multiphase microstructures. There are cases, however, where the interest is in switching back and forth between glassy and crystalline states, the reversibility of the transformation itself being the exploitable property. In this subsection, we examine the reversible transformation in chalcogenide alloys, exploited in various types of nonvolatile computer memory. There are two main types of behavior, distinguished by the role of nucleation in the crystallization stage.

Switching between glassy and crystalline states is used in phase-change optical recording in rewritable media such as CDs and DVDs. The disks have a stack of layers within which the active layer, 10–30-nm thick, is a chalcogenide alloy. Chalcogenides combine elements in group 16 of the periodic table (particularly sulfur, selenium, or tellurium) with more electropositive elements such as silver. They are chosen for this application because they show significant differences in local order and bond type in the two states, the glassy state being more semiconducting and the crystalline state being more metallic [56]. The differences in reflectivity allow the written data to be read easily with a low-power laser beam. In a disk with no data, the chalcogenide layer is polycrystalline. Writing consists of melting a submicrometer diameter patch using a high-power laser pulse and through the subsequent rapid quench obtaining a glassy "mark." Erasure of a mark is achieved by using the same laser at lower power to heat sufficiently for crystallization to occur.

The rates at which data can be written and erased are key performance indicators for phase-change media, and the crystallization rate, characterized as the *complete erasure time* (CET), is the limiting factor [57]. There have consequently been many studies to try to minimize the CET. When CET is measured as a function of the diameter of the glassy mark, two types of behavior are observed, corresponding to different crystallization modes. There are many chalcogenide alloys, but extensive development has led to widespread use of two compositions that happen to show the different modes. In the chalcogenide $Ge_2Sb_2Te_5$, glassy marks crystallize as shown in Figure 10a. The crystallization, essentially progressing in two dimensions, starts at nucleation sites dispersed across the glassy mark. In such a case, the CET is independent of mark size, but may include an undesirable incubation time. In the chalcogenide $Ag_{5.5}In_{6.5}Sb_{59}Te_{29}$ (composition in at.%), there is no nucleation, only growth from the surrounding crystalline layer (Figure 10b). In this case, the CET has no contribution from an incubation time, but it increases linearly with mark diameter. The two erasure modes have been termed *nucleation-driven* and *growth-driven*; which is preferable (to reproducibly obtain the shortest CET) depends on mark size.

The active chalcogenide layer is sandwiched between two dielectric layers in the optical disk. Given the limited thickness of the active layer, it is not surprising that the crystallization behavior of a glassy mark is influenced by the thickness of the layer and by the nature of the neighboring layers. In particular, the introduction of carbide or nitride interface layers can greatly accelerate the crystallization in nucleation-driven erasure, apparently by increasing the nucleation rate at the interfaces [57].

The conditions relevant for phase-change optical recording are extreme: melt volume $\approx 10^{-20}\,m^3$; cooling rate $\approx 10^{10}\,K\,s^{-1}$ [58]. It is difficult or impossible to make quantitative measurements of nucleation kinetics in timescales directly relevant for the CET. Measurements made under very different conditions, for example isothermal annealing

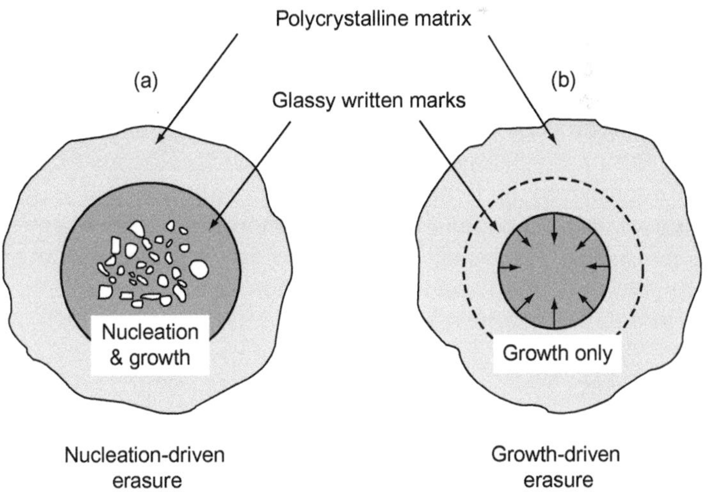

Fig. 10 Schematic comparison of (a) nucleation-driven and (b) growth-driven erasure (crystallization) of glassy marks in a chalcogenide thin film within an optical disk. (Reprinted from Ref. [57], copyright (2001), with permission from Elsevier.)

around 150°C, do show differences in nucleation rates consistent with the different modes shown in Figure 10 [59]. Measurements of critical supercoolings for nucleation in liquid droplets of various chalcogenides (fluxed in B_2O_3) permit a lower bound to be set on the crystal–liquid interfacial energy (assuming homogeneous nucleation for the largest supercoolings observed, ~9% of the absolute liquidus temperature) [60]. Subsequent calculations of steady-state nucleation rates [60], supported by TEM of partially crystallized thin films [61], show that in steady state, glass formation would not be possible even at the cooling rate in actual disks. Glass formation is possible only because of transient effects of the kind highlighted in Figures 2 and 3 in Chapter 8 and expected at high cooling rates. From laser-pulse experiments, direct estimates have been made of nucleation transient times [62]. The chalcogenide compositions of interest for rewritable memory are clearly marginal glass formers, consistent with need for rapid crystallization (short CET).

While optical disks remain under active development, mainly to use shorter laser wavelengths and thereby achieve higher data-storage density, a key research focus is on the application of chalcogenides in solid-state computer memory [63]. In this application, phase-change switching is achieved by electrical heating, passing a current through a memory cell. A higher-power electrical pulse causes melting and gives a glassy volume, and a lower-power pulse reverses the state back to crystalline. Detection of the state of the memory cell is through its resistance, two to three orders of magnitude higher in the glassy state than when crystalline. The electrical switching has potential complications — an electronic transition may be involved at the threshold of switching from the glassy state, and there may be roles for ionic transport [64] — but the operation of practical PC-RAMs (phase-change random-access memories) can be taken to be thermally driven. The very small memory cell volumes, down to $10^{-23}\,m^3$, and the importance of the pattern of the phase change within the volume (analogous to, but likely to be different from [65] those shown in Figure 10) will ensure that nucleation of crystallization remains of interest for optimizing device operation. In our discussion of switching in chalcogenides, we have implicitly assumed that there is no nucleation barrier for melting, a topic covered in detail in the next section.

3. MELTING

We begin our coverage of nucleation in *crystalline* solids by considering a transition that brings the crystalline state to an end, namely melting. As discussed extensively so far in this book, there is a distinct nucleation barrier for freezing of the liquid, associated with obvious supercooling.

In contrast, there is normally virtually no barrier to melting of the solid and no obvious superheating. This marked asymmetry is but one reason why the melting transition has received much attention, both theoretical and experimental, and why it remains an area of active study. As will be discussed further, nucleation analyses are important in understanding melting.

3.1 Theories of melting

The theory of melting has been rigorously developed for two-dimensional systems, where between the crystal and liquid there is an intermediate *hexatic phase* with orientational but not translational order [66, 67]. In three dimensions, however, theories of melting have remained less clearly established. In a key early approach, Lindemann [68] suggested that melting would occur when the root-mean-square amplitude of atomic vibration reaches a critical fraction of the interatomic distance. Noting that the shear moduli of a crystal decrease on heating, Born [69] suggested that melting would occur when one of these moduli reaches zero. This can be considered as a *rigidity catastrophe* or *mechanical melting* in which the entire crystal lattice uniformly makes a continuous transition into the liquid state. Lindemann and Born intended to describe melting as normally observed, but actual melting is not uniform throughout the solid. It is common experience that it begins at the surface of a crystalline solid and propagates inward. At the melting temperature, solid and liquid coexist with a well-defined interface between them, and the solid is rigid with non-zero elastic moduli. Surface melting normally starts without a nucleation barrier and, therefore, accounts for the usual lack of superheating for melting. Surface melting can be suppressed, however, for example by coating with a material of higher melting point, and then it is found that the interior of a solid can be very substantially superheated. It is of interest to know what then limits the degree of superheating that can be achieved. The degree of superheating and the consequences of it are of interest in fields as diverse as the partial melting of rock-producing magma bodies [70], the effects of intense laser irradiation [71], and the electrical explosion of wires [72].

The Born rigidity catastrophe sets one possible limit to the superheating of solids. Other types of catastrophe that would lead to uniform or homogeneous melting have been proposed, based on the enthalpy, entropy, or volume of the superheated crystal becoming equal to that of the liquid (outlined in [73, 74]). These catastrophes occur at temperatures as high as twice the equilibrium melting temperature. Lu and Li [73] first analyzed the kinetics of internal nucleation of a melt in a crystal and showed that nucleation can occur on heating well before any of the catastrophes is reached. It is now well accepted that it is the nucleation of the melt within the crystal, and not a catastrophic collapse of the crystal, that sets the limit to superheating [74]. Since the onset of melting is thus kinetically limited, the degree of superheating that can be achieved is very dependent on heating rate [75].

Lu and Li's original analysis used the classical theory for homogeneous nucleation [73]. While correctly indicating that the volume change on melting gives a substantial nucleation barrier, this is far from being a complete analysis of melt nucleation. As outlined in Chapter 6, the defects in crystalline solids provide possibilities for heterogeneous nucleation. It is also of interest to look beyond the classical theory. Because the superheated state is difficult to maintain for long times, quantitative observations of superheating are rarer than for supercooling, and many studies are based on atomistic simulations. These topics are covered in Section 3.3. Before dealing with them, we examine surface melting and how it can be suppressed.

3.2 Surface melting and its suppression

Surface melting is a topic that has long been of interest. Faraday [76], for example, suggested that some of the distinctive properties of ice could be attributed to a surface layer of water. Atomic vibration amplitudes are greater at surfaces, so the Lindemann criterion would indeed suggest that the melting point at surfaces might be less than for the bulk. The first direct evidence came from studies of lead that showed surface melting beginning at 500 K, well below the bulk melting temperature (T_m) of 601 K [77]. Some surfaces, however, notably the close-packed {111} planes of ccp metals, do not show melting below T_m [78]. Surface melting can be simply interpreted in terms of interfacial energies. If the solid–vapor energy exceeds the sum of the liquid–solid and liquid–vapor energies:

$$\sigma_{sv} > \sigma_{ls} + \sigma_{lv}, \tag{3}$$

then the solid is wetted by its own melt and a thin layer of liquid should coat the solid even below T_m. The layer must remain thin below T_m (when the bulk liquid is not thermodynamically stable), but its thickness diverges as T_m is approached on heating. For a given crystal, σ_{sv} may vary significantly from one type of crystal facet to another, explaining why surface melting occurs on some and not others [79]. As outlined by Olson et al. [75], such differences can also be expressed in terms of the Hamaker constant. In most cases, this constant is positive, corresponding to repulsion between the solid–liquid interface and the liquid–vapor interface. In the relatively unusual cases where the liquid phase has a higher electrical conductivity than the solid (e.g., as in bismuth and germanium), the Hamaker constant is negative and the attraction of the two interfaces prohibits surface melting. In such cases, when the surface is the only likely origin of melting, for example in nanoparticles, the solid can be significantly superheated. For example, free-standing particles of bismuth, <14 nm in diameter, can be superheated by ~ 50 K [75].

Superheating can be achieved for larger particles, even for materials (unlike bismuth) that would naturally show surface melting, by coating with a material of higher melting point. For example, Daeges et al. coated monocrystalline particles of silver with gold, and found that a superheating of at least 25 K was possible [80]. There have been many studies of melting of small particles embedded in a solid matrix, as reviewed by Jin et al. [81]. Substantial superheatings have been found, for example, up to 72 K for lead particles in an aluminum matrix. Melting of embedded particles has been analyzed in terms of heterogeneous nucleation [82]. An extreme confinement is that within fullerene shells; tin ($T_m = 505$ K) and lead ($T_m = 601$ K) particles ~ 10 nm in diameter were still solid at 770 and 740 K, respectively [83]. Confinement, and substantial superheating, can also be achieved in multilayered structures [84]. For example, tin layers sandwiched between silicon are still solid at ~ 720 K [85].

Superheating can also be achieved by arranging that the interior of the crystal is hotter than its surface, for example by focusing radiation inside a sample, or by forced cooling of the surface of an electrically heated sample. In single crystals, large superheats (e.g., 450 K for quartz [70]) can be sustained without any evidence for internal nucleation of the melt, demonstrating the difficulty of such nucleation. Internal melt nucleation can be observed, however, probably associated with impurities; it was first analyzed by Tyndall [86], who noted the appearance of "flower-shaped figures" each with six petals forming inside ice exposed to sunlight. These *Tyndall figures* (Figure 11) are dendrites of water formed at small superheating, and they continue to be a subject of active study [87].

A superheated solid can also survive, as discussed in connection with Figure 14 in Chapter 6, if a suitable substrate provides cavities in which the solid can be stabilized by the curvature of the solid–liquid interface.

Since the liquid phase takes time to nucleate and grow, superheating can be achieved by a sufficiently rapid change of the thermodynamic conditions to favor the liquid phase. This can be by shock-wave loading, intense laser irradiation, or electrical pulse heating, with effective heating rates as high as $\sim 10^{12}\,\mathrm{K\,s^{-1}}$ [71, 72, 90, 91]. With laser irradiation, while there can be nonthermal melting through the excitation of an electron-hole plasma, in most cases the melting is thermal. With laser pulses of only femtosecond to picosecond duration, it still takes several picoseconds for melting to start; the dominant contribution to this delay is the time for electron-lattice equilibration rather than the nucleation induction time [71]. With rapid heating, the transient superheatings are

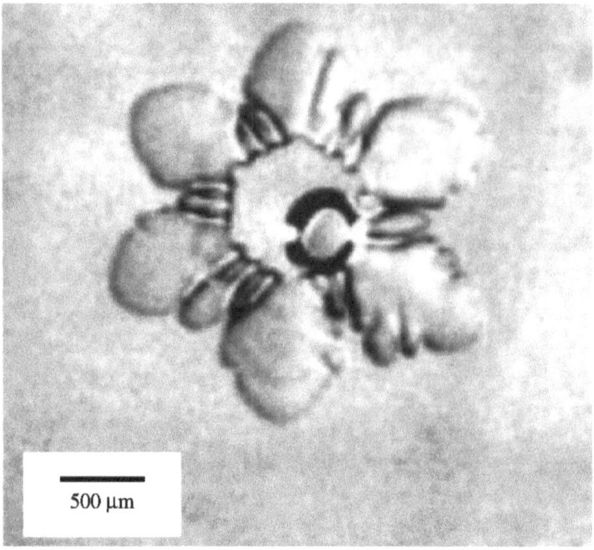

Fig. 11 A dendrite of water inside ice superheated by 0.15 K. (Reprinted with permission from Ref. [87], copyright (2006), American Institute of Physics.)

particularly large. With shock-wave loading, the largest normalized superheat is that of Mg_2SiO_4, which starts to melt at 7000 K; relative to a pressure-modified equilibrium melting temperature of $T_m = 4300\,K$, this represents a superheat of $0.63T_m$. With laser irradiation, the largest normalized superheat is that of aluminum ($T_m = 933\,K$) starting to melt at 1300 K, a superheat of $0.39T_m$ [90]. Some examples of superheats are collected in Table 1.

3.3 Nucleation of melting — The limit to superheating

Lu and Li [73] applied classical theory to the homogeneous nucleation of the melt within a crystalline solid. They took the simplest case of a spherical cluster of radius r, for which the work of formation $W(r)$ is given by

$$W(r) = \frac{4\pi}{3} r^3(\Delta g + \Delta e) + 4\pi r^2 \sigma_{ls}, \tag{4}$$

where Δg is the free energy change (per unit volume) on melting, σ_{ls} is the liquid–solid interfacial free energy (per unit area), and Δe is the elastic strain energy (per unit volume) associated with the volume change on melting. This strain energy plays an important role in suppressing internal melting [70]. It can be estimated using the analysis of Eshelby [92, 93]. We take the stress-free volume change on melting to be ΔV, the shear modulus of the solid to be μ_s, and the bulk modulus of solid and

Table 1 Measured superheats for crystalline solids achieved in a range of experiments

Material	Melting temperature T_m (K)	Maximum temperature T_{max} (K)	Superheat $\Delta T = T_{max} - T_m$ (K)	Relative superheat $\Delta T / T_m$	Ref.
Bulk single crystals					
SiO_2 (quartz)	1698[a]	>2148	>450	>0.27	[70]
Nanoparticles: free-standing					
Bi (<14 nm diameter)	545	595	50	0.09	[75]
Bi (>100 nm diameter)	545	553	8	0.01	[88]
Nanoparticles: embedded in a matrix, coating, or shell					
Ag (coated with Au)	1234	1259	25	0.02	[80]
In (in Al matrix)	430	459	29	0.07	[89]
Sn (in fullerene)	505	>770	>265	>0.52	[83]
Pb (in fullerene)	601	>740	>139	>0.23	[83]
Sn (in Sn/Si multilayer)	601	>720	>215	>0.43	[85]
Pb (in Al matrix)	601	673	72	0.12	[81]
Ultra-fast heating (~10^{12} K s^{-1}): shock-wave loading					
Fe	5800[b]	7250	1450	0.25	[90]
V	6150[b]	7600	1450	0.24	[90]
KBr	3500[b]	4200	700	0.20	[90]
SiO_2 (quartz)	4800[b]	6100	1300	0.27	[90]
Mg_2SiO_4	4300[b]	7000	2700	0.63	[90]
Ultra-fast heating (~10^{12} K s^{-1}): laser irradiation					
H_2O (ice)	273	330	57	0.21	[91]
Al	933	1300	367	0.39	[90]
Pb(111)	601	721	120	0.20	[90]
GaAs	1511	2061	550	0.36	[90]

Further values of superheat can be found in the reviews by Luo et al. [90] and Olson et al. [75].
[a] A metastable melting point [70].
[b] These equilibrium melting temperatures are not those for ambient pressure, but are the elevated values estimated for the pressures under shock-wave loading [90].

liquid to be K. The influence of the free surface of the solid is taken to be negligible. In the usual case ΔV is positive, and the liquid constrained within the solid is under hydrostatic compression; its volume change ΔV_l is given by

$$\Delta V_l = \left(\frac{4\mu_s}{4\mu_s + 3K} \right) \Delta V. \tag{5}$$

The volume change of the solid ΔV_s is given by

$$\Delta V_s = \left(\frac{3K}{4\mu_s + 3K}\right)\Delta V. \tag{6}$$

The elastic strain energy per unit volume in the liquid Δe_l is given by

$$\Delta e_l = \frac{1}{2}K\left(\frac{\Delta V_l}{V_l}\right)^2, \tag{7}$$

and the corresponding energy in the solid Δe_s by

$$\Delta e_s = \frac{2}{3}\mu_s\left(\frac{\Delta V_s}{V_s}\right)^2. \tag{8}$$

Making the approximation that $V_l \approx V_s = V = (4\pi/3)r^3$, the overall strain energy per unit volume is obtained by summing

$$\Delta e = \Delta e_l + \Delta e_s = \left(\frac{2\mu_s K}{4\mu_s + 3K}\right)\left(\frac{\Delta V}{V}\right)^2. \tag{9}$$

The critical work of nucleus formation W^* is given by

$$W^* = \frac{16\pi\sigma_{ls}^3}{3(\Delta g + \Delta e)^2}. \tag{10}$$

Lu and Li related the driving force for melting to the superheating ΔT in the simplest way, that is, $\Delta g = \Delta s \Delta T$. They took the homogeneous nucleation rate I_{hom} of the liquid in the solid to be

$$I_{hom} = I_0 \exp\left(-\frac{W^*}{k_B T}\right)\exp\left(-\frac{Q}{k_B T}\right), \tag{11}$$

where Q is the activation energy for atomic diffusion in the crystalline lattice. They approximated the prefactor I_0 as nk_BT/h, where n is the number of atoms or molecules per unit volume and h is the Planck constant.

Taking values of the materials constants for aluminum, Lu and Li [73] found that, on heating, the homogeneous nucleation rate of the liquid within the solid increases very rapidly at ~1127 K, which is approximately $1.21T_m$. This is shown in Figure 12, where it is noted that this critical temperature is lower than those for the various proposed catastrophes based on rigidity, volume, or entropy. Using similar calculations for a range of metals, Lu and Li found that, with the exception of those with particularly low melting temperature T_m, typical superheatings for melting are $\sim 0.20T_m$, and thus comparable in magnitude to typical supercoolings for freezing (as shown in Figure 12). If the strain energy Δe were ignored, a typical superheating for melting would be $\sim 0.13T_m$; Lu and Li note that this condition might be approached at higher temperatures as elastic moduli decrease.

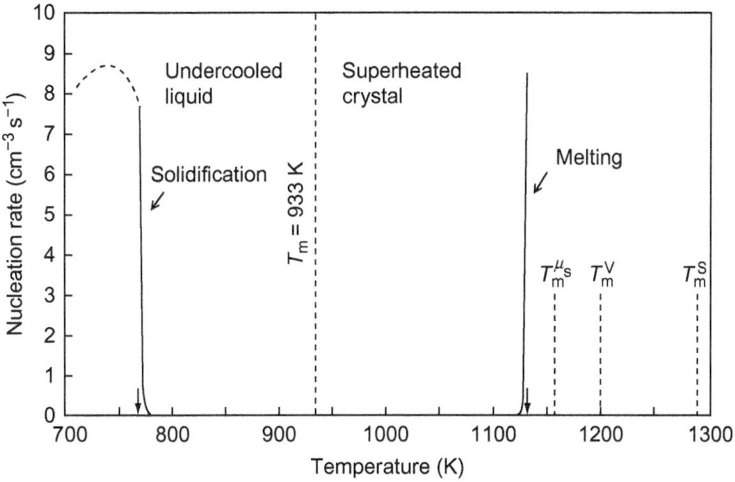

Fig. 12 Nucleation rates for solidification or melting of aluminum, calculated using classical theory. The arrows show the effective critical onset temperatures for the homogeneous nucleation of solidification and melting. Estimated temperatures for uniform melting associated with a catastrophe in rigidity, volume, or entropy ($T_m^{\mu_s}$, T_m^V, T_m^S, respectively) are also shown. (Reprinted with permission from Ref. [73], copyright (1998) by the American Physical Society.)

Figure 13 is a time–temperature–transformation (TTT) diagram for freezing and melting in aluminum. Freezing in the supercooled liquid is described by a characteristic C-curve, reflecting that, as the temperature is lowered, there are competing effects of increasing driving force for solidification and decreased atomic mobility. For superheating, in contrast, the nucleation rate increases monotonically as the temperature is raised; both the driving force for melting and the atomic mobility increase on heating. Lu and Li note that the TTT curve for melting provides a way of estimating the conditions necessary to reach, for example, the entropy catastrophe. In essence, sufficient rapid heating at $\sim 10^{13}\,\mathrm{K\,s^{-1}}$ would bypass homogeneous nucleation allowing the solid to reach the temperature at which the catastrophe is expected. While classical nucleation calculations can be performed for such heating rates, it has already been noted that other factors such as the need for electron-lattice equilibration in laser heating [71] may control the onset of melting.

The classical nucleation analysis is well supported by molecular-dynamics simulations [94, 95]. Bai and Li extended these analyses by considering the effect of curvature of the liquid–solid interface in reducing σ_{ls} [93]. Molecular-dynamics simulations have also been applied to analyze superheating at high heating rates [96].

The classical nucleation approach suggests that the critical liquid nucleus at the onset of uniform melting has approximately 120 atoms [95].

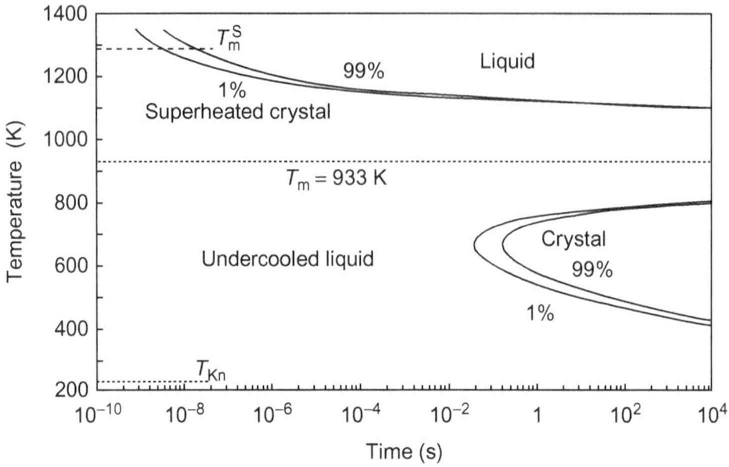

Fig. 13 A TTT diagram for melting of a superheated crystal or freezing of a supercooled liquid, calculated for aluminum assuming classical nucleation and polymorphic diffusion-controlled growth. The entropy catastrophe point T_m^S (at which the superheated crystal and the equilibrium liquid have the same entropy) and the Kauzmann temperature T_{Kn} (at which the supercooled liquid and the equilibrium crystal have the same entropy) are also shown. (Reprinted with permission from Ref. [73], copyright (1998) by the American Physical Society.)

Some molecular-dynamics simulations suggest, however, that fewer atoms may be involved. In aluminum, for example, Forsblom and Grimvall suggest that the thermal fluctuation initiating melting is an aggregate with three to four vacancies and six to seven interstitials [97]. Through nucleation of the liquid on such aggregates, melting starts at ~1190 K. It has also been suggested that melting can nucleate on vacancies alone [98]. The detailed processes by which dislocation-free single crystals undergo uniform melting remain under study, with more to be done to elucidate the effects of thermally activated point defects [74]. Such studies will help to unify the approaches to the superheating limit, based on instability catastrophes on the one hand and homogeneous nucleation of the liquid on the other [99]. In nonideal systems, however, other heterogeneities are likely to initiate melting.

Lu and Li noted that their calculated superheat should be an upper limit, as the nucleation of the liquid is likely to be heterogeneous, not homogeneous. Studies of colloidal systems indeed show that there can be premelting (i.e., melting below T_m) at vacancies, dislocations, and grain boundaries [100]. We consider first the role of grain boundaries. In some cases, simulations suggest that the free energy of the solid–solid boundary can be larger than that of the two liquid–solid interfaces formed when the boundary is wetted by a layer of liquid, thus favoring premelting [101, 102].

But, as reviewed by Fan and Gong [103], both microscopical studies and molecular-dynamics simulations find that there are grain boundaries that show no significant premelting. Indeed, simulations suggest that high-symmetry tilt boundaries can even be superheated [103]. Lu et al. have shown that polycrystalline selenium with a grain size of 10 μm can be superheated by ~5 K [104]. We conclude that grain boundaries are not always potent nucleation catalysts for melting.

We next consider nucleation on dislocations, for which the relevant theory has been covered in Chapter 6, Section 3. The liquid forming along a dislocation line is free from any shear stresses (its shear modulus being zero), and thus dislocations are expected to be powerful catalysts for the nucleation of melting. Huang et al. [105] performed molecular-dynamics simulations on copper containing Shockley partial dislocations. They found that the core structure of the dislocations remains solid-like up to T_m, but that above T_m liquid always appears first in the dislocation cores. This is a qualitative indication that the nucleation barrier is lower. More quantitatively, Ma and Li [106] performed continuum-mechanics calculations, taking an arbitrary dislocation density of $10^6\,m^{-2}$. Their estimate of the work of nucleation is based on that given by Eq. (44) in Chapter 6, but with account taken of the elastic strain energy due to the volume change on melting. They estimated the superheat for the onset of melting in aluminum to be $\sim 0.068T_m$ in the presence of screw dislocations, and $\sim 0.095T_m$ in the presence of edge dislocations. Clearly these superheats are much lower than the $\sim 0.21T_m$ estimated by Lu and Li [73] in the dislocation-free case. The temperature dependence of the nucleation rate is so sharp that the onset superheat is not strongly dependent on the dislocation density. For a variety of metals, Ma and Li estimated onset superheats in the range $(0.068-0.133)T_m$. Such estimates have been extended to misfit dislocations at semicoherent interfaces, of particular relevance for comparison with measurements of superheating in embedded droplets and thin films [107].

4. METALLURGICAL CONTROL OF NUCLEATION

In the manufacture of engineering materials, the exploitation and control of phase transformations, to obtain desired microstructures and properties, are most widespread in the heat treatment and thermomechanical processing of alloys. Often the objective is to strengthen the alloy by increasing its plastic yield stress. The toughness of a material, that is, its resistance to crack propagation, is however lower if plastic flow at a crack tip is more difficult. As the microstructure of a given material is adjusted, then, there is a clear correlation between higher strength and lower toughness. Much of the control of microstructure in engineering alloys is aimed at optimizing the compromise between these two desirable properties of strength and toughness. That control naturally involves nucleation. There is a vast literature on metallurgical processing, often focusing more on property optimization than on fundamental studies of kinetics and, in particular, there has been little quantitative analysis of nucleation. This is largely because of the complexity of engineering alloys. They may have many chemical components, and even trace additions or impurities may exert a large influence. In addition, as noted in Section 1, there is a great range of microstructural features that may catalyze heterogeneous nucleation. Even for the model precipitation system analyzed in Chapter 9, Section 2, where the focus was on homogeneous nucleation in the absence of complicating factors such as elastic strain, it was not possible to achieve a fully quantitative match of experiment and theory. Although a quantitative analysis is still further from being achieved in engineering alloys, there are many examples of a successful exploitation of the understanding of nucleation phenomena. The underlying principles of nucleation on microstructural features such as grain boundaries and dislocations are set out in Chapter 6. Further examples of dislocation-mediated microstructural changes, such as recrystallization, twinning, and martensitic transformations, have been analyzed in Chapter 12. In this section, there is insufficient space to attempt a comprehensive survey of the many ways in which nucleation is exploited in the optimization of engineering alloys. We focus on just two areas of particular importance: precipitation, of particular interest for the age-hardening of aluminum alloys (Section 4.1), and changes of structure in the matrix phase of an alloy, of particular importance in steels (Section 4.2). Classical studies of well-known metallurgical phenomena are in some cases being superseded by new work exploiting superior microstructural characterization (using techniques such as high-resolution TEM and 3DAP) and atomistic simulations; where possible, such developments are identified.

4.1 Precipitation: Aging of aluminum alloys

We are concerned with the precipitation of a new phase from a supersaturated solid solution; some examples are cited in Table 2 [108–110]. Precipitation is of very broad metallurgical significance, and the subject is covered in several definitive reviews [111–117]. In Chapter 9, Section 2, the model system chosen for the study of precipitation was Cu–Co, in which the two atomic species are identical in size. Also, the equilibrium precipitate phase has the same crystal structure and lattice parameter as the matrix. Thus the precipitate particles are fully coherent with the matrix (Chapter 6, Figure 20a) and the elastic strain energy associated with their formation is negligible. In this case, the equilibrium

Table 2 Precipitate-phase sequences in age-hardening [108–110]

Base metal	Alloy system	Intermediate precipitate sequence	Equilibrium precipitate phase
Aluminum	Al–Ag	GP (sphere) → γ′ (plate)[a]	γ (Ag_2Al)
	Al–Cu	GP → θ″ (plate)[b] → θ′ (plate)[a]	θ (Al_2Cu)
	Al–Mg	GP (sphere) → β′ (plate)[a]	β (Al_2Mg_3)
	Al–Mg–Cu	GP (needle) → S′ (lath)[a]	S (Al_2CuMg, lath)
	Al–Mg–Si	GP (needle) → β′ (rod)[b]	β (Mg_2Si, plate)
	Al–Zn	GP (sphere) → α′ (plate)[b]	Zn
	Al–Zn–Mg	GP (sphere) → η′ (plate)[b]	η ($MgZn_2$, plate or rod)
		GP (sphere) → T′ (cuboid)[b]	T ($(Al, Zn)_{48}Mg_{32}$)
Copper	Cu–Be	GP (disk) → γ′	γ (CuBe)
	Cu–Co	GP (sphere)	β-Co (plate)
Iron	Fe–C	ε-carbide (disk) → Fe_3C	(graphite)
	Fe–N	α″ (disk)	Fe_4N
Nickel	Ni–Al–Ti–Cr		γ′ (cuboid or spheroid)

[a]Nucleated on dislocations.
[b]Nucleated on GP zones.

precipitate phase nucleates directly from the matrix, apparently homogeneously, with a comparatively low nucleation barrier.

As Table 2 makes clear, precipitation in engineering alloys is typically much more complex. The equilibrium precipitate phase can be preceded by a sequence of intermediate metastable precipitates. The sequence is of practical importance, as optimum alloy properties are often obtained with intermediate precipitates. The sequence of phases is controlled by nucleation. The phase that appears first has the structure most closely resembling that of the solid solution within which it forms. Indeed, commonly the precipitate is fully coherent with the matrix and thus has a low interfacial energy with it. Also, the structural similarity may be associated with a low volume change, reducing the elastic strain energy. Low interfacial energy and low strain energy clearly facilitate nucleation, and can ensure that a metastable precipitate forms first, despite the thermodynamic driving force for its formation being lower than for the equilibrium precipitate.

The coherent precipitates that form first are termed Guinier-Preston (GP) zones. These were first analyzed for Al–Cu alloys. The nucleation of such precipitate phases far from equilibrium may be catalyzed by, for example, small vacancy clusters, but in essence it appears to be homogeneous. In contrast the equilibrium phase normally nucleates only on high-angle grain boundaries. Intermediate phases may nucleate on dislocations. In a precipitation sequence, the interface between preexisting precipitates and the matrix may be a favored site for nucleation of the next phase in the sequence. Nucleation sites are noted in Table 2.

The treatment of elastic strain energy and nucleation in Section 3.3 can readily be applied to the nucleation of one solid phase in another. A full treatment of the problem lies beyond our present scope, but it is of interest to consider the shapes of the GP zones in different alloy systems. In essence, the zones are solute-rich clusters forming on matrix lattice sites. Table 3 presents data on Al-based systems. The solute atoms (Ag, Cu, Zn) are not identical in size to the Al solvent. The given values of zone misfit are the differences in stress-free lattice parameter between zone and matrix if the zone were pure solute. For low values of misfit, as in Al–Ag alloys, the zones tend to be spherical, as expected from a minimization of interfacial area. For large misfits, as in Al–Cu alloys, the elastic strain energy for a spherical nucleus would be large. The aluminum matrix is elastically anisotropic and has significantly lower Young modulus in $\langle 1\,0\,0 \rangle$ directions than in $\langle 1\,1\,1 \rangle$ directions. It is then favorable for the nuclei to be disk-shaped aligned on $\{1\,0\,0\}$ planes so that the misfit is mainly accommodated in a compliant direction normal to the disk. The critical misfit above which zones tend to be disk-shaped appears to be $\sim 5\%$. It can also be favorable for zones to be needle-shaped, and examples are noted in Table 2.

While exploited in alloys as diverse as nickel-based superalloys hardened by ordered phases and tempered martensitic steels, precipitation has been most intensively studied in aluminum alloys, as reviewed by Lorimer [118]. The matrix solid solution in aluminum-based alloys shows only the cubic close-packed (ccp) structure, and so does not permit the types of microstructural control made possible by the transformations between face-centered and body-centered phases that are so important in steels (Section 4.2). Furthermore, aluminum alloys do not show particularly large Hall-Petch hardening (the impeding of dislocation motion by grain boundaries). Strengthening of aluminum alloys is therefore mainly by *age-hardening* in which an annealing treatment (*aging*) gives a dispersion of precipitates that impede dislocation motion.

Table 3 Guinier-Preston zones in Al–Ag, Al–Cu, and Al–Zn alloys

Atomic radius (Å)	Al: 1.43	Ag: 1.44	Cu: 1.28	Zn: 1.38
Zone misfit (%)	–	+0.7	−10.5	−3.5
Zone shape	–	Sphere	Disk	Sphere

From Ref. [110], copyright (1992), with kind permission of Springer Science and Business Media.

Age-hardening was discovered first in Al–Cu, and we focus on this system that is the most widely studied and is still of importance as the basis for commercial high-strength alloys. It also has a particularly complex precipitation sequence (Table 2). The equilibrium phase diagram for aluminum with small additions of copper (Figure 14; [118]) shows the liquid phase at high temperature, and a eutectic tie-line at 548°C linking the ccp solid solution of Cu in Al (α phase) and the compound phase Al_2Cu (θ phase). The most-studied alloy is that with 4 wt% Cu (i.e., 1.7 at.% Cu). If such an alloy is *solution-treated*, that is, held sufficiently long in the α-phase field, for example at 540°C, then a polycrystalline, chemically uniform solid solution is obtained. When this solid solution is rapidly quenched, for example by immersion in water, it becomes supersaturated, as there is insufficient time for precipitation and the copper is retained in solution. Precipitation occurs within the supersaturated solution even at room temperature, but is accelerated by heating. In isothermal heat treatments below roughly 180°C, the full sequence of intermediate precipitate phases is observed.

The sequence begins with GP zones, first inferred from streaks in X-ray diffraction maxima [119, 120], indicating very thin platelets

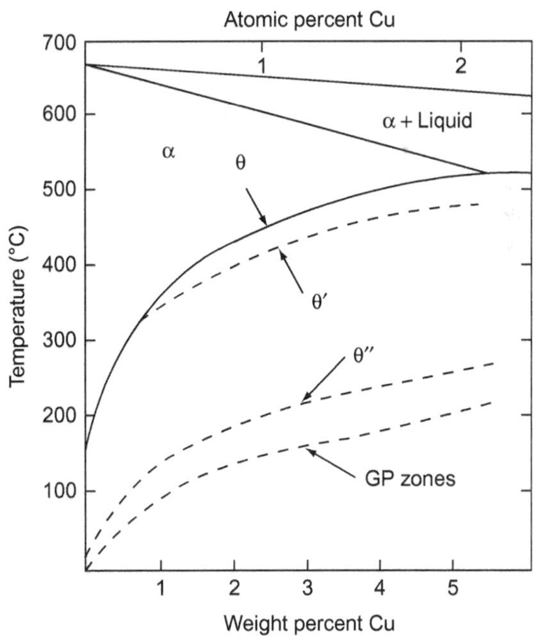

Fig. 14 Al-rich side of the equilibrium phase diagram for Al–Cu, in addition showing the solvus lines for the metastable precipitates: GP zones, θ″ and θ′. (From Ref. [110], copyright (1992), with kind permission of Springer Science and Business Media.)

(presumably copper-rich) parallel to {1 0 0} planes in the matrix. Early conventional TEM studies did not resolve the platelets themselves, but showed contrast arising from the strain fields in the matrix [121]. The images did confirm the existence of platelets up to $\sim$10 nm in diameter. The very limited thickness of the zones has impeded direct characterization. In particular, the composition of the zones and the strain field around them has remained the focus of active study [122–124]. Techniques such as synchrotron X-ray diffraction [125, 126], high-resolution electron microscopy (HREM) [127, 128], 3DAP [123], and nuclear-magnetic resonance [129] have all been applied. Recent work [130] has combined HREM with 3DAP to provide a secure interpretation of the observed phenomena. The GP zones are copper-rich disks, the majority just one monolayer thick. Their diameters are in the range 1–10 nm. A zone is imaged at A in Figure 15, and the schematic structure is shown in Figure 16a, as inferred from 3DAP studies. The copper content of the monolayer disks is in the range 40–100 at.%, higher than that in the equilibrium precipitate phase Al_2Cu. There is evidence from HREM [127] and from 3DAP [131] that a minority of zones can be thicker than a single monolayer (Figure 16b).

The θ'' phase consists of a layering of copper atoms on {1 0 0} planes within the Al matrix. It appears to evolve from GP zones by the formation of a second Cu-rich monolayer separated by three Al layers from the first. This structure is illustrated in Figure 16c, and can be seen partially and fully formed (at B and C, respectively) in the image in Figure 15. The θ'' crystal structure (Figure 17) is tetragonal. It matches the α matrix perfectly on (0 0 1), and there is only a little distortion on (1 0 0) and (0 1 0). The θ'' precipitates remain fully coherent with the matrix and take the form of plates parallel to $(0\,0\,1)_{\theta''}$ and $\{1\,0\,0\}_{\alpha}$ up to 10 nm thick and 100 nm in lateral dimension.

The θ' phase is also tetragonal (Figure 17) matching the α matrix well on (0 0 1). There is no good match with the matrix on (1 0 0) and (0 1 0); the corresponding interfaces with the matrix are incoherent (Chapter 6, Figure 20b) or semicoherent (Chapter 6, Figure 20c) even for the smallest precipitate size. The θ' precipitates nucleate on dislocations because these can reduce the degree of misfit. Their nucleation is also greatly facilitated by low contents of solutes such as cadmium, indium, or tin. The nucleation appears to be associated with prior clustering of these elements that have rather high diffusivities in the α matrix [123]. With different degrees of matching on different faces, the θ' precipitates again take the form of plates parallel to $(0\,0\,1)_{\theta''}$ and $\{1\,0\,0\}_{\alpha}$. The (0 0 1) faces of the precipitates are initially fully coherent with the matrix, but they become semicoherent as they grow. As shown in Figure 20c in Chapter 6, semicoherent interfaces contain misfit dislocations, and these are evident on the predominant (0 0 1) faces of the θ' particles in Figure 18 [132].

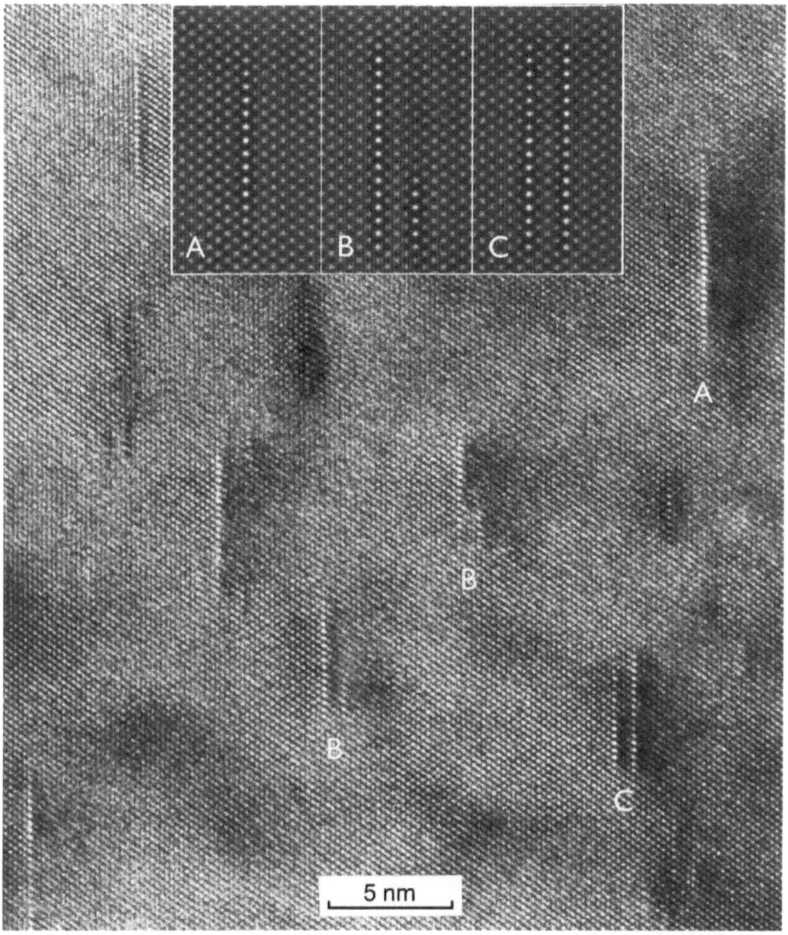

Fig. 15 Transmission electron micrograph of Al–1.54 at.% Cu quenched into ice-water and aged at 100°C for 30 h, showing: (A) monolayer GP zones, (C) θ″ precipitates, and (B) an intermediate stage between A and C. The inset shows image simulations. (Reprinted from Ref. [130], copyright (2004), with permission from Elsevier.)

The equilibrium θ phase has a body-centered tetragonal structure (Figure 17) and has no planes that provide good crystallographic matching with the matrix. The precipitates of θ phase grow with essentially incoherent interfaces. They nucleate either on grain boundaries or on the interfaces between the α matrix and θ′.

The θ″, θ′, and θ phases have crystal structures quite distinct from that of the α matrix and they have compositions at or close to Al_2Cu. In contrast, GP zones have a structure matching that of the matrix, but are much richer in copper. Porter and Easterling's [110] schematic free-energy composition diagram (Figure 19) is drawn accordingly and shows

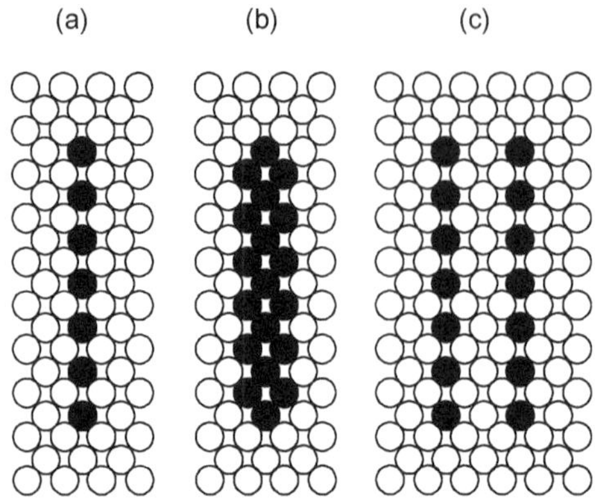

Fig. 16 Schematic drawings of (a) a monolayer GP zone, (b) a multilayer GP zone, (c) a θ″ precipitate. The open and filled circles represent Al and Cu atoms respectively. (From Ref. [131], with permission; Taylor & Francis Ltd. http://www.informaworld.com.)

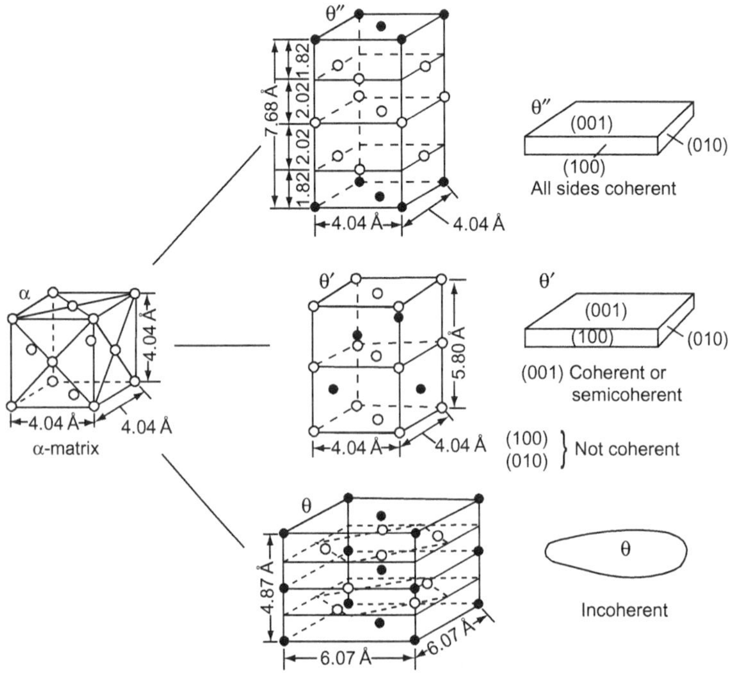

Fig. 17 Crystal structures of the θ″, θ′ and θ phases in Al–Cu, and their morphology as precipitates. (From Ref. [110], copyright (1992), with kind permission of Springer Science and Business Media.)

the increasing thermodynamic stability of the phases in the precipitation sequence. For each precipitate phase, a common tangent can be drawn, showing that as more-stable phases are formed, the copper content of the

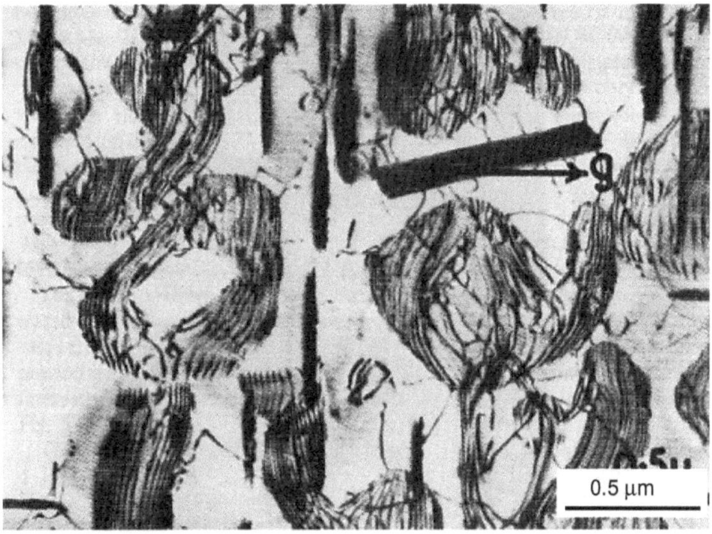

Fig. 18 Transmission electron micrograph of Al–4 wt.% Cu quenched and aged at 300°C for 1 day, showing arrays of dislocations on the large (0 0 1) faces of θ′ precipitates. (From Ref. [132], with permission; Taylor & Francis Ltd. http://www.informaworld.com.)

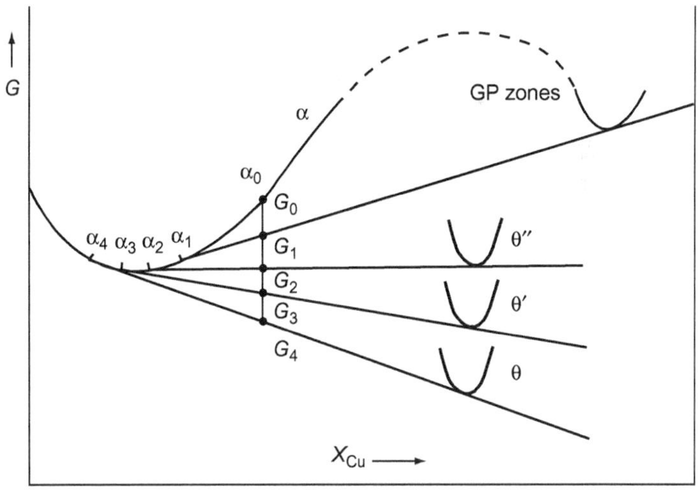

Fig. 19 A schematic molar free energy diagram for the Al–Cu system. (From Ref. [110], copyright (1992), with kind permission of Springer Science and Business Media.)

matrix is lowered (α_0 to α_4) and the free energy of the system is lowered from G_0 (for the supersaturated matrix α_0) to G_4 (for the final equilibrium state of the system with the α matrix and θ precipitate). The precipitation sequence can thus be represented as [110]

$$\alpha_0 \to \alpha_1 + \mathrm{GP} \to \alpha_2 + \theta'' \to \alpha_3 + \theta' \to \alpha_4 + \theta. \tag{12}$$

It is clear that the α_1 matrix in equilibrium with GP zones is much richer in copper than the α_4 matrix in equilibrium with θ phase. Correspondingly, the solvus line for GP zones is to the right of the equilibrium θ solvus in Figure 14. It follows that the formation of a given phase will lead to dissolution of a prior less-stable phase as the copper content of the matrix is lowered, a process known as *reversion*. Also, for a given copper content, an aging temperature could, for example, be chosen that is above the θ'' solvus but below the θ' solvus; then GP zones and θ'' cannot form, and the precipitation sequence must start with θ'. In this way, control of nucleation through the choice of aging treatment can lead to profoundly different microstructures. The initial nucleation stage and its dependence on aging treatment can be represented on a nucleation map (Figure 20; [133]).

We turn now to examine in more detail the nucleation of the various phases in the Al–Cu precipitation sequence. Attempts have been made to evaluate the work of formation of GP zones from macroscopic thermodynamic descriptions of the Al–Cu solid solution, the GP-α interfacial energy and the associated elastic strain [124]; given that the zones are only one monolayer thick, such approaches are at best approximate. Rather, the small size of GP zones favors atomistic

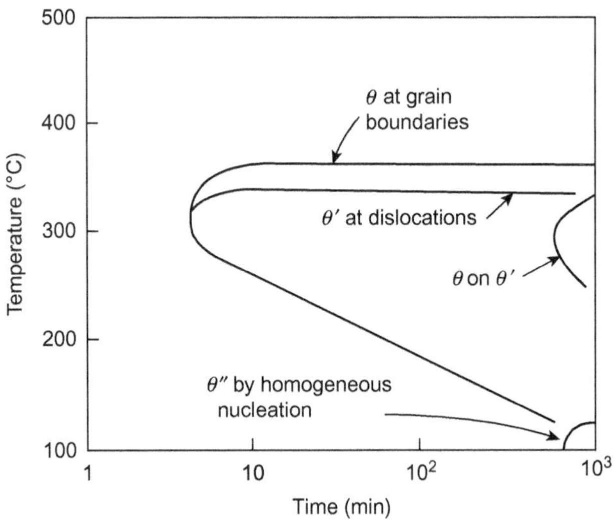

Fig. 20 A nucleation map for precipitation during isothermal annealing of an Al–0.99 wt% Cu alloy. (Reprinted from Ref. [133], copyright (1986), with permission from Elsevier.)

modeling of their formation. Wolverton showed that first-principles calculations combined with large-scale Monte-Carlo simulations can yield useful results for coherent precipitates: the atomic structure and composition of precipitates, their equilibrium shape and their work of formation can all be determined [134]. He showed that GP zones are disk-shaped monolayers composed entirely of copper atoms; thicker disks and those of <100% Cu have very high energies. Monolayer GP zones are observed only up to a diameter of 15 nm. The other favored configuration is that discussed above for θ'', namely a disk shape with two monolayers of copper atoms separated by three layers of aluminum atoms; these are observed only down to a diameter of 10 nm. Wolverton suggests that there is a comparatively large interfacial energy around the perimeter of a θ'' disk. Thus GP zones are initially favored despite the lower thermodynamic driving force for their formation, but θ'' becomes favored at larger sizes. The structures of GP zones and θ'', the maximum size for GP, and the minimum size for θ'' predicted by these simulations are all in remarkably close agreement with HREM observations (Figure 15) [127, 130]. The interaction energies in this atomistic modeling suggest that the compliant elastic response of copper along $\langle 1\,0\,0 \rangle$ is a factor underlying the favored disk shape of GP zones. Elastic interactions are similarly important in favoring the three-plane separation of copper monolayers in θ''.

The purely thermodynamic analysis of Wolverton was extended to kinetics by Wang et al. [135]. They followed the number of GP zones, their size, and size distribution as a function of aging time (Figure 21). They concluded that the precipitation of GP zones is a classical nucleation and growth process. Hu et al. have used molecular-dynamics simulations on Al–Cu to study a precipitate artificially placed in the center of the simulation cell [136]. The energy of the system was compared with the homogeneous alloy, holding the overall copper content fixed. Figure 22 shows the work of formation of a GP zone as a function of zone radius.

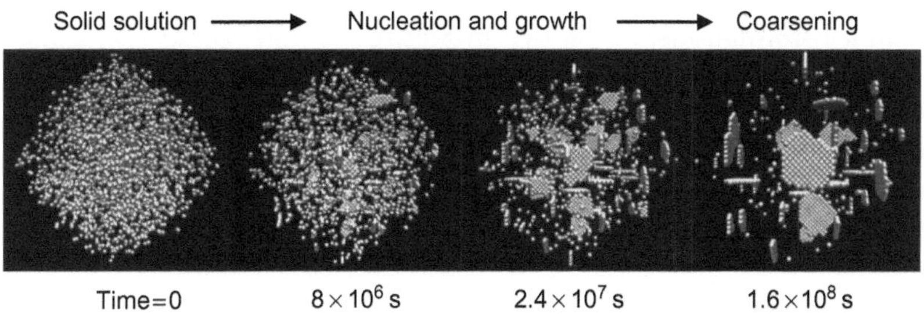

Fig. 21 First-principles atomistic simulation of the evolution of GP zones in an Al–1.0 at.% Cu alloy at 373 K. Only the copper atoms are shown; {1 0 0} monolayer disks form by nucleation and growth and then coarsen. (Reprinted from Ref. [135], copyright (2005), with permission from Elsevier.)

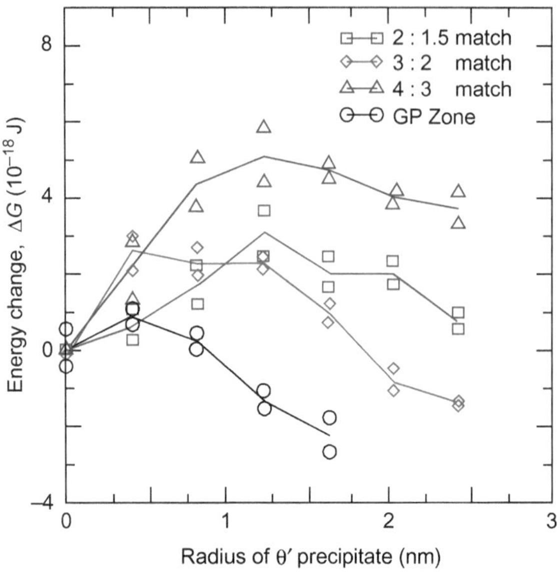

Fig. 22 Works of formation of GP zones and θ′ particles in Al–Cu, from atomistic simulations. Each data point is the result of a molecular-dynamics calculation on an artificially embedded particle. For θ′ particles, three different matches on the (1 0 0) and (0 1 0) interfaces with the matrix are tested. A match of 3 unit cells of α to 2 unit cells of θ′ gives the lowest barrier for homogeneous nucleation. (Reprinted from Ref. [136], copyright (2006), with permission from Elsevier.)

The critical radius is ~0.4 nm, and the nucleation barrier is $\sim 8 \times 10^{-19}$ J, a very low value.

As noted earlier, θ″ appears to evolve from GP zones, so the next distinct nucleation stage to consider is that of θ′. This analysis can also benefit from an atomistic approach, and has been analyzed by Hu et al. [136]. For small particles, the (1 0 0) and (0 1 0) interfaces with the matrix are semicoherent. Hu et al. considered a number of possible interface matches for the artificially embedded particles in their molecular-dynamics simulations. As shown in Figure 22, they found that a match of three unit cells of α to two unit cells of θ′ gives the lowest work of formation. They estimate the critical radius to be 0.8–1.2 nm and the critical work of formation to be $2.3\text{–}3.0 \times 10^{-18}$ J, much higher than the values for GP zones [136]. These values are, however, for homogeneous nucleation. The nucleation of θ′ on dislocations, as is found in practice, has not yet been simulated.

It is clear from the above that atomistic simulations are very useful, arguably indispensable, in understanding the complexities of precipitation in aluminum alloys. The simulations considered so far have been for Al–Cu. First-principles calculations have been applied to a wide range of alloys. The particular interest is in commercial alloys with compositions of much greater complexity, such as Al–Mg–Si–Cu, and these have been widely studied [137].

It has been noted earlier (Table 2) that GP zones often act as the nucleation sites for later phases in a precipitation sequence. As pointed out by Marth et al. [138], this could be regarded as surprising, because the solute supersaturation adjacent to a zone must be lower than in the far-field matrix. They analyze, in essentially qualitative terms, why the GP zones are indeed likely to be favored sites. These are chiefly as follows:

(i) The work of nucleation is reduced when the nucleation is heterogeneous on the GP zone as a substrate.
(ii) With the very high population density of GP zones (up to $10^{24}\,\mathrm{m}^{-3}$), there is a very large substrate area per unit volume.

In crystalline solids, there is an equilibrium fraction of vacant sites (*vacancies*) $X_{\mathrm{vac}}^{\mathrm{eq}}$, given by

$$X_{\mathrm{vac}}^{\mathrm{eq}} = A \exp\left(-\frac{\Delta H_{\mathrm{vac}}}{k_{\mathrm{B}}T}\right), \tag{13}$$

where A is a constant of order one and ΔH_{vac} is the enthalpy of vacancy formation (the change in enthalpy on taking an atom from a lattice site and transferring it to the surface). Typical values of ΔH_{vac} ($\approx 1.3 \times 10^{-19}\,\mathrm{J}$, for aluminum) are such that the vacancy concentration is strongly temperature-dependent. When the temperature is changed, the vacancy concentration changes to adopt the equilibrium value for the new temperature, through the operation of vacancy *sources* and *sinks* (predominantly at the free surface, grain boundaries and dislocations). As vacancies have to diffuse from the sources or to the sinks, it takes time for an equilibrium vacancy concentration to be established. This is important for age-hardening. When a solid solution is rapidly quenched, the vacancy concentration characteristic of the annealing temperature is initially retained at lower temperature. Taking aluminum alloys as an example, the vacancy concentration at room temperature can exceed the equilibrium concentration by a factor of 10^6 to 10^{17}. This accelerates precipitation rates, and, as expected, the acceleration is more evident if the quench has been from a higher annealing temperature. There are several ways in which an excess vacancy concentration of this kind can accelerate precipitation, and specifically nucleation:

(i) The atomic diffusivities of substitutionally diffusing species (e.g., Cu in Al) are directly proportional to the vacancy concentration, and are therefore greatly enhanced after a quench.
(ii) The precipitate–matrix interface, if incoherent, can act as a vacancy sink; this reduces the strain energy if the volume per atom of the precipitates exceeds that of the matrix; in this way the vacancies act to increase the driving force for precipitation [139].

(iii) The vacancies can condense to form clusters, prismatic dislocation loops, and stacking fault tetrahedra (SFTs, as discussed later in Section 5.2); these defects, notably dislocations (Chapter 6, Section 3), are sites for heterogeneous nucleation.

(iv) Preexisting screw dislocations act as sinks for vacancies and thereby evolve into helical edge dislocations of greater length, facilitating nucleation (Figure 23; [112]).

Quenched Al–Cu alloys form GP zones at room temperature. That would not happen in reasonable time without the quenched-in excess vacancy concentration. The acceleration in that case is attributable entirely to the enhanced diffusion of copper.

The importance of vacancy concentration is dramatically illustrated at grain boundaries. The microstructures of age-hardened alloys commonly show a precipitate-free zone (PFZ) at each boundary (Figure 24) [140, 141]. Such zones are a problem, as they are mechanically weak (in particular showing decreased fatigue resistance) and the electrochemical difference between the PFZ and the matrix may facilitate corrosion. The study of PFZs reveals much about nucleation mechanisms. The grain boundary acts as a sink for vacancies, and thus the vacancy mole fraction at a boundary in a quenched alloy is likely to have the form shown in Figure 25a. As noted earlier, the vacancy supersaturation resulting from a rapid quench is necessary for precipitation of GP zones in the short timescales that are observed. We can consider that there is a critical vacancy concentration for zone formation, and we then expect that a PFZ

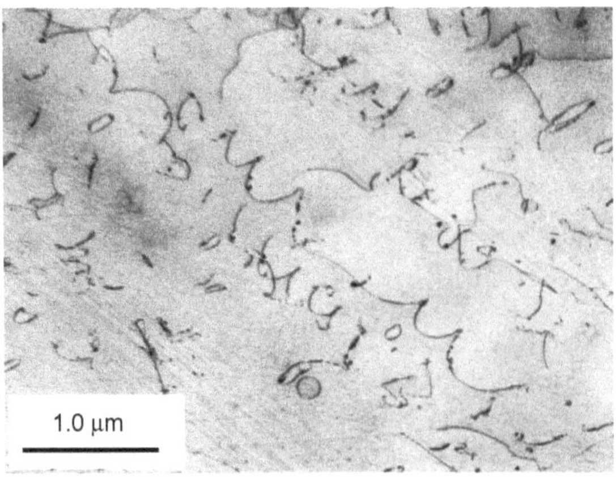

Fig. 23 Transmission electron micrograph of the early stages of nucleation of θ′ precipitates on helical dislocations in Al–0.85 at.% Cu quenched and then aged at 200°C for 1 min. (Reprinted from Ref. [112], copyright (1969), with permission from Marcel Dekker.)

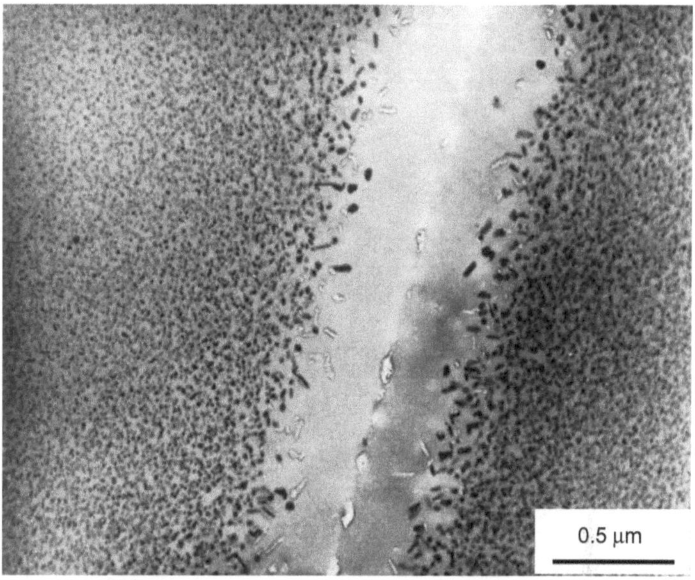

Fig. 24 Distribution of η precipitates in Al–Zn–Mg water-quenched and immediately aged at 180°C for 3 h. (Reprinted from Ref. [141], copyright (1969), with permission from Elsevier.)

will arise, with a width related to the quench rate. Slower quenches give more time for diffusion of vacancies to the boundary and a broader PFZ should result (Figure 25b). It is indeed found that PFZs are wider if the quench is interrupted or is at a lower rate. In an Al–6 wt% Zn–2.2 wt% Mg alloy, the PFZ width is approximately inversely proportional to the square root of the quenching rate, as is expected for vacancy depletion [142].

The alloy shown in Figure 24 is from the commercially important Al–Zn–Mg system, with the complex precipitation sequence:

$$GP \rightarrow \eta' \rightarrow \eta(MgZn_2) \rightarrow T((Al,Zn)_{48}Mg_{32}). \tag{14}$$

We now consider in more detail how the microstructure in such an alloy can be controlled. The aging treatment at 180°C used for the Al–5.9 wt% Zn–2.9 wt% Mg alloy in Figure 24 leads to η-phase precipitates. Much finer microstructures are obtained if such an aging treatment is preceded by quenching to below the GP zone solvus (~155°C for this alloy) and holding for some time below the solvus. In that case, the alloy contains a dispersion of GP zones that then act as nuclei for η′ precipitates on subsequent aging. Lorimer and Nicholson [140] suggest that there is a critical zone size above which the zone can act as nuclei. PFZs then arise because near a grain boundary, the GP zones are less developed, are not big enough to act as nuclei, and dissolve during the aging treatment.

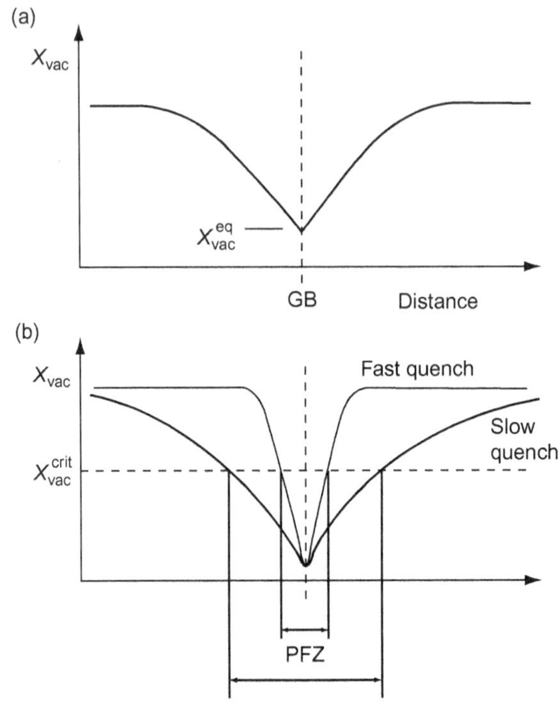

Fig. 25 Schematic diagrams of (a) the vacancy concentration profile near a grain boundary in a quenched polycrystalline alloy; (b) profiles for fast and slow quenches, relating to the observed widths of PFZ. (From Ref. [110], copyright (1992), with kind permission of Springer Science and Business Media.)

An alloy that has been aged and shows a PFZ can later be held for a significant time below the GP solvus to develop a dispersion of GP zones. A second aging treatment then gives a second precipitation within the PFZ (Figure 26), and a third precipitation has similarly been demonstrated [141]. The success of such treatments in "filling in" PFZs shows that the zone near the boundary cannot be attributed to depletion of solute, and supports the analysis of PFZ formation based on vacancy depletion. All the cases of PFZs considered so far have been in rapidly quenched alloys, in which no precipitation occurred during the quench. If an alloy is more slowly cooled after solution treatment, precipitates can nucleate and grow on grain boundaries during cooling. Subsequent aging to develop a fine precipitate dispersion then does show solute depletion effects, for example the widened PFZ around an equilibrium T-phase precipitate in Figure 27.

Clearly there has been much detailed study of precipitation in alloys, in aluminum alloys in particular. Despite this, it appears that there remains significant potential for improving alloy performance through

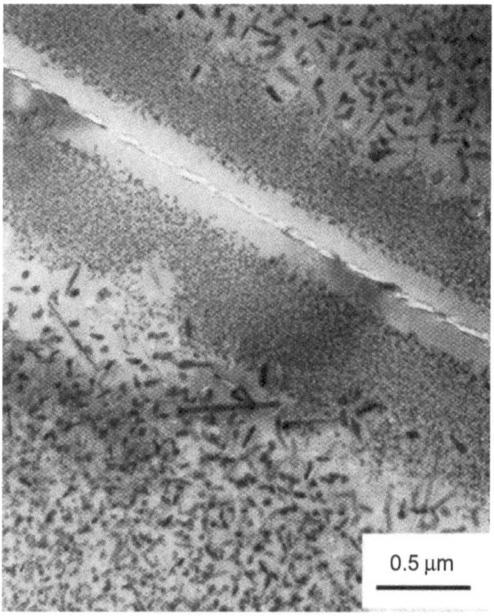

Fig. 26 The distribution of η′ precipitates in Al–5.9 wt% Zn–2.9 wt% Mg quenched into acetone at −95°C, aged at 180°C for 2 h, at room temperature for 8 days, and finally at 180°C for 2 h to give a filled-in PFZ. (Reproduced in part from Ref. [118], copyright (1978), by permission of AIME.)

more carefully designed heat treatments to control nucleation. More complex annealing treatments can give highly desirable simultaneous increases in strength and toughness [143].

4.2 Matrix polymorphism: Decomposition of austenite

During precipitation of the kind considered in the previous section, the matrix phase becomes depleted in solute, but does not change structure. There are, however, elements that, unlike aluminum, do show more than one polymorph. Transformations from one polymorph to another can be induced by temperature or pressure change, or by alloying. Important examples are found in commercial titanium alloys and, above all, in steels. In this section, we restrict our coverage to steels, and specifically to the decomposition of the cubic close-packed (ccp) (fcc) γ phase (*austenite*). Figure 28 shows the iron-rich end of the Fe–C phase diagram relevant for steels [144]. On cooling, the γ phase undergoes a eutectoid decomposition at 727°C to α-phase body-centered cubic (bcc) iron (*ferrite*) and Fe_3C (*cementite*). This phase diagram, although reflecting equilibrium for most practical purposes, depicts metastable

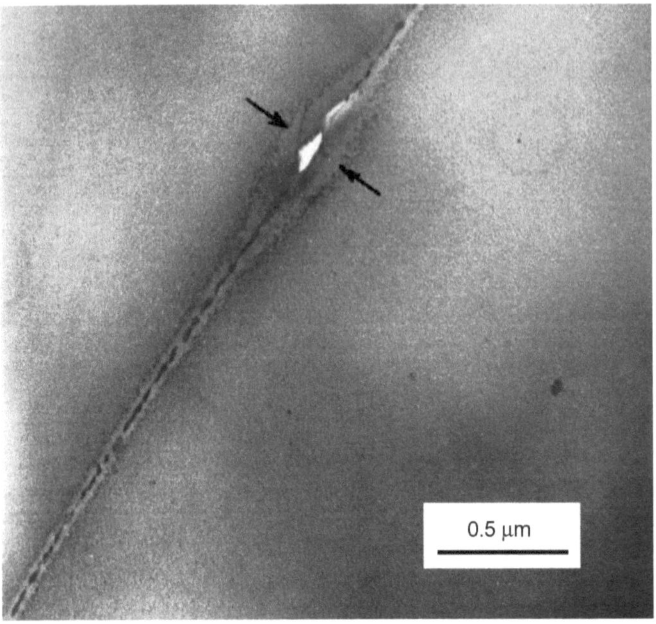

Fig. 27 Distribution of precipitates in an Al–Zn–Mg alloy oil-quenched to 330°C, held at 330°C for 2 min, water-quenched to 100°C and held at 100°C for 5 min, and finally aged at 180°C for 3 h. The PFZ width is increased (as arrowed) around a T-phase precipitate nucleated on the grain boundary causing local solute depletion. (Reprinted from Ref. [141], copyright (1969), with permission from Elsevier.)

equilibria, as cementite is itself metastable. The equilibrium carbon-rich phase is graphite, but in the solid state its nucleation and growth are effectively hindered by the large volume increase that would be involved. The eutectoid mixture of $\alpha+Fe_3C$ is known as *pearlite*, and the eutectoid composition is 0.8 wt% C. On cooling *hypoeutectoid* steels (with <0.8 wt% C), the austenite should first start to transform to ferrite (*proeutectoid* ferrite). As we will see below, this can take many forms, with different nucleation characteristics.

One reason for the complexity of transformations in steels is that different solutes in iron diffuse at very different rates. Important additions in alloy steels, such as nickel and manganese, are *substitutional* solutes, occupying lattice sites that would otherwise be occupied by iron. These solutes diffuse at rates comparable with iron itself. In contrast, carbon in particular is an *interstitial* solute, occupying interstitial sites and able to diffuse many orders of magnitude faster than the substitutional solutes. In this way, it is possible for there to be partitioning of carbon during the transformation from original to new phase without partitioning of substitutional solutes. In that case, only

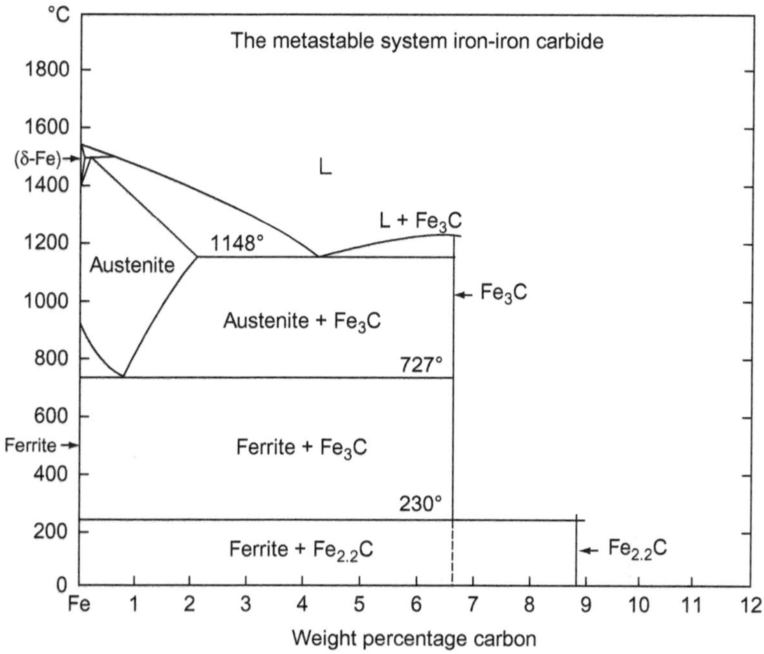

Fig. 28 The metastable phase diagram for Fe–Fe_3C equilibria. For steels, the key reaction is the eutectoid decomposition of austenite on cooling at 727°C. (Reprinted from [144], copyright (1973), with permission from ASM International.)

the carbon shows equilibrium behavior, a condition known as *para-equilibrium.* The diverse behavior of solutes is evident even at the nucleation stage of transformations in steels.

Another reason for complexity is that transformations can be *reconstructive* or *displacive* (Chapter 12, Section 4.1). Transformations such as the precipitation of distinct new phases considered in Section 4.1 in this chapter are reconstructive. All the atoms involved make diffusive rearrangements to reconstruct the local coordination. A solid-state transformation can, however, also be achieved by a shear of the lattice. In this way, there is a one-to-one correspondence of lattice sites in the original and new structures. Displacive transformations are often termed diffusionless, but this is misleading. In steels, for example, a shear of the lattice sites occupied by the iron atoms may be accompanied by diffusion of interstitial carbon. There may also be some rearrangement of substitutional solutes on the lattice sites. A key characteristic of displacive transformations is that there is a shape change with a large shear component. This takes the form of an invariant plane strain, and is readily detected, for example, by the development of surface relief on polished samples taken through the transformation. The lattice-site correspondence of the original and new phases means that in principle the shear can be reversed to take the structure back to its original condition. This underlies a range of phenomena such as the *shape memory effect* and *superelasticity*. A new phase forming by a displacive transformation typically adopts a plate-like or lath shape to minimize the strain energy in the surrounding matrix. Furthermore, there is typically a well-defined crystallographic orientation relationship between the original and new phases; for the austenite-to-ferrite transformation, the most common are the Kurdjumov–Sachs and the Nishiyama–Wasserman relationships. The crystallography of displacive transformations lies beyond our present scope, but is well covered in standard texts [116, 117, 145].

The decomposition kinetics of austenite are commonly presented on isothermal-transformation (TTT) diagrams. These indicate the diverse decomposition reactions, and the times for their commencement or completion, as a function of the temperature at which the steel, just cooled from the austenite phase field, is held. TTT diagrams may be very complex. For a given reaction, the kinetics typically shows a C-curve on the diagram; the transformation rate is maximum at intermediate temperatures, lowered at higher temperatures by the small thermodynamic driving force for transformation and at lower temperatures by low atomic mobility. Bhadeshia [146] has suggested a generic diagram (Figure 29), showing reconstructive transformations at higher

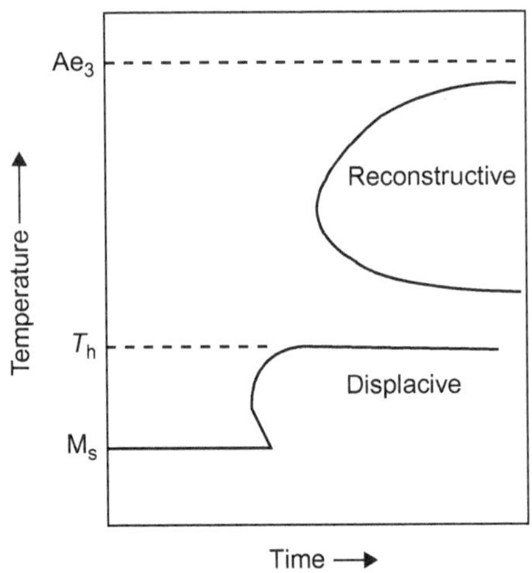

Fig. 29 A schematic TTT diagram showing curves for the onset of the decomposition of austenite. Ae_3 is the standard designation for equilibrium lower temperature limit to the austenite phase field at which ferrite first appears on cooling; T_h is the highest temperature at which ferrite can form by a displacive mechanism; M_s is the martensite-start temperature. (Reprinted with permission from Ref. [146], copyright (1992), Maney Publishing.)

temperature, able to proceed at low driving force and favored by high atomic mobility. Reconstructive transformations are sluggish below ~850 K. Displacive transformations, requiring a higher driving force because of the strain energy associated with them, but not dependent on substitutional diffusion, predominate at lower temperature. The C-curve for displacive transformations in Figure 29 has a clear horizontal top, representing the highest temperature at which ferrite can form by a displacive mechanism. Rapid cooling of austenite leads to athermal, displacive transformation to *martensite*, as discussed in Chapter 12, Section 4.4. Table 4 summarizes the nucleation characteristics of many possible decomposition products of austenite. We now examine these in turn.

Allotriomorphic ferrite is the name given to carbon-depleted ferrite forming with no preferred shape on grain boundaries in the austenite. It forms at high temperatures, with low thermodynamic driving force for transformation. Carbon is partitioned into the remaining austenite, and at some point the growth of allotriomorphic ferrite may stop and the remaining austenite may decompose to pearlite, giving microstructures such as those seen in Figure 30. The subject of nucleation on grain boundaries has been discussed in general terms in Chapter 6, Section 2.6. The particular case of the nucleation of ferrite on austenite boundaries has been studied in detail by Aaronson and co-workers [147–149]. They applied classical nucleation theory and concluded that the high nucleation rate of ferrite and its low temperature dependence were consistent only with a low work of nucleation, such as would be obtained if the nucleus had low-energy, fully or partially coherent interfaces with *both* austenite grains. It would take the form of a pill-box (Chapter 6, Figure 22f) [149], a disk sitting in the plane of the boundary with planar faces on both sides. Since the grains are in random crystallographic orientations, it must be rare that the grain-boundary orientation is nearly parallel to a favored *habit plane* (i.e., preferred orientation of the interphase interface) in both grains. Only a very small fraction of the total austenite grain area is capable of catalyzing nucleation. Only a minority of grain boundaries (approximately one third) have any sites for ferrite nucleation [149]. It was noted in Chapter 6, Section 5 that nucleation may be more favored on grain edges and grain corners, and that the driving force for transformation determines which nucleation sites are dominant (Chapter 6, Figure 32). Enomoto and Aaronson analyzed nucleation of ferrite on grain edges and concluded that only a small fraction of the edges is suitable for nucleation [150]. The studies of Militzer et al. suggest that as the temperature is lowered, the predominant nucleation sites evolve from grain corners to grain edges and then to grain boundaries [151]. Increasing the cooling rate limits the time for growth of ferrite from early nucleation events, leaving more potential nucleation sites at austenite boundaries for later nucleation.

Table 4 Characteristics of the nucleation of the decomposition of austenite in steels

	Allotriomorphic ferrite	Idiomorphic ferrite	Pearlite	Widmanstätten ferrite	Acicular ferrite	Upper bainite	Lower bainite	Martensite
Reconstructive transformation	✓	✓	✓					
Displacive transformation				✓	✓	✓	✓	✓
Plate shape, invariant plane strain shape change with shear				✓	✓	✓	✓	✓
Diffusionless nucleation								✓
Diffusion of carbon only during nucleation				✓	✓	✓	✓	
Reconstructive diffusion during nucleation	✓	✓	✓					
Nucleation typically on grain boundaries	✓		✓	✓		✓	✓	
Nucleation typically intragranular, on inclusions		✓			✓			
Nucleation typically intragranular on dislocations								✓

Reprinted with permission from Ref. [146], copyright (1992), Maney Publishing.

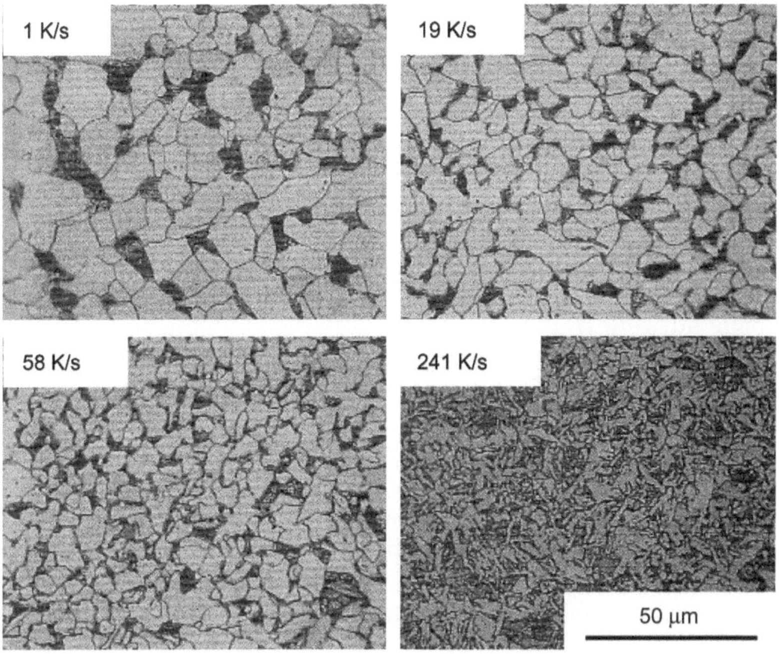

Fig. 30 Optical micrographs of the microstructures of allotriomorphic ferrite (light) and pearlite (dark) obtained in an Fe–Mn–C steel, with an initial austenite grain size of 18 μm, cooled at the rates shown. (From Ref. [151], copyright (1996), with kind permission of Springer Science and Business Media.)

Faster cooling, thus, gives grain refinement in the ferritic microstructure (Figure 30), and this desirable outcome is achievable under practical rolling mill conditions.

Idiomorphic ferrite nucleates inside austenite grains at rather higher driving force than Widmanstätten ferrite, and has an equiaxed shape. It nucleates mainly on inclusion particles [152]. Its formation is again reconstructive and its growth also occurs at an incoherent interface with the austenite.

In austenite with very low carbon content, there can be a *massive* transformation to ferrite with identical composition to the austenite. The characteristics of such nonpartitioning, yet reconstructive, transformations have been reviewed [153]. For present purposes, however, we note that massive ferrite nucleates in just the same way as allotriomorphic ferrite.

Pearlite can be nucleated on austenite grain boundaries, with either ferrite or cementite appearing first. The nucleating phase has a crystallographic orientation relationship with the austenite grain on one side of the boundary. The semicoherent interface with that grain is effectively the nucleation substrate, and growth occurs into the other austenite grain at a mobile incoherent boundary. Ferrite (α) nucleating on austenite (γ) adopts the Kurdjumov–Sachs orientation relationship

$$\{1\,1\,0\}_\alpha//\{1\,1\,1\}_\gamma \quad \langle 1\,1\,1\rangle_\alpha//\langle 1\,1\,0\rangle_\gamma, \tag{15}$$

under which the close-packed planes and directions in the two phases are parallel [110]. The orthorhombic phase cementite nucleating on austenite adopts a relationship close to:

$$(1\,0\,0)_{Fe_3C}//\{1\bar{1}1\}_\gamma \quad (0\,1\,0)_{Fe_3C}//\{1\,1\,0\}_\gamma \quad (0\,0\,1)_{Fe_3C}//\{\bar{1}1\,2\}_\gamma. \tag{16}$$

Figure 31a shows schematically the nucleation and growth of pearlite when cementite is the phase that nucleates first. Carbon depletion in the austenite adjacent to the cementite favors the subsequent nucleation of ferrite on the same austenite grain, also with a semicoherent interface and the usual Kurdjumov–Sachs relationship. The microstructure of interleaved plates of ferrite and cementite in the pearlite forms partly by branching of existing plates, and partly by nucleation of new plates. The pearlite *nodule* or *colony* is effectively two interpenetrating single crystals. Both phases grow cooperatively, sharing a common transformation front with the austenite, and in the absence of impingement with other nodules, the shape is hemispherical (Figure 31c).

Pearlite can also be nucleated on ferrite coating grain boundaries in the case of hypoeutectoid steels, or on cementite coating the boundaries in hypereutectoid steels; in those cases, there is a continuous crystal lattice from the proeutectoid phase into the same phase in the pearlite (Figure 31b).

So far, our consideration of the decomposition of austenite has been restricted to reconstructive transformations. The formation of *martensite* on rapid cooling (quenching) of austenite is, in contrast, displacive. The nucleation of martensite has been treated in Chapter 12, Section 4, where it was noted that it has several characteristic features. The nucleation, on preexisting dislocation arrays, is largely athermal, and is without solute partitioning of any kind. At lower temperatures, isothermal nucleation can be observed because of the need for thermal activation of dislocation motion; the effective activation energy for nucleation (describing the temperature dependence of the nucleation rate) is then proportional to the chemical driving force for transformation (Chapter 12, Section 4.4). As noted in Chapter 12, Section 4.5, the nucleation of the decomposition products to be discussed below has similarities with that of martensite.

Widmanstätten ferrite grows from austenite grain boundaries in the form of plates or laths that are formed by a displacive transformation and have semicoherent interfaces with the grain into which they are growing. Data on the highest temperature at which Widmanstätten ferrite can form suggest that it nucleates in the same way as martensite, except that carbon is mobile, diffusing out of the ferrite as it forms [154]. The maintenance of this paraequilibrium partitioning of the carbon means that the growth rate of the ferrite plates is controlled by the diffusion of carbon in the austenite. The

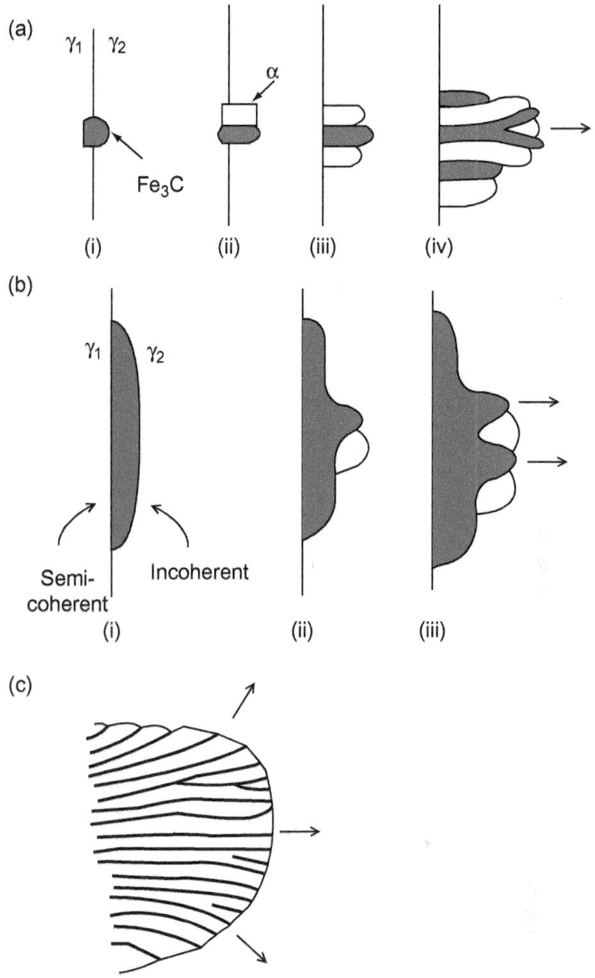

Fig. 31 Nucleation and growth of pearlite from an austenite grain boundary. (a) Nucleation of cementite with a coherent interface on the grain γ_1 (i) is followed by nucleation of ferrite also on γ_1, and the pearlite colony grows by further nucleation (iii) and branching of plates (iv). (b) With a proeutectoid phase on an austenite grain boundary (i), pearlite can nucleate on the incoherent boundary with the austenite, giving a different α:Fe_3C orientation relationship from that in (a). (c) A pearlite colony with a characteristic hemispherical form. (From Ref. [110], copyright (1992), with kind permission of Springer Science and Business Media.)

subsequent growth continues in the same way, with displacive transformation of the iron lattice accompanied by partitioning of the carbon solute. Widmanstätten ferrite nucleates with a low energy barrier on austenite grain boundaries. Its nucleation at a comparatively high temperature, at which the driving force for transformation is low, leads to a distinctive growth mode in which two plates with mirror-image shears grow cooperatively in an arrowhead shape to reduce the overall strain energy.

The nucleation of *bainite* is essentially identical to that of Widmanstätten ferrite. It forms at higher driving force, and its growth is diffusionless. Soon after growth, however, the carbon does come out of solution in the ferrite. At higher temperatures, the carbon diffuses out into the remaining austenite and forms carbides there; the resulting microstructure is *upper bainite*. At lower temperatures, the carbon forms carbides within the ferrite, giving *lower bainite*. The burst of transformation that is usual for martensite (Chapter 12, Section 4.1) does not arise for bainite, which forms at lower driving force and with carbon diffusing during nucleation. The growth of a bainite plate is limited by the accumulation of lattice damage around it. Transformation proceeds by the nucleation of new plates in the undamaged matrix under shear stress near the tips of existing plates. In this way, a characteristic sheaf microstructure develops consisting of clusters of ferrite plates (subunits) with near-identical crystallographic orientation and size (Figure 32). The parallel plates arising from this autocatalytic nucleation reduce the toughness of the steel.

Plates of ferrite formed by a displacive transformation can also nucleate *inside* grains; this is *acicular ferrite*. It differs from bainite only in that its intragranular nucleation allows the ferrite plates to grow outward

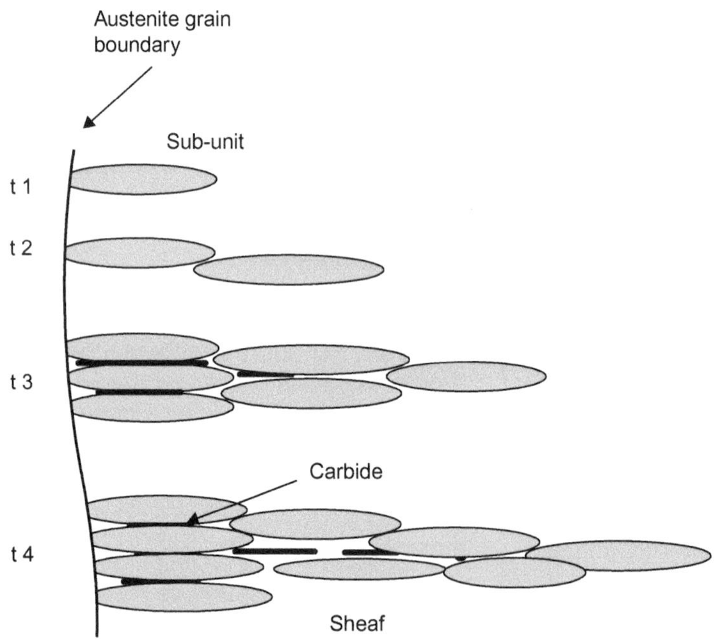

Fig. 32 A schematic depiction of the development of upper bainite at successive times. (Reprinted with permission from Ref. [146], copyright (1992), Maney Publishing.)

in many orientations. The resulting microstructure shows much improved toughness compared to a bainitic steel. Acicular ferrite nucleates in the same way as bainite, but on nonmetallic inclusions in the steel rather than at grain boundaries. It is often the last decomposition product to form, after allotriomorphic and Widmanstätten ferrite or bainite. It is favored by the presence of potent inclusions and by a large austenite grain size. For given transformation conditions, the content of inclusions can determine whether the decomposition is mainly to acicular ferrite or to bainite [155]. In the presence of some alloying elements (e.g., Cr, Mn, Mo, Ni), the formation of allotriomorphic ferrite is hindered. As the solute content is increased (Figure 33), this at first facilitates the formation of acicular ferrite. At still higher solute contents, however, the suppression of allotriomorphic ferrite allows bainite to be

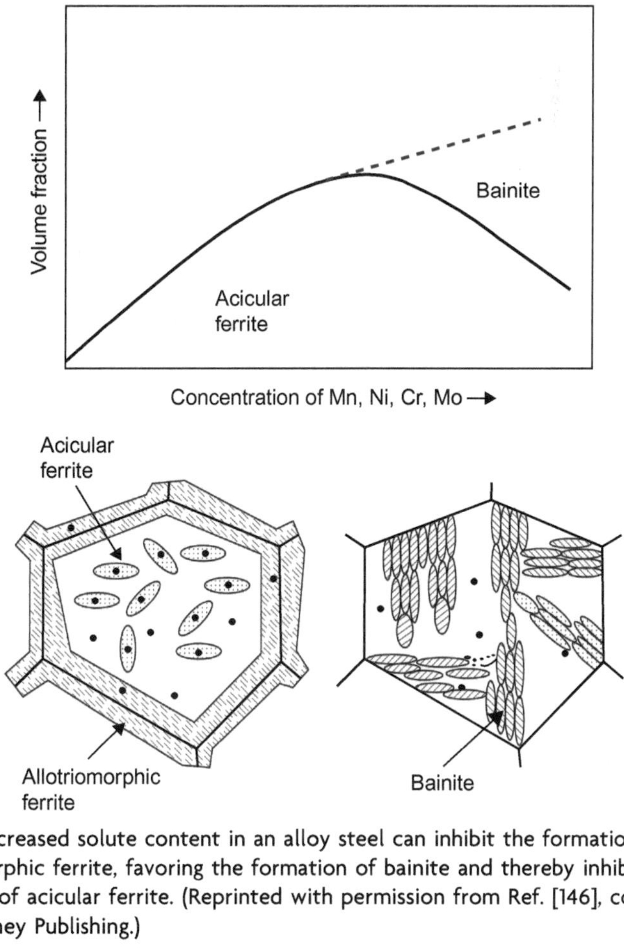

Fig. 33 Increased solute content in an alloy steel can inhibit the formation of allotriomorphic ferrite, favoring the formation of bainite and thereby inhibiting the formation of acicular ferrite. (Reprinted with permission from Ref. [146], copyright (1992), Maney Publishing.)

nucleated on the grain boundaries and this reduces the opportunities for acicular ferrite to nucleate [156].

Because acicular ferrite has such a beneficial effect on the toughness of the steel, particularly at welds, its intragranular nucleation on inclusions has been very widely studied. The inclusions themselves form in the liquid steel. Although it may be possible in some cases for ferrite to nucleate on the inclusions in the melt [157], it is usually considered that the ferrite forms by decomposition of austenite. There has been extensive work on the likely nucleation potency of different inclusion phases, based on their degree of lattice matching with ferrite [158]. However, the ferrite forms with a well-defined orientation relationship with the austenite. Since the inclusions are embedded in the austenite in random crystallographic orientation, it is unlikely that the nucleating ferrite could have an epitaxial relationship with the inclusion as well as the required relationship with the austenite. In any case, templated nucleation may be unlikely for a displacive, nonreconstructive transformation. Experiments do suggest that there is no crystallographic orientation relationship between inclusion particles and the ferrite plates nucleated on them [159]. Inclusions may be favored sites because of local strains (associated with cooling) and reactions giving changes in matrix composition. Studies of the nucleation potency of different inclusion phases show that there may be different mechanisms on different phases, and in some cases the mechanism remains quite unclear [146, 160]. Titanium oxide particles are particularly effective nucleants, leading to their exploitation in inoculated steels [161]. Figure 34 shows the dramatic microstructural effect in the heat-affected zone of welds. Greater heat input leads to more

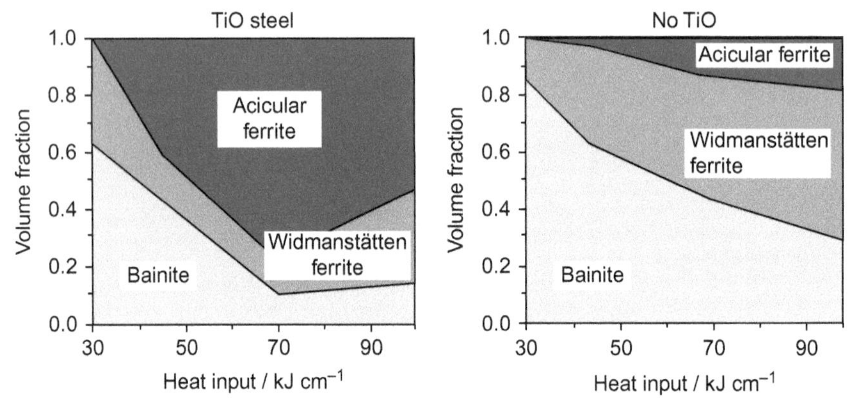

Fig. 34 Data from Ref. [161] on the nature of the microstructure in the heat-affected zone of steel welds as a function of the heat input during welding: (a) a steel inoculated with titanium oxide particles, and (b) the same steel without such inoculation. (Reprinted with permission from Ref. [146], copyright (1992), Maney Publishing.)

grain growth in the austenite. The reduced area available for grain-boundary nucleation of Widmanstätten ferrite or bainite favors acicular ferrite, but the presence of titanium oxide inclusions has a particularly strong beneficial effect [161].

In this section on the decomposition of austenite, we have focused on the nucleation of ferrite. The nucleation of carbides is also of interest, and it can occur at the austenite–ferrite interface; this is considered in Chapter 15, Section 3.

5. RADIATION DAMAGE AND VOIDS

5.1 Introduction to radiation damage

Radiation damage has been widely studied because of its importance in structural components in and near the cores of nuclear reactors. Such components are subject to extreme conditions of various kinds — high temperature, high stress, corrosive environment — but neutron irradiation leads to specific types of damage, the nucleation aspects of which are the subject of this section. The essence of radiation damage is that the incident radiation is sufficiently energetic to displace atoms from their equilibrium sites in the crystal structure, creating vacancies and self-interstitials. This can occur not only under neutron irradiation but also under incident electrons and ions, and the latter species have often been used in studies simulating neutron-induced radiation damage. Incident electrons with energies greater than $\sim$1 MeV are capable of creating single vacancy-interstitial pairs (*Frenkel pairs*). The much greater energy transfer possible with ions and neutrons gives *displacement cascades* in which many pairs are created. An atom struck by an incident neutron or ion receives enough energy in the primary recoil that it in turn can displace many atoms. The initial mean free path between collisions is of the order of 1 cm for fast neutrons, but decreases as energy is dissipated in successive collisions; the result is that displacement cascades are concentrated in volumes 1–10 nm in diameter. Typical damage rates (measured as displacements per atom per second, dpa s^{-1}) range from 10^{-9} dpa s^{-1} in thermal reactors to 10^{-5} dpa s^{-1} in the first wall of proposed fusion reactors. Over the lifetime of a component, each atom could be displaced as many as 100 times. Clearly such extreme conditions can have profound effects on the microstructure of the alloys involved. These effects have been extensively reviewed [162–164], and include dissolution of precipitates, changes in their morphology, and appearance of nonequilibrium phases. We focus on nucleation phenomena associated with radiation damage, especially on voiding, which is among the most important types of damage in reactor components, notably in stainless steel cladding.

In principle, the vacancies and interstitials generated in displacement cascades might recombine to restore the equilibrium structure. In practice, several factors give a substantial supersaturation of vacancies [165]. Self-interstitial atoms form stable clusters and are ultimately removed by processes of glide and dislocation sweeping; in effect radiation damage has a bias toward the production of vacancies [166]. Typically the vacancies and interstitials are created at a temperature high enough for them to be mobile. The interstitials are the more mobile defect, especially when processes not requiring thermal activation, such as collision sequences along close-packed rows of atoms [167], are taken into account. The greater mobility of the interstitials is an additional factor leaving the centers of displacement cascades vacancy-rich. Dislocations, through the process known as *climb* [168], can act as sinks for vacancies and interstitials, but the greater elastic strain around the latter again leads to their preferential removal.

The supersaturation of vacancies leads to the appearance of new microstructural features. At low temperatures ($T < 0.2T_m$, where T_m is the absolute melting temperature of the irradiated alloy), the vacancies condense mainly to form stacking faults and dislocation loops, ultimately leading to the development of a dislocation network. These changes, in particular the nucleation of dislocation loops, are considered in Section 5.2.

Irradiation at higher temperatures ($T > 0.2T_m$), leads to cavities, the nucleation of which is related to supersaturations not only of vacancies but also of dissolved helium atoms. The helium arises from irradiation-induced transmutation reactions accompanied by the emission of alpha particles. The rate of helium production is in the range 0.5–12 atomic parts per million per dpa [164]. While the condensation of vacancies gives *voids*, the precipitation of helium atoms gives *bubbles*. Of course these represent extremes, and observed cavities involve both vacancies and helium atoms. Indeed, small helium bubbles can subsequently act as sinks for more vacancies than helium atoms, thereby evolving from bubbles into voids. The analysis of general cavity nucleation is outlined in Section 5.3. The coupled fluxes of vacancies, self-interstitial atoms, helium atoms, and indeed other impurities in the irradiated material, necessarily lead to complexity, though some simple cases can be distinguished. Methods of analyzing coupled fluxes have been outlined in Chapter 5, Section 5. Cavity nucleation under irradiation has the additional feature, not often encountered in nucleation problems that the principal species (vacancies and helium atoms) are under continual production; without this, cavities already produced would largely disappear on annealing.

5.2 Nucleation of dislocation loops

In a metal with the ccp structure (also known as fcc), vacancies can agglomerate to form SFTs. These are defects with four {111} faces each bounded by partial dislocations with Burgers vectors of the type

$(a/6)\ \langle 110 \rangle$, where a is the lattice parameter. These Burgers vectors are shorter than any of the other types commonly found in ccp metals: $(a/2)\langle 110 \rangle$ for a perfect dislocation, $(a/6)\langle 112 \rangle$ for partial dislocations formed by dissociation on the glide plane, or $(a/3)\langle 111 \rangle$ for a prismatic loop formed by local collapse after vacancies agglomerate as a disk on a close-packed plane (next paragraph). There is a competition between the formation of SFTs and dislocation loops, with a low stacking-fault energy favoring the former. In irradiated copper, for example, which has a particularly low stacking fault energy, SFTs of diameter ~2 nm outnumber dislocation loops by 9 to 1 [169]. SFTs can interact with interstitials and transform into dislocation loops. The role of SFTs in damage accumulation has been reviewed [170], but overall dislocation loops are more important and are the focus in this section.

The nucleation of dislocation loops on application of a shear stress was discussed in Chapter 12, Section 2.1. In that case, the loop itself, and its Burgers vector, lie on a glide plane and the loop is glissile. We now consider the nucleation of a quite distinct type of loop. If a disk-shaped monolayer of self-interstitials forms in a stack of crystal planes (Figure 35a), the disrupted stacking sequence can be described as an otherwise perfect lattice containing a *prismatic* dislocation loop bounding the disk. A similar prismatic loop is formed if vacancies aggregate to form a disk-shaped hole in one of the planes (Figure 35b), and the planes on either side come together to close up the hole (Figure 35c). In these cases, in contrast to the loops considered in Chapter 12, Section 2.1, the Burgers vector is perpendicular to the plane of the disk. If the planes are the close-packed {111} planes of a ccp crystal, the Burgers vector is $\pm(a/3)\langle 111 \rangle$ and the dislocation is clearly *sessile* (unable to glide). The sequence of the planes normal to the disk is disrupted so that the dislocation loop, known as a *Frank* loop, encloses a *stacking fault*, said to be *intrinsic* for a disk formed via the condensation of vacancies (Figure 35c) and *extrinsic* for the interstitial case (Figure 35a). The distinction can be seen by considering the dislocation reactions necessary to convert the Frank loops into perfect loops (i.e., loops bounded by perfect dislocations, for which the Burgers vector is a lattice vector) thus eliminating the stacking fault and rendering the loop glissile. The unfaulting reaction is simpler for intrinsic loops (Eq. (17)) than for extrinsic loops (Eq. (18)):

$$\frac{a}{3}[1\,1\,1] + \frac{a}{6}[1\,1\,\bar{2}] \rightarrow \frac{a}{2}[1\,1\,0] \tag{17}$$

$$\frac{a}{3}[\bar{1}\,\bar{1}\,\bar{1}] + \frac{a}{6}[1\,1\,\bar{2}] + \frac{a}{6}[1\,\bar{2}\,1] \rightarrow \frac{a}{2}[0\,\bar{1}\,\bar{1}]. \tag{18}$$

The partial dislocations involved in these reactions may be freshly nucleated, or may be preexisting. Prismatic dislocation loops are not restricted to ccp metals: one of the equivalents of Eq. (17) for a

Fig. 35 End-on view of a stack of close-packed planes in a crystalline metal: (a) with an interleaved disk of self-interstitial atoms; (b) with a disk of vacancies; and (c) after collapse of the vacancy disk. Cases (a) and (c) can be described as a *prismatic* dislocation loop with Burgers vector normal to the planes and of magnitude equal to the interplanar spacing. The stacking sequence is disrupted differently in the two cases: (a) is an *extrinsic* fault, (c) *intrinsic*.

bcc metal is [162, 171]

$$\frac{a}{2}[1\,1\,0] + \frac{a}{2}[0\,0\,1] \rightarrow \frac{a}{2}[1\,1\,1]. \tag{19}$$

A classical analysis of the homogeneous nucleation of dislocation loops has been presented by Davis and Hirth [172]. The elastic strain energy $E_{\text{loop}}(n)$ of a circular prismatic dislocation loop arising from the condensation of n vacancies or self-interstitials is [168, 173]

$$E_{\text{loop}}(n) = \frac{\mu_{\text{S}} b^{3/2} \bar{v}^{1/2} n^{1/2}}{2\pi^{1/2}(1-\nu_{\text{P}})} \ln\left(\frac{16n^{1/2}}{\pi^{1/2}\text{e}}\right), \tag{20}$$

where μ_{S} is the shear modulus, b the magnitude of the Burgers vector, $\bar{v}$ the atomic volume, ν_{P} the Poisson ratio, and e is the natural logarithm base. This strain energy contributes to the work of formation of the loop, which must also take account of the reduction in the number of vacancies or interstitials, and of the formation of the stacking fault. For n vacancies condensing to form a circular intrinsic fault, the work of formation $W_{\text{loop}}(n)$ is given by

$$W_{\text{loop}}(n) = -nk_{\text{B}}T \ln\left(\frac{C_{\text{vac}}}{C_{\text{vac}}^{\text{eq}}}\right) + \frac{\bar{v}n}{b}\gamma_{\text{I}} + E_{\text{loop}}(n), \tag{21}$$

where C_{vac} is the vacancy concentration, which in thermal equilibrium would have the value C_{vac}^{eq}, and γ_I is the energy per unit area of the intrinsic fault. The maximum value W_{loop}^* of this work is then the critical barrier in a classical analysis expressing the nucleation rate I of the loops in the form:

$$I = kZN_V \exp\left(\frac{-W_{loop}^*}{k_B T}\right), \tag{22}$$

where N_V is the number of atomic sites per unit volume, Z is the Zeldovich factor (Eq. (51) in Chapter 2 and [172]), and k is the frequency with which vacancies join or leave the loop, given by

$$k = 10\pi R^* \frac{X_{vac}}{a}\omega, \tag{23}$$

where R^* is the critical radius of the loop, X_{vac} is the mole fraction of vacancies, a is the interatomic spacing, and ω is the jump frequency of a vacancy. This treatment predicts that the nucleation of dislocation loops would be affected by an applied stress, and stresses do arise from the swelling and distortion of irradiated specimens (Section 5.3). The nucleation of an interstitial type Frank loop is favored by a tensile stress acting normal to the plane of the loop, a phenomenon suggested also by later modeling and now confirmed in irradiation experiments [174]. The nucleation of a perfect loop is favored by a shear stress acting on its plane.

This analysis of loop nucleation, based on Eq. (20) or a similar expression for the elastic strain energy of the loop, has problems. Davis and Hirth [172] noted that actual nucleation rates are much higher than predicted by this approach, and they concluded that observed loops have not nucleated homogeneously. A dislocation reaction of the type in Eq. (17) removes the term in γ_I from Eq. (21) but increases $E_{loop}(n)$ by increasing b. Overall, the work of formation of the loop is lowered, but only beyond a size much greater than can be relevant for its nucleation. The origin of a nucleation barrier lower than given by Eq. (21) must be sought elsewhere.

In ccp metals, the binding energy of two vacancies is strong enough for the divacancy to be a stable defect [175]. Divacancies can be highly mobile and make a significant contribution to diffusivities [175]. Trivacancies are relatively immobile and mark the start of a vacancy cluster. For a small planar cluster, it is energetically preferable for the fault to remain open as in Figure 35b, and thus its work of formation is not well modeled, and is overestimated, by assuming a collapse of the planes as in Figure 35c. The work of formation of a disk of vacancies has a term from the surface energy of what is in effect a thin void, but this energy is less than for a free surface of the same area because of interactions between the two surfaces that are so close. Davis [176] noted that a nucleation analysis based on uncollapsed vacancy disks gives a much better fit to observed nucleation rates of loops. The formation of a dislocation loop thus appears to go through several stages:

$$\begin{gathered}\text{(i) irreversible aggregation of vacancies}\\ \downarrow\\ \text{(ii) stochastic growth of uncollapsed vacancy disk}\\ \downarrow\\ \text{(iii) collapse of disk to form Frank loop}\\ \downarrow\\ \text{(iv) reaction with partial dislocation to give a perfect loop.}\end{gathered} \tag{24}$$

According to Davis [176], the critical barrier for loop nucleation is in stage (ii) of this sequence (24). With a lower vacancy supersaturation than is usual under irradiation, the critical disk or loop radius would be larger, and then the critical barrier might be in a later stage. In that case, however, the homogeneous nucleation rate is likely to be negligibly small. If gas atoms enter a vacancy disk, they may impede its collapse, thereby favoring the formation of a void rather than a dislocation loop [177]. The formation of voids is considered in Section 5.3.

If the critical work of nucleation were that of a dislocation loop according to Eq. (21), it would be difficult to interpret the significant enhancements of nucleation rate found in the presence of some solute elements [178]. But, given the importance of early stages in the sequence in Eq. (24), it is possible that, for example, solute atoms acting as centers for vacancy aggregation can facilitate nucleation. In austenitic stainless steels, for example, silicon atoms in solution act as centers for heterogeneous nucleation, and their effect in stimulating loop nucleation can be stifled by addition of molybdenum, which forms Mo–Si clusters [179]. In concentrated alloys, the kinetics of loop nucleation can be affected by bond energies and relative atomic sizes [180, 181].

Our focus has been on loops formed from condensation of vacancies rather than interstitials. As outlined in Section 5.1, there are a number of reasons for vacancies to predominate over interstitials under irradiation. However, loops can be formed by condensation of interstitials, and these contribute to reducing the excess concentration of interstitials. Di-interstitials can play an analogous role to that of divacancies in the early stages of formation of the loops [182]. Nucleation of interstitial-type loops is relatively rare, however, suppressed not only by a low population of interstitials, but also by the high mobility of interstitial clusters making their way to sinks and by relative instability of interstitial disks [183].

So far there has been separate consideration of the nucleation of vacancy-type and interstitial-type dislocation loops. Russell and Powell [173] noted that in, for example, the nucleation of vacancy-type loops, both vacancies and self-interstitials can be important. The presence of self-interstitials very strongly retards the nucleation of vacancy-type loops.

Throughout the coverage of irradiation effects, very little attention has been paid to any possible anisotropy in the damage pattern, that is, whether the pattern might reflect not only an isotropic supersaturation of point defects but also the directionality of the irradiation. At least for ion irradiation of a ceramic, it has been shown that the relative populations of dislocation loops formed on different types of plane ({1 1 0} and {1 1 1} in $MgAl_2O_4$) depend strongly on how the planes are oriented relative to the ion-beam direction [184]. As loop size is independent of orientation, it appears that the anisotropy is in loop nucleation rather than growth, but this has not yet been fully analyzed.

The formation of vacancy-type and interstitial-type loops under irradiation is a major contribution to the observed increase in dislocation density. The continuing evolution of dislocation density under irradiation, however, involves many processes and is not susceptible to analysis in terms of loop nucleation [185].

5.3 Nucleation of voids

From the first observations of cavity formation in austenitic stainless steel [186], it was recognized that this is a particularly important form of radiation damage, leading to swelling and distortion of irradiated components. Particularly at higher temperatures, it also causes hardening of irradiated alloys and associated embrittlement and loss of ductility [164]. The development of voids on grain boundaries from initial helium bubbles shortens creep life, and affects the frequency dependence of the number of cycles to failure in fatigue [187]. The rate of swelling, typically $\sim 1\%$ per dpa in steady state [188], can give total volume changes of several tens of percent [164], is much greater than can be accounted for by the helium production rate, and is mainly due to the condensation of vacancies. Although the presence of helium may be important, and there are cases where cavities do not form without it [189], the cavities can often be regarded as voids and are usually labeled as such.

Once voids are nucleated, they continue to grow; the interstitials formed in displacement cascades are absorbed mainly at dislocations and the vacancies mainly join the voids. Modeling of the evolution of the void size distribution has received considerable attention, taking account of the various species involved and the formation of vacancy and interstitial clusters [190]. The latest modeling takes account of cascade-induced fluctuations [191] and of the role of the evolving dislocation structure [188]. Measurements of microstructural evolution under irradiation all show that void nucleation occurs very early and has ended before the onset of steady-state swelling or any change in dislocation structure [188]. Even though void nucleation occurs before the complications of microstructural evolution, it has proved difficult to analyze.

One problem is the range of species involved: vacancies, interstitials, gaseous impurities (notably He), and nongaseous impurities. The fluxes of these species into and out of embryos are likely to be linked. The general form of nucleation analysis required in a multicomponent system is described in Chapter 5, and a particular example (oxygen precipitation in silicon) is discussed in Chapter 9, Section 3. Early analyses of void nucleation in irradiated metals were reviewed by Russell [189]. The earliest treatments considered vacancies and interstitials only [192], but it is important to include impurities, and we focus on the role of gaseous impurities. Voids can be characterized in terms of the number n of vacant lattice sites they occupy and the number x of gas atoms they contain. As sketched in Figure 36, the transitions of an embryo are normally assumed

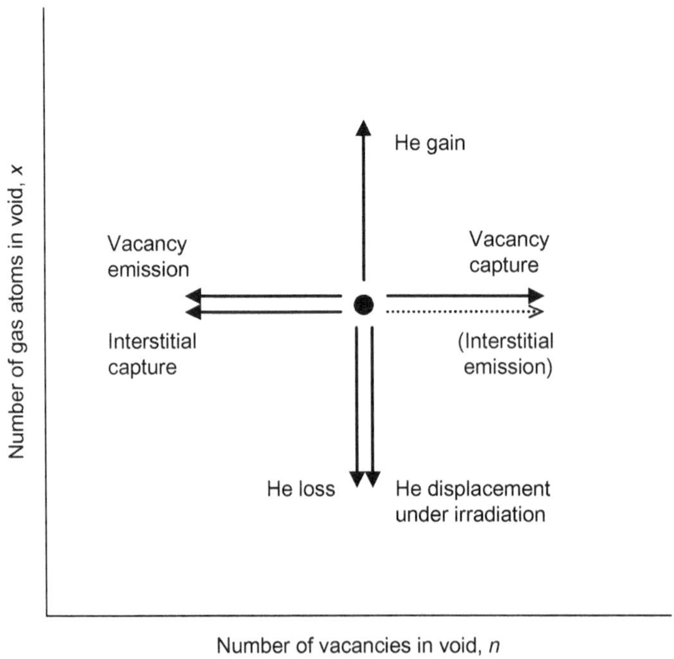

Fig. 36 A cavity embryo in an irradiated metal can be represented by its size (the number n of vacancy sites occupied) and by the number x of gas atoms (assumed to be helium) it contains. Nucleation involves transitions of ±1 in n and x values. (Adapted from Ref. [189], copyright (1978), with permission from Elsevier.)

to involve a change of no more than one in the value of n or x. The void can grow by gain of a vacancy, or shrink by loss (emission) of a vacancy or gain (capture) of an interstitial; thermal emission of an interstitial is energetically improbable. The void can also lose or gain a gas atom. For the usual case of helium, its solubility is so low that loss of a gas atom is very unlikely by a thermal fluctuation; re-solution of helium atoms does occur, however, by displacement under irradiation [189]. In some metals, divacancies are stable and highly mobile. The inclusion of divacancy migration and capture in cluster dynamics leads to an acceleration of predicted void nucleation rate by three to six orders of magnitude [193].

The work W_{void} of forming a void embryo, characterized by n and x, is given by Russell [189] as

$$W_{\mathrm{void}}(n,x) = -nk_{\mathrm{B}}T\ \ln(s_{\mathrm{vac}}) + (36\pi\bar{v}^2)^{1/3}\sigma n^{2/3} + xk_{\mathrm{B}}T\ \ln(x/x_{\mathrm{eq}}) - xk_{\mathrm{B}}T, \tag{25}$$

where $\bar{v}$ is the atomic volume, σ is the energy per unit area of the void–matrix interface, and the void gas constant x_{eq} is the number of gas atoms in the void that would be in equilibrium with concentration of gas dissolved in the matrix. The vacancy supersaturation s_{vac} is given by

$$s_{\mathrm{vac}} = \frac{C_{\mathrm{vac}}}{C_{\mathrm{vac}}^{\mathrm{eq}}}, \tag{26}$$

where C_{vac} is the concentration of vacancies and $C_{\mathrm{vac}}^{\mathrm{eq}}$ is the equilibrium concentration at the given temperature. Equation (25) was derived assuming ideal gas behavior, but at high gas pressure, such as may be found in some cases, it is better to assume van der Waals behavior. As shown by Parker and Russell [194], this reduces $W_{\mathrm{void}}(n, x)$. Recent work has noted that the pressure in helium-filled bubbles may greatly exceed the value $2\sigma/r$ expected from the interfacial tension at a bubble radius r. The mechanical stability limit beyond which the matrix flows plastically is at a pressure of approximately 20% of the shear modulus of the matrix [187].

As presented by Russell [189], the fluxes of the reactions shown in Figure 36 depend on the form of $W_{\mathrm{void}}(n, x)$, and on rate constants. The existence and nature of possible critical embryos can be determined from nodal lines. In the n-x plane, a locus of $\dot{n} = 0$ can be found along which the embryo is equally likely to lose or gain a vacancy while keeping x constant. Equivalently, the locus of $\dot{x} = 0$ shows where loss or gain of a gas atom is equally likely at a given n. Figure 37 shows possible combinations of loci. In each case, embryos with x above the $\dot{n} = 0$ line would gain vacancies spontaneously and vice versa. Embryos with x above the $\dot{x} = 0$ line would lose gas atoms spontaneously and vice versa. Where the loci cross, the embryo is in equilibrium with regard to both vacancies and gas atoms. The arrows on Figure 37 show the directions for

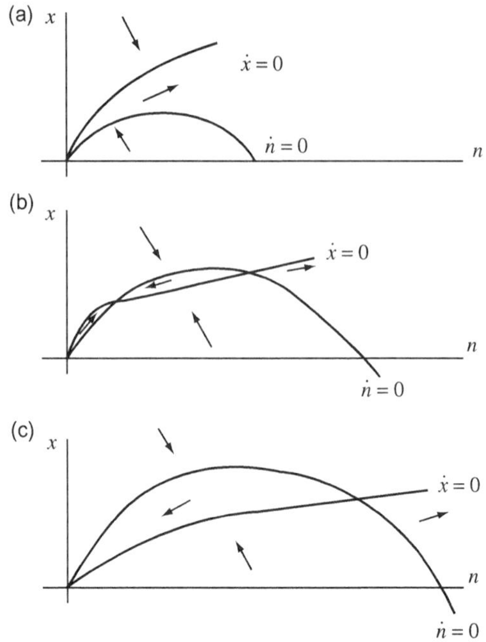

Fig. 37 The variation in work of formation of a cavity embryo represented by the nodal line $\dot{n} = 0$ along which, at given x, n is equally likely to increase or decrease, and by its counterpart $\dot{x} = 0$. The cases are: (a) spontaneous nucleation when $\psi > 1$; (b) gas-assisted nucleation when $\psi < 1$; (c) homogeneous nucleation when $\psi \ll 1$. (Reprinted from [189], copyright (1978), with permission from Elsevier.)

spontaneous evolution and show whether the equilibria are metastable or unstable, the latter corresponding to a nucleation barrier (in the manner shown for a one-component system in Chapter 6, Figure 16). The cases in Figure 37 correspond to different values of the dimensionless parameter ψ, given by

$$\psi = \frac{9k_x^+ k_B^2 T^2 \ln(s_{\text{eff}})}{(36\pi\bar{v}^2)^{2/3}\sigma^2 K_x^c}, \tag{27}$$

where k_x^+ is the jump frequency for addition of a gas atom, K_x^c is the probability per unit time of displacement-induced re-solution of a gas atom into the matrix, and k_B and T have their usual meanings. The *effective vacancy supersaturation* s_{eff} is given by

$$s_{\text{eff}} = s_{\text{vac}}\left(1 - \frac{k_{\text{int}}^+}{k_{\text{vac}}^+}\right), \tag{28}$$

where k_{int}^+ and k_{vac}^+ are, respectively, the jump frequencies for self-interstitial atoms and vacancies joining the embryo.

When ψ>1 (Figure 37a), there is no nucleation barrier. Nucleation is spontaneous following a path between the nodal lines. In the early stages, the void is built by simultaneous gain of vacancies and gas atoms. At low displacement rates and low vacancy supersaturations nucleation, though spontaneous, may be slow because of the need for gas atoms. The $\psi > 1$ regime is most likely to be found at low temperature when s_{eff} is large. As noted in early work [195] and confirmed by recent computer simulations, divacancies containing two helium atoms are very stable, and nucleation in this regime can be modeled as an irreversible aggregation of helium atoms each bound in a vacancy [196].

When $\psi < 1$ (Figure 37b), the nodal lines cross twice. As shown readily by the arrows, the first intersection (at lower n and x) represents the favored size and gas content of metastable voids. The second intersection shows the unstable equilibrium representing the nucleation barrier. In this case, termed *gas-assisted nucleation* [189], the work of nucleation of a freely growing void is clearly reduced by the prior existence of the metastable void.

When $\psi \ll 1$ (Figure 37c), the first intersection of the nodal lines is at such low values of n and x that a standard homogeneous nucleation analysis can be applied. In each of the three cases in Figure 37, an operational definition is that a void is nucleated when it can grow freely by acquisition of vacancies without the need for more gas atoms [189].

We have so far considered the role of impurities only when they are species, such as helium, with very low solubility in the matrix and forming a gas in the voids. Other impurities may affect void nucleation by segregating to the void/matrix interface and lowering the interfacial energy [189, 193]. In some cases, for example, irradiation of aluminum in which there is some transmutation to silicon, which then segregates to the interface, the adsorbed layer becomes a hard shell coating the void [197]; this shell is likely to inhibit further growth by preventing the void from acting as a sink for vacancies.

Under irradiation, the nucleation of voids is discernable after an incubation time and then quickly stops, giving a well-defined final population density. This is interpreted as related to the supersaturation of vacancies, which initially rises and then falls back to a lower steady-state value in the presence of voids [187]. Differences in behavior between ccp and bcc metals, and between pure metals and alloys, are largely attributable to different clustering tendencies affecting the degrees of vacancy supersaturation [165]. At a given irradiation flux, and therefore given rate of production of helium, the density of voids shows a clear dependence on temperature with two regimes (Figure 38). At a given temperature, an increased helium production rate leads to a higher density of smaller voids. These effects arise from the competition between void nucleation and void growth. This competition has been

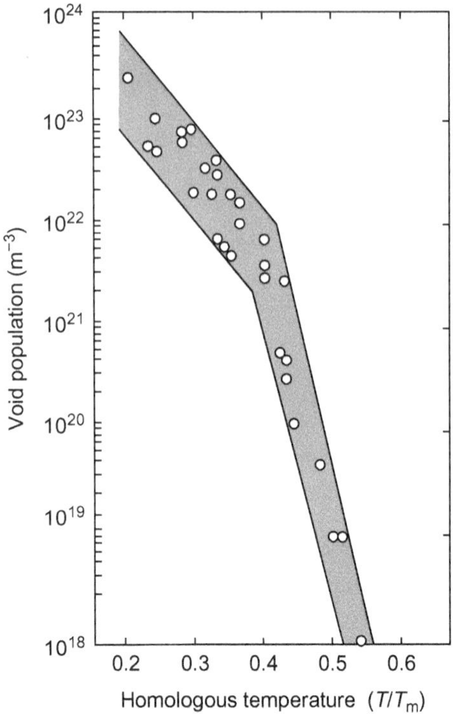

Fig. 38 Temperature dependence of void densities observed in neutron-irradiated molybdenum. (Adapted from Ref. [165], copyright (2002), with permission from Elsevier.)

analyzed by Trinkaus and Singh [165, 187], and some qualitative aspects of their analysis are summarized here. In the low-temperature regime, as discussed earlier in connection with Figure 37a, the key mobile species are helium atoms in solution in vacancies. In steady state, the production rate $\mathscr{P}_{He}$ of helium atoms must balance their rate of capture by voids, giving

$$\mathscr{P}_{He} \propto D_{He} C_{He} N_{void}, \tag{29}$$

where D_{He} and C_{He} are the diffusivity and concentration of the dissolved helium, and N_{void} is the number density of voids. Void nucleation requires only the aggregation of two helium atoms. It must start to saturate when the population of nucleated voids is sufficiently high that a helium atom has as much chance of meeting a void as another helium atom, that is, when $C_{He} \propto N_{void}$. Thus nucleation is self-limiting, with

$$N_{void} \propto \left(\frac{\mathscr{P}_{He}}{D_{He}} \right)^{1/2}. \tag{30}$$

The temperature dependence of N_{void} should, therefore, show an effective activation energy one-half that for diffusion of helium, and this is the case [187]. The relationship between N_{void} and $\mathscr{P}_{\text{He}}$ has also been roughly verified, the larger population of voids at higher fluxes forcing individual voids to be smaller. The high-temperature regime in Figure 38 is one in which small helium clusters are no longer thermally stable. Consistent with Figure 37c, there is now a well-defined nucleation barrier and nucleation tends to occur at a critical value of C_{He}. Therefore, we have

$$N_{\text{void}} \propto \frac{\mathscr{P}_{\text{He}}}{D_{\text{He}} C_{\text{He}}} \tag{31}$$

giving a much stronger temperature dependence.

The early attainment of a well-defined void density under steady-state conditions facilitates the development of *void lattices*. In this remarkable phenomenon, voids originally in disordered arrays align to form superlattices analogous to the lattice of the crystal structure of the metallic matrix. In irradiated molybdenum, for example, the voids form a bcc lattice like that of the host metal, but with a lattice parameter some 50 times larger [198].

All the earlier treatments of void nucleation have ignored the possible catalytic effects of extended microstructural features such as grain boundaries and precipitates. Such effects do not appear to have been analyzed quantitatively [187]. It is known, however, that preferential nucleation and growth of voids at grain boundaries can give rise to void-free zones, analogous to the PFZs discussed in Section 4.1 [187].

6. SUMMARY

Transformations in the solid state show in several examples just how complex nucleation can be. In particular, the nucleation often occurs in multiple stages and may involve coupled fluxes. A simple application of classical nucleation theory is then rarely, if ever, appropriate. While quantitative analysis of nucleation kinetics has been achieved for some complex cases, it has to be recognized that the absence of such analysis has not prevented very wide-ranging manipulation of nucleation in solid-state processing of advanced materials.

REFERENCES

[1] S.D. Stookey, Corning develops new ceramic material, Am. Ceram. Soc. Bull. **36** (1957) 279–280.
[2] G.H. Beall, Design and properties of glass-ceramics, Ann. Rev. Mater. Sci. **22** (1992) 91–119.

[3] W. Höland, V. Rheinberger, M. Schweiger, Control of nucleation in glass ceramics, Phil. Trans. Roy. Soc. Lond. A **361** (2003) 575–589.
[4] S.D. Stookey, Photosensitive glass, Ind. Eng. Chem. **41** (1949) 856–861.
[5] G.H. Beall, B.R. Karstetter, H.L. Rittler, Crystallization and chemical strengthening of stuffed β-quartz glass-ceramics, J. Am. Ceram. Soc. **50** (1967) 181–190.
[6] V. Maier, G. Müller, Mechanism of oxide nucleation in lithium aluminosilicate glass-ceramics, J. Am. Ceram. Soc. **70** (1987) C176–C178.
[7] D.R. Uhlmann, A.G. Kolbeck, Phase separation and the revolution in concepts of glass structure, Phys. Chem. Glasses **17** (1976) 146–158.
[8] V. Gerold, G. Kostorz, Small-angle scattering applications to materials science, J. Appl. Cryst. **11** (1978) 376–404.
[9] E.D. Zanotto, A.F. Craievich, P.F. James, SAXS and TEM studies of phase separation in BaO-SiO_2 glasses, J. Physique Coll. **43**(9), (1982) 107–110.
[10] S. Bras, A. Craievich, J.M. Sanchez, C. Williams, E.D. Zanotto, Small-angle X-ray-scattering study of phase-separation in glasses using a new position-sensitive detector, Nucl. Instrum. Meth. Phys. Res. **208** (1983) 489–494.
[11] E.D. Zanotto, P.F. James, A.F. Craievich, The effects of amorphous phase separation on crystal nucleation kinetics in BaO-SiO_2 glasses: III. Isothermal treatments at 718°C to 760°C — small-angle X-ray scattering results, J. Mater. Sci. **21** (1986) 3050–3064.
[12] K. Hono, K. Hirano, Early stages of decomposition of alloys (spinodal or nucleation), Phase Trans. B **10** (1987) 223–255.
[13] S. Schneider, U. Geyer, P. Thiyagarajan, W.L. Johnson, Time and temperature dependence of decomposition and crystallization in a multicomponent bulk metallic glass forming alloy, Mater. Sci. Forum **235** (1997) 337–342.
[14] T. Takei, Y. Kameshima, A. Yasumori, K. Okada, N. Kumada, N. Kinomura, SAXS analysis of textures formed by phase separation and crystallization of Al_2O_3-SiO_2 glasses, J. Non-Cryst. Solids **282** (2001) 265–277.
[15] K. Hono, T. Sakurai, Atom-probe studies of nanostructured alloys, Appl. Surf. Sci. **87–88** (1995) 166–178.
[16] R. Busch, S. Schneider, A. Peker, W.L. Johnson, Decomposition and primary crystallization in undercooled $Zr_{41.2}Ti_{13.8}Cu_{12.5}Ni_{10.0}Be_{22.5}$ melts, Appl. Phys. Lett. **67** (1995) 1544–1546.
[17] M.K. Miller, R.B. Schwarz, Y. He, Decomposition in $Pd_{40}Ni_{40}P_{20}$ bulk metallic glass, Mater. Res. Soc. Symp. Proc. **554** (1999) 9–14.
[18] H.G. Read, K. Hono, A.P. Tsai, A. Inoue, Preliminary atom probe studies of PdNi(Cu)P supercooled liquids, Mater. Sci. Eng. A **226** (1997) 453–457.
[19] J.W. Cahn, On spinodal decomposition, Acta Metall. **9** (1961) 795–801.
[20] J.W. Cahn, On spinodal decomposition in cubic crystals, Acta Metall. **10** (1962) 179–183.
[21] W. Höland, V. Rheinberger, M. Frank, Mechanism of nucleation and controlled crystallization of needle-like apatite in glass ceramics of the SiO_2-Al_2O_3-K_2O-CaO-P_2O_5 system, J. Non-Cryst. Solids **253** (1999) 170–177.
[22] J.F. MacDowell, G.H. Beall, Immiscibility and crystallization in Al_2O_3–SiO_2 glasses, J. Am. Ceram. Soc. **52** (1969) 17–25.
[23] T. Kokubo, Bioactive glass ceramics: properties and applications, Biomaterials **12** (1991) 155–163.
[24] W. Vogel, W. Höland, The development of bioglass ceramics for medical application, Angew. Chem. Int. Ed. Engl. **26** (1987) 527–544.
[25] S.K. Das, J.H. Perepezko, R.I. Wu, G. Wilde, Undercooling and glass formation in Al-based alloys, Mater. Sci. Eng. A **304-306** (2001) 159–165.
[26] Y. Yoshizawa, K. Yamauchi, Fe-based soft magnetic alloys composed of ultrafine grain structure, Mater. Trans. JIM **31** (1990) 307–314.

[27] E. Cini, B. Vinet, P.J. Desré, A thermodynamic approach to homogeneous nucleation via fluctuations of concentration in binary liquid alloys, Philos. Mag. A **80** (2000) 955–966.
[28] P.J. Desré, E. Cini, B. Vinet, Homophase-fluctuation-mediated mechanism of nucleation in multicomponent liquid alloys and glass-forming ability, J. Non-Cryst. Solids **288** (2001) 210–217.
[29] K.F. Kelton, A new model for nucleation in bulk metallic glasses, Philos. Mag. Lett. **77** (1998) 337–343.
[30] A.L. Greer, Stress effects on interdiffusion in amorphous multilayers, Defect Diffusion Forum **129–130** (1996) 163–180.
[31] D.R. Allen, J.C. Foley, J.H. Perepezko, Nanocrystal development during primary crystallization of amorphous alloys, Acta Mater. **46** (1998) 431–440.
[32] K. Hono, K. Hiraga, Q. Wang, A. Inoue, T. Sakurai, The microstructure evolution of a $Fe_{73.5}Si_{13.5}B_9Nb_3Cu_1$ nanocrystalline soft magnetic material, Acta Metall. Mater. **40** (1992) 2137–2147.
[33] M. Ohnuma, K. Hono, S. Linderoth, J.S. Pedersen, Y. Yoshizawa, H. Onodera, Small-angle neutron scattering and differential scanning calorimetry studies of the copper clustering stage of Fe-Si-B-Nb-Cu nanocrystalline Alloys, Acta Mater. **48** (2000) 4783–4790.
[34] S.H. Kim, M. Matsuura, M. Sakurai, K. Suzuki, Local structure changes around Cu atoms in the early-stage of nanocrystalline formation of amorphous $Fe_{73.5}Cu_1Nb_3Si_{13.5}B_9$, Jap. J. Appl. Phys. **32** (1993) 676–678.
[35] J.D. Ayers, V.G. Harris, J.A. Sprague, W.T. Elam, On the role of Cu and Nb in the formation of nanocrystals in amorphous $Fe_{73.5}Nb_3Cu_1Si_{13.5}B_9$, Appl. Phys. Lett. **64** (1994) 974–976.
[36] J.D. Ayers, V.G. Harris, J.A. Sprague, W.T. Elam, H.N. Jones, On the formation of nanocrystals in the soft magnetic alloy $Fe_{73.5}Nb_3Cu_1Si_{13.5}B_9$, Acta Mater. **46** (1998) 1861–1874.
[37] Y. Yoshizawa, S. Oguma, K. Yamaguchi, New Fe-based soft magnetic alloys composed of ultrafine grain structure, J. Appl. Phys. **64** (1988) 6044–6046.
[38] K. Hono, D.H. Ping, M. Ohnuma, H. Onodera, Cu clustering and Si partitioning in the early crystallization stage of an $Fe_{73.5}Si_{13.5}B_9Nb_3Cu_1$ amorphous alloy, Acta Mater. **47** (1999) 997–1006.
[39] X.L. Wang, J. Almer, C.T. Liu, Y.D. Wang, J.K. Zhao, A.D. Stoica, D.R. Haeffner, W.H. Wang, In situ synchrotron study of phase transformation behaviors in bulk metallic glass by simultaneous diffraction and small angle scattering, Phys. Rev. Lett. **91** (2003) 265501/1–4.
[40] A.K. Gangopadhyay, T.K. Croat, K.F. Kelton, The effect of phase separation on subsequent crystallization in $Al_{88}Gd_6La_2Ni_4$, Acta Mater. **48** (2000) 4035–4043.
[41] K.F. Kelton, T.K. Croat, A.K. Gangopadhyay, L.-Q. Xing, A.L. Greer, M. Weyland, X. Li, K. Rajan, Mechanisms for nanocrystal formation in metallic glasses, J. Non-Cryst. Solids **317** (2003) 71–77.
[42] J. Antonowicz, A.R. Yavari, W.J. Botta, P. Panine, Phase separation and nanocrystallization in $Al_{92}Sm_8$ metallic glass, Philos. Mag. **46** (2006) 4235–4242.
[43] B. Radiguet, D. Blavette, N. Wanderka, J. Banhart, K.L. Sahoo, Segregation controlled nanocrystallization in Al-Ni-La metallic glass, Appl. Phys. Lett. **92** (2008) 103126/1–4.
[44] L. Yang, M.K. Miller, X.L. Wang, C.T. Liu, A.D. Stoica, D. Ma, J. Almer, D. Shi, Nanoscale solute partitioning in bulk metallic glasses, Adv. Mater. **21** (2009) 305–308.
[45] N. Tian, M. Ohnuma, T. Ohkubo, K. Hono, Primary crystallization of an $Al_{88}Gd_6Er_2Ni_4$ metallic glass, Mater. Trans. **46** (2005) 2880–2885.
[46] A.L. Greer, Crystallisation kinetics of $Fe_{80}B_{20}$ glass, Acta Metall. **30** (1982) 171–192.

[47] K.F. Kelton, A.L. Greer, Transient nucleation effects in glass formation, J. Non-Cryst. Solids **79** (1986) 295–309.
[48] A. Inoue, T. Zhang, N. Nishiyama, K. Ohba, T. Masumoto, Preparation of 16 mm diameter rod of amorphous $Zr_{65}Al_{7.5}Ni_{10}Cu_{17.5}$ alloy, Mater. Trans. JIM **33** (1993) 1234–1237.
[49] T. Egami, Y. Waseda, Atomic size effect on the formability of metallic glasses, J. Non-Cryst. Solids **64** (1984) 113–134.
[50] Z. Altounian, J.O. Strom-Olsen, Superconductivity and spin fluctuations in *M*-Zr metallic glasses (M = Cu, Ni, Co, and Fe), Phys. Rev. B **27** (1983) 4149–4156.
[51] F.L. Yang, R.E. Somekh, W.C. Shih, A.L. Greer, Measurements and simulation of asymmetric interdiffusion in amorphous Ni-Zr multilayers, Mater. Sci. Forum **179–181** (1995) 807–812.
[52] G.B. Stephenson, Deformation during interdiffusion, Acta Metall. **36** (1988) 2663–2683.
[53] A.L. Greer, N. Karpe, J. Bøttiger, Diffusional aspects of the solid state amorphizing reaction, J. Alloys Comp. **194** (1993) 199–211.
[54] M. Baricco, L. Battezzati, Nanocrystals through crystallization of amorphous alloys, Mater. Sci. Forum **179–181** (1995) 597–602.
[55] U. Köster, M. Blank-Bewersdorff, Crystallization of zirconium-based metallic glasses, J. Less-Common Metals **140** (1988) 7–16.
[56] W. Welnic, A. Pamungkas, R. Detemple, C. Steimer, S. Blugel, M. Wuttig, Unravelling the interplay of local structure and physical properties in phase-change materials, Nature Mater. **5** (2006) 56–62.
[57] G.F. Zhou, Materials aspects in phase change optical recording, Mater. Sci. Eng. A **304–306** (2001) 73–80.
[58] C.B. Peng, L. Cheng, M. Mansuripur, Experimental and theoretical investigations of laser-induced crystallization and amorphization in phase-change optical recording media, J. Appl. Phys. **82** (1997) 4183–4191.
[59] J. Kalb, F. Spaepen, M. Wuttig, Atomic force microscopy measurements of crystal nucleation and growth rates in thin films of amorphous Te alloys, Appl. Phys. Lett. **84** (2004) 5240–5242.
[60] J.A. Kalb, F. Spaepen, M. Wuttig, Kinetics of crystal nucleation in undercooled droplets of Sb- and Te-based alloys used for phase change recording, J. Appl. Phys. **98** (2005) 054910/1–7.
[61] J.A. Kalb, C.Y. Wen, F. Spaepen, H. Dieker, M. Wuttig, Crystal morphology and nucleation in thin films of amorphous Te alloys used for phase change recording, J. Appl. Phys. **98** (2005) 054902/1–6.
[62] S. Ziegler, M. Wuttig, Nucleation of AgInSbTe films employed in phase-change media, J. Appl. Phys. **99** (2006) 064907/1–4.
[63] M. Wuttig, Phase-change materials — Towards a universal memory?, Nat. Mater. **4** (2005) 265–266.
[64] A.L. Greer, N.D. Mathur, Materials science — Changing face of the chameleon, Nature **437** (2005) 1246–1247.
[65] D.H. Kim, F. Merget, M. Laurenzis, P.H. Bolivar, H. Kurz, Electrical percolation characteristics of $Ge_2Sb_2Te_5$ and Sn doped $Ge_2Sb_2Te_5$ thin films during the amorphous to crystalline phase transition, J. Appl. Phys. **97** (2005) 083538/1–5.
[66] D.R. Nelson, B.I. Halperin, Dislocation-mediated melting in two dimensions, Phys. Rev. B **19** (1979) 2457–2484.
[67] A.P. Young, Melting and the vector Coulomb gas in two dimensions, Phys. Rev. B **19** (1979) 1855–1866.
[68] F.A. Lindemann, Uber die Berechnung molekylarer Eigenfrequenzen, Z. Phys. **11** (1910) 609–615.
[69] M. Born, Thermodynamics of crystals and melting, J. Chem. Phys. **7** (1939) 591–603.

[70] D.R. Uhlmann, On the internal nucleation of melting, J. Non-Cryst. Solids **41** (1980) 347–357.
[71] B. Rethfeld, K. Sokolowski-Tinten, D. von der Linde, Ultrafast thermal melting of laser-excited solids by homogeneous nucleation, Phys. Rev. B **65** (2002) 092103/1–4.
[72] M.N. Krivoguz, G.E. Norman, V.V. Stegailov, A.A. Valuev, Superheating of solid metal prior to electric explosion of wires at fast energy deposition, J. Phys. A **36** (2003) 6041–6048.
[73] K. Lu, Y. Li, Homogeneous nucleation catastrophe as a kinetic stability limit for superheated crystal, Phys. Rev. Lett. **80** (1998) 4474–4477.
[74] F. Delogu, Homogeneous melting of metals with different crystalline structure, J. Phys. Cond. Mat. **18** (2006) 5639–5653.
[75] E.A. Olson, M. Yu. Efremov, M. Zhang, Z. Zhang, L.H. Allen, Size-dependent melting of Bi nanoparticles, J. Appl. Phys. **97** (2005) 034304/1–9.
[76] M. Faraday, Note on regelation, Proc. Roy. Soc. Lond. **10** (1860) 440–450.
[77] J.W.M. Frenken, J.F. van der Veen, Observation of surface melting, Phys. Rev. Lett. **54** (1985) 134–137.
[78] B. Pluis, A.W. Denier van der Gon, J.W.M. Frenken, J.F. van der Veen, Crystal-face dependence of surface melting, Phys. Rev. Lett. **59** (1987) 2678–2681.
[79] D.W. Oxtoby, New perspectives on freezing and melting, Nature **347** (1990) 725–730.
[80] J. Daeges, H. Gleiter, J.H. Perepezko, Superheating of metal crystals, Phys. Rev. A **119** (1986) 79–82.
[81] Z.H. Jin, H.W. Sheng, K. Lu, Melting of Pb clusters without free surfaces, Phys. Rev. B **60** (1999) 141–149.
[82] J. Zhong, Z.H. Jin, K. Lu, Melting, superheating and freezing behaviour of indium interpreted using a nucleation-and-growth model, J. Phys. Cond. Matt. **13** (2001) 11443–11452.
[83] F. Banhart, E. Hernández, M. Terrones, Extreme superheating and supercooling of encapsulated metals in fullerenelike shells, Phys. Rev. Lett. **90** (2003) 185502/1–4.
[84] L. Zhang, Z.H. Jin, L.H. Zhang, M.L. Sui, K. Lu, Superheating of confined Pb thin films, Phys. Rev. Lett. **85** (2000) 1484–1487.
[85] B.R. Cuenya, W. Keune, W.A. Adeagbo, P. Entel, Thermal stability, atomic vibrational dynamics, and superheating of confined interfacial Sn layers in Sn/Si multilayers, Phys. Rev. B **73** (2006) 045311/1–11.
[86] J. Tyndall, On some physical properties of ice, Phil. Trans. Roy. Soc. Lond. **148** (1858) 211–229.
[87] S. Takeya, Growth of internal melt figures in superheated ice, Appl. Phys. Lett. **88** (2006) 074103/1–3.
[88] G.L. Allen, R.A. Bayles, W.W. Gile, W.A. Jesser, Small particle melting of pure metals, Thin Solid Films **144** (1986) 297–308.
[89] D.L. Zhang, B. Cantor, Melting behaviour of In and Pb particles embedded in an Al matrix, Acta Metall. Mater. **39** (1991) 1595–1602.
[90] S.N. Luo, T.J. Ahrens, T. Çağın, A. Strachan, W.A. Goddard, III, D.C. Swift, Maximum superheating and undercooling: systematics, molecular-dynamics simulations, and dynamic experiments, Phys. Rev. B **68** (2003) 134206/1–11.
[91] M. Schmeisser, H. Iglev, A. Laubereau, Maximum superheating of bulk ice, Chem. Phys. Lett. **442** (2007) 171–175.
[92] J.D. Eshelby, The continuum theory of lattice defects, Solid State Phys. **3** (1956) 79–144.
[93] X.M. Bai, M. Li, Nature and extent of melting in superheated solids: liquid-solid coexistence model, Phys. Rev. B **72** (2005) 052108/1–4.
[94] Z.H. Jin, K. Lu, Melting of surface-free bulk single crystals, Philos. Mag. Lett. **78** (1998) 29–35.
[95] S.N. Luo, L. Zheng, A. Strachan, D.C. Swift, Melting dynamics of superheated argon: nucleation and growth, J. Chem. Phys. **126** (2007) 034505/1–7.

[96] X. Liu, C.G. Meng, C.H. Liu, Molecular-dynamics study on superheating of Pd at high heating rates, Phase Trans. **79** (2006) 249–259.
[97] M. Forsblom, G. Grimvall, How superheated crystals melt, Nat. Mater. **4** (2005) 388–390.
[98] L.W. Wang, Q. Wang, K.Q. Lu, Melting of superheated crystals initiated on vacancies, Philos. Mag. Lett. **87** (2007) 19–24.
[99] Z.H. Jin, P. Gumbsch, K. Lu, E. Ma, Melting mechanisms at the limit of superheating, Phys. Rev. Lett. **87** (2001) 055703/1–4.
[100] A.M. Alsayed, M.F. Islam, J. Zhang, P.J. Collings, A.G. Yodh, Premelting at defects within bulk colloidal crystals, Science **309** (2005) 1207–1210.
[101] J.Q. Broughton, G.H. Gilmer, Thermodynamic criteria for grain-boundary melting: a molecular-dynamics study, Phys. Rev. Lett. **56** (1986) 2692–2695.
[102] S.R. Phillpot, J.F. Lutsko, D. Wolf, S. Yip, Molecular-dynamics study of lattice-defect-nucleated melting in silicon, Phys. Rev. B **40** (1989) 2831–2840.
[103] W. Fan, X.G. Gong, Superheated melting of grain boundaries, Phys. Rev. B **72** (2005) 064121/1–5.
[104] H. Lu, N.X. Sun, Y.C. Zhou, Y.F. Wang, Superheating of bulk polycrystalline selenium, Philos. Mag. Lett. **76** (1997) 49–56.
[105] J. Huang, M. Meyer, V. Pontikis, Core structure of a dissociated easy-glide dislocation in copper investigated by molecular-dynamics, Phys. Rev. B **42** (1990) 5495–5504.
[106] D. Ma, Y. Li, Heterogeneous nucleation catastrophe on dislocations in superheated crystals, J. Phys. Cond. Matt. **12** (2000) 9123–9128.
[107] Q.S. Mei, Z.H. Jin, K. Lu, The kinetic limit of superheating induced by semicoherent interfaces, Philos. Mag. Lett. **85** (2005) 203–211.
[108] J.W. Martin, *Precipitation Hardening*, Pergamon, Oxford (1986).
[109] G.W. Lorimer, Nucleation and growth transformations in the solid state, in: *Encyclopedia of Materials Science and Engineering*, Ed. M.B. Bever, Vol. 4, Pergamon, Oxford (1986), pp. 3255–3260.
[110] D.A. Porter, K.E. Easterling, *Phase Transformations in Metals and Alloys*, 2nd edn, Chapman & Hall, London (1992).
[111] A. Kelly, R.B. Nicholson, Precipitation hardening, Prog. Mater. Sci. **10** (1963) 149–391.
[112] E. Hornbogen, Nucleation of precipitates in defect solid solutions, in: *Nucleation*, Ed. A.C. Zettlemoyer, Marcel Dekker, New York (1969), pp. 309–378.
[113] K.C. Russell, H.I. Aaronson, Sequences of precipitate nucleation, J. Mater. Sci. **10** (1975) 1991–1999.
[114] K.C. Russell, H.I. Aaronson (eds.), *Precipitation Processes in Solids*, Met. Soc. AIME, Warrendale PA, 1978.
[115] K.C. Russell, Nucleation in solids: the induction and steady state effects, Adv. Colloid Inter. Sci. **13** (1980) 205–318.
[116] R.D. Doherty, Diffusive phase transformations in the solid state, in: *Physical Metallurgy*, 4th edn, Eds. R.W. Cahn, P. Haasen, Vol. 2, Elsevier, Amsterdam (1996), pp. 1363–1505.
[117] J.W. Christian, *The Theory of Transformations in Metals and Alloys, Parts I ands II*, 3rd edn., Elsevier, Oxford (2002).
[118] G.W. Lorimer, Precipitation in aluminium alloys, in: *Precipitation Processes in Solids*, Eds. K.C. Russell, H.I. Aaronson, Met. Soc. AIME, Warrendale PA (1978), pp. 87–119.
[119] A. Guinier, Structure of age-hardened aluminium-copper alloys, Nature **142** (1938) 569–570.
[120] G.D. Preston, The diffraction of X-rays by age-hardening aluminium copper alloys, Proc. Roy. Soc. Lond. A **167** (1938) 526–538; Pl. 14–18.

[121] R.B. Nicholson, J. Nutting, Direct observation of the strain field produced by coherent precipitated particles in an age-hardened alloy, Philos. Mag. **3** (1958) 531–535; Pl. 22–24.
[122] K. Hirano, K. Hono, Field-ion studies of structure of GP zones, Ann. Rev. Mater. Sci. **18** (1988) 351–380.
[123] S.P. Ringer, K. Hono, Microstructural evolution and age hardening in aluminium alloys: atom probe field-ion microscopy and transmission electron microscopy studies, Mater. Char. **44** (2000) 101–131.
[124] Y. Jing, C. Li, Z. Du, F. Wang, Y. Song, The thermodynamic analysis of Guinier-Preston zones in aged supersaturated Al-Cu alloys, Comp. Coup. Phase Diag. Thermochem. **32** (2008) 164–170.
[125] E. Matsubara, J.B. Cohen, The G.P. zones in Al-Cu alloys—I, Acta Metall. **33** (1985) 1945–1955.
[126] E. Matsubara, J.B. Cohen, The G.P. zones in Al-Cu alloys—II, Acta Metall. **33** (1985) 1957–1969.
[127] T. Sato, T. Takahashi, High resolution electron microscopy on the layered structures of GP zones in an Al-1.7at.%Cu alloy, Scripta Metall. **22** (1988) 941–946.
[128] M. Karlík, B. Jouffrey, High resolution electron microscopy study of Guinier-Preston (GP1) zones in Al-Cu based alloys, Acta Mater. **45** (1997) 3251–3263.
[129] T.J. Bastow, A.J. Hill. Guinier-Preston, Guinier-Preston-Bagaryatsky zone reversion in Al-Cu-Mg alloys studied by NMR, Mater. Sci. Forum **519–521** (2006) 1385–1390.
[130] M. Karlík, A. Bigot, B. Jouffrey, P. Auger, S. Belliot, HREM, FIM and tomographic atom probe investigation of Guinier-Preston zones in an Al-1.54at.% Cu alloy, Ultramicroscopy **98** (2004) 219–230.
[131] K. Hono, T. Satoh, K. Hirano, Evidence of multi-layer GP zones in Al-1.7at.%Cu alloy, Philos. Mag. A **53** (1986) 495–504.
[132] G.C. Weatherly, R.B. Nicholson, An electron microscope investigation of the interfacial structure of semi-coherent precipitates, Philos. Mag. **17** (1968) 801–831.
[133] K.C. Russell, Nucleation in the solid state, in: *Encyclopedia of Materials Science and Engineering*, Ed. M.B. Bever, Vol. 4, Pergamon, Oxford (1986), pp. 3263–3267.
[134] C. Wolverton, First-principles prediction of equilibrium precipitate shapes in Al-Cu alloys, Philos. Mag. Lett. **79** (1999) 683–690.
[135] J. Wang, C. Wolverton, S. Müller, Z.K. Liu, L.Q. Chen, First-principles growth kinetics and morphological evolution of Cu nanoscale particles in Al, Acta Mater. **53** (2005) 2759–2764.
[136] S.Y. Hu, M.I. Baskes, M. Stan, L.Q. Chen, Atomistic calculations of interfacial energies, nucleus shape and size of θ' precipitates in Al-Cu alloys, Acta Mater. **54** (2006) 4699–4707.
[137] C. Ravi, C. Wolverton, First-principles study of crystal structure and stability of Al–Mg–Si–(Cu) precipitates, Acta Mater. **52** (2004) 4213–4227.
[138] P.E. Marth, H.I. Aaronson, G.W. Lorimer, T.L. Bartel, K.C. Russell, Application of heterogeneous nucleation theory to precipitate nucleation at GP zones, Metall. Trans. **7A** (1976) 1519–1528.
[139] K.C. Russell, The role of excess vacancies in precipitation, Scripta Metall. **3** (1969) 313–316.
[140] G.W. Lorimer, R.B. Nicholson, Further results on the nucleation of precipitates in the Al-Zn-Mg system, Acta Metall. **14** (1966) 1009–1013.
[141] P.N.T. Unwin, G.W. Lorimer, R.B. Nicholson, The origin of the grain boundary precipitate free zone, Acta Metall. **17** (1969) 1363–1377.
[142] S. Chang, J.E. Morral, The influence of quenching rate on precipitate-free-zones in an Al-Zn-Mg alloy, Acta Metall. **23** (1975) 685–689.

[143] R.N. Lumley, I.J. Polmear, A.J. Morton, Control of secondary precipitation to improve the performance of aluminium alloys, Mater. Sci. Forum **396–402** (2002) 893–898.
[144] *Metals Handbook* (8th edn.), Vol. 8; Metals Park, OH: ASM (1973).
[145] C.M. Wayman, H.K.D.H. Bhadeshia, Phase transformations, non-diffusive, in: *Physical Metallurgy*, 4th edn, Eds. R.W. Cahn, P. Haasen, Vol. 2, Elsevier, Amsterdam (1996), pp. 1507–1554.
[146] H.K.D.H. Bhadeshia, *Bainite in Steels*, Inst. of Materials, London (1992).
[147] M. Enomoto, H.I. Aaronson, On the critical nucleus composition of ferrite in an Fe-C-Mn alloy, Metall. Trans A **17A** (1986) 1381–1384.
[148] M. Enomoto, H.I. Aaronson, Nucleation kinetics of proeutectoid ferrite at austenite grain boundaries in Fe-C-X alloys, Metall. Trans A **17A** (1986) 1385–1397.
[149] W.F. Lange, III, M. Enomoto, H.I. Aaronson, The kinetics of ferrite nucleation at austenite grain boundaries in Fe-C alloys, Metall. Trans. A **19A** (1988) 427–440.
[150] M. Enomoto, H.I. Aaronson, The kinetics of ferrite nucleation at austenite grain edges in Fe-C and Fe-C-X alloys, Metall. Trans A **17A** (1986) 1399–1407.
[151] M. Militzer, R. Pandi, E.B. Hawbolt, Ferrite nucleation and growth during continuous cooling, Metall. Mater. Trans. A **27A** (1996) 1547–1556.
[152] T. Furuhara, J. Yamaguchi, N. Sugita, G. Miyamoto, T. Maki, Nucleation of proeutectoid ferrite on complex precipitates in austenite, ISIJ Int. **43** (2003) 1630–1639.
[153] H.I. Aaronson, Mechanisms of the massive transformation, Metall. Mater. Trans. A **33A** (2002) 2285–2297.
[154] A. Ali, H.D.K.H. Bhadeshia, Nucleation of Widmanstätten ferrite, Mater. Sci. Technol. **6** (1990) 781–784.
[155] P.L. Harrison, R.A. Farrar, Application of continuous cooling transformation diagrams for welding of steels, Int. Mater. Rev. **34** (1989) 35–51.
[156] G.M. Evans, "The effect of molybdenum on the microstructure and properties of C-Mn all-weld metal deposits," Int. Inst. of Welding, Document II-A-666-86 (1986).
[157] A.O. Kluken, Ø. Grong, J. Hjelen, The origin of transformation textures in steel weld metals containing acicular ferrite, Metall. Trans. A **22A** (1991) 657–663.
[158] A.R. Mills, G. Thewlis, J.A. Whiteman, Nature of inclusions in steel weld metals and their influence on the formation of acicular ferrite, Mater. Sci. Technol. **3** (1987) 1051–1061.
[159] J.M. Dowling, J.M. Corbett, H.W. Kerr, Inclusion phases and the nucleation of acicular ferrite in submerged arc welds in high strength low alloy steels, Metall. Trans. A **17A** (1986) 1611–1623.
[160] J.M. Gregg, H.K.D.H. Bhadeshia, Titanium-rich mineral phases and the nucleation of bainite, Metall. Mater. Trans. A **25A** (1994) 1603–1611.
[161] R. Chijiiwa, H. Tamehiro, M. Iría, H. Matsuda, H. Mimura, Extra high toughness titanium-oxide steel plates for offshore structures and line pipe, in *Proc. 7th Int. Conf. on Offshore Mechanics and Arctic Engineering*, Houston, TX, pp. 1–8.
[162] J. Gittus, *Irradiation Effects in Crystalline Solids*, Applied Science Publishers, London (1978).
[163] J.A. Hudson, Precipitation under irradiation, in: *Precipitation Processes in Solids*, Eds. K.C. Russell, H.I. Aaronson, Met. Soc. AIME, Warrendale PA (1978), pp. 284–313.
[164] J.O. Stiegler, L.K. Mansur, Radiation effects in structural materials, Ann. Rev. Mater. Sci. **9** (1979) 405–454.
[165] H. Trinkaus, B.N. Singh, Modeling of void nucleation under cascade damage conditions, J. Nucl. Mater. **307–311** (2002) 900–906.
[166] C.H. Woo, B.N. Singh, Production bias due to clustering of point defects in irradiation-induced cascades, Philos. Mag. A **65** (1992) 889–912.
[167] R.H. Silsbee, Focusing in collision problems in solids, J. Appl. Phys. **28** (1957) 1246–1250.

[168] J.P. Hirth, J. Lothe, *Theory of Dislocations*, 2nd edn, Wiley, New York (1982).
[169] R. Schäublin, Z. Yao, N. Baluc, M. Victoria, Irradiation-induced stacking fault tetrahedra in fcc metals, Philos. Mag. **85** (2005) 769–777.
[170] B.N. Singh, S.I. Golubov, H. Trinkaus, D.J. Edwards, M. Eldrup, Review: Evolution of stacking fault tetrahedra and its role in defect accumulation under cascade damage conditions, J. Nucl. Mater. **328** (2004) 77–87.
[171] B.L. Eyre, R. Bullough, On the formation of interstitial loops in B.C.C. metals, Philos. Mag. **12** (1965) 31–39.
[172] T.L. Davis, J.P. Hirth, Nucleation rate of vacancy clusters in crystals, J. Appl. Phys. **37** (1966) 2112–2116.
[173] K.C. Russell, R.W. Powell, Dislocation loop nucleation in irradiated metals, Acta Metall. **21** (1973) 187–193.
[174] H. Tanigawa, A. Kohyama, Y. Katoh, A modeling of radiation induced microstructural evolution under applied stress in austenitic alloys, J. Nucl. Mater. **239** (1996) 80–84.
[175] V.G. Weizer, L.A. Girifalco, Vacancy-vacancy interaction in copper, Phys. Rev. **120** (1960) 837–839.
[176] T.L. Davis, Nucleation rate of vacancy clusters in aluminum, J. Appl. Phys. **38** (1967) 3756–3760.
[177] M.F. Wehner, W.G. Wolfer, Vacancy cluster evolution in metals under irradiation, Philos. Mag. A **52** (1985) 189–205.
[178] S.N. Buckley, S.A. Manthorpe, Dislocation loop nucleation and growth in zirconium-2.5wt.% niobium alloy during 1 MeV electron irradiation, J. Nucl. Mater. **90** (1980) 169–174.
[179] N. Shigenaka, T. Hashimoto, M. Fuse, Effects of alloying elements (Mo, Si) in an austenitic stainless-steel on dislocation loop nucleation under ion irradiation, J. Nucl. Mater. **207** (1993) 46–52.
[180] T. Hashimoto, N. Shigenaka, A theory of dislocation loop formation in binary alloys under irradiation, J. Nucl. Mater. **187** (1992) 161–169.
[181] T. Hashimoto, N. Shigenaka, M. Fuse, Nucleation of dislocation loops during irradiation in binary fcc alloys with different alloy compositions, Mater. Trans. JIM **33** (1992) 366–373.
[182] H. Nakamichi, C. Kinoshita, K. Yasuda, S. Fukada, Formation and growth process of dislocation loops in Zircaloys under electron irradiation, J. Nucl. Sci. Technol. **34** (1997) 1079–1086.
[183] T. Yoshiie, S. Kojima, M. Kiritani, Nucleation of interstitial-type dislocation loops in metals under neutron irradiation, Sci. Rep. Res. Inst. Tohoku Univ. Ser. A **40** (1994) 77–84.
[184] S.J. Zinkle, Anisotropic dislocation loop nucleation in ion-irradiated $MgAl_2O_4$, J. Nucl. Mater. **191** (1992) 645–649.
[185] W.G. Wolfer, B.B. Glasgow, Dislocation evolution in metals under irradiation, Acta Metall. **33** (1985) 1997–2004.
[186] C. Cawthorne, E.J. Fulton, Voids in irradiated stainless steel, Nature **216** (1967) 575–576.
[187] H. Trinkaus, B.N. Singh, Helium accumulation in metals during irradiation — where do we stand? J. Nucl. Mater. **323** (2003) 229–242.
[188] M.P. Surh, J.B. Sturgeon, W.G. Wolfer, Radiation swelling behavior and its dependence on temperature, dose rate, and dislocation structure evolution, J. Nucl. Mater. **336** (2005) 217–224.
[189] K.C. Russell, The theory of void nucleation in metals, Acta Metall. **26** (1978) 1615–1630.
[190] M.P. Surh, J.B. Sturgeon, W.G. Wolfer, Master equation and Fokker-Planck methods for void nucleation and growth in irradiation swelling, J. Nucl. Mater. **325** (2004) 44–52.

[191] A.A. Semenov, C.H. Woo, Classical nucleation theory of microstructure development under cascade-damage irradiation, J. Nucl. Mater. **323** (2003) 192–204.
[192] J.L. Katz, H. Wiedersich, Nucleation of voids in materials supersaturated with vacancies and interstitials, J. Chem. Phys. **55** (1971) 1414–1425.
[193] W.G. Wolfer, A. Si-Ahmed, Nucleation of voids in materials supersaturated with mobile interstitials, vacancies and divacancies, Philos. Mag. A **46** (1982) 723–736.
[194] C.A. Parker, K.C. Russell, Void nucleation in metals assisted by non-ideal inert gas, Scripta Metall. **15** (1981) 643–647.
[195] G.W. Greenwood, A.J.E. Foreman, E.A. Rimmer, The role of vacancies and dislocations in the nucleation and growth of gas bubbles in irradiated fissile material, J. Nucl. Mater. **1** (1959) 305–324.
[196] A.J. Schwartz, M.A. Wall, T.G. Zocco, W.G. Wolfer, Characterization and modelling of helium bubbles in self-irradiated plutonium alloys, Philos. Mag. **85** (2005) 479–488.
[197] K. Farrell, J. Bentley, D.N. Braski, Direct observation of radiation-induced coated cavities, Scripta Metall. **11** (1977) 243–248.
[198] J.H. Evans, Observations of a regular void array in high purity molybdenum irradiated with 2 MeV nitrogen ions, Nature **229** (1971) 403–404.

CHAPTER 15

Interfacial and Thin-Film Reactions

Contents

1. INTRODUCTION

When two phases are brought into contact, there may be atomic or molecular interchange between them. At the interface, the system may be far from thermodynamic equilibrium, the chemical potentials of the component species differing substantially from one phase to the other. Furthermore, with access to the components in both original phases, it may be possible to form other, more stable phases. Thus the interaction between two phases in contact, sometimes called *reactive diffusion*, may include *interdiffusion* and *interfacial reactions* to form a new phase or phases. Nucleation of interfacial reactions is important, but the dynamic environment at an interface poses special problems in analysis as discussed in this chapter.

Reactions can in principle occur at an interface or surface between fluid phases or between a solid and a fluid phase; a prominent example

Pergamon Materials Series, Volume 15
ISSN 1470-1804, DOI 10.1016/S1470-1804(09)01515-6

of the latter is oxidation occurring at a solid–gas interface [1]. Our focus, however, is on reactions at interfaces between two solid phases. Such reactions can be important in bulk systems, for example, in diffusion bonding [2], and in composites where reaction between the matrix and reinforcing phase can have both beneficial and detrimental effects [3, 4]. Nevertheless, most interest in interfacial reactions centers on thin-film systems, in which the reactions are naturally more prominent because of the greater ratio of interfacial area to volume. An extreme case is found in multilayered materials (sometimes called *superlattices*) made by thin-film deposition; these consist of 10 to 100 alternating thin layers (each of a thickness from an atomic monolayer up to 100 nm) of two phases and are of interest for a wide range of fundamental and practical reasons [5, 6]. Interfacial reactions are not only of concern in analyzing the stability of multilayered materials. They can also be exploited for materials synthesis [7], and the heat released by reaction in multilayers can be used in joining, soldering, and ignition [8–10]. Above all, however, interfacial reactions in thin films have attracted attention because of their importance in electronic devices, and in particular at metal–semiconductor interfaces [11, 12]. The interest lies not only in degradation of as-deposited structures limiting device reliability, but also in controlled reaction to produce selected silicide layers [13, 14].

After long annealing of an interface between two effectively semi-infinite phases, for example, blocks of two elemental phases, all intermediate phases expected across the range of composition in the equilibrium diagram are observed. Ideally there is a set of planar layers, with the layer of each phase continuing to thicken at a rate governed by atomic diffusion through it. For example, in the system Ni–Si, there are six equilibrium crystalline silicides: Ni_3Si, Ni_5Si_2, Ni_2Si, Ni_3Si_2, NiSi, and $NiSi_2$. In bulk diffusion couples (total reacted layer thickness $\approx 10\,\mu m$), most or all of these phases grow simultaneously [11]. The two elements are progressively consumed by the extended interfacial reaction, but are far from being exhausted. In such a case, the kinetics of the continuing reactions is governed by diffusion in each phase, and each interphase interface is close to equilibrium. The path to such a state can be very complex, however, and it cannot be attained if one or other of the original phases is completely consumed. It is in the early stages of reaction, obviously of particular relevance in thin films, that nucleation plays an important role. Correspondingly, thin-film reactions show features not found in bulk diffusion couples. Typically there is only one product phase at any time. Different intermediate phases may appear sequentially in time, rather than simultaneously as a set of layers. Thus a phase once formed may subsequently disappear. The nature of the first product phase varies from case to case and is difficult to predict. In thin-film Ni–Si, for example, the first product phase is Ni_2Si and it continues growing until one or the other of the initial elemental phases is exhausted. At higher temperature, a second product phase forms at the interface between the Ni_2Si and the remaining element, and the new phase then consumes the Ni_2Si. At still higher temperature, the second product phase is itself consumed in a similar way by a third phase [11]. In many systems the first product phase is metastable (not appearing on the equilibrium phase diagram), and there has been particular interest in the formation of *amorphous phases* by interfacial reaction [15, 16].

The dynamic nature of early-stage interfacial reactions, far from equilibrium, poses problems for the analysis of nucleation because many of the assumptions, explicit or implicit, made in earlier sections of this book are invalid. This chapter considers key aspects in turn. In Chapter 6, Section 2, there was extensive discussion of heterogeneous nucleation on a stationary interface. In an interfacial reaction, however, the relevant interface may migrate, generally inhibiting nucleation on it, as explored in Section 3. When a new phase forms within an original phase of uniform chemical composition, the driving force for transformation is readily calculated (Chapter 5, Section 1), according to principles that are clear even for systems far from equilibrium. However, if a nucleus forms within a phase of steeply graded composition, or at an interface between two phases of different composition, the matrix can show significant composition variation even over the short scale of the nucleus diameter. In that case, as explored in Section 4, the evaluation of the driving force for transformation is far from straightforward. The same variations in composition also drive diffusional fluxes through the original phases, and there are likely to be fluxes within any newly formed nucleus as well. As analyzed in Section 5, these fluxes exert an effective pressure on the interphase interfaces that can control phase stability. First, however, we examine briefly the evidence that a distinct nucleation stage can be important for interfacial reactions.

2. EVIDENCE FOR NUCLEATION

In thin-film interfacial reactions, new product phases can appear after some time. Such behavior could be nucleation-controlled, but it could alternatively be due to a barrierless adsorption or to a process of wetting, as analyzed in Chapter 6, Section 2.1. In the latter case, the product phase would completely coat the interface, and an exactly planar geometry could be maintained as the layer of the new phase thickens. In some cases, the silicides formed from metallic films deposited on silicon wafers maintain the mirror finish of the wafer and original film [13]; the lack of distinct nucleation centers on the interface could arise from adsorption or wetting, or from very profuse (possibly barrierless) nucleation. In other cases, the interfacial reaction leads to surface modulations, showing the pattern of growth from distinct centers (Figure 1). In the latter case, it is expected that the growth from the centers coating the interface would be more rapid than the subsequent thickening of the product layer; only at the interface are both component species of the product phase immediately available. The kinetics has been investigated in detail using calorimetry to measure the rate of release of heat of reaction. This technique has good sensitivity when used on multilayered samples (in

Fig. 1 Surface modulations showing the nucleation and growth pattern of Rh_4Si_5 formed on annealing a thin film of Rh deposited on single-crystal Si. (From Ref. [13], with permission.)

contrast to the bilayer sample in Figure 1). As first shown by Coffey et al. [17], and followed by much further work, the rate of reaction can show two distinct maxima on heating (Figure 2). The first peak shows the rapid growth of the product phase in the plane of the interface. Coffey et al. [17] modeled this as the growth of circular disks of constant thickness. When the disks impinge, the overall rate of reaction decreases dramatically. As heating continues, the thickening of the product layer accelerates (because of faster interdiffusion) and then stops when one or other original phase is exhausted, giving the second peak. The two-stage

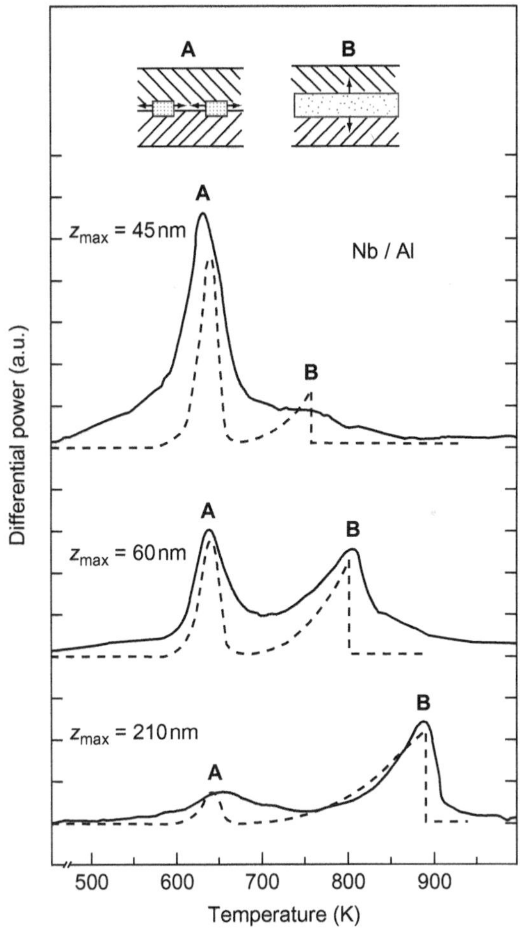

Fig. 2 Comparison of measured (solid lines) and calculated (dashed lines) differential scanning calorimetry traces on heating Al–Nb multilayer thin films at 50°C min^{-1}. The product layer thickness is z_{max}. (Reprinted with permission from Ref. [17], copyright (1989), American Institute of Physics.)

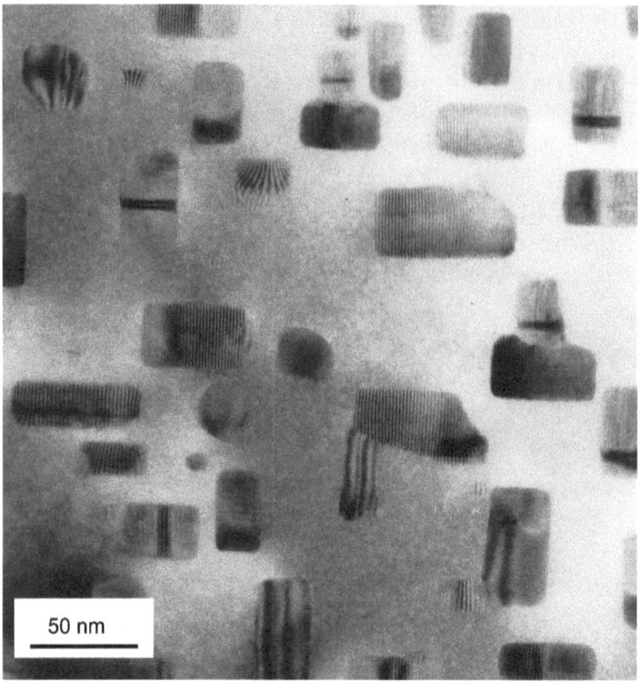

Fig. 3 Plan-view transmission electron micrograph of Pd_2Si particles forming epitaxially on single-crystal silicon at its interface with an amorphous Pd–Er alloy. (Reprinted with permission from Ref. [20], copyright (1987), American Institute of Physics.)

growth of the product phase is reflected also in the development of stresses arising from the volume change accompanying the reaction [18]. Analysis of the position and shape of the first peak in calorimetry traces (Figure 2) can give information on the nucleation stage. The nucleation is affected by the exact structure and composition of the original interfaces, determined for example by the conditions of thin-film deposition. It is important to note that interfaces between the same two original phases can behave quite differently from case to case. This can be seen even in a single multilayered sample, where the interfaces are not necessarily symmetric. That is, the interface formed when B is deposited on A, may differ from that when A is deposited on B. This can be revealed, for example, in calorimetry traces in which the first peak of the kind seen in Figure 2, is split into two peaks, one for the nucleation stage at each type of interface [19].

While the sequence of phases formed in interfacial reactions can be revealed by X-ray diffraction, the distribution of nucleation events can be revealed only by imaging techniques such as transmission electron microscopy (TEM) [20–23] or atom-probe field-ion microscopy (three-dimensional atom probe, 3DAP) [24]. There have, in particular, been

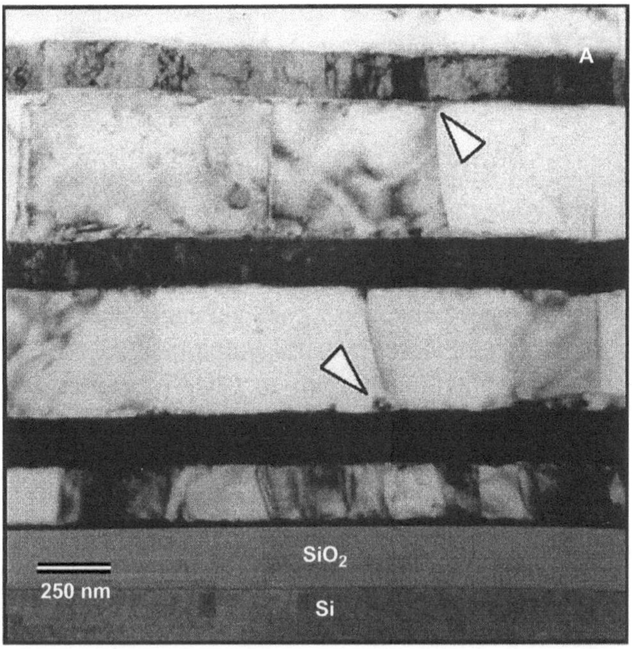

Fig. 4 Transmission electron micrograph of a cross-section of a multilayer of Al (thicker) and Nb (thinner) layers. The arrows indicate where Al_3Nb has nucleated at the Al-Nb interfaces and is growing along grain boundaries in the Al layers. (Adapted from Ref. [23], copyright (2001), with permission from Elsevier.)

many studies using TEM to show both plan (Figure 3) and cross-sectional (Figure 4) views of interfaces. Studies of this kind show a range of kinetics analogous to nucleation within uniform phases. For example, Ma et al. [21] found very different nucleation kinetics of VSi_2 and Al_3Ni in the corresponding multilayers of the alternating elements. Multilayers of polycrystalline vanadium and amorphous silicon (a-Si) in their as-deposited state show thin layers of an amorphous vanadium silicide at the interfaces. On subsequent annealing, crystalline VSi_2 nucleates at the interfaces between this silicide and the a-Si. The nucleation appears to be at random sites, shows an effective time lag and then attains a steady-state rate. This is in marked contrast to the nucleation of Al_3Ni at the interfaces in multilayers of polycrystalline aluminum and polycrystalline nickel; in this case, there is no time lag and the population of nuclei appears independent of time, showing that the nuclei existed before the reaction anneal, or that there was early site saturation during the anneal. The increase of the nucleus population with decreasing grain size in the polycrystalline layers suggested that the preferred sites for nucleation are grain-boundary triple junctions (Chapter 6, Figure 24a). This has been confirmed by many studies that show initial phase formation at grain boundaries (Figure 4) with protrusion of the new phase along the boundary as well as along the interphase interface [21, 22, 25], as reviewed by Lucadamo et al. [23].

In deposited multilayer samples, the selection of the first product phase at the interfaces appears to be independent of the periodicity of the multilayer (i.e. the thickness of two adjacent layers) and its composition (i.e. the ratio of thickness of the two types of layer) [22]. But the nucleation kinetics is clearly dependent on the detailed structure, as affected by deposition and annealing. For example, the kinetics can be different for otherwise similar multilayers deposited by evaporation and sputtering [22]. While the differences are qualitatively explained by differing grain structures in the layers of original phases, detailed analysis is complicated by the fact that these grain structures can themselves evolve during annealing. Grain growth and recrystallization in the original phases, and in the product phase give complex kinetics and can lead to distinct final grain structures, including equiaxed grain structures in which the initial layering is completely destroyed [23].

We have seen that there is clear evidence for the nucleation of interfacial reactions. It is tempting to analyze the selection of the first product phase and the sequence of subsequent phases in terms of *competitive nucleation*. It is important to note, however, that *competitive growth* may also play a key role in phase selection [26], a topic that will examined more closely in Section 5. For the moment we note that in a diffusion couple in which there is a stack of layers of intermediate phases, the competitive growth governed by the interdiffusion fluxes within each phase can lead to a negative thickening rate for a given phase. Such a phase would be incapable of forming, or if already formed it would disappear. As reviewed and analyzed by Gösele and Tu [27], there are many examples of these competitive growth effects in thin-film interfacial reactions. Such effects cannot of themselves explain, however, the appearance of metastable product phases. A competitive growth analysis shows that a metastable phase in a stack of phases would always be consumed by the neighboring equilibrium intermediate phases. The appearance of the metastable phase must therefore depend on stifling the nucleation of the equilibrium product phases.

This has been examined in detail for the cases in which the metastable product phase in amorphous. The *solid-state amorphization reaction* (SSAR) has been extensively reviewed [15, 16]. SSAR has been found at the interfaces between many combinations of metals, but for convenience of discussion here we focus on the most studied system: Ni/Zr. This represents a large family of reactions between an early transition metal (e.g., Zr) and a late transition metal (e.g., Ni). When deposited under normal conditions, thin films of nickel or zirconium are polycrystalline, and the reaction between the metals has mostly been studied in multilayers. When a multilayer stack of polycrystalline layers is heated, planar layers of amorphous Ni–Zr (spanning the approximate composition range $Ni_{47}Zr_{53}$ to $Ni_{68}Zr_{32}$, at.% [28]) form at each original Ni/Zr interface and grow at a diffusion-limited rate until they impinge [29]. In the deposition of such multilayers, however, the conditions are such that some amorphous phase is already present in the as-deposited state [30]. The SSAR in such samples is therefore not limited by nucleation of the amorphous phase; the reaction is detectable below 200°C, and proceeds rapidly at 300°C (as shown by calorimetry [29, 31]). In contrast, when polycrystalline Ni is deposited onto *single-crystal* Zr, there is no interfacial reaction of any kind at 300°C [32]. Similarly, there is no reaction at 300°C between polycrystalline Ni and a recrystallized Zr foil, which was polycrystalline but with only low-energy grain boundaries [33]. Thus, there is clear evidence that high-energy grain boundaries in Zr are required for nucleation of the amorphous phase. Grain boundaries in the Ni are not required. SSAR is found to occur readily at interfaces between single-crystal Ni and polycrystalline Zr [33, 34]. This asymmetry in the nucleation of SSAR is not unexpected in view of other asymmetries. Ni is a fast, interstitial diffuser in Zr, while Zr (having twice the atomic volume of Ni) is a slow diffuser in Ni [35]. Finally, it should be noted that the observed barrier to the formation of the amorphous phase is indeed for nucleation not growth. SSAR proceeds readily in contact with single-crystal zirconium, provided the initial interface has some amorphous phase nucleated at it by ion-beam mixing [32] or by initial deposition of an amorphous layer [36].

As reviewed by Highmore [37] and Thompson [38], the observed barrier to the nucleation of the amorphous phase, in particular Ni–Zr, has been regarded as paradoxical. It is equally unexpected that there is any barrier to nucleating the stable crystalline intermetallic compounds. The free energy of reaction between Ni and Zr (whether to form an amorphous or crystalline product phase) is large and negative. For any reasonable interfacial energies, even a hypothetical atomic monolayer of product phase would have a negative work of formation [39]. There is a volume decrease on reaction, but the associated strain energy is not sufficient to change the conclusion that there should be no barrier to the formation of the amorphous or other possible product phases. We show later (Section 4), however, that the apparent paradox arises only because of an incorrect calculation of the driving force for the formation of a phase at an interface.

The importance of grain boundaries (in one of the original phases) has been noted above for a number of interfacial reactions. In Chapter 6, Section 2, the preference for nucleation on grain boundaries was analyzed in terms of the energetic benefit of removal of part of the grain-boundary area. Figure 5 shows schematically the nucleation of a product γ phase between original phases α and β. While the nucleation at a grain boundary, for example as shown in the β phase, does remove some grain-boundary area, it also increases the area of the γ–β interface compared to the planar configuration if nucleation occurred elsewhere on the α–β interface. Overall, if only grain-boundary energy is considered, there is likely to be little, if any, reduction of the work of nucleation on a grain boundary. This is in contrast to the cases considered in Chapter 6, Figure 22.

Another factor that must be considered is that a grain boundary is a fast path (sometimes termed a *short circuit*) for atomic diffusion. Nucleation, and subsequent growth, may be favored on a grain boundary because the boundary facilitates the partitioning of solute necessary for the transformation. In the particular case of nucleation of amorphous Ni–Zr at grain boundaries in the original Zr phase, this is unlikely to be a significant factor as there is fast interstitial transport of Ni through the Zr grains. Short-circuit diffusion, while it may play a role in some cases, has also been discounted for other types of interfacial reaction not so far considered. For example, in conventional aluminum alloys the intermetallic particles dispersed in the α-Al matrix may be induced by annealing to transform in-situ from one phase to another. It is found that the transformation of a given particle nucleates at its surface where a matrix grain boundary meets the surface. The detailed study of such transformations in model systems [40, 41] shows that the grain boundaries remain preferred sites even when the boundaries are no different in composition from the surrounding grains. Instead in this case [41], and others, grain boundaries may be preferred nucleation sites because plastic flow at boundaries can occur more easily than within a grain [42]. This flow can accommodate shear and volume changes accompanying nucleation, and so reduces the contribution of elastic strain energy to the work of formation of the critical nucleus.

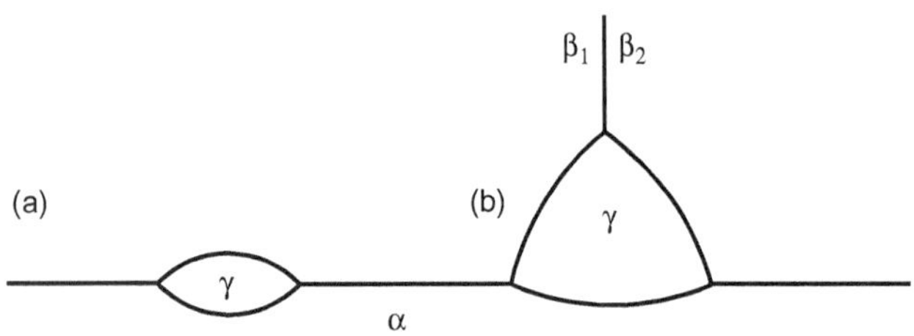

Fig. 5 Sketch of a γ-phase nucleus forming at the interface between α and β phases: (a) in contact only with a single grain on either side of the interface; (b) at a grain boundary in the β phase.

3. NUCLEATION ON A MOVING INTERFACE

The first attempt at a quantitative analysis of nucleation on a *moving* interphase boundary was by Aaronson et al. [43], considering the isothermal decomposition of the ccp-iron phase (*austenite*, γ) in some steels. The transformation of austenite to the bcc-iron phase (*ferrite*, α), discussed in Chapter 14, Section 4.2, can be accompanied by precipitation of carbide particles at the γ–α interface. The formation of the carbides requires long-range solute transport, while the γ-to-α transformation does not. The conditions are, therefore, such that migration of the γ–α interface can readily be sufficiently fast to affect the nucleation of the carbides. The austenite transforms to the ferrite by a ledge-growth mechanism. The broad faces of the ledges (A in Figure 6) are low-energy interfaces and are essentially immobile. Transformation occurs by lateral motion of the risers (B) of the ledges. These risers are high-energy interfaces. Given the considerations of Section 2.6 in Chapter 6, nucleation of carbides should be more likely on the risers, favored by the higher interface energy there. In practice, however, nucleation is

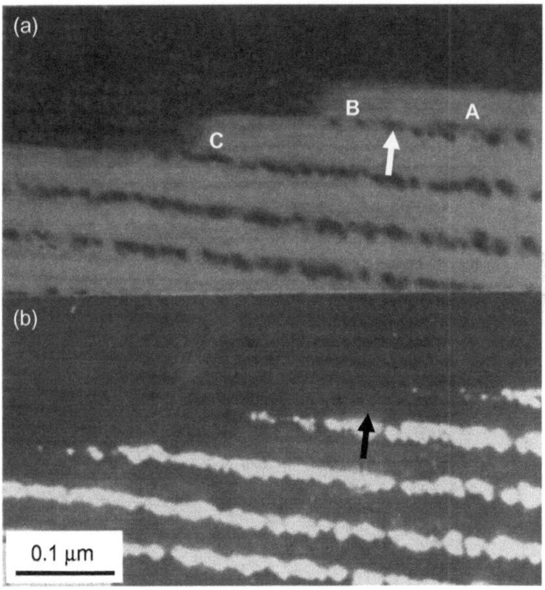

Fig. 6 Matching (the arrow shows corresponding locations) bright-field (a) and dark-field (b) transmission electron micrographs showing precipitation of $Cr_{23}C_6$ on the broad faces (A), but not the migrating risers (B), of ledges as ferrite grows into austenite in a Fe–Cr–C steel annealed at 650°C. In (b), at the point corresponding to C in (a), it can be seen that precipitates nucleate and grow just behind an advancing riser. (Adapted with permission from Ref. [44], copyright (1974), Maney Publishing.)

exclusively on the stationary ledge faces, giving rise to a very distinctive layered pattern of precipitation. In Figure 6, at C for example, it can be seen that precipitates nucleate and grow on the fresh ledge just behind an advancing riser. As postulated in the original work [44], the suppression of nucleation on the risers themselves is likely to be due to their migration.

Aaronson et al. [43] considered that the increased work of formation would prohibit any significant deviation of carbide nuclei from their equilibrium shape. They therefore assumed that carbide nucleation would be prohibited if the interphase interface upon which it would occur were to migrate more than a negligible distance in the time required for a nucleus to form. Arbitrarily, they took the critical distance to be one lattice parameter a. The time to form a nucleus, on average, can be taken to be the effective time lag θ (Chapter 3, Eq. (2)), modified for nucleation on an interface according to the discussion of transient times in Chapter 6, Section 2 (Eq. (15)). Accordingly, for nucleation on a migrating γ–α interface to be possible, the velocity of the interface $\upsilon_{\gamma\alpha}$ must be such that

$$\theta\upsilon_{\gamma\alpha} \leq a. \tag{1}$$

Aaronson et al. used an expression for θ reviewed by Feder et al. [45] (Chapter 3, Section 5, Eq. (23)), giving results largely similar to Chapter 3 later approaches. They found that at typical levels of driving force for austenite decomposition, the velocity of the risers is at least an order of magnitude too great to permit any carbide nucleation on them. They also noted that in other cases, the nucleation at the interphase interface may be controlled by interfacial diffusion, rather than long-range transport through the volume. In that case, the faster kinetics, though not precluding inhibition of nucleation on moving interfaces, does make it less likely.

For interfacial reactions at a diffusion couple, it has already been noted that a characteristic of a thin-film geometry is that there can be a succession of product phases. We now consider the role of nucleation in the appearance of the second product phase. For convenience, we again focus on the reaction between elemental nickel and zirconium. As noted in Section 2, the first product phase is often amorphous Ni–Zr. On isothermal annealing, the amorphous phase grows to a critical thickness (of the order of 100 nm, decreasing slightly with increasing temperature) and then the crystalline intermetallic phase NiZr appears at the interface between the amorphous phase and the original zirconium [33, 46, 47]. Under diffusion control, the thickening of the amorphous phase layer continuously decelerates during isothermal annealing. It is likely that nucleation of the second product phase can occur only when the rate of thickening has slowed below a critical value. As the new phase has the same composition as the amorphous phase at that interface, it is reasonable to assume that its nuclei form predominantly within the amorphous phase, as shown schematically in Figure 7. As suggested by Highmore [37], the appropriate analysis of the appearance of this second product phase depends on the catalytic potency of the interface, represented by the contact angle ϕ. For small ϕ, and in particular if the interface is wetted by the new phase ($\phi = 0$), there is no nucleation barrier, and the appearance of the new phase is best analyzed in terms of a growth competition between amorphous Ni–Zr and crystalline NiZr [37]. For larger ϕ, however, a competitive nucleation is appropriate. The crystalline NiZr, when it appears, grows rapidly to consume the amorphous alloy, suggesting that the barrier to the nucleation of the second product phase is significant [46, 47]. Highmore et al. [37, 48] analyzed this in terms of a critical velocity of the a-Ni–Zr/α-Zr interface below which NiZr can nucleate. Their general approach was the same as that of Aaronson et al. [43] described above, but the condition for nucleation to be possible was much less stringent than that described by Eq. (1). Highmore et al. took a nucleus with ϕ tending towards 180° to represent the nucleation-dominated case, and considered that nucleation could occur provided the developing nucleus of NiZr is not overrun by

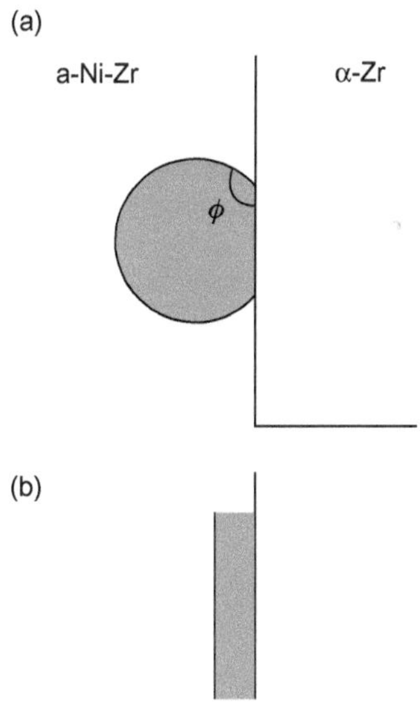

Fig. 7 Sketch of possible modes of formation of crystalline NiZr (shaded) within amorphous Ni–Zr. In the interfacial reaction between elemental Ni and Zr, a-Ni–Zr is the first product phase and NiZr is the second. In case (a) with a large contact angle ϕ a *competitive-nucleation* analysis is more appropriate; in case (b) with $\phi = 0$ a *competitive-growth* analysis is more appropriate. (Adapted from Ref. [37], with permission; Taylor & Francis Ltd. http://www.informaworld.com.)

the a-Ni–Zr/α-Zr interface, i.e. provided the NiZr nucleus can stay in contact with the α-Zr. The effective velocity of the nucleus surface, as considered by Aaronson et al., is related to the effective time lag for nucleation θ. For a near-spherical nucleus, the velocity can be approximated as θ/r^*, where r^* is the critical nucleation radius. Highmore et al. considered that nucleation could occur when the velocity of the a-Ni–Zr/α-Zr interface decreased to be below θ/r^*. Combining the Kashchiev expression for θ (discussed in Chapter 3, Section 4.2) and r^* (Chapter 6, Eq. (12)), they derived the critical interface velocity v^* to be of the form:

$$v^* = -\left(\frac{3\bar{v}^4}{4\pi}\right)^{1/3} \frac{\gamma}{k_B T} \Delta g, \tag{2}$$

where $\bar{v}$ is the molecular volume, γ is the unbiased molecular jump frequency, and Δg is the free-energy driving force for the transformation per unit volume. Since both θ and r^* are proportional to the liquid–solid interfacial energy σ_{ls}, this quantity does not appear in Eq. (2), a useful feature since σ_{ls} is not well known. Using Eq. (2), combined with a model for the diffusion-limited thickening of the first product layer of amorphous phase, Highmore [37] derived an expression for the critical thickness of that phase before the appearance of the second product phase. It was found that with reasonable parameters, measurements on the temperature-dependent critical thickness could be matched. Thin-film reaction maps were calculated to show the amorphous layer thickness as a function of processing conditions.

The two approaches outlined above, while having some success in qualitative matching of experimental behavior, are far from quantitative and fail to take account explicitly of the atomic interdiffusion fluxes through which interface migration and nucleation of a new phase must be linked. As we will see in Section 5, these fluxes are important in determining phase stability.

4. DRIVING FORCE FOR TRANSFORMATION IN NONUNIFORM COMPOSITION

4.1 In a composition gradient

When the composition of a new phase differs from that of the original phase within which it is nucleating, a full kinetic analysis involves the factors discussed in Chapter 5, Section 3; in particular, the atomic fluxes involved in nucleation and in solute transport are coupled (Chapter 5, Section 5). For the moment, we set these kinetic considerations aside and focus only on the thermodynamics of nucleation. The full basis for calculating the work of nucleus formation in a multicomponent system has been set out in Chapter 5, Section 2, but for present purposes, we restrict our consideration to a two-component system. In the expression for the work of nucleation, the term scaling with nucleus volume is calculated from the chemical potential change for each species joining the nucleus. In the treatments considered so far, it is assumed that the composition outside the nucleus is uniform and is not changed by the nucleation, so that the original phase acts as a reservoir of atoms, each type of which has a fixed chemical potential. For a two-component system, the parallel-tangent construction on a free-energy vs composition plot gives the composition of the nucleating phase that maximizes the driving force for transformation and the value of the chemical potential change for each species.

The calculation of the driving force for transformation is obviously more complex when nucleation occurs in a region of nonuniform composition. In this section, we consider nucleation within a single phase of graded composition, and in the next section nucleation at an interphase interface. In the early stages of an interfacial reaction, it is expected that there may be steep composition gradients in the phases. This is especially so for a new phase formed at the original interface. A first product phase of this kind may be metastable, and therefore may itself transform. For example, a newly formed amorphous product phase has a driving force to crystallize, and there has been much analysis of this and similar cases. For convenience of discussion, we focus on the crystallization of an amorphous phase to a stable intermetallic phase, though the treatment is equally valid for precipitation of an intermetallic within a crystalline solid solution.

The driving force for transformation is maximized when the compositions of the original and new phases are the same. In an original phase of graded composition, this matching can occur only on one plane ($y = 0$ in the schematic in Figure 8, where the composition is graded along the y-axis). Assuming that nucleation is centered on that plane, the nucleus grows outward into regions of composition ($y>0$ and $y<0$), which are increasingly unfavorable for transformation. If the composition of the original matrix is graded significantly on the length scale of the critical nucleus diameter, then the work of nucleation can be affected by the composition gradient. The deviation of the matrix composition from that maximizing the driving force for transformation constitutes a *thermodynamic suppression* of nucleation. As has been reviewed [49–51], the calculation of the work of nucleation depends on the assumptions made about the solute partitioning when the nucleus is formed. The simplest assumption is that the original phase (in our particular example the amorphous phase) transforms *polymorphically*, i.e. by changing its

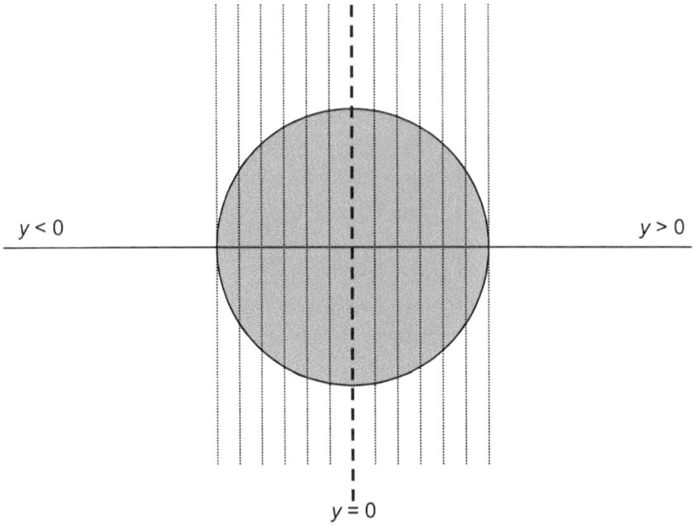

Fig. 8 A schematic diagram of a nucleus in a matrix of composition that is graded along the y axis. The nucleus is centered on the plane $y = 0$, at which the matrix composition matches that of the nucleus phase. Away from that plane ($y>0$ or $y<0$) three modes have been considered for how that the composition of the nucleus is graded: *polymorphic* in which each slice remains of uniform composition throughout both phases, *transversal* in which there is solute partitioning between the phases only within each slice, and *longitudinal* in which there is complete mixing within the nucleus only with no net partitioning with the matrix.

structure, but not its composition. Alternatively, in what has been termed the *tranversal mode,* solute transport is permitted perpendicular to, but not parallel to the y-axis in Figure 8. Within each of the slices sketched in the figure, the nucleus forms with the composition maximizing the driving force for the transformation within that slice (found using the parallel-tangent construction). The nucleus thereby has a composition that is graded, but (assuming that its free-energy curve has a sharper minimum than that of the original phase) less so than if it had formed polymorphically. In both cases, the critical work of nucleation $W^*_{\rm grad}$ takes the form [51]:

$$W^*_{\rm grad} = W^*_{\rm hom} + \xi\left(\frac{{\rm d}X}{{\rm d}y}\right)r^5, \tag{3}$$

where $W^*_{\rm hom}$ is the usual work of homogeneous nucleation, ξ is a positive constant, X is the mole fraction of one of the components (i.e. ${\rm d}X/{\rm d}y$ is the composition gradient), and r is the radius (or other characteristic dimension) of the embryo. The work of formation then takes the forms shown in Figure 9, showing that as the composition gradient is increased, the new phase is limited to a decreasing thickness centered on the

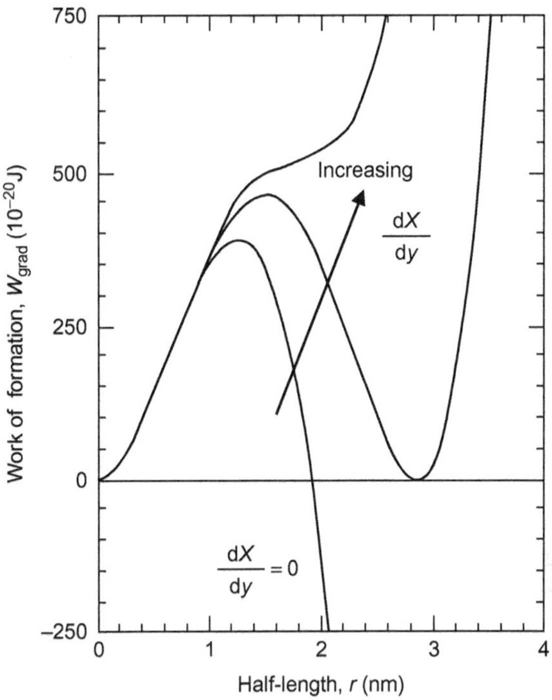

Fig. 9 The work of formation of a cube-shaped nucleus (of side length $2r$) forming in an original phase with composition gradient dX/dy. For a finite gradient, the growth of the nucleus is limited, and as the gradient increases and the limiting thickness decreases, the new phase becomes metastable rather than stable, and ultimately nucleation of the new phase is forbidden. (Adapted from Ref. [53], copyright (1996), with permission from Elsevier.)

composition plane most favorable for transformation. The layer of the new phase is stable for low gradients, then metastable at higher gradients, and ultimately forbidden at yet higher gradients, i.e. nucleation can be completely suppressed. In the presence of a composition gradient, the nucleus shape is altered from that of the simple homogeneous case; the work of nucleation is lowered if the nucleus is shortened parallel to the y-axis and extended normal to y. Taking account of such effects does not significantly alter the thermodynamic suppression, however [52].

A third possibility has also been considered for solute transport accompanying nucleation: in this *longitudinal mode*, it is assumed that there is solute redistribution only within the nucleus so that it adopts a uniform composition equal to the average composition of the matrix it has consumed [53]. It has generally been assumed that the composition gradient in the matrix in this case *increases* the driving force for transformation, but, for reasons to be discussed in the next section, this is incorrect.

The three modes of solute redistribution outlined in this section are evidently idealized, and have not been linked to the general atomic fluxes expected in an interfacial region. While it may be expected that the tranversal mode would be a better approximation than the longitudinal mode when the atomic mobility in the original phase exceeds that in the new phase, and vice versa [50], each fails to provide a fully self-consistent solution to the nucleation kinetics taking account of the full spatial variation of the chemical potentials.

4.2 At an interphase interface

Consider the nucleation of an equiatomic phase γ at the interface between α phase initially composed only of A atoms and β phase initially composed only of B atoms. At a particular temperature, the free energies of the phases vary with composition as shown schematically in Figure 10. When placed in contact there is a thermodynamic driving force for interdiffusion of A and B atoms between the α and β phases, which at the interface should tend toward the equilibrium compositions (shown in Figure 10 as X, the mole fraction of B) $X_{\alpha\beta}^{eq}$ and $X_{\beta\alpha}^{eq}$. Before any interdiffusion occurs, however, the chemical potential of A atoms in the α phase is indicated by point a on the figure, the chemical potential of B atoms in the β phase by point b, and the apparent driving force for the reaction to form γ phase at the interface is indicated by the energy change c. Driving forces for transformation, evaluated in this way, were used to conclude (as reviewed in Section 2 for examples such as Ni/Zr) that the existence of a nucleation barrier is paradoxical: the calculated critical thickness of a reacted layer is less than one monolayer. It is likely that some interdiffusion between α and β phases would precede nucleation, changing the compositions of the phases as shown by the arrows at points a and b. This interdiffusion apparently reduces the driving force for formation of γ-phase, and has been cited as the basis of an otherwise unexpected nucleation barrier.

The apparently straightforward evaluation of driving force given above is in error. As always for nucleation of a new phase, we need to consider the interface surrounding it: unstable equilibrium of that interface defines the critical condition for nucleation. Consider now the interface between a newly formed γ phase and the original α phase. If the α phase has remained pure A, the chemical potential of the A atoms within it easily exceeds that of the A atoms in γ phase of roughly stoichiometric composition. Thus it would be favorable for A atoms to migrate from α to γ. This does not mean, however, that it is thermodynamically possible for the γ phase to grow at the expense of the α phase. As shown by point d in Figure 10, the chemical potential of B atoms in the α-phase is much less than that of B atoms in γ phase, thus

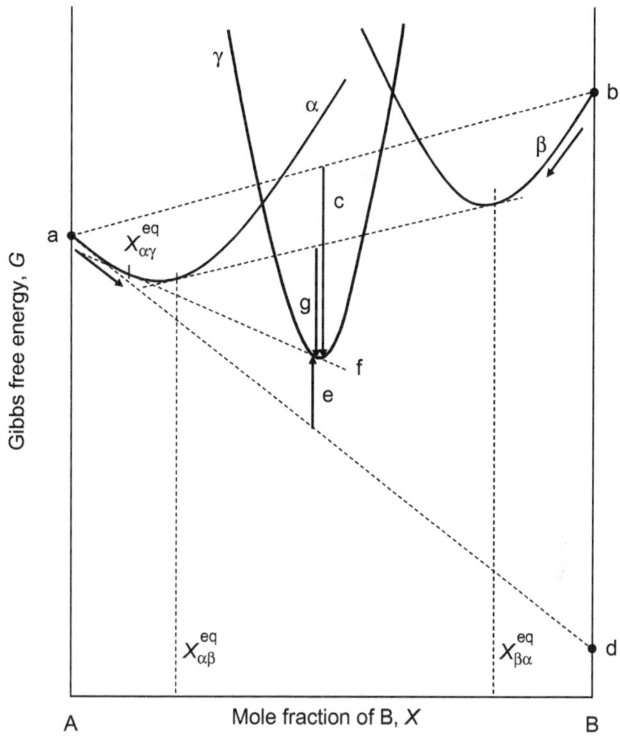

Fig. 10 A free-energy composition diagram for a given temperature at which there is a reaction at the interface between α phase (an A-rich solid solution) and β phase (a B-rich solid solution) to form γ phase (Figure 12a). The compositions are given in mole fraction of component B. Equilibria are given by common tangents; driving forces for formation of γ phase by parallel tangents. Arrow c indicates the apparent, but fictitious, driving force to form γ phase at the interface between pure α and β phases. There is no driving force for formation of γ until some interdiffusion has taken place between α and β. Arrow g indicates the maximum possible driving force for formation of γ, obtained when α and β have attained diffusional equilibrium with each other.

driving migration of B from γ to α. Overall the driving force for growth of γ phase into the pure α phase is given by the energy change e on the figure. Notably, this change is positive — the γ phase should dissolve at this interface.

If we now consider that there is progressive interdiffusion between the α and β phases, the changing composition of, for example, the α phase at the interface causes the chemical potentials of both A and B species within it to change. When the composition of the α phase reaches $X_{\alpha\gamma}^{eq}$ it is in equilibrium with the γ phase as indicated by the common tangent f. Only if the B content of α exceeds $X_{\alpha\gamma}^{eq}$ is there a driving force for growth of γ phase into α. The same argument applies, of course, at the γ–β interface. Thus, as first pointed out clearly by Thompson [38], interdiffusion between the two original phases not only *may* precede nucleation of a new phase at the interface between them, but *must* precede nucleation. Without interdiffusion there is no driving force for growth of the new phase, let alone its nucleation, which is further opposed by interfacial energy terms not so far considered in this discussion. The maximum possible driving force for formation of γ phase is when the α and β phases have attained their equilibrium compositions with respect to each other (i.e. $X^{eq}_{\alpha\beta}$ and $X^{eq}_{\beta\alpha}$); this is indicated by arrow g. The quite different initial assumption, that the driving force would be given by arrow c, is in error because it does not correctly account for all the chemical potential changes at the interface around a nucleus forming at an interphase interface. For similar reasons, it is wrong to conclude (as noted briefly in Section 4.1 in this chapter) that a composition gradient in a single phase promotes nucleation if the new phase is of uniform composition. Whether at an interphase interface, or within a single phase of graded composition, the nucleus interface makes contact with nonuniform chemical potentials, and the stability of the nucleus is then difficult to analyze fully.

The interdiffusion necessary to have a driving force for formation of the new phase is not necessarily symmetric. Consider that the diffusion of B into the α phase, already discussed for Figure 10, could greatly exceed the diffusion of A into the β phase. If so, the γ phase would nucleate and grow preferentially into the α phase. An example is provided by the SSAR, highlighted in Section 2. In the particular case of the reaction between nickel and zirconium, Ni diffuses much faster into Zr than vice versa [35]. Thus the amorphous phase forms mainly by growth into the zirconium rather than the nickel, explaining why (as noted in Section 2) the grain structure of the zirconium is important, but not the structure of the nickel. According to the arguments made above, the key factor favoring growth into the zirconium is the increase in its nickel content. In addition to all the reasons already discussed why grain boundaries may be important sites for nucleation, they are sites where interstitial nickel in zirconium (a fast diffuser but with low solubility) can be converted into substitutional nickel capable of reaching a higher mole fraction in solution. The difficulties of forming a substitutional solid solution of Ni in crystalline Zr have been discussed by Highmore [37].

As emphasized by Thompson [38], and easily envisaged by considering the evolution of the tangents in Figure 10, the relative diffusion rates into the original phases can determine the delay in nucleating a given phase, and thereby the selection of which of many potential product phases forms first. Also, the relative diffusion rates of the species within a given original phase may be important. For example, in the zirconium phase in contact with nickel, the diffusivity of the nickel is much higher than that of the host zirconium. This may favor the formation of an amorphous phase, as the nickel mobility permits changes in composition but the immobility of the zirconium may inhibit the structural changes necessary to form the equilibrium crystalline intermetallic phase [29].

Having established that interdiffusion must precede the nucleation of a product phase, it is clear that the shape of its nucleus could be affected by the composition profiles in the original phases. Hoyt and Brush [54] considered error-function composition profiles on either side of the interface. Assuming a static interface, they calculated the shape of embryo to minimize the work of nucleation. As would be expected, this shape is flatter than the double spherical-cap shape (Chapter 6, Figure 22b) that would apply for phases of uniform composition. In the case considered, however, the shape change was not great. In the non-equilibrium environment of an interphase interface, kinetic factors — specifically the distribution of interdiffusion fluxes — can influence the nucleation and growth of the product phase.

It is known that grain boundaries are fast-transport paths, and in some cases, the same can be expected of interphase interfaces. It can then be useful to define chemical potentials of the species within the interfaces themselves, as proposed by Coffey and Barmak [55]. When the original α and β phases have interdiffused to mutual equilibrium, the chemical potential of each atomic species has the same value in each phase at the interface, and this value can also be assigned to the more mobile atoms of that species in the interface itself. When equilibrium has not been reached, the value of the interfacial chemical potential can be taken to be a weighted average of the values in α and β [55]. Dramatic confirmation of the importance of transport along interphase interfaces is provided by the atom-probe field-ion microscopy observations of Vovk et al. [24] on the interfacial reaction in Co–Al bilayers. This high-resolution technique permits evaluation of composition profiles on fine length scales relevant for analysis of the nucleation of the first product phase. Direct measurement of the composition profiles in the original phases confirms that substantial interdiffusion, giving an interfacial width of ~3.5 nm, precedes nucleation. The first product phase is Al_9Co_2, which forms as globular nuclei and quickly thickens to at least 10 nm. This thickness greatly exceeds the interfacial width, which, according to an analysis such as that of Hoyt and Brush [54], should limit the growth of the phase. It is concluded that the nucleation of the Al_9Co_2 phase opens up a new fast-transport path. The new phase grows asymmetrically, mostly into the original aluminum layer, as shown schematically in Figure 11. This suggests that the interfaces between Co and Al and between Al_9Co_2 and Al (but not between Co and Al_9Co_2) are the fast-transport paths. Since

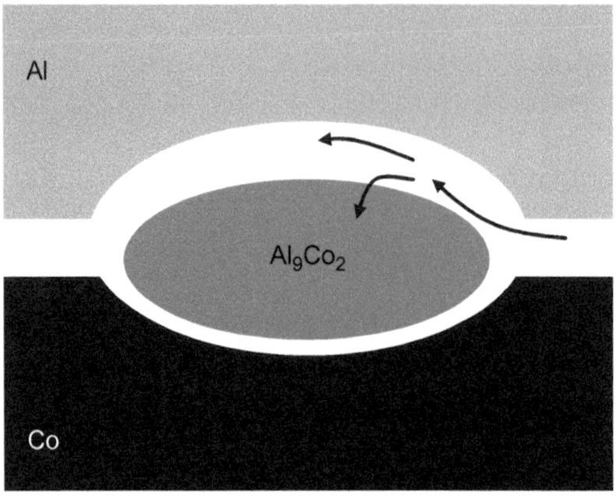

Fig. 11 The model of Vovk et al. [24] for diffusion along interphase interfaces permitting asymmetric growth of the Al_9Co_2 compound at the interface between Al and Co. (Reprinted with permission from Ref. [24], copyright (2004) by the American Physical Society.)

growth of the new phase occurs, as shown in Figure 11, by the supply of reactants from the original Co–Al interface, rather than by interdiffusional exchange with the bulk phases on either side, it can grow with a uniform composition, rather than the graded composition otherwise expected in a diffusion couple. The new phase also grows rapidly to cover the original interface and when, after impingement, the interface is completely covered, the rate of reaction slows down. Thus these high-resolution studies confirm the two-stage model of Coffey et al. [17] illustrated in Figure 2.

The predominant theme of this section and Section 4.1 has been the *thermodynamic suppression* of nucleation at an interface. Repeatedly, however, has been noted that the evaluation of nucleus stability is complicated by the diffusional fluxes in the original and product phases. In a diffusion couple, these fluxes exert pressures on the interphase interfaces. As has been noted extensively [27, 50], these fluxes can result in a *kinetic suppression* of nucleation, as considered next.

5. PHASE GROWTH AND STABILITY INFLUENCED BY INTERDIFFUSION FLUXES

Nucleation of a phase can occur only when there is an effective supersaturation, i.e. when the conditions are favorable for growth of the phase. We turn now to examine how the growth of a phase in a diffusion couple is influenced by the interdiffusion fluxes within it and within its neighboring phases. We will find that fluxes drive the migration of the interphase interfaces, and can force a phase to shrink rather than to grow. In such a case nucleation of the phase would clearly be prohibited. Analysis of growth kinetics is thus relevant for consideration of phase nucleation. We will also see that the competitive growth of different phases may mimic effects of nucleation.

5.1 Growth kinetics

The growth of product phase layers in interfacial reactions has been analyzed by several authors, as reviewed, e.g., by Gusak et al. [50]; the analysis given here follows closely that of Gösele and Tu [27]. We consider (Figure 12a) two elemental phases α and β reacting at the

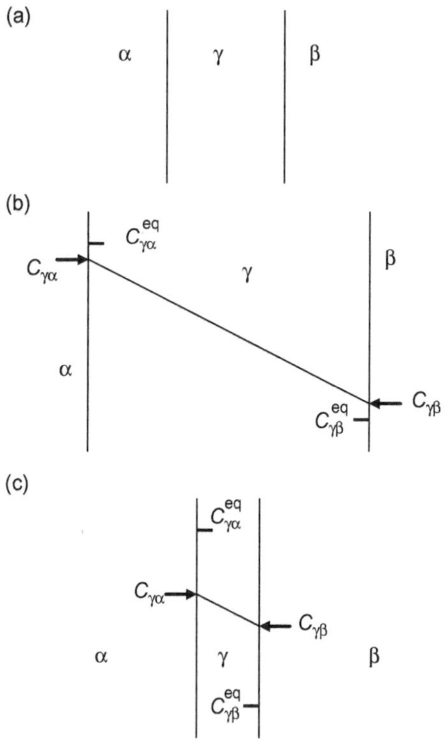

Fig. 12 (a) A γ-phase layer forms by reaction at the interface between α and β. When the γ-phase layer is thick, its rate of growth is controlled by the interdiffusion of A and B species through it; in that case (b) the actual interface compositions, i.e. concentrations of species A, in the γ phase $C_{\gamma\alpha}$ and $C_{\gamma\beta}$ deviate only slightly from their equilibrium values $C_{\gamma\alpha}^{eq}$ and $C_{\gamma\beta}^{eq}$. When the γ-phase layer is thin, its rate of growth is controlled by the reactions at its interfaces with α and β; in that case (c) the interface compositions $C_{\gamma\alpha}$ and $C_{\gamma\beta}$ deviate substantially from their equilibrium values $C_{\gamma\alpha}^{eq}$ and $C_{\gamma\beta}^{eq}$.

interface between them to form a planar layer of reaction product, a compound phase γ. Initially we take the case where the α phase (a solid solution rich in component A) and the β phase (solid solution rich in B) are both saturated, i.e. interdiffusion of A and B has taken place so that each phase has a uniform composition in equilibrium with the γ phase, the chemical potentials of both A and B then being continuous across each interface. The α and β phases then have the compositions $C_{\alpha\gamma}^{eq}$ and $C_{\beta\gamma}^{eq}$ given by the common tangents on the free-energy vs composition plot (Figure 13) at the annealing temperature. These compositions remain uniform and time-invariant throughout the interfacial reaction. All compositions are specified in terms of the concentration (number of atoms per unit volume) of species A. For simplicity, we take the atomic volume to be the same for A and B and the same in all phases; thus volume changes and consequent stresses, which can arise as a result of interdiffusion in real systems, are ignored here.

The uniform compositions of the α and β phases ensure that there are no interdiffusion fluxes (i.e., net exchanges of A and B atoms) within them. In the γ phase, if there were local equilibrium at the interfaces, the compositions in contact with α and β would be $C_{\gamma\alpha}^{eq}$ and $C_{\gamma\beta}^{eq}$ respectively.

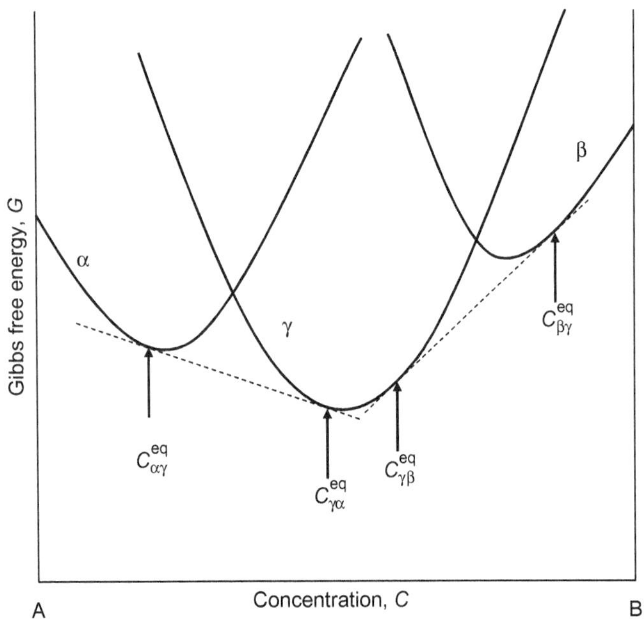

Fig. 13 Schematic free-energy vs composition diagram at a given temperature for the reaction depicted in Figure 12. The compositions are expressed in terms of concentration (atoms m^{-3}) of component A. Interfacial equilibrium compositions are given by the common tangents as shown.

However, the difference between these concentrations would drive diffusive fluxes through the γ phase, showing that (as expected from the phase rule) the three-phase configuration is not stable. The actual composition variation can be approximated by assuming a steady state in which the diffusive fluxes are uniform throughout the γ phase. Taking the A-B interdiffusivity in the γ phase, $\tilde{D}_\gamma$, to be independent of composition, the concentration profile in γ is linear. The frame of reference for measuring the atomic fluxes is chosen so the net flux is zero. (This is the laboratory frame, relative to which the lattice of the γ phase may exhibit drift [56].) The fluxes of A and B atoms are then equal and opposite ($j^{\mathrm{A}} = -j^{\mathrm{B}}$), and j^{A} alone (henceforth written simply as j) can be used to characterize the rate of reaction. The flux in γ is related to the concentration gradient in the phase $(\mathrm{d}\,C/\mathrm{d}\,x)_\gamma$ by

$$j_\gamma = -\tilde{D}_\gamma \left(\frac{\mathrm{d}\,C}{\mathrm{d}\,x}\right)_\gamma . \tag{4}$$

For this flux to be maintained, both α and β phases must dissolve, α acting as a source and β as a sink for A atoms. The migration of each interphase interface is driven by the deviation of the interfacial concentrations within the γ phase, $C_{\gamma\alpha}$ and $C_{\gamma\beta}$ from their equilibrium values. The resulting flux is taken to be linearly proportional to the deviations:

$$j_\gamma = \kappa_{\gamma\alpha}(C^{\mathrm{eq}}_{\gamma\alpha} - C_{\gamma\alpha}) = \kappa_{\gamma\beta}(C^{\mathrm{eq}}_{\gamma\beta} - C_{\gamma\beta}), \tag{5}$$

where $\kappa_{\gamma\alpha}$ and $\kappa_{\gamma\beta}$ are (positive) rate constants for the motion of (i.e. atomic rearrangement at) the γ–α and γ–β interfaces. Denoting the thickness of the γ phase as x_γ, we have

$$\left(\frac{\mathrm{d}\,C}{\mathrm{d}\,x}\right)_\gamma = \frac{C_{\gamma\beta} - C_{\gamma\alpha}}{x_\gamma} . \tag{6}$$

Eliminating $C_{\gamma\alpha}$ and $C_{\gamma\beta}$ from Eqs. (4) to (6), and rearranging, we have

$$j_\gamma = \frac{\kappa_\gamma(C^{\mathrm{eq}}_{\gamma\alpha} - C^{\mathrm{eq}}_{\gamma\beta})}{1 + (x_\gamma \kappa_\gamma / \tilde{D}_\gamma)}, \tag{7}$$

where κ_γ is an effective rate constant for thickening of the γ layer, taking account of the mobility of both the γ–α and γ–β interfaces:

$$\frac{1}{\kappa_\gamma} = \frac{1}{\kappa_{\gamma\alpha}} + \frac{1}{\kappa_{\gamma\beta}} . \tag{8}$$

At the α–γ interface, mass conservation gives the interface velocity, $\mathrm{d}\,x_{\alpha\gamma}/\mathrm{d}\,t$, to be:

$$\frac{\mathrm{d}\,x_{\alpha\gamma}}{\mathrm{d}\,t} = \frac{-j_\gamma}{C^{\mathrm{eq}}_{\alpha\gamma} - C_{\gamma\alpha}} . \tag{9}$$

Similarly,

$$\frac{\mathrm{d}\,x_{\gamma\beta}}{\mathrm{d}\,t} = \frac{j_\gamma}{C_{\gamma\beta} - C^{\mathrm{eq}}_{\beta\gamma}}. \tag{10}$$

The rate of thickening of the γ layer, $\mathrm{d}\,x_\gamma/\mathrm{d}\,t$, is given by

$$\frac{\mathrm{d}\,x_\gamma}{\mathrm{d}\,t} = \frac{\mathrm{d}\,x_{\gamma\beta}}{\mathrm{d}\,t} - \frac{\mathrm{d}\,x_{\alpha\gamma}}{\mathrm{d}\,t} \tag{11a}$$

$$= j_\gamma \left[\frac{1}{C^{\mathrm{eq}}_{\alpha\gamma} - C_{\gamma\alpha}} + \frac{1}{C_{\gamma\beta} - C^{\mathrm{eq}}_{\beta\gamma}}\right] \tag{11b}$$

$$= H_\gamma j_\gamma, \tag{11c}$$

where H_γ is, to a good approximation, independent of time and of layer thickness. (The interfacial concentrations $C_{\gamma\alpha}$ and $C_{\gamma\beta}$ are time-dependent, but for compounds with a narrow composition range their variation is negligible.)

We now have the basic description of the thickening of the γ layer:

$$\frac{\mathrm{d}\,x_\gamma}{\mathrm{d}\,t} = \frac{H_\gamma \kappa_\gamma (C^{\mathrm{eq}}_{\gamma\alpha} - C^{\mathrm{eq}}_{\gamma\beta})}{1 + (x_\gamma \kappa_\gamma / \tilde{D}_\gamma)}. \tag{12}$$

After significant reaction, the layer of γ has thickened so that $x_\gamma \gg (\tilde{D}_\gamma/\kappa_\gamma)$. The flux through the layer and its rate of thickening are then inversely proportional to the layer thickness:

$$j_\gamma = \frac{(C^{\mathrm{eq}}_{\gamma\alpha} - C^{\mathrm{eq}}_{\gamma\beta})\tilde{D}_\gamma}{x_\gamma}, \tag{13}$$

and

$$\frac{\mathrm{d}\,x_\gamma}{\mathrm{d}\,t} = \frac{H_\gamma (C^{\mathrm{eq}}_{\gamma\alpha} - C^{\mathrm{eq}}_{\gamma\beta})\tilde{D}_\gamma}{x_\gamma}. \tag{14}$$

It follows that the thickness is proportional to $t^{1/2}$, the behavior expected under diffusion control, and observed in macroscopic diffusion couples. Figure 12b shows the corresponding concentration profile. There is a substantial variation of concentration across the γ layer and negligible deviations of $C_{\gamma\alpha}$ and $C_{\gamma\beta}$ from their equilibrium values $C^{\mathrm{eq}}_{\gamma\alpha}$ and $C^{\mathrm{eq}}_{\gamma\beta}$.

In early stages of reaction, however, $x_\gamma \ll (\tilde{D}_\gamma/\kappa_\gamma)$; this is the regime relevant for nucleation of interfacial reactions and for the subsequent growth in many thin-film reactions. In this case the flux through the γ layer and its thickening rate have the limiting values:

$$j_\gamma = \kappa_\gamma (C^{\mathrm{eq}}_{\gamma\alpha} - C^{\mathrm{eq}}_{\gamma\beta}), \tag{15}$$

and

$$\frac{d x_\gamma}{d t} = H_\gamma \kappa_\gamma (C^{eq}_{\gamma\alpha} - C^{eq}_{\gamma\beta}). \tag{16}$$

In this case, the rate of thickening is controlled by the rate of the reactions at the interphase interfaces, and not by diffusion through the product layer. There is a small concentration variation across the γ layer and large deviations of $C_{\gamma\alpha}$ and $C_{\gamma\beta}$ from their equilibrium values $C^{eq}_{\gamma\alpha}$ and $C^{eq}_{\gamma\beta}$ (Figure 12c).

We now extend this treatment to consider the case where there are interdiffusion fluxes j_α and j_β in the α and β phases, which could themselves be compound phases rather than the terminal solid solutions. Thus, in the following treatment, the three adjacent layers of α, γ, and β phases could be part of a larger set of layers in a diffusion couple. Focusing on the central γ phase, its thickening rate now depends not only on j_γ (Eq. 11c), but also on j_α and j_β. Equations (9) and (10) become

$$\frac{d x_{\alpha\gamma}}{d t} = \frac{j_\alpha - j_\gamma}{C_{\alpha\gamma} - C_{\gamma\alpha}}, \tag{17}$$

and

$$\frac{d x_{\gamma\beta}}{d t} = \frac{j_\gamma - j_\beta}{C_{\gamma\beta} - C_{\beta\gamma}}. \tag{18}$$

In contrast to Eqs. (9) and (10), the concentrations $C_{\alpha\gamma}$ and $C_{\beta\gamma}$ can no longer be taken to have their equilibrium values. Using Eq. (11a), the thickening rate of the γ layer is found to be

$$\frac{d x_\gamma}{d t} = j_\gamma \left(\frac{1}{C^{eq}_{\alpha\gamma} - C_{\gamma\alpha}} + \frac{1}{C_{\gamma\beta} - C^{eq}_{\beta\gamma}} \right) - \frac{j_\alpha}{C_{\alpha\gamma} - C_{\gamma\alpha}} - \frac{j_\beta}{C_{\gamma\beta} - C_{\beta\gamma}} \tag{19a}$$

$$= H_\gamma j_\gamma - H_{\gamma\alpha} j_\alpha - H_{\gamma\beta} j_\beta, \tag{19b}$$

where H_γ (apart from a relaxation of the interfacial equilibria) is the same as in Eq. (11c). The coefficients H_γ, $H_{\gamma\alpha}$, and $H_{\gamma\beta}$ are all positive (as are the fluxes), and are approximately constant because the stoichiometry range of typical compounds limits the possible variation in interfacial concentrations. Equation (19b) shows that for any intermediate layer (γ phase in this case) to thicken, the flux through it must exceed a critical value determined by the fluxes in both adjacent layers, i.e. it must satisfy:

$$j_\gamma > \frac{H_{\gamma\alpha} j_\alpha + H_{\gamma\beta} j_\beta}{H_\gamma}. \tag{20}$$

The flux within any layer is still governed by Eq. (7).

The foregoing analysis in principle permits quantitative study of interfacial reactions. In practice, however, while the thermodynamic parameters are mostly known, the interfacial mobilities κ are completely unknown. The interdiffusivities $\tilde{D}$ are mostly known for solid-solution phases, but not for compounds. A further problem is that, the effective value of $\tilde{D}$ for a given phase depends on its grain structure. In principle, values of an effective interdiffusivity, $\tilde{D}_{\mathrm{eff}}$, can be obtained from independent experiments, but the values of κ can be obtained only from reaction kinetics, and may be impossible to obtain for many combinations of phases in which one of the phases is kinetically unstable. Nevertheless the analysis is useful in predicting the basic forms of behavior. In particular, we can understand the suppression of phases in thin-film diffusion couples. Equation (20) shows that the kinetic stability of a phase relative to its neighbors in a stack of layers is determined by the interdiffusion fluxes. If the flux for each layer is limited by the diffusion through that layer according to Eq. (13), then a low value of interdiffusivity $\tilde{D}$ for the phase can be balanced by reduced layer thickness, without any need for the phase to disappear. This is the case in macroscopic diffusion couples, when at an interface between elemental phases, all intermediate phases should be present. In thin-film couples, however, the interdiffusion flux within a phase has an absolute limit given by Eq. (15). It is possible that the limiting value of, for example, flux j_γ in the γ phase is too small for Eq. (20) to be satisfied; the γ phase would in that case be suppressed. As suggested by Gösele and Tu [27], this is the reason why typically there is only one product phase layer growing at any instant in a thin-film diffusion couple.

We have concluded that diffusion control of layer thickening leads to the complete set of intermediate phases forming in an interfacial reaction region, and that interface mobility (such as is found in the early stages of reactions, for example in thin films) must be rate-controlling for there to be phase suppression. However, we have been considering what may be called a *series competition* between phases, in which the phases are present in order of composition as a stack of planar layers. If different phases form side-by-side on the original interface between two phases, this can be regarded as a *parallel competition*. In that case, as shown by Williams et al. [57], there can be suppression of phases, even under diffusion control. However, Williams et al. considered only atomic transport through the bulk phases, and not interfacial transport of the kind shown in Figure 11. The interfacial transport is more likely to be relevant for the nucleation stage of a reaction.

5.2 Nucleation of the first product phase

Consider the nucleation of γ phase at the interface between α and β phases. For growth of the phase, and therefore its nucleation, to be possible, Eq. (20) must be satisfied, favored by higher interdiffusion flux in γ and lower fluxes in α and β. While a higher value of H_γ would favor nucleation of γ, the key factor is j_γ. When the new γ phase is thin, the interfacial reactions are rate-controlling, and j_γ is given by Eq. (15); this seems likely to be applicable for a nucleus. In this case, nucleation is favored by a high value of the effective interfacial mobility κ_γ. As the nucleus thickens, it is possible that interdiffusion through it could become rate-controlling; in that case, nucleation is favored by a high value of the interdiffusivity $\tilde{D}_\gamma$ (Eq. (13)). As shown in Figure 11, however, diffusion control of this kind is highly unlikely for an isolated nucleus. When the new phase does not completely coat the interface, diffusion along the interphase interfaces is likely to be faster than within the nucleus, and, as observed by Vovk et al. [24], can permit growth even when there is no perceptible concentration gradient in the nucleus. Whether interfacial mobility or interdiffusion is rate-controlling, Eqs. (13) and (15) show that nucleation of γ is favored by a high value of $(C^{eq}_{\gamma\alpha} - C^{eq}_{\gamma\beta})$, the composition range of the phase in equilibrium with α and β. An amorphous phase (or in general a solid-solution phase) shows a much wider composition range than a compound (which is likely to stay close to stoichiometry, represented by a free-energy vs composition curve with a sharply defined minimum). The wider composition range is the most general factor favoring nucleation (and subsequent growth) of an amorphous phase, though such phases could also be favored by higher κ_γ or $\tilde{D}_\gamma$.

The nucleation of γ is impeded (Eq. (20)) by the interdiffusion fluxes in α and β. As pointed out by Thompson [38], for pure elemental phases brought into contact, the initial fluxes would be very large, of the order of $6\tilde{D}_\gamma/\lambda^4$, where λ is the diffusion jump distance ($\approx$ atomic diameter). The initial fluxes (which are, course, positive) would clearly prohibit nucleation of γ. The fluxes in α and β phases are reduced if $\tilde{D}_\alpha$ or $\tilde{D}_\beta$ is small, or by interdiffusion between the phases. Thus as interdiffusion between α and β proceeds, the increase in chemical potential of B in α and of A in β, and the reduction of j_α and j_β all act to progressively favor the nucleation and subsequent growth of γ. If, as already noted for the SSAR, the interdiffusion is asymmetric, these thermodynamic and kinetic factors favor asymmetric consumption of the same original phase (Zr in the case of the SSAR between Ni and Zr).

5.3 Critical thickness of the first product phase

Adopting the treatment of Gösele and Tu [27], we now consider the case of two intermediate layers between saturated terminal solid solutions. In terms of a growth competition, it is of interest to derive the critical thickness of the first product phase that must be exceeded before the second product phase can appear. We consider the set of layers α-γ-δ-β, in which α and β are the terminal, saturated, solid solutions (in which $j_\alpha = j_\beta = 0$), and γ and δ are two product phases assumed already present. Analogous to Eq. (19b), the thickening rates of the γ and β phases are

$$\frac{\mathrm{d}x_\gamma}{\mathrm{d}t} = H_\gamma j_\gamma - H_{\gamma\delta} j_\delta, \tag{21}$$

and

$$\frac{\mathrm{d}x_\delta}{\mathrm{d}t} = H_\delta j_\delta - H_{\delta\gamma} j_\gamma. \tag{22}$$

We take γ to be the first product phase and assume that it has thickened to the extent that the flux through it is diffusion-controlled (Eq. (13)). The initial appearance of the second product phase is, in contrast, controlled by the interfacial reactions (Eq. (15)). From Eq. (22), we then have:

$$\frac{\mathrm{d}x_\delta}{\mathrm{d}t} = H_\delta \kappa_\delta (C_{\delta\gamma}^{\mathrm{eq}} - C_{\delta\beta}^{\mathrm{eq}}) - H_{\delta\gamma} \frac{(C_{\gamma\alpha}^{\mathrm{eq}} - C_{\gamma\beta}^{\mathrm{eq}})\tilde{D}_\gamma}{x_\gamma}, \tag{23}$$

from which it is readily seen that a layer of δ would shrink (i.e. could not appear) unless the first product phase γ exceeds a critical thickness given by

$$x_\gamma^{\mathrm{crit}} = \frac{H_{\delta\gamma}(C_{\gamma\alpha}^{\mathrm{eq}} - C_{\gamma\delta}^{\mathrm{eq}})\tilde{D}_\gamma}{H_\delta \kappa_\delta (C_{\delta\gamma}^{\mathrm{eq}} - C_{\delta\beta}^{\mathrm{eq}})}. \tag{24}$$

Once the critical thickness of γ is exceeded, the δ phase can appear, and if it does nucleate both phases can thicken. In actual thin-film reactions, simultaneous growth of two phases is rarely observed because one of the original phases is often exhausted before the first product phase reaches its critical thickness.

From Eq. (24) it is seen that, even if there were strictly no nucleation barrier (i.e. no critical work requiring thermal activation), the growth competition gives a critical thickness of the first product phase before the second product phase can appear. In practice, it is likely to be unclear whether the critical thickness arises only from the growth competition or is influenced by the difficulty of nucleating the second product phase on a moving interface (as discussed in Section 3).

5.4 Electromigration

In Chapter 6, Section 4, it was noted that an applied electric field is unlikely to have a significant effect on nucleation kinetics; this conclusion was reached because the field has a negligible effect on the free energies of the original and new phases. The nucleation of an interfacial reaction, however, is an exceptional case when an electric field can have a significant effect. An applied field drives an electron current density through a conductor, and there is an associated force on the atoms in the conductor. In a good conductor, the main contribution to this force is the *electron wind*, resulting from the scattering that gives rise to the electrical resistance; as a result, through a slight biasing of their diffusion, the atoms migrate in the same direction as the electrons, thus displaying a negative effective charge [58]. This *electromigration* is a significant reliability problem in metallization lines in integrated circuits where the small cross-sections of the conductors lead to high current densities of the order of $10^{10}\,\mathrm{A\,m^{-2}}$. Electromigration damage arises from nonuniform drift of the atoms [58]. For currents normal to an interphase interface the natural nonuniformity of the system leads to significant electromigration effects at much lower current densities, of the order of only $10^{6}\,\mathrm{A\,m^{-2}}$ [59]. Consequently the effects may be significant at a wide variety of junctions in electronic devices. The effects of applied current on interfacial reactions have now been observed for a wide range of bimetallic couples, with a particular interest in contacts with lead-free solders [60]. Sandwich structures of two metals A and B, i.e. the configuration A:B:A, allow the effects of current direction to be detected. A direct current flowing normal to interfaces in the sandwich can induce acceleration of the interfacial reaction between A and B at one interface and deceleration at the other [61–64]. It has been suggested that there could be a direct effect on product phase nucleation [60, 65, 66]; we now examine the possible basis of this effect.

We consider, again, the formation of a product phase γ at the interface between original α and β phases. In an electric field E, the interdiffusion of A and B atoms within γ is influenced by the current flowing through the phase. The electromigration force on A atoms is $E|e|z_A^*$, where z_A^* is their effective charge and e is the electronic charge. The mobility of the A atoms is $D_A^*/k_B T$, where D_A^* is the tracer diffusivity (i.e. the diffusivity in the absence of a chemical driving force for mixing [56]). The force on B atoms and their mobility are similarly determined. The electromigration of both species when combined makes a contribution $j_{\gamma,\mathrm{EM}}$ to the intermixing flux [67, 68], given by:

$$j_{\gamma,\mathrm{EM}} = \frac{C_\gamma^{\mathrm{A}} C_\gamma^{\mathrm{B}}}{C_0} Z, \tag{25}$$

where $C_\gamma^{\mathrm{A,B}}$ is the concentration of A, B atoms (no. $\mathrm{m^{-3}}$) in the γ phase and C_0 is the concentration of atomic sites (inverse of the atomic volume). The quantity Z, defined by

$$Z = \frac{E|e|}{k_B T}(z_A^* D_A^* - z_B^* D_B^*), \tag{26}$$

can be of either sign, depending on the relative values of z_A^* and z_B^*, or D_A^* and D_B^*. Accordingly, the electromigration flux can augment or oppose the interdiffusion flux, which between saturated phases is given by Eq. (7). With the electromigration flux included, Eq. (7) becomes:

$$j_\gamma = \frac{\kappa_\gamma(C^{eq}_{\gamma\alpha} - C^{eq}_{\gamma\beta}) + (x_\gamma\kappa_\gamma/\tilde{D}_\gamma)(C^A_\gamma C^B_\gamma/C_0)Z}{1 + (x_\gamma\kappa_\gamma/\tilde{D}_\gamma)}. \tag{27}$$

When $j_{\gamma,EM}$ is positive (i.e. the interchange of A and B driven by electromigration is in the same sense as interdiffusion), the overall flux j_γ is increased, as is the thickening rate of the γ layer (Eq. 11c). When $j_{\gamma,EM}$ is negative, j_γ is still positive for small thickness, but decreases through zero. As noted by Orchard and Greer [68], the condition $j_\gamma = 0$ defines a limiting thickness of the γ phase $x_{\gamma,\text{lim}} \propto -1/j_{\gamma,EM}$. A layer thickening under a fixed applied current must stop at $x_{\gamma,\text{lim}}$, and if the current is then increased the layer must get thinner, a prediction consistent with monitoring of product phase thickness via resistance measurements [69]. The inverse proportionality of $x_{\gamma,\text{lim}}$ and $j_{\gamma,EM}$ shows, though, that an applied current, however large, cannot eliminate a single product phase; this is in agreement with observations that even when there are clear electromigration effects on the kinetics of interfacial reactions, there is no change in the identity of the phase(s) formed [60, 62]. When there are two product phases, however, e.g. γ and δ as considered in Section 5.3, it is possible for electromigration to promote the growth of one and inhibit the growth of the other (i.e. for a given current polarity, Z can be of opposite sign in the two phases); in that case the inhibited phase can be completely suppressed [70], changing the phase selection at the interface.

6. SUMMARY

There is ample evidence from microstructural, and other, measurements that there can be a distinct nucleation stage in interfacial reactions. However, the interface between two phases is a particularly complex environment in which to analyze nucleation. In particular, the phases in contact may be very far from equilibrium with each other when they have been made by thin-film deposition. The reactions at such nonequilibrium interfaces are of wide relevance, but are especially notable in microelectronic devices. In the interfacial region, the phases may have steep composition gradients, there can be interdiffusion fluxes within the phases, and the interphase interface can move.

The materials parameters in the analyses of nucleation and growth at interfaces are poorly known, if at all. The effective values of the parameters may also depend on the grain structure of the reactant and product phases. Quantitative analyses of nucleation kinetics are in general not possible. Nevertheless, some important points are clear.

The driving force for reaction at an interface is relevant for the nucleation stage of the reaction. The correct evaluation of the driving force requires careful accounting for the chemical potentials of the reacting species in each of the original phases at the interface. Interdiffusion between the phases initially in contact is expected to precede, and indeed to be a prerequisite for, nucleation of a product phase at the interface. Composition gradients in the reacting phases contribute to the thermodynamic suppression of nucleation, while interdiffusion fluxes within the reacting phases and a moving interface contribute to the kinetic suppression of nucleation. Both thermodynamic and kinetic suppression weaken as the two original phases approach diffusional equilibrium. Eventually the growth of one or more product phases becomes possible. Which product phase appears first may depend not only on the growth competition, but also on the influence of phase nucleation. In thin-film diffusion couples, in particular, continued interdiffusion is likely to lead to a succession of product phases. The appearance of an amorphous phase as the first reaction product is evidence for suppression of equilibrium phases by a nucleation barrier; a growth competition alone cannot explain the appearance of metastable phases.

At the nucleation stage, the formation of a new phase at the interface between reacting phases can be facilitated by fast atomic transport along one or more interphase interfaces. Nucleation is followed by rapid lateral growth to cover the interface. Once the product layer completely covers the interface between the original phases, its further growth depends on interdiffusional transport through it. Thus interfacial reactions can have two quite distinct stages of nucleation and growth.

The effects of an electrical current normal to an interphase interface remain to be fully explored. Electromigration within product phases at the interface can augment or oppose their growth, and in this way a particular polarity of current can suppress the nucleation and growth of a product phase.

REFERENCES

[1] B.E. Deal, A.S. Grove, General relationship for the thermal oxidation of silicon, J. Appl. Phys. **36** (1965) 3770–3778.

[2] J.T. Klomp, Chemistry and kinetic aspects of metal–ceramic diffusion bonding, in: *Joining Ceramics, Glass and Metal*, Ed. W. Kraft, DGM, New York (1989), pp. 55–64.

[3] S.G. Fishman, Ramifications of recent micromechanical innovations on understanding in composite interfaces, in: *Interfaces in Metal–Ceramics Composites*, Eds. R.Y. Lin, R.J. Arsenault, G.P. Martins, S.G. Fishman, TMS, Warrendale, PA (1990), pp. 3–29.

[4] J.H. Perepezko, M.H.D. Bassani, J.S. Park, A.S. Edelstein, R.K. Everett, Diffusional reactions in composite synthesis, Mater. Sci. Eng. A **195** (1995) 1–11.
[5] L. Chang, B.C. Giessen (Eds), *Synthetic Modulated Structures*, Academic Press, New York (1985), pp. 419–486.
[6] T.W. Barbee, F. Spaepen, A.L. Greer (Eds.), *Multilayers: Synthesis, Properties and Non-Electronic Applications*, Materials Research Society Symposium Proceedings, Vol. 103 (1988).
[7] L. Fister, D.C. Johnson, Controlling solid-state reaction mechanisms using diffusion length in ultrathin-film superlattice composites, J. Am. Chem. Soc. **114** (1992) 4639–4644.
[8] E. Besnoin, S. Cerutti, O.M. Knio, T.P. Weihs, Effect of reactant and product melting on self-propagating reactions in multilayer foils, J. Appl. Phys. **92** (2002) 5474–5481.
[9] J. Wang, E. Besnoin, A. Duckham, S.J. Spey, M.E. Reiss, O.M. Knio, M. Powers, M. Whitener, T.P. Weihs, Room-temperature soldering with nanostructured foils, Appl. Phys. Lett. **83** (2003) 3987–3989.
[10] J. Wang, E. Besnoin, O.M. Knio, T.P. Weihs, Investigating the effect of applied pressure on reactive multilayer foil joining, Acta Mater. **52** (2004) 5265–5274.
[11] K.N. Tu, Interdiffusion in thin films, Ann. Rev. Mater. Sci. **15** (1985) 147–176.
[12] J.W. Mayer, S.S. Lau, *Electronic Materials Science for Integrated Circuits in Si and GaAs*, Macmillan, New York (1990).
[13] F.M. d'Heurle, Nucleation of a new phase from the interaction of two adjacent phases: Some silicides, J. Mater. Res. **3** (1988) 167–195.
[14] R. Pretorius, C.C. Theron, A. Vantomme, J.W. Mayer, Compound phase formation in thin film structures, Crit. Rev. Solid State Mater. Sci. **24** (1999) 1–62.
[15] W.L. Johnson, Thermodynamic and kinetic aspects of the crystal to glass transformation in metallic materials, Prog. Mater. Sci. **30** (1986) 81–134.
[16] B.X. Liu, W.S. Lai, Z.J. Zhang, Solid-state crystal-to-amorphous transition in metal-metal multilayers and its thermodynamic and atomistic modelling, Adv. Phys. **50** (2002) 367–429.
[17] K.R. Coffey, L.A. Clevenger, K. Barmak, D.A. Rudman, C.V. Thompson, Experimental evidence for nucleation during thin-film reactions, Appl. Phys. Lett. **55** (1989) 852–854.
[18] G. Lucadamo, K. Barmak, Stress evolution in polycrystalline thin film reactions, Thin Solid Films **389** (2001) 8–11.
[19] M. Gonzalez-Silveira, M.T. Clavaguera-Mora, F. Pi, J. Rodriguez-Viejo, Calorimetric evidence of asymmetry in the nucleation of $CuMg_2$ in Cu/Mg multilayers, Phys. Rev. B **69** (2004) 113411/1–4.
[20] C.B. Boothroyd, W.M. Stobbs, K.N. Tu, Formation of submicron epitaxial islands of Pd_2Si on silicon, Appl. Phys. Lett. **50** (1987) 577–579.
[21] E. Ma, L.A. Clevenger, C.V. Thompson, Nucleation of an intermetallic at thin-film interfaces: VSi_2 contrasted with Al_3Ni, J. Mater. Res. **7** (1992) 1350–1355.
[22] K. Barmak, C. Michaelsen, S. Vivekanand, F. Ma, Formation of the first phase in sputter-deposited Nb/Al multilayer thin films, Philos. Mag. A **77** (1998) 167–185.
[23] G. Lucadamo, K. Barmak, D.T. Carpenter, J.M. Rickman, Microstructure evolution during solid-state reactions of Nb/Al multilayers, Acta Mater. **49** (2001) 2813–2826.
[24] V. Vovk, G. Schmitz, R. Kirchheim, Nucleation of product phase in reactive diffusion of Al/Co, Phys. Rev. B **69** (2004) 104102/1–9.
[25] Z. Ma, L.H. Allen, D.D. Allman, Microstructural aspects and mechanism of the C49-to-C54 polymorphic transformation in titanium disilicide, J. Appl. Phys. **77** (1995) 4384–4388.
[26] S. Wohlert, R. Bormann, Phase selection governed by different growth velocities in the early stages of the Ti/Al phase reaction, J. Appl. Phys. **85** (1999) 825–832.
[27] U. Gösele, K.N. Tu, Growth kinetics of planar binary diffusion couples — thin-film versus bulk cases, J. Appl. Phys. **53** (1982) 3252–3260.

[28] J.C. Barbour, Diffusivity of Ni in an amorphous Ni–Zr alloy, Phys. Rev. Lett. **55** (1985) 2872–2875.
[29] R.J. Highmore, J.E. Evetts, A.L. Greer, R.E. Somekh, Differential scanning calorimetry study of solid-state amorphization in multilayer thin-film Ni/Zr, Appl. Phys. Lett. **50** (1987) 566–568.
[30] R.J. Highmore, R.E. Somekh, A.L. Greer, J.E. Evetts, A differential scanning calorimetry study of solid state amorphisation in multilayer Ni/Zr, Mater. Sci. Eng. **97** (1988) 83–86.
[31] E.J. Cotts, W.J. Meng, W.L. Johnson, Calorimetric study of amorphization in planar, binary, multilayer thin-film diffusion couples of Ni and Zr, Phys. Rev. Lett. **57** (1986) 2295–2298.
[32] A.M. Vredenberg, J.F.M. Westendorp, F.W. Saris, N.M. van der Pers, Th.H. de Keijser, Evidence for a nucleation barrier in the amorphous phase formation by solid-state reaction of Ni and single-crystal Zr, J. Mater. Res. **1** (1986) 774–780.
[33] W.J. Meng, C.W. Nieh, E. Ma, B. Fultz, W.L. Johnson, Solid-state interdiffusion reactions of Ni/Zr diffusion couples, Mater. Sci. Eng. **97** (1988) 87–91.
[34] K. Pampus, K. Samwer, J. Bøttiger, Amorphization by solid-state reaction of a Ni single-crystal and Zr, Europhys. Lett. **3** (1987) 581–585.
[35] A.L. Greer, N. Karpe, J. Bøttiger, Diffusional aspects of the solid state amorphizing reaction, J. Alloys Compounds **194** (1993) 199–211.
[36] P. Ehrhart, R.S. Averback, H. Hahn, S. Yadavalli, C.P. Flynn, Fast diffusion and nucleation of the amorphous phase in Ni–Zr films, J. Mater. Res. **3** (1988) 1276–1278.
[37] R.J. Highmore, Nucleation issues in solid state amorphization, Philos. Mag. B **62** (1990) 455–467.
[38] C.V. Thompson, On the role of diffusion in phase selection during reactions at interfaces, J. Mater. Res. **7** (1992) 367–373.
[39] W.L. Johnson, B. Dolgin, M. Van Rossum, Glass formation and growth in solids, in: A. F. Wright, L. Dupuy (Eds.), *Glass: Current Issues*, NATO ASI Series, Martinus Nijhoff, Dordrecht (1985), pp. 172–187.
[40] D.T.L. Alexander, A.L. Greer, Nucleation of the Al_6(Fe,Mn)-to-α-Al–(Fe,Mn)–Si transformation in 3XXX aluminium alloys. I: Roll-bonded diffusion couples, Philos. Mag. **84** (2004) 3051–3070.
[41] D.T.L. Alexander, A.L. Greer, Nucleation of the Al_6(Fe,Mn)-to-α-Al–(Fe,Mn)–Si transformation in 3XXX aluminium alloys. II: transformation in cast aluminium alloys, Philos. Mag. **84** (2004) 3071–3083.
[42] J.W. Christian, *The Theory of Transformations in Metals and Alloys, Part I*, 3rd ed., Elsevier: Oxford. (2002), p. 463.
[43] H.I. Aaronson, M.R. Plichta, G.W. Franti, K.C. Russell, Precipitation at interphase boundaries, Metall. Trans. A **9A** (1978) 363–371.
[44] K. Campbell, R.W.K. Honeycombe, The isothermal decomposition of austenite in simple chromium steels, Metal. Sci. **8** (1974) 197–203.
[45] J. Feder, K.C. Russell, J. Lothe, G.M. Pound, Homogeneous nucleation and growth of droplets in vapours, Adv. Phys. **15** (1966) 111–178.
[46] S.B. Newcomb, K.N. Tu, Transmission electron-microscopic observations of amorphous NiZr alloy formation by solid-state reaction, Appl. Phys. Lett. **48** (1986) 1436–1438.
[47] W.J. Meng, C.W. Nieh, W.L. Johnson, Maximum thickness of amorphous NiZr interlayers formed by a solid-state reaction technique, Appl. Phys. Lett. **51** (1987) 1693–1695.
[48] R.J. Highmore, A.L. Greer, J.A. Leake, J.E. Evetts, A transient nucleation model for solid state amorphisation, Mater. Lett. **6** (1988) 401–405.
[49] P.J. Desré, A.R. Yavari, Suppression of crystal nucleation in amorphous layers with sharp concentration gradients, Phys. Rev. Lett. **64** (1990) 1533–1536.
[50] A.M. Gusak, F. Hodaj, A.O. Bogatyrev, Kinetics of nucleation in the concentration gradient, J. Phys. Cond. Matt. **13** (2001) 2767–2787.

[51] F. Hodaj, A.M. Gusak, Suppression of intermediate phase nucleation in binary couples with metastable solubility, Acta Mater. **52** (2004) 4305–4315.
[52] F. Hodaj, A.M. Gusak, P.J. Desré, Effect of sharp concentration gradients on the nucleation of intermetallics in disordered solids: influence of the embryo shape, Philos. Mag. A **77** (1998) 1471–1479.
[53] F. Hodaj, P.J. Desré, Effect of a sharp gradient of concentration on nucleation of intermetallics at interfaces between polycrystalline layers, Acta Mater. **44** (1996) 4485–4490.
[54] J.J. Hoyt, L.N. Brush, On the nucleation of an intermediate phase at an interface in the presence of a concentration gradient, J. Appl. Phys. **78** (1995) 1589–1594.
[55] K.R. Coffey, K. Barmak, A new model for grain boundary diffusion and nucleation in thin film reactions, Acta Metall. Mater. **42** (1994) 2905–2911.
[56] M.E. Glicksman, *Diffusion in Solids: Field Theory, Solid-State Principles, and Applications*, Wiley, New York (2000).
[57] D.S. Williams, R.A. Rapp, J.P. Hirth, Phase suppression in the transient stages of interdiffusion in thin films, Thin Solid Films **142** (1986) 47–64.
[58] R.E. Hummel, Electromigration and related failure mechanisms in integrated-circuit interconnects, Int. Mater. Rev. **39** (1994) 97–111.
[59] N. Bertolino, J. Garay, U. Anselmi-Tamburini, Z.A. Munir, High-flux current effects in interfacial reactions in Au–Al multilayers, Philos. Mag. B **82** (2002) 969–985.
[60] S.-W. Chen, C.-M. Chen, Electromigration effects upon interfacial reactions, JOM **55**(2), (2003) 62–67.
[61] L. De Schepper, W. De Ceuninck, G. Lekens, L. Stals, B. Vanhecke, J. Roggen, E. Beyne, L. Tielemans, Accelerated ageing with in situ electrical testing: a powerful tool for the building-in approach to quality and reliability in electronics, Qual. Rel. Eng. Int. **10** (1994) 15–26.
[62] S.-W. Chen, C.-M. Chen, W.-C. Liu, Electric current effects upon the Sn/Cu and Sn/Ni interfacial reactions, J. Electron. Mater. **27** (1998) 1193–1198.
[63] C.-M. Chen, S.-W. Chen, Electric current effects on Sn/Ag interfacial reactions, J. Electron. Mater. **28** (1999) 902–906.
[64] C. Passagrilli, L. Gobbato, R. Tiziani, Reliability of Au/Al bonding in plastic packages for high temperature (200°C) and high current applications, Microelectron. Rel. **42** (2002) 1523–1528.
[65] M. Braunovic, N. Alexandrov, Intermetallic compounds at aluminum-to-copper electrical interfaces — Effect of temperature and electric current, IEEE Trans. Comp. Pack. Man. Technol. Pt A **17** (1994) 78–85.
[66] W.-C. Liu, S.-W. Chen, C.-M. Chen, The Al/Ni interfacial reactions under the influence of electric current, J. Electron. Mater. **27** (1998) L5–L8.
[67] K.P. Gurov, A.M. Gusak, Theory of phase growth in the diffusion zone during mutual diffusion in external electric field, Phys. Met. Metall. **52**(4), (1981) 75–81; (transl. of Fiz. Metal. Metalloved. **52** (1981) 767–773).
[68] H.T. Orchard, A.L. Greer, Electromigration effects on compound growth at interfaces, Appl. Phys. Lett. **86** (2005) 231906/1–3.
[69] H.T. Orchard, A.L. Greer, Electromigration effects on intermetallic growth at wire–bond interfaces, J. Electron. Mater. **35** (2006) 1961–1968.
[70] H.T. Orchard, A.L. Greer, Electromigration effects on competitive compound growth at interfaces, unpublished.

CHAPTER 16

Biology and Medicine

Contents

1. INTRODUCTION

Relating to biology and medicine, one area of interest for nucleation is its role in producing crystals to permit determination of the structure of large molecules, notably proteins, nucleic acids, and multimacromolecular assemblies. The crystals grown from solution must be of sufficient size and quality for diffraction studies. Spontaneous nucleation may be irreproducible and requires careful control of conditions, as the supersaturation required for nucleation is higher than that for growth. Spontaneous nucleation centers are often polycrystalline and unsuitable for growing single crystals. For proteins, nucleation of crystals may compete with the formation of amorphous aggregates in the form of

Pergamon Materials Series, Volume 15
ISSN 1470-1804, DOI 10.1016/S1470-1804(09)01516-8

precipitates or protein skins. For all these reasons, seeding at small supersaturation is widely applied. The seeds can be the best crystals previously grown, or crystals formed from a related molecule, or extraneous material that happens to be a suitable template, such as cellulose fibers. There is a wealth of experience in this area, sometimes involving complex sequences of heteroepitaxial seeding. Such matters are covered elsewhere [1], and lie beyond our scope. Instead, our focus is on nucleation occurring within living systems.

Terms such as *nucleus* and *nucleation* can have specific meanings in describing biological systems and their functioning. Confusion in the terminology of physical and biological processes is not confined to English; in French, for example, nucleation as a physical process is *germination*, and nuclei are *germes*. Specifically biological structures and processes — for example, the nucleus of a cell, the germination of seeds — are excluded from the coverage in this chapter, which is concerned only with physical nucleation of phase changes of the kind considered throughout this book. It will be seen that in biological systems, there can be strong effects inhibiting or promoting nucleation. These effects can be apparently accidental, or may serve a clear function, for the nucleation of a new phase can be detrimental or beneficial for the health or even the survival of living species.

The control exercised by biological systems is illustrated in Section 2 for the nucleation of ice, for which there are clear examples of both catalysis and inhibition. Another example of nucleation of a phase that is not itself part of the organism is the formation of gas bubbles (Section 3), important for decompression disease (*the bends*) in divers and for gas-bubble trauma in fish. *Biomineralization* (Section 4) in the formation of shells, bones, and teeth shows that nucleation may be important in forming the organism itself, and may be relevant for new types of ceramic processing. When biomineralization goes wrong — usually through the failure to control nucleation in a supersaturated solution — there are pathological developments such as gout and kidney stones (Section 5). Finally, this chapter includes discussion of the role of nucleation of protein aggregates in neurodegenerative disease (Section 6).

2. NUCLEATION OF ICE

2.1 Water in living systems

Water, as reagent, solvent, transport medium, or environment, is crucial for life processes. A typical mammalian cell, for example, is 70% water by weight. The equilibrium temperature between ice and water at atmospheric pressure is, interestingly, in the middle of the temperature range tolerated by living species; consequently, the formation of ice is a major, arguably the major, opponent of life. This is particularly direct for species with a temperature matching that of their environment — plants and *ectothermic* (cold-blooded) animals. Freezing inside a cell is, almost without exception, fatal, and mostly associated with perforation of the cell membrane. Although the growing ice crystals can penetrate cell membranes [2], the expansion as water transforms to ice is of itself insufficient to rupture the cell membrane and cause fatal damage. Ice, despite its open structure, has very low solubilities for typical solutes in the cellular liquid (*cytosol*). The formation of ice, freezing, therefore leads to a concentration of solutes in the remaining liquid; this concentration, including such effects as changes in pH, can be toxic [3]. Importantly, the increased solute concentration disturbs the osmotic balance between the intracellular and extracellular liquid. Water is drawn into cells in which the ice forms, resulting in swelling and ultimately membrane rupture [2]. In Section 2.2 we examine the methods used in living systems to prevent the formation of ice. Organisms can also find it beneficial to nucleate ice, and this is the focus of Section 2.3. The strategies for avoiding ice formation discussed in Section 2.2 fail at large supercooling, but organisms can still survive by controlling where the ice forms; this type of nucleation control is the basis for freeze tolerance outlined in Section 2.4. Finally, in Section 2.5, we note that the absence of freezing can also be important in long-term storage of organic materials, ranging from blood to (potentially) organs, and including foodstuffs.

2.2 Inhibition of freezing

The largest detected supercooling for the onset of freezing of water is 40 K [4], measured in droplet-emulsion experiments of the kind discussed in Chapter 7, Section 2.1.1. Such large supercoolings are outside the normal range of interest for living systems, within which the nucleation can be taken to be heterogeneous. For nucleation of ice in living systems, it is relevant that the water is not pure, is in contact with other matter, and is subdivided. Of prime importance is the cellular nature of the tissue. Liquid within the cells is separated from that outside by the cell membrane, which permits osmotic but not chemical equilibrium between the liquids. On slow cooling, it is often the extracellular liquid that freezes first. Nucleation within the cells can be studied, however, by using the droplet-emulsion technique to suppress freezing of the external fluid. An emulsion can be prepared that has cells within aqueous droplets suspended in oil [5]. In this way the extracellular water can be very substantially supercooled, down to the same limit as in the absence of the cells. In every case, the intracellular liquid attains substantially greater supercoolings than would be the case in the presence of prior extracellular ice. Intracellular freezing does nucleate, however, before the supercooling limit of the extracellular water, showing that the structures of the cell can act as heterogeneous catalysts. Mostly, however, the catalytic action is far from potent; in human red blood cells, for example, heterogeneous nucleation, apparently on the membrane, occurs only at an supercooling of 39 K [5], very close to that observed in pure water droplets in oil.

Whole animals can be supercooled. In smaller animals, there are smaller probabilities of forming an ice nucleus or of containing an effective seed for ice growth, and consequently greater supercoolings can be reached (Figure 1) [6]. The lowest temperatures observed for the onset of freezing appear to be lower than that for homogeneous nucleation of ice, but the liquidus temperature in body fluids can be reduced because of the solute present. Small animals, such as turtle hatchlings, improve their resistance to freezing by altering their skin to become more resistant to the penetration of external ice and by expelling potential heterogeneous nucleants for ice [7].

In plants, and in particular trees, even macroscopic tissues can show large supercoolings of around 40 K, presumably reflecting the subdivision of water into cellular volumes, which can promote supercooling in the same manner as in the droplet-emulsion technique. The lowest nucleation temperatures are found in *cold-acclimated* (*hardened*) plants, that is, those that have adapted to cold by altering the chemistry of their tissue water to inhibit ice formation, thermodynamically by lowering the liquidus temperature (freezing-point depression) and kinetically by increasing the viscosity. In general, though, freezing-point depression is

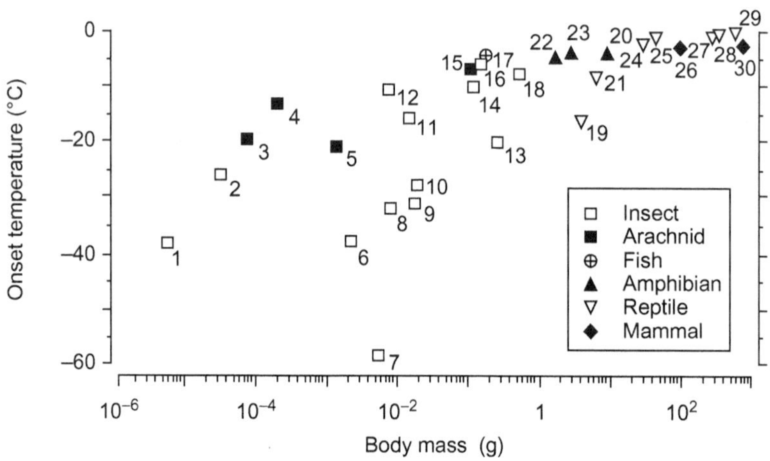

Fig. 1 For animals, the onset temperature for freezing is correlated with body mass. (Reprinted, with permission, from Ref. [6] by Annual Reviews. www.annualreviews.org.)

not a major contributor to freeze avoidance in plants [8]. Additional adaptations are to reduce the number of heterogeneous nucleants and to have more effective barriers against the growth of ice through the tissue [8]. Figure 2 shows thermal analysis data on the freezing of water in hardened plant tissue [8]. It is most remarkable that the freezing in

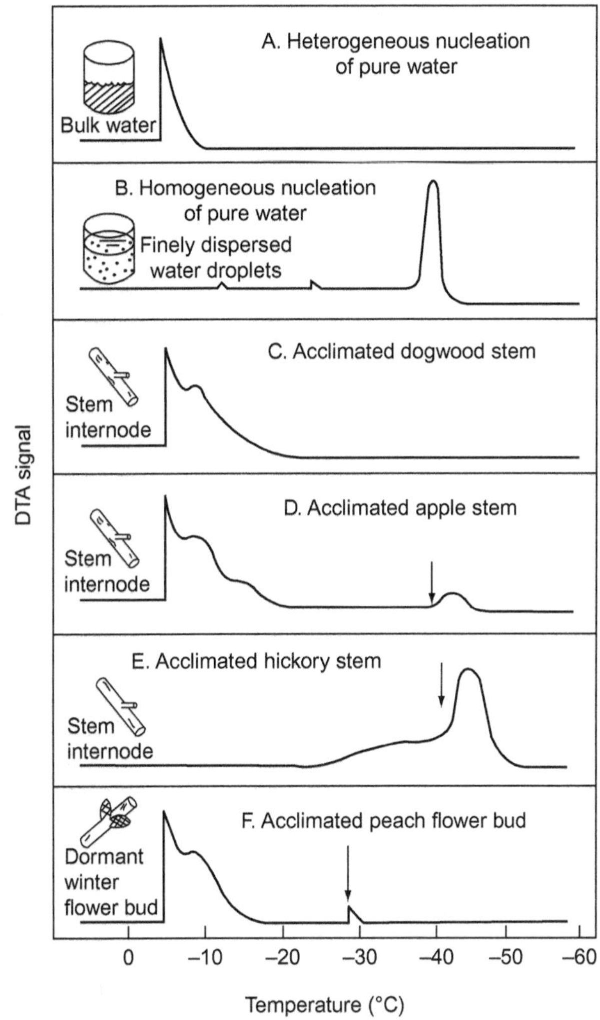

Fig. 2 Differential thermal analysis (DTA) exotherms on cooling: (A) bulk water; (B) water dispersed in the form of a droplet emulsion in an inert oil; (C)–(E) tissue water in acclimated or hardened woods; and (F) tissue water in hardened peach-flower buds. The vertical displacement from the baseline is proportional to the rate of heat output from the sample and the temperature scale shows the supercooling. (Reprinted, with permission, from Ref. [8] by Annual Reviews www.annualreviews.org.)

hardened hickory stem (curve E) occurs at very similar temperatures to the freezing of water in an oil emulsion. Curve F shows more common behavior in which the bulk of the tissue water (e.g., in the bark and nonliving xylem cells) freezes at rather small supercooling and only a small proportion, 5% in this case of peach buds, attains large supercooling. This small proportion is in the living cells, and its large supercooling provides an example where ice formed elsewhere does not appear to act as an efficient seed. The arrows in the figure show the temperatures below which death occurs, clearly correlated with freezing in the living cells. That a living system can survive partial freezing is considered further in Section 2.4.

In plant tissues, subdivision of the liquid can clearly be effective in inhibiting freezing, but this strategy is more difficult to apply in complex animal systems with such features as large vessels and circulating blood. A different strategy is very common, based on freezing inhibitors [9–12]. The best known are termed *antifreeze peptides* or *antifreeze proteins* in fish [13, 14] and *thermal hysteresis proteins* in insects [15–17]. In this chapter, these substances are termed antifreeze proteins (AFPs), and we note that they also have been found in plants [18, 19]. These proteins have now been isolated from many species to permit measurements of the onset supercooling for freezing under controlled in vitro conditions (Figure 3) [9]. When AFPs are present in water, the liquidus temperature is slightly depressed, as can be measured from the point at which ice crystals in the water start to melt (as noted in Chapter 14, Section 3, there is in most cases no barrier to melting). DeVries [20] was the first to show clearly that just below the liquidus temperature ice crystals did not grow, even though the system was clearly supercooled. A critical supercooling had to be exceeded for the ice to grow. Thus, for coexisting ice and water, the temperature for the onset of melting is higher than the temperature for the onset of freezing, a phenomenon termed *freezing hysteresis* attributed to adsorption of AFPs on the ice blocking its growth. As shown in Figure 3, the freezing hysteresis, in effect the supercooling required for the onset of freezing, greatly exceeds the depression of the liquidus.

The freezing hysteresis is small, less than 2 K, but it can be crucial for some species, as we now examine for polar fish. A mixture of seawater and ice has a temperature of 271.3 K (–1.9°C), and this is the temperature of fish in polar waters. Some species have high concentrations of solutes such as chloride, glucose, or glycerol in their body fluid and thereby depress the liquidus temperature to avoid supercooling. For *polar teleost fish*, however, their blood has a liquidus temperature of –0.6°C, so the supercooling, from birth to death, is 1.3 K [14]. Special measures are required to maintain this supercooling throughout the lives of the fish, which have a large body mass and which can be in contact with external ice. They achieve this by having AFPs in their blood plasma, and in other

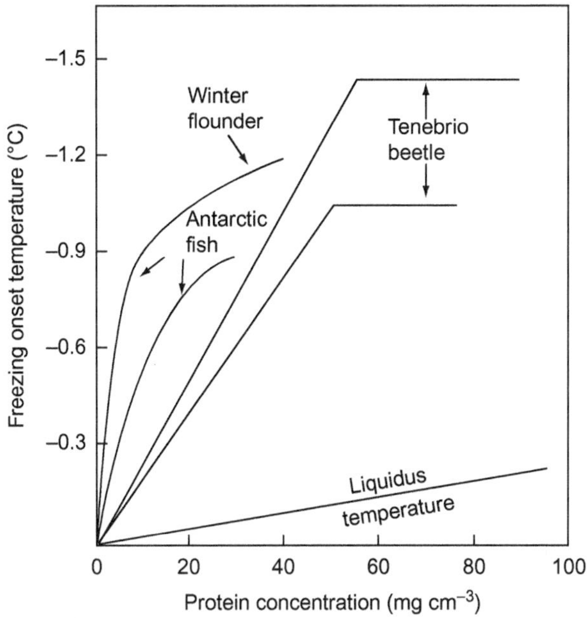

Fig. 3 In vitro supercooling potential of antifreeze proteins isolated from various species. (Adapted from Ref. [9], copyright (1985), with permission from Cambridge University Press.)

body fluids, for example in the intestine and bile [21]. The AFPs are not present within the cells.

The critical supercooling for the onset of ice growth is dependent on the concentration of AFPs present (Figure 3). When the critical supercooling is exceeded, the ice crystal morphology is greatly altered, to needles parallel to $\langle 0001 \rangle$, instead of the usual tabular forms with predominant $\{0001\}$ basal faces [12]. This is consistent with AFPs adsorbed on prismatic $\{10\bar{1}0\}$ and other nonbasal faces, blocking the usual growth mechanisms, though in some cases AFPs do adsorb on $\{0001\}$ [12]. AFPs isolated from different species are quite varied. They are glycopeptides with molecular masses in the range 2–25 kDa, but there is no clear correlation between the antifreeze action and the proportions of different amino acids. They can have a relatively hydrophobic side and a relatively hydrophilic one, the latter being that which binds to the ice surface. Insect AFPs confer the greatest freezing hysteresis, and some insect AFPs show very clear structural matching to ice. Liou et al. [22], for example, show that a protein in a β-helix structure has an array of oxygen and bound water molecules on its surface that may be the most regular protein structure yet observed. It matches the $(10\bar{1}0)$ plane of ice very well (Figure 4) and shows strong hydrogen bonding. In general, AFPs should have a planar [23] array of polar groups, matching the ice structure.

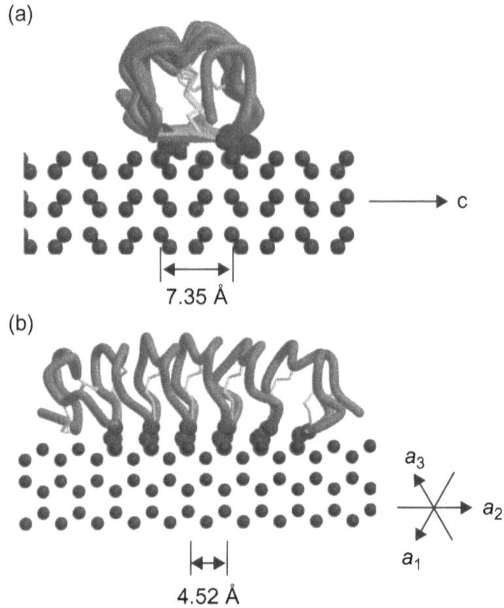

Fig. 4 Lattice matching of an insect antifreeze protein (AFP) (from the beetle *Tenebrio molitor*) on to the $(10\bar{1}0)$ prismatic face of ice. (a) The β-helix structure of the AFP viewed along its axis, parallel to the $[\bar{1}2\bar{1}0]$ axis of ice; (b) the same configuration viewed orthogonally along [0001] of ice. The crystallographic axes of the hexagonal ice structure are shown. (Reprinted by permission from Macmillan Publishers Ltd: *Nature* Ref. [22], copyright 2000.)

AFPs appear to act through the poisoning of the usual sites for molecular addition at the ice–water interface. Between the points where an AFP molecule blocks growth, the ice–water interface may still advance, but it becomes curved, hindering further growth of ice through the Gibbs–Thomson effect (Chapter 4, Section 5.3 and Chapter 6, Section 2.5) [11]. Observed onset supercoolings for freezing can then be interpreted in a similar way to those for the onset of free growth considered earlier (Chapter 6, Section 2.5 and Chapter 13, Section 3.4). From the onset supercoolings observed in fish, the spacing between adsorbed AFPs has been estimated to be 16–22 nm; for insects, with their greater supercoolings, the spacing must be smaller, perhaps as little as 5 nm [24].

2.3 Ice-nucleating agents

The afro-alpine plant *Lobelia teleki* grows on Mount Kenya in temperatures showing a daily cycle from 263 to 283 K. In water in a central vessel, it uses a nucleating agent to ensure that there is no significant supercooling below the liquidus temperature of 272.66 K [25]. The latent heat released on ice formation in this water inhibits the fall of the plant temperature during the night. The agent has been extracted from the plant and shown to prevent any significant supercooling, even of microdroplets.

The most potent ice-nucleating agents (INAs) are those found in bacteria, for example *Pseudomonas syringae* [26, 27], and fungi that cause frost damage to crops. The damage releases nutrients and the bacteria benefit if this happens at smaller supercoolings. These INAs have been closely studied because of their detrimental effects on agricultural production [28, 29], and controversially genetic modification has been used to render some microorganisms ineffective as nucleants [30]. They are also likely to be important in nucleating the freezing of water droplets in the atmosphere, an important contributor to the processes of precipitation [31–33].

The INAs have a complex proteinaceous structure, localized in the outer membrane of the bacteria. The structures have hydrophilic active groups in a regular array matching that of ice. Freezing catalyzed by INAs shows a wide range of onset supercoolings, and the associated INAs have been classified as follows: type 1 giving nucleation temperatures of −5°C, type 2 giving −5 to −7°C, and type 3 giving −7 to −10°C [34, 35]. Govindarajan and Lindow [31] showed clearly that more potent nucleation (smaller supercooling) is associated with larger mass of the nucleating site. Burke and Lindow [36] examined this correlation in more detail. They showed that the basic protein with a mass of ~150 kDa would give a nucleation temperature of −12 to −15°C, but that a large aggregate of mass, ~8700 kDa, would give a temperature of −3°C. They used the analysis of Fletcher (Chapter 6, Section 2.4, Eqs. (19)–(21)) to calculate nucleation rates on spherical, disk-shaped, and cylindrical substrates. They adjusted the water–ice interfacial energy σ_{ls} and the contact angle ϕ for an ice nucleus on the substrate to obtain best fits to observed onset supercoolings. They obtained the best fits for disk-shaped substrates with $\sigma_{ls} \approx 30\,\mathrm{mJ\,m^{-2}}$ and $\phi = 0$. Since the effective contact angle is zero, the correct analysis to use, however, is probably not that of Fletcher based on the temperature dependence of nucleation rates, but rather the much simpler athermal nucleation model outlined in Chapter 6, Section 2.5, and applied specifically to freezing in Chapter 13, Section 3.4. Figure 5 is based on INA disk diameters quoted by Burke and Lindow, and the supercooling is fitted to Eq. (4), Chapter 13. A reasonable fit is obtained, suggesting that the free-growth model can be applied. Furthermore, the value of $\sigma_{ls} \approx 31.3\,\mathrm{mJ\,m^{-2}}$ obtained from the fit is well within the range of values quoted for the ice–water interface, 20–44 $\mathrm{mJ\,m^{-2}}$ [37].

Cold-conditioning improves the potency of nucleation by INAs. The fitted value of $\phi = 0$ rules out any changing surface structure as an explanation. Burke and Lindow conclude that cold-conditioning can be

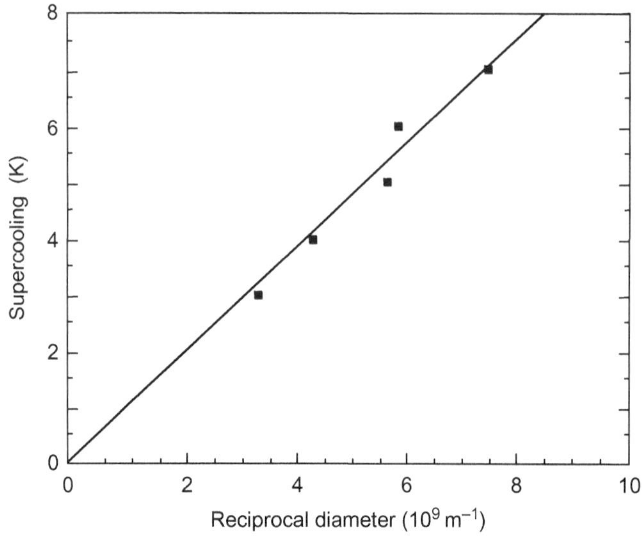

Fig. 5 Measured supercoolings for ice nucleation in water catalyzed by the bacterium *P. syringae* as a function of the reciprocal of the diameter of disk-shaped ice-nucleating agent substrate. The straight line is a fit to Eq. (4), Chapter 13, consistent with the heterogeneous nucleation being athermal (Chapter 6, Section 2.5). The gradient of the line corresponds to $\sigma_{ls} = 31.3\,\mathrm{mJ\,m^{-2}}$. (Data from [36].)

explained only by increasing the size of the INAs. It is generally accepted that the size can increase by aggregation [38], associated with decreasing temperature of the microorganism. This aggregation can also be reversed by heating. Figure 6 shows a detailed model for these effects [35].

We are used to expecting that the evolutionary processes in nature can optimize performance beyond that possible in man-made systems. There is, for example, nothing in conventional solidification processing (Chapter 13) to match the nucleation- and growth-inhibiting effects of biogenic AFPs. Further examples of the superior performance of biological systems will be found in Section 4. But the action of INAs is not so impressive. In Chapter 13, Eq. (4), the ratio of σ_{ls} to Δs for water is $\sim$17% of that for the solidification of aluminum. For a disk-shaped nucleant substrate of a given radius, then, the free-growth supercooling for the freezing of water should be $\sim$17% of that for aluminum. In metallurgical processing, the supercoolings obtainable by adding inoculant particles are of the order of 0.2 K, whereas in the biological systems, the supercoolings are large, at least 3 K and possibly 10 K or more. The origin of the poor performance of the biogenic nucleants lies in their small size. Whereas the inoculant particles in metallurgical processing have radii of a few micrometers, the INA proteins have radii of only at most a few tens of nanometers. Flat surfaces of even these radii

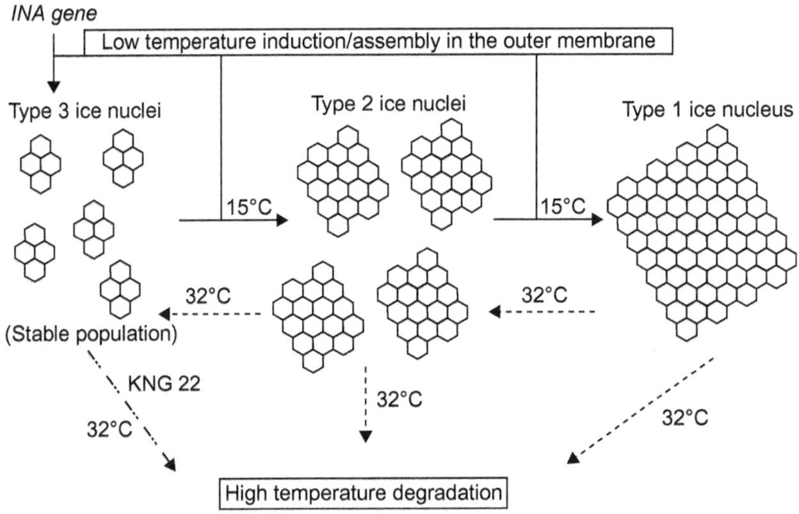

Fig. 6 Conceptual model for the thermal control of aggregation to form, and of disassembly to remove, type-1 ice nuclei on the outer membrane of *P. syringae*. The hexagons represent individual ice-nucleating agent protein subunits. (Reprinted with permission from Ref. [35], copyright (1993) by the American Society for Microbiology.)

are anomalously large for proteins that would more usually fold into compact spheroidal structures.

It is intriguing that the arrays of sites for hydrogen bonding on the surfaces of AFPs (Figure 4) and INAs must be similar; in each case the requirement on a local scale is to bind firmly to water molecules in an arrangement matching that in ice. Most simply this means that the array must be flat [23]. AFPs mostly bind onto prismatic or other nonbasal faces [12], while INAs provide templates for the basal plane. The key difference, though, is that the AFPs are isolated molecules, whereas the INAs must have an area large enough to act as a template and even more stringently must be large enough to permit the onset of freezing through free growth at sufficiently small supercooling.

As many substrates may catalyze heterogeneous nucleation of ice, there may well be INAs in a system that are *incidental*, rather than *adaptive* (i.e., developed to fulfill a function) [39].

2.4 Freeze tolerance

The avoidance of any ice formation is the most obvious way to promote survival in cold conditions. As noted earlier, avoidance can be favored by subdivision of the tissue water, by expulsion of nucleant substrates, by the actions of AFPs, and by increasing and changing solute content to depress the liquidus temperature and increase viscosity.

A contrasting survival strategy, *freeze tolerance*, is necessary when temperatures are so low that ice formation is inevitable. With only a few exceptions, intracellular ice is lethal, so the essence of freeze tolerance is that the ice forms outside the cells [6]. The growth of ice in extracellular fluid may assist survival by release of latent heat [21]. The more important effect, however, is that the freeze concentration of solutes in the residual extracellular liquid draws water from within the cells as the system tries to maintain the osmotic equilibrium. In this way, the water content of the intracellular liquid is so decreased that formation of ice is prohibited. The cells can become very dehydrated and shrunken. As discussed further in Section 2.5, cells can remain viable under dehydration if the cellular fluid (*cytosol*) vitrifies, avoiding the crystallization of salts from the supersaturated cytosol, limiting further dehydration and stabilizing the structures within the cells. This stabilization of the cells is promoted by the presence of *cryoprotectant* molecules. Freeze tolerance relies on osmotically driven permeation of water from the cells through the cell membrane to the extracellular fluid. For survival, the cooling and ice formation must be slow. Rapid cooling is almost always lethal for plants and for animals [6, 8].

Pure water cannot readily be vitrified, but addition of solutes in many biological systems does permit glass formation. The addition of solute can depress the ice liquidus and raise the glass-transition temperature of the solution. Cooling, and the concomitant freeze concentration of solute in the remaining liquid, can bring the liquidus and the glass-transition temperatures to equal values. For example, the addition of 64 wt.% sucrose to water brings both temperatures to 241 K, as explored further in Chapter 17, Section 2 (Figure 1) [40, 41]. Further cooling then leads to vitrification of the remaining liquid. There is not a large risk of ice nucleation, because the solute addition makes this more difficult, as can be monitored from the ultimate temperature for nucleation onset in droplet-emulsion experiments. A number of studies indicate that the depression of the nucleation onset is about twice that of the liquidus [42, 43]. The vitrification occurs when the solution is supersaturated in the solute, but many solutes are themselves very reluctant to nucleate. Notably, polyhydroxy compounds (carbohydrates and polyols, including sugars) are resistant to unaided crystallization from solution (Chapter 17, Section 2). The vitrification sets a kinetic limit to the freezing of ice (or any other crystal), and leads to the phenomenon of *unfreezable water*. An additional factor inhibiting complete freezing is the subdivision of the liquid into cells and into small pores and capillaries. The formation of glass, rather than continued freezing, makes an important contribution to preservative action on cooling (Section 2.5).

Most plants that survive significantly sub-zero temperatures do so by being tolerant of extracellular freezing, and by developing barriers to the propagation of ice into cells. Some cold-acclimated woody plants survive all of their freezable water (~70% of the total) transforming to ice extracellularly [8]. Survival is then essentially a matter of coping with extreme cellular dehydration. Freezing of the extracellular fluids in plants is readily nucleated by external ice (itself likely to have been nucleated by bacterial INAs, Section 2.3) through various entry sites and then propagates rapidly through the plant.

The same mechanisms of freeze tolerance are found in animals, in particular insects, marine invertebrates, and hibernating amphibians and reptiles [6, 44]. Insects are among the most cold-resistant organisms. The two strategies of freeze avoidance and freeze tolerance are illustrated schematically in Figure 7. In each case, the *cold-hardiness* is promoted by the accumulation of polyols such as glycerol (or exceptionally ethylene glycol) in the autumn and early winter [45]. In summer, the insects show supercoolings of 8–12 K, compared to supercoolings of 20 K measured in artificial body fluid samples of the same volume (~15 μL) as the insects. This suggests that there are incidental INAs in the cells or intestine. In freeze-avoiding insects, these INAs are removed in winter and there is production of AFPs. In contrast, in freeze-tolerant insects, there is production of INAs, which stimulate freezing in the extracellular fluid. The adaptive INAs in overwintering insects have been extensively studied [45, 46].

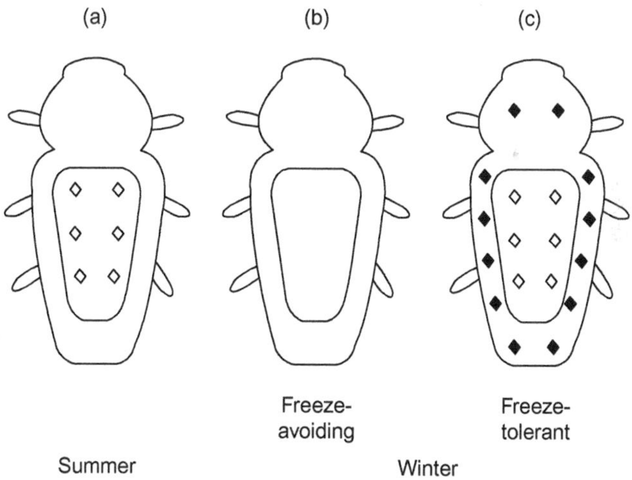

Fig. 7 Alternative strategies for avoiding death by freezing, illustrated for an insect. Ice-nucleating agents (INAs) in blood plasma (◆) can be distinguished from those (◇) in the intestine. In summer (A) there is no risk of freezing. In winter, *freeze avoidance* (B) requires removal of all INAs, but *freeze tolerance* (C) may be easier as the requirement is only for active INAs in extracellular regions where ice formation will do no damage. (From Ref. [45], copyright (1980), with kind permission of Springer Science and Business Media.)

Other species as diverse as lizards [47] and earthworms [48] can survive extracellular freezing of 50% of their total body water and use glucose as a cryoprotectant for their cells. The most studied case is probably that of the wood frog *Rana sylvatica*, whose hibernation involves extracellular freezing of up to 70% of its body water, and survival in this frozen state for periods of more than 1 month [6, 49]. Augmenting (by injection) their natural glucose levels improves their survival rates for even longer periods of freezing [50]. As survival depends on the initial freezing being slow, it is important that freezing begins at supercoolings that are as small as possible. For invertebrates, INAs play an important role. Bacterial INAs have been isolated from vertebrates such as the wood frog [51] and they can enhance freeze tolerance, but more usually freezing is nucleated by external ice (itself often nucleated by bacterial INAs in soil or leaf mold). The moist, highly permeable skin of the frog facilitates this *inoculative freezing* triggered by external ice [52, 53].

2.5 Glass formation and preservation

Nucleation of ice, its nature or avoidance, is important throughout the field of *cryopreservation*. This is based on the same principles as the cold resistance of species in the natural environment, but there is control of heating and cooling, and much higher rates can be achieved. It is not appropriate to review the large field of cryopreservation here (the interested reader is referred to the book by Franks [9]), but the range of applications will indicate its importance: preservation of plant tissue, spermatozoa, microorganisms, and insects. Notable success has been achieved with blood storage and with the freezing of mammalian embryos. A related topic is the use of freezing in the processing and preservation of foodstuffs.

In a glass, degradation reactions that might progress in an aqueous system are stifled. Vitrification, promoting survival in living systems or preservation generally, can be induced not only by freeze concentration on cooling, but also by direct dehydration. Survival depends on the preservation of the structures within the cells, and in particular it is necessary to avoid the nucleation and growth of salt or other crystals from the increasingly supersaturated cytosol during drying. Seeds are a good example; they are substantially dehydrated, but glassy, and in consequence can survive long periods in a dormant state [54]. Whole plants, particularly the so-called *resurrection plants*, can survive extreme dehydration, also by vitrification [55]. The intracellular glasses are based on a mixture of proteins and sugars, often sucrose; crystallization is retarded primarily due to their high viscosity [55]. Primitive animals, for example, various types of worm [56, 57], can also survive extreme dehydration by the same mechanism.

3. NUCLEATION OF GAS BUBBLES

Gas bubbles can nucleate and grow in the blood within the tissues, organs, or vascular systems of animals. A well-known example is in the *decompression sickness* suffered by divers. Another example is *gas-bubble trauma* in fish. In all cases, the formation of gas bubbles in animals causes major injury and mortality by a wide variety of mechanisms. It is subject to research to optimize the decompression sequences for divers and to protect commercial fisheries.

For gas bubbles to form, the blood must be supersaturated in dissolved gas. Chapter 13 Eqs. (8) and (9) apply. Taking the H_2O vapor pressure over blood to be negligible compared to the other terms, the critical radius for bubble nucleation is given by

$$r^* = \frac{2\sigma}{(p_g - p_l)}, \tag{1}$$

where σ is the surface energy (surface tension) of the blood, p_g is the external pressure of gas that would be in equilibrium with the actual concentration dissolved in the blood, and p_l is the actual pressure of the liquid blood. Bubble nucleation can be driven by increasing p_g (as in gas-bubble trauma in fish) or by decreasing p_l (as in decompression sickness).

We do not consider the separate components of the gas, taking there to be a single dominant gas species that is nondissociating. (The example of a dissociating species, hydrogen, has already been encountered in Chapter 13, Section 6, when treating porosity formation in solidification of metals.) Most diving is performed breathing air, and the nitrogen acts as an inert gas. It is the gas of interest in decompression sickness. For nondissociating species, the concentration of dissolved gas molecules C_g is related to the equilibrium pressure by Henry's law:

$$C_g = Hp_g, \tag{2}$$

where H is an empirically determined constant. The *supersaturation ratio* α_s is the factor by which C_g exceeds the concentration C_0 that would be in equilibrium with the actual pressure of the liquid p_l. Unlike the previous case considered in Chapter 13, Eq. (12), the supersaturation ratio is now given by the ratio of pressures:

$$\alpha_s = \frac{C_g}{C_0} = \frac{p_g}{p_l}. \tag{3}$$

As noted earlier, the *supersaturation*, *s*, is $(\alpha_s - 1)$. Equation (1) can be expressed in terms of supersaturation as

$$r^* = \frac{2\sigma}{p_l s}. \tag{4}$$

For a bubble of radius r, the pressure within it, p_b, is given by

$$p_b = p_l + \frac{2\sigma}{r}. \tag{5}$$

At the critical radius, the pressure of the gas in the bubble is in equilibrium for the supersaturation of dissolved gas ($p_b = p_g$). For larger bubbles $p_b < p_g$ and growth is driven by more gas coming out of solution. Conversely, smaller bubbles have $p_b > p_g$ and shrink.

In the case of divers, the supersaturation rises by approximately 1 for each 10.5 m of water depth because of the high pressure of the gas needed to permit breathing. Final supersaturations on subsequent decompression may exceed 9, the resulting bubble formation leading to *the bends*. Similar effects can be suffered by aircraft pilots when decompression takes the ambient pressure below 1 atmosphere. Less well known is gas-bubble trauma (also known as *gas-bubble disease*) in fish. In this case, the supersaturation can arise from an increase in temperature, a decrease in barometric pressure, or a decrease in immersion depth, but the problem became significant only when fish encountered significantly supersaturated water in the courses from hydroelectric generators and in shallow hatcheries. The supersaturations in these cases are around 0.5, but can be as high as 1. In young fish, however, gas-bubble trauma appears at supersaturations as low as 0.03. Levels such as this, and even the significantly higher levels found in mammalian decompression, are far below the supersaturation needed to induce homogeneous nucleation.

Consistent with Eq. (4), the critical work W^* for homogeneous nucleation is given by

$$W^* = \frac{16\pi\sigma^3}{3p_l^2 s^2}. \tag{6}$$

As discussed in Chapter 7, Section 2.2.2, at the threshold of homogeneous nucleation, $W^* \approx 60k_BT$. The surface tension of blood has been determined to be 55–72 mN m^{-1} [58]. The critical supersaturation for the onset of homogeneous nucleation is then found to be greater than 1000. It is clear that heterogeneous nucleation is controlling the supersaturation at which bubbles appear, and indeed it is found that these threshold values are sensitive to environmental and physiological characteristics.

There has been much effort to characterize the nucleation sites. Inevitably, in living systems there is a complex array of possible nucleant substrates. Nevertheless, it seems clear that nucleation is favored on hydrophobic surfaces. By having a relatively high interfacial energy with water/blood, these would promote the formation of gas-filled spherical-cap nuclei with low contact angle (Chapter 6,

Section 2.1). Lipids, including human fat, are hydrophobic, and it is known that subjects with greater fat content do suffer more from decompression sickness [59]. The same correlation is found for gas-bubble trauma in fish [60]. Lipids occur inside cell membranes, but expose hydrophobic surfaces to the liquid only if the membrane is damaged. Gas-bubble nucleation is promoted by local injury where such damage would occur.

Planar hydrophobic surfaces are not potent catalysts, however, requiring supersaturations of approximately 9 to become active [61]. This is well beyond observed threshold levels. Harvey et al. [62] proposed that crevices in hydrophobic surfaces could act as potent nucleation sites by stabilizing prior bubbles, even in the case of undersaturation. Figure 8 illustrates the main features of this model (related to that discussed in Chapter 6, Section 2.4, Figure 14). Even in an undersaturated solution, a small bubble in a crevice can be stabilized by the curvature of the gas/liquid interface. With supersaturation, the bubble can grow. Typically the crevice defines a lateral dimension. When the supersaturation becomes great enough for the critical nucleus diameter to fall below this dimension, free growth of the bubble can occur (as in Chapter 6, Figure 12c), analogous to the processes in inoculation of melts (Chapter 13, Section 3.4). Evidence for the crevice theory is provided by memory effects. Without changing dissolved gas content, application of high pressure raises the supersaturation threshold for bubble formation on subsequent decompression [62, 63].

At the lowest observed onset supersaturations, the critical diameter of a crevice that would permit free bubble growth is as large as 10 μm. This is comparable with the diameter of smaller capillaries, and suggests that such crevices are most likely to be found in the walls of large structures such as arteries.

Bubbles formed in blood are found to become coated by organic skins. These skins have been observed, for example, by Philp et al. [64]; they appear to have an inner layer (approximately 0.2 μm thick) of fatty acids and an outer layer of cellular material (red blood cells and endothelial cells). It is proposed that bubbles that become subcritical could be stabilized by this skin; they would, therefore, survive to act as nuclei in case of supersaturation [65, 66]. Quite hard surfaces can form on bubbles, as the blood begins to clot. The resulting rigidity causes pain when recompression is used to alleviate decompression sickness.

The above account has focused on bubble formation in the vascular system (for more details, such as the preference for arteries or veins, see Fidler [67]). Bubble formation within tissues has been much less studied. Further examples of bubble nucleation are considered in Chapter 17, Section 4.

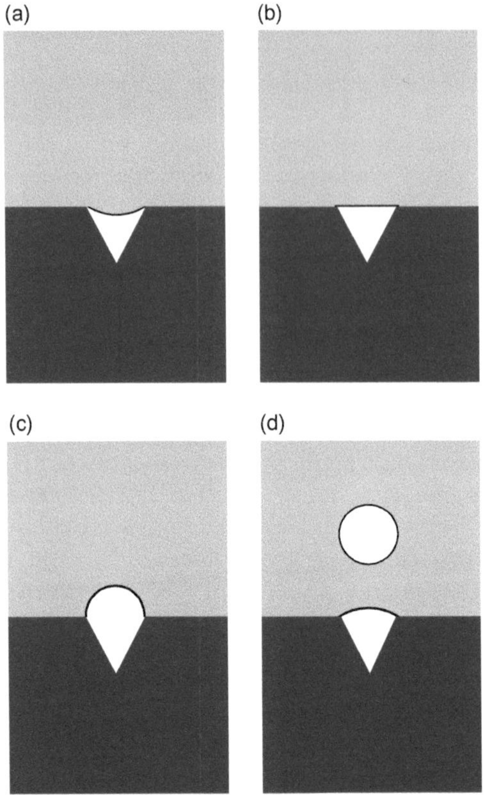

Fig. 8 Gas bubbles in a liquid take different forms in a crevice, depending on the relative values of p_g (the external pressure of gas that would be in equilibrium with the actual concentration dissolved in the liquid) and p_l (the actual pressure of the liquid). (a) The curvature of the gas–liquid interface (Eq. 1) can stabilize a bubble even when the liquid is undersaturated ($p_g < p_l$); (b) the curvature is zero at equilibrium ($p_g = p_l$); (c) the curvature reverses with supersaturation ($p_g > p_l$), and if the critical hemispherical shape can be reached free growth and subsequent bubble detachment (d) can occur.

4. BIOMINERALIZATION

4.1 Introduction

Biomineralization, the formation of inorganic solids in biological systems, is a focus of much current research and has been extensively reviewed [68]. In contrast to the formation of ice considered in Section 2, the solids formed may be integral, structural components of the organism, such as shells, bones, and teeth. Biominerals can also serve as optical lenses, as mineral stores, or as sensors for magnetic or gravitational fields. Roughly 60 biogenic minerals are known, of which easily the most common are calcium phosphate (in bones and teeth), calcium carbonate (in shells), and amorphous silica (in the shells of unicellular organisms such as diatoms). Calcium carbonate has six polymorphs: calcite, aragonite, vaterite, two hydrated phases, and an amorphous phase. Calcite is thermodynamically the most stable, the amorphous phase the least stable. Only the two most stable phases, calcite and aragonite, are commonly found in shells.

The processes of evolution have led to biogenic materials that are remarkably well adapted for their function [69], sometimes vastly outperforming their synthetic counterparts. A good example is the inner shell (*nacre, mother of pearl*) of mollusks; this is composed of aragonite crystals, but has a toughness roughly 3000× that of the nonbiogenic monolithic mineral [70]. It is possible to analyze evolutionary developments in terms of design strategies, one being to produce mineralized tissues with greater isotropy of properties [71]. Of course, biomineralization is itself vital in charting evolution, as mineralized biological materials are far better preserved in the fossil record than organic tissues [71].

The excellent properties of biominerals are all the more remarkable considering that the organisms can craft them in very intricate shapes, that the materials have a range of compositions that is very limited when compared to the range of man-made engineering materials, and that the materials are formed under mild conditions including ambient temperature and pressure. Producing advanced ceramic materials without high-temperature processing, as nature evidently can, is of great environmental interest. This is among the reasons for natural processes to be taken as inspiration, in the fast-growing area of biomimetic materials synthesis. The excellent properties of biominerals have their origins in complex, hierarchical composite structures combining the mineral with an organic matrix. Although the matrix is typically less than 2 wt.% of the material, it determines the structural organization and the properties. The organic matrix is intimately mixed with the mineral crystals, often even down to a nanometer scale. The living system producing the tissue controls the composition of the mineral crystals, their phase selection, their crystallographic orientation, their size, shape, and distribution, and their aggregation into complex composite structures.

Nucleation of the mineral phase is central to achieving such control. Biomineralization never involves spontaneous crystallization; rather, nucleating substrates are used to achieve dispersions of crystals of uniform size and shape, often aligned [72]. Calcium carbonate, the focus of our coverage, provides a comparison of biogenic and nonbiogenic processes. Precipitation of calcium carbonate from a supersaturated aqueous solution will often yield a mixture of polymorphs. But living systems only ever select one polymorph in a given type of location. While overall biomineralization processes are very diverse and complex, it appears that there are common mechanisms underlying nucleation whether derived from a common ancestor or independently evolved [73].

Analysis of nucleation in vivo has not been achieved, but there are many studies of related systems in vitro. While it is important to recognize the huge differences from the living environment, the in vitro studies have led to many insights on nucleation processes. Much remains to be done, however. There are as yet no direct studies of nucleation rates. Analyses are commonly based on the variation of nucleation induction time with supersaturation, and the kinetic analyses applied are typically primitive, taking no account of the transient effects discussed in Chapter 3 [74].

The basic mechanisms of nucleation in biomineralization are considered in Section 4.2. Selection between the calcium-carbonate polymorphs calcite and aragonite in mollusk shells is particularly interesting for what it reveals about nucleation control; this selectivity has been much studied (Section 4.3). Finally, the great progress in biomimetic materials synthesis is briefly outlined in Section 4.4.

4.2 Nucleation mechanisms

The importance of the organic matrix in biominerals has already been noted. In this section, we focus on the role of the matrix in mediating the nucleation of the crystalline inorganic phase. This nucleation is assumed to be from a supersaturated aqueous solution, given the hydrated conditions in a living system [75]. Both structural and functional aspects of the matrix need to be considered [68]. The major fraction of the organic matrix is an insoluble, hydrophobic framework, defining the structure within which the mineral phase forms. In most cases, the framework stays in place in the final material, though tooth enamel provides an example where the framework proteins are broken down and progressively removed as the mineral grows [71]. While there is some evidence that the framework proteins can play a role in mineral nucleation [76], this would not be expected. To form a good template for a mineral, a defined pattern of ions or hydrated species is sought, such as would be found on a hydrophilic surface. Such a surface is provided by attaching on to the framework macromolecules that are hydrophilic, relatively soluble, and predominantly acidic. These macromolecules control nucleation, but are difficult to study. They are so closely associated with the mineral phase that they can be difficult to extract from mineralized tissues. They are diverse, but have common chemical attributes, such as being rich in carboxylate groups.

As reviewed by Mann and others [68], a successful template, favoring heterogeneous nucleation, may have several characteristic features. The first is that the nucleant substrate should have an electrostatic interaction with the solution favoring an accumulation of relevant ions. For example, the nucleation of calcium phosphate on silica is enhanced by coating the silica with a cationic or anionic polyelectrolyte [77]. Studies of Langmuir monolayers as model substrates suggest that the surface charge density can be a dominant factor controlling the selection of mineral polymorph and crystallographic orientation [78]. One way of building a three-dimensional crystal nucleus through attraction of ions would be to have a patterned substrate with a local pocket (Figure 9) [68].

Of course, lattice matching to give heteroepitaxial growth on the substrate could explain both phase selection and crystallographic orientation. There is surprisingly little direct evidence for matching [68]. One of the clearest cases is the mineralization in nacre. Nacre is the inner layer of mollusk shell and consists of thin platelets of aragonite around 0.5 μm thick (Figure 10 [75]) in a matrix of β-chitin. The platelets are parallel to the shell surface and to (0 0 1) of the aragonite. The evidence for lattice matching comes from the alignment of the [1 0 0] and [0 1 0] axes of the aragonite (orthorhombic system) with the corresponding axes in the

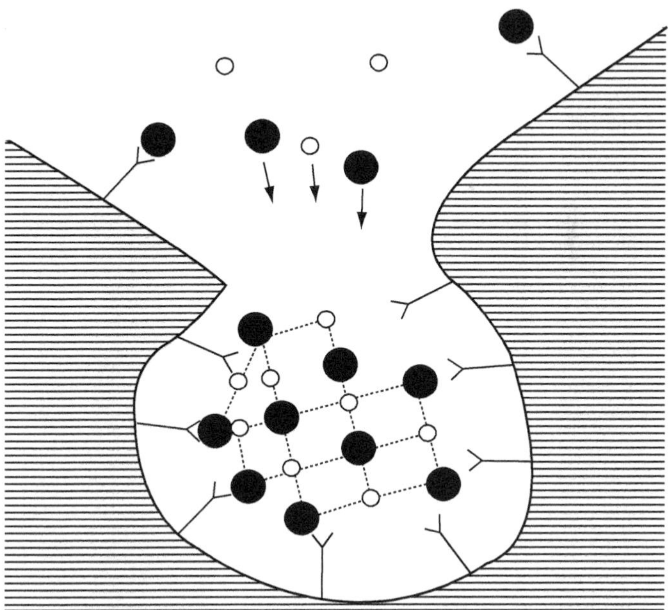

Fig. 9 A schematic representation of a pocket in an organic surface. The distribution of functional groups can direct nucleation of a particular inorganic phase. (Reprinted from Ref. [68], copyright (2001), with permission from Oxford University Press.)

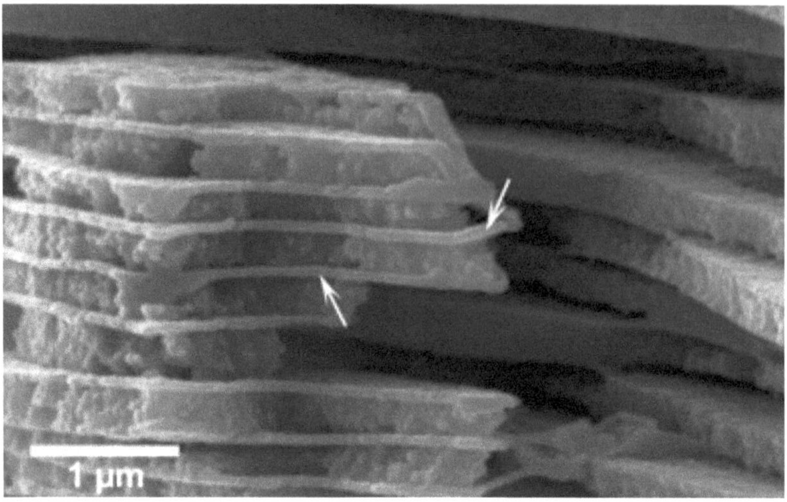

Fig. 10 Scanning electron micrograph of a fracture section through nacre from *Nautilus pompilius*. The aragonite platelets are separated by thin layers (arrowed) of organic matrix, swollen in this case by the specimen preparation for microscopy. (Adapted from Ref. [75], copyright (2006), with permission from Elsevier.)

chitin and from the approximate matching of lattice parameters (Figure 11) [68]. It should be noted, though, that the degree of in-plane orientation of the aragonite may exceed the degree of order in the chitin. This suggests that factors such as preferred growth directions can also play a role in defining the in-plane orientation [79]. The organic matrices supporting templates for nucleation in biomineralization are much more flexible than would be possible for the inorganic substrates in conventional studies of heteroepitaxy. Stretching of the templates can lead to reorientation of functional groups and can promote the formation of different crystalline phases [80].

It would be expected that a combination of electrostatic and lattice-matching effects could give a strong nucleating effect. This was the basis for the model system developed by Addadi et al. [81], in which polystyrene films were sulfonated and then treated so that β-sheet poly (aspartate) was adsorbed on to them. The adsorbed carboxylates are weak binders to calcium, but have an appropriate ordered structure. The sulfonated surface generates a strong field to which calcium ions are drawn, without being attracted to specific groups. The combination does enhance heterogeneous nucleation of calcite (trigonal system) with (0 0 0 1) parallel to the substrate as is typical in vivo. The same cooperative effect is implied the functional groups found on organic matrix surfaces from nacre [82]. These are the surfaces of the β-chitin interlamellar sheets. Before the mineral forms, these sheets are separated by a gel. Each aragonite platelet appears to nucleate at a specific location

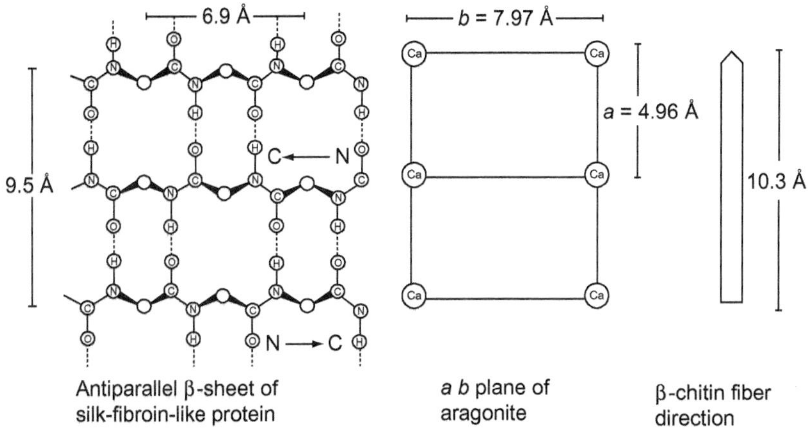

Fig. 11 Lattice matching between the aragonite (0 0 1) plane and the β-chitin sheet in nacre. (Reprinted from Ref. [68], copyright (2001), with permission from Oxford University Press.)

on the sheet surface, grow normal to the surface until stopped by the next chitin sheet, and then grow laterally until it impinges on its neighbors, ending up with a polygonal perimeter detectable on the sheet after the mineral has been extracted. Nudelman et al. used an ion-exchange resin to demineralize the shell, and by staining were able to detect the nucleation site on the sheet at the center of each platelet and to map the distribution of functional groups (Figure 12) [82]. Corresponding to the center of an aragonite platelet 10–15 μm in diameter, they found sulfates concentrated in a ring (outer diameter 3 μm, inner diameter 600 nm) surrounding a spot rich in carboxylates. Both the inner spot and the ring also showed a raised concentration of acidic macromolecules known to selectively nucleate aragonite. This arrangement thus seems to have everything needed to act as a nucleation site.

The earlier discussion has included both calcite and aragonite. A mollusk shell typically includes both, platelets of aragonite in the inner layer of nacre as just described, and prismatic crystals of calcite elongated perpendicular to the shell surface in an outer layer (Figure 13). It is not clear that there is a particular reason for these phases to be selected, but it is clear that the phase selection is rigorously controlled. These polymorphs of calcium carbonate form with (0 0 0 1) of calcite and (0 0 1) of aragonite parallel to the nucleating substrate. The arrangement of calcium ions on these planes in the two structures is practically identical (Figure 14), so lattice matching on these faces fails to explain how one or other phase can be selected. Just how phase selection might work has been the focus of many studies, as outlined in Section 4.3.

A possible mechanism for selection between calcite and aragonite is that the arrangement of carbonate ions around the calcium ions is

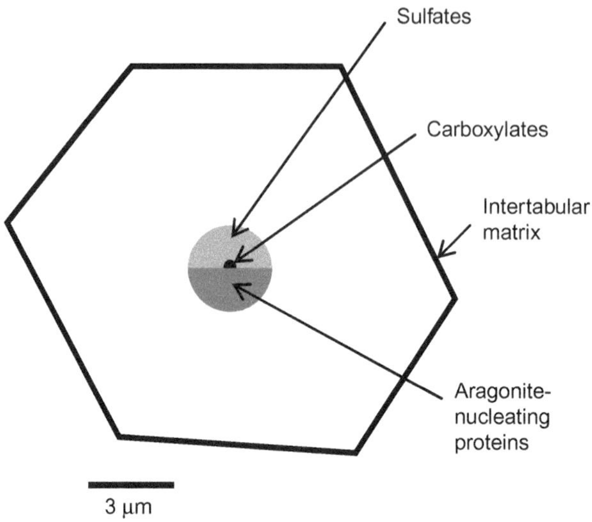

Fig. 12 The distribution of active species on the organic matrix covering the (0 0 1) face of an aragonitic tablet in nacre from *N. pompilius*. (Adapted from Ref. [82], copyright (2006), with permission from Elsevier.)

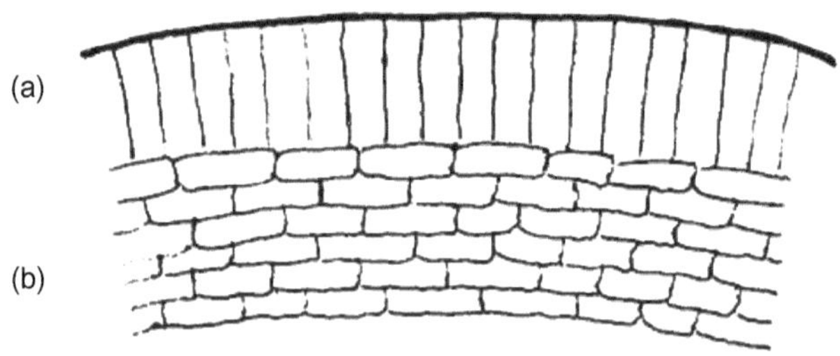

Fig. 13 Schematic section through a mollusk shell. In the outer layer (a), the phase is calcite and the crystals are columnar with their axis perpendicular to the shell surface. In the inner layer (b), the phase is aragonite and the crystals are in the form of platelets parallel to the shell surface.

different in the two phases. A particular feature is that the CO_3^{2-} lying in the (0 0 0 1) plane in calcite are all aligned in the same way, while those in (0 0 1) aragonite are in two distinct orientations. Thus some more complete structural matching with the substrate, a stereochemical correspondence [68], could provide a basis for phase selection. That such a degree of matching might be possible is made clear by the observation that when calcite nucleates on $(01\bar{1}2)$ to form sea-urchin spicules, it forms in an orientation that is chirally defined [71].

The account so far can be taken to be the *standard model* for nucleation in biomineralization in which an organic template acts to select a crystal

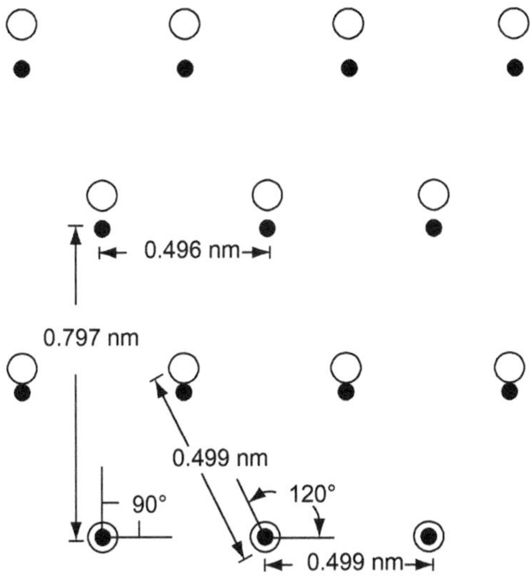

Fig. 14 The arrangement of Ca^{2+} ions (filled circles) on the (0 0 1) plane of aragonite closely matches the arrangement on (0 0 0 1) calcite (open circles). (Reprinted from Ref. [68], copyright (2001), with permission from Oxford University Press.)

phase and orientation for precipitation from supersaturated solution. Several pieces of work show that it is necessary to look beyond this standard model. In the standard model, a switch in phase from calcite to aragonite during mineral growth would require the deposition of a protein sheet on which fresh nucleation could occur. In vitro, Belcher et al. first grew a layer of $(1 0 \bar{1} 4)$ calcite. The solution was then removed and replaced with a fresh solution containing the same inorganic components, but with soluble proteins extracted from either the inner or outer layers of mollusk shell. With proteins from the inner layer associated with aragonite, the growth switched to aragonite and switched back to calcite when the soluble proteins were depleted. The presence or absence of soluble aragonite-specific proteins can thus be sufficient to switch phase to and fro without elaborate templates [83, 84].

It has become clearer that the nucleation and growth of crystalline mineral phases can occur not only from a supersaturated aqueous solution, but also from solid amorphous calcium carbonate. For some time, this possibility was overlooked as amorphous calcium carbonate dissolves easily, is difficult to detect in the presence of crystalline phases, and is often a transient precursor in the formation of calcite or aragonite [85]. Sea urchins have spicules composed of single crystals of calcite. By examining their larvae [86], it is known that the spicule first consists of amorphous calcium carbonate that then crystallizes. This may be attractive because of difficulties in transporting carbonate to the mineralization site. If the transport were by supersaturated solution, the solubility of calcium carbonate is so low that the volume of solution needed to deposit a given volume of mineral is around $10^5\times$ that of the mineral [75]. This is alleviated by transporting amorphous calcium carbonate formed elsewhere to the site in small vesicles and then depositing it. The nucleation aspects of this may not be much different from the conventionally considered case, as the first crystalline phase is thought to be deposited before the transport of the amorphous phase. Nevertheless, formation of crystalline mineral from an amorphous precursor rather than from solution may limit the relevance of many in vitro studies of mineral nucleation. The formation of initially amorphous structures that later crystallize may be an attractive alternative to direct crystallization from solution because in this way elaborate shapes can be formed, not restricted or influenced by crystal-growth mechanisms. It has been suggested that this would be attractive in inorganic materials synthesis [86], in some sense analogous to the exploitation of devitrification in glass-ceramics (Chapter 14, Section 2.1), but without any requirement for melting. The attractions of forming crystalline structures by devitrification rather than precipitation from solution are such that some systems appear to use particular proteins to poison the growth of crystal nuclei in solution [85], their action being analogous to that of the AFPs described in Section 2.2.

Spicules, as just noted, have often been described as single crystals. Yet it is now clear that many are in fact a composite of nanocrystals [87]. In such a *mesocrystal*, the individual nanocrystals are all in crystallographic alignment so that the mesocrystal is faceted and diffracts like a single crystal. It is likely that mesocrystalline structures are much more common than has been assumed. Such a structure helps to explain why an apparently single-crystal calcite spicule shows brittle fracture not along the $\{10\bar{1}4\}$ cleavage planes as in the monolithic mineral, but in a conchoidal pattern more characteristic of glass. The division into nanocrystalline domains is likely to have toughening effect. It has been discussed that even the much-studied platelets of aragonite in nacre may be mesocrystals [88].

In such structures, and also in larger-scale structures such as stacks of aligned aragonite platelets in nacre, there is the question whether neighboring aligned units nucleate independently or are linked by growth through some as yet unobserved nanobridges [89]. It is likely that both mechanisms can operate, an example of the redundancy often found in mechanisms in living systems [82].

4.3 Calcite and aragonite

The occurrence of calcite and aragonite in mollusk shells has already been noted (Figure 13). There have been many studies aiming to elucidate the mechanism of phase selection by nucleation. An early study

[90] used compressed Langmuir monolayers as the substrate on top of an in vitro solution. The precipitation of aragonite rather than calcite is favored by the addition of magnesium ions. Both polymorphs nucleate in the orientation found in vivo, with the plane parallel to the substrate being (0 0 0 1) for calcite and (0 0 1) for aragonite. As already noted, the atomic arrangements on these two planes are very similar and, as expected in this study, did not offer any basis for phase selection.

In nacre, the matrix sheets have a central layer of β-chitin, sandwiched between layers with a structure very similar to that of silk fibroin and predominantly also in the β-sheet conformation. These sheets are coated with acidic macromolecules rich in aspartic acid. Attempts to study the influence of the framework components, when extracted from the shell itself, are complicated by the potential influence of residual seeds of crystalline mineral. The seminal study, by Falini et al. [91], used artificial substrates composed of β-chitin from squid and silkworm fibroin, these nonmollusk sources being devoid of any crystalline mineral. The aspartate-rich soluble macromolecules were extracted from individual calcite or aragonite layers. When crystallization occurred, it always yielded the polymorph, calcite, or aragonite, corresponding to the layer from which the soluble macromolecule had been extracted. Without any added macromolecules, no crystallization occurred inside the substrate. When β-chitin was used as the substrate without any silk fibroin, vaterite formed inside the substrate in the presence of 'aragonitic' macromolecules and calcite formed in the presence of 'calcitic' macromolecules. The same macromolecules adsorbed on polystyrene induced the formation of calcite only, whether fibroin was present or not.

Thus the acidic macromolecules can give complete control over phase selection, and in particular can promote the formation of aragonite, as seen in vivo, but only when they are in the appropriate microenvironment including both β-chitin and silk fibroin. The influence of the macromolecules must be to promote nucleation, and cannot be growth inhibition, as when they are absent there is no crystallization in their artificial substrate. Falini et al. conclude that polymorph selection must involve the three-dimensional structure of the nucleation site. Subsequent work has shown that the order of assembly of the substrate is important. When the silk fibroin is added after adsorption of the macromolecules onto the β-chitin, there is never any nucleation of aragonite [92]. The full subtleties of the microenvironment, in particular the juxtaposition of proteins, needed for aragonite nucleation remain to be elucidated.

4.4 Biomimetic materials synthesis

The examples of nucleation control exercised in living systems, and to some extent mimicked in vitro, offer inspiration for developing new synthesis routes. The rapidly growing and important field of *biomimetic mineralization* has been extensively reviewed [68, 88]. Clearly the complexities of aragonite nucleation just discussed are going to be difficult to reproduce, yet enormous progress has been made in transferring biomineralization principles into synthetic materials chemistry.

Living systems are restricted to relatively narrow ranges of temperature and pressure, and biomineralization is of interest at least partly because it demonstrates that sophisticated materials synthesis is possible under ambient conditions. Nevertheless, it is of interest to explore a wider range of conditions. It has been found that even mildly nonambient conditions ($T < 100^{\circ}C$) can exercise control over polymorph selection and crystal morphology [93].

The importance of surfactants in biological systems has led to their exploitation in various ways. For example, they can be used to stabilize single molecules in solution, thereby delaying crystal nucleation [94]. In this way, it has been possible to greatly narrow the size distribution of nanocrystals formed from supersaturated solution.

The templating evident in biological systems has led to much work on artificial soft organic templates for the nucleation of inorganic structures. As reviewed by Qi et al. [95], many approaches have been taken. These include molecular-level crystal imprinting of polymer surfaces to nucleate a chosen polymorph [96] and chemical treatments to provide a high surface density of functional groups [97]. The substrates of interest are not all planar, but include fibers [98].

A notable example of a bio-inspired approach is the work of Aizenberg et al. [99] on the direct fabrication of large (~ 1 mm), controlled-orientation single crystals of calcite patterned on a sub-10-μm scale. This exploited the mineral-nucleating ability of a self-assembled monolayer (SAM) that can be fixed to a substrate (also the basis for the work shown in Chapter 6, Figure 13). Standard photolithography was used to define a pattern of features such as a square array of posts. This micropatterned glass substrate was coated with gold or silver and a patch of SAM was deposited on top from the tip of an atomic-force microscope (Figure 15). The SAM was chosen to induce calcite in a selected orientation. The rest of the surface was treated to create a disordered organic surface suppressing the nucleation calcite. When calcium carbonate is then deposited from solution, it is amorphous, but calcite nucleates on the SAM patch and grows laterally to transform the patterned thin film into a single crystal of calcite. Without the SAM patch, the thin film transformed to a polycrystalline mixture of calcite and vaterite. This work not only suggests novel synthesis routes, but helps to understand the mechanisms of in vivo mineralization.

(a)

(b)

(c)

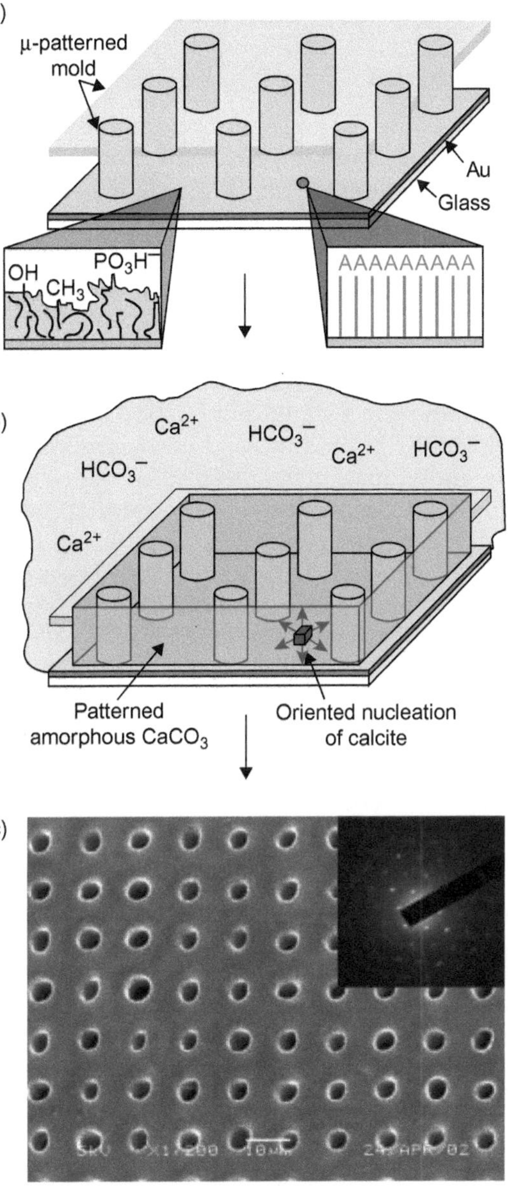

Fig. 15 The experiment of Aizenberg et al. [99] in which ~1 mm crystals of calcite are produced in controlled orientation with patterning on a sub-10-µm scale: (a) template preparation; (b) deposition of amorphous $CaCO_3$; (c) scanning electron micrograph of the product, a patterned single crystal. (From Ref. [99]. Reprinted with permission from AAAS.)

5. PATHOLOGICAL MINERALIZATION

5.1 Survey of examples

Medicine provides many examples of undesirable mineralization in the formation of crystals or stones (*calculi*) in a variety of organs and species. In humans, stone formation is the origin of many painful conditions. Gout can lead to the deposition of monosodium urate monohydrate crystals around joints, often causing devastating damage. Gallstones, composed mainly of cholesterol, form in the gallbladder or bile duct. Stones mainly of calcium carbonate or phosphate can form in the prostate. There are many other examples, including stones in the intestine, pancreas, and tonsils. Stone formation is precipitation from a supersaturated solution; in a significant fraction of cases, supersaturation is usual even in healthy individuals and is not abnormal. The absence of stones in healthy individuals is then associated with a nucleation barrier. The breakdown of the barrier through the development of a substrate suitable for heterogeneous nucleation can be an important disease mechanism. It is a characteristic that stone formation can be favored in damaged tissues, including those damaged by previous stone formation. We focus only on the formation of kidney stones; this has been particularly widely studied, and some aspects of the nucleation have been revealed.

5.2 Kidney stones

Once rare, kidney stones now occur in a large and increasing fraction of the population in developed countries. Up to 20% of Caucasian men in the USA will have kidney stones at some time. The incidence rate in women was as little as one-third that in men, but is climbing rapidly [100]. Even with modern, minimally invasive techniques for breaking up stones, their treatment is a significant call on medical resources. Kidney stones show a wide variety of compositions, from a variety of causes as surveyed by Kavanagh [101]. For example, some drugs can be excreted in sufficient concentration in urine to form specific stones. The majority of stones ($\sim$75%), however, are based on calcium oxalate (written here as CaOx) and these will be the focus of the present discussion. The increasing incidence of kidney stones may be associated with several factors, but the most prominent are increased consumption of animal protein and salt (NaCl). Animal protein favors the precipitation of CaOx by increasing excretion of calcium and oxalate ions and by lowering the pH of urine. Salt increases calcium excretion. The complexity of human physiology thwarts simple remedies. For example, reduction of calcium consumption, once recommended, is now recognized to increase the risk of stone formation by inhibiting oxalate absorption in the gut [102].

Up to 40% of individuals can have crystals (of calcium phosphate or calcium oxalate) in their urine. In most cases, the presence of these crystals (*crystalluria*) is harmless and can be regarded as normal. The incidence of crystalluria is somewhat higher in those suffering from stone formation and the crystals themselves tend to be larger than in the urine of healthy subjects [103]. Somewhat analogous to the two strategies outlined in Section 2 for the prevention of damage by freezing, prevention of kidney stones may involve either inhibition of nucleation altogether, or reduction of supersaturation by promotion of nucleation to form fine crystals in the urine.

Supersaturations for the precipitation of a given salt from a solution involve the concentrations of the relevant ions, their dissociation constants, and the solubility of the salt. The calculation of supersaturations in this way for precipitation from urine has been described by Finlayson [104, 105]. As summarized in Figure 16 [101], for stones not involving calcium, supersaturation of the relevant species is mostly associated with an abnormal condition and leads directly to stone formation. For calcium ions, however, supersaturation is common and calcium-based stones could be expected to be much more prevalent than they are [106]. Furthermore, the supersaturation ranges in both healthy

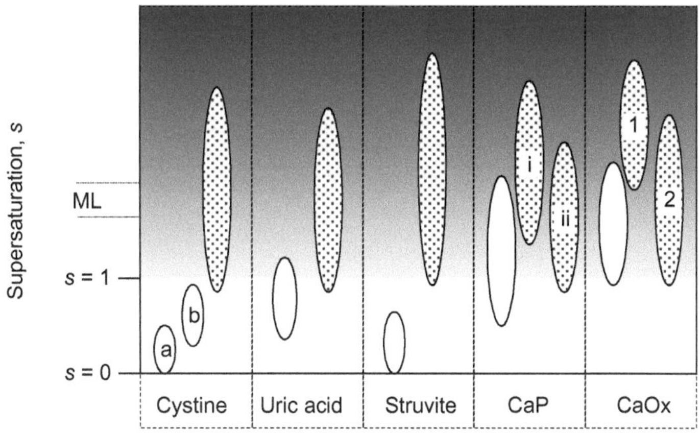

Fig. 16 Supersaturations in urine for the main salts forming kidney stones. The open ellipses are for healthy individuals: (a) normal, and (b) showing hyperexcretion of cystine because of a gene defect. Shaded ellipses are for stone-forming individuals: (i) associated with urinary infection, (1) showing hyperexcretion of oxalate because of a gene defect, (ii, 2) stone formation without well-defined cause (idiopathic). The metastable limit (ML) is the supersaturation range above which precipitation occurs. (From [101], with permission.)

and stone-forming individuals are broad and show substantial overlap [107]. As surveyed by Kavanagh, many methods have been used to measure the supersaturation beyond which nucleation occurs, the metastable limit [108].

Comparisons of inorganic artificial urine and natural whole urine reveal that natural urines contain several inhibitors and promoters of crystallization. The overall effect is illustrated in Figure 17, which shows the interdependence of calcium and oxalate ion concentrations for which nucleation is observed [109]. Inorganic, artificial urine shows a well-defined metastable limit (supersaturation ratio ≈ 13.6) for the precipitation of CaOx. The critical supersaturation ratio is lower (~ 9.9) in natural urine; it is still lower in the urine of stone formers.

In addition to measurements of the metastable limit, the nucleation of crystals in urine has been characterized by measurements of rate. As in virtually all the cases considered in this book, the nucleation events themselves are not observable, so the measurement of their rate is indirect. Mixed-suspension, mixed-product-removal continuous crystallizers are particularly useful, especially when miniaturized [110]. The particle-size distribution from a continuous crystallizer can be analyzed to yield nucleation and growth rates [111]. In artificial urine, the nucleation rate is simply related to the supersaturation ratio [112], but in natural urine this is not the case. When natural urine is diluted, but the

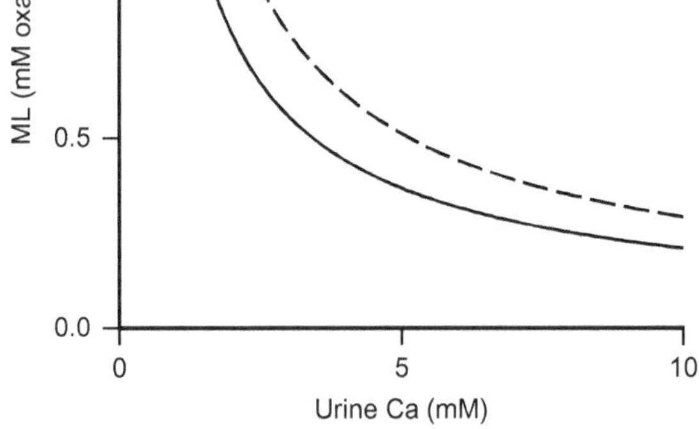

Fig. 17 Metastable limits for CaOx precipitation in artificial urine (dashed line) and natural urine (solid line), determined for additions of oxalate at different calcium concentrations. (Reproduced with permission, from Ref. [109], copyright (2000) the Biochemical Society. http://www.clinsci.org)

concentrations of calcium and oxalate ions are held constant, the CaOx nucleation rate falls (Figure 18), presumably reflecting the influence of a nucleation catalyst. It is well known that CaOx nucleation in urine must be heterogeneous. Over a wide range of studies, the maximum observed supersaturation is ~20, well below the level required for homogeneous nucleation [104, 113].

Surprisingly, there is evidence that the nucleation rate in urine from healthy individuals is higher (when compared for a given supersaturation ratio) than that in stone-formers' urine [110]. Although a higher supersaturation is required to start precipitation in normal urine, once it does start there may be an advantage in generating many small crystals.

What are the nucleant substrates in natural urine? The production of urine has several stages during which its composition changes. It is still not certain where stones start to form, but for CaOx stones at least, it is most likely that they form in the tubules and ducts, in the final stages in which water is extracted [114]. Several studies consistently point to the role of the cell membranes that are naturally shed into the urine as a result of cell turnover in the kidney. Typical kidney stones contain 2 to 5 wt.% of biological material. When this material is extracted, it does promote precipitation from supersaturated solution [115]. This appears to be

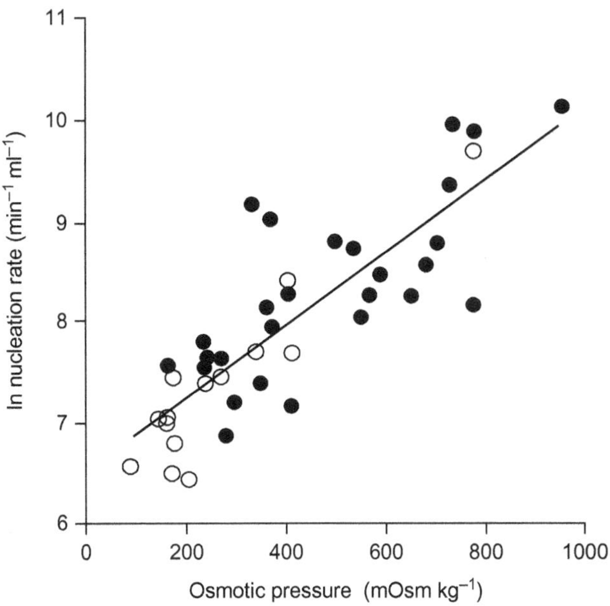

Fig. 18 CaOx crystal nucleation rate in normal urine samples diluted while holding the calcium and oxalate concentrations fixed. The concentration of the urine is represented by its osmotic pressure. (From Ref. [112], copyright (1999), with kind permission of Springer Science and Business Media.)

associated with the lipid fraction of the material, and in particular with acidic phospholipids. These are found at a higher level in the urine of CaOx stone formers [116]. Furthermore, artificial Langmuir monolayers of phospholipids clearly promote nucleation [116], giving the same phase, calcium oxalate monohydrate (COM), as added membranes. The Langmuir monolayer studies have permitted investigation of the roles of lipid chemistry and packing. The COM crystals mostly have their $\{1 0 \bar{1}\}$ face parallel to the substrate, but epitaxy does not appear to be an important factor. Rather the chemistry may be important, as suggested by Khan et al. [116] in their model shown in Figure 19. Acidic phospholipids

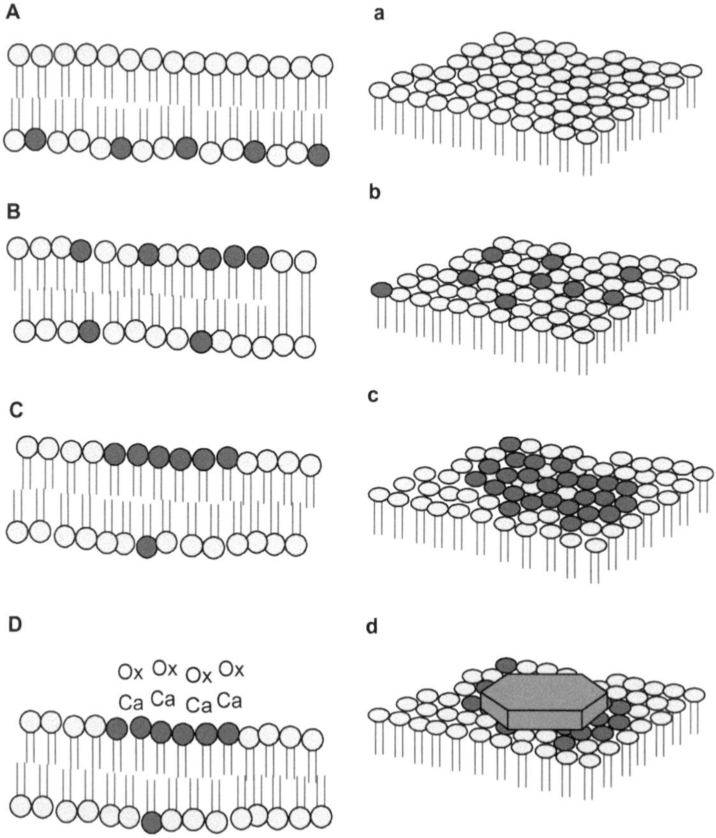

Fig. 19 The model of Khan et al. [116] for heterogeneous nucleation of CaOx on a cell membrane, shown in two (A–D) and three dimensions (a–d). A normal membrane (A, a) with only neutral phospholipids (open circles) on the outer surface. Migration of acidic phospholipids (filled circles) from inside to outside (B, b), and aggregation into anionic domains (C, c). Interaction (D, d) between calcium ions and an anionic domain, and nucleation of calcium oxalate monohydrate crystal on the $(10\bar{1})$ face. (Reprinted by permission from Macmillan Publishers Ltd: *Kidney International* Ref. [116], copyright 2002.)

in the cell membrane migrate to the outer surface and aggregate into domains of anionic character attractive for the calcium-rich $\{10\bar{1}\}$ face of COM.

Studies of nucleation in urine have included measurements and analysis of incubation time [117]. These show that the incubation time is dramatically reduced on addition of albumin, which may be significant in promoting beneficial nucleation in natural urine [118].

Kidney stones are mostly CaOx, forming for no well-defined reason. The present understanding of their nucleation has been reviewed by Kavanagh [101]. While it is clear that the nucleation is heterogeneous, the presence of the cell-membrane substrates is inevitable and unlikely to be part of a well-defined nucleation control mechanism. The potency and population of the substrates have not yet been characterized so that prediction of nucleation rates has not yet been approached. The incidence of kidney stones is likely to be associated with an abnormal nucleation mechanism, but the understanding of this is as yet insufficient to assist in the development of therapies. The best simple way to reduce the risk of stone formation is to drink plenty of water, thereby reducing the supersaturation of salts in the urine.

6. NEURODEGENERATIVE DISEASE

There is a very wide range of neurodegenerative diseases, fatal to animals and humans, in which a key step is the formation of protein aggregates in the brain. Public concern about these diseases has increased enormously with the recognition that bovine spongiform encephalopathy (BSE, *mad cow disease*) can lead to variant Creutzfeldt–Jacob disease (vCJD) in humans [119, 120]. These and other diseases relying on *prions* as the infectious agent are considered in Section 6.1. Prions do not include nucleic acids, so infection may be by a physical seeding. There are other neurodegenerative diseases that do not involve prions as the infectious agent, but where nucleation concepts may again be relevant in analyzing the formation of protein aggregates. The example of Huntington's disease is considered in Section 6.2. In both prion-related diseases and Huntington's disease, the key phenomenon is abnormal folding of protein molecules. The appropriate forms of analysis inevitably build on those used for crystallization of polymers (Chapter 11).

6.1 Prion diseases

The disease *scrapie* is infectious among sheep, but there is no evidence of transmission to humans despite long exposure (in Britain, e.g., scrapie has been endemic for over 250 years). BSE in cattle, however, has contaminated foodstuffs and is implicated not only in vCJD in humans but in feline spongiform encephalopathy (FSE) in domestic and large cats. Prion diseases are unique in medicine in that they can arise sporadically, be inherited, or be infectious. Although public concern is largely with infection crossing from other species to humans, the majority of cases of human prion disease arise sporadically as Creutzfeldt–Jacob disease (sCJD); the rate is roughly 1 in 10^6 worldwide, and can be influenced by genetic susceptibility [121]. Other human prion diseases include Gerstmann–Straussler–Scheinker disease (GSS) and fatal familial insomnia (FFI); roughly 15% are inherited [122]. Transmission of prion disease from one human to another is restricted to special cases such as contamination from transplanted tissue or infected medical instruments, or cannibalism. The disease *kuru* is of particular interest [123]. The mortuary-feast cannibalism in Papua New Guinea that gave rise to kuru was last practiced around 1956; although the disease has largely been eliminated, a few cases still appear in those who participated in the feasts as children. This indicates that the incubation period for the disease can be more than 40 years.

The common feature of prion diseases is the conversion of the prion protein (a sialoglycoprotein, designated PrP, found anchored to cell surfaces) to an abnormal form. The normal or *cellular* form is PrP^C; the abnormal *scrapie* form is PrP^{Sc}. Amino acid sequencing shows no covalent-bond differences between the PrP^C and PrP^{Sc} molecules; the conversion involves conformation only [124]. PrP^{Sc} is self-replicating and deposited as stable aggregates. The exact state of aggregation that constitutes the infectious agent remains to be determined. The formation of evident amyloid plaques (insoluble PrP^{Sc} aggregates) in cerebral gray matter is a characteristic of vCJD, GSS, and kuru, but is rare in sCJD. Microscopic examination of diseased brains typically reveals micro-vacuolation giving a spongiform appearance. The link between PrP^{Sc} and this neurodegeneration has not been established. It is possible that there is a direct neurotoxic effect of PrP^{Sc} [125], or that the problem is the depletion of PrP^C the normal function of which is not well understood. For example, PrP^C may have important roles as an antioxidant [126] and in regulating programmed cell death (*apoptosis*) [127]. It is also possible that there is a toxic intermediate in the conversion of PrP^C to PrP^{Sc} [128].

Inoculation by transfer of infected tissue can transmit prion disease within a species and from one species to another. For example, chimpanzees inoculated with human tissue infected with kuru [129] or CJD [130] develop fatal neurodegeneration. Inoculation is most effective when it is intracerebral. Contact with infected tissue outside the central nervous system can also lead to infection, but with much longer incubation times (up to a few decades, as noted earlier) before evident onset of disease. In such cases, prion replication can proceed in lymphoreticular tissue with no clinical effect until neuroinvasion occurs much later [131]. Incubation times are also longer when transmission is from one species to another, the so-called *species barrier* [132]. Several factors influence this barrier, as briefly reviewed by Jackson et al. [128]. The remarkable feature of inoculation of these diseases is that the infectious agent is devoid of nucleic acids. It is evident from the progression of disease after infection that there are distinct prion strains, the characteristics of which are maintained on transmission. Nevertheless, several studies have suggested that the strain specificity is encoded by the PrP protein itself and not by a nucleic acid [128]. Overall, there is strong support for the *protein-only hypothesis* for prion replication [133, 134]. This holds that PrP^{Sc} acts as a conformational template driving the conversion of PrP^{C} to PrP^{Sc}. The ability of PrP^{Sc} to impart an abnormal conformation to PrP^{C} has been demonstrated [135], though the material produced was noninfectious [136]. The essentially physical mechanism of templating a new conformation is the focus of our analysis.

Although the structure of the PrP^{C} protein has not been determined crystallographically, it is known to have essentially the same conformation in different species [137]. The protein has an N-terminal region of $\sim$100 amino acids and is unstructured in solution. The key structural features are in the C-terminus of $\sim$110 amino acids. This consists of three α-helices and a small two-strand β-sheet. This conformation is stabilized by a single disulfide bond. In contrast, PrP^{Sc} is predominantly a β-sheet structure. The ordered domain of the protein shows rapid folding and unfolding; there is no significant presence of partially unfolded or intermediate states [137]. It is considered that PrP^{Sc} is unlikely to arise from an intermediate structure, but rather that it is formed from the unfolded molecule. There are alternative pathways of folding and unfolding, dependent on the pH of the medium [138]. Ignoring these distinctions, Figure 20 (from [128]) shows schematically the possible mechanisms of prion infection. The normal PrP^{C} molecule is in equilibrium with the unfolded molecule. On reduction of the disulfide bond, β-sheet PrP can form from the unfolded molecules. The β-sheet is rare, but can be stabilized by aggregation to from a stable PrP^{Sc} seed. The formation and further growth of this seed can be by aggregation with protein molecules already folded in β-sheet form or by acting as a template for unfolded molecules to fold into the PrP^{Sc} conformation. Large plaques of PrP^{Sc} may then break up into smaller infectious units.

In terms of a physical nucleation theory, it is important to note that aggregated PrP^{Sc} is significantly more stable biochemically than PrP^{C}. Despite being identical in its covalent bonds, it is relatively insoluble and is partially resistant to being broken down by standard proteolysis techniques [139]. Although the stability of various PrP aggregates has not yet been investigated in detail, it is clear that in all likely conditions there

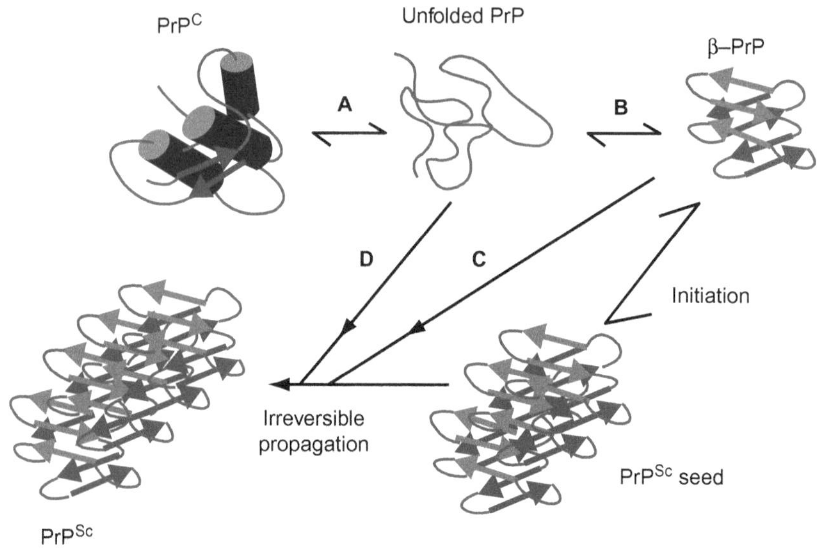

Fig. 20 Pathways to prion infection. Equilibrium A is between PrP^C and unfolded protein in solution. Equilibrium B is between the unfolded protein and β-sheet PrP. Infection may be initiated by the aggregation of β-sheet PrP to form a stable seed. The seed may grow by acquisition of unfolded protein or further β-sheet to form an insoluble plaque. (From [128], with permission.)

is a critical size for a PrP^{Sc} aggregate above which it is stable. The question is how this stable size may be reached. In a healthy animal without inherited prion disease, PrP^C is relatively stable, and the nature of equilibrium A in Figure 20 is such that the concentration of unfolded protein is very low. In that case, formation of β-PrP and ultimately a stable PrP^{Sc} seed or nucleus is very unlikely. It may still occur, however, and is a possible origin of sporadic disease. Inherited disease arises from mutations that destabilize the PrP^C conformation and thereby facilitate formation of β-sheet and PrP^{Sc}. Uninherited, somatic mutation is another possible origin of sporadic disease. It is particularly clear, however, that inoculation with already formed seeds of PrP^{Sc} can lead to straightforward transformation of the protein folding to be of the PrP^{Sc} type. In this way, the general model presented in Figure 20 can account for basic features of the onset of prion diseases.

The protein-only hypothesis that prion replication can be explained in terms of only differently folded states is well accepted. But it should be noted that the dissemination of prions may involve active biological processes [140]. It is notable that infective prions have not yet been produced from purified material. Also, the energies of key protein conformations and aggregates have not been calculated. Thus the molecular mechanisms of the formation of the key PrP^{Sc} conformation are far from being understood in detail. In the next section, it will be seen that there is more detailed understanding of the narrower range of phenomena in Huntington's disease.

6.2 Huntington's disease

In the previous section, it was noted that prion diseases, while there may be a genetic contribution to susceptibility, are mainly of concern because of infection. In contrast, there is a range of neurodegenerative diseases that are clearly inherited, although showing a late onset. In animals or patients with the inherited disorder, neurons function normally for periods up to tens of years while a few, genetically identical, cells die prematurely. This remarkable behavior was thought to be a result of damage accumulation. However, Clarke et al. [141] have shown clearly for 16 diseases that the probability of cell death is constant over time, or in some cases decreasing, and that this is incompatible with the increasing probability expected for damage accumulation. A particularly well-studied case is the human disease *Huntington's chorea*. Perutz and Windle [142] suggested that for this and other cases, the constant probability of neuron death is most likely to arise from steady-state nucleation of protein aggregates. The protein huntingtin (Ht) is of uncertain function, but appears to be essential in embryonic development [143]. It contains a glutamine amino acid sequence, the length of which is clearly correlated with the disease. In healthy individuals, the number of structural repeat units in the poly(l-glutamine) sequence is 6–36. Those with Huntington's disease have more units, as many as 180 [144, 145], the greater length arising from a mutation in the gene that acts as a template for Ht. Those with more than 40 units never remain free from the disease. The age of onset of the disease shows an inverse exponential dependence on the number of repeats in the polyglutamine sequence. There appears to be no direct toxic effect of the sequence length; rather it leads to a misfolding of the protein [146]. This in turn leads to aggregation into insoluble amyloid fibers that accumulate particularly in neural cells [147] and are associated with neurodegeneration [148], as for the prion diseases. A nucleation barrier of some kind is likely, because seeding with already aggregated proteins shortens the time for the onset of symptoms [149].

The importance of sequence length is not seen for the prion diseases; its crucial role for diseases such as Huntington's makes it the focus for study of nature of the self-aggregation of the polyglutamine sequence. The structural repeat units are linked to form a chain (the molecular backbone) that adopts conformations favoring low steric hindrance of the side groups and intramolecular hydrogen bonding. These are commonly analyzed using a Ramachandran diagram [150] that shows the dihedral angles between particular neighboring bonds on the backbone. There are two main conformations of natural proteins, the α-helix and the β-sheet. In the case of polyglutamines, two helical structures have been proposed. Monoi [151] proposed the μ-helix, while the structure proposed by Perutz et al. [152] is less tightly coiled. The hydrogen bonding between the backbone and the side-groups in the Perutz structure is such that it can be considered to be a coiled β-sheet. While there is considerable interest in modeling the nucleation of such structures from more random configurations in aqueous solution, the simulation of realistic conditions is not practicable because of the extreme improbability of the nucleation event in Huntington's disease (onset time of the order of a human lifetime). Instead the relative or absolute stabilities of imposed conformations can be assessed, as demonstrated in the work of Starikov et al. [153]. A small reduction in the stability of correctly folded proteins can dramatically accelerate nucleation of the disordered aggregates from which amyloid plaques can grow [154].

7. SUMMARY

The many processes of life include phase nucleation of the kinds familiar in physical systems. Biological systems are complex, and evolved to an often-high degree of optimization. In particular, proteins are the most complicated of all molecules, with the most sophisticated functions. They feature in all of the broad categories of nucleation considered in this chapter. There are clear examples where the control of nucleation, either its promotion at favored sites or its suppression, is important for the health or even the survival of the organism. The relevance of nucleation studies extends, therefore, to such areas as agriculture, food science, and medicine. In this chapter, it has been possible to cover only a few topics from a great range of possibilities. Furthermore, progress in biological sciences and medicine is such that the coverage is often just a snapshot of an understanding that is fast-changing and improving.

Water is essential to life, and living systems exploit interface-active species with hydrophilic and hydrophobic functionalities. The dominant effects of such species in biological systems are not so evident for the different functionalities that might be relevant for physical systems. Nevertheless, the role of interface-active species in physical nucleation may deserve more attention.

The formation of ice inside a cell is almost always fatal for the cell. There is more than one strategy for avoidance of this problem, and it is typical of the redundancy found in biological systems that different strategies may be deployed at the same time. Plants and animals can prevent intracellular ice formation using AFPs, which are interface-active and adapted to block the growth of ice. Their action can be relevant even at the nucleation stage. Often of bacterial origin, INAs have planar surfaces that act as a template for ice. According to the analysis of athermal nucleation, the larger the surface the more effective the nucleant, and indeed INAs, with radii up to a few tens of nanometers, are unusually large for protein structures. Organisms can exploit INAs to promote the formation of ice outside their cells and in this way can develop freeze tolerance.

Athermal nucleation features again in the formation of bubbles in living systems, notably decompression sickness in divers and gas-bubble trauma in fish. The supersaturations of dissolved gas necessary to cause bubble formation are so small that growth is always likely to be from some form of preexisting bubble. A proteinaceous skin can form on bubble nuclei, stabilizing them. Analogous effects in nucleation in physical systems are essentially unexplored.

Biomineralization offers many examples in which nucleation is important in the control of crystal orientation, location, and distribution. The nucleants can select between structurally very similar polymorphs and may do so by sophisticated templating in three dimensions. The processes involved are under active investigation for exploitation in biomimetic materials synthesis.

Pathological mineralization, examined in the one example of kidney stones, again shows alternative strategies for avoidance. Nucleation of stones may be directly inhibited, or the supersaturation of stone-forming species in the urine can be reduced by the promotion of nucleation to form finer crystals.

Many neurodegenerative disorders have their origin in abnormal protein folding. A physical nucleation of such folding may explain the long incubation times sometimes observed. The nucleation of harmful plaques can involve several stages, key details of which remain to be elucidated.

REFERENCES

[1] A. Ducruix, R. Giegé (eds), *Crystallization of Nucleic Acids and Proteins: A Practical Approach*, 2nd edn, Oxford University Press, Oxford (1999).

[2] A.M. Karow, Jr., W.R. Webb, A theory for injury and survival, Cryobiology **2** (1965) 99–108.

[3] J.E. Lovelock, The mechanism of the cryoprotective effect of glycerol against freezing and thawing, Biochim. Biophys. Acta **11** (1953) 28–36.

[4] C.A. Angell, Supercooled water, in: Water — A Comprehensive Treatise Ed. F. Franks, vol. 7, Plenum Press, New York (1982), pp. 1–82.

[5] F. Franks, S.F. Mathias, P. Galfre, S.D. Webster, D. Brown, Ice nucleation and freezing in undercooled cells, Cryobiology **20** (1983) 298–309.

[6] R.E. Lee, Jr., J.P. Costanzo, Biological ice nucleation and ice distribution in cold-hardy ectothermic animals, Ann. Rev. Physiol. **60** (1998) 55–72.
[7] G.C. Packard, M.J. Packard, Natural freeze-tolerance in hatchling painted turtles?, Comp. Biochem. Physiol. A **134** (2003) 233–246.
[8] M.J. Burke, L.V. Gusta, H.A. Quamme, C.J. Weiser, P.H. Li, Freezing and injury in plants, Ann. Rev. Plant Physiol. **27** (1976) 507–528.
[9] F. Franks, *Biophysics and Biochemistry at Low Temperatures*, Cambridge University Press, Cambridge (1985).
[10] J. Duman, Insect antifreezes and ice-nucleating agents, Cryobiology **19** (1982) 613–617.
[11] K.E. Zachariassen, R. Lundheim, Applications of antifreeze proteins, in: *Biotechnological Applications of Cold-Adapted Organisms*, Eds. R. Margesin, F. Schinner, Springer Verlag, Berlin (1999), pp. 319–332.
[12] C.A. Knight, Adding to the antifreeze agenda, Nature **406** (2000) 249–251.
[13] J.A. Raymond, A.L. DeVries, Adsorption inhibition as a mechanism of freezing resistance in polar fishes, Proc. Natl. Acad. Sci. USA **74** (1977) 2589–2593.
[14] A.L. DeVries, Biological antifreeze agents in coldwater fishes, Comp. Biochem. Physiol. **73A** (1982) 627–640.
[15] S. Bremdal, K.E. Zacharaissen, Thermal hysteresis factors and supercooling of hibernating *Rhagium inquisitor* beetles, in: F. Sehnal, A. Zabza, D.L. Denlinger (Eds.), *Endocrinological Frontiers in Physiological Insect Ecology*, Proc. Int. Conf., Technical University of Wroclaw, 1990, pp. 187–191.
[16] J.G. Duman, D.W. Wu, K.L. Yeung, E.E. Wolf, Hemolymph proteins involved in the cold tolerance of terrestrial arthropods: antifreeze and ice nucleator proteins, in: *Water and Life*, Eds. G.N. Somero, C.B. Osmond, C.L. Bolis, Springer Verlag, Berlin (1992), pp. 282–300.
[17] M.G. Thyshenko, D. Doucet, P.L. Davies, V.K. Walker, The antifreeze potential of the spruce budworm thermal hysteresis protein, Nat. Biotechnol. **15** (1997) 887–890.
[18] D. Worrall, L. Elias, D. Ashford, M. Smallwood, C. Sidebottom, P. Lillford, J. Telford, C. Holt, D. Bowles, A carrot leucine-rich-repeat protein that inhibits ice recrystallization, Science **282** (1998) 115–117.
[19] C. Sidebottom, S. Buckley, P. Pudney, S. Twigg, C. Jarman, C. Jolt, J. Telford, A. McArthur, D. Worrall, R. Hubbard, P. Lillford, Heat-stable antifreeze protein from grass, Nature **406** (2000) 256.
[20] A.L. DeVries, Glycoproteins as biological antifreeze agents in Antarctic fishes, Science **172** (1971) 1153–1155.
[21] J.A. Ahlgren, C.C. Cheng, J.D. Schrag, A.L. DeVries, Freezing avoidance and the distribution of antifreeze glycopeptides in body fluids and tissues of Antarctic fish, J. Exp. Biol. **137** (1988) 549–563.
[22] Y.C. Liou, A. Tocilj, P.L. Davies, Z. Jia, Mimicry of ice structure by surface hydroxyls and water of a β-helix antifreeze protein, Nature **406** (2000) 322–324.
[23] D.S.C. Yang, W.C. Hon, S. Bubanko, Y. Xue, J. Seetharaman, C.L. Hew, F. Sicheri, Identification of the ice-binding surface on a type III antifreeze protein with a 'flatness function' algorithm, Biophys. J. **74** (1998) 2142–2151.
[24] P. Wilson, Explaining thermal hysteresis by the Kelvin effect, Cryo-Lett. **14** (1993) 31–36.
[25] J. Krog, K.E. Zachariassen, B. Larsen, O. Smidsrød, Thermal buffering in Afro-alpine plants due to nucleating agent-induced water freezing, Nature **282** (1979) 300–301.
[26] R.C. Schnell, G. Vali, Atmospheric ice nucleation from decomposing vegetation, Nature **236** (1972) 163–165.
[27] L.R. Maki, E.L. Gaylan, M. Chang-Chien, D.R. Caldwell, Ice nucleation induced by *Pseudomonas syringae*, Appl. Microbiol. **28** (1974) 456–459.
[28] C.D. Upper, G. Vali, The discovery of ice nucleation and its role in the injury of plants by frost, in: *Biological Ice Nucleation and its Applications*, Eds. R.E. Lee, G.J. Warren, L.V. Gusta, APS Press, St Paul (1995), pp. 29–39.

[29] G. Vali, Principles of ice nucleation, in: *Biological Ice Nucleation and its Applications*, Eds. R.E. Lee, G.J. Warren, L.V. Gusta, APS Press, St. Paul (1995), pp. 1–28.
[30] S.E. Lindow, The role of bacterial ice nucleation in frost injury to plants, Ann. Rev. Phytopathol. **21** (1983) 363–384.
[31] A.G. Govindarajan, S.E. Lindow, Size of bacterial ice-nucleation sites measured in situ by radiation inactivation analysis, Proc. Nat. Acad. Sci. USA **85** (1988) 1334–1338.
[32] W. Szyrmer, I. Zawadzki, Biogenic and anthopogenic sources of ice-forming nuclei: a review, Bull. Am. Meteorol. Soc. **78** (1997) 209–228.
[33] C.E. Morris, D.G. Georgakopoulos, D.C. Sands, Ice nucleation active bacteria and their potential role in precipitation, J. Phys. IV **121** (2004) 87–103.
[34] S.A. Yankofsky, Z. Levin, T. Bertold, N. Sandlerman, Some basic characteristics of bacterial freezing nuclei, J. Appl. Meteorol. **20** (1981) 1013–1019.
[35] J.A. Ruggles, M. Nemecek-Marshall, R. Fall, Kinetics of appearance and disappearance of classes of bacterial ice nuclei support an aggregation model for ice nucleus assembly, J. Bacteriol. **175** (1993) 7216–7221.
[36] M.J. Burke, S.E. Lindow, Surface properties and size of the ice nucleation site in ice nucleation active bacteria: theoretical considerations, Cryobiology **27** (1990) 80–84.
[37] F. Franks, The properties of aqueous solutions at subzero temperatures, in: Water — A Comprehensive Treatise Ed. F. Franks, Vol. 7, Plenum Press, New York (1982), pp. 215–238.
[38] G.M. Mueller, P.K. Wolber, G.J. Warren, Clustering of ice nucleation protein correlates with ice nucleation activity, Cryobiology **27** (1990) 416–422.
[39] R. Lundheim, *Adaptive and Incidental Biological Ice Nucleators*. Doctoral thesis, NTNU Trondheim, 1996.
[40] A.P. MacKenzie, Non-equilibrium freezing behaviour of aqueous systems, Phil. Trans. Roy. Soc. Lond. B **278** (1977) 167–188.
[41] T. Soesanto, M.C. Williams, Volumetric interpretation of viscosity for concentrated and dilute sugar solutions, J. Phys. Chem. **85** (1981) 3338–3341.
[42] D.H. Rasmussen, A.P. Mackenzie, Effects of solutes on ice-solution interfacial free energy; calculation from measured homogeneous nucleation temperatures, in: *Water Structure at the Water–Polymer Interface*, Ed. H.H.G. Jellinek, Plenum Press, New York (1972), pp. 126–145.
[43] F. Franks, The nucleation of ice in undercooled aqueous solutions, Cryo-Lett. **2** (1981) 27–31.
[44] K.B. Storey, J.M. Storey, Natural freezing survival in animals, Annu. Rev. Ecol. Syst. **27** (1996) 365–386.
[45] K.E. Zachariassen, The role of polyols and nucleating agents in cold-hardy beetles, J. Comp. Physiol. **140** (1980) 227–234.
[46] J.G. Duman, J.L. Patterson, The role of ice nucleators in the frost tolerance of overwintering queens in the bald faced hornet, Comp. Biochem. Physiol. A **59** (1978) 69–72.
[47] Y. Voituron, J.M. Storey, C. Grenot, K.B. Storey, Freezing survival, body ice content and blood composition of the freeze-tolerant European common lizard, *Lacerta vivipara*, J. Comp. Physiol. B **172** (2002) 71–76.
[48] M. Holmstrup, Overwintering adaptations in earthworms, Pedobiologia **47** (2003) 504–510.
[49] K.B. Storey, J.M. Storey, Biochemical adaption for freezing tolerance in the wood frog, *Rana sylvatica*, J. Comp. Physiol. B **155** (1984) 29–36.
[50] J.R. Layne, J.P. Costanzo, R.E. Lee, Freeze duration influences postfreeze survival in the frog *Rana sylvatica*, J. Exp. Zool. **280** (1998) 197–201.
[51] M.R. Lee, R.E. Lee, Jr., J.M. Strong-Gunderson, S.R. Minges, Isolation of ice-nucleating active bacteria from the freeze-tolerant frog, *Rana sylvatica*, Cryobiology **32** (1995) 358–365.

[52] J.P. Costanzo, R.E. Lee, Ice nucleation in freeze-tolerant vertebrates, Cryo-Lett. **17** (1996) 111–118.

[53] J.P. Costanzo, J.M. Bayuk, R.E. Lee, Inoculative freezing by environmental ice nuclei in the freeze-tolerant wood frog, *Rana sylvatica*, J. Exp. Zool. **284** (1999) 7–14.

[54] J. Buitink, O. Leprince, Glass formation in plant anhydrobiotes: survival in the dry state, Cryobiology **48** (2004) 215–228.

[55] M. Vicré, J.M. Farrant, A. Driouich, Insights into the cellular mechanisms of desiccation tolerance among angiosperm resurrection plant species, Plant Cell Environ. **27** (2004) 1329–1340.

[56] A. Solomon, P. Ilan, G. Itamar, Desiccation tolerance of *Muellerius* cf. *capillaris* (Nematoda: Protostrongylidae) first-stage larvae, J. Parasitol. **84** (1998) 802–805.

[57] M. Holmstrup, V. Loeschcke, Genetic variation in desiccation tolerance of *Dendrobaena octaedra* cocoons originating from different climatic regions, Soil Biol. Biochem. **35** (2003) 119–124.

[58] P.L. Altmann, D.S. Ditmer, *Respiration and Circulation*, Fed. of American Soc. for Experimental Biology, 1971.

[59] B.A. Hills, *Decompression Sickness*, John Wiley & Sons, New York (1977).

[60] S. Shirahata, Experiments on nitrogen gas disease with rainbow trout fry, Bull. Fresh. Res. Lab. (Tokyo) **15** (1966) 197–211.

[61] E.A. Hemmingsen, Supersaturation of gases in water: absences of cavitation on decompression from high pressures, Science **167** (1970) 1493–1494.

[62] E.N. Harvey, D.K. Barnes, W.D. McElroy, A.H. Whiteley, D.C. Pease, K.W. Cooper, Bubble formation in animals: I. Physical factors, J. Cell. Comp. Physiol. **24** (1944) 1–22; E.N. Harvey, A.H. Whiteley, W.D. McElroy, D.C. Pease, D.K. Barnes, Bubble formation in animals: II. Gas nuclei and their distribution in blood and tissues, J. Cell. Comp. Physiol. **24** (1944) 23–34.

[63] E.N. Harvey, Physical factors in bubble formation, in: *Decompression Sickness*, Ed. J.F. Fulton, Saunders, Philadelphia (1951).

[64] R.B. Philp, M.J. Inwood, B.A. Warren, Interactions between gas bubbles and components of the blood: Implications in decompression sickness, Aerospace Med. **43** (1972) 946–953.

[65] F.E. Fox, K.F. Herzfeld, Gas bubbles with organic skin as cavitation nuclei, J. Acoust. Soc. Am. **26** (1954) 984–989.

[66] D.E. Yount, A model for microbubble fission in surfactant solutions, J. Colloid Interface Sci. **91** (1983) 349–360.

[67] L.E. Fidler, *A study of biophysical phenomena associated with gas bubble trauma in fish.* Report of Penny Applied Sciences Ltd, Penny, British Columbia, July 1984.

[68] S. Mann, *Biomineralization: Principles and Concepts in Bioinorganic Materials Chemistry,* Oxford University Press, Oxford (2001).

[69] U.G.K. Wegst, M.F. Ashby, The mechanical efficiency of natural materials, Philos. Mag. **84** (2004) 2167–2181.

[70] J.D. Currey, Mechanical properties of mother of pearl in tension, Proc. Roy. Soc. Lond. B **196** (1977) 443–463.

[71] S. Weiner, L. Addadi, Design strategies in mineralized biological materials, J. Mater. Chem. **7** (1997) 689–702.

[72] S. Weiner, I. Sagi, L. Addadi, Choosing the crystallization path less traveled, Science **309** (2005) 1027–1028.

[73] A. Pérez-Huerta, M. Cusack, Common crystal nucleation mechanism in shell formation of two morphologically distinct calcite brachiopods, Zoology **111** (2008) 9–15.

[74] H.D. Jiang, X.Y. Liu, G. Zhang, Y. Li, Kinetics and template nucleation of self-assembled hydroxyapatite nanocrystallites by chondroitin sulfate, J. Biol. Chem. **280** (2005) 42061–42066.

[75] L. Addadi, D. Joester, F. Nudelman, S. Weiner, Mollusk shell formation: A source of new concepts for understanding biomineralization processes, Chem. Eur. J. **12** (2006) 980–987.
[76] Z. Yan, G. Jing, N. Gong, C. Li, Y. Zhou, L. Xie, R. Zhang, N40, a novel nonacidic matrix protein from pearl oyster nacre, facilitates nucleation of aragonite *in vitro*, Biomacromolecules **8** (2007) 3597–3601.
[77] P.A. Ngankam, P. Lavalle, J.C. Voegel, L. Szyk, G. Decher, P. Schaaf, F.J.G. Cuisinier, Influence of polyelectrolyte multilayer films on calcium phosphate nucleation, J. Am. Chem. Soc. **122** (2000) 8998–9005.
[78] M. Fricke, D. Volkmer, C.E. Krill III, M. Kellermann, A. Hirsch, Vaterite polymorph switching controlled by surface charge density of an amphiphilic dendron-calix[4] arene, Cryst. Growth Des. **6** (2006) 1120–1123.
[79] A.G. Checa, A.B. Rodríguez-Navarro, Self-organisation of nacre in the shells of Pterioida (Bivalvia: Mollusca), Biomaterials **26** (2005) 1071–1079.
[80] S. Cavali, D.C. Popescu, E.E. Tellers, M.R.J. Vos, B.P. Pichon, M. Overhand, H. Rapaport, N.A.J.M. Sommerdijk, A. Kros, Self-organizing β-sheet lipopeptide monolayers as template for the mineralization of $CaCO_3$, Angew. Chem. Int. Ed. **45** (2006) 739–744.
[81] L. Addadi, J. Moradian, E. Shay, N.G. Maroudas, S. Weiner, A chemical model for the cooperation of sulfates and carboxylates in calcite crystal nucleation: relevance to biomineralization, Proc. Natl. Acad. Sci. USA **84** (1987) 2732–2736.
[82] F. Nudelman, B.A. Gotliv, L. Addadi, S. Weiner, Mollusk shell formation: mapping the distribution of organic matrix components underlying a single aragonitic tablet in nacre, J. Struct. Biol. **153** (2006) 176–187.
[83] A.M. Belcher, X.H. Wu, R.J. Christensen, P.K. Hansma, G.D. Stucky, D.E. Morse, Control of crystal phase switching and orientation by soluble mollusk-shell proteins, Nature **381** (1996) 56–58.
[84] J.B. Thompson, G.T. Paloczi, J.H. Kindt, M. Michenfelder, B.L. Smith, G. Stucky, D.E. Morse, P.K. Hansma, Direct observation of the transition from calcite to aragonite growth as induced by abalone shell proteins, Biophys. J. **79** (2000) 3307–3312.
[85] L. Addadi, S. Raz, S. Weiner, Taking advantage of disorder: amorphous calcium carbonate and its roles in biomineralization, Adv. Mat. **15** (2003) 959–970.
[86] Y. Politi, T. Arad, E. Klein, S. Weiner, L. Addadi, Sea urchin spine calcite forms via a transient amorphous calcium carbonate phase, Science **306** (2004) 1161–1164.
[87] I. Sethmann, R. Hinrichs, G. Wörheide, A. Putnis, Nano-cluster composite structure of calcitic sponge spicules — a case study of basic characteristics of biominerals, J. Inorg. Biochem. **100** (2006) 88–96.
[88] A.W. Xu, Y. Ma, H. Cölfen, Biomimetic mineralization, J. Mater. Chem. **17** (2007) 415–449.
[89] Y. Oaki, H. Imai, Nanoengineering in echinoderms: the emergence of morphology from nanobricks, Small **2** (2006) 66–70.
[90] B.R. Heywood, S. Mann, Molecular construction of oriented inorganic materials: controlled nucleation on calcite and aragonite under compressed Langmuir monolayers, Chem. Mater. **6** (1994) 311–318.
[91] G. Falini, S. Albeck, S. Weiner, L. Addadi, Control of aragonite or calcite polymorphism by mollusk shell macromolcules, Science **271** (1996) 67–69.
[92] Y. Levi, S. Albeck, A. Brack, S. Weiner, L. Addadi, Control over aragonite crystal nucleation and growth: an *in vitro* study of biomineralization, Chem. Eur. J. **4** (1998) 389–396.
[93] S.B. Mukkamala, C.E. Anson, A.K. Powell, Modelling calcium carbonate biomineralization processes, J. Inorg. Biochem. **100** (2006) 1128–1138.
[94] M.F. Casula, Y.W. Jun, D.J. Zaziski, E.M. Chan, A. Corrias, A.P. Alivisatos, The concept of delayed nucleation in nanocrystal growth demonstrated for the case of iron oxide nanodisks, J. Am. Chem. Soc. **128** (2006) 1675–1682.

[95] L. Qi, J. Li, J. Ma, Biomimetic morphogenesis of calcium carbonate in mixed solutions of surfactants and double-hydrophilic block copolymers, Adv. Mater. **14** (2002) 300–303.

[96] S.M. D'Souza, C. Alexander, S.W. Carr, A.M. Waller, M.J. Whitcombe, E.N. Vulfson, Directed nucleation of calcite at a crystal-imprinted polymer surface, Nature **398** (1999) 312–316.

[97] W.L. Murphy, D.J. Mooney, Bioinspired growth of crystalline carbonate apatite on biodegradable polymer substrata, J. Am. Chem. Soc. **124** (2002) 1910–1917.

[98] R. Lakshminarayan, S. Valiyaveettil, G.L. Loy, Selective nucleation of calcium carbonate polymorphs: role of surface functionalization and poly(vinyl alcohol) additive, Cryst. Growth Design **3** (2003) 953–958.

[99] J. Aizenberg, D.A. Muller, J.L. Grazul, D.R. Hamann, Direct fabrication of large micropatterned single crystals, Science **299** (2003) 1205–1208.

[100] W.G. Robertson, The changing pattern of urolithiasis in the UK and its causes, in: D.J. Kok, H.C. Romijn, P.C. M.S. Verhagne, C.F. Verkoelen (Eds.), *Eurolithiasis* (Proc. 9th Eur. Symp. on Urolithiasis), Shaker Pub., Maastricht, 2001, pp. 9–11.

[101] J.P. Kavanagh, Kidney stones and renal nucleation of calcium salts, unpublished preprint.

[102] D.G. Assimos, R.P. Holmes, Role of diet in the therapy of urolithiasis, Urol. Clin. North. Am. **27** (2000) 255–268.

[103] M. Menon, B.G. Parulkar, G.W. Drach, Urinary lithiasis: etiology, diagnosis and medical management, in: *Campbell's Urology,* Eds. P.C. Walsh, A.B. Retik, A.J. Vaughan, 7th edn, W.B. Saunders Co., Philadelphia (1988), pp. 2661–2733.

[104] B. Finlayson, Physicochemical aspects of urolithiasis, Kidney Int. **13** (1978) 344–360.

[105] P.G. Werness, C.M. Brown, L.H. Smith, B. Finlayson, Equil2: A basic computer program for the calculation of urinary saturation, J. Urol. **134** (1985) 1242–1244.

[106] H. Fleisch, Inhibitors and promoters of stone formation, Kidney Int. **13** (1978) 361–371.

[107] J.P. Kavanagh, A critical appraisal of the hypothesis that urine is a saturated equilibrium with respect to stone-forming calcium salts, BJU Int. **87** (2001) 589–598.

[108] J.P. Kavanagh, Calcium oxalate crystallisation *in vitro,* in: *Calcium Oxalate in Biological Systems,* Ed. S.R. Khan, Boca Raton, CRC Press (1996), pp. 1–21.

[109] J.P. Kavanagh, L. Jones, P.N. Rao, Calcium oxalate induced crystallization kinetics studied by oxalate-induced turbidity in fresh human urine and artificial urine, Clin. Sci. **98** (2000) 151–158.

[110] J.P. Kavanagh, S. Nishio, J. Garside, N.J. Blacklock, Crystallization kinetics of calcium oxalate in fresh minimally diluted urine: comparison of recurrent stone formers and healthy controls in a continuous mixed suspension mixed product removal crystallizer, J. Urol. **149** (1993) 614–617.

[111] O. Söhnel, J. Garside, *Precipitation: Basic Principles and Industrial Applications,* Butterworth-Heinemann, Oxford (1992).

[112] J.P. Kavanagh, L. Jones, P.N. Rao, Calcium oxalate crystallization kinetics at different concentrations of human and artificial urine, with a constant calcium to oxalate ratio, Urol. Res. **27** (1999) 231–237.

[113] L. Borghi, A. Guerra, T. Meschi, A. Briganti, T. Scianchi, F. Allegri, A. Novarini, Relationship between supersaturation and calcium oxalate crystallization in normals and idiopathic calcium oxalate stone formers, Kidney Int. **55** (1999) 1041–1050.

[114] J.M. Fasano, S.R. Khan, Intratubular crystallization of calcium oxalate in the presence of membrane vesicles: an *in vitro* study, Kidney Int. **59** (2001) 169–178.

[115] S.R. Khan, P.S. Shevock, R.L. Hackett, *In vitro* precipitation of calcium oxalate in the presence of whole matrix or lipid components of the urinary stones, J. Urol. **139** (1988) 418–422.

[116] S.R. Khan, P.A. Glenton, R. Backov, D.R. Talham, Presence of lipids in urine, crystals and stones: Implications for the formation of kidney stones, Kidney Int. **62** (2002) 2062–2072.

[117] W. Wu, D.E. Gerard, G.H. Nancollas, Nucleation at surfaces: the importance of interfacial energy, J. Am. Soc. Nephrol. **10**(suppl 14), (1999) S355–S358.

[118] C. Cerini, S. Geider, B. Dussol, C. Hennequin, M. Daudon, S. Veesler, S. Nitsche, R. Boistelle, P. Berthezene, P. Dupuy, A. Vazi, Y. Berland, J.C. Dagorn, J.M. Verdier, Nucleation of calcium oxalate crystals by albumin: involvement in the prevention of stone formation, Kidney Int. **55** (1999) 1776–1786.

[119] M.E. Bruce, R.G. Will, J.W. Ironside, I. McConnell, D. Drummond, A. Suttie, L. McCardle, A. Chree, J. Hope, C. Birkett, S. Cousens, H. Fraser, C.J. Bostock, Transmissions to mice indicate that 'new variant' CJD is caused by the BSE agent, Nature **389** (1997) 498–501.

[120] J. Collinge, K.C.L. Sidle, J. Meads, J. Ironside, A.F. Hill, Molecular analysis of prion strain variation and the aetiology of 'new variant' CJD, Nature **383** (1996) 685–690.

[121] S. Mead, S.P. Mahal, J. Beck, T. Campbell, M. Farrall, E. Fisher, J. Collinge, Sporadic — but not variant — Creutzfeldt–Jakob disease is associated with polymorphisms upstream of *PRNP* Exon 1, Am. J. Hum. Genet. **69** (2001) 1225–1235.

[122] J. Collinge, Human prion diseases and bovine spongiform encephalopathy (BSE), Hum. Mol. Genet. **6** (1997) 1699–1705.

[123] D.C. Gajdusek, Unconventional viruses and the origin and disappearance of kuru, Science **197** (1977) 943–960.

[124] N. Stahl, M.A. Baldwin, D.B. Teplow, L. Hood, B.W. Gibson, A.L. Burlingame, S.B. Prusiner, Structural Studies of the scrapie prion protein using mass spectrometry and amino acid sequencing, Biochemistry **32** (1993) 1991–2002.

[125] G. Forloni, N. Angeretti, R. Chiesa, E. Monzani, M. Salmona, O. Bugiani, F. Tagliavini, Neurotoxicity of a prion protein fragment, Nature **362** (1993) 543–546.

[126] D.R. Brown, K. Qin, J.W. Herms, A. Madlung, J. Manson, R. Strome, P.E. Fraser, T. Kruck, A. von Bohlen, W. Schulz-Schaeffer, A. Giese, D. Westaway, H. Kretzschmar: the cellular prion protein binds copper *in vivo*,, Nature **390** (1997) 684–687.

[127] C. Kurschner, J.I. Morgan, Analysis of interaction sites in homo- and heteromeric complexes containing Bcl-2 family members and the cellular prion protein, Mol. Brain Res. **37** (1996) 249–258.

[128] G.S. Jackson S.J. Tabrizi, J. Collinge, Protein aggregation and the replication of prion infectivity, unpublished preprint.

[129] D.C. Gajdusek, C.J. Gibbs Jr, M.P. Alpers, Experimental transmission of a kuru-like syndrome to chimpanzees, Nature **209** (1966) 794–796.

[130] C.J. Gibbs Jr, D.C. Gajdusek, D.M. Asher, M.P. Alpers, E. Beck, P.M. Daniel, W.B. Matthews, Creutzfeldt–Jakob disease (Spongiform Encephalopathy): transmission to the Chimpanzee, Science **161** (1968) 388–389.

[131] C. Weissmann, A.J. Raeber, F. Montrasio, I. Hegyi, R. Frigg, M.A. Klein, A. Aguzzi, Prions and the lymphoreticular system, Phil. Trans. Roy. Soc. Lond. B **356** (2001) 177–184.

[132] I.H. Pattison, Experiments with scrapie with special reference to the nature of the agent and the pathology of the disease, in: *Slow, Latent and Temperate Virus Infections, NINDB Monograph 2*, Eds. C.J. Gajdusek, C.J. Gibbs, M.P. Alpers, US Government Printing, Washington DC (1965), pp. 249–257.

[133] J.S. Griffith, Self replication and scrapie, Nature **215** (1967) 1043–1044.

[134] S.B. Prusiner, Novel proteinaceous infectious particles cause scrapie, Science **216** (1982) 136–144.

[135] D.A. Kocisko, J.H. Come, S.A. Priola, B. Chesebro, G.J. Raymond, P.T. Lansbury, B. Caughey, Cell-free formation of protease-resistant prion protein, Nature **370** (1994) 471–474.

[136] A.F. Hill, M. Antoniou, J. Collinge, Protease-resistant prion protein produced *in vitro* lacks detectable infectivity, J. Gen. Virol. **80** (1999) 11–14.
[137] L.L.P. Hosszu, N.J. Baxter, G.S. Jackson, A. Power, A.R. Clarke, J.P. Waltho, C.J. Craven, J. Collinge, Structural mobility of the human prion protein probed by backbone hydrogen exchange, Nat. Struct. Biol. **6** (1999) 740–743.
[138] S. Hornemann, R. Glockshuber, A scrapie-like unfolding intermediate of the prion protein domain PrP(121-231) induced by acidic pH, Proc. Natl. Acad. Sci. USA **95** (1998) 6010–6014.
[139] M.P. McKinley, D.C. Bolton, S.B. Prusiner, A protease-resistant protein is a structural component of the scrapie prion, Cell **35** (1983) 57–62.
[140] S. Paquet, C. Langevin, J. Chapuis, G.S. Jackson, H. Lande, D. Vilette, Efficient dissemination of prions through preferential transmission to nearby cells, J. Gen. Virol. **88** (2007) 706–713.
[141] G. Clarke, R.A. Collins, B.R. Leavitt, D.F. Andrews, M.R. Hayden, C.J. Lumsden, R.R. McInnes, A one-hit model of cell death in inherited neuronal degenerations, Nature **406** (2000) 195–199.
[142] M.F. Perutz, A.H. Windle, Cause of neural death in neurodegenerative diseases attributable to expansion of glutamine repeats, Nature **412** (2001) 143–144.
[143] M.P. Duyao, A.B. Auerbach, A. Ryan, F. Persichetti, G.T. Barnes, S.M. McNeil, P. Ge, J. P. Vonsattel, J.F. Gusella, A.L. Joyner, M.E. MacDonald, Inactivation of the mouse Huntington's disease gene homolog Hdh, Science **269** (1995) 407–410.
[144] D.C. Rubinsztein, J. Leggo, R. Coles, E. Almqvist, V. Biancalana, J.J. Cassiman, K. Chotai, M. Connarty, D. Craufurd, A. Curtis, D. Curtis, M.J. Davidson, A.M. Differ, C. Dode, A. Dodge, M. Frontali, N.G. Ranen, O.C. Stine, M. Sherr, M.H. Abbott, M.L. Franz, C.A. Graham, P.S. Harper, J.C. Hedreen, A. Jackson, J.C. Kaplan, M. Losekoot, J. C. MacMillan, P. Morrison, Y. Trottier, A. Novelletto, S.A. Simpson, J. Theilmann, J.L. Whittaker, S.E. Folstein, C.A. Ross, M.R. Hayden, Phenotypic characterization of individuals with 30-40 CAG repeats in the Huntington disease (HD) gene reveals HD cases with 36 repeats and apparently normal elderly individuals with 36-39 repeats, Am. J. Hum. Genet. **59** (1996) 16–22.
[145] K. Sathasivam, I. Amaechi, L. Mangiarini, G. Bates, Identification of an HD patient with a (CAG)(1 8 0) repeat expansion and the propagation of highly expanded CAG repeats in lambda phage, Hum. Genet. **99** (1997) 692–695.
[146] C. Soto, Protein misfolding and disease; protein refolding and therapy, FEBS Lett. **498** (2001) 204–207.
[147] M. DiFiglia, E. Sapp, K.O. Chase, S.W. Davies, G.P. Bates, J.P. Vonsattel, N. Aronin, Aggregation of huntingtin in neuronal intranuclear inclusions and dystrophic neurites in brain, Science **277** (1997) 1990–1993.
[148] J.S. Steffan, L. Bodai, J. Pallos, M. Poelman, A. McCampbell, B.L. Apostol, A. Kazantsev, E. Schmidt, Y.Z. Zhu, M. Greenwald, R. Kurokawa, D.E. Housman, G.R. Jackson, J.L. Marsh, L.M. Thompson, Histone deacetylase inhibitors arrest polyglutamine-dependent neurodegeneration in Drosophila, Nature **413** (2001) 739–743.
[149] Y. Narain, A. Wyttenbach, J. Rankin, R.A. Furlong, D.C. Rubenstein, A molecular investigation of true dominance in Huntington's disease, J. Med. Genet. **36** (1999) 739–746.
[150] G.N. Ramachrandran, V. Sasiekharan, Conformation of polypeptides and proteins, Adv. Protein Chem. **23** (1968) 283–437.
[151] H. Monoi, New tubular single-stranded helix of poly-l-amino acids suggested by molecular mechanics calculations: 1. Homopolypeptides in isolated environments, Biophys. J. **69** (1995) 1130–1141.

[152] M.F. Perutz, J.T. Finch, J. Berriman, A. Lesk, Amyloid fibers are water-filled nanotubes, Proc. Natl. Acad. Sci. USA **99** (2002) 5591–5595.
[153] E.B. Starikov, H. Lehrach, E.E. Wanker, Folding of oligoglutamines: a theoretical approach based upon thermodynamics and molecular dynamics, J. Biomol. Struct. Dyn. **17** (1999) 409–427.
[154] M.F. Perutz, B.J. Pope, D. Owen, E.E. Wanker, E. Scherzinger, Aggregation of proteins with expanded glutamine and alanine repeats of the glutamine-rich and asparagine-rich domains of Sup35 and of the amyloid β-peptide of amyloid plaques, Proc. Natl. Acad Sci. USA **99** (2002) 5596–5600.

CHAPTER 17

Food and Drink

Contents

1. PHASE TRANSFORMATIONS IN FOODS

The production of foods, their cooking, and their degradation all involve changes in structure. In some cases, these changes require the nucleation of a phase transformation; such cases are the focus of this chapter. Foods have four main components: carbohydrates, lipids, proteins, and water.

Carbohydrates are sugars and long-chain molecules composed of sugars. The most common in foods is starch, but the category also includes cellulose. Starch and sugars often have an amorphous or glassy form that can crystallize. The glass transition and crystallization are greatly affected by water content. Bread going stale involves the partial crystallization of starch [1], and the sandiness that can develop in badly kept ice cream is the result of crystallization of amorphous lactose [2]; these are but two examples of many in which food degradation is associated with crystallization. The crystallization of sugars in confectionery can be either detrimental or beneficial. It has been the subject of much work on nucleation control, and is considered in Section 2.

Lipids include *fats* (solid at room temperature) and *oils* (liquid). Fats and oils are mixtures of *triglycerides,* which are esters formed from glycerol and three fatty acid molecules. They are immiscible with water and their phase transformations are, therefore, not affected by water content. When fats solidify, they typically do so in stages, through several crystalline forms of increasing stability. Cocoa butter is a particularly simple fat, and its crystalline polymorphs are relevant for the quality of chocolate; nucleation control in this context is the subject of Section 3.

Pergamon Materials Series, Volume 15
ISSN 1470-1804, DOI 10.1016/S1470-1804(09)01517-X

Proteins are polymers of amino acids. Although some proteins show a glass transition, the change of most interest in food and cooking is the *denaturing* [3] of their molecular structure by heating or chemical means, and does not involve nucleation. (We have seen in the previous chapter, however, that the conformation of protein molecules can play a central role in the nucleation of neurodegenerative disease.)

Water is the ubiquitous plasticizer of food solids. Removal of water often results in the formation of an amorphous or glassy structure, and this has received much attention. Avoidance of crystallization is often an important part of the preserving action of dehydration, familiar in dried foods such as legume seeds (beans and peas) and pasta. The glassy phase plays a vital role in governing the texture and controlling the shelf life of food products such as cookies (in USA, biscuits in UK) and potato chips (crisps) [4, 5]. The glassy phase is also important more broadly in fields such as cryopreservation [6, 7], including its application to complex biological systems.

It is clear, then, that phase transformations, and in particular the avoidance or promotion of crystallization, are important in food production and storage. They have been subject to widespread thermodynamic and kinetic study [1]. All such studies, including those of nucleation, are hindered by the variability and complexity of natural raw materials.

Phase transformations occasionally contribute to the pleasure of eating itself. Examples include the progressive softening of taffy/toffee at its glass transition as it is heated in the mouth, and in contrast the sharp melting of chocolate (accompanied by a slight coolness as the latent heat of melting is absorbed). A further example of a phase transformation contributing to the gaiety of nations is the effervescence in certain beverages, analyzed in Section 4.

2. SUGARS

The production of sugar typically involves at least two stages in which it is crystallized from solution (a process known as *graining*) in order to purify it. For purification by graining to work, it is important that the impurities in the residual liquid are not trapped in a fine equiaxed mush. Such a solidification pattern (*false grain*) can arise if primary crystals are broken up under turbulent conditions (similar to the *big bang* phenomenon in alloy casting, Chapter 13, Section 2.1). The most common sugar is sucrose, for which the state diagram with water is shown in Figure 1 (from [8], incorporating data from [9–13]). Although metastable sucrose hydrates have been reported [13], they rarely form spontaneously and only the anhydrous form of sucrose is considered in the

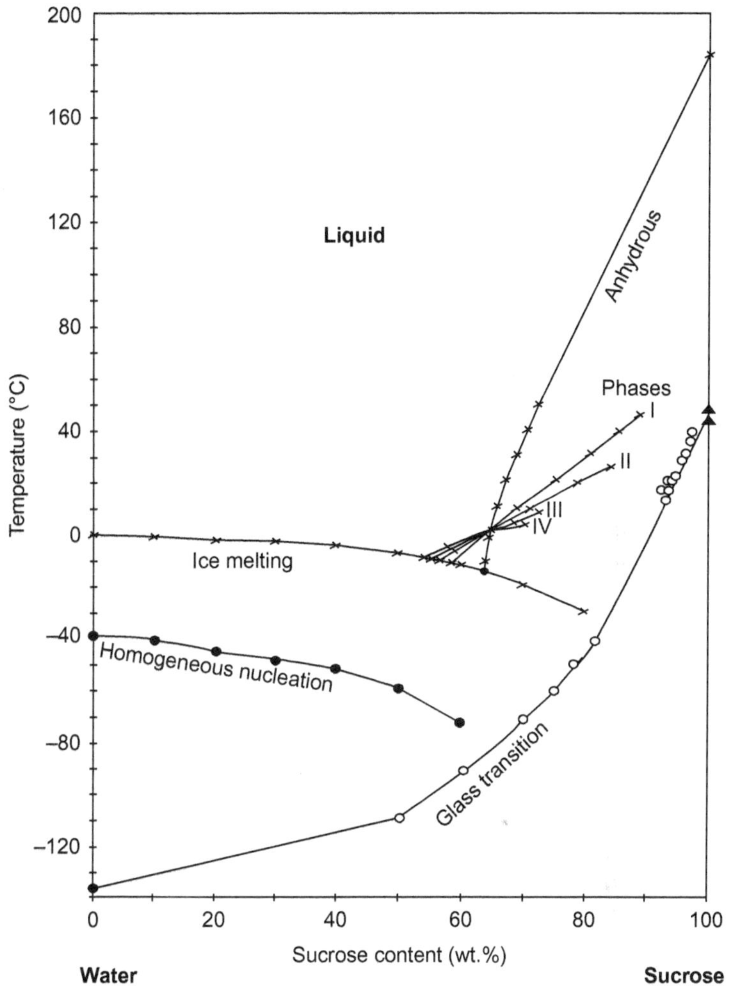

Fig. 1 The sucrose–water state diagram, incorporating data from [8–13]. (Adapted from Ref. [8], with permission.)

diagram. In this eutectic system, freezing can start with ice or with sucrose and the liquidus lines for both are shown. For the compositions at which cooling is expected to give primary crystallization of sucrose, its nucleation is hindered by the viscosity of the system so that, in practice, the ice liquidus is still often the point at which freezing starts. Solutions with more than 87% sucrose are very difficult to crystallize [14]. The homogeneous nucleation temperature reported for ice is shown, but none is available for sucrose. The difficulty of obtaining spontaneous nucleation of sucrose has led to a variety of graining methods, outlined in [15]. Well-controlled sugar processing typically involves seeding. In the production of granulated sugar, for example, the concentrated sucrose solution is seeded with milled icing sugar [16]. (Similarly, the crystallization of glucose in set honey is seeded with set honey that has been chilled and ground into a fine powder.)

Sugar is a significant ingredient in a wide range of foods — baked goods (bread, cakes, cookies/biscuits), confectionery, and dairy products. While crystalline phases may be preferred in some cases, mostly the glassy state is desirable and the importance of this state for preservation has already been noted. The importance of controlling crystallization of glassy sugars in foods is well recognized [17, 18]. Even so, data on crystal nucleation rates are not available, acquisition and interpretation being hindered by the complexity of food structure. When sugar is the predominant component in confectionery, there is the best chance of quantitative interpretation of crystal nucleation phenomena.

Clear hard candies (boiled sweets) are a rare example of a single-phase food; they consist of a glassy sugar solution, with occasional inclusions of flavoring or coloring agents and possibly air [19]. Crystallization is almost always considered undesirable. The glass-transition temperature T_g (Figure 1) increases sufficiently in near-anhydrous compositions for the sucrose–water system to be glassy at room temperature. A typical *high-boiled* sweet has a water content of 2–5%. The resistance of a sucrose–water glass to devitrification is insufficient for long storage, and it is necessary to add an *inhibitor* (also known as a *doctor*), which is usually glucose. Glucose acts by decreasing both the mobility in the system (by increasing the viscosity) and the thermodynamic driving force for crystallization (by increasing the solubility of sucrose). With too much glucose, however, the candy can be soft and sticky. The final boiled candy has a T_g of 43–45°C, and is very resistant to crystallization, unless there is an increase in mobility (by raising the water content) or seeding. Unfortunately, the glassy phase is hygroscopic; increased water content at the surface of the candy lowers T_g below room temperature and crystallization ultimately progresses inward [20].

Taffy (toffee) consists of fat droplets in a sugar glass matrix. (Caramels have essentially the same composition, but the matrix is syrupy, not glassy.) The fat can be vegetable fats and butter, and it is now usual for taffy also to contain milk protein, often from skimmed condensed milk [21]. The water content in taffy is 8–9% [21], and the most common defect in toffee is devitrification. This is hindered, as for boiled sweets, by including glucose. During processing, it is essential that all the original sucrose crystals are dissolved to avoid residual crystals acting as seeds later. Also, it is important during cooling to avoid moisture on the surface of the taffy. Internal devitrification is nucleated at the interfaces between the fat droplets and the sugar glass matrix [22]. In taffy containing milk, this interface is where whey protein forms a membrane at which casein micelles congregate [19].

A *fondant* is a smooth sugar paste, in which the crystal phase fraction should be high (typically 60%) and near equilibrium in a syrup matrix. Further crystallization is undesirable, as it would lead to interlocking of the crystals and hardening of the paste. For a fondant to be smooth, the crystals must be small, typically 8–10 μm in diameter [23]. The required high population of fine grains can be attained only by vigorous stirring (*beating*). The syrup is cooled rapidly to ~50 °C to attain a high supersaturation and stimulate nucleation. Stirring at that temperature gives seeding by broken-up crystals (a process known as *secondary nucleation*). Fondant is of interest not only in its own right, but also as a nucleating agent, for example, in fudge.

Fudge attains its desired texture by crystallization of sugar during processing. It typically contains more sugar and milk than taffy does and the water content is 7–10%. The solid content in fudge (including the fat) is ~60%. The sucrose crystals in fudge are ~30 μm in diameter. Their formation is seeded by the addition of about 5% fondant to the mix. The mix must be cooled to ~105°C when the fondant is added to avoid dissolution of the small seed crystals [23]. The population of seed crystals in fondant is $\sim 6 \times 10^{14}\,m^{-3}$, while in the final fudge the population of sucrose grains is $\sim 1.5 \times 10^{13}\,m^{-3}$. As has been noted [21], this means that roughly half of the added seeds initiate fudge grains, with the others presumably dissolving.

3. CHOCOLATE

The processing of cocoa beans into the various forms of chocolate available as foods is complex, with many steps, as reviewed, for example, by Beckett [24]. We consider only the last of these in which molten chocolate is molded into final form. In this solidification, it is necessary to select the cocoa-butter polymorph that crystallizes, and this is achieved through control of nucleation. The selection of a stable polymorph ensures that the chocolate is solid but melts rapidly in the mouth, and it also hinders the formation of unsightly surface *bloom* during storage.

The various types of chocolate (milk, dark, bittersweet) contain ingredients in different proportions [25], but all have a minimum solid fat content of 45% by weight. Molten chocolate consists of a suspension of particles of sugar, cocoa solids, and possibly milk solids, in liquid fat. The fat is predominantly cocoa butter, with 12–32% milk fat in milk chocolate [26], and with possible additions of compatible vegetable fats (*cocoa-butter equivalents* [27]). The other fats are fully miscible with cocoa butter and do not change its crystalline polymorphs, though they do somewhat affect their melting temperatures. Cocoa butter itself is relatively simple for a natural product, consisting of a mixture of fatty acids, of which three — palmitic, stearic, and oleic — account for more than 95%. The exact mixture of fatty acids varies with the origin of the cocoa butter. Fats in general show three main polymorphs, α, β', and β, all of which have the alkyl (fatty acid) chains stacked parallel to each other. In the α and β' polymorphs, the repeat length parallel to the chains corresponds to two fatty acid chains, while in β it corresponds to three chains. These polymorphs may have subforms, and in the usual classification schemes cocoa butter is considered to exhibit six forms (Table 1) [25, 28–31]. The melting point indicates the relative stability, and the more stable forms are also significantly more dense. The least stable phase, Form I, is obtained upon rapid cooling, but rapidly transforms to Form II, which then changes more slowly to Forms III and IV. Molten chocolate is carefully treated (*tempered*) to ensure that the molded product is mainly in Form V. Without tempering, it would be mainly in Form IV. The most stable phase, Form VI, is found when cocoa butter is crystallized from solution. Although it is difficult to produce directly from the melt, it does arise from transformation of Form V during long storage, and thereby contributes to bloom. While a start has been made on the determination of time–temperature–transformation diagrams for solidification of model systems based on cocoa-butter components [32, 33], there has been little work on the quantification of nucleation and growth rates.

It is evident from its melting point (Table 1) that Form V is well suited to give a sharp melting in the mouth (taken to be at 37°C), deemed to be an attractive characteristic. A well-tempered chocolate, solidifying to Form V, sets quickly with a glossy finish. If solidification is to a less stable

Table 1 Cocoa-butter polymorphs

Polymorph classification		Melting point (°C)	Chain packing
Wille and Lutton [28]	Larsson [29]		
I	β'_1	17.3	double
II	α	23.3	double
III	mixed	25.5	double
IV	β'_1	27.3	double
V	β_2	33.8	triple
VI	β_1	36.3	triple

In the Larsson classification, the subscripts 1 and 2 denote sub-forms of greater and lesser melting point. (From Refs. [30, 31], see also Ref. [25]).

polymorph, subsequent transformation to more stable forms is accompanied by contraction and unsightly surface roughening. When such contraction leads to microscopic cracks, the liquid fat that reaches the surface along them can ultimately crystallize to Form VI, giving the phenomenon of bloom [34]. The process is accelerated by the migration, through the chocolate itself, of lighter (e.g., nut) oils from a filling material. The process can be inhibited by the addition of other proprietary oils [27].

The process of tempering is designed then to ensure that the solidifying polymorph is the most stable that can be formed directly from the melt. As less stable polymorphs melt at lower temperatures, the principle of eliminating them is straightforward. In practice, the thermal program is as illustrated in Figure 2. The exact temperatures depend on chocolate composition; the figure shows values for a typical milk chocolate. After complete melting of the cocoa butter (ensured above 45°C), cooling is used to nucleate crystals (*strike seed*). Rapid needle-like growth is followed by reheating to remove the less stable polymorph (Form IV). Stirring of the partially solidified mix, including scraping of container surfaces, gives a uniform dispersion of small crystals [35]. On subsequent solidification, cooling curves (Figure 3) correlate well with other characteristics and give a useful indication of the quality of the temper [31, 35]. If insufficient stable seeds of Form V have been produced, the liquid shows clear supercooling and recalescence (Figure 3a, similar to Chapter 13, Figure 10). If perfectly tempered, there is no supercooling and the thermal plateau is as expected for nonpartitioning

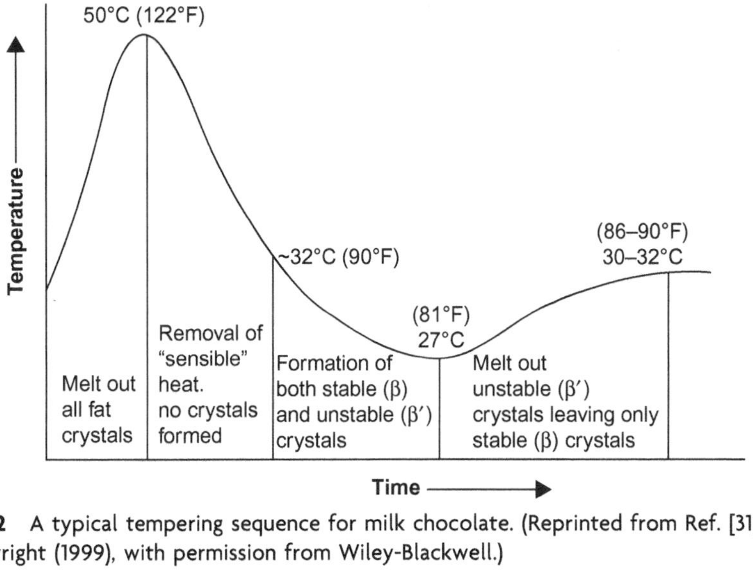

Fig. 2 A typical tempering sequence for milk chocolate. (Reprinted from Ref. [31], copyright (1999), with permission from Wiley-Blackwell.)

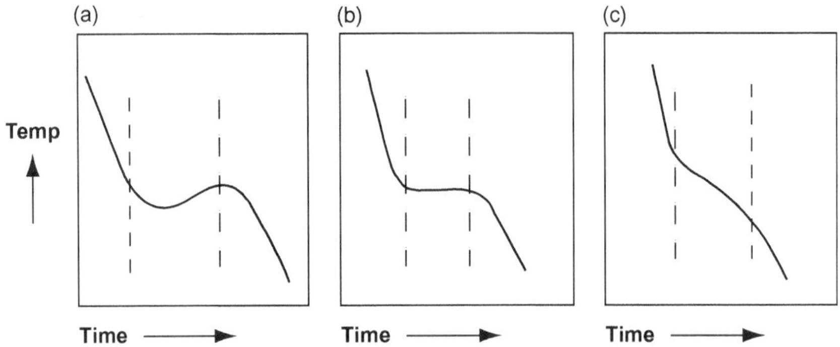

Fig. 3 Schematic cooling curves showing solidification of molten chocolate in three starting conditions: (a) undertemper, (b) perfect temper, (c) overtemper. (Reprinted from Ref. [31], copyright (1999), with permission from Wiley-Blackwell.)

solidification (Figure 3b). Overtempering leads to excessive crystallization before molding and a blurring of the thermal plateau (Figure 3c). On the other hand, long tempering times (up to 6 hours) can be beneficial in allowing crystals to grow to a size that enables them to survive a subsequent raising of the temperature, this being useful to have greater fluidity for casting, particularly for thin coatings [35].

Processing of chocolate would be simpler, were it possible to implement direct seeding with a stable cocoa-butter polymorph [31]. This has been achieved by introducing crystals dispersed in an oil carrier into untempered chocolate [36], but is not yet a widespread practice. There is also some interest in introducing seeds that are more stable than the cocoa-butter forms. The triglyceride BOB, containing two behenic acid groups, has the β_1 structure with a melting point of 53°C, well above that of chocolate. In this case, the chocolate raised above its melting point still solidifies readily to Form V of cocoa butter [37]. The continuing survival of the seeds greatly reduces the chance of bloom arising subsequently.

Conventional tempering is achieved through variation of temperature (Figure 2), but similar effects can be achieved through variation of pressure [38]. The melting point of cocoa butter rises linearly with pressure, so applying pressure at constant temperature has an effect equivalent to cooling at constant pressure. If cocoa butter at 30°C is subjected to a pressure of 150 MPa, the effective temperature drop is 20°C, and all main polymorphs form. On release of the pressure, the less stable ones melt away. Successful tempering requires two pressure pulses held for 5 minutes [39]. The main difficulty appears to be in nucleating Form V and not in growth to a suitable size. This can be seen when successfully tempered chocolate is recycled as 10% of the original mix. For this seeded mix, no extended hold at pressure is required [39].

4. CARBONATED DRINKS

The formation and growth of bubbles occurs when a liquid is supersaturated with dissolved gas, is superheated, or is subjected to hydrostatic tension. Bubble formation is important in many contexts; metallurgical and medical examples have already been considered in Chapter 13, Section 6 and Chapter 16, Section 3. Other examples, ranging from explosive degassing of volcanic magma to foaming of plastics, have been collected in the review by Jones et al. [40]. The familiar and much appreciated example of bubbling in carbonated drinks [41, 42] is treated in this section. Quantitative studies of bubble nucleation and growth in drinks are of fundamental interest, but also reflect practical concerns of the drinks industry.

The gas in carbonated drinks is carbon dioxide, CO_2. In champagne, other sparkling wines, and beer, the CO_2 is a by-product of the fermentation of sugars into alcohol. In carbonated soft drinks (in the USA, *soda*, from the sometime use of sodium bicarbonate), the CO_2 is added as a gas. The bubbles have effects not only in the mouth. The bouquet of champagne and other carbonated drinks arises largely from the *jet drops* that are projected a few centimeters above the liquid when bubbles burst at its surface. The aromatic molecules in the drink are often surfactants, concentrated at the bubble surfaces and thereby in the jet drops. Bubble size is often a characteristic, being smallest in champagne, intermediate in beer, and largest in carbonated soft drinks. For champagne, the fineness of the bubbles is often taken to be a measure of quality.

In a sealed container, the dissolved concentration of CO_2 in the liquid is in equilibrium with the internal pressure of gas. As CO_2 does not dissociate on dissolution, Henry's law is obeyed and Chapter 16, Eqs. (1)–(6) can be applied. For a typical champagne, the pressure in the corked bottle is 6 atm. On opening the bottle, the supersaturation s is thus ~ 5 (from Chapter 16, Eq. (3), and noting that $s = \alpha_s - 1$). The surface tension $\sigma = 46.8 \pm 0.6\,\mathrm{mN\,m^{-1}}$, and Chapter 16, Eq. (4) then gives a critical radius $r^* \approx 0.2\,\mu\mathrm{m}$. In beer, the supersaturation is about one-third of this value, and the critical radius is correspondingly higher.

Even the critical radius in fully supersaturated champagne is far above molecular dimensions and corresponds to a vanishingly low homogeneous nucleation rate. For homogeneous nucleation to be possible, the supersaturation would need to be greater than 100. Given typical contact angles, heterogeneous nucleation of a spherical cap of gas also requires a supersaturation far greater than that typically found in carbonated drinks. These points have been extensively reviewed by Jones et al. [40]; they classify homogeneous and spherical-cap heterogeneous nucleation as Types I and II. They show that the nucleation of bubbles in lightly supersaturated liquids, such as carbonated drinks, must be from preexisting gas-filled cavities. (This is the same conclusion as seen earlier for bubble nucleation in the metallurgical and medical examples in Chapter 13, Section 6 and Chapter 16, Section 3, Figure 8.) If the mouth of the cavity has a radius less than r^* (Chapter 16, Eq. (4)), then the gas pocket in the cavity will grow until the radius of curvature of its interface with the liquid decreases to r^*. Growth beyond that point involves surmounting an energy barrier, and relies on thermal fluctuation; Jones et al. classify this as Type III nucleation. If the mouth of the cavity has a radius greater than r^*, there is growth of the bubble without an energy barrier, classified as Type IV. A transition from Type III to Type IV would be observed on increasing the supersaturation and it would be closely analogous to the athermal nucleation illustrated in (Chapter 6, Section 2.5, Figure 15). Type IV nucleation is dominant at the low supersaturations typical of carbonated drinks, and in such a case the sequence of events generating a bubble rising in a glass of liquid is as illustrated in Figure 4.

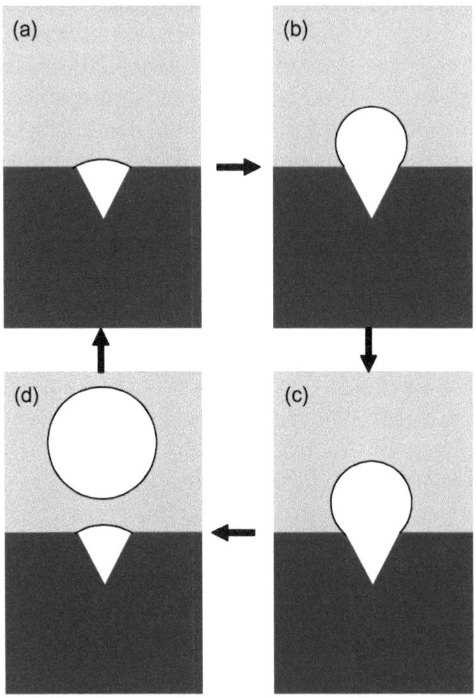

Fig. 4 Schematic illustration of bubble production from an initial gas-filled cavity: (a) cavity growth with decreasing radius of curvature of the gas–liquid interface, (b) continued growth with increase in the radius of curvature, (c) detachment when buoyancy and capillary forces balance, (d) the growing bubble is released, leaving gas in the cavity for the cycle to be repeated. (Reprinted with permission from Ref. [44], copyright (2002) American Chemical Society.)

In stage (a), CO_2 molecules diffuse from the liquid into the gas-filled cavity, decreasing the radius of curvature of the gas–liquid interface. As the mouth of the cavity has a radius greater than the critical radius for the supersaturation, the growth continues unimpeded through stage (b), but the bubble remains anchored by capillary forces. At stage (c), the buoyancy force on the bubble matches the capillary forces, and as growth continues to stage (d) the bubble breaks free and rises in the glass. The radius of a bubble when it detaches is $\leq 100\,\mu m$. Free bubbles are produced, that is, they break free from capillary forces at a regular frequency dependent on the supersaturation. In champagne, a typical production rate at a given site is 30 bubbles per second, and in beer 10 per second. The regular production of bubbles is particularly evident in champagne, where the clear bubble trains can be subjected to quantitative analysis as described further.

What provides the gas-filled cavities? With critical radii of $0.2\,\mu m$ in champagne, and larger in other carbonated drinks, defects in the glass surface are not sufficiently large to be the basis for Type IV nucleation [43]. In the study of Liger-Belair et al. [44], nucleation sites were observed to be the gas cavities in hollow cylindrical fibers, as shown in Figure 5. These are likely to be cellulose fibers left from glass drying. The survival of gas pockets even in liquids undersaturated in gas is analogous to the survival of crystalline seeds in cavities, even in melts that have been superheated (Chapter 6, Section 2.4, Figure 14).

After a bubble has been nucleated and has broken free, it rises in the glass and grows. Surfactant molecules (proteins and glycoproteins) have water-soluble and water-insoluble ends and, therefore, congregate at the bubble surface. The effect is to stiffen the bubble and retard its ascent through the liquid. In beer, there is a higher concentration of surfactants than in sparkling wine; there is also a lower supersaturation, leading to lower bubble growth rates. Bubbles in beer are typically stiffened by surfactant. In contrast, in champagne the lower concentration of surfactants combined with faster growth arising from the greater supersaturation keeps the bubble surfaces with less surfactant than in beer and, therefore, relatively supple. Analysis of the growth of bubbles as they rise in a supersaturated liquid is complex, and beyond the scope of the present treatment. However, the growth rate (i.e., rate of increase of radius) is directly proportional to the supersaturation. In freshly opened champagne, bubbles grow at up to $400\,\mu m\,s^{-1}$.

The frequency of bubble production also varies linearly with the supersaturation s and has been widely studied [44, 45]. As bubbling continues, s decreases and bubble production slows. Eventually, for a given cavity with mouth of radius r, the supersaturation decreases until it satisfies Eq. (4), Chapter 16. At that point bubble production stops; the critical radius has become equal to or larger than the radius of the cavity

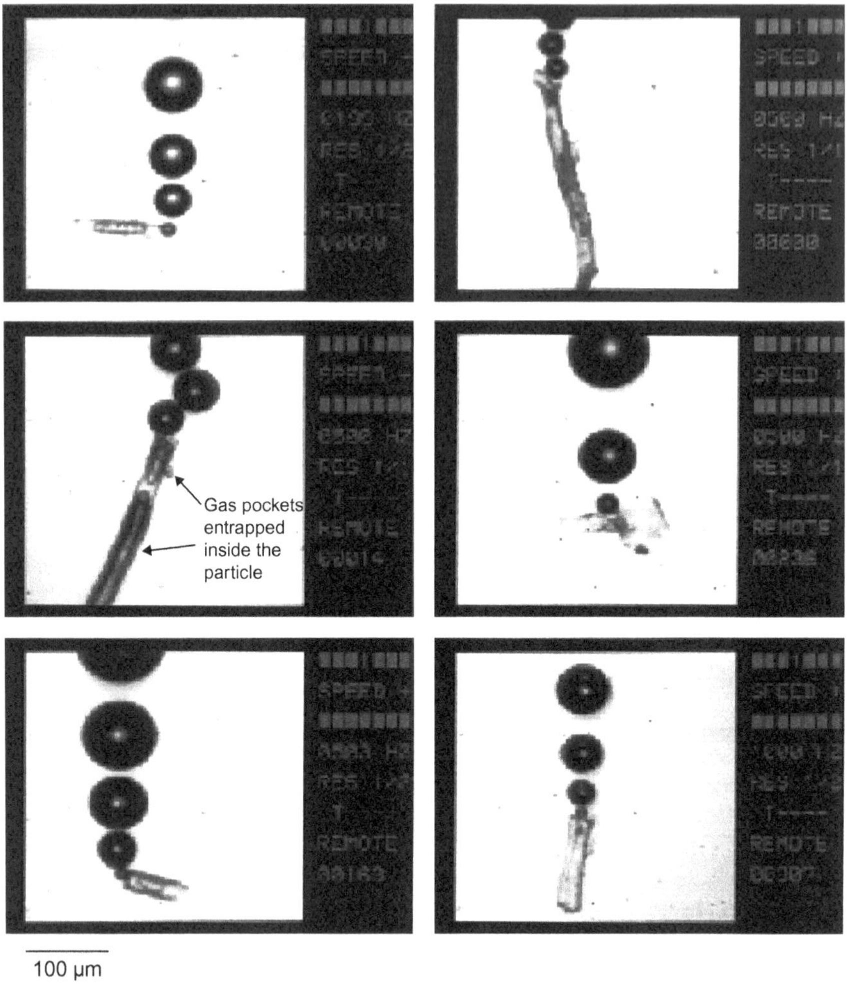

Fig. 5 Six different bubble sources observed in a glass of champagne. (Reprinted with permission from Ref. [44], copyright (2002) American Chemical Society.)

mouth and Type IV nucleation is no longer possible. As *s* decreases, the sources of bubbles shut down in sequence, the smallest first. This effect of cavity size distribution is analogous to that of inoculant particle size distribution discussed in Chapter 13, Section 3. In that case, crystal initiation in the liquid starts first on the largest particles and progresses to smaller particles as the supercooling (analogous to supersaturation in the present case) is increased.

The relationship between bubble production frequency and bubble growth rate once released has been measured and is shown in Figure 6 [44]. Similar data have been acquired in a range of studies, for example,

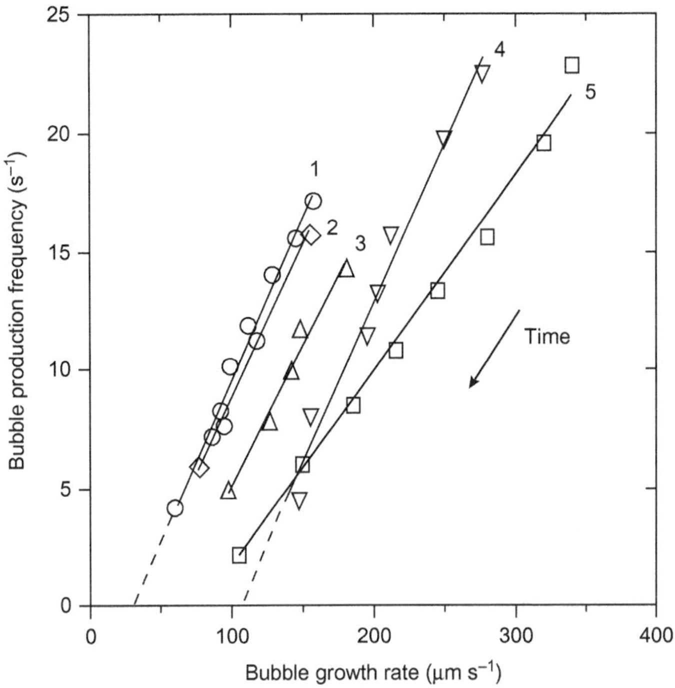

Fig. 6 For a given source, the frequency of bubble production varies linearly with the bubble growth rate as it rises through the liquid; both decrease with the supersaturation as bubbling continues. These data are for five sources in a typical champagne. (Reprinted with permission from Ref. [44], copyright (2002) American Chemical Society.)

[45]. For each bubble source, the data can readily be extrapolated to yield a growth rate and, therefore, the supersaturation at which bubble production would cease. From this value of s, via Eq. (4), Chapter 16, the radius of the cavity mouth can be determined. In this study of sources (of the kind in Figure 5) in a typical glass of champagne, the source radii were in the range 0.6–2.3 μm. These are all significantly greater than the critical nucleation radius of 0.2 μm in freshly opened champagne, confirming the basic model of Type IV nucleation [40].

When champagne is first poured into a glass, there is less than perfect wetting of the inner surface of the glass, and the gas cavities and bubbles give explosive foaming. Once the glass is wetted and coated with a layer of liquid in which the supersaturation has been reduced, bubbles are produced mainly at sources of the kind identified in Figure 5; the drink is then quiescent. Waiters exploit these phenomena. The fastest way to fill several glasses is first to pour a little champagne into each. The resulting foaming achieves the wetting that inhibits further formation of bubbles on the glass itself. Then champagne can be poured rapidly into the glass without foaming and wastage. Wastage is, of course, exactly the intention of Formula 1 victors; shaking the bottle introduces large pockets of gas into the liquid, which then expand rapidly when the cork is expelled.

In this section, and in earlier sections (Chapter 13, Section 6 and Chapter 16, Section 3) concerned with pore or bubble nucleation, we have taken a rather simple approach. There have been more complete analyses taking account of mixtures of species in liquid and gas phases [46–49]. As in other cases, there are queries about the validity of the classical nucleation theory for bubble nucleation. It is expected that the classical theory may break down for small critical radii, and this has been studied, for example, using density-functional methods [50, 51]. In all the cases of practical interest analyzed in this section and earlier, bubble nucleation was at very low supersaturation, corresponding to critical radii easily large enough for the classical theory to be valid.

5. SUMMARY

The nucleation of phase transformations can be important in foodstuffs. Its importance may lie in the production and processing of the food, in its shelf life, in cooking, or in the enjoyment we derive from eating it. Most of the transformations occur close to equilibrium, ruling out homogeneous nucleation. Indeed, in many cases, ranging from the formation of gas bubbles in carbonated drinks to the controlled crystallizing of sugars, the initiation is from seeds (gas pockets or crystals) of the new phase itself. The relatively simple nature of drinks has permitted fully quantitative studies to be made of gas nucleation. This has yet to be achieved for transformations in the more complex environment of foodstuffs, but opportunities for progress are numerous.

REFERENCES

[1] Y.H. Roos, *Phase Transitions in Foods*, Academic Press, San Diego (1995).

[2] T.A. Nickerson, Lactose crystallization in ice cream IV: factors responsible for reduced incidence of sandiness, J. Dairy Sci. **45** (1962) 354–359.

[3] W. Kauzmann, Some factors in the interpretation of protein denaturation, Adv. Protein Chem. **14** (1959) 1–63.

[4] T.W. Schenz, Glass transition and product stability: an overview, Food Hydrocolloids **9** (4), (1995) 307–315.

[5] L. Slade, H. Levine, J. Ievolella, M. Wang, The glassy state phenomenon in applications for the food industry: application of the food polymer science approach to structure-function relationships of sucrose in cookie and cracker systems, J. Sci. Food Agric. **63** (1993) 133–176.

[6] F. Franks, R.H.M. Hatley, S.F. Mathias, Materials science and the production of shelf-stable biologicals, BioPharm **4**(9), (1991) 38–55.
[7] F. Franks, The importance of being glassy, Cryo Lett. **13** (1992) 349–350.
[8] A. Gough, *A Physicochemical Study of the Role of Sugar in Baked Products*, Ph.D. Thesis, Univ. of Nottingham (1995).
[9] L. Finegold, F. Franks, R.H.M. Hatley, Glass/rubber transitions and heat capacities of binary sugar blends, J. Chem. Soc. Faraday Trans. I **85** (1989) 2945–2951.
[10] A.P. Mackenzie, Non-equilibrium freezing behaviour of aqueous systems, Phil. Trans. Roy. Soc. Lond. B **278** (1977) 167–189.
[11] H. Levine, L. Slade, Water as a plasticizer: physicochemical aspects of low-moisture polymeric systems, in: *Water Science Reviews*, Ed. F. Franks, Cambridge University Press, Cambridge (1987), pp. 79–185.
[12] B. Luyet, D. Rasmussen, Study by differential thermal analysis of the temperatures of instability of rapidly cooled solutions of glycerol, ethylene glycol, sucrose and glucose, Biodynamica **211**(10), (1968) 167–191.
[13] F.E. Young, F.T. Jones, Sucrose hydrates: the sucrose-water phase diagram, J. Phys. Chem. **53** (1949) 1334–1350.
[14] R. Lees, General technical aspects of industrial sugar confectionery manufacture, Chapter 6, in: *Sugar Confectionery Manufacture*, Ed. E.B. Jackson, 2nd edn, Blackie, London (1995), pp. 106–128.
[15] A. Vanhook, Graining in sugar boiling, in: *Nucleation*, Ed. A.C. Zettlemoyer, Marcel Dekker, New York (1969), pp. 573–588.
[16] J.C. Abram, J.T. Ramage, Sugar refining: present technology and future developments, Chapter 3, in: Sugar Science and Technology, Eds. G.G. Birch, K.J. Parker, Appl. Sci. Pub., London (1979), pp. 49–95.
[17] R.W. Hartel, A.V. Shastry, Sugar crystallization in food products, Crit. Rev. Food Sci. Nutrition **30** (1991) 49–112.
[18] R.W. Hartel, Controlling sugar crystallization in food products, Food Technol. **47**(11), (1993) 99–107.
[19] D.F. Lewis, Structure of sugar confectionery, Chapter 15, in: *Sugar Confectionery Manufacture*, Ed. E.B. Jackson, 2nd edn, Blackie, London (1995), pp. 312–333.
[20] B. Makower, W.B. Dye, Equilibrium moisture content and crystallization of amorphous sucrose and glucose, J. Agric. Food Chem. **4**(1), (1956) 72–77.
[21] D. Stansell, Caramel, toffee and fudge, Chapter 8, in: *Sugar Confectionery Manufacture*, Ed. E.B. Jackson, 2nd edn, Blackie, London (1995), pp. 170–188.
[22] E.J. DeBruin, P.G. Keeney, in: *Structural Features of Milk Caramel Revealed through Mechanical Measurements and Scanning Electron Microscopy*, Proc. 26th PMCA Production Conf., Penn. Manufacturing Confectioners' Assoc., Bethlehem, PA (1972), pp. 12–17.
[23] D. Stansell, The composition and structure of confectionery, Chapter 14, in: *Sugar Confectionery Manufacture*, Ed. E.B. Jackson, 2nd edn, Blackie, London (1995), pp. 298–311.
[24] S.T. Beckett, Traditional chocolate making, Chapter 1, in: *Industrial Chocolate Manufacture and Use*, Ed. S.T. Beckett, 3rd edn, Blackwell Science, Oxford (1999), pp. 1–7.
[25] P. Fryer, K. Pinschower, The materials science of chocolate, Mater. Res. Soc. Bull. **25**(12), (2000) 25–29.
[26] S.J. Haylock, T.M. Dodds, Ingredients from milk, Chapter 4, in: *Industrial Chocolate Manufacture and Use*, Ed. S.T. Beckett, 3rd edn, Blackwell Science, Oxford (1999), pp. 57–77.
[27] G. Talbot, Vegetable fats, Chapter 17, in: *Industrial Chocolate Manufacture and Use*, Ed. S. T. Beckett, 3rd edn, Blackwell Science, Oxford (1999), pp. 307–322.
[28] R.L. Wille, E.S. Lutton, Polymorphism of cocoa butter, J. Am. Oil Chem. Soc. **43** (1966) 491–496.

[29] K. Larsson, Classification of glyceride crystal forms, Acta Chem. Scand. **20** (1966) 2255–2260.
[30] J. Schlichter-Aronhime, S. Sarig, N. Garti, Reconsideration of polymorphic transformations in cocoa butter using the DSC, J. Amer. Oil Chem. Soc. **65** (1988) 1140–1143.
[31] G. Talbot, Chocolate temper, Chapter 12, in: *Industrial Chocolate Manufacture and Use*, Ed. S.T. Beckett, 3rd edn, Blackwell Science, Oxford (1999), pp. 218–230.
[32] P. Rousset, M. Rappaz, E. Minner, Polymorphism and solidification kinetics of the binary system POS-SOS, J. Amer. Oil Chem. Soc. **75** (1998) 857–864.
[33] H. Schenk, R. Peschar, Understanding the structure of chocolate, Rad. Phys. Chem. **71** (2004) 829–835.
[34] J. Kleinert, in: *Studies on the Formation of Fat Bloom and Methods of Delaying It*, Proc. 15th PMCA Production Conf., Sect. 14, Penn. Manufacturing Confectioners' Assoc., Bethlehem, PA (1961), pp. 1–14.
[35] R.B. Nelson, Tempering, Chapter 13, in: *Industrial Chocolate Manufacture and Use*, Ed. S. T. Beckett, 3rd edn, Blackwell Science, Oxford (1999), pp. 231–258.
[36] F.W. Cain, N.G. Hargreaves, D.J. Cebula, Production of tempered confectionery, European Patent EP0521205 (1993).
[37] T. Koyano, I. Hachiya, K. Sato, Fat polymorphism and crystal seeding effects on fat bloom stability of dark chocolate, Food Struct. **9** (1990) 231–240.
[38] S.T. Beckett, Non-conventional machines and processes, Chapter 22, in: *Industrial Chocolate Manufacture and Use*, Ed. S.T. Beckett, 3rd edn, Blackwell Science, Oxford (1999), pp. 405–428.
[39] A. Yasuda, K. Mochizuki, The behaviour of triglycerides under high pressure, in: *High Pressure and Biotechnology*, Eds. C. Balny, R. Hayashi, K. Heremans, P. Masson, Colloque INSERM (1992), pp. 255–259.
[40] S.F. Jones, G.M. Evans, K.P. Galvin, Bubble nucleation from gas cavities — a review, Adv. Coll. Inter. Sci. **80** (1999) 27–50.
[41] G. Liger-Belair, The science of bubbly, Sci. American **288**(1), (Jan. 2003) 68–73.
[42] N. Shafer, R. Zare, Through a beer glass darkly, Phys. Today **44** (1991) 48–52.
[43] M.W. Carr, A.R. Hillman, S.D. Lubetkin, Nucleation rate dispersion in bubble evolution kinetics, J. Colloid Inter. Sci. **169** (1995) 137–142.
[44] G. Liger-Belair, M. Vignes-Adler, C. Voisin, B. Robillard, P. Jeandet, Kinetics of gas discharging in a glass of champagne: the role of nucleation sites, Langmuir **18** (2002) 1294–1301.
[45] S.F. Jones, G.M. Evans, K.P. Galvin, The cycle of bubble production from a gas cavity in a supersaturated solution, Adv. Colloid Interface Sci. **80** (1999) 51–84.
[46] M. Blander, J.L. Katz, Bubble nucleation in liquids, AIChE J. **21** (1975) 833–848.
[47] C.A. Ward, A. Balakrishnan, F.C. Hooper, On thermodynamics of nucleation in weak gas-liquid solutions, Mech. Eng. **92** (1970) 71–80.
[48] A.S. Tucker, C.A. Ward, Critical state of bubbles in liquid-gas solutions, J. Appl. Phys. **46** (1975) 4801–4808.
[49] B.S. Holden, J.L. Katz, The homogeneous nucleation of bubbles in superheated binary liquid mixtures, AIChE J. **24** (1978) 260–267.
[50] V. Talanquer, D.W. Oxtoby, Nucleation of bubbles in binary fluids, J. Chem. Phys. **102** (1995) 2156–2164.
[51] V. Talanquer, C. Cunningham, D.W. Oxtoby, Bubble nucleation in binary mixtures: a semiempirical approach, J. Chem. Phys. **114** (2001) 6759–6762.

CHAPTER 18

Key Themes and Prospects

Contents

1. EMERGING THEMES

A key aim of this book is to enable the reader to analyze nucleation in condensed matter. Nucleation is relevant in cases as diverse as precipitation in alloys, the production of glass, the survival of Antarctic fish with supercooled blood, and the appearance of champagne. It is important, of course, to recognize when nucleation is occurring and when it is not, and to apply the correct analysis to the problem. There are phenomena such as aggregation that can mimic some of the features of nucleation, but these have not been covered in the book.

From the wide range of nucleation examples that are examined, it becomes clear that even when nucleation is strictly defined, it comprises a range of phenomena of great diversity and potential complexity. Nucleation of a new phase is achieved by rearrangement of some species. When the kinetics of rearrangement is examined, even the relevant type of species shows a great range. In most of the cases considered, the relevant species are atoms or small spheroidal molecules. But we have also analyzed cases in which the relevant species are long-chain molecules or even dislocations, with, of course, distinctive behavior. The essence of nucleation is that there is an energetic barrier to the appearance of the new phase. Yet the ways of surmounting the barrier are also diverse. Mostly nucleation is regarded as a stochastic process found to proceed sporadically under fixed conditions. Yet there are other cases where nucleation is deterministic, proceeding only when the conditions are changed. The concept of nucleation has been deployed in topics as diverse as the birth of the universe and in the dynamics of traffic. But even with our

Pergamon Materials Series, Volume 15
ISSN 1470-1804, DOI 10.1016/S1470-1804(09)01518-1

narrow focus on condensed matter, it is clear that one simple model cannot be applied. The key goal must be to apply the correct model in each case.

From the cases examined, two themes emerge:

Nucleation is often a multistage process. In making glass ceramics, the nucleation of devitrification can start with the nucleation of noble metal precipitates, followed by nucleation of the main crystallization process on those precipitates. In irradiated alloys, the nucleation of loops of perfect dislocation proceeds through four stages (Chapter 14, Eq. (24)) involving aggregation of vacancies and dislocation reactions. Also under irradiation, voids nucleate by evolving from helium bubbles. These are just some examples, and they emphasize the role of heterogeneities. Even in cases of homogeneous nucleation, however, fluctuations and general inhomogeneities in the original phase undoubtedly influence the process. Icosahedral ordering in super-cooled liquids can influence the phase selection when they solidify. And there are many other possible types of order in an original phase that may be significant, for example, chemical order and magnetic order. One message is that the distinction between homogeneous and heterogeneous nucleation may not always be as clear as one would like. In some cases, such as the precipitation of noble metals in glass ceramics, noted earlier, the multistage nature of the nucleation has been designed into the material processing. But there are also cases where the early stages, involving small and possibly nonobservable heterogeneities, have not been noticed, let alone analyzed. Unless all stages of the nucleation process are understood to some degree, it is unlikely that the rate-limiting stage can be securely identified or the overall kinetics quantitatively analyzed.

Nucleation often involves coupled fluxes. The stages noted earlier are largely considered to be sequential. Yet in many cases of nucleation, distinct processes proceed in parallel. When a nucleus is formed, there is a need for rearrangement of species at the interface between the original and new phases. The new phase differs from the original phase in having a different enthalpy per unit volume and possibly a different composition. Thus nucleation also involves fluxes of heat and possibly of solute. There may be also fluxes of defects such as vacancies and interstitials. The key point is that these fluxes proceeding in parallel are all, to some degree, coupled. The coupling can be complex and affects the analysis of the nucleation.

In identifying the most appropriate form of nucleation analysis, a good start can be made by considering various length scales and time scales that can be involved.

2. LENGTH SCALES IN NUCLEATION

Three important length scales can be identified.

Radius of the critical nucleus. A central concept in nucleation is that of the critical nucleus. As described in Chapter 2, Figure 1, and noted repeatedly, a nucleus of critical size is in unstable equilibrium, its work of formation as a function of size being at a maximum. In the simplest geometry of a spherical nucleus, we can describe this condition in terms of the critical radius r^*. In the discussion that follows, we use r^* more generally to represent any characteristic dimension of a critical nucleus. The critical radius is inversely proportional to the free energy change driving the transformation and, therefore, diverges to infinity as the nucleating phase and the original phase approach equilibrium with each other. At the other extreme, when the driving force for transformation is high, r^* can approach the scale of an atomic diameter.

Width of the interface between nucleus and original phase. When a nucleus is formed within the original phase, there is an interface separating the two phases, and the energy of the interface is the key barrier to nucleation. The structures of the two phases and the interactions between the atomic species within them determine an interfacial width δ. While such a width may diverge to large values near a thermodynamic critical point, it mostly remains within one order of magnitude larger than the atomic diameter.

Dimension of heterogeneities in the original phase. Heterogeneities in the original phase can act as substrates catalyzing nucleation. The parameter r^*, representing the radius of curvature of the interface between original and nucleating phases, is the same for homogeneous and heterogeneous nucleation. But the action of the substrates is to reduce the work of forming a critical nucleus such that nucleation can still be observed at much larger r^* than would be possible in the homogeneous case. The size of the heterogeneities is important and can be characterized by a linear dimension or radius R. While our focus is mostly on heterogeneities such as foreign particles or crystal defects, there are cases where fluctuations within the original phase can have similar effects. For such homophase fluctuations, the time-scale over which they are stable is, of course, also an issue.

We now examine the interactions between these three length scales. In introducing the length scales above and in the following discussion, we focus on systems where the relevant species are atoms. Yet many of the key concepts arising from comparison of r^*, δ, and R remain valid for other cases, for example, in dislocation-mediated nucleation.

Classical nucleation theory (Chapter 2) rests on being able to treat the work of nucleation as a sum of two independent terms, one relating to the bulk free energy difference between the nucleating and original phases, the other to the interface between the phases. Given that $r^* \gg \delta$, the interface can be treated as sharp, and the classical approach is valid. As the driving force for the transformation is increased, r^* shrinks, and when $r^* \leq \delta$, the classical approach breaks down. Not only does the thermodynamics of nucleus formation deviate from the classical model, but the interfacial kinetics may also be affected. Of course, a high driving force is needed for homogeneous nucleation to be observable in reasonable time.

We next consider a system with heterogeneities in the original phase. Even if these heterogeneities are perfectly potent, that is they are thermodynamically wetted by the nucleating phase, they do not necessarily have a catalytic effect. If $R = r^*$, they can have little or no effect on the work of nucleation. The condition $R = r^*$ is where the heterogeneities become active nucleating centers (Chapter 6). For many cases of practical interest, R is so large that there is no possibility of thermal fluctuations generating active nuclei before the condition $R = r^*$ is reached; nucleation is then athermal and deterministic, not stochastic. If the heterogeneities are not perfectly potent, the nucleation is stochastic. The driving force necessary to have nucleation on a particle is greater than in the case of perfectly potent particles, but still less than that for homogeneous nucleation.

The length scales r^*, δ, and R are compared in Figure 1. Plotted on the vertical axis, the value of r^* in a given system varies with the driving force for the transformation. Thus moving in the direction of the arrow might, for example, correspond to increasing the supercooling of a liquid about to freeze. At low enough driving force, the original phase is stable on an observable time-scale, but as the driving force is increased there is a detectable onset of homogeneous nucleation. The interfacial width δ is shown and marks the transition into a nonclassical regime. In practice, the transition from classical to nonclassical is likely to occur at a driving force not much greater than that for the onset of detectable homogeneous nucleation. When heterogeneities are present, other possibilities open up. The catalytic potency of the heterogeneities can be characterized by the contact angle ϕ. The lower the value of ϕ, the smaller the driving force necessary for nucleation, still a stochastic process, to be observed on the substrates. When $R \gg r^*$, the driving force necessary for observable heterogeneous nucleation is independent of R. As ϕ is lowered to zero and the heterogeneities become perfectly potent, the behavior tends to a clear limit in which nucleation occurs deterministically at a driving force that decreases with increasing R. At small R, when the driving force is comparable with that for observable homogeneous nucleation, the heterogeneous nucleation becomes stochastic.

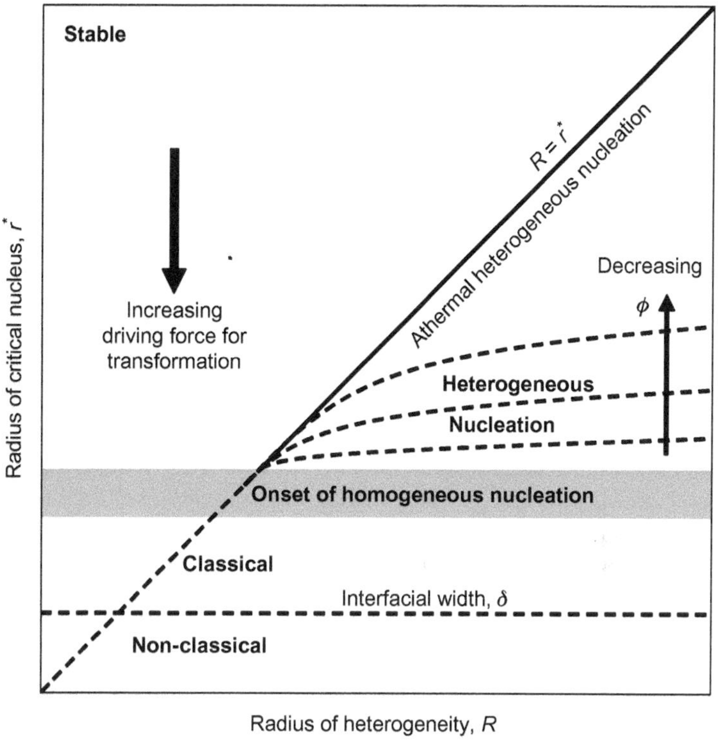

Fig. 1 Schematic comparison of length scales in nucleation: the radius of the critical nucleus r^*, the width δ of the interface between original and new phases, and the radius R of heterogeneities in the original phase. As the driving force for transformation is increased, r^* decreases until homogeneous nucleation becomes observable. In the presence of heterogeneities on which nuclei can form with a contact angle ϕ, nucleation occurs at larger r^* (smaller driving force), and for perfect nucleating substrates, heterogeneous nucleation becomes an athermal process.

In analyzing nucleation, the regimes indicated in Figure 1 can be helpful in identifying the correct approach. The regimes are not intended to indicate predominance of one process over another. For example, the importance of heterogeneous nucleation in an overall transformation clearly depends on the population density of heterogeneities, a factor not considered in this discussion.

3. TIME SCALES IN NUCLEATION

Even for given mobilities of the relevant species, steady-state nucleation rates, and therefore the times required to generate a given population density of nuclei, vary enormously. A more fundamental measure of nucleation kinetics is the rate of relaxation of the cluster size distribution to the steady-state populations, characterized by the nucleation transient time τ. This is the time for a cluster to diffuse through the critical region, and for a one-component system it scales inversely with the rate of the rearrangement processes at the interface between the original and new phases. The nature of the analysis required for a given case of nucleation depends on how τ (evaluated as though for a one-component, nonpartitioning system as in Chapter 3) compares with other characteristic times:

Supersaturation relaxation time. In practice, nucleation is often induced by changing the conditions to increase the driving force for transformation (the supersaturation). For example, a liquid can be cooled to nucleate freezing, or pressure can be decreased to induce nucleation of gas bubbles. The characteristic relaxation time τ_{sup} over which the thermodynamic state is changed is relevant for the analysis of the nucleation. If $\tau_{sup} \gg \tau$, then it is adequate to consider only steady-state nucleation kinetics (Chapter 2). If $\tau_{sup} \approx \tau$, then a full transient analysis has to be used. If $\tau_{sup} \ll \tau$, nucleation is avoided. Cooling in such cases leads to a kinetically frozen state in which the cluster size distribution is very far from steady state. On heating to accelerate the cluster dynamics, transient nucleation behavior can be observed. Examples are found in the nucleation of precipitation in a supersaturated solid solution, or of devitrification in a glass.

Thermal relaxation time. In addition to effects of changing system temperature, there are local effects of the release of heat from forming nuclei. The temperature of the nucleus is raised relative to the surrounding original phase, but whether the temperature increase is significant or not depends on a thermal relaxation time τ_{th}. A coupling of nucleation with thermal fluxes can be significant in vapor condensation, when nucleating droplets have difficulty in dissipating heat. In condensed systems, however, significant coupling with thermal fluxes is rare — mostly thermal conductivities are high enough to ensure that $\tau_{th} \ll \tau$, and there is then no effect.

Solutal relaxation time. It is common for a newly nucleated phase to have a composition different from the original phase. The transport rate of solute to or from a nucleus is characterized by a solutal relaxation time τ_{sol}. If $\tau_{sol} \ll \tau$, then partitioning occurs readily and the cluster dynamics underlying nucleation can be analyzed very similarly to those for a one-component system. It may, though, be necessary to account for changes in the average composition of the original phase as nucleation proceeds. If $\tau_{sol} \gg \tau$, then there is no solute transport on a time-scale relevant for nucleation, and the original phase, if it transforms at all, is forced to do so polymorphically, without changes in composition. Importantly, it is most likely that $\tau_{sol} \approx \tau$, in which case nucleation and solute fluxes are strongly coupled, greatly complicating the kinetic analysis (Chapter 5).

Homophase fluctuation lifetime. Equilibrium fluctuations of order or composition in the original phase can strongly affect the nucleation of a new phase, as frozen-in fluctuations can do. In each case, the deviation of order or composition has a characteristic lifetime τ_{flu}. If $\tau_{flu} < \tau$, then there is no effect on nucleation kinetics, but otherwise there may be strong coupling.

4. PROSPECTS

From the early studies by Fahrenheit, interest in nucleation has a long history. Great progress has been made, but many questions remain. Research on nucleation is very active, and we can note from the cases examined in the book some of the drivers of current advances in understanding:

Improved experimental techniques. The characterization of structures and degrees of order in condensed matter continues to make great progress. Transmission electron microscopy gives ever better information on structures and compositions, while atom-probe field-ion microscopy can now characterize samples with around one billion atoms. Improved processing techniques, such as electromagnetic and electrostatic levitation to stabilize supercooled liquids, allow more quantitative investigations of metastable states as well as increasing the accessible departures from equilibrium. The range of possible experimental studies will continue to be extended by the development of new technologies.

Developments in theory. Nucleation theories are still, for the most part, based on classical principles that are not appropriate for small clusters of a few atoms. They do not adequately handle coupling effects. These include not only coupling with heat and solute fluxes, but also coupling with other phase transitions (first-order or not), and with homophase fluctuations in the original phase. Complexities of these kinds may be better dealt with using a formal structure different from that of the liquid drop model of the classical theory of nucleation. Density-functional models and similar approaches can make a major impact here.

Improved computational methods. The advances in computing power now enable molecular-dynamics and Monte-Carlo methods to be applied to larger systems over longer time-scales, so much so that direct comparison with experimental systems is starting to be possible. The derivation of better interatomic potentials has been crucial to the useful exploitation of modeling techniques. Computer simulations will play an increasing role in guiding nucleation studies. In particular, they can allow regions of parameter space to be more easily explored than is possible for real systems with significant experimental difficulties.

Studies of a wider range of systems. Colloids have emerged as key systems for fundamental studies, providing insights that may be applicable more broadly. The range of phenomena being analyzed continues to widen, notably in biological and medical systems where many of the most exciting future directions will be found. Processes in these systems can arise from nucleation-initiated first-order phase transformations, even at the subcellular level. Nature's control, as in biomineralization, highlights new aspects of nucleation, and can teach us lessons for the tailoring of desired microstructures.

The most exciting advances certainly are not predictable. What is clear, however, is that almost three centuries after Fahrenheit was first fascinated by the ability of water to remain liquid through the cold nights of an Amsterdam winter, nucleation remains an open and exciting topic.

APPENDIX

MODELS USED FOR ANALYSIS OF LIQUID AND GLASS NUCLEATION DATA

In this appendix are collected expressions for the work of cluster formation from selected extended classical, diffuse-interface, and density functional (DF) theories. Many, but not all, were discussed in Chapters 2 and 4. The models listed here are used to fit data for homogeneous nucleation of polymorphic crystallization in liquids and glasses in Chapters 7 and 8. Following the terminology of Gránásy [1, 2], the models and their acronyms are listed in Table 1.

The reversible work of critical cluster formation in the classical nucleation theory is (Chapter 2, Eq. (14))

$$W^*_{\mathrm{CNT}} = \left(\frac{16\pi}{3}\right)\frac{\sigma^3_{\mathrm{CNT}}}{(\Delta g)^2}. \tag{1}$$

where σ_{CNT} is the classical-theory value for the interfacial free energy between the cluster of the new phase and the original phase and Δg is the free energy decrease per unit volume for the formation of the new phase. This expression suggests that a non-zero work $W(1)$ is required to form a single-molecule cluster. In the simplest extension [3] of the CNT to get round this problem, $W(1)$ is subtracted from $W(n)$ in the self-consistent classical theory (SCCT) giving for the critical cluster (a variation of Chapter 2, Eq. (37)),

$$W^*_{\mathrm{SCCT}} = W^*_{\mathrm{CNT}} - W(1) = \left(\frac{16\pi}{3}\right)\frac{\sigma^3_{\mathrm{CNT}}}{(\Delta g)^2} - \left(\bar{v}\Delta g + (36\pi)^{1/3}\bar{v}^{2/3}\sigma_{\mathrm{CNT}}\right), \tag{2}$$

where $\bar{v}$ is the molecular volume (see Chapter 2, Section 5). The diffuse-interface theory (DIT) [1], or phenomenological DIT (see Chapter 4, Section 3), parameterizes the enthalpy and entropy of the interface yielding (from Ref. [1])

$$W^*_{\mathrm{DIT}} = \frac{4\pi}{3}\left(\frac{\sigma_{\mathrm{CNT}}}{\Delta h_{\mathrm{f}}}\right)^3 \Delta g_{\mathrm{ct}}\psi, \tag{3}$$

Table 1 Some non-classical models of nucleation

Model	Acronym	Equation
Self-consistent CNT[a]	SCCT	2
Phenomenological DIT[c]	DIT	3
Perturbative DF[b]	PDFA	4
Semi-empirical DF[b]	SDFA	5
Modified-weighted DF[b]	MWDA	6
Ginzburg–Landau free energy used in Cahn–Hilliard model	GLCH	7

[a]CNT – classical nucleation theory.
[b]DF – density-functional.
[c]DIT – diffuse-interface theory.

where $\Delta g_{ct} = \Delta h_{ct} - T\Delta s_{ct}, \psi = 2(1+q)\xi^{-3} - (3+2q)\xi^{-3} + \xi^{-1}, \xi = \Delta g_{ct}/\Delta h_{ct}$, $q = \sqrt{(1-\xi)}$, Δh_f is the heat of fusion per unit volume, and Δg_{ct}, Δh_{ct}, and Δs_{ct} are the free energy, enthalpy, and entropy differences per unit volume, evaluated at the center of the cluster.

The next four methods, listed in Table 1, are derived by the DF methods discussed in Chapter 4. In all cases, expressions for the work of cluster formation are obtained that must be solved analytically or numerically to determine the interfacial profile, and then the work of critical cluster formation. Based on the theory of freezing proposed by Ramakrishnan & Yussouff [4], within the perturbative density-functional approximation (PDFA), the work of cluster formation [4] can be obtained in terms of the excess local grand potential density, $\Delta\omega = (c_0 - 1)\eta + \frac{1}{2}\, c_0\, \eta^2 + \frac{1}{2}\sum_i c_i m_i^2$, where η is the fractional density change on freezing and m_i are the Fourier coefficients corresponding to the set of reciprocal lattice vectors of the solid cluster, $\{\boldsymbol{q}_i\}$. The c_i are the Fourier components of the direct correlation function, which can be directly related to the structure factor of the liquid, $S(q)$, $S(\boldsymbol{q}) = 1 - (c(\boldsymbol{q}))^{-1}$. This link between nucleation and diffraction measurements has not been well explored. The critical work is (Chapter 4, (Eq. (81)),

$$W^*_{PDFA} = 4\pi\rho_1 k_B T \int_0^\infty r^2 \left\{ \Delta\omega(\eta(\boldsymbol{r}), m_i(r)) - \frac{c_0''}{4}\left(\frac{d\eta(r)}{dr}\right)^2 - \sum_i \frac{c_i''}{12}\left(\frac{dm_i(r)}{dr}\right)^2 \right\} dr, \tag{4}$$

where ρ_1 is the uniform density of the original phase (liquid or glass in this case) and T is the temperature; $c_i'' = d^2c/dq^2$ is evaluated at the ith reciprocal lattice vector, q_i. To fit the measured nucleation data presented in Chapters 7 and 8, only η and one structural order parameter were used to obtain the values for W^*_{PDFA} [5]. Also, in Chapter 8, these expressions are used to fit data taken only from congruently melting glasses, so the assumption of a unique melting temperature remains valid.

The semi-empirical density functional approximation (SDFA) is a one-order-parameter DF theory [6] that was developed in Chapter 4, Section 4.3. The work of cluster formation, scaled to the density of the liquid phase[1] and the melting temperature T_m can be written as (Chapter 4, Eq. (59))

$$W^*_{\mathrm{SDFA}} = 4\pi\rho_1 k_B T_m \int_0^\infty \left(\omega(M) + \frac{1}{2}K_M^2\left(\frac{\mathrm{d}M}{\mathrm{d}r}\right)^2\right) r^2 \mathrm{d}r, \tag{5}$$

where M is a structural order parameter and K_M^2 is coefficient of the square-gradient term divided by $\rho_1 k_B T_m$ (see Chapter 4, Section 4.3). The densities of the initial and final phases are assumed equal, which is a good approximation for the liquids and glasses of interest.

In the modified-weighted DF approximation (MWDA), particularized to a Lennard–Jones system [7], the square-gradient approximation is used to compute the work of cluster formation (Chapter 4, Eq. (94)),

$$W^*_{\mathrm{MWDA}} = 4\pi\rho_1 k_B T_m \int_0^\infty r^2 \left\{ \begin{array}{c} \omega[\rho_0(r), m_1(r)] - \omega_m + \frac{1}{2}K^2_{m_1 m_1}\left(\frac{\partial m_1(r)}{\partial r}\right)^2 \\ + \frac{1}{2\rho_1^2}K^2_{\rho_0\rho_0}\left(\frac{\partial \rho_0(r)}{\partial r}\right)^2 + \frac{1}{\rho_1}K^2_{m_1\rho_0}\left(\frac{\partial m_1(r)}{\partial r}\right)\left(\frac{\partial \rho_0(r)}{\partial r}\right) \end{array} \right\} \mathrm{d}r, \tag{6}$$

where ω_m is the local grand canonical potential of the metastable liquid. Equation (6) is analogous to Eq. (5), except with two order parameters. The correlation lengths, $K^2_{m_1 m_1}$, $K^2_{\rho_0\rho_0}$, and $K^2_{m_1\rho_0}$, are functionals of ρ_0 and m_1, are related to $c^{(2)}$, the direct correlation function of the liquid.

A model not discussed in Chapter 4 is the Ginzburg–Landau Cahn–Hilliard (GLCH) model. Iwamatsu & Horii [8] proposed to use a Ginzburg–Landau-type expansion of the free-energy functional originally proposed by Shih et al. for bcc phases [9]. For the work of cluster formation this gives

$$W^*_{\mathrm{GLCH}} = c_0 \int_0^\infty r^2 \left\{ \Delta\omega[m(r)] + \tfrac{1}{2}\left(\frac{\mathrm{d}m(r)}{\mathrm{d}r}\right)^2 \right\} \mathrm{d}r, \tag{7}$$

where $c_0 = -\Delta h_f (2b/a_{2f})^{3/2}(T_r/a)$; a_{2f} and b are related to the liquid structure factor and the direct correlation function, $a \approx (4.8\alpha)^{-1}$, $\Delta\omega = a(T_r - 1)m^2 + m^2(1-m)^2$, T_r is the reduced temperature ($T_r = T/T_m$), and $\alpha = 0.45$.

REFERENCES

[1] L. Gránásy, Diffuse interface theory of nucleation, J. Non-Cryst. Solids **162** (1993) 301–303; Diffuse interface model of crystal nucleation, J. Non-Cryst. Solids **219** (1997) 49–56.

[2] L. Gránásy, P.F. James, Nucleation in oxide glasses: comparison of theory and experiment, Proc. Roy. Soc. Lond. A **454** (1998) 1745–1766.

[1] Note that Gránásy scaled W^*_{SDFA} (Eq. 5) and W^*_{MWDA} (Eq. 6) to the liquid density; in Chapter 4 they were scaled to the solid density.

[3] D.T. Wu, Nucleation theory, in: *Solid State Physics*, Eds. H. Ehrenreich, F. Spaepen, Academic, Boston (1997), pp. 37–187 .
[4] T.V. Ramakrishnan, M. Yussouff, First-principles order-parameter theory of freezing, Phys. Rev. B **19** (1979) 2775–2794.
[5] P. Harrowell, D.W. Oxtoby, A molecular theory of crystal nucleation from the melt, J. Chem. Phys. **80** (1984) 1639–1646.
[6] C.K. Bagdassarian, D.W. Oxtoby, Crystal nucleation and growth from the undercooled liquid: a nonclassical piecewise parabolic free-energy model, J. Chem. Phys. **100** (1994) 2139–2148.
[7] Y.C. Shen, D.W. Oxtoby, Nucleation of Lennard-Jones fluids: a density functional approach, J. Chem. Phys. **105** (1996) 6517–6524.
[8] M. Iwamatsu, K. Horii, A one order parameter theory of crystal-melt nucleation of spherical clusters, J. Phys. Soc. Jpn. **65** (1996) 2311–2316; Temperature-dependent surface tension of critical nucleus in crystal-melt nucleation, J. Phys. Soc. Jpn. **66** (1997) 1210–1218.
[9] W.H. Shih, Z.Q. Wang, X.X. Sheng, D. Stroud, Ginzburg-Landau theory for the solid-solid interface of bcc elements, Phys. Rev. A **35** (1987) 2611–2618.

AUTHOR INDEX

Index to authors' names mentioned in the text.

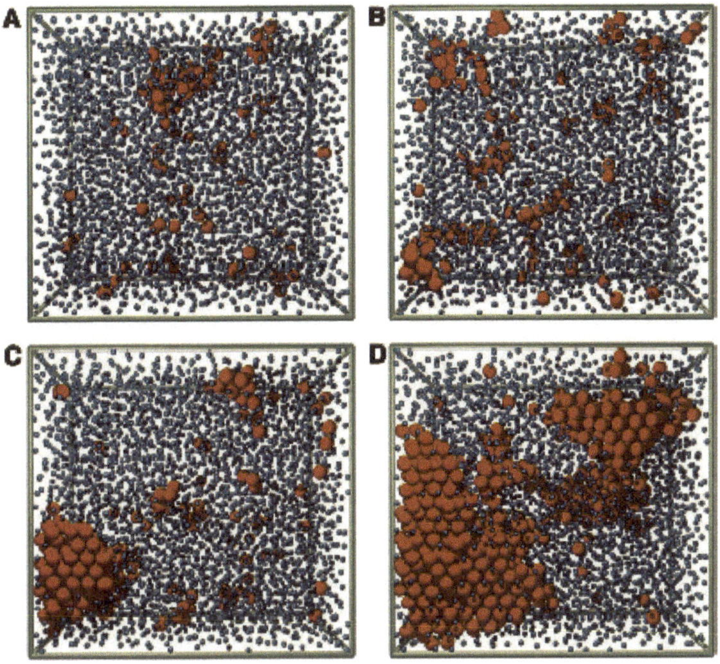

Fig. 3 Scanning confocal microscopy observations of particle positions during the crystallization of a colloid of polymethyl methacrylate spheres that were stabilized with poly-12-hydroxystearic acid. Particles in a crystal-like configuration are red, while those in the liquid-like phase are blue. Time after shear melting: (a) 20 min; (b) 43 min; (c) 66 min; (d) 89 min. (From Ref. [19], reprinted with permission from AAAS.)

Ref. 19 U. Gasser, E.R. Weeks, A. Schofield, P.N. Pusey & D.A. Weitz, "Real-space imaging of nucleation and growth in colloidal crystallization," *Science*, **292**, 258–262 (2003).

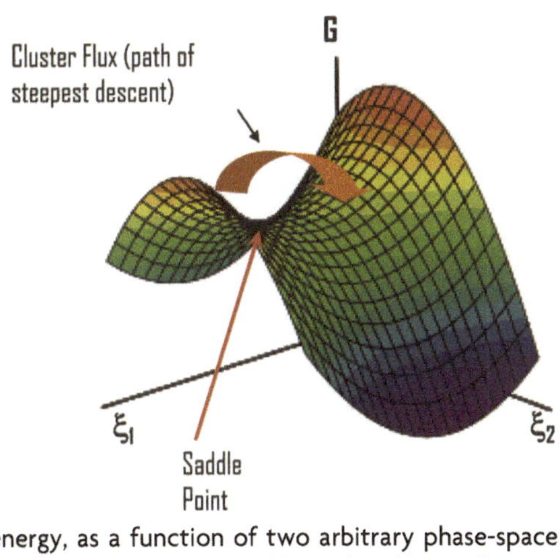

Fig. 3 The free energy, as a function of two arbitrary phase-space variables, in the saddle transition region between metastable and stable configurations.

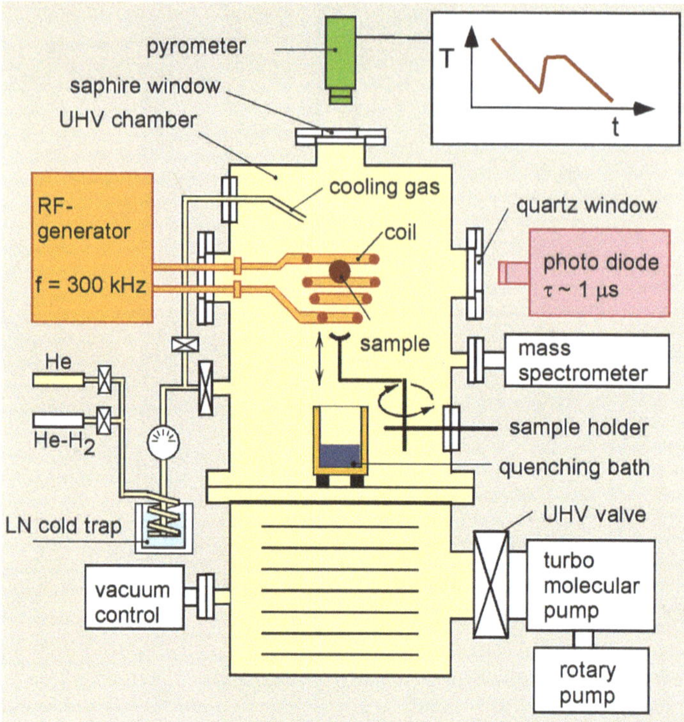

Fig. 5 Electromagnetic levitation (EML) facility for studies of supercooled liquid metallic samples. (Reprinted with permission from Ref. [12], copyright (1993), Maney Publishing.)

Ref. 12 D.M. Herlach, R.F. Cochrane, I. Egry, H.J. Fecht & A.L. Greer, "Containerless processing in the study of metallic melts and their solidification," *Int. Mater. Rev.* **38**, 273–347 (1993)

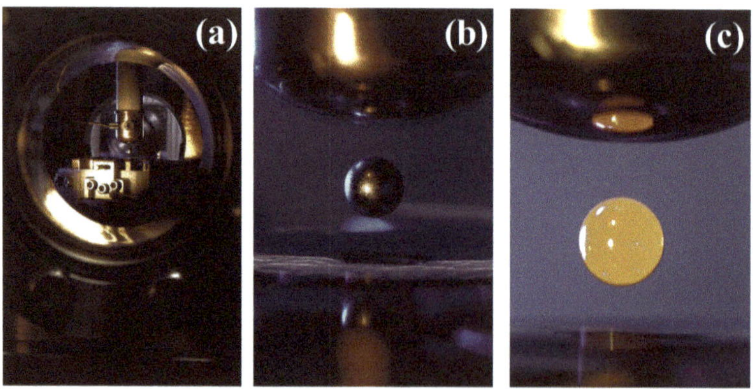

Fig. 7 (a) The vacuum chamber and levitation electrodes of the ESL facility at NASA Marshall Space Flight Center; (b) a solid $Ti_{39.5}Zr_{39.5}Ni_{21}$ sphere suspended between the top and bottom electrostatic electrodes; (c) liquid $Ti_{39.5}Zr_{39.5}Ni_{21}$ after laser melting. (Courtesy Jan Rogers, NASA Marshall Space Flight Center, Huntsville, AL.)

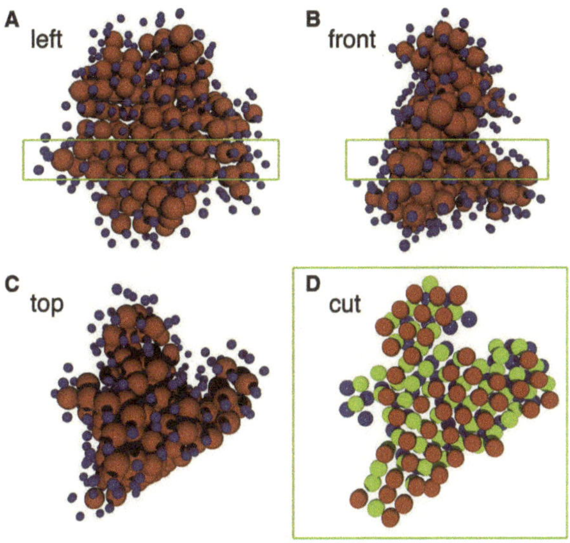

Fig. 18 (a) to (c) Crystal colloidal clusters that are larger than the critical size (observed by confocal microscopy). The larger lighter gray spheres represent colloidal particles that have crystal-type bonds and are part of crystal clusters; the smaller darker gray spheres represent those particles with at least one crystal-like bond with a larger gray particle. (d) A slice through the center of a cluster showing three stacked hexagonal layers (the shading of the spheres represents their layer location, i.e. dark gray spheres on one layer, medium gray spheres on the next and light gray spheres on the final layer, in order from front to rear). (From Ref. [72], Reprinted with permission from AAAS.)

Ref. 72 U. Gasser, E.R. Weeks, A. Schofield, P.N. Pusey & D.A. Weitz, "Real-space imaging of nucleation and growth in colloidal crystallization," *Science*, **292**, 258–262 (2003).

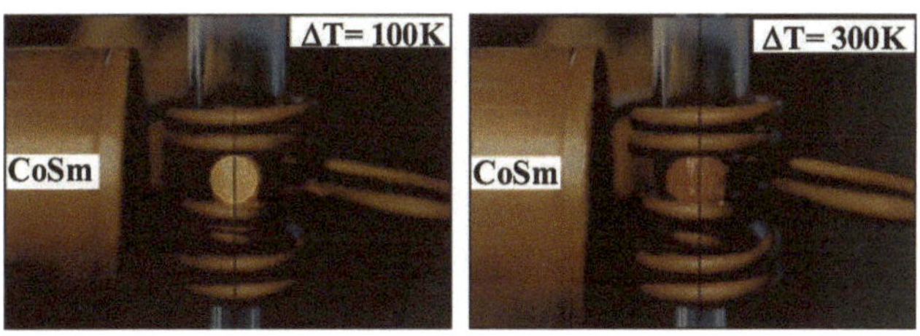

Fig. 19 Left: EML levitated $Co_{80}Pd_{20}$ sample at an supercooling of 100 K, far above the Curie temperature, T_C. The center of the sample coincides with the symmetry axis of the levitation coil, shown by the vertical line. Right: The sample at a deeper supercooling near T_C. The liquid has become ferromagnetic and is attracted by the cobalt–samarium magnet, moving the sample center to the left of the coil's symmetry axis. (Reprinted with permission from Ref. [91], copyright (2003), The Royal Society.)

Ref. 91 D.M. Herlach, D. Holland-Moritz, R. Willnecker, D. Herlach & K. Maier, "Magnetic ordering and crystal nucleation in undercooled Co-based melts," *Phil. Trans. Roy. Soc. Lond. A*, **361**, 497–515 (2003)

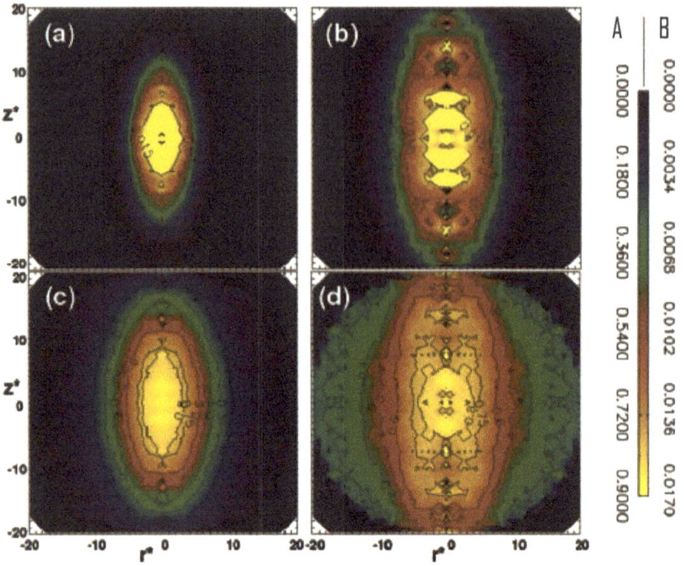

Fig. 16 Monte Carlo simulations of a nematic cluster of rigid rods (length-to-diameter ratio 15) forming in an isotropic phase. The contour plots show the concentration of rods (a, b, scale B) and the degree of their nematic alignment (c, d, scale A) for a subcritical embryo (a, c) with $n = 20$ and for a critical nucleus with $n^* = 95$ (b, d). (Reprinted with permission from [87], copyright (2007) by the American Physical Society.)

Ref. 87 A. Cuetos, M. Dijkstra, Kinetic pathways for the isotropic-nematic phase transition in a system of colloidal hard rods: a simulation study, Phys. Rev. Lett. **98** (2007) 095701/1–4

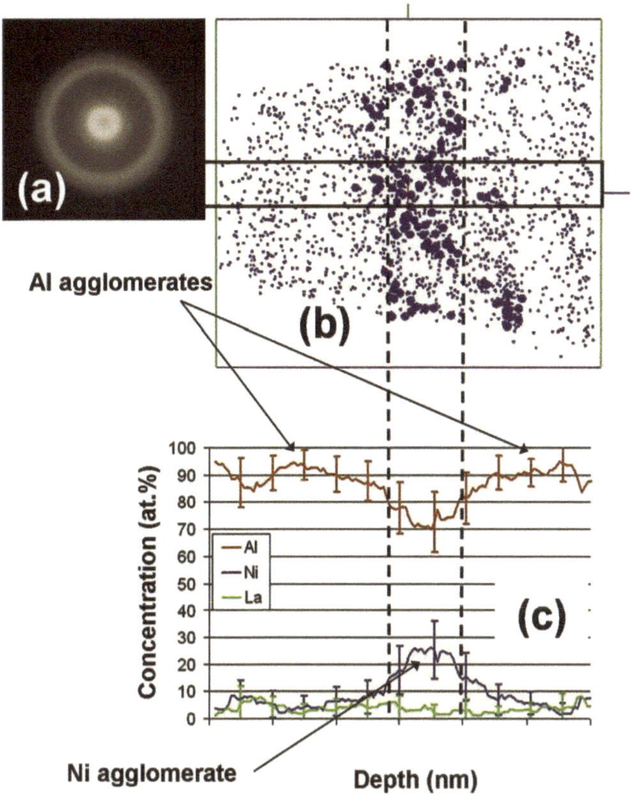

Fig. 6 (a) TEM selected-area diffraction pattern from an as-quenched $Al_{89}Ni_6La_5$ metallic glass; (b) 3DAP of the Ni atom distribution within an 11 × 11 × 12 nm volume — the bold dots show regions where the Ni concentration is greater than 10 at.%; (c) concentration profiles for the Al, Ni, and La atoms measured by 3DAP along the axis of the cylindrical region shown in (b). The arrows indicate the regions where the Al content exceeds the average of 89 at.%. (Reprinted with permission from Ref. [43], copyright (2008), American Institute of Physics.)

Ref. 43 B. Radiguet, D. Blavette, N. Wanderka, J. Banhart, K.L. Sahoo, Segregation controlled nanocrystallization in Al-Ni-La metallic glass, Appl. Phys. Lett. **92**(200) 103126/1–4

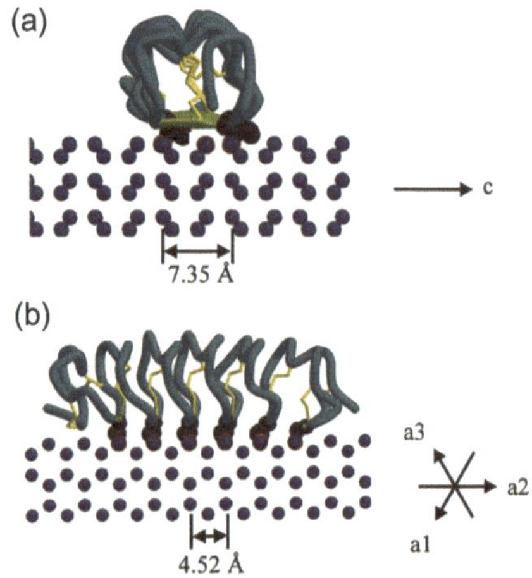

Fig. 4 Lattice matching of an insect antifreeze protein (AFP) (from the beetle *Tenebrio molitor*) on to the $(10\bar{1}0)$ prismatic face of ice. (a) The β-helix structure of the AFP viewed along its axis, parallel to the $[\bar{1}2\bar{1}0]$ axis of ice; (b) the same configuration viewed orthogonally along [0 0 0 1] of ice. The crystallographic axes of the hexagonal ice structure are shown. (Reprinted by permission from Macmillan Publishers Ltd: *Nature* Ref. [22], copyright 2000.)

Ref. 22 Y.C. Liou, A. Tocilj, P.L. Davies, Z. Jia, Mimicry of ice structure by surface hydroxyls and water of a b-helix antifreeze protein, Nature **406** (2000) 322–324.

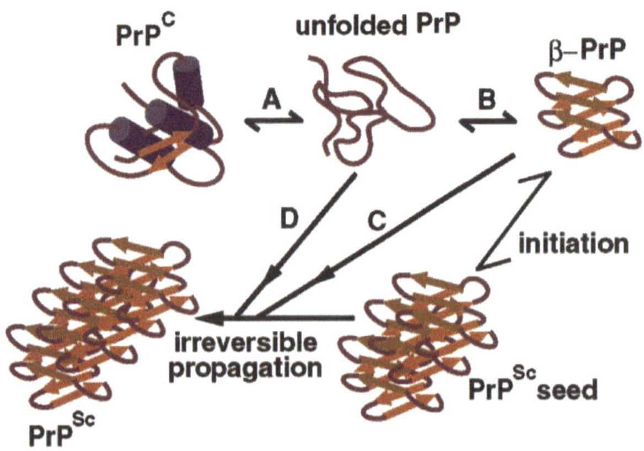

Fig. 20 Pathways to prion infection. Equilibrium A is between PrP^C and unfolded protein in solution. Equilibrium B is between the unfolded protein and β-sheet PrP. Infection may be initiated by the aggregation of β-sheet PrP to form a stable seed. The seed may grow by acquisition of unfolded protein or further β-sheet to form an insoluble plaque. (From [128], with permission.)

Ref. 128 G.S. Jackson S.J. Tabrizi, J. Collinge, Protein aggregation and the replication of prion infectivity, unpublished preprint.

SUBJECT INDEX

FSC
www.fsc.org
MIX
Papier | Fördert gute Waldnutzung
FSC® C083411

Kolibri 360 GmbH
Ferdinand-Jühlke-Straße 7
99095 Erfurt, Deutschland
produktsicherheit@kolibri360.de